ALLE · ZEIT · WACH · 1842

Solomon G. Mikhlin   Siegfried Prössdorf

Singular Integral Operators

With 11 Figures

Springer-Verlag
Berlin Heidelberg New York Tokyo

Prof. Dr. Solomon G. Mikhlin
University of Leningrad
pr. Schaumjana 33, kw. 47
195112 Leningrad
USSR

Prof. Dr. Siegfried Prössdorf
Karl-Weierstraß-Institut für Mathematik
der Akademie der Wissenschaften der DDR
Mohrenstr. 39
DDR-1086 Berlin
GDR

This book is an extended and partly modified
English version of the authors' book
„Singuläre Integraloperatoren“

Translated from the German by
Dr. Albrecht Böttcher (Chapters I–VII, XVI–XVII)
Doz. Dr. Reinhard Lehmann (Chapters VIII–XV, XVIII)

Sole distribution rights for all non-socialist countries:
Springer-Verlag Berlin Heidelberg New York Tokyo

ISBN 3-540-15967-3
Springer-Verlag Berlin Heidelberg New York Tokyo
ISBN 0-387-15967-3
Springer-Verlag New York Heidelberg Berlin Tokyo

Library of Congress Cataloging-in-Publication Data
Mikhlin, S. G. (Solomon Grigorevich), 1908–
Singular integral operators.
Translation of: Singuläre Integraloperatoren.
Bibliography: p.
Includes indexes.
1. Integral operators. 2. Integral equations. I. Prössdorf, Siegfried.
II. Title.
QA329.6.M5413 1986 515.7′23 85-26159
ISBN 0-387-15967-3 (U.S.)

Printed in the German Democratic Republic
Typesetting and printing: VEB Druckhaus Köthen
Bookbinding: Konrad Triltsch, Würzburg
2141/3140-543210

Preface

The present edition differs from the original German one mainly in the following additional material: weighted norm inequalities for maximal functions and singular operators (§ 12, Chap. XI), polysingular integral operators and pseudo-differential operators (§§ 7, 8, Chap. XII), and spline approximation methods for solving singular integral equations (§ 4, Chap. XVII). Furthermore, we added two subsections on polynomial approximation methods for singular integral equations over an interval or with discontinuous coefficients (Nos. 3.6 and 3.7, Chap. XVII). In many places we incorporated new results which, in the vast majority, are from the last five years after publishing the German edition (note that the references are enlarged by about 150 new titles).

S. G. Mikhlin wrote §§ 7, 8, Chap. XII, and the other additions were drawn up by S. Prössdorf.

We wish to express our deepest gratitude to Dr. A. Böttcher and Dr. R. Lehmann who together translated the text into English carefully and with remarkable expertise.

Leningrad and Berlin, July 1986

S. G. Mikhlin and S. Prössdorf

Preface to the German Edition

In mathematical literature, there is a number of books devoted to theory and applications of singular integral equations. We have in mind the books by MUSKHELISHVILI, VEKUA, GAKHOV, GOHBERG and KRUPNIK, and PRÖSSDORF on one-dimensional equations, and of the books by MIKHLIN and STEIN on higher-dimensional equations as well. Each of those monographs deals with a certain section from the general theory of singular operators only. It appears to us to be reasonable to develop the fundamentals of the theory of singular integral operators and its applications in one book in a unified framework using extensively ideas and methods of functional analysis. Furthermore, it should be mentioned that a large number of new results has been published recently and they are either entirely missing in the books mentioned above or, at least, presented incompletely.

A global view of the topics covered by the present book is given by a glance at the table of contents. Here, we merely point out some particularities of it which seem to be important.

Special emphasis is given to one-dimensional equations and systems of such equations with discontinuous (particularly piecewise continuous) coefficients on piecewise Lyapunov curves (Chaps. IV and V). In Chap. VI these equations are investigated when the symbol degenerates. In Chap. VII, under analogous assumptions singular integro-differential equations, singular integral equations with Carleman shift, Tricomi equations, boundary value problems for elliptic systems of partial differential equations in the plane with violated Shapiro-Lopatinski condition, etc. are considered. An outstanding topic in Chap. IX is the generalization of the Zygmund inequality to higher dimensions. The exact correlation of the differentiability properties between symbol and characteristic is given in Chap. X. Furthermore, fundamental theorems on singular operators in weighted spaces are proved (Chaps. II, XI). Chap. XIII contains the modern theory of higher-dimensional singular integrals and integral equations in Lipschitz-Hölder spaces. Basic results on the index of multi-dimensional singular systems are given (Chap. XIV). The theory of locally Fredholm operators due to Simonenko and its applications to singular operators is discussed in detail (Chap. XV). Chap. XVI contains several new results from the theory of multi-dimensional singular operators with degenerate symbol. The last two chapters are devoted to methods for the approximate solution of one- and multi-dimensional singular integral equations. Representing the material in Chap. IV as well as in Chap. II, §§ 2, 3, 5, we essentially followed the book by GOHBERG and KRUPNIK [4].

Most of the results mentioned above have not yet been published in monographs, in part they are known from mathematical journals. Several results are new, and some of them are published here for the first time.

We have endeavoured to present the material in this monograph, on the one hand, according to the current scientific level in order to lead the reader to modern research in that field. On the other hand, it was our aim to write as elementary an account as possible to render it accessible to a large spectrum of readers. At the same time we were anxious to keep the volume of the book within a reasonable limit. This, in turn, inevitably forced us to leave some problems aside. Thus, we do not consider, for example pseudo-differential operators (apart from Chap. I, §§ 5, 6, and Chap. XV, § 10) or Fourier integral operators (concerning these operators, the reader is referred to Duistermaat [1], Eskin [1], Hörmander [5], Shubin [1]). We do not consider either Wiener-Hopf operators (see Gohberg and Feldman [1], Prössdorf [17]).

Also nonlinear singular equations could not be included. We do not deal with maximal functions which are in connection to singular operators. Finally, the index theorems for general multi-dimensional systems are formulated without proof.

S. Prössdorf wrote Chaps. I–VII (except §§ 5, 6 of Chap. I) and Chaps. XVI, XVII. The remaining chapters (as well as §§ 5, 6 of Chap. I) were drawn up by S. G. Mikhlin.

A few words concerning the organization of the material are in order. Formulas are numbered in each section separately. Referring to a formula in the current section we indicate only the corresponding number. When a reference is made to a formula in another section of the current chapter the numbers of the corresponding section and formula are given. In case of a reference to a formula in another chapter the number of the corresponding chapter in Roman numerals is placed following the formula number.

We assume the reader to be familiar with elements of analysis and functional analysis to the extent of usual university courses.

When compiling the final voluminous manuscript we were supported considerably by many colleagues and collaborators. We wish to express our gratitude to Professor V. G. Mazya and to Professor B. Silbermann, who also read the entire manuscript, for many valuable discussions and suggestions. Dr. W. Sprössig communicated some new unpublished results which we used when writing § 3, Chap. XVI. Dr. J. Nagel translated Mikhlin's part from Russian into German, and Dr. J. Elschner helped us in proofreading. Mrs. Ch. Huber typed the whole manuscript extremely carefully, and Mr. W. Ley, Mr. A. Pomp, and Mr. G. Schmidt helped in preparing the text for print. We wish to thank sincerely all these persons.

Finally, we are pleased to express our gratitude to the Akademie-Verlag Publishing House, especially to the Deputy Chief Editor, Miss R. Helle, and to the Editor, Dr. R. Höppner, for the careful performance of the book and for their patience when confronted with various changes in the project, and–last not least–to the Printing Office VEB Druckerei "Druckhaus Köthen" for the careful formula type setting.

Autumn 1979

S. G. Mikhlin and S. Prössdorf

Contents

Chapter I
Basic facts from functional analysis

§ 1. Basic concepts

1.1. Let X and Y be Banach spaces and let A be a linear operator acting from X into Y. We denote by $D(A) \subset X$ the domain and by $\operatorname{Im} A \subset Y$ the image space (range) of A. The linear set of all $x \in D(A)$ such that $Ax = 0$ is called the *kernel* of the operator A and is denoted by $\operatorname{Ker} A$. The quotient space $Y / \overline{\operatorname{Im} A}$ (or any direct complement of $\overline{\operatorname{Im} A}$ in Y (cf. Section 1.4)) is referred to as the *cokernel* of A and denoted by $\operatorname{Coker} A$. The dimensions[1])

$$\alpha(A) = \dim \operatorname{Ker} A, \qquad \beta(A) = \dim \operatorname{Coker} A \tag{1}$$

are called the *nullity* and the *deficiency* of A, respectively. If at least one of the integers (1) is finite, their difference is called the *index* of A and denoted by $\operatorname{Ind} A$:

$$\operatorname{Ind} A = \alpha(A) - \beta(A).$$

Clearly, $\operatorname{Ind} A$ is finite if and only if both $\alpha(A)$ and $\beta(A)$ are finite.

In what follows we always assume that $D(A)$ is dense in X. Then one can define the adjoint operator A^*, acting from Y^* into X^*. Here X^* and Y^* are the conjugate spaces of X and Y, respectively. Note that A^* is always a closed operator. Furthermore, if A itself is closed, then $\overline{D(A^*)} = Y$ and $A^{**} = A$. It is easy to see that $\alpha(A^*) = \beta(A)$. If A is a bounded operator (in which case we shall always suppose that $D(A) = X$), then $\|A^*\| = \|A\|$. The space of all bounded linear operators from X into Y is denoted by $\mathscr{L}(X, Y)$ ($\mathscr{L}(X)$ in case $X = Y$).

1.2. An operator $P \in \mathscr{L}(X)$ is called a *projection* if it is idempotent, that is, $P^2 = P$. Clearly, if P is a projection then the operator $Q = I - P$ (I the identity operator) is also a projection. It is referred to as the *complementary projection* of P. Since $\operatorname{Im} P = \operatorname{Ker} Q$, the range of a projection is always closed. Note that $PQ = QP = 0$ for any complementary projections.

The norm of a projection $P \neq 0$ is always no less than 1. The norm of an orthogonal projection on a Hilbert space is equal to 1.

1.3. Let M_1 and M_2 be (in general, not necessarily closed) subspaces of X such that $M_1 \cap M_2 = \{0\}$. The *direct sum* $L = M_1 \dotplus M_2$ is defined as the subspace consisting of all elements of the form $x = x_1 + x_2$ with $x_1 \in M_1$ and $x_2 \in M_2$. The operator P defined by $Px = x_1$ is obviously idempotent and one has $\operatorname{Im} P = M_1$ and $\operatorname{Ker} P = M_2$.

---

[1]) The dimension of a subspace $L \subset X$ is defined as the maximal number of linearly independent vectors in L. It is denoted by $\dim L$. We put $\dim L = \infty$ if L contains infinitely many linearly independent vectors.

It is well-known that the direct sum $L = M_1 \dotplus M_2$ of two closed subspaces M_1 and M_2 is closed if and only if the operator P is bounded and is therefore a projection (see, e.g., A. P. Robertson and W. Robertson [1]). In that case P is called the *projection of L onto M_1 parallel to M_2*. The direct sum $M_1 \dotplus M_2$ of a closed subspace M_1 and a finite-dimensional subspace M_2 is always closed.

Obviously, every projection $P \in \mathscr{L}(X)$ generates a direct sum $X = M_1 \dotplus M_2$ with $M_1 = \operatorname{Im} P$ and $M_2 = \operatorname{Ker} P$.

1.4. Let M be a closed subspace of a Banach space X. Then each closed subspace $N \subset X$ such that $X = M \dotplus N$ is called a *direct complement* of M in X. Each direct complement of M in X, if it exists at all[1]), is isomorphic to the quotient space X/M (see A. P. Robertson and W. Robertson [1], V.7). The dimension of the quotient space X/M is called the *codimension* of the subspace M and denoted by codim M.

Of course, the direct complement is not determined uniquely. Hilbert spaces have the property that every closed subspace possesses a direct complement (take, for instance, the orthogonal complement). In a Banach space every closed subspace M having finite dimension or finite codimension possesses a direct complement.

Indeed, let first $m = \dim M$ be finite and let $e_1, \ldots, e_m$ be a basis in M. By the Hahn-Banach theorem there exists a corresponding biorthogonal system of functionals $f_1, \ldots, f_m$ in X^* (i.e. $(e_i, f_j) = \delta_{ij}$, $i, j = 1, \ldots, m$).[2]) Put $N = \bigcap_{i=1}^{m} \operatorname{Ker} f_i$. Then N is a closed subspace of X and it is easy to see that $X = M \dotplus N$.

If, on the other hand, $n = \dim X/M$ is finite and if $\xi_1, \ldots, \xi_n$ denotes a basis in the quotient space X/M, we choose from each coset ξ_k $(k = 1, \ldots, n)$ an arbitrary element $x_k \in X$. Then $X = M \dotplus N$, with N the subspace spanned by the linearly independent elements $x_1, \ldots, x_n$.

1.5. Let A be a linear operator from X into Y with dense domain $D(A)$. Then for the equation

$$Ax = y \qquad (y \in Y) \tag{2}$$

to have a solution $x \in D(A)$ it is necessary that y be orthogonal to $\operatorname{Ker} A^*$ (that is, $(y, f) = 0$ for all $f \in \operatorname{Ker} A^*$). Indeed, if x is a solution of (2), then, for arbitrary $f \in \operatorname{Ker} A^*$,

$$(y, f) = (Ax, f) = (x, A^*f) = (x, 0) = 0.$$

The operator A is said to be *normally solvable* (in the sense of Hausdorff [1]), if the orthogonality condition just mentioned is also sufficient for (2) to have a solution $x \in X$.

Hausdorff's lemma. *The operator A is normally solvable if and only if its range* $\operatorname{Im} A$ *is closed.*

Proof. Obviously, it suffices to show that A is normally solvable if $\operatorname{Im} A$ is closed. Fix an arbitrary element $y_0 \in Y$ such that $y_0 \notin \operatorname{Im} A$. The assertion will follow if we have proved that there exists a functional $f_0 \in \operatorname{Ker} A^*$ with $(y_0, f_0) \neq 0$.

To this end define first f_0 by $(y + \lambda y_0, f_0) = \lambda$ on the closed subspace $Y_0 = \operatorname{Im} A \dotplus \{\lambda y_0\}$. Then f_0 is a continuous linear functional on Y_0, and by the Hahn-Banach theorem

---

[1]) In every Banach space which is not isomorphic to a Hilbert space there exist closed subspaces which have no direct complement (see Lindenstrauss and Tzafriri [1]).

[2]) Here and in what follows we denote by (e, f) the value of the functional f at the element e and by δ_{ij} the Kronecker symbol.

it can be extended continuously to the whole space Y. This extension is the functional sought. ∎

Banach-Hausdorff theorem. *For a closed linear operator* $A : X \to Y$ *with dense domain the following statements are equivalent:*

(a) $\operatorname{Im} A$ *is closed.*
(b) A *is normally solvable, that is,* $\operatorname{Im} A = \{y \in Y : (y, f) = 0$ *for all* $f \in \operatorname{Ker} A^*\}$.
(c) A^* *is normally solvable, that is,*

$$\operatorname{Im} A^* = \{f \in X^* : (x, f) = 0 \text{ for all } x \in \operatorname{Ker} A\}.$$

(d) $\operatorname{Im} A^*$ *is closed in* X^*.

For a proof of this theorem we refer the reader to Yosida [1].

§ 2. Regularization of operators

Throughout this section let X, Y, Z denote arbitrary Banach spaces.

2.1. Let A be a linear operator from X into Y. In this section we shall always suppose that the domain $D(A)$ is dense in X.

The operator A is said to admit a *left regularization* if there exists an operator $R_l \in \mathscr{L}(Y, X)$ such that

$$R_l A = I_X + T_X, \tag{1}$$

where I_X is the identity operator and T_X a compact operator on X. The operator R_l is called a *left regularizer* of A.

We say the operator A admits a *right regularization* if there is an operator $R_r \in \mathscr{L}(Y, X)$ such that $\operatorname{Im} R_r \subset D(A)$ and

$$A R_r = I_Y + T_Y, \tag{2}$$

with I_Y the identity operator and T_Y a compact operator on Y. In that case R_r is referred to as a *right regularizer* of A. If an operator A admits both a left and right regularization, then A is said to admit a *two-sided regularization.*

Notice the following simple properties of operators which admit a regularization:

1. *If A admits a left (right) regularization, then A^* admits a right (left) regularization.* This is immediate from

$$A^* R_l^* = I_X^* + T_X^*, \qquad R_r^* A^* = I_Y^* + T_Y^*.$$

2. *If R_l and R_r, respectively, are left and right regularizers of A, then $R_l - R_r$ is compact.*

Indeed, on multiplying (1) from the right by R_r, (2) from the left by R_l, and then subtracting the equalities obtained, we get $R_r - R_l = R_l T_Y - T_X R_r$.

3. *If R is a left (right) regularizer of an operator $A \in \mathscr{L}(Y, X)$, then, for every compact operator T, the operator $R + T$ is a left (right) regularizer of A.*

This implies, in particular, that, if a bounded operator admits a two-sided regularization, then it can be assumed that $R_l = R_r$.

Theorem 2.1. *If an operator admits a left (right) regularization, then its nullity (deficiency) is finite.*

Proof. Let R_l be a left regularizer of the operator A. Since each solution of the equation

$Ax = 0$ ($x \in D(A)$) is also a solution of the equation $R_l A x = 0$, we have $\alpha(A) \leqq \alpha(R_l A)$. But $\alpha(R_l A) = \alpha(I_X + T_X) < \infty$ and thus $\alpha(A) < \infty$.

On the other hand, if A admits a right regularization, then A^* admits a left one and by what has just been proved, $\beta(A) = \alpha(A^*) < \infty$. ▮

Theorem 2.2. *If a closed operator admits a left or right regularization, then it is normally solvable.*

We prove this theorem with the help of the following test, which will be frequently used later on, too.

Lemma 2.1. *For a closed operator $A: X \to Y$ to be normally solvable and to have finite nullity it is necessary and sufficient that there exist a compact operator $T: X \to Z$ and a constant $C > 0$ such that*

$$\|x\| \leqq C(\|Ax\| + \|Tx\|), \qquad \forall x \in D(A). \tag{3}$$

Proof. *Sufficiency.* Suppose (3) holds. We first show that then the closed subspace Ker A is finite-dimensional. Because of (3) we have

$$\|x\| \leqq C\,\|Tx\|$$

for all $x \in \operatorname{Ker} A$. From this it is easily seen that the unit ball $\|x\| \leqq 1$ is a compact subset of Ker A. A well-known theorem of F. Riesz then implies that Ker A has finite dimension.

We now prove that the range Im A is closed. Let $y_n = Ax_n \to y$. We decompose X into a direct sum

$$X = X_1 \dotplus \operatorname{Ker} A. \tag{4}$$

Clearly, x_n can be chosen from X_1. Let us first show that the norms $\|x_n\|$ are uniformly bounded.

Assume the contrary. Then (take a subsequence, if necessary) $\|x_n\| \to \infty$ as $n \to \infty$. Putting $x_n' = \dfrac{x_n}{\|x_n\|}$ we get $Ax_n' \to 0$. Since $\|x_n'\| = 1$, there exists a subsequence $\{x_{n_k}'\}$ such that $\{Tx_{n_k}'\}$ is a Cauchy sequence and has, consequently, a limit $x \in X_1$. Obviously, $\|x\| = 1$. But on the other hand, $x \in D(A)$ and $Ax = 0$, since A is closed. Thus, taking into account that $X_1 \cap \operatorname{Ker} A = \{0\}$, we get $x = 0$. This contradiction shows that $\|x_n\| \leqq M$ ($n = 1, 2, \ldots$) with some constant $M > 0$.

By virtue of (3) and the compactness of T, we can again choose a convergent subsequence $\{x_{nk}\} \subset \{x_n\}$. But if $x_{nk} \to x$ and $Ax_{nk} \to y$, then $x \in D(A)$ and $y = Ax \in \operatorname{Im} A$. This implies that Im A is closed and so we have proved that A is normally solvable.

Necessity. Suppose A is normally solvable and has a finite-dimensional kernel. We decompose X into the direct sum (4), denote by P the projection of X onto Ker A parallel to X_1, and define A_1 as the restriction of the operator A to X_1. Then obviously $\operatorname{Ker} A_1 = \{0\}$ and $\operatorname{Im} A_1 = \operatorname{Im} A$. The inverse A_1^{-1} is a closed operator on the Banach space Im A and therefore it is bounded. Thus, we can find a constant $C' > 0$ such that

$$\|x_1\| \leqq C'\,\|A_1 x_1\|, \qquad \forall x_1 \in X_1.$$

Now, given an arbitrary $x \in D(A)$ we have, by (4), $x = x_1 + x_2$ with $x_1 = (I - P)\,x \in X_1$ and $x_2 = Px \in \operatorname{Ker} A$. The operator $T = P$ has a finite rank and is therefore compact. Since obviously $Ax = A_1 x_1$, we obtain

$$\|x\| \leqq \|x_1\| + \|x_2\| \leqq C(\|Ax\| + \|Tx\|)$$

with $C = \max(1, C')$. ▮

Remark. It follows from Lemma 2.1 that a closed operator $A: X \to Y$ is normally solvable and has finite nullity if, for all $x \in D(A)$,

$$\|x\| \leqq C\left(\|Ax\| + \sum_{j=1}^{r} \|T_j x\|\right) \tag{5}$$

with certain compact operators $T_j: X \to Z_j$ ($j = 1, \ldots, r$). To see this, define Z as the space of all r-tuples $z = (z_1, \ldots, z_r)$ with $z_j \in Z_j$ and $\|z\| = \sum_{j=1}^{r} \|z_j\|$, and define the operator $T: X \to Z$ by $Tx = (T_1 x, \ldots, T_r x)$. It is clear that then (5) takes the form (3).

Proof of Theorem 2.2. Let R_l be a left regularizer of the closed operator A. By virtue of Lemma 2.1 we have

$$\|x\| \leqq C'(\|(I_X + T_X)\, x\| + \|Tx\|), \qquad \forall x \in X, \tag{6}$$

where T is a compact operator on X. But (1) and (6) give (3) with $C = C' \max(1, \|R_l\|)$, and this proves the normal solvability of A.

If A admits a right regularization, then A^* is left regularizable. Consequently, A^* is normally solvable, and the Banach-Hausdorff theorem implies the normal solvability of A. The proof of Theorem 2.2 is complete. ∎

Remark. For a bounded operator $A \in \mathscr{L}(X, Y)$ one has the following necessary and sufficient conditions for the existence of a left or right regularizer (see Yood [1]):

1° *The operator A admits a left regularization if and only if it is normally solvable, has finite nullity, and there exists a projection of Y onto* Im A.

2° *The operator A admits a right regularization if and only if it is normally solvable, has finite deficiency, and there exists a projection of X onto* Ker A.

2.2. In Section 2.1 we saw that regularizable operators have a series of important properties. If the regularizer is known, the operator equation

$$Ax = y \tag{7}$$

can be transformed into an equation of the form

$$x + Tx = R_l y \tag{8}$$

(for a left regularizer) or of the form

$$z + Tz = AR_r z = y \tag{9}$$

(for a right regularizer), where T is a compact operator. This transformation is called a *regularization of equation* (7), and the equations (8) and (9) are called *Riesz-Schauder equations* or *equations of the second kind*.

It is obvious that for arbitrary $y \in Y$ every solution of (7) is also a solution of (8). If, conversely, for arbitrary $y \in Y$ every solution $x \in X$ of (8) is also a solution of (7), then R_l is said to be an *equivalent left regularizer* of A.[1])

It is easy to see that a left regularizer R_l of an operator A is an equivalent left regularizer if and only if it has the following two properties: 1. Ker $R_l = \{0\}$; 2. if $x \in X$ is any solution of (8), then $x \in D(A)$.

With regard to equation (9) note first of all the following: If z is a solution of this equation, then

$$x = R_r z \tag{10}$$

[1]) Note that for a left regularizer R_l the equality $R_l A x = (I + T)\, x$ holds for all $x \in D(A)$. Thus, if R_l is an equivalent left regularizer, then each solution $x \in X$ of equation (8) must belong to $D(A)$.

is a solution of the original equation (7). However, in general (7) can have still other solutions, that is, solutions which are not of the form (10). If for arbitrary $y \in \operatorname{Im} A$ every solution of (7) can be obtained through (10) (i.e., $\operatorname{Im} R_r = D(A)$), then R_r is called an *equivalent right regularizer* of A.

Thus, if R_r is an equivalent right regularizer, then either (7) and (9) are simultaneously solvable or none of them is solvable. In the first case we obtain the solutions of one of these equations from those of the other equation by means of the substitution (10).

In the following two theorems we state criteria for the existence of an equivalent regularizer.

Theorem 2.3. *For a closed operator to admit an equivalent left regularization it is necessary and sufficient that it be normally solvable and that it have a finite and non-negative index.*

Proof. *Necessity.* Let R_l be an equivalent left regularizer of the closed operator $A: X \to Y$. By the Theorems 2.1 and 2.2, A has finite nullity and is normally solvable. Since (7) and (8) are equivalent, (7) is solvable if and only if

$$(y, R_l^* \varphi_j) = 0, \qquad j = 1, \dots, n,$$

where $n = \beta(I + T) = \alpha(A)$ and $\varphi_1, \dots, \varphi_n$ denote linearly independent solutions of the adjoint homogeneous equation $\varphi + T^*\varphi = 0$. But then the functionals $\psi_j = R_l^*\varphi_j$ $(j = 1, \dots, n)$ belong to $\operatorname{Ker} A^*$ and $\operatorname{Ker} A^*$ is spanned by them. Thus,

$$\alpha(A^*) \leqq n = \alpha(A)$$

and therefore $\operatorname{Ind} A \geqq 0$.

Sufficiency. Let now A be a closed and normally solvable operator with finite and non-negative index. Decompose X into the direct sum (4) and Y into the direct sum $Y = \operatorname{Im} A \dotplus Y_1$, where Y_1 is a subspace of the dimension $n = \beta(A)$. Let $y_1, \dots, y_n$ be a basis in Y_1 and $x_1, \dots, x_m$ be a basis in $\operatorname{Ker} A$. By assumption, $m \geqq n$.

Let P denote the finite-dimensional projection of X onto $\operatorname{Ker} A$ parallel to X_1 and define A_1 as the restriction of the operator A to X_1. Then A_1^{-1} is defined and bounded on the closed subspace $\operatorname{Im} A$ (cf. also the proof of Lemma 2.1), and we have $A_1^{-1}A = I - P$ and $AA_1^{-1} = I$.

If $n = 0$, i.e., $\operatorname{Im} A = Y$, then $A_1^{-1} = A^{-1}$ is the inverse of A and is therefore an equivalent regularizer.

In case $n \geqq 1$ we extend A_1^{-1} to an operator $B \in \mathscr{L}(Y, X)$ by setting

$$By_j = x_j \qquad (j = 1, \dots, n).$$

The equalities $BA = A_1^{-1}A = I - P$ imply that B is a left regularizer of A. Our objective is to show that B is even an equivalent regularizer, i.e., that every solution of the equation

$$x - Px = By \tag{11}$$

also satisfies (7). Notice first that $\operatorname{Im} B \subset D(A)$, and hence every solution $x \in X$ of (11) belongs to $D(A)$.

Thus, let x be a solution of (11) and write $y = y' + y''$ with $y' \in \operatorname{Im} A$ and $y'' \in Y_1$. Then

$$x - Px - A_1^{-1}y' = By'' \tag{12}$$

with the left-hand side belonging to $\operatorname{Ker} P$ and the right-hand side being in $\operatorname{Im} P$. Because $\operatorname{Ker} P \cap \operatorname{Im} P = \{0\}$, we get $By'' = 0$ and so $y'' = 0$. The equality (12) now implies that

$$Ax = y' = y$$

and this is what was wanted. ∎

Theorem 2.4. *For a closed operator to admit an equivalent right regularization it is necessary and sufficient that it be normally solvable and that it have a finite and non-positive index.*

Proof. *Necessity.* Let R_r be an equivalent right regularizer of the closed operator $A : X \to Y$. Due to the Theorems 2.1 and 2.2, A has finite deficiency and is normally solvable. Since $\operatorname{Im} R_r = D(A)$, we deduce that

$$\alpha(A) \leqq \alpha(AR_r), \qquad \operatorname{Im} A = \operatorname{Im} AR_r$$

and thus $\beta(AR_r) = \beta(A)$. Because $AR_r = I + T$, we get finally that $\alpha(AR_r) = \beta(AR_r)$. Hence $\alpha(A) \leqq \beta(A)$, that is, $\operatorname{Ind} A \leqq 0$.

Sufficiency. Now let A be a closed and normally solvable operator with finite and non-positive index. We use the notations introduced in the proof of the sufficiency part of the preceding theorem.

If $n = \beta(A) = 0$, then $\operatorname{Ker} A = \{0\}$, since $\alpha(A) \leqq \beta(A)$. Thus, $\operatorname{Im} A_1^{-1} = D(A)$, and $A^{-1} = A_1^{-1}$ is therefore an equivalent right regularizer of A.

Now assume $n \geqq 1$. The operator A_1^{-1} extends to an operator $B \in \mathscr{L}(Y, X)$ by putting

$$By_j = \begin{cases} x_j & \text{for } j = 1, \ldots, m-1, \\ x_m & \text{for } j = m, \ldots, n. \end{cases}$$

We claim that B is an equivalent right regularizer of A.

Let K denote the projection of Y onto Y_1 parallel to $\operatorname{Im} A$. Then

$$B = A_1^{-1}(I - K) + BK,$$

and since $\operatorname{Im} BK \subset \operatorname{Ker} A$, this gives

$$AB = AA_1^{-1}(I - K) = I - K.$$

Thus, B is a right regularizer of A, since K (as an operator with finite rank) is compact.

It remains to show that $\operatorname{Im} B = D(A)$. Every element $x \in D(A)$ has a unique representation as a sum $x = x' + x''$ with $x' \in \operatorname{Ker} A$ and $x'' \in D(A) \cap \operatorname{Ker} P = D(A_1)$. Thereby, $x' = \sum_{j=1}^{m} \alpha_j x_j$. Put

$$y' = \sum_{j=1}^{m} \alpha_j y_j, \qquad y'' = A_1 x'', \qquad y = y' + y''.$$

Then $y' \in Y_1$, $y'' \in \operatorname{Im} A$, and thus

$$By = \sum_{j=1}^{m} \alpha_j x_j + A_1^{-1} A_1 x'' = x.$$

Consequently, $D(A) \subset \operatorname{Im} B$. Since the reverse inclusion is obvious, the proof is complete. ∎

Remark. The Theorems 2.1 and 2.3 are due to Mikhlin [22], Theorem 2.4 is due to Prössdorf [9] (see also Prössdorf [1]). Lemma 2.1 was established by U. Köhler in his diploma paper in 1971; for a special case it had already been proved by Agranovich [1] (see also Köhler and Silbermann [1, 2]).

§ 3. Fredholm and semi-Fredholm operators on Banach Spaces

Throughout this section, X, Y, Z again denote arbitrary Banach spaces. All operators occuring are supposed to be bounded (and thus to be defined on the whole space).

3.1. A normally solvable operator A is called a *Fredholm operator* or, concisely, a *Φ-operator*[1]) if it has a finite index. It is said to be a *semi-Fredholm operator* if at least one of the numbers $\alpha(A)$ and $\beta(A)$ is finite. A semi-Fredholm operator is referred to as a Φ_+-operator if $\alpha(A) < \infty$ and as a Φ_--operator if $\beta(A) < \infty$.

The collection of all Φ (resp. $\Phi_\pm$)-operators $A \in \mathcal{L}(X, Y)$ is denoted by $\Phi(X, Y)$ (resp. $\Phi_\pm(X, Y)$).

The class $\Phi(X) = \Phi(X, X)$ contains, for instance, all operators of the form $I + T$ with T a compact operator. Note that then $\operatorname{Ind}(I + T) = 0$.

By the Banach-Hausdorff theorem, $A \in \Phi_\pm(X, Y)$ if and only if $A^* \in \Phi_\pm(Y^*, X^*)$. Furthermore, in that case $\operatorname{Ind} A = -\operatorname{Ind} A^*$. Notice that Lemma 2.1 yields a necessary and sufficient condition for an operator to be a Φ_+-operator.

By virtue of the Theorems 2.1 and 2.2 every operator that admits a two-sided regularization is a Φ-operator. The reverse is also true.

Indeed, let $A \in \Phi(X, Y)$. With the operators A_1, P, K introduced in the proof of the Theorems 2.3 and 2.4, we have $B = A_1^{-1}(I_Y - K) \in \mathcal{L}(Y, X)$ and

$$BA = I_X - P, \qquad AB = I_Y - K. \tag{1}$$

Thus, B is both a left and right regularizer of A.

The just constructed operator B is also a *generalized inverse* of the operator A. This term is used for every operator $A^{(-1)} \in \mathcal{L}(Y, X)$ such that

$$AA^{(-1)}A = A.$$

If $A^{(-1)}$ is a generalized inverse of A, then, obviously,

$$\operatorname{Ker} A = \operatorname{Ker} A^{(-1)}A, \qquad \operatorname{Im} A = \operatorname{Im} AA^{(-1)}.$$

The following theorem, due to Atkinson [1], summarizes the above results.

Theorem 3.1. *For an operator $A \in \mathcal{L}(X, Y)$ the following properties are equivalent:*
(i) *A is a Φ-operator.*
(ii) *A admits a two-sided regularization.*
(iii) *There is a regularizer $B \in \mathcal{L}(Y, X)$ such that both the operators $BA - I_X$ and $AB - I_Y$ are finite-dimensional.*

For what follows we need the definition of the trace of an operator. Let $K \in \mathcal{L}(X, Y)$ be a finite-dimensional operator and denote by $\lambda_1, \ldots, \lambda_n$ its non-zero eigenvalues (counted up to their multiplicity). Similarly as this is done for matrices, one calls the sum of the eigenvalues $\lambda_1, \ldots, \lambda_n$ the *trace* of the operator K and denotes it by sp K:

$$\operatorname{sp} K = \lambda_1 + \lambda_2 + \ldots + \lambda_n.$$

The trace of an n-dimensional projection is easily seen to be n.

---

[1]) The term F-operator is also in use. In German and Russian Fredholm operators are usually called *Noether operators*. Such operators with non-zero index first appeared in F. Noether's paper [1] in connection with the investigation of singular integral operators.

Two important properties of the trace can be stated as follows:

1° *If $K_1, K_2 \in \mathscr{L}(X, Y)$ are any finite-dimensional operators, then*

$$\operatorname{sp}(K_1 + K_2) = \operatorname{sp} K_1 + \operatorname{sp} K_2 .$$

2° *If $K \in \mathscr{L}(X, Y)$ is finite-dimensional and if $A \in \mathscr{L}(Y, X)$ is an arbitrary operator, then*

$$\operatorname{sp} AK = \operatorname{sp} KA .$$

These two assertions are well-known for matrices and their generalizations to the situation considered here is not difficult.

Theorem 3.2. *Let $A \in \Phi(X, Y)$ and suppose $R \in \mathscr{L}(Y, X)$ is a regularizer of A such that both the operators $RA - I_X$ and $AR - I_Y$ have finite rank. Then*

$$\operatorname{Ind} A = \operatorname{sp}(I_X - RA) - \operatorname{sp}(I_Y - AR). \tag{2}$$

Proof. In the case where $R = A^{(-1)}$ is a generalized inverse of A we have in view of (1)

$$\operatorname{sp}(I_X - A^{(-1)}A) = \dim P = \alpha(A),$$
$$\operatorname{sp}(I_Y - AA^{(-1)}) = \dim K = \beta(A),$$

and this proves (2).

If R is an arbitrary regularizer then

$$R - A^{(-1)} = RK - (I_X - RA)\, A^{(-1)},$$

which implies that $R - A^{(-1)}$ is finite-dimensional. Because of 2° we have

$$\operatorname{sp}(R - A^{(-1)})\, A = \operatorname{sp} A(R - A^{(-1)}),$$

and this combined with the property 1° gives

$$\operatorname{sp}(I_X - A^{(-1)}A) - \operatorname{sp}(I_X - RA) = \operatorname{sp}(I_Y - AA^{(-1)}) - \operatorname{sp}(I_Y - AR)$$

and hence

$$\operatorname{sp}(I_X - RA) - \operatorname{sp}(I_Y - AR) = \operatorname{sp}(I_X - A^{(-1)}A) - \operatorname{sp}(I_Y - AA^{(-1)}) = \operatorname{Ind} A . \blacksquare$$

Remark. In the case where $X = Y$, formula (2) can be simplified to

$$\operatorname{Ind} A = \operatorname{sp}(AR - RA).$$

Formula (2) goes back to FEDOSOV [2] (see also Theorem 3.10). The proof of Theorem 3.2 presented here, as well as that of the following theorem, was communicated to the authors by A. S. Markus.

Theorem 3.3. (*Atkinson's theorem*). *If $A \in \Phi(X, Y)$ and $B \in \Phi(Y, Z)$, then $BA \in \Phi(X, Z)$ and*

$$\operatorname{Ind} BA = \operatorname{Ind} A + \operatorname{Ind} B .$$

Proof. Let $A^{(-1)}$ and $B^{(-1)}$ be generalized inverses of A and B, respectively (or any other regularizers such that the operators $A^{(-1)}A - I_X$, $AA^{(-1)} - I_Y$, $B^{(-1)}B - I_Y$, $BB^{(-1)} - I_Z$ have finite rank). Then the operators $A^{(-1)}B^{(-1)}BA - I_X$ and $BAA^{(-1)}B^{(-1)} - I_Z$ are finite-dimensional, and so Theorem 3.1 implies that $BA \in \Phi(X, Z)$.

From formula (2) and from property 1° we obtain that $d = \operatorname{Ind} BA - \operatorname{Ind} A - \operatorname{Ind} B$ equals

$$d = \operatorname{sp}(A^{(-1)}A - A^{(-1)}B^{(-1)}BA) + \operatorname{sp}(BAA^{(-1)}B^{(-1)} - BB^{(-1)})$$
$$+ \operatorname{sp}(B^{(-1)}B - AA^{(-1)}).$$

By virtue of property 2°

$$\operatorname{sp}(A^{(-1)}A - A^{(-1)}B^{(-1)}BA) = \operatorname{sp} A^{(-1)}(I_Y - B^{(-1)}B)\,A = \operatorname{sp}(I_Y - B^{(-1)}B)\,AA^{(-1)}$$

and analogously

$$\operatorname{sp}(BAA^{(-1)}B^{(-1)} - BB^{(-1)}) = \operatorname{sp} B(AA^{(-1)} - I_Y)\,B^{(-1)} = \operatorname{sp} B^{(-1)}B(AA^{(-1)} - I_Y).$$

Thus, once more taking into account property 1°, we get

$$d = \operatorname{sp}[(I_Y - B^{(-1)}B)\,AA^{(-1)} + B^{(-1)}B(AA^{(-1)} - I_Y) + B^{(-1)}B - AA^{(-1)}] = \operatorname{sp} 0 = 0. \blacksquare$$

Theorem 3.3 was proved by Atkinson [1] for the first time, however, by other methods. The following theorem goes back to Mikhlin [7], Atkinson [1], Gohberg [3], and Yood [1].

Theorem 3.4. *Let* $A \in \Phi(X, Y)$ *and suppose* $T: X \to Y$ *is compact. Then* $A + T \in \Phi(X, Y)$ *and* $\operatorname{Ind}(A + T) = \operatorname{Ind} A$.

Proof. Theorem 3.1 implies the existence of a two-sided regularizer R of the operator A. Since A can also be viewed as a two-sided regularizer of R, we deduce that $R \in \Phi(Y, X)$, and Theorem 3.3 then shows that $\operatorname{Ind} A + \operatorname{Ind} R = \operatorname{Ind} RA = 0$, that is,

$$\operatorname{Ind} A = -\operatorname{Ind} R. \tag{3}$$

But R is also a two-sided regularizer of $A + T$, and so, analogously, $\operatorname{Ind}(A + T) = -\operatorname{Ind} R$. Putting this and (3) together, we obtain the equality $\operatorname{Ind}(A + T) = \operatorname{Ind} A$. $\blacksquare$

Theorem 3.5. *For every operator* $A \in \Phi(X, Y)$ *there exists a number* $\varrho > 0$ *such that* $C \in \mathscr{L}(X, Y)$ *and* $\|C\| < \varrho$ *imply that* $A + C \in \Phi(X, Y)$ *and* $\operatorname{Ind}(A + C) = \operatorname{Ind} A$.

Proof. Again let R be a two-sided regularizer of A (cf. Theorem 3.1), that is,

$$RA = I + T_1, \qquad AR = I + T_2$$

with compact operators T_1, T_2. We claim that the assertion is true with $\varrho = \|R\|^{-1}$.

If $\|C\| < \varrho$, then $\|RC\| \leqq \|R\|\,\|C\| < 1$. Consequently, by a well-known theorem of Banach, the inverse $(I + RC)^{-1} \in \mathscr{L}(X)$ exists. Hence

$$R(A + C) = I + T_1 + RC = (I + RC)\,(I + T), \tag{4}$$

where $T = (I + RC)^{-1}\,T_1$ is compact on X. Now, with $R_1 := (I + RC)^{-1}\,R$, we obtain

$$R_1(A + C) = I + T. \tag{5}$$

Thus, R_1 is a left regularizer of $A + C$.

It can be shown analogously that $R_2 = R(I + CR)^{-1}$ is a right regularizer of $A + C$. Theorem 3.1 then gives that $A + C \in \Phi(X, Y)$.

Using Theorem 3.3, we get from (4) the equality

$$\operatorname{Ind}(A + C) + \operatorname{Ind} R = \operatorname{Ind}(I + RC) + \operatorname{Ind}(I + T) = 0.$$

This combined with (3) proves the equality $\operatorname{Ind}(A + C) = \operatorname{Ind} A$. $\blacksquare$

Theorem 3.5 was stated by Dieudonné [1] and Atkinson [1].

3.2. Our next objective is to extend the results of Section 3.1 to semi-Fredholm operators. The following two theorems are due to Yood [1].

Theorem 3.6. *If $A \in \Phi_{\pm}(X, Y)$ and $B \in \Phi_{\pm}(Y, Z)$ then $BA \in \Phi_{\pm}(X, Z)$.*

Proof. We prove the theorem for Φ_+-operators. For Φ_--operators the assertion then follows by taking adjoints.

Thus, let $A \in \Phi_+(X, Y)$ and $B \in \Phi_+(Y, Z)$. It follows from Lemma 2.1 that there exist compact operators T_j and constants $C_j > 0$ $(j = 1, 2)$ such that

$$\begin{aligned} \|x\| &\leqq C_1(\|Ax\| + \|T_1x\|), \qquad \forall x \in X, \\ \|y\| &\leqq C_2(\|By\| + \|T_2y\|), \qquad \forall y \in Y. \end{aligned} \tag{6}$$

Therefore, with $y = Ax$,

$$\|x\| \leqq C_1C_2(\|BAx\| + \|C_2^{-1}T_1x\| + \|T_2Ax\|), \qquad \forall x \in X.$$

But the operators $C_2^{-1}T_1$ and T_2A are compact and so, taking into consideration Lemma 2.1 and the remark made to it, we see that $BA \in \Phi_+(X, Z)$. ▮

The following theorem can be viewed as a reverse of the just proved Theorem 3.6.

Theorem 3.7. *Let $A \in \mathscr{L}(X, Y)$, $B \in \mathscr{L}(Y, Z)$, and $BA \in \Phi_+(X, Z)$ (resp. $BA \in \Phi_-(X, Z)$). Then $A \in \Phi_+(X, Y)$ (resp. $B \in \Phi_-(Y, Z)$).*

Proof. Let $BA \in \Phi_+(X, Z)$. Because of Lemma 2.1 there exist a compact operator T and a positive constant C such that

$$\|x\| \leqq C(\|BAx\| + \|Tx\|) \leqq C'(\|Ax\| + \|Tx\|)$$

for all $x \in X$. Here $C' = C \max(1, \|B\|)$. The last inequality combined with Lemma 2.1 now implies that $A \in \Phi_+(X, Y)$.

The assertion for Φ_--operators again follows by taking adjoints. ▮

Theorem 3.8. *Let $A \in \Phi_{\pm}(X, Y)$ and suppose $T: X \to Y$ is compact. Then $A + T \in \Phi_{\pm}(X, Y)$ and* $\operatorname{Ind}(A + T) = \operatorname{Ind} A$.

Proof. Let $A \in \Phi_+(X, Y)$. Then the first inequality in (6) holds and it implies that

$$\|x\| \leqq C_1(\|(A + T)x\| + \|T_1x\| + \|Tx\|), \qquad \forall x \in X.$$

From Lemma 2.1 we deduce that $A + T \in \Phi_+(X, Y)$. If $\beta(A) = \infty$ then, by Theorem 3.4, $\beta(A + T) = \infty$ and thus $\operatorname{Ind}(A + T) = \operatorname{Ind} A$. If $\beta(A) < \infty$, then that equality results from Theorem 3.4.

As above, to prove the assertion for Φ_--operators, it suffices to take adjoints. ▮

Theorem 3.8 was established by Köthe [1], Pietsch [2], and also by Gohberg and M. G. Krein [1].

Theorem 3.9. *For every operator $A \in \Phi_{\pm}(X, Y)$ there exists a number $\varrho > 0$ such that $C \in \mathscr{L}(X, Y)$ and $\|C\| < \varrho$ imply that $A + C \in \Phi_{\pm}(X, Y)$ and*

$$\alpha(A + C) \leqq \alpha(A), \qquad \beta(A + C) \leqq \beta(A), \qquad \operatorname{Ind}(A + C) = \operatorname{Ind} A$$

Proof. Assume first $A \in \Phi_+(X, Y)$. From the proof of Lemma 2.1 it follows that there exists a finite-dimensional operator $T \in \mathscr{L}(X)$ such that $\dim \operatorname{Im} T = \alpha(A)$ and

$$\|x\| \leqq \gamma(\|Ax\| + \|Tx\|), \qquad \forall x \in X,$$

where $\gamma > 0$ is some constant. Put $\varrho = 1/\gamma$. Then, if $\|C\| < \varrho$, the last inequality gives the estimate

$$\|x\| \leqq \gamma'(\|(A + C)x\| + \|Tx\|), \qquad \forall x \in X,$$

where $\gamma' = \gamma/(1 - \gamma \|C\|)$. Hence $A + C \in \Phi_+(X, Y)$ (cf. Lemma 2.1) and, furthermore, the finite-dimensional spaces $X_0 = \operatorname{Ker}(A + C)$ and $\operatorname{Im}(T \mid X_0)$ are isomorphic. Consequently, $\alpha(A + C) \leqq \alpha(A)$.

Let now $A \in \Phi_-(X, Y)$. Again as in the proofs of the preceding theorems, passage to the adjoint operators shows that $A + C \in \Phi_-(X, Y)$ and that $\beta(A + C) \leqq \beta(A)$ if only $\|C\| < \varrho$.

Thus, taking into account Theorem 3.5, what remains to be proved is that $\operatorname{Ind} A = \infty$ implies that also $\operatorname{Ind}(A + C) = \infty$ whenever $\|C\| < \varrho$ for a sufficiently small ϱ. We shall prove this for the case where $X = Y = H$ are Hilbert spaces.[1]) Thus, let $A \in \Phi_+(H)$. Since H is a Hilbert space, the subspace $\operatorname{Im} A$ has an orthogonal complement in H. The arguments preceding Theorem 3.1 imply that A possesses a left regularizer R. Now choose $\varrho = \frac{1}{2} \|R\|^{-1}$ and let $\|C\| < \varrho$. Exactly as in the proof of Theorem 3.5 we can derive the equality (5), and hence $A + C \in \Phi_+(H)$ (cf. Theorem 3.7). We now assume that $\beta(A + C)$ is finite and claim that then $\beta(A)$ must be finite, too.

Thus, $A + C \in \Phi(H)$. Recalling Theorem 3.1 and property 3 stated in Section 2.1 for regularizers we conclude that the operator $R_1 = (I + RC)^{-1} R$ (cf. (5)) is also a right regularizer of $A + C$, that is,

$$AR_1 = (I - CR_1) + T_1$$

with a compact operator T_1. Since $\|R_1\| < 2 \|R\|$, the operator $I - CR_1$ is invertible and therefore $R_1(I - CR_1)^{-1}$ is a right regularizer of A. From Theorem 2.1 we now deduce that $\beta(A) < \infty$.

Analogously it can be proved that $\operatorname{Ind}(A + C) = \infty$ if $A \in \Phi_-(H)$ and $\operatorname{Ind} A = \infty$. ∎

Theorem 3.9 is due to Gohberg and M. G. Krein [1].

3.3. If we have to do with (separable) Hilbert spaces, the trace can be defined not only for finite-dimensional operators, but also for the (essentially larger) class of nuclear operators. An operator $A \in \mathscr{L}(H)$ is said to be *nuclear* if the series

$$\sum_{k=1}^{\infty} (Ae_k, e_k) \tag{7}$$

converges for all orthonormal bases $e_1, e_2, \ldots$ of H. Note that a nuclear operator is necessarily compact, and also that for a nuclear operator A the sum of the series (7) does not depend on the choice of $e_1, e_2, \ldots$ and coincides with the sum of the eigenvalues of A (counted up to algebraic multiplicity). The sum (7) is referred to as the *trace* of the nuclear operator A and is denoted by $\operatorname{sp} A$. The trace of nuclear operators possesses the properties 1° and 2° stated in Section 3.1. By repeating the arguments of the proof of Theorem 3.2 one gets the following theorem, which is due to Fedosov [2].

Theorem 3.10. *Let $A \in \Phi(H_1, H_2)$. Suppose $R \in \mathscr{L}(H_2, H_1)$ is a regularizer of A such that both $RA - I_1$ and $AR - I_2$ are nuclear. Then*

$$\operatorname{Ind} A = \operatorname{sp}(I_1 - RA) - \operatorname{sp}(I_2 - AR). \tag{8}$$

Remark. For $H_1 = H_2 = H$ this simplifies to

$$\operatorname{Ind} A = \operatorname{sp}(AR - RA).$$

3.4. From Theorem 3.4 we can derive a further important property of the index, namely, its homotopic invariance.

---

[1]) A proof for the general case can be found, for instance, in Kato's book [1].

Definition. Two operators $A_0, A_1 \in \mathscr{L}(X, Y)$ are said to be *homotopic* (in the class $\Phi(X, Y)$) if there exists a continuous operator function $A: [0, 1] \to \mathscr{L}(X, Y)$ that possesses the following two properties:

(i) $A(t) \in \Phi(X, Y), \qquad \forall t \in [0, 1]$.

(ii) $A(0) = A_0, A(1) = A_1$.

Theorem 3.11. *Homotopic operators have equal index.*

Proof. It follows from Theorem 3.5 that for each fixed point $t \in [0, 1]$ there exists a number $\varrho_t > 0$ such that $\operatorname{Ind} B = \operatorname{Ind} A(t)$ whenever $B \in \mathscr{L}(X, Y)$ and $\|B - A(t)\| < \varrho_t$. The continuity of the operator function A implies that there is a number $\delta_t > 0$ such that $\|A(t') - A(t)\| < \varrho_t$ if only $|t' - t| < \delta_t$ and $t' \in [0, 1]$. Thus, for these t' we have $\operatorname{Ind} A(t') = \operatorname{Ind} A(t)$. Consequently, each point $t \in [0, 1]$ belongs to an open interval on which $\operatorname{Ind} A(t)$ is constant. By the Heine-Borel theorem, the closed interval $[0, 1]$ can be covered by finitely many of those open intervals, say $I_0, I_1, \ldots, I_k$, where $0 \in I_0$, $1 \in I_k$, and $I_j \cap I_{j+1} \neq \emptyset$ $(j = 0, 1, \ldots, k - 1)$. Then the index $\operatorname{Ind} A(t)$ is constant on $I_j \cap I_{j+1}$ $(j = 0, 1, \ldots, k - 1)$ and therefore it is constant on the whole interval $[0, 1]$. In particular, $\operatorname{Ind} A(0) = \operatorname{Ind} A(1)$. ∎

Analogously it can be shown that if two operators are homotopic in the class of one-sided invertible operators, then either both they are invertible or none of them is invertible.

An operator $A \in \mathscr{L}(X, Y)$ is said to be *left* (resp. *right*) *invertible* if there exists an operator $A^{(-1)} \in \mathscr{L}(Y, X)$ such that

$$A^{(-1)}Ax = x, \qquad \forall x \in X \qquad (AA^{(-1)}y = y, \forall y \in Y).$$

The operator $A^{(-1)}$ is then called a *left* (resp. *right*) *inverse* of A. If an operator A is both left and right invertible, then all left and right inverses are equal to each other and coincide with the inverse A^{-1}. In that case A is called *invertible* (or two-sided invertible).

It is well-known that the collection of all invertible operators forms an open subset of the space $\mathscr{L}(X, Y)$. The same can be said about the collection of all one-sided invertible operators.

Lemma 3.1. *If an operator $A \in \mathscr{L}(X, Y)$ is only left (resp. only right) invertible, then all operators $B \in \mathscr{L}(X, Y)$ with $\|B - A\| < \|A^{(-1)}\|^{-1}$ are only left (resp. only right) invertible and*

$$\dim \operatorname{Ker} B = \dim \operatorname{Ker} A, \dim \operatorname{Coker} B = \dim \operatorname{Coker} A.$$

Proof. To see this, represent B as

$$B = [I - (A - B)\, A^{(-1)}]\, A$$

if A is left invertible and as

$$B = A[I - A^{(-1)}(A - B)]$$

if A is right invertible, and note that the operators in square brackets are invertible. ∎

It can be shown analogously that the set of all one-sided invertible elements of a Banach algebra with identity element is open. This and the arguments of the proof of Theorem 3.11 without difficulty give the following lemma.

Lemma 3.2. *Let $\mathfrak{R}$ be a Banach algebra with identity element and let a be a continuous mapping of the interval $[0, 1]$ into the algebra $\mathfrak{R}$. If the element $a(t)$ is at least one-sided* *invertible for all* $t \in [0, 1]$ *and if it is (two-sided) invertible for at least one* $t_0 \in [0, 1]$, *then* $a(t)$ *is (two-sided) invertible for all* $t \in [0, 1]$.

Remark. A comprehensive treatment of the theory of (closed) Φ- and $\Phi_\pm$-operators can be found in the surveys of GOHBERG and M. G. KREIN [1], of GOHBERG, MARKUS, and FELDMAN [1], and of KRACHKOVSKI and DIKANSKI [1]. For various aspects of the theory and the application of Fredholm operators we also refer to the following books: BREUER [1], DOUGLAS [1], GOHBERG and KRUPNIK [4], GOLDBERG [1], KATO [1], S. G. KREIN [1], PALAIS [1], PRÖSSDORF [17], D. PRZEWORSKA-ROLEWICZ [1], D. PRZEWORSKA-ROLEWICZ and S. ROLEWICZ [1], SCHECHTER [1].

3.5. In applications one is frequently concerned with operator equations the right-hand side of which depends on certain parameters and whose solutions are subject to some additional conditions. This problem will be precisely formulated in a sufficiently general form in the present subsection. The material of this subsection is taken from the paper PRÖSSDORF and v. WOLFERSDORF [1].

Let X and Y be Banach spaces. Furthermore, suppose we are given fixed linearly independent elements $z_i \in Y$ $(i = 1, \dots, p)$ and fixed linearly independent functionals $f_j \in X^*$ $(j = 1, \dots, q)$. Finally, let $A \in \mathcal{L}(X, Y)$. We consider the equation

$$Ax = y + \sum_{i=1}^{p} c_i z_i \tag{9}$$

with the restrictions

$$(x, f_j) = 0 \qquad (j = 1, \dots, q). \tag{10}$$

Here $y \in Y$ is a given element and the c_i are scalar-valued parameters. An element $x \in X$ is said to be a *solution* of the problem (9), (10) if x satisfies (10) and if there exist c_i $(i = 1, \dots, p)$ that make x satisfy (9).

The problem (9), (10) is said to be *normally solvable* if the following orthogonality condition is sufficient for the problem to have a solution:

$$(y, g) = 0 \tag{11}$$

for every $g \in Y^*$ such that

$$\text{a)}\ \ gA \in \mathfrak{L}(\{f_1, \dots, f_q\})^{1)}, \qquad \text{b)}\ \ (z_i, g) = 0 \qquad (i = 1, \dots, p).$$

Note that this condition is obviously necessary for the problem (9), (10) to possess a solution.

Denote by $\tilde{\alpha}(A)$ the number of linearly independent solutions of the problem (9), (10) for $y = 0$, and let $\tilde{\beta}(A)$ denote the number of linearly independent solvability conditions (11). A normally solvable problem (9), (10) is called a *Φ_+-problem* (resp. a *Φ_--problem*) if $\tilde{\alpha}(A)$ (resp. $\tilde{\beta}(A)$) is finite. It is called a Φ-problem or a *Fredholm problem* if both $\tilde{\alpha}(A)$ and $\tilde{\beta}(A)$ are finite. In all those cases the difference $\tilde{\varkappa}(A) = \tilde{\alpha}(A) - \tilde{\beta}(A)$ is referred to as the *index* of the problem (9), (10).

Theorem 3.12. *The problem* (9), (10) *is a* Φ- *(resp.* $\Phi_\pm$-*)problem if and only if* A *is a* Φ- *(resp.* $\Phi_\pm$-*) operator. In that case*

$$\tilde{\varkappa}(A) = \operatorname{Ind} A + p - q \tag{12}$$

[1]) By $\mathfrak{L}(\{f_1, \dots, f_q\})$ we denote the subspace of X^* that has the basis $\{f_1, \dots, f_q\}$, that is, the subspace spanned by the elements $f_1, \dots, f_q$.

and

$$\alpha(A) - q \leqq \tilde{\alpha}(A) \leqq \alpha(A) + p, \tag{13}$$

$$\beta(A) - p \leqq \tilde{\beta}(A) \leqq \beta(A) + q. \tag{14}$$

Proof. Let $x_i \in X$ $(i = 1, \ldots, q)$ be linearly independent elements satisfying $(x_i, f_j) = \delta_{ij}$ $(i, j = 1, \ldots, q)$, define $X_0 = \mathfrak{L}(\{x_1, \ldots, x_q\})$, and put $X_1 = \bigcap_{j=1}^{q} \operatorname{Ker} f_j$. Then X decomposes into the direct sum $X = X_0 \dotplus X_1$ (see § 1). Let P_1 denote the projection of X onto X_1 parallel to X_0, and let A_1 be the restriction of the operator A to X_1. Obviously, $P_1 \in \Phi(X)$ and

$$\alpha(P_1) = \beta(P_1) = q. \tag{15}$$

We claim that A_1 is a $\Phi_\pm$-operator if and only if A is a $\Phi_\pm$-operator.

Indeed, the obvious equality

$$\operatorname{Im} A_1 = \operatorname{Im} AP_1 \tag{16}$$

implies that A_1 and AP_1 are simultaneously $\Phi_\pm$-operators, and since $I - P_1$ is a finite-dimensional projection (on X_0), we deduce from the equality $A = AP_1 + A(I - P_1)$ and Theorem 3.8 that AP_1 and A are simultaneously $\Phi_\pm$-operators, too.

Notice that (16) also implies

$$\beta(A_1) = \beta(AP_1), \qquad \alpha(A_1) = \alpha(AP_1) - q,$$

and making use of (15) and the Atkinson theorem, we get

$$\operatorname{Ind} A_1 = \operatorname{Ind} AP_1 - q = \operatorname{Ind} A - q. \tag{17}$$

It is obvious that the problem (9), (10) is equivalent to the equation

$$A_1 x = y + \sum_{i=1}^{p} c_i z_i \qquad (x \in X_1). \tag{18}$$

Therefore the problem (9), (10) is normally solvable if and only if (18) has a solution and if (11) is satisfied for every $g \in Y^*$ such that

$$\text{a}')\quad g \perp \operatorname{Im} A_1, \qquad \text{b})\quad (z_i, g) = 0 \qquad (i = 1, \ldots, p).$$

Let now Z be the subspace of Y spanned by $z_1, \ldots, z_p$ and put $Z_0 = Z \cap \operatorname{Im} A_1$. Without loss of generality assume that $\{z_1, \ldots, z_r\}$ $(1 \leqq r \leqq p)$ forms a basis of Z_0 (in case $Z_0 = \{0\}$ put $r = 0$). Then $Z = Z_0 \dotplus Z_1$ with Z_1 the $(p - r)$-dimensional subspace spanned by $z_{r+1}, \ldots z_p$. Thus, for $i = 1, \ldots, r$, $z_i = A_1 \tilde{x}_i$, where $\tilde{x}_i \in X_1$ are linearly independent elements that do not belong to $\operatorname{Ker} A_1$. Denote by M the subspace of X_1 spanned by the elements $\tilde{x}_1, \ldots, \tilde{x}_r$. Then $M \cap \operatorname{Ker} A_1 = \{0\}$ and, consequently, (18) is equivalent to the equation

$$A_1 \tilde{x} = y + z, \tag{19}$$

where

$$\tilde{x} = x - \sum_{i=1}^{r} c_i \tilde{x}_i, \qquad z = \sum_{i=r+1}^{p} c_i z_i.$$

Equation (19) has a solution if and only if $y \in \operatorname{Im} A_1 \dotplus Z =: L$.

On the other hand, for any $g \in Y^*$ satisfying a′) and b), the condition (11) is equivalent to the condition $y \in \bar{L}$ (see Kantorovich and Akilov [1], Theorem 4 (2.IV)). Thus, the problem (9), (10) is a $\Phi_\pm$-problem if and only if A_1 or, equivalently, A is a $\Phi_\pm$-operator.

In that case

$$\tilde{\alpha}(A) = \alpha(A_1) + r, \tag{20}$$

since x is a solution of the problem (9), (10) for $y = 0$ if and only if (19) is fulfilled with $\tilde{x} \in \operatorname{Ker} A_1$ and $z = 0$, i.e., if $x \in \operatorname{Ker} A_1 \dotplus M$.

Furthermore, it results that

$$\tilde{\beta}(A) = \dim Y/L = \dim Y/\operatorname{Im} A_1 - \dim Z_1 = \beta(A_1) - (p - r). \tag{21}$$

Now (17), (20), and (21) imply (12). Finally, it is easy to verify that

$$\alpha(A) \leqq \alpha(A_1) + q, \quad \beta(A) \geqq \beta(A_1) - q,$$

and this combined with (20) and (21) proves (13) and (14). ∎

The following facts result immediately from the just given proof.

Corollary 3.1. *Let $A \in \mathscr{L}(X, Y)$, let $X_1 \subset X$ be a closed subspace with* codim $X_1 = q < \infty$, *and put $A_1 = A \mid X_1$. Then either both A and A_1 are Φ-operators or none of them is a Φ-operator. In the first case,* Ind A = Ind $A_1 + q$.

Corollary 3.2. *Let $A \in \mathscr{L}(X, Y)$, $B \in \mathscr{L}(Z, Y)$, and suppose* $\dim Z = n < \infty$. *Define $C \in \mathscr{L}(X \times Z, Y)$ on $X \times Z$*[1]) *by*

$$C(x, z) = Ax + Bz \qquad (x \in X, z \in Z).$$

Then $C \in \Phi(X \times Z, Y)$ if and only if $A \in \Phi(X, Y)$; in this case Ind C = Ind $A + n$.

Remark. It follows from (13), (14) that the integers $\alpha(A)$ and $\tilde{\alpha}(A)$ (resp. $\beta(A)$ and $\tilde{\beta}(A)$) are either both finite or both infinite. If one of them is infinite, then $\tilde{\varkappa}(A) = \operatorname{Ind} A = \infty$ and so (12) holds obviously.

Finally, we are interested in conditions ensuring that the problem (9), (10) is uniquely solvable for each $y \in Y$, i.e., that $\tilde{\alpha}(A) = \tilde{\beta}(A) = 0$. From (20) we deduce immediately that $\tilde{\alpha}(A) = 0$ if and only if $\alpha(A_1) = r = 0$, i.e., if and only if the following two conditions are satisfied:

1. $z_i \notin \operatorname{Im} A_1 = \operatorname{Im} AP_1$ $(i = 1, \ldots, p)$;
2. $Ax = 0$ and (10) imply that $x = 0$.

Furthermore, (21) shows that $\tilde{\beta}(A) = 0$ if and only if

$$\beta(A_1) = \beta(AP_1) = p - r.$$

From (13) we see that $\alpha(A) \leqq q$ is necessary for $\tilde{\alpha}(A)$ to be 0 and, analogously, (14) implies that $\beta(A) \leqq p$ is a necessary condition for $\tilde{\beta}(A) = 0$.

Let us still consider two special cases:

(a) $\alpha(A) = 0, q = 0$.

Then: $\tilde{\alpha}(A) = 0 \Leftrightarrow z_i \notin \operatorname{Im} A$ $(i = 1, \ldots, p)$;

$$\tilde{\varkappa}(A) = 0 \Leftrightarrow \beta(A) = p.$$

(b) $\beta(A) = 0, p = 0$.

In this case: $\tilde{\alpha}(A) = 0 \Leftrightarrow$ condition 2 is fulfilled;

$$\tilde{\varkappa}(A) = 0 \Leftrightarrow \alpha(A) = q.$$

---

[1]) The cartesian product $X \times Z$ can be endowed e.g. with the norm $\|(x, z)\| = \|x\|_X + \|z\|_Z$.

3.6. Being concerned with concrete operator equations one has sometimes to distinguish between spaces of real- and complex-valued functions and, accordingly, between the operators on those spaces. In this connection consider the following general situation.

Let X be a Banach space over the field $\mathbb{C}$ of complex numbers, suppose $X_{\mathbb{R}}$ is a closed subspace of X which itself is a Banach space over the field $\mathbb{R}$ of real numbers, and assume

$$X = X_{\mathbb{R}} + iX_{\mathbb{R}}. \tag{22}$$

The last in particular involves that, for $x = x_1 + ix_2$ ($x_1, x_2 \in X_{\mathbb{R}}$),

$$c_1[\|x_1\|_X + \|x_2\|_X] \leqq \|x\|_X \leqq c_2[\|x_1\|_X + \|x_2\|_X] \tag{23}$$

with some positive constants c_1, c_2. Analogously, let $Y = Y_{\mathbb{R}} + iY_{\mathbb{R}}$.

If an operator $A \in \mathscr{L}(X, Y)$ has the property $A(X_{\mathbb{R}}) \subset Y_{\mathbb{R}}$, then its restriction $A_{\mathbb{R}} = A \mid X_{\mathbb{R}}$ is obviously in $\mathscr{L}(X_{\mathbb{R}}, Y_{\mathbb{R}})$. Vice versa, if $A_{\mathbb{R}} : X_{\mathbb{R}} \to Y_{\mathbb{R}}$ is an $\mathbb{R}$-linear bounded operator, then through

$$A(x_1 + ix_2) = A_{\mathbb{R}}x_1 + iA_{\mathbb{R}}x_2, \qquad x_1, x_2 \in X_{\mathbb{R}} \tag{24}$$

a $\mathbb{C}$-linear bounded operator $A \in \mathscr{L}(X, Y)$ is defined.

Theorem 3.13. *Let $A_{\mathbb{R}} \in \mathscr{L}(X_{\mathbb{R}}, Y_{\mathbb{R}})$ and define $A \in \mathscr{L}(X, Y)$ by* (24). *Then:*

(i) *A is normally solvable $\Leftrightarrow$ $A_{\mathbb{R}}$ is normally solvable.*

(ii) *$\alpha(A) < \infty \Leftrightarrow \alpha(A_{\mathbb{R}}) < \infty$. If the nullities are finite, then $\alpha(A) = \alpha(A_{\mathbb{R}})$.*

(iii) *$\beta(A) < \infty \Leftrightarrow \beta(A_{\mathbb{R}}) < \infty$. If the deficiencies are finite, then $\beta(A) = \beta(A_{\mathbb{R}})$.*

Proof. (i) If A is normally solvable, then using (23) and (24) it is easy to see that $\operatorname{Im} A_{\mathbb{R}}$ is closed in $Y_{\mathbb{R}}$, and thus $A_{\mathbb{R}}$ is normally solvable. Conversely, if $A_{\mathbb{R}}$ is normally solvable, then $\operatorname{Im} A$ is closed in Y as the sum of the disjoint closed sets $\operatorname{Im} A_{\mathbb{R}}$ and $i \operatorname{Im} A_{\mathbb{R}}$ (cf. (22), (23)) and, hence, A is normally solvable.

(ii) Obviously, $x = x_1 + ix_2 \in \operatorname{Ker} A$ if and only if $x_1, x_2 \in \operatorname{Ker} A_{\mathbb{R}}$. Let $m = \alpha(A_R) < \infty$ and let $\{e_1, \ldots, e_m\}$ be a basis in $\operatorname{Ker} A_{\mathbb{R}}$. Choose an arbitrary $x = x_1 + ix_2 \in \operatorname{Ker} A$. Then

$$x_1 = \sum_{j=1}^{m} a_j e_j, \qquad x_2 = \sum_{j=1}^{m} b_j e_j \qquad (a_j, b_j \in \mathbb{R}),$$

consequently, $x = \sum_{j=1}^{m} c_j e_j$ with $c_j = a_j + ib_j$ and hence $\operatorname{Ker} A$ is spanned by $\{e_1, \ldots, e_m\}$. But, if $0 = \sum_{j=1}^{m} d_j e_j = \sum_{j=1}^{m} \operatorname{Re} d_j e_j + i \sum_{j=1}^{m} \operatorname{Im} d_j e_j$, then necessarily $\operatorname{Re} d_j = \operatorname{Im} d_j = 0$ for $j = 1, \ldots, m$. Thus, $\{e_1, \ldots, e_m\}$ forms a basis in $\operatorname{Ker} A$ and $\alpha(A) = \alpha(A_{\mathbb{R}})$.

(iii) The same arguments as in (i) can be applied to deduce that $f = f_1 + if_2 \in \operatorname{Im} A$ if and only if $f_1, f_2 \in \operatorname{Im} A_{\mathbb{R}}$. Hence $f \in \overline{\operatorname{Im} A}$ if and only if $f_1, f_2 \in \overline{\operatorname{Im} A_{\mathbb{R}}}$. Put $\hat{Y} = Y/\overline{\operatorname{Im} A}$ and $\hat{Y}_{\mathbb{R}} = Y_{\mathbb{R}}/\overline{\operatorname{Im} A_{\mathbb{R}}}$. Then, for any two elements $f, g \in Y$, the cosets $\hat{f}$ and $\hat{g}$ coincide iff $\hat{f_1} = \hat{g}_1$ and $\hat{f_2} = \hat{g}_2$ (here $\hat{f_1}, \hat{f_2}, \hat{g}_1, \hat{g}_2 \in \hat{Y}_{\mathbb{R}}$). Thus, to each coset $\hat{f}$ in Y there corresponds a pair of cosets in $\hat{Y}_{\mathbb{R}}$. So we have for $\hat{Y}$ and $\hat{Y}_{\mathbb{R}}$ the same situation as in (ii) for $\operatorname{Ker} A$ and $\operatorname{Ker} A_{\mathbb{R}}$, and the proof can now be completed analogously to (ii). ∎

Remark. Theorem 3.13 generalizes some results of Muskhelishvili ([1], § 54) about real integral equations (see also Kopp [1], Theorem 16).

§ 4. Fredholm and semi-Fredholm operators on linear topological spaces

The results of the last two sections can, of course in modified form, be carried over to locally convex linear topological spaces. This is important e.g. for the study of singular integral equations in distribution spaces. We first recall some definitions.

4.1. A *linear topological space* is a linear space X over the field $\mathbb{C}$ of complex (or the field $\mathbb{R}$ of real) numbers which is endowed with a topology such that the mappings $(x, y) \mapsto x + y$ of $X \times X$ into X and $(\lambda, x) \mapsto \lambda x$ of $\mathbb{C} \times X$ into X are continuous. A set $E \subset X$ is said to be *convex*, if $x \in E$ and $y \in E$ always implies that $\mu x + \nu y \in E$ for all non-negative numbers μ, ν with $\mu + \nu = 1$. A subset E is said to be *bounded* if to every neighborhood $U \subset X$ of 0 corresponds a number $\lambda > 0$ such that $\lambda E \subset U$. Finally, a linear Hausdorff topological space is said to be *locally convex* if it possesses a local base $\mathfrak{U}$ consisting of convex neighborhhoods of 0.

Given a locally convex space X, one has many possibilities to define a locally convex topology in the conjugate (or dual) space X^* of all continuous linear functionals on X. Among them the most important topologies are the strong and the weak topology. A strong neighborhood of the zero functional $0 \in X^*$ is any set of functionals $f \in X^*$ with the property

$$\sup_{x \in E} |(x, f)| < \varepsilon, \tag{1}$$

where E is any bounded subset of X and $\varepsilon > 0$ any real number. The weak neighborhoods of $0 \in X^*$ are obtained when the E in (1) runs through the finite subsets of X. In what follows we shall always assume that X^* is equipped with one of these two topologies. Locally convex spaces have the remarkable property that the Hahn-Banach theorem about the extension of continuous linear functionals remains true for them.

If X and Y are locally convex spaces, $\mathcal{L}(X, Y)$ will denote the collection of all continuous linear operators from X into Y. An operator $T \in \mathcal{L}(X, Y)$ is said to be *compact* (or *completely continuous*), if for a neighborhood $U \subset X$ of zero the image set $T(U)$ is relatively compact in Y (i.e., the closure $\overline{T(U)}$ is compact in Y). An operator $A: X \to Y$ is called a *homomorphism* if for each open set $U \subset D(A)$ ($D(A)$ equipped with the topology inherited from X) the image set $A(U)$ is open in $\operatorname{Im} A$ (with respect to the topology in $\operatorname{Im} A$ inherited from Y). Thus a one-to-one operator A has a continuous inverse A^{-1} (from $\operatorname{Im} A$ onto $D(A)$) if and only if it is a homormophism.

4.2. Locally convex spaces which are metrizable and complete are of particular importance in applications to analysis. These are called *Fréchet spaces*. They are characterized by the property that their topology can be generated by a countable system of seminorms such that

$$\|x\|_1 \leqq \|x\|_2 \leqq \cdots \leqq \|x\|_n \leqq \cdots \tag{2}$$

Therefore system (2) is also referred to as the *generating system of seminorms* of the Fréchet space at hand. Given such a system, a local base of convex neighborhoods of 0 can be defined by

$$U_{n,\varepsilon} = \{x \in X: \|x\|_n < \varepsilon\} \qquad (n = 1, 2, \ldots; \varepsilon > 0).$$

A *seminorm* on a linear space X is a real-valued function $\|x\|$ with the following properties: 1. $\|x + y\| \leqq \|x\| + \|y\|$ and 2. $\|\lambda x\| = |\lambda| \, \|x\|$ for all $x, y \in X$ and $\lambda \in \mathbb{C}$.

It is obvious that $\|x\| \geqq 0$ for all x and that $\|0\| = 0$. A seminorm is called a *norm* if $\|x\| = 0$ implies $x = 0$.

A linear space on which a norm is defined is called *normed* or *normable*. Note that every Banach space is a normed space and thus also a Fréchet space.

An example of a Fréchet space which is not normable is the space $\mathbf{C}^\infty(\mathbb{R})$ of all infinitely differentiable functions f on the real axis with the generating system of seminorms

$$\|f\|_n = \sum_{k=0}^{n} \max_{-n \leqq t \leqq n} |f^{(k)}(t)| \qquad (n = 0, 1, 2, \ldots).$$

If X and Y are Fréchet spaces, then a linear operator $A : X \to Y$ with $D(A) = X$ is continuous (i.e., $A \in \mathscr{L}(X, Y)$) iff it is bounded, that is, if to each positive integer n there correspond a positive integer m_n and a constant $C_n > 0$ such that

$$\|Ax\|_n \leqq C_n \|x\|_{m_n}, \qquad \forall x \in X.$$

Furthermore, if X and Y are Fréchet spaces, if $A \in \mathscr{L}(X, Y)$, and if $\operatorname{Im} A = Y$, then A is a homomorphism (Banach's theorem).

4.3. In applications one is frequently concerned with Fréchet spaces whose generating system of seminorms (2) actually consists of norms, which, moreover, are *pairwise coordinated*. The latter means that each sequence $\{x_n\} \subset X$ which is a Cauchy sequence in the p-th norm and converges to zero in the $(p - 1)$st norm also converges to zero in the p-th norm.

Let X_p $(p = 1, 2, \ldots)$ denote the Banach space obtained from the linear space X by completion with respect to the p-th norm. It is easy to see that

$$X_1 \supset X_2 \supset \ldots \supset X_p \supset \ldots \supset X,$$

where the inclusions are continuous embeddings. One can show that the Fréchet space X whose topology is given by a countable system (2) of pairwise coordinated norms is complete if and only if

$$X = \bigcap_{p=1}^{\infty} X_p \tag{3}$$

(cf. Gelfand and Shilov [2], p. 14). If the condition (3) is satisfied, X is called a *countably normed space.*[1])

An example of a countably normed space is the space $\mathbf{C}^\infty[a, b]$ of all infinitely differentiable functions f on the closed interval $[a, b]$ with the system of norms

$$\|f\|_n = \sum_{k=0}^{n} \max_{a \leqq t \leqq b} |f^{(k)}(t)| \qquad (n = 0, 1, 2, \ldots).$$

For countably normed spaces one has the following simple criterion for the compactness of an operator.

Lemma 4.1. *Let X be a countably normed space and suppose T is a linear operator defined on the whole space X_1 with values in X. Then, if the operator $T : X_1 \to X_p$ is compact for each $p = 1, 2, \ldots$, the restriction $T \mid X$ is compact on X.*

Proof. It suffices to show that $T(U) \subset X$ is relatively compact for every neighborhood of zero of the form $U = \{x \in X : \|x\|_1 < \varepsilon\}$.

1) Countably normed spaces were introduced by Gelfand and Shilov (cf. [2]) for the first time, who took them as the basis of a general theory of distributions.

Since U is a bounded subset of X_1 and since by our hypotheses T is compact on X_1, the image set $T(U)$ contains a sequence $\{y_{1n}\}_{n=1}^{\infty}$ which is a Cauchy sequence in X_1. This sequence in turn contains a subsequence which is a Cauchy sequence with respect to the norm of X_2. Repeating this argument we obtain a family of sequences $\{y_{mn}\}_{m,n=1}^{\infty}$ the m-th sequence of which is a Cauchy sequence in X_m. Thus, the diagonal sequence $\{y_{mm}\}$ is a Cauchy sequence with respect to each of the norms $\|x\|_p$ ($p = 1, 2, \ldots$) and this means that it is a Cauchy sequence in X. But since X is complete, the sequence $\{y_{mm}\}$ must converge, which shows that $T(U)$ is relatively compact. ∎

4.4. All definitions given in the first three sections can be literally carried over to the case of Fréchet spaces.

Lemma 2.1. admits the following generalization.

Lemma 4.2. *Let X, Y, Z be Fréchet spaces. Then for a closed operator $A: X \to Y$ to be normally solvable and to have a finite nullity it is necessary and sufficient that there exists a compact operator $T: X \to Z$ with the following property: to every n there correspond a number $m = m_n$ ($n, m = 1, 2, \ldots$) and a constant $C_n > 0$ such that*

$$\|x\|_n \leq C_n(\|Ax\|_m + \|Tx\|_m), \qquad \forall x \in D(A).$$

The proof is completely analogous to that of Lemma 2.1 (cf. Prössdorf [17], pp. 17–18).

Repeating the arguments of the Sections 2 and 3 and taking into consideration Lemma 4.2 it is easy to see that the results of these sections, with the exception of the Theorems 3.5, 3.9, 3.10, 3.11, and the Lemma 3.1, remain true for Fréchet spaces. The point is that a continuous linear operator which maps a Fréchet space one-to-one onto another Fréchet space always possesses a continuous inverse (Banach's theorem). Notice that in the case of arbitrary locally convex spaces only homomorphisms have the latter property.

4.5. The preceding remark also implies that in the case of arbitrary locally convex spaces results analogous to those of the Sections 2 and 3 can only be derived if the operator A is not only required to be normally solvable (i.e., to have a closed image space $\operatorname{Im} A$), but also to be a homomorphism. In particular, a normally solvable continuous operator A is called a Φ_+ (resp. Φ_-)-operator, if it is a homomorphism and if the number $\alpha(A)$ (resp. $\beta(A)$) is finite. An operator is called a *Φ-operator*, if it is both a Φ_+- and a Φ_--operator.

After having adopted this definition the results of the Section 3 can be carried over to the case of arbitrary locally convex spaces (of course, with the exception of the Theorems 3.5, 3.9, 3.10, 3.11, and the Lemma 3.1, whose formulation makes sense only for normed spaces).

Remark. In connection with the results of the Sections 4.1–4.4 we refer to Schaefer [1, 2], Pietsch [1–3], and Prössdorf ([17], Chapter 1 and the references listed there).

4.6. We finally record some facts from the theory of distributions.

In what follows suppose X is a Fréchet space which is continuously and densely embedded into a Banach space E. Put $H = E^*$.

Every element $\varphi \in H$ can be viewed as a functional on X. Indeed, for all $x \in X$ we have

$$|(x, \varphi)| \leq \|\varphi\|_H \|x\|_E, \tag{4}$$

and if $x_n \to 0$ (in the metric of X), then $\|x_n\|_E \to 0$ and therefore $(x_n, \varphi) \to 0$. Furthermore, since X is dense in E, the element $\varphi \in H$ is uniquely determined by its values on X. Consequently, $H \subset X^*$.

Lemma 4.3. *Let $\{\varphi_n\}_{n=1}^{\infty} \subset H$ and suppose $\|\varphi_n\|_H \to 0$. Then $\varphi_n \to 0$ in the strong topology of the space X^*.*

Proof. Recall that convergence in the strong topology of X means uniform convergence on every bounded set $G \subset X$.

Thus, let $G \subset X$ be an arbitrary bounded set. It is easy to see that then G is a bounded subset of the Banach space E, too. Hence, by virtue of (4)

$$|(x, \varphi_n)| \leqq C \|\varphi_n\|_H \qquad \forall x \in G,$$

which immediately implies that (x, φ_n) converges to zero uniformly with respect to $x \in G$. ∎

In analogy to the theory of distributions we call X a *test function space* and the functionals $f \in X^*$ *generalized functions* or *distributions* on the test function space X. To the distributions $\varphi \in H$ we refer to as *regular functionals* (cf. Gelfand and Shilov [2]).

If E is a Hilbert space, then $E = H = E^*$ and therefore

$$X \subset H \subset X^*. \tag{5}$$

4.7. Now let $A \in \mathscr{L}(H)$ be a continuous linear operator on the space H. Suppose there is an operator $A' \in \mathscr{L}(E)$ such that $A = (A')^*$. Then obviously

$$(x, Af) = (A'x, f) \qquad (x \in X) \tag{6}$$

for all $f \in H$.

Further, assume that $A'(X) \subset X$ and that the restriction A'_X of A' to X is continuous on X, that is $A'_X \in \mathscr{L}(X)$. Under this hypothesis the operator A can be defined for all $f \in X^*$ via the formula (6). Clearly, this extension of the operator A to the whole space X^* ($\supset H$) coincides with $(A'_X)^*$ and it is therefore a continuous linear operator on the distribution space X^* (with respect to both the strong and the weak topology of X^*).

The collection of all operators A with the above properties will henceforth be denoted by $\Pi(X; H)$.

Definition. Let $A \in \Pi(X; H)$ and $f \in X^*$. An element $u \in X^*$ is called a *generalized solution* of the equation

$$Au = f \tag{7}$$

if[1])

$$(A'x, u) = (x, f) \qquad \forall x \in X.$$

If $f \in H$ and if the element u which satisfies (7) also belongs to H, then u will be called a *classical solution* of this equation.

Note that because of (5) in the case of a Hilbert space $H = E$ the right-hand side of (7) may, in particular, be taken from X.

Let now $A \in \Pi(X; H)$. Then, obviously, if equation (7) with $f \in X^*$ possesses a generalized solution, we have

$$(x, f) = 0 \qquad \forall x \in \operatorname{Ker} A'_X. \tag{8}$$

[1]) Note that this definition is an accordance with the definition of the generalized (or weak) solution of boundary value problems for partial differential equations of elliptic type.

In accordance with the definition given in Section 1 we call the operator A (or (7)) *normally solvable in generalized functions*, if condition (8) is also sufficient for (7) to have a solution $u \in X^*$. The operator A will be called a *Fredholm operator on the distribution space* X^*, if it is normally solvable in generalized functions and if the following two numbers are finite:

$$\dim \operatorname{Ker}_* A = \beta(A'_X), \qquad \dim \operatorname{Coker}_* A = \alpha(A'_X).$$

The difference of these two numbers,

$$\operatorname{Ind}^* A = -\operatorname{Ind} A'_X, \tag{9}$$

is referred to as the *index* of the operator A on the distribution space X^*.

If $A'_X \in \Phi(X)$, a generalized solution of (7) can be easily constructed in the following way. Let Q be the projection of X onto $\operatorname{Im} A'_X$ and let A_1 be the restriction of the operator A'_X to a direct complement of $\operatorname{Ker} A'_X$ in X (cf. § 1). Then, if the functional $f \in X^*$ satisfies (8), $u_0 = (A_1^{-1} Q)^* f$ is a generalized solution of the equation (7).

To see this, note that by virtue of (8)

$$(A'x, u_0) = (A_1^{-1}QA'x, f) = (A_1^{-1}A'x, f) = (x, f)$$

for all $x \in X$.

Combining this with the Banach-Hausdorff theorem we get the following result.

Theorem 4.1. *An operator $A \in \Pi(X; H)$ is Fredholm on the distribution space X^* if and only if $A'_X \in \Phi(X)$.*

§ 5. The symbol

5.1. Let $\mathfrak{R}$ be a ring of linear operators acting on a linear space X and assume we are given a homomorphism of the ring $\mathfrak{R}$ onto a certain other ring $\mathfrak{r}$. Thus, to each element from $\mathfrak{R}$ corresponds exactly one element of the ring $\mathfrak{r}$ and to each element of $\mathfrak{r}$ corresponds at least one element of $\mathfrak{R}$ such that, if a and b in $\mathfrak{r}$ correspond to the operators A and B from $\mathfrak{R}$, then the elements $a + b$ and ab correspond to the operators $A + B$ and AB, respectively. In this situation we call $\mathfrak{r}$ a *symbol ring* of the ring $\mathfrak{R}$. If $a \in \mathfrak{r}$ corresponds to the operator $A \in \mathfrak{R}$, then a is called the *symbol* of the operator A and we write

$$a = \operatorname{Smb} A \tag{1}$$

in this case.

The symbol ring is not determined uniquely: if $\mathfrak{r}_1$ is any ring such that $\mathfrak{r}$ can be mapped onto $\mathfrak{r}_1$ homomorphically, then $\mathfrak{r}_1$ can be taken as a symbol ring of $\mathfrak{R}$, too. In general, the class of all operators having the same symbol will then become larger, but this class remains the same as it was if the mapping of $\mathfrak{r}$ onto $\mathfrak{r}_1$ is an isomorphism.

There are two trivial cases which will be excluded from our further considerations:

a) $\mathfrak{r} = \mathfrak{R}$ and the above mentioned homomorphism is the identity mapping. In this case the symbol of an arbitrary operator coincides with the operator itself.

b) $\mathfrak{r} = \{0\}$. Then the symbol of an arbitrary operator is the zero element.

It is obvious that to each ideal (resp. maximal ideal) in the ring of operators corresponds an ideal (resp. maximal ideal) in the symbol ring and vice versa. In particular, the collection of all operators in $\mathfrak{R}$ whose symbol is zero forms an ideal in the ring $\mathfrak{R}$. We call this ideal the *zero ideal* of the ring $\mathfrak{R}$. Clearly, if the correspondence between operators and symbols is one-to-one, then the zero ideal is the trivial one, that is, it contains the zero element of the ring $\mathfrak{R}$ only.

5.2. Now suppose that $\mathfrak{R}$ contains the identity operator I. We denote its symbol by $\boldsymbol{i}$. If a is an arbitrary element of the symbol ring and if A is any operator with the symbol a, then the identities $A = AI = IA$ imply that $a = ai = ia$ and, consequently, i is the identity element of the symbol ring. Thus, the symbol of the identity operator is always the identity element of the symbol ring.

If an element a in the symbol ring is not invertible and if $A \in \mathfrak{R}$ is any operator the symbol of which is a, then A cannot be invertible. Indeed, suppose there is an operator $B \in \mathfrak{R}$ such that $AB = BA = I$ and let b denote the symbol of B. Then $ab = ba = i$, and this is a contradiction to our assumption that a is not invertible.

Now let $a \in \mathfrak{r}$ be an invertible element and let A and R be operators from $\mathfrak{R}$ whose symbols are a and a^{-1}, respectively. The identities $aa^{-1} = a^{-1}a = i$ imply that then $AR = I + T$ and $RA = I + T_1$ with certain operators T and T_1 from the zero ideal in $\mathfrak{R}$.

We mention a special case of particular importance. Let X be a Banach space, $\mathfrak{R}$ a ring of bounded operators on X, and assume the zero ideal of $\mathfrak{R}$ contains only compact operators. Then, if $A \in \mathfrak{R}$ is an operator with invertible symbol a and if R is any operator with the symbol a^{-1}, R is a two-sided regularizer of A (cf. § 2).

5.3. Let us consider some examples.

1. Put $X = \mathbf{C}^\infty[a, b]$, where a and b are finite or infinite real numbers and let $\mathfrak{R}$ be the ring of linear differential operators of the form

$$\sum_{k=0}^{n} a_k \frac{\mathrm{d}^k}{\mathrm{d}x^k} \tag{2}$$

with constant coefficients a_k and with n a positive integer. The coefficients a_k are not necessarily required to be scalars but are also allowed to be matrices. As the ring $\mathfrak{r}$ one can take the ring of polynomials in a new variable ξ with the ring operations being the usual addition and multiplication for polynomials. Then the symbol of the operator (2) can be defined as its characteristic polynomial

$$\sum_{k=1}^{n} a_k \xi^k . \tag{3}$$

In this case the zero ideal is trivial.

2. On the same set $\mathbf{C}^\infty[a, b]$ consider the larger ring of operators of the form

$$\sum_{k=0}^{n} a_k(x) \frac{\mathrm{d}^k}{\mathrm{d}x^k} \tag{4}$$

with $a_k \in \mathbf{C}^\infty[a, b]$. As the symbol ring we choose the set of polynomials in ξ whose coefficients depend on x, with the addition given in the usual way, but with the multiplication defined by

$$\sum_{k=0}^{n} a_k(x)\, \xi^k \sum_{l=0}^{m} b_l(x)\, \xi^l = \sum_{k=0}^{n+m} c_k(x)\, \xi^k , \tag{5}$$

where

$$c_k(x) = \sum_{p=0}^{n} \sum_{r=0}^{p} \binom{p}{r} a_p(x)\, b_{k-r}^{(p-r)}(x) \tag{6}$$

with $b_{k-r}(x) = 0$ if $k - r > m$ or $k - r < 0$. The symbol of the operator (4) is defined as the polynomial

$$\sum_{k=1}^{n} a_k(x)\, \xi_k. \tag{7}$$

Note that the multiplication given through (5) and (6) is associative and distributive, but not commutative. As in example 1, the zero ideal is trivial.

In an analogous fashion symbols can be constructed for linear partial differential operators. Clearly, this will lead to the consideration of polynomials in several variables.

3. Let $\mathfrak{R}$ be the ring of operators of the form

$$(Au)(x) = a(x)\, u(x) + (Tu)(x) \tag{8}$$

considered on a certain Banach space X of functions defined on a certain measurable subset G of the m-dimensional Euclidean space $\mathbb{R}^m$. Here $a(x)$ runs through a certain ring $\mathfrak{a}$ of functions defined on G and generating bounded multiplication operators on X, and T runs through the compact operators on X. It is obvious that the operators (8) form a ring. As a symbol ring one can take the ring $\mathfrak{a}$ itself and as the symbol of the operator (8) one can then take the function $a(x)$. Thus, in the case under consideration the symbol of a compact operator is always zero and the zero ideal consists of the compact operators on X.

5.4. Let $\mathfrak{R}$ be a commutative ring of bounded operators on a certain Banach space X and suppose $\mathfrak{R}$ is a closed subset of $\mathscr{L}(X)$. Then $\mathfrak{R}$ is a commutative normed ring[1]) (see Gelfand, Raikov, and Shilov [1]). Denote by $\mathfrak{M}$ the set of all maximal ideals of $\mathfrak{R}$ and for $M \in \mathfrak{M}$ let π_M be the canonical projection of $\mathfrak{R}$ onto $\mathfrak{R}/\mathfrak{M}$ and ϱ_M the canonical isomorphism of $\mathfrak{R}/\mathfrak{M}$ onto the complex field. Then the collection of all complex-valued functions on $\mathfrak{M}$ which are of the form

$$\varphi_A(M) := (\varrho_M \circ \pi_M)(A) \qquad (M \in \mathfrak{M}) \tag{9}$$

with some $A \in \mathfrak{R}$ can serve as a symbol ring of $\mathfrak{R}$. Then the function φ_A defined by (9) is just the symbol of the operator $A \in \mathfrak{R}$.

A symbol ring can also be constructed in the more general case where $\mathfrak{R}$ is a non-commutative ring which, however, possesses a (non-trivial) ideal $\mathfrak{N}$ such that for arbitrary $A_1, A_2 \in \mathfrak{R}$ the commutator $[A_1, A_2] = A_1A_2 - A_2A_1$ is in $\mathfrak{N}$. Indeed, in that case the quotient ring $\mathfrak{R}/\mathfrak{N}$ is commutative and so its symbol ring and the corresponding symbol map can be defined as above. Thus, the symbols are complex-valued functions on the maximal ideal space of $\mathfrak{R}/\mathfrak{N}$ and the symbol of an operator from $\mathfrak{N}$ is zero. An example of such a ring is the ring of all operators of the form (8) with $\mathfrak{a}$ the ring of measurable and essentially bounded functions on G and $X = \mathbf{L}_p(G)$; here $\mathfrak{N}$ is the ideal of compact operators on X.

5.5. The concept of the symbol was in 1936 introduced by Mikhlin for the first time. In his works [4, 5] it was defined for singular integral operators on the two-dimensional plane and in [6] for singular integral operators on a sufficiently smooth two-dimensional manifold. Also in 1936, Giraud [1] published a formula for the symbol of the singular integral operator on an Euclidean space of arbitrary dimension. Giraud's formula was a generalization of Mikhlin's formula for the two-dimensional case. Giraud has never

[1]) *Normed rings* are more currently called *Banach algebras*. In the following chapters we shall define the symbol of operators which belong to a (in general, non-commutative) Banach algebra (over the complex field). In this context the symbol ring is also referred to as the *symbol algebra*.

published a proof of his formula. A proof was published by MIKHLIN [8] in 1955, some years after Giraud's death, for the first time. In MIKHLIN's article [9] the symbol of the singular integral on a sufficiently smooth manifold of arbitrary dimension was defined, and in his paper [7] it was observed that the symbol can in a natural fashion also be defined for the one-dimensional singular integral operator.

The concept of the symbol was extended to pseudodifferential operators in the work of KOHN and NIRENBERG [1] and HÖRMANDER [2], and to pseudomultiplication operators, which also involve the multidimensional Wiener-Hopf integral operators, by PRÖSSDORF [18]. The general definition of the symbol given in the present paragraph is a natural generalization of the definitions contained in the works cited above.

§ 6. The symbol of the convolution operator

6.1. In this section we consider an important class of operator rings the symbol rings of which exist and can be constructed in a relatively simple way. This class of rings is of great importance for what follows.

We denote by $\mathscr{S}$ the countably normed space of all infinitely differentiable functions which are defined on the Euclidean space $\mathbb{R}^m$ and which as well as all their derivatives converge to zero as $|x| \to \infty$ more rapidly than any power of $|x|^{-1}$. By $\mathscr{S}'$ we denote the space of continuous linear functionals on $\mathscr{S}$; these are called *tempered distributions* (see L. SCHWARTZ [1]). We consider the class $\mathfrak{F}$ of convolution operators over $\mathbb{R}^m$ which are of the form

$$(Ku)(x) = (K * u)(x) = \int_{\mathbb{R}^m} K(x - y)\, u(y)\, \mathrm{d}y, \tag{1}$$

where K and u belong to a certain subclasses of $\mathscr{S}'$. We shall now give an example of such a subclass. The Fourier transform, defined for a function $u \in \mathscr{S}$ by the formula

$$(Fu)(\xi) = \hat{u}(\xi) = \int_{\mathbb{R}^m} \mathrm{e}^{-2\pi \mathrm{i}(\xi, x)}\, u(x)\, \mathrm{d}x$$

is a continuous operator on $\mathscr{S}$ and its inverse operator (on $\mathscr{S}$) is given by

$$(F^{-1}v)(x) = \int_{\mathbb{R}^m} \mathrm{e}^{2\pi \mathrm{i}(\xi, x)}\, v(\xi)\, \mathrm{d}\xi .$$

For a distribution $u \in \mathscr{S}'$ we define Fu and $F^{-1} u$ by

$$(\varphi, Fu) = (F\varphi, u), \qquad (\varphi \in \mathscr{S}),$$
$$(\varphi, F^{-1} u) = (F^{-1} \varphi, u) \qquad (\varphi \in \mathscr{S}).$$

Then Fu, $F^{-1} u \in \mathscr{S}'$ for every $u \in \mathscr{S}'$.

We denote by Φ the collection of all distributions $u \in \mathscr{S}'$ whose Fourier transform is an ordinary function which does not increase at infinity more rapidly than a polynomial. Given arbitrary distributions K and u from Φ, the convolution (1) can be defined by the formula

$$(\varphi(x), K * u) = ((\varphi(x + y), u_y), K_x) \qquad (\varphi \in \mathscr{S}).$$

Then $K * u \in \mathscr{S}'$ and

$$F(K * u) = FKFu, \tag{2}$$
$$K * u = F^{-1}(FKFu). \tag{3}$$

Indeed, for sufficiently large n we have

$$w(\xi) := \hat{u}(\xi)/(1 + |\xi|^2)^n \in \mathbf{L}_2 = \mathbf{L}_2(\mathbb{R}^m).$$

Consequently,

$$u(x) = \left(1 - \frac{\Delta}{4\pi^2}\right)^n v(x) \qquad (v \in \mathbf{L}_2), \qquad \hat{v}(\xi) = w(\xi),$$

where Δ denotes the Laplace operator. Let now $\{v_j\}$ be a sequence of infinitely differentiable functions with compact support such that $v_j \xrightarrow{\mathbf{L}_2} v$ as $j \to \infty$. Put

$$u_j(x) = \left(1 - \frac{\Delta}{4\pi^2}\right)^n v_j(x).$$

In the same way construct a sequence $\{K_j\}$ for the distribution $K \in \Phi$. Then it is easily seen that $u_j \xrightarrow{\mathcal{S}'} u$, $K_j \xrightarrow{\mathcal{S}'} K$, and $K_j * u_j \xrightarrow{\mathcal{S}'} K * u$. From the continuity of the operator F on $\mathcal{S}'$ we conclude that $F(K_j * u_j) \xrightarrow{\mathcal{S}'} F(K * u)$. Taking into account the relation

$$F\left(\frac{\partial}{\partial x_k} u(x)\right) = 2\pi \mathrm{i} \xi_k \hat{u}(\xi)$$

we readily deduce that $FK_j \to FK$ and $Fu_j \to Fu$ in the norm of $\mathbf{L}_2(\Omega)$, where $\Omega \subset \mathbb{R}^m$ is an arbitrary bounded region. On passing to the limit $j \to \infty$ in the well-known equality

$$F(K_j * u_j) = FK_j Fu_j$$

we obtain formula (2).

It follows immediately from formula (2) that for $K \in \Phi$ the operator (1) (or, equivalently, the operator (3)) maps the set Φ into itself. Thus, the collection $\mathfrak{F}$ of convolution operators forms a commutative ring and the symbol of an operator from this ring can be defined as the function $\hat{K}(\xi)$, the Fourier transform of its kernel. Note that this construction also applies if u is a vector function and K is a square matrix of corresponding dimension. Then, however, the ring of operators is not commutative.

From formula (2) and the Plancherel theorem we deduce that the operator (1) is bounded on the space $\mathbf{L}_2(\mathbb{R}^m)$ if and only if $\hat{K}(\xi)$ is bounded.

6.2. Example. Consider the operator

$$(Au)(x) = au(x) + \frac{b}{\pi \mathrm{i}} \int_{-\infty}^{\infty} \frac{u(y)}{y - x} \mathrm{d}y, \qquad -\infty < x < \infty, \tag{4}$$

where a and b are constants. The integral in (4) is understood in the sense of the Cauchy principal value:

$$\int_{-\infty}^{\infty} \frac{u(y)}{y - x} \mathrm{d}y = \lim_{\varepsilon \to 0} \int_{|y-x| > \varepsilon} \frac{u(y)}{y - x} \mathrm{d}y;$$

more about this will be said in § 1 of Chapter II. The operator (4) can be written as a convolution, namely,

$$(Au)(x) = \int_{-\infty}^{+\infty} \left[a\delta(x - y) - \frac{b}{\pi \mathrm{i}(x - y)}\right] u(y)\, \mathrm{d}y, \tag{5}$$

where δ denotes the Dirac δ-function.

The Fourier transform of the kernel of the integral (5) is

$$\left(F\left(a\delta(x)-\frac{b}{\pi i x}\right)\right)(\xi)=a(F\delta(x))(\xi)-\frac{b}{\pi i}\left(F\frac{1}{x}\right)(\xi)$$

and we have

$$(F\delta(x))(\xi)=\int_{-\infty}^{\infty}\delta(x)\,e^{-2\pi i x\xi}\,dx=1;$$

$$\left(F\frac{1}{x}\right)(\xi)=\int_{-\infty}^{\infty}\frac{e^{-2\pi i x\xi}}{x}\,dx=\lim_{\varepsilon\to 0}\left[\int_{-\infty}^{-\varepsilon}\frac{e^{-2\pi i x\xi}}{x}\,dx+\int_{\varepsilon}^{\infty}\frac{e^{-2\pi i x\xi}}{x}\,dx\right]$$

$$=\lim_{\varepsilon\to 0}\int_{\varepsilon}^{\infty}\frac{e^{-2\pi i x\xi}-e^{2\pi i x\xi}}{x}\,dx=-2i\int_{0}^{\infty}\frac{\sin 2\pi x\xi}{x}\,dx=-\pi i\ \mathrm{sign}\,\xi.$$

Thus, as the symbol of the operator (4) we can take the function

$$a+b\ \mathrm{sign}\,\xi, \tag{6}$$

where ξ is a real variable that can assume all values with the exception of zero.

The symbol (6) can be simplified. Put $\mathrm{sign}\,\xi=\theta$. Thus θ only assumes the values $\theta=+1$ and $\theta=-1$. So the function of the variable θ given by

$$\Phi_A(\theta)=a+b\theta;\qquad \theta=\pm 1 \tag{7}$$

can be viewed as a symbol of the operator (4).

Note that through (6) and (7) a symbol can also be defined for the case where $u(x)$ is a vector function and a and b are constant square matrices.

The correspondence between the operators (4) and the symbols (7) is one-to-one. This combined with the facts stated in § 5 has the following consequences.

1. The symbol of the operator S defined by

$$(Su)(x)=\frac{1}{\pi i}\int_{-\infty}^{\infty}\frac{u(y)}{y-x}\,dy$$

is θ. Since then $\mathrm{Smb}\,S^2=\theta^2=1$ and since in the case at hand to each symbol corresponds exactly one operator, we get

$$S^2=I, \tag{8}$$

or, written down in full,

$$-\frac{1}{\pi^2}\int_{-\infty}^{\infty}\frac{dt}{t-x}\int_{-\infty}^{\infty}\frac{u(y)}{y-t}\,dy=u(x). \tag{9}$$

(9) is a special case of the well-known Poincaré-Bertrand commutation formula (see below, § 4, Chapter II).

2. The ring of the symbols (7) has two maximal ideals: the sets of functions

$$a\cdot(1-\theta)\ \text{and}\ a\cdot(1+\theta)\qquad (a=\text{const.}), \tag{10}$$

respectively. Hence, the ring of the operators (7) also possesses two maximal ideals: the set of all operators of the form $a(I-S)$ on the one hand and the set of all operators $a(I+S)$ on the other hand. Here a denotes a constant and I the identity operator.

3. Every element of the symbol ring which does not belong to the ideals (10) is invertible. Indeed, if $a \neq \pm b$, then

$$(a + b\theta)\left(\frac{a}{a^2 - b^2} - \frac{b\theta}{a^2 - b^2}\right) = \frac{a^2 - b^2\theta^2}{a^2 - b^2} = 1.$$

Consequently, if $a \neq \pm b$, then the operator A given by (4) is invertible and we have

$$(A^{-1}u)(x) = \frac{a}{a^2 - b^2} u(x) - \frac{b}{(a^2 - b^2)\pi \mathrm{i}} \int_{-\infty}^{\infty} \frac{u(y)}{y - x} \mathrm{d}y. \tag{11}$$

Chapter II
The one-dimensional singular integral

§ 1. The singular integral and its simplest properties

1.1. Definition 1. 1. An oriented Jordan curve Γ in the complex plane $\mathbb{C}$ is called a *Lyapunov curve* if it satisfies a *Lyapunov condition*, i.e., if at each point $t \in \Gamma$ there exists a tangent to the curve and if the angle $\theta_\Gamma(t)$ between the tangent and the x-axis, measured from the latter counter-clockwise, satisfies a Hölder condition:

$$|\theta_\Gamma(t_1) - \theta_\Gamma(t_2)| \leqq c\,|t_1 - t_2|^\alpha.$$

Here t_1 and t_2 are any points on Γ, c is a positive constant, and $0 < \alpha < 1$. Notice that a Lyapunov curve is always simple (i.e., without points of self-intersection) and rectifiable.

2. A collection of finitely many closed or open Lyapunov curves which have no common points is referred to as a *Lyapunov curve system.*

3. A collection of finitely many open Lyapunov curves $\Gamma_1, \Gamma_2, \ldots, \Gamma_n$ is called a *piecewise Lyapunov curve system* (and is simply denoted by Γ) if the curves $\Gamma_1, \Gamma_2, \ldots, \Gamma_n$ have only finitely many common points and if, in addition, the following condition is fulfilled: if Γ_j and Γ_k have a point t_0 in common then either $\Gamma_j \cup \Gamma_k$ is a Lyapunov curve or the tangents to Γ_j and Γ_k at t_0 do not coincide. Clearly, the orientations of the curves Γ_j define an orientation of the curve system Γ.

The points which are end points of one or of several curves Γ_j ($j = 1, \ldots, n$) are called the *nodes* of the curve system Γ. All other points are referred to as *regular points*. Finally, a node which is end point of only one curve Γ_j is called an *end point of the curve system* Γ, while a node which is end point of at least two curves Γ_j is called a *corner point.*

Note that every Lyapunov curve system is obviously a piecewise Lyapunov curve system.

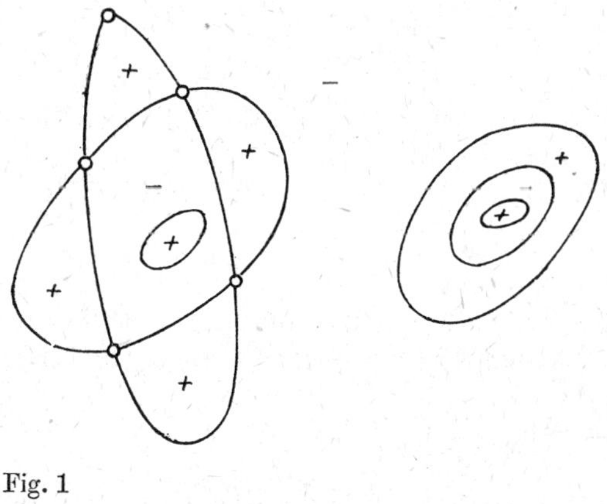

Fig. 1

4. A piecewise Lyapunov curve system Γ is said to be a *closed curve system* if it divides the closed complex plane into two non-empty open sets D_Γ^+ and D_Γ^- such that Γ is the boundary of either of these sets. The positive direction on a closed curve system Γ will be chosen so that, if Γ is traced out in this direction, D_Γ^+ is always on the left. For the sake of simplicity we shall henceforth assume that D_Γ^+ contains the origin $z = 0$ and that the point at infinity, $z = \infty$, belongs to D_Γ^- (see Fig. 1; the regions designated by "$\pm$" form $D_\Gamma^\pm$).

In the case where Γ is constituted by only one closed curve the latter agreements mean that Γ is oriented counter-clockwise.

In applications one is frequently concerned with the case where D_Γ^+ is a connected bounded region; then one of the curves Γ_j encloses the others, while the latter curves do not enclose one another (see Fig. 2).

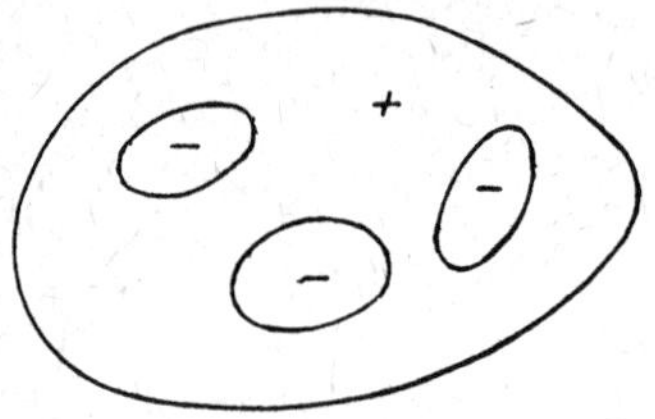

Fig. 2

1.2. Let Γ and Γ' be piecewise Lyapunov curve systems. As usual, we denote by $\mathbf{C}(\Gamma)$ the Banach algebra of all (complex-valued) continuous functions φ on Γ with the norm

$$\|\varphi\|_{\mathbf{C}(\Gamma)} = \max_{t\in\Gamma} |\varphi(t)|. \tag{1}$$

$\mathbf{H}^\mu(\Gamma)$ (or $\mathbf{Lip}_\mu(\Gamma)$), where $0 < \mu \leqq 1$, denotes the collection of all functions φ defined on Γ and there satisfying a *Hölder condition* with exponent μ, that is, for all $t_1, t_2 \in \Gamma$

$$|\varphi(t_1) - \varphi(t_2)| \leqq A\,|t_1 - t_2|^\mu$$

with some constant $A > 0$ (depending on φ). $\mathbf{H}^{\mu,\nu}(\Gamma\times\Gamma')$, where $0 < \mu, \nu \leqq 1$, is defined as the collection of all functions of two variables, $\varphi(t, \tau)$, which are given on $\Gamma\times\Gamma'$ and satisfy the Hölder condition

$$|\varphi(t_1, \tau_1) - \varphi(t_2, \tau_2)| \leqq A[|t_1 - t_2|^\mu + |\tau_1 - \tau_2|^\nu]$$

for all pairs $(t_j, \tau_j) \in \Gamma\times\Gamma'$ $(j = 1, 2)$.

To a function $\varphi \in \mathbf{H}^\mu(\Gamma)$ we assign the norm

$$\|\varphi\|_{\mathbf{H}^\mu(\Gamma)} = \|\varphi\|_{\mathbf{C}(\Gamma)} + \sup_{t_1,t_2\in\Gamma} \frac{|\varphi(t_1) - \varphi(t_2)|}{|t_1 - t_2|^\mu}. \tag{2}$$

It is well known that $\mathbf{H}^\mu(\Gamma)$ provided with the norm (2) forms a Banach algebra.

Finally, we denote by $\mathbf{R}(\Gamma)$ the set of all rational functions without poles on Γ. Note that $\mathbf{R}(\Gamma)$ is a dense subset of $\mathbf{C}(\Gamma)$. For a closed curve system Γ we let $\mathbf{R}^\pm(\Gamma)$ denote the set of all functions from $\mathbf{R}(\Gamma)$ with poles outside $D_\Gamma^\pm$.

1.3. Definition 2. Let Γ be a piecewise Lyapunov curve system and φ a function defined and integrable on Γ. For a sufficiently small positive number ε and for a fixed point $t \in \Gamma$, put $\Gamma_\varepsilon = \{\tau \in \Gamma : |\tau - t| \geqq \varepsilon\}$. The Cauchy principal value of the integral

$$(S_\Gamma\varphi)(t) := \frac{1}{\pi i} \int_\Gamma \frac{\varphi(\tau)\,d\tau}{\tau - t} = \lim_{\varepsilon\to 0} \frac{1}{\pi i} \int_{\Gamma_\varepsilon} \frac{\varphi(\tau)\,d\tau}{\tau - t} \qquad (t \in \Gamma), \tag{3}$$

if it exists, is called the *singular integral* (or the *Cauchy singular integral*) of the function φ taken along the curve system Γ. This value will henceforth be denoted by $(S_\Gamma \varphi)(t)$.

The function φ is usually referred to as the *density* of the singular integral and the expression $\frac{d\tau}{\tau - t}$ as the *Cauchy kernel*.

In what follows we shall see that the singular integral exists for a sufficiently large class of functions φ.

Lemma 1.1. *Let Γ be a piecewise Lyapunov curve system and suppose $\varphi \in \mathbf{H}^\mu(\Gamma)$ $(0 < \mu \leqq 1)$. Then the singular integral* (3) *exists at each regular point $t \in \Gamma$.*

Proof. Obviously, it suffices to prove the assertion for the case where Γ is an open Lyapunov curve and where t is not an end point of Γ. Then

$$\int_{\Gamma_\varepsilon} \frac{\varphi(\tau)\, d\tau}{\tau - t} = \int_{\Gamma_\varepsilon} \frac{\varphi(\tau) - \varphi(t)}{\tau - t}\, d\tau + \varphi(t) \int_{\Gamma_\varepsilon} \frac{d\tau}{\tau - t}.$$

Since

$$\left| \frac{\varphi(\tau) - \varphi(t)}{\tau - t} \right| \leqq A\, |\tau - t|^{\mu - 1},$$

the limit of the first integral on the right-hand side exists and is equal to the improper integral

$$\int_\Gamma \frac{\varphi(\tau) - \varphi(t)}{\tau - t}\, d\tau.$$

Thus let us consider the second integral. To this end, cut the complex plane along a line joining the points $\tau = t$ and $\tau = \infty$ and lying on the right of the curve Γ. Then each (fixed) branch of the function $\ln(\tau - t)$ in the cut plane is a primitive of $(\tau - t)^{-1}$. Choose ε small enough, so that the circle $|\tau - t| = \varepsilon$ intersects the curve Γ at exactly two points t_1 and t_2 (see Fig. 3). Then, with a and b the end points of Γ,

$$\int_{\Gamma_\varepsilon} \frac{d\tau}{\tau - t} = \ln(\tau - t)\Big|_{\tau=a}^{\tau=t_1} + \ln(\tau - t)\Big|_{\tau=t_2}^{\tau=b} = \ln \frac{b - t}{a - t} + \ln \frac{t_1 - t}{t_2 - t}.$$

Because $|t_1 - t| = |t_2 - t| = \varepsilon$, we get

$$\ln \frac{t_1 - t}{t_2 - t} = i\,[\arg(t_1 - t) - \arg(t_2 - t)] \xrightarrow[\varepsilon \to 0]{} i\pi.$$

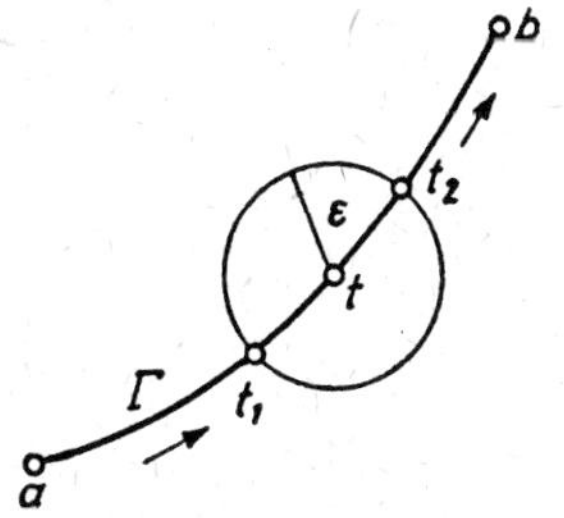

Fig. 3

This proves the existence of the limit (3). Moreover, we have actually shown that

$$\int_\Gamma \frac{\varphi(\tau)\,d\tau}{\tau - t} = \int_\Gamma \frac{\varphi(\tau) - \varphi(t)}{\tau - t}\,d\tau + \varphi(t)\left[\ln\frac{b-t}{a-t} + \pi i\right]. \blacksquare \tag{4}$$

Remark. The preceding arguments give rise to the following slightly more general definition of the singular integral. Choose points t_1, t_2 on Γ in a neighborhood of the regular point $t \in \Gamma$ so that the arc $(t_1, t_2) \subset \Gamma$ with the end points t_1 and t_2 contains t in its interior and that, in addition,

$$\lim_{\substack{t_1 \to t \\ t_2 \to t}} \left|\frac{t_1 - t}{t_2 - t}\right| = 1. \tag{*}$$

Then, under the hypotheses of Lemma 1.1,

$$\int_\Gamma \frac{\varphi(\tau)\,d\tau}{\tau - t} = \lim_{\substack{t_1 \to t \\ t_2 \to t}} \int_{\Gamma \setminus (t_1,t_2)} \frac{\varphi(\tau)\,d\tau}{\tau - t}.$$

Corollary 1.1. *Let Γ be a closed piecewise Lyapunov curve system and $t \in \Gamma$ an arbitrary regular point. Then*

$$\int_\Gamma \frac{\varphi(\tau)\,d\tau}{\tau - t} = \int_\Gamma \frac{\varphi(\tau) - \varphi(t)}{\tau - t}\,d\tau + \pi i\varphi(t). \tag{5}$$

In particular,

$$\int_\Gamma \frac{d\tau}{\tau - t} = \pi i. \tag{6}$$

Theorem 1.1. *Let Γ be a closed piecewise Lyapunov curve system, $r_+ \in \mathbf{R}^+(\Gamma)$, $r_- \in \mathbf{R}^-(\Gamma)$, and $r_-(\infty) = 0$. Then, for an arbitrary regular point $t \in \Gamma$,*

$$(S_\Gamma r_+)(t) = r_+(t), \qquad (S_\Gamma r_-)(t) = -r_-(t). \tag{7}$$

Proof. Since $(r_+(\tau) - r_+(t))(\tau - t)^{-1}$ is an analytic function of τ in D_Γ^+, the first equality in (7) immediately results from (5) by the Cauchy integral theorem.

Let now $\varphi(t) = (t - \alpha)^{-m}$, where α belongs to D_Γ^+ and m is a positive integer. Then

$$\frac{\varphi(\tau) - \varphi(t)}{\tau - t} = -\sum_{k=0}^{m-1} (t - \alpha)^{k-m}(\tau - \alpha)^{-k-1},$$

and thus the Cauchy residue theorem gives

$$\frac{1}{2\pi i}\int_\Gamma \frac{\varphi(\tau) - \varphi(t)}{\tau - t}\,d\tau = -(t - \alpha)^{-m} = -\varphi(t).$$

This combined with (5) implies that $(S_\Gamma\varphi)(t) = -\varphi(t)$. The second equality in (7) now follows from the fact that every function $r_- \in \mathbf{R}^-(\Gamma)$ with $r_-(\infty) = 0$ is representable in the form

$$r_-(t) = \sum_{k=1}^{n} \frac{A_k}{(t - \alpha_k)^{m_k}} \qquad (\alpha_k \in D_\Gamma^+, A_k = \text{const.}). \blacksquare$$

Corollary 1.2. *Let Γ be a closed piecewise Lyapunov curve system and $r \in \mathbf{R}(\Gamma)$. Then, for every regular point $t \in \Gamma$,*

$$(S_\Gamma^2 r)(t) = r(t), \tag{8}$$

where $(S_\Gamma^2 r)(t) = (S_\Gamma(S_\Gamma r))(t)$.

Proof. It is easy to see that every function $r \in \mathbf{R}(\Gamma)$ has a unique representation as a sum $r = r_+ + r_-$ with $r_\pm \in \mathbf{R}^\pm(\Gamma)$ and $r_-(\infty) = 0$. Consequently, on applying (7) we obtain

$$(S_\Gamma r)(t) = r_+(t) - r_-(t), \qquad (S_\Gamma^2 r)(t) = r_+(t) + r_-(t) = r(t). \blacksquare$$

Theorem 1.2. *Let Γ and Γ' be Lyapunov curves and $\tau = \alpha(\zeta)$ a function which maps Γ' one-to-one onto Γ. Furthermore, suppose the derivative $\alpha'(\zeta)$ exists, satisfies a Hölder condition on Γ', and vanishes nowhere on Γ'. If $\varphi \in \mathbf{H}^\mu(\Gamma)$ $(0 < \mu \leq 1)$, then*

$$\int_\Gamma \frac{\varphi(\tau)}{\tau - t}\,d\tau = \int_{\Gamma'} \frac{\varphi[\alpha(\zeta)]\,\alpha'(\zeta)}{\alpha(\zeta) - \alpha(\xi)}\,d\zeta, \tag{9}$$

where $t = \alpha(\xi)$.

Proof. By Lemma 1.1 and the definition of the singular integral we have

$$\int_{\Gamma'} \frac{\varphi[\alpha(\zeta)]\,\alpha'(\zeta)}{\alpha(\zeta) - \alpha(\xi)}\,d\zeta = \lim_{\varepsilon\to 0} \int_{\Gamma'_\varepsilon} \frac{\varphi[\alpha(\zeta)]\,\alpha'(\zeta)}{\alpha(\zeta) - \alpha(\xi)}\,d\zeta, \tag{10}$$

where $\Gamma'_\varepsilon = \{\zeta \in \Gamma' : |\zeta - \xi| \geqq \varepsilon\}$. Upon the substitution $\tau = \alpha(\zeta)$ in the latter integral, one sees that the right-hand side of (10) is equal to the limit

$$\lim_{\substack{t_1\to t\\ t_2\to t}} \int_{\Gamma\setminus(t_1,t_2)} \frac{\varphi(\tau)}{\tau - t}\,d\tau,$$

where $t_j = \alpha(\xi_j)$ $(j = 1, 2)$ and ξ_1, ξ_2 are the points of intersection of the curve Γ' and the circle $|\zeta - \xi| = \varepsilon$. A few application of Taylor's formula shows that (*) holds. This and the remark made before Corollary 1.1 give the assertion. $\blacksquare$

Remark. Theorem 1.1 was first proved by Sokhotski [1] and later again by Plemelj [1] (in this connection see also Muskhelishvili [1], p. 52). Theorem 1.2 is due to Mikhlin [7].

§ 2. The boundedness of the singular integral operator on the space $\mathbf{L}_p(\Gamma)$

For Γ a piecewise Lyapunov curve system, let $\mathbf{L}_p(\Gamma)$ $(1 \leqq p < \infty)$ be the Banach space of all measurable functions φ on Γ which are absolutely integrable in the p-th power. The norm in $\mathbf{L}_p(\Gamma)$ is given by

$$\|\varphi\|_{\mathbf{L}_p(\Gamma)} = \Big(\int_\Gamma |\varphi(t)|^p\,|dt|\Big)^{1/p},$$

where $|dt|$ is the differential of the arc measure. By $\mathbf{L}_\infty(\Gamma)$ we denote the set of all measurable and essentially bounded functions on Γ. We define

$$\|\varphi\|_{\mathbf{L}_\infty} = \operatorname*{vrai\,max}_{t\in\Gamma} |\varphi(t)|.$$

If there is no fear of misunderstandings, we simply write $\|\varphi\|_p$ in place of $\|\varphi\|_{\mathbf{L}_p(\Gamma)}$.

In this section we extend the singular integral operator S_Γ to a linear bounded operator defined on the whole space $\mathbf{L}_p(\Gamma)$ for every p such that $1 < p < \infty$. This will be done using the theorem about the continuous extension of an operator and the following interpolation theorem, due to M. Riesz (see, e.g., Dunford and Schwartz [1] or Zygmund [3], Vol. II).

Theorem 2.1. *If a linear operator A is bounded on the spaces $\mathbf{L}_{p_0}(\Gamma)$ and $\mathbf{L}_{p_1}(\Gamma)$ $(1 \leqq p_0 < p_1 < \infty)$ and if $p_0 < p < p_1$, then A is also bounded on $\mathbf{L}_p(\Gamma)$. Moreover, for the norms of the operator A one has the estimate*

$$\| A \|_p \leqq \| A \|_{p_0}^{1-t} \| A \|_{p_1}^{t},$$

where $1/p = (1-t)/p_0 + t/p_1$ $(0 < t < 1)$.

2.1. We first suppose that Γ is a closed Lyapunov curve. In the preceding section, we showed that the singular integral operator S_Γ is defined on the set $\mathbf{R}(\Gamma)$, which is dense in $\mathbf{L}_p(\Gamma)$, and that it maps this set onto itself (Corollary 1.2). We now prove the following theorem.

Theorem 2.2. *If Γ is a closed Lyapunov curve, then the singular integral operator S_Γ is bounded on the spaces $\mathbf{L}_p(\Gamma)$ for every p such that $1 < p < \infty$.*
The proof of this theorem will be based on the following lemma.

Lemma 2.1. *Let Γ_0 be the unit circle $|t| = 1$ and suppose $1 < p < \infty$. Then the operator $S_0 = S_{\Gamma_0}$ is bounded on the space $\mathbf{L}_p(\Gamma)$. Moreover, one has the following estimates*

$$\| S_0 \|_p \leqq \begin{cases} \cot \dfrac{\pi}{2p} & \text{for} \quad p = 2^n \qquad (n = 1, 2, \ldots), \\[2ex] \tan \dfrac{\pi}{2p} & \text{for} \quad p = \dfrac{2^n}{2^n - 1} \quad (n = 1, 2, \ldots). \end{cases} \tag{1}$$

Proof. Let $|t| = 1$ and $m = 0, \pm 1, \ldots$ By Theorem 1.1, $S_0 t^m = t^m$ for $m \geqq 0$ and $S_0 t^m = -t^m$ for $m < 0$. But the system of functions $\{t^m\}_{m=-\infty}^{\infty}$ forms an orthogonal basis in the Hilbert space $\mathbf{L}_2(\Gamma_0)$, and so the operator S_0 defined on the linear hull of this basis is bounded on $\mathbf{L}_2(\Gamma_0)$ and $\| S_0 \|_2 = 1$.

Let now $\varphi(t) = \sum_{k=-N}^{N} a_k t^k$ be an arbitrary trigonometric polynomial. Put $\varphi_+(t) = \sum_{k=0}^{N} a_k t^k$ and $\varphi_-(t) = \sum_{k=-N}^{-1} a_k t^k$. Then $\varphi = \varphi_+ + \varphi_-$, $S_0 \varphi = \varphi_+ - \varphi_-$, and thus Theorem 1.1 gives that

$$\varphi^2 + (S_0 \varphi)^2 = 2(\varphi_+^2 + \varphi_-^2) = 2 S_0(\varphi_+^2 - \varphi_-^2) = 2 S_0(\varphi S_0 \varphi).$$

This implies that

$$\| (S_0 \varphi)^2 \|_p \leqq \| 2 S_0 (\varphi S_0 \varphi) \|_p + \| \varphi^2 \|_p,$$

and taking into account the relations

$$\| \varphi^2 \|_p = \| \varphi \|_{2p}^2, \qquad \| \varphi \psi \|_p \leqq \| \varphi \|_{2p} \| \psi \|_{2p}$$

we obtain that

$$\| S_0 \varphi \|_{2p}^2 \leqq 2 \| S_0 \|_p \| \varphi \|_{2p} \| S_0 \varphi \|_{2p} + \| \varphi \|_{2p}^2.$$

Thus, if S_0 is bounded on the space $\mathbf{L}_p(\Gamma_0)$, then

$$\| S_0 \varphi \|_{2p} \leqq (\| S_0 \|_p + \sqrt{1 + \| S_0 \|_p^2}) \| \varphi \|_{2p}$$

and, consequently, S_0 is bounded on the space $\mathbf{L}_{2p}(\Gamma_0)$ and one has

$$\| S_0 \|_{2p} \leqq \| S_0 \|_p + \sqrt{1 + \| S_0 \|_p^2}.$$

From the last estimate we deduce that S_0 is bounded on all the spaces $\mathbf{L}_{2^n}(\Gamma_0)$ $(n = 1, 2, \ldots)$ and that $\| S_0 \|_{2^{n+1}} \leqq \| S_0 \|_{2^n} + \sqrt{1 + \| S_0 \|_{2^n}^2}$. Using this, it is an easy matter to verify the first inequality in (1). Finally, Theorem 2.1 now gives the boundedness of the operator S_0 on the space $\mathbf{L}_p(\Gamma_0)$ for all p such that $2 \leqq p < \infty$.

Thus we are left with the case $1 < p < 2$. Then $q = p(p-1)^{-1} > 2$. Given two trigonometric polynomials

$$\varphi(t) = \sum_{k=-N}^{N} a_k t^k, \qquad \psi(t) = \sum_{k=-N}^{N} b_k t^k,$$

we have the equality

$$\int_{\Gamma_0} \varphi(t) \overline{(S_0\psi)(t)} \, |dt| = \sum_{k=-N}^{N} \varepsilon_k a_k \bar{b}_k = \int_{\Gamma_0} (S_0\varphi)(t) \overline{\psi(t)} \, |dt|, \tag{2}$$

where $\varepsilon_k = 2\pi$ if $k \geqq 0$ and $\varepsilon_k = -2\pi$ if $k < 0$. The identity (2) shows that the operator $S_0 \in \mathscr{L}(\mathbf{L}_q(\Gamma_0))$ and its adjoint operator S_0^* takes the same values on the dense subset $\mathbf{R}(\Gamma_0)$ of $\mathbf{L}_p(\Gamma_0)$. Consequently, the operator S_0 is bounded on the space $\mathbf{L}_p(\Gamma_0)$ if $1 < p < 2$. Furthermore, for $p = 2^n/(2^n - 1)$ the first inequality in (1) implies that

$$\|S_0\|_p = \|S_0\|_q \leqq \cot \frac{\pi}{2q} = \tan \frac{\pi}{2p}$$

and this completes the proof. ∎

Proof of Theorem 2.2. Since Γ is a closed Lyapunov curve, there exists a one-to-one and differentiable function $\tau = \alpha(\zeta)$ mapping the unit circle Γ_0 onto Γ such that its derivative satisfies a Hölder condition and does not vanish on Γ_0. Put

$$k(\xi, \zeta) = \frac{1}{\pi i} \left(\frac{\alpha'(\zeta)}{\alpha(\zeta) - \alpha(\xi)} - \frac{1}{\zeta - \xi} \right) \qquad (\xi, \zeta \in \Gamma_0).$$

It is easily seen that

$$|k(\xi, \zeta)| \leqq \frac{c}{|\xi - \zeta|^\lambda}, \qquad 0 \leqq \lambda < 1, \tag{3}$$

with some positive constant c (see, e.g., Muskhelishvili [1], § 7). Therefore, the operator K defined on the space $\mathbf{L}_p(\Gamma_0)$ $(1 < p < \infty)$ by

$$(K\psi)(\xi) = \int_{\Gamma_0} k(\xi, \zeta) \, \psi(\zeta) \, d\zeta$$

is an integral operator with weak singularity and, hence, compact (see, e.g., Kantorovich and Akilov [1]).

Now denote by A the linear bounded operator from $\mathbf{L}_p(\Gamma)$ onto $\mathbf{L}_p(\Gamma_0)$ given by $(A\varphi)(\zeta) = \varphi(\alpha(\zeta))$. Clearly, A is one-to-one. Thus, by Theorem 1.2, we have $S_\Gamma\varphi = A^{-1}(K + S_0) A\varphi$ for every $\varphi \in \mathbf{R}(\Gamma)$. But S_0 is bounded on $\mathbf{L}_p(\Gamma_0)$ by virtue of Lemma 2.1, and thus S_Γ is so on $\mathbf{L}_p(\Gamma)$. ∎

Corollary 2.1. *Let Γ be a closed Lyapunov curve and let $\varphi \in \mathbf{L}_p(\Gamma)$ $(1 < p < \infty)$. Then the limit*

$$\lim_{\varepsilon \to 0} \frac{1}{\pi i} \int_{\Gamma_\varepsilon} \frac{\varphi(\tau) \, d\tau}{\tau - t} \tag{4}$$

exists for almost all $t \in \Gamma$ and is equal to $(S_\Gamma\varphi)(t)$. Here $\Gamma_\varepsilon = \{\tau \in \Gamma: |\tau - t| \geqq \varepsilon\}$.

Proof. Choose $r_n \in \mathbf{R}(\Gamma)$ $(n = 1, 2, \ldots)$ so that $\|r_n - \varphi\|_p \to 0$ as $n \to \infty$. Then $\|S_\Gamma r_n - S_\Gamma\varphi\|_p \to 0$, by Theorem 2.2. Decompose r_n into a sum $r_n = r_n^+ + r_n^-$ with

$r_n^+ \in \mathbf{R}^+(\Gamma)$, $r_n^- \in \mathbf{R}^-(\Gamma)$, and $r_n^-(\infty) = 0$, and, furthermore, put $\varphi^+ = P_\Gamma\varphi$ and $\varphi^- = Q_\Gamma\varphi$, where $P_\Gamma = \frac{1}{2}(I + S_\Gamma)$, $Q_\Gamma = \frac{1}{2}(I - S_\Gamma)$.

By virtue of Theorem 1.1, $P_\Gamma r_n = r_n^+$, $Q_\Gamma r_n = r_n^-$, and hence $\|r_n^\pm - \varphi^\pm\|_p \to 0$. Thus, the almost obvious relations

$$\int_\Gamma r_n^+(t)\, t^k\, \mathrm{d}t = 0 \quad (k = 0, 1, \ldots), \qquad \int_\Gamma r_n^-(t)\, t^{-k}\, \mathrm{d}t = 0 \quad (k = 1, 2, \ldots)$$

imply that

$$\int_\Gamma \varphi^+(t)\, t^k\, \mathrm{d}t = 0 \quad (k = 0, 1, \ldots), \qquad \int_\Gamma \varphi^-(t)\, t^{-k}\, \mathrm{d}t = 0 \quad (k = 1, 2, \ldots). \tag{5}$$

From (5) we deduce that for $\varphi = \varphi^+$ (resp. $\varphi = \varphi^-$) the limit (4) exists at almost all points $t \in \Gamma$ and equals $\varphi^+(t)$ (resp. $-\varphi^-(t)$) (cf. Privalov [2], p. 136 and p. 145). To complete the proof it remains to notice that $\varphi = \varphi^+ + \varphi^-$ and $S_\Gamma\varphi = \varphi^+ - \varphi^-$. ∎

Corollary 2.2. *Let Γ be an open Lyapunov curve. Then the singular integral operator S_Γ is bounded on the space $\mathbf{L}_p(\Gamma)$ $(1 < p < \infty)$.*

Proof. Let $\tilde{\Gamma}$ be any closed Lyapunov curve a portion of which is formed by Γ. Then, by Theorem 2.2, the operator $S_{\tilde{\Gamma}}$ is bounded on the space $\mathbf{L}_p(\tilde{\Gamma})$ $(1 < p < \infty)$. Thus, with $\chi(t)$ $(t \in \tilde{\Gamma})$ the characteristic function of the set $\tilde{\Gamma}$, we conclude that

$$\|S_\Gamma r\|_{\mathbf{L}_p(\Gamma)} = \|\chi S_{\tilde{\Gamma}} \chi r\|_{\mathbf{L}_p(\tilde{\Gamma})} \leqq \|S_{\tilde{\Gamma}}\|_{\mathbf{L}_p(\tilde{\Gamma})} \|r\|_{\mathbf{L}_p(\Gamma)}$$

for every function $r \in \mathbf{R}(\Gamma)$ and this implies the assertion at once. ∎

2.2. We now consider the case of an arbitrary piecewise Lyapunov curve system. Our objective is to prove the following theorem.

Theorem 2.3. *Let Γ be a piecewise Lyapunov curve system. Then the operator S_Γ is bounded on the space $\mathbf{L}_p(\Gamma)$ for every p such that $1 < p < \infty$.*

We first prove an auxiliary fact.

Lemma 2.2. *Let Γ_1 and Γ_2 be two open Lyapunov curves which have the point z_0 in common. Suppose the tangents to these curves at z_0 do not coincide. Then the operator $S_{\Gamma_1\Gamma_2}$ defined as*

$$(S_{\Gamma_1\Gamma_2}\varphi)(t) = \frac{1}{\pi \mathrm{i}} \int_{\Gamma_1} \frac{\varphi(\tau)}{\tau - t}\, \mathrm{d}\tau \qquad (t \in \Gamma_2) \tag{6}$$

is bounded from $\mathbf{L}_p(\Gamma_1)$ into $\mathbf{L}_p(\Gamma_2)$ $(1 < p < \infty)$.

Proof. 1. First suppose Γ_1 and Γ_2 are straight line segments. Without loss of generality assume $z_0 = 0$ and $\Gamma_1 \subset [0, \infty)$. Choose a complex number ζ such that Γ_2 lies on the ray ζy $(0 \leqq y < \infty)$. Let $\varphi \in \mathbf{L}_p(\Gamma_1)$ be an arbitrary function and $\psi \in \mathbf{L}_q(\Gamma_2)$ $(1/p + 1/q = 1)$ a function vanishing identically in a neighborhood of the origin. Extend the functions by zero to the whole rays $[0, \infty)$ and ζy $(0 \leqq y < \infty)$, respectively. Then

$$\left| \int_{\Gamma_2} \overline{\psi(t)}\, |\mathrm{d}t| \int_{\Gamma_1} \frac{\varphi(\tau)\, \mathrm{d}\tau}{\tau - t} \right| \leqq |\zeta| \int_0^\infty \mathrm{d}y \int_0^\infty \frac{|\psi(\zeta y)\, \varphi(x)|}{|x - \zeta y|}\, \mathrm{d}x$$

$$= |\zeta| \int_0^\infty \mathrm{d}y \int_0^\infty \frac{|\psi(\zeta y)\, \varphi(sy)|}{|s - \zeta|}\, \mathrm{d}s = |\zeta| \int_0^\infty \frac{\mathrm{d}s}{|s - \zeta|} \int_0^\infty |\psi(\zeta y)\, \varphi(sy)|\, \mathrm{d}y.$$

On estimating the inner integral by the Hölder inequality we readily obtain that

$$\left|\int_{\Gamma_2} \overline{\psi(t)}\,(S_{\Gamma_1\Gamma_2}\varphi)(t)\,|\mathrm{d}t|\right| \leqq |\zeta|^{1/p} \int_0^\infty \frac{\mathrm{d}s}{s^{1/p}\,|s-\zeta|}\,\|\psi\|_{\mathbf{L}_q(\Gamma_2)}\,\|\varphi\|_{\mathbf{L}_p(\Gamma_1)}.$$

Since the above functions ψ form a dense subset of $\mathbf{L}_q(\Gamma_2)$, the last inequality shows that the operator $S_{\Gamma_1\Gamma_2}: \mathbf{L}_p(\Gamma_1) \to \mathbf{L}_p(\Gamma_2)$ is bounded.

2. Now consider the general case. Denote by G_1 and G_2 any two segments on the tangents to the curves Γ_1 and Γ_2 at the point z_0, respectively. Let $t = \alpha(\xi)$ be a one-to-one mapping of $G = G_1 \cup G_2$ onto $\Gamma = \Gamma_1 \cup \Gamma_2$. Since Γ_1 and Γ_2 are Lyapunov curves, the function $\alpha(\xi)$ can be chosen in such a manner that its derivative satisfies a Hölder condition and, furthermore, does not vanish. Consider the operators $A_k: \mathbf{L}_p(\Gamma_k) \to \mathbf{L}_p(\Gamma_k)$ $(k = 1, 2)$ defined as $(A_k\varphi)(\xi) = \varphi(\alpha(\xi))$. It is easily seen that $K = A_2 S_{\Gamma_1\Gamma_2} A_1^{-1} - S_{G_1G_2}$ is an integral operator with the kernel

$$k(\xi, \zeta) = \frac{1}{\pi\mathrm{i}}\left(\frac{\alpha'(\zeta)}{\alpha(\zeta)-\alpha(\xi)} - \frac{1}{\zeta-\xi}\right) \qquad (\zeta \in G_1, \xi \in G_2).$$

Thus, in view of the estimate (3), the operator $K: \mathbf{L}_p(G_1) \to \mathbf{L}_p(G_2)$ is bounded. The assertion is now an immediate consequence of the equality $S_{\Gamma_1\Gamma_2} = A_2^{-1}(K + S_{G_1G_2})A_1$. ∎

Proof of Theorem 2.3. Our piecewise Lyapunov curve system is the union of finitely many open Lyapunov curves $\Gamma_1, \ldots, \Gamma_n$ which satisfy the following conditions: any two curves Γ_j and Γ_k $(j, k = 1, \ldots, n)$ have at most one point in common, namely, a possible common end point t_0; if Γ_j and Γ_k have a common point t_0, then either $\Gamma_j \cup \Gamma_k$ is a Lyapunov curve or the tangents to the curves Γ_j and Γ_k at the point t_0 do not coincide. We shall prove the boundedness of the operator $S_{\Gamma_j\Gamma_k}: \mathbf{L}_p(\Gamma_j) \to \mathbf{L}_p(\Gamma_k)$ defined by (6). For $j = k$ this results from Corollary 2.2. Thus let $j \neq k$.

If Γ_j and Γ_k have a common point and if $\Gamma_{jk} = \Gamma_j \cup \Gamma_k$ is not a Lyapunov curve, then $S_{\Gamma_j\Gamma_k}$ is bounded, due to Lemma 2.2. Thus, suppose Γ_{jk} is a Lyapunov curve. By Corollary 2.2, $S_{\Gamma_{jk}}$ is a bounded operator on $\mathbf{L}_p(\Gamma_{jk})$ and we have the estimate

$$\|S_{\Gamma_j\Gamma_k}\varphi\|_{\mathbf{L}_p(\Gamma_k)} \leqq \|S_{\Gamma_{jk}}\tilde{\varphi}\|_{\mathbf{L}_p(\Gamma_{jk})} \leqq \|S_{\Gamma_{jk}}\|\,\|\varphi\|_{\mathbf{L}_p(\Gamma_j)},$$

where $\tilde{\varphi}$ denotes the function obtained from a function $\varphi \in \mathbf{L}_p(\Gamma_j)$ by extending it by zero to the whole curve Γ_{jk}. But the last inequality implies the boundedness of the operator $S_{\Gamma_j\Gamma_k}$. Finally, if Γ_j and Γ_k have no point in common then $S_{\Gamma_j\Gamma_k}$ is an integral operator with continuous kernel and, therefore, it is bounded.

Now denote by χ_j the characteristic function of the set $\Gamma_j \subset \Gamma$. Then

$$S_\Gamma = \sum_{j,k=1}^{n} \chi_j S_\Gamma \chi_k I,$$

$$\|\chi_j S_\Gamma \chi_k \varphi\|_{\mathbf{L}_p(\Gamma)} \leqq \|S_{\Gamma_j\Gamma_k}\|\,\|\varphi\|_{\mathbf{L}_p(\Gamma)},$$

which proves the boundedness of the operator S_Γ on the space $\mathbf{L}_p(\Gamma)$. ∎

Notice the following immediate consequence of Theorem 2.3 and formula (1.8).

Theorem 2.4. *If Γ is an arbitrary closed piecewise Lyapunov curve system then $S_\Gamma^2\varphi = \varphi$ for every function $\varphi \in \mathbf{L}_p(\Gamma)$ $(1 < p < \infty)$.*

Remark 1. If $p = 1$ or $p = \infty$, then Theorem 2.3 is not true even in the case where Γ is the unit circle Γ_0. This follows from the fact that e.g.

$$\sum_{n=2}^{\infty} \frac{1}{n \ln n}(t^n - t^{-n}) \qquad (|t| = 1)$$

is a continuous function on Γ_0, whereas the conjugate function

$$\sum_{n=2}^{\infty} \frac{1}{n \ln n} (t^n + t^{-n}) \qquad (|t| = 1)$$

is continuous for $t \neq 1$ and unbounded for $t = 1$ (see ZYGMUND [3], Chapter VII.2).

Remark 2. Lemma 2.1 (without the estimates (1)) goes back to LUZIN [1] ($p = 2$) and M. RIESZ [1] (general p, $1 < p < \infty$). The proof of this lemma given here is due to COTLAR [1]. For $p = 2$, Theorem 2.2 was proved by MIKHLIN [7] in a different way. The proof of Theorem 2.3 (for general p, $1 < p < \infty$) is due to KHVEDELIDZE [1, 3]. Here we essentially followed GOHBERG and KRUPNIK [4] (Chap. I, §§ 2 and 3).

Remark 3. A Jordan curve Γ is called a *Radon curve* or a curve of bounded variation if the angle $\theta_\Gamma(t)$ (see Section 1.1) is a function of bounded variation. In the case of a Radon curve Γ, the boundedness of the singular integral operator S_Γ on the space $\mathbf{L}_p(\Gamma)$ (or even on $\mathbf{L}_p(\Gamma, \varrho)$, cf. § 3) was proved and a corresponding theory of singular integral equations was developed by HAVIN [1, 2], DANILYUK, SHELEPOV [1], and others (cf. to this DANILYUK [1]). GORDADZE [1–3] studied the case of Radon curves with cusps.

Recently CALDERON [5] (cf. also his article [3]) has established the following remarkable result:

Theorem (CALDERON [5]). *Let Γ be a curve in the complex plane given as $z(t) = t + i\varphi(t)$, where φ is a real-valued function which is defined on the real line* R *and has a bounded and derivative φ'. Furthermore, let*

$$(A_{\varphi,\varepsilon} f)(t) = \frac{1}{\pi i} \int_{|s-t|>\varepsilon} \frac{f(s)}{z(s) - z(t)} \, dz(s), \qquad \varepsilon > 0.$$

Then there exists a positive number a such that if $\|\varphi'\|_\infty < a$:

(i) *the operator $A_\varphi f = \lim_{\varepsilon \to 0} A_{\varphi,\varepsilon} f$ is bounded on the space $\mathbf{L}_p(\Gamma)$ $(1 < p < \infty)$;*

(ii) *the limit*

$$(A_\varphi f)(t) = \lim_{\varepsilon \to 0} (A_{\varphi,\varepsilon} f)(t)$$

exists for almost all $t \in \Gamma$ if $f \in \mathbf{L}_p(\Gamma)$ $(1 \leq p < \infty)$;

(iii) *the operator A_φ is of the weak type* (1, 1) (*see the inequality* (5.7), *Chapter XI.*).

Note that both smooth curves and Radon curves satisfy the conditions of this theorem.

For several interesting applications of this result, in particular, for its application to the Dirichlet or Neumann problem in C^1 domains in the Euclidean space $\mathbb{R}^m$, we refer to the work [1] of A. CALDERON, C. CALDERON, FABES, JODEIT, and RIVIERE.

A version of Calderon's proof which is due to Y. Meyer can be found in the book of JOURNÉ [1] (Chap. 7). A. Calderon also conjectured that the restriction $\|\varphi'\|_\infty < a$ in the preceding theorem can be replaced by $\|\varphi'\|_\infty < \infty$. This was then proved by COIFMAN, MCINTOSH, and MEYER [1] using a method completely different from that of Calderon. A bit later, DAVID [1] developed a real variable technique (based on some perturbation arguments) to remove the restriction $\|\varphi'\|_\infty < a$ (see also JOURNÉ [1], Chap. 8, BONY [1], and T. MURAI [1]).

§ 3. The boundedness of the singular integral operator on the space $\mathbf{L}_p$ with weight

3.1. Let Γ be a piecewise Lyapunov curve system, let $t_1, \ldots, t_m$ be pairwise distinct points on Γ, let $\beta_1, \ldots, \beta_m$ be real numbers, and let

$$\varrho(t) = \prod_{k=1}^{m} |t - t_k|^{\beta_k}. \tag{1}$$

We denote by $\mathbf{L}_p(\Gamma, \varrho)$ $(1 \leq p < \infty)$ the Banach space of all measurable functions φ on Γ for which $\varphi\varrho \in \mathbf{L}_p(\Gamma)$; the norm in $\mathbf{L}_p(\Gamma, \varrho)$ is defined as

$$\|\varphi\|_{\mathbf{L}_p(\Gamma,\varrho)} = \|\varrho\varphi\|_{\mathbf{L}_p(\Gamma)} = \Big(\int_\Gamma |\varphi(t)|^p \varrho^p(t) \,|dt|\Big)^{1/p}.$$

Let us first state some simple properties of the spaces $\mathbf{L}_p(\Gamma, \varrho)$.

Obviously,

$$\mathbf{L}_\infty(\Gamma) \subset \mathbf{L}_p(\Gamma, \varrho), \quad \text{if} \quad \beta_k > -1/p \qquad (k = 1, \ldots, m),$$
$$\mathbf{L}_p(\Gamma, \varrho) \subset \mathbf{L}_1(\Gamma), \quad \text{if} \quad \beta_k < 1 - 1/p \qquad (k = 1, \ldots, m).$$

Lemma 3.1. *If $\beta_k > -1/p$ $(k = 1, \ldots, m)$ then the set of continuous functions, $\mathbf{C}(\Gamma)$, is dense in the space $\mathbf{L}_p(\Gamma, \varrho)$.*

Proof. If $\beta_k > -1/p$ $(k = 1, \ldots, m)$ then $\varrho \in \mathbf{L}_1(\Gamma)$ and therefore $\mathbf{C}(\Gamma) \subset \mathbf{L}_p(\Gamma, \varrho)$.

Now, given a function $\varphi \in \mathbf{L}_p(\Gamma, \varrho)$ we can approximate the function $\varphi\varrho \in \mathbf{L}_p(\Gamma)$ by a sequence of continuous functions ψ_n which, moreover, vanish identically in certain neighborhoods of the points $t_1, \ldots, t_m$. Then the functions $\varphi_n = \psi_n\varrho^{-1}$ are obviously continuous on Γ and we have

$$\|\varphi - \varphi_n\|_{\mathbf{L}_p(\Gamma,\varrho)} = \|\varphi\varrho - \psi_n\|_{\mathbf{L}_p(\Gamma)} \xrightarrow[n\to\infty]{} 0. \quad \blacksquare$$

Corollary 3.1. *If $\beta_k > -1/p$ $(k = 1, \ldots, m)$ then the set $\mathbf{R}(\Gamma)$ is dense in the space $\mathbf{L}_p(\Gamma, \varrho)$.*

To see this, note that $\mathbf{C}(\Gamma)$ is densely and continuously embedded in $\mathbf{L}_p(\Gamma, \varrho)$ and that every continuous function can be uniformly approximated by rational functions as closely as desired.

Corollary 3.2. *If $\beta_k > -1/p$ $(k = 1, \ldots, m)$ and if Γ_0 is the unit circle, then the linear hull of the system of functions $\{t^k\}_{k=-\infty}^{\infty}$ $(|t| = 1)$ is dense in the space $\mathbf{L}_p(\Gamma_0, \varrho)$.*

It is easy to describe the continuous linear functionals on the space $\mathbf{L}_p(\Gamma, \varrho)$ $(1 < p < \infty)$. Let $\varphi \in \mathbf{L}_p(\Gamma, \varrho)$ and $\psi \in \mathbf{L}_q(\Gamma, \varrho^{-1})$, where $1 < p < \infty$, $p^{-1} + q^{-1} = 1$. Then Hölder's inequality gives

$$\int_\Gamma \varphi(t)\,\overline{\psi(t)}\,|dt| = \int_\Gamma \varphi(t)\,\varrho(t)\,\overline{\psi(t)}\,\varrho^{-1}(t)\,|dt| \leq \|\varphi\|_{\mathbf{L}_p(\Gamma,\varrho)} \|\psi\|_{\mathbf{L}_q(\Gamma,\varrho^{-1})}, \tag{*}$$

and so the theorem of F. Riesz on the general form of the continuous linear functionals on the space $\mathbf{L}_p(\Gamma)$ implies that every continuous linear functional on the space $\mathbf{L}_p(\Gamma, \varrho)$ has the form

$$\Psi(\varphi) = \int_\Gamma \varphi(t)\,\overline{\psi(t)}\,|dt| =: (\varphi, \psi) \qquad (\varphi \in \mathbf{L}_p(\Gamma, \varrho)),$$

where $\psi \in \mathbf{L}_q(\Gamma, \varrho^{-1})$ and, moreover, $\|\Psi\|_{\mathbf{L}_p^*(\Gamma,\varrho)} = \|\psi\|_{\mathbf{L}_q(\Gamma,\varrho^{-1})}$. In other words, $\mathbf{L}_p^*(\Gamma, \varrho) = \mathbf{L}_q(\Gamma, \varrho^{-1})$.

3.2. We now prove the boundedness of the singular integral operator S_Γ on the space $\mathbf{L}_p(\Gamma, \varrho)$, where the weight is given by (1).

Theorem 3.1. *Let Γ be a piecewise Lyapunov curve system and suppose $p, \beta_1, \ldots, \beta_m$ are real numbers satisfying*

$$1 < p < \infty, \qquad -1/p < \beta_k < 1 - 1/p \qquad (k = 1, \ldots, m). \tag{2}$$

Then the singular integral operator S_Γ is bounded on the space $\mathbf{L}_p(\Gamma, \varrho)$. If Γ is a closed curve system, then $S_\Gamma^2 = I$.

We first prove two lemmas.

Lemma 3.2. *Suppose $t_0 \in \Gamma$, $1 < p < \infty$, and let α be a real number such that $-q^{-1} < \alpha < p^{-1}$ $(p^{-1} + q^{-1} = 1)$. Then the operator B defined as*

$$(B\varphi)(t) = \int_\Gamma \frac{|\tau - t_0|^\alpha - |t - t_0|^\alpha}{|\tau - t|\,|t - t_0|^\alpha}\,\varphi(\tau)\,|d\tau| \qquad (t \in \Gamma)$$

is bounded on the space $\mathbf{L}_p(\Gamma)$.

Proof. 1. Assume first that $0 \leqq \alpha < p^{-1}$. Let δ be a real number satisfying the inequality $\alpha(p-1) < p\,\delta < \min(p-1, 1-\alpha)$. Then, for any function $\varphi \in \mathbf{C}(\Gamma)$, Hölder's inequality implies that

$$|(B\varphi)(t)| \leqq c \int_\Gamma \frac{|\varphi(\tau)|\,|d\tau|}{|t - t_0|^\alpha\,|\tau - t|^{1-\alpha}} = \frac{c}{|t - t_0|^\alpha} \int_\Gamma \frac{|\tau - t_0|^\delta\,|\varphi(\tau)|\,|d\tau|}{|\tau - t|^{1-\alpha}\,|\tau - t_0|^\delta}$$

$$\leqq \frac{c}{|t - t_0|^\alpha} \left(\int_\Gamma \frac{|\tau - t_0|^{\delta p}\,|\varphi(\tau)|^p\,|d\tau|}{|\tau - t|^{1-\alpha}}\right)^{1/p} \left(\int_\Gamma \frac{|d\tau|}{|\tau - t|^{1-\alpha}\,|\tau - t_0|^{\delta q}}\right)^{1/q}.$$

Next, if β and γ are any real numbers such that $\beta < 1, \gamma < 1, \beta + \gamma > 1$, then through the substitution $\tau - t = \zeta(t - t_0)$ one easily gets the estimate

$$\int_\Gamma \frac{|d\tau|}{|\tau - t|^\beta\,|\tau - t_0|^\gamma} \leqq \frac{c}{|t - t_0|^{\beta + \gamma - 1}}. \tag{3}$$

Thus,

$$|(B\varphi)(t)| \leqq \frac{c}{|t - t_0|^{\delta + \alpha/p}} \left(\int_\Gamma \frac{|\tau - t_0|^{\delta p}\,|\varphi(\tau)|^p\,|d\tau|}{|\tau - t|^{1-\alpha}}\right)^{1/p}$$

and therefore

$$\|B\varphi\|_p^p \leqq c \int_\Gamma \frac{|dt|}{|t - t_0|^{\alpha + \delta/p}} \int_\Gamma \frac{|\tau - t_0|^{\delta p}\,|\varphi(\tau)|^p\,|d\tau|}{|\tau - t|^{1-\alpha}}$$

$$= c \int_\Gamma \frac{|\varphi(\tau)|^p\,|d\tau|}{|\tau - t_0|^{-\delta p}} \int_\Gamma \frac{|dt|}{|t - t_0|^{\alpha + \delta/p}\,|\tau - t|^{1-\alpha}} \leqq c\,\|\varphi\|_p^p.$$

To obtain the last inequality we used the estimate (3).

2. Let now $-q^{-1} < \alpha < 0$. Put $\beta = -\alpha$. By what was just proved, the operator C defined as

$$(C\psi)(t) = \int_\Gamma \frac{|\tau - t_0|^\beta - |t - t_0|^\beta}{|\tau - t|\,|t - t_0|^\beta}\,\psi(\tau)\,|d\tau|$$

is bounded on the space $\mathbf{L}_q(\Gamma)$. But it is readily seen that $C^* = B$ and, consequently, B is bounded on the space $\mathbf{L}_p(\Gamma)$. ∎

Lemma 3.3. *Let $t_1, \ldots, t_m \in \Gamma$ and suppose $p, \alpha_1, \ldots, \alpha_m$ are real numbers satisfying*

$$1 < p < \infty, \qquad -q^{-1} < \alpha_k < p^{-1} \qquad (k = 1, \ldots, m;\ p^{-1} + q^{-1} = 1).$$

Define $h(t) = \prod_{k=1}^{m} |t - t_k|^{\alpha_k}$ $(t \in \Gamma)$. Then the operator $A = h^{-1} S_\Gamma h I$ is bounded on the space $\mathbf{L}_p(\Gamma)$.

Proof. Choose open arcs $\Gamma_k \subset \Gamma$ $(k = 1, \ldots, m)$ so that $t_k \in \Gamma_k$, but $t_j \notin \Gamma_k$ if $j \neq k$. Furthermore, let the distance between any two of these arcs be positive. Put $\Gamma_0 = \Gamma \setminus \bigcup_{k=1}^{m} \Gamma_k$ and denote by R_k $(k = 0, 1, \ldots, m)$ the operator defined as $(R_k\varphi)(t) = \chi_k(t)\,\varphi(t)$, where χ_k is the characteristic function of the set $\Gamma_k \subset \Gamma$. Then the operator A can be represented in the form

$$A = \left(\sum_{j=0}^{m} R_j\right) A \left(\sum_{k=0}^{m} R_k\right) = \sum_{j,k=0}^{m} R_jAR_k$$

and thus it remains to prove the boundedness of each of the operators R_jAR_k. To do this we distinguish four cases.

1. $j = k \neq 0$. Then $R_kAR_k = gA_kfI$, where

$$g(t) = \chi_k(t)\, h^{-1}(t)\, |t - t_k|^{\alpha_k}, \qquad f(t) = \chi_k(t)\, h(t)\, |t - t_k|^{-\alpha_k},$$
$$A_k = |t - t_k|^{-\alpha_k}\, S_\Gamma\, |t - t_k|^{\alpha_k}\, I.$$

The operator $A_k - S_\Gamma$ is bounded on $\mathbf{L}_p(\Gamma)$ by Lemma 3.2 and S_Γ is so due to Theorem 2.3. Consequently, A_k is a bounded operator on $\mathbf{L}_p(\Gamma)$, too. This and the boundedness of the functions f and g on Γ imply that the operator R_kAR_k is bounded on $\mathbf{L}_p(\Gamma)$.

2. $k \neq j$, $k \neq 0$, $j \neq 0$. In this case we have

$$|(R_jAR_k\varphi)(t)| = \frac{\chi_j(t)}{\pi h(t)} \left| \int_{\Gamma_k} \frac{\varphi(\tau)\, h(\tau)}{\tau - t}\, \mathrm{d}\tau \right| \leqq \frac{\|\varphi\|_p\, \|h\|_q}{\pi d h(t)},$$

where $d > 0$ denotes the distance between Γ_j and Γ_k. Since, obviously, $h \in \mathbf{L}_q(\Gamma)$ and $h^{-1} \in \mathbf{L}_p(\Gamma)$, we get the estimate

$$\|R_jAR_k\varphi\|_p \leqq (\pi d)^{-1}\, \|h^{-1}\|_p\, \|h\|_q\, \|\varphi\|_p$$

and thus the boundedness of R_jAR_k on $\mathbf{L}_p(\Gamma)$.

3. $k \neq j$, $k = 0$ or $j = 0$. Then $R_jAR_k = R_jRARR_k$ with $R = R_j + R_k$. The boundedness of the operator RAR results from case 1 and, hence, R_jAR_k is bounded on $\mathbf{L}_p(\Gamma)$.

4. $k = j = 0$. The boundedness of the operator R_0AR_0 is an immediate consequence of Theorem 2.3. ∎

Proof of Theorem 3.1. Notice that the operator ϱI is an isometric isomorphism of $\mathbf{L}_p(\Gamma, \varrho)$ onto $\mathbf{L}_p(\Gamma)$. Thus, the operator S_Γ is bounded on $\mathbf{L}_p(\Gamma, \varrho)$ if and only if the operator $\varrho S_\Gamma \varrho^{-1} I$ is bounded on $\mathbf{L}_p(\Gamma)$. But the latter immediately follows from Lemma 3.3, since the function $h = \varrho^{-1}$ in view of the restrictions (2) satisfies the hypotheses of Lemma 3.3. ∎

Remark. Conditions (2) are not only sufficient, but also necessary for the operator S_Γ to be bounded on the space $\mathbf{L}_p(\Gamma, \varrho)$ (see Gohberg and Krupnik [4], Chapter IX, § 8).

3.3. The remainder of this section is devoted to the proof of the boundedness of the singular integral operator S_Γ on the space $\mathbf{L}_p(\Gamma, \varrho)$ for the case where $\Gamma = \mathbb{R}$. The weight is now given as

$$\varrho_\infty(t) = |t - \mathrm{i}|^{\beta} \prod_{k=1}^{m} |t - t_k|^{\beta_k},$$

where $t_1, \ldots, t_m$ are pairwise distinct points on the real line and where $\beta, \beta_1, \ldots, \beta_m$ are real numbers.

Theorem 3.2. *Let $p, \beta, \beta_1, \ldots, \beta_m$ be real numbers satisfying*

$$1 < p < \infty, \qquad -1/p < \beta_k < 1 - 1/p \qquad (k = 1, \ldots, m),$$

$$-1/p < \beta + \sum_{k=1}^{m} \beta_k < 1 - 1/p. \tag{4}$$

Then the singular integral operator S_∞ defined as

$$(S_\infty \varphi)(t) = \frac{1}{\pi i} \int_{-\infty}^{\infty} \frac{\varphi(\tau)}{\tau - t} \, d\tau \qquad (-\infty < t < \infty) \tag{5}$$

is bounded on the space $\mathbf{L}_p(\mathbb{R}, \varrho_\infty)$.

The singular integral (5) is referred to as the *Hilbert transform*.

Proof. Let Γ_0 be the unit circle and put $\zeta_k = (t_k + i)(t_k - i)^{-1}$ $(k = 1, \ldots, m)$, $\zeta_0 = 1$, $\beta_0 = 1 - 2/p - \beta - \sum_{k=1}^{m} \beta_k$, $\quad \varrho_0(\zeta) = \prod_{k=0}^{m} |\zeta - \zeta_k|^{\beta_k}$.

The operator defined by

$$(B\varphi)(\zeta) = \frac{1}{\zeta - 1} \varphi\left(i \frac{\zeta + 1}{\zeta - 1}\right)$$

is a linear bounded operator from $\mathbf{L}_p(\mathbb{R}, \varrho_\infty)$ into the space $\mathbf{L}_p(\Gamma_0, \varrho_0)$. Indeed, given any $\varphi \in \mathbf{L}_p(\mathbb{R}, \varrho_\infty)$ we have

$$\| B\varphi \|^p_{\mathbf{L}_p(\Gamma_0, \varrho_0)} = \int_{\Gamma_0} \left| \varphi\left(i \frac{\zeta + 1}{\zeta - 1}\right) \right|^p \varrho_0^p(\zeta) \, |\zeta - \zeta_0|^{-p} \, |d\zeta| = c \int_{-\infty}^{\infty} |\varphi(\tau)|^p \, \varrho_\infty^p(\tau) \, d\tau$$

with some positive constant c. The operator B is obviously invertible and

$$(B^{-1} \psi)(t) = \frac{2i}{t - i} \psi\left(\frac{t + i}{t - i}\right).$$

The restrictions (4) guarantee that the numbers p and β_k $(k = 0, 1, \ldots, m)$ satisfy conditions (2). Thus, by Theorem 3.1, the operator $S_0 = S_{\Gamma_0}$ is bounded on the space $\mathbf{L}_p(\Gamma_0, \varrho_0)$ and consequently, $B^{-1} S_0 B$ is a bounded operator on $\mathbf{L}_p(\mathbb{R}, \varrho_\infty)$.

Let φ be an arbitrary continuously differentiable function with finite support defined on the real line. Application of Theorem 1.2 gives

$$(B^{-1} S_0 B\varphi)(t) = \frac{2}{\pi(t - i)} \int_{\Gamma_0} \frac{\varphi\left(i \frac{\zeta + 1}{\zeta - 1}\right) d\zeta}{(\zeta - 1)\left(\zeta - \frac{t + i}{t - i}\right)} = \frac{1}{\pi i} \int_{-\infty}^{\infty} \frac{\varphi(\tau) \, d\tau}{\tau - t} = (S_\infty \varphi)(t). \tag{6}$$

Note that the function $\varphi\left(i \frac{\zeta + 1}{\zeta - 1}\right)$ vanishes identically in some neighborhood of the point $\zeta = 1$. Since the set of our functions φ is dense in $\mathbf{L}_p(\mathbb{R}, \varrho_\infty)$, the equality (6) implies the boundedness of the operator S_∞ on the space $\mathbf{L}_p(\mathbb{R}, \varrho_\infty)$. ∎

The just proved theorem in particular implies the boundedness of the Hilbert transform on the space $\mathbf{L}_p(\mathbb{R})$ $(1 < p < \infty)$ as well as the boundedness of the singular integral operator S_Γ on the space $\mathbf{L}_p(\Gamma, \varrho_\infty)$, where $\Gamma = [a, b]$ is an arbitrary finite or infinite interval. Moreover, from (6) we deduce that $S_\infty^2 = I$.

Remark. Theorem 3.2 was proved by Hardy and Littlewood for the first time. Theorem 3.1 for arbitrary Lyapunov curves (different from the line and from line segments) is due to Khvede-

LIDZE [3]. The essence of the proof presented here is taken from KHVEDELIDZE's papers [1, 3] (also compare with GOHBERG and KRUPNIK [4], Chapter I, §§ 4 and 5).

Moreover, if $S_\Gamma \in \mathcal{L}(\mathbf{L}_2(\Gamma))$ for a rectifiable Jordan curve Γ then Γ is necessarily a *Carleson curve*, i.e., the length of the part $\Gamma(z, r) := \cap B(z, r)$ is less than $C \cdot r$ for each disk $B(z, r)$ of radius r centered at a point $z \in \mathbb{C}$. The following generalization of Theorem 3.1 has been recently proved by DAVID [1], using Calderon's theorem.

Theorem 3.3. *Let Γ be a Carleson curve and let $\varrho \geqq 0$ be an arbitrary measurable function on Γ. Then S_Γ is bounded on $\mathbf{L}_p(\Gamma, \varrho)$ $(1 < p < \infty)$ if and only if*

$$\|\varrho\|_{\mathbf{L}_p(\Gamma(z,r))} \|\varrho^{-1}\|_{\mathbf{L}_q(\Gamma(z,r))} \leqq C_p \cdot r \; (p^{-1} + q^{-1} = 1)$$

for all $z \in \Gamma$ and all $r > 0$, where C_p is a constant independent of z and r.

Theorem 3.3. was proved by HUNT, MUCKENHOUPT, and WHEEDEN [1] for the case of the unit circle $\Gamma = \Gamma_0$ and then generalized to the case of a larger class of curves Γ by KOKILASHVILI [1] and KRUPNIK [6] (cf. also § 12, Chap. XI).

§ 4. Further properties of the singular integral operator

In this section we prove some properties of the singular integral operator which can be derived from its boundedness on the space $\mathbf{L}_p(\Gamma, \varrho)$. Here Γ denotes a piecewise Lyapunov curve system, ϱ is the weight (3.1), and the numbers $p, \beta_1, \ldots, \beta_m$ are subject to the conditions (3.2).

4.1. Integral operators with weak singularity

Let $k(t, \tau)$ be a function which is measurable on $\Gamma \times \Gamma$ and satisfies the estimate

$$|k(t, \tau)| \leqq c\, |t - \tau|^{-\mu} \qquad (c = \text{const}, 0 \leqq \mu < 1).$$

Such a function is called a *weakly singular kernel*. It is well known that the *integral operator with weak singularity* defined by

$$(T\varphi)(t) = \int_\Gamma k(t, \tau)\, \varphi(\tau)\, d\tau \qquad (t \in \Gamma) \tag{1}$$

is compact on the space $\mathbf{L}_p(\Gamma)$, $1 < p < \infty$ (see KRASNOSELSKI, ZABREIKO et al. [1], for instance). With the help of Theorem 3.1 it is easy to show that T is also compact on the space $\mathbf{L}_p(\Gamma, \varrho)$.

Theorem 4.1. *If the numbers $p, \beta_1, \ldots, \beta_m$ satisfy the conditions* (3.2), *then the integral operator with weak singularity* (1) *is compact on the space* $\mathbf{L}_p(\Gamma, \varrho)$.

Proof. Similarly to the proof of Theorem 3.1, it is sufficient to prove the compactness of the operator $K = h^{-1} ThI$ on the space $\mathbf{L}_p(\Gamma)$, where the function h satisfies the hypotheses of Lemma 3.3. For a positive integer n set

$$k_n(t, \tau) = \begin{cases} k(t, \tau), & \text{if } |t - \tau| \geqq 1/n, \\ 0, & \text{if } |t - \tau| < 1/n. \end{cases}$$

Denote by A, A_n, and K_n the integral operators whose kernels are $h^{-1}(t)\, k(t, \tau)\, h(\tau) - k(t, \tau)$, $h^{-1}(t)\, k_n(t, \tau)\, h(\tau) - k_n(t, \tau)$, and $h^{-1}(t)\, k_n(t, \tau)\, h(\tau)$, respectively. Because $h \in \mathbf{L}_q(\Gamma)$, $h^{-1} \in \mathbf{L}_p(\Gamma)$ $(1/p + 1/q = 1)$ and since $k_n(t, \tau)$ is a bounded kernel, we have

$$\int_\Gamma |dt| \left(\int_\Gamma |h^{-1}(t)\, k_n(t, \tau)\, h(\tau)|^q\, |d\tau| \right)^{p-1} < \infty,$$

and consequently, K_n and thus also A_n are compact on $\mathbf{L}_p(\Gamma)$. We shall show that $\lim_{n\to\infty} \|A - A_n\|_p = 0$. Clearly, this will imply the compactness of A and, hence, that of K on $\mathbf{L}_p(\Gamma)$.

Put $M_n = A - A_n$ and let $m_n(t, \tau)$ denote the kernel of the operator M_n. From the proof of Lemma 3.3 it is easy to deduce that the operator B defined as

$$(B\varphi)(t) = \int_\Gamma \left|\frac{h^{-1}(t)\, h(\tau)}{\tau - t} - \frac{1}{\tau - t}\right| \varphi(\tau)\, |\mathrm{d}\tau|$$

is bounded on $\mathbf{L}_p(\Gamma)$. With $b(t, \tau)$ the kernel of B, we have

$$|m_n(t, \tau)| \leqq cn^{\mu-1} b(t, \tau), \qquad \|M_n\|_p \leqq cn^{\mu-1} \|B\|_p$$

and thus $\|M_n\|_p \to 0$ as $n \to \infty$. ▮

4.2. Two theorems on commutators

Theorem 4.2. *If a is a continuous function on Γ then the commutator[1]) $T = aS_\Gamma - S_\Gamma a$ is compact on the space $\mathbf{L}_p(\Gamma, \varrho)$.*

Proof. If $a \in \mathbf{R}(\Gamma)$ then T is an integral operator with continuous kernel and thus compact.

If $a \in \mathbf{C}(\Gamma)$ is an arbitrary continuous function, choose a sequence of functions $a_n \in \mathbf{R}(\Gamma)$ so that $\|a - a_n\|_{\mathbf{C}(\Gamma)} \to 0$ as $n \to \infty$. Then the commutators $T_n = a_n S_\Gamma - S_\Gamma a_n$ are compact and we have

$$\|T - T_n\| \leqq 2\, \|S_\Gamma\|\, \|a - a_n\|_{\mathbf{C}(\Gamma)}.$$

Consequently, $\|T - T_n\| \to 0$ and this implies that the operator T is compact on the space $\mathbf{L}_p(\Gamma, \varrho)$. ▮

Theorem 4.3. *Let Γ_1 and Γ_2 be two piecewise Lyapunov curve systems and let $\tau = \alpha(\zeta)$ be a one-to-one mapping of Γ_2 onto Γ_1 possessing a derivative α' which satisfies a Hölder condition and vanishes nowhere on Γ_2. Define*

$$\varrho_1(t) = \prod_{k=1}^m |t - t_k|^{\beta_k}, \qquad \varrho_2(\xi) = \prod_{k=1}^m |\xi - \xi_k|^{\beta_k}, \qquad t_k = \alpha(\xi_k)$$

and denote by S_{Γ_1} and S_{Γ_2} the singular integral operators on the spaces $\mathbf{L}_p(\Gamma_1, \varrho_1)$ and $\mathbf{L}_p(\Gamma_2, \varrho_2)$, respectively. Then

$$S_{\Gamma_2} = BS_{\Gamma_1}B^{-1} + T, \tag{2}$$

where T is a compact operator on $\mathbf{L}_p(\Gamma_2, \varrho_2)$ and B denotes the linear bounded and invertible operator from $\mathbf{L}_p(\Gamma_1, \varrho_1)$ onto $\mathbf{L}_p(\Gamma_2, \varrho_2)$ defined by $(B\varphi)(\zeta) = \varphi(\alpha(\zeta))$.

Proof. Let $r \in \mathbf{R}(\Gamma_2)$, $\varphi = B^{-1} r$, and $T = S_{\Gamma_2} - BS_{\Gamma_1}B^{-1}$. Then, by (1.9),

$$(Tr)(\xi) = \frac{1}{\pi i} \int_{\Gamma_2} \frac{r(\zeta)\, \mathrm{d}\zeta}{\zeta - \xi} - \frac{1}{\pi i} \int_{\Gamma_1} \frac{\varphi(\tau)\, \mathrm{d}\tau}{\tau - \alpha(\xi)} = \frac{1}{\pi i} \int_{\Gamma_2} \left(\frac{1}{\zeta - \xi} - \frac{\alpha'(\zeta)}{\alpha(\zeta) - \alpha(\xi)}\right) r(\zeta)\, \mathrm{d}\zeta$$

and since the kernel of this operator satisfies the estimate (2.3), T is an integral operator with weak singularity and therefore compact on $\mathbf{L}_p(\Gamma_2, \varrho_2)$ (Theorem 4.1). ▮

Remark. The Theorems 4.2 and 4.3 are due to Mikhlin [7].

---

[1]) As for notation, here and in what follows (if there is no fear of misunderstandings) we do not distinguish between a function and the corresponding operator of multiplication by this function.

4.3. The Poincaré-Bertrand commutation formula

From the boundedness of the operator S_Γ on the space $\mathbf{L}_p(\Gamma, \varrho)$ it is easy to deduce that the singular integral commutes with the ordinary integral.

Theorem 4.4. *Let* $\varphi \in \mathbf{L}_p(\Gamma, \varrho)$ *and* $\psi \in \mathbf{L}_q(\Gamma, \varrho^{-1})$ $(1 < p < \infty,\ 1/p + 1/q = 1)$. *Then*

$$\int_\Gamma \psi(t)\,dt \int_\Gamma \frac{\varphi(\tau)}{\tau - t}\,d\tau = \int_\Gamma \varphi(\tau)\,d\tau \int_\Gamma \frac{\psi(t)}{\tau - t}\,dt. \tag{3}$$

Proof. Recall that the numbers $p, \beta_1, \ldots, \beta_m$ are assumed to satisfy (3.2). Thus, S_Γ is bounded on $\mathbf{L}_p(\Gamma, \varrho)$. Furthermore, $\varrho^{-1}(t) = \prod_{k=1}^{m} |t - t_k|^{-\beta_k}$ and we have

$$-1/q < -\beta_k < 1 - 1/q. \tag{4}$$

Therefore S_Γ is also bounded on $\mathbf{L}_q(\Gamma, \varrho^{-1})$.

For fixed $t \in \Gamma$ and variable $\tau \in \Gamma$ put

$$\varphi_n(\tau) = \begin{cases} \varphi(\tau), & \text{if } |\tau - t| \geqq 1/n, \\ 0, & \text{if } |\tau - t| < 1/n. \end{cases}$$

Then $\|\varphi - \varphi_n\|_{\mathbf{L}_p(\Gamma,\varrho)} \to 0$ and thus $\|f - f_n\|_{\mathbf{L}_p(\Gamma,\varrho)} \to 0$ as $n \to \infty$, where

$$f(t) = \int_\Gamma \frac{\varphi(\tau)}{\tau - t}\,d\tau, \qquad f_n(t) = \int_\Gamma \frac{\varphi_n(\tau)}{\tau - t}\,d\tau \qquad (t \in \Gamma).$$

This and the estimate (*) in § 3 give

$$\int_\Gamma \psi(t) f(t)\,dt = \lim_{n\to\infty} \int_\Gamma \psi(t) f_n(t)\,dt = \lim_{n\to\infty} \int_\Gamma \psi(t)\,dt \int_\Gamma \frac{\varphi_n(\tau)}{\tau - t}\,d\tau.$$

But in the last integral the order of integration may be reversed and since obviously $\psi(t)\,\varphi_n(t) = \varphi(\tau)\,\psi_n(t)$, we get

$$\int_\Gamma \psi(t) f(t)\,dt = \lim_{n\to\infty} \int_\Gamma \varphi(\tau)\,d\tau \int_\Gamma \frac{\psi_n(t)}{\tau - t}\,dt = \int_\Gamma \varphi(\tau)\,d\tau \int_\Gamma \frac{\psi(t)\,dt}{\tau - t}$$

and this proves (3). ∎

The following theorem can be easily proved using Theorem 4.1 and the definition of the singular integral.

Theorem 4.5. *Suppose* $K(t, \tau) = k(t, \tau)\,|t - \tau|^{-\mu}$, *where* $k(t, \tau) \in \mathbf{H}^\lambda(\Gamma \times \Gamma)$ $(0 \leqq \mu < 1,$ $0 < \lambda \leqq 1)$. *Then, if* $\varphi \in \mathbf{L}_p(\Gamma, \varrho)$ $(1 < p < \infty)$,

$$\int_\Gamma \frac{dt}{t - t_0} \int_\Gamma K(t, \tau)\,\varphi(\tau)\,d\tau = \int_\Gamma \varphi(\tau)\,d\tau \int_\Gamma \frac{K(t, \tau)}{t - t_0}\,dt. \tag{5}$$

Note that two singular integrals do not commute. Indeed, if Γ is a closed Lyapunov curve and φ an arbitrary function in $\mathbf{L}_p(\Gamma, \varrho)$, then by Theorem 3.1

$$\int_\Gamma \frac{dt}{t - t_0} \int_\Gamma \frac{\varphi(\tau)}{\tau - t}\,d\tau = -\pi^2 \varphi(t_0) \qquad (t_0 \in \Gamma),$$

whereas

$$\int_\Gamma \varphi(\tau)\,d\tau \int_\Gamma \frac{dt}{(t-t_0)(\tau-t)} = 0,$$

since the inner integral is equal to

$$\frac{1}{\tau-t_0}\left\{\int_\Gamma \frac{dt}{t-t_0} - \int_\Gamma \frac{dt}{t-\tau}\right\} = 0$$

(cf. (1.6)).

However, we have the following extremely important *commutation formula*, due to H. Poincaré and G. Bertrand.

Theorem 4.6. *Let $\varphi(t,\tau) \in \mathbf{H}^\lambda(\Gamma\times\Gamma)$ $(0 < \lambda \le 1)$ and let t_0 be a regular point of the curve system Γ. Then the Poincaré-Bertrand formula holds:*

$$\int_\Gamma \frac{dt}{t-t_0}\int_\Gamma \frac{\varphi(t,\tau)}{\tau-t}\,d\tau = \int_\Gamma d\tau \int_\Gamma \frac{\varphi(t,\tau)\,dt}{(t-t_0)(\tau-t)} - \pi^2\varphi(t_0,t_0). \tag{6}$$

Proof. For the integral on the left-hand side of (6) we have

$$\int_\Gamma \frac{dt}{t-t_0}\int_\Gamma \frac{\varphi(t,\tau)}{\tau-t}\,d\tau = \int_\Gamma \frac{dt}{t-t_0}\int_\Gamma \frac{\varphi(t,\tau)-\varphi(t,t)}{\tau-t}\,d\tau$$

$$+ \int_\Gamma \frac{\varphi(t,t)-\varphi(t_0,t_0)}{t-t_0}\,dt \int_\Gamma \frac{d\tau}{\tau-t} + \varphi(t_0,t_0)\int_\Gamma \frac{dt}{t-t_0}\int_\Gamma \frac{d\tau}{\tau-t}.$$

In the first two integrals on the right-hand side the inversion of the order of integration is legitimated by (3) and (5). Hence,

$$\int_\Gamma \frac{dt}{t-t_0}\int_\Gamma \frac{\varphi(t,\tau)}{\tau-t}\,d\tau = \int_\Gamma d\tau \int_\Gamma \frac{\varphi(t,\tau)\,dt}{(t-t_0)(\tau-t)}$$

$$+ \varphi(t_0,t_0)\left[\int_\Gamma \frac{dt}{t-t_0}\int_\Gamma \frac{d\tau}{\tau-t} - \int_\Gamma d\tau \int_\Gamma \frac{dt}{(t-t_0)(\tau-t)}\right].$$

If Γ is a closed curve system, the first integral in square brackets is $-\pi^2$ (cf. (1.6)), while the second one is zero. Thus, in this case the proof is complete.

Now suppose the curve system Γ is not closed. Clearly, it suffices to consider the case of a single open curve. Let $\tilde\Gamma$ be a closed curve such that $\Gamma \subset \tilde\Gamma$ and put $\Gamma' = \tilde\Gamma \setminus \Gamma$. It is easy to see that in the integrals

$$\int_\Gamma \frac{dt}{t-t_0}\int_{\Gamma'} \frac{d\tau}{\tau-t},\quad \int_{\Gamma'} \cdots \int_\Gamma \cdots,\quad \int_{\Gamma'} \cdots \int_{\Gamma'} \cdots$$

the order of integration may be inverted. Consequently, the term in square brackets is equal to

$$\int_{\tilde\Gamma} \frac{dt}{t-t_0}\int_{\tilde\Gamma} \frac{d\tau}{\tau-t} - \int_{\tilde\Gamma} d\tau \int_{\tilde\Gamma} \frac{dt}{(t-t_0)(\tau-t)} = -\pi^2$$

and (6) is proved. ▮

Remark. In this section we essentially followed Mikhlin [7], § 3.

4.4. The singular integral operator on the space $\mathbf{H}^\mu(\Gamma)$

Theorem 4.7. *Let Γ be a closed Lyapunov curve system. Then the singular integral operator S_Γ is bounded on the space $\mathbf{H}^\mu(\Gamma)$ for every μ such that $0 < \mu < 1$.*

Proof. By a well-known theorem of Plemelj and Privalov, $S_\Gamma\varphi \in \mathbf{H}^\mu(\Gamma)$ for every $\varphi \in \mathbf{H}^\mu(\Gamma)$ (see Privalov [3] or Muskhelishvili [1])[1]. The boundedness of the operator S_Γ on $\mathbf{L}_2(\Gamma)$ (Theorem 2.3) implies that S_Γ is a closed operator on the space $\mathbf{H}^\mu(\Gamma)$. Finally, since S_Γ is defined on the whole space $\mathbf{H}^\mu(\Gamma)$, we deduce from a well-known theorem of Banach that the operator S_Γ is actually bounded on the space $\mathbf{H}^\mu(\Gamma)$. ∎

A direct proof of Theorem 4.7 can be found, for instance, in Prössdorf [17] (see Section 3.4.1 there).

Furthermore, the following important analogue of Theorem 4.2 holds.

Theorem 4.8. *Let Γ be a closed Lyapunov curve system and let $a \in \mathbf{H}^\mu(\Gamma)$ $(0 < \mu < 1)$. Then the commutator $T = aS_\Gamma - S_\Gamma a$ is compact from $\mathbf{H}^\nu(\Gamma)$ $(0 < \nu \leqq 1)$ into $\mathbf{H}^\mu(\Gamma)$.*

For a proof see Prössdorf [17] (Corollary 4.4, Chapter 3).

Remark. Theorem 4.7 is, in principle, due to Plemelj (in this connection see also Muskhelishvili [1] §§ 16–18). Theorem 4.8 for $\mu = \nu$ was first stated by I. A. Feldman (compare Budyanu and Gohberg [2]). For generalizations of the Theorems 4.7 and 4.8 to the case of piecewise Lyapunov curve systems and Hölder spaces with weight we refer to Duduchava [1].

§ 5. Operators related to the Cauchy singular integral

5.1. The adjoint singular integral operator

Let $t \in \Gamma$. It is easily seen that

$$\mathrm{d}t = h_\Gamma(t)\,|\mathrm{d}t|, \qquad h_\Gamma(t) = \mathrm{e}^{\mathrm{i}\theta_\Gamma(t)}, \tag{1}$$

where $\theta_\Gamma(t)$ denotes the angle between the tangent to Γ at the point t and the positive direction of the x-axis. Obviously, the function $h_\Gamma(t)$ is defined at all regular points $t \in \Gamma$ and is a bounded piecewise continuous function.

Further, recall that under the conditions (3.2) the singular integral operator S_Γ is bounded on both $\mathbf{L}_p(\Gamma, \varrho)$ and $\mathbf{L}_q(\Gamma, \varrho^{-1}) = \mathbf{L}_p^*(\Gamma, \varrho)$ $(1/p + 1/q = 1)$ (see Theorem 3.1 and the proof of Theorem 4.4).

Theorem 5.1. *For the adjoint operator S_Γ^* on the space $\mathbf{L}_q(\Gamma, \varrho^{-1})$ we have*

$$S_\Gamma^* = -H_\Gamma S_\Gamma H_\Gamma, \tag{2}$$

where H_Γ is the operator acting on the space $\mathbf{L}_q(\Gamma, \varrho^{-1})$ by the rule

$$(H_\Gamma\varphi)(t) = \overline{h_\Gamma(t)\,\varphi(t)}.$$

Proof. Let $\varphi \in \mathbf{L}_p(\Gamma, \varrho)$ and $\psi \in \mathbf{L}_q(\Gamma, \varrho^{-1})$. Application of formula (4.3) gives

$$(S_\Gamma\varphi, \psi) = \frac{1}{\pi\mathrm{i}}\int_\Gamma \overline{\psi(t)}\,|\mathrm{d}t| \int_\Gamma \frac{\varphi(\tau)}{\tau - t}\,\mathrm{d}\tau$$

$$= \frac{1}{\pi\mathrm{i}}\int_\Gamma \varphi(\tau)\,h_\Gamma(\tau)\,|\mathrm{d}\tau| \int_\Gamma \frac{\overline{\psi(t)}\,h_\Gamma^{-1}(t)}{\tau - t}\,\mathrm{d}t = (\varphi, -H_\Gamma S_\Gamma H_\Gamma\psi)$$

and this implies (2) at once. ∎

---

[1]) See also Chapter IX (Theorem 4.2). There the theorem of Plemelj and Privalov will be proved for the multidimensional case.

Corollary 5.1. *Let Γ be a circle, a circular arc, or an interval. Then $S_\Gamma^* = S_\Gamma$.*

Theorem 5.2. *Let Γ be a Lyapunov curve system and S_Γ^* the adjoint of the singular integral operator $S_\Gamma \in \mathscr{L}(\mathbf{L}_p(\Gamma, \varrho))$. Then the operator $T = S_\Gamma^* - S_\Gamma$ is compact on $\mathbf{L}_q(\Gamma, \varrho^{-1})$.*

Proof. We split the proof into three steps.

1. First suppose Γ is a closed Lyapunov curve. Then there exists a function $\tau = \alpha(\zeta)$ mapping the unit circle Γ_0 one-to-one onto Γ and having a derivative α' which satisfies a Hölder condition and vanishes nowhere on Γ_0. From Theorem 4.3 we know that $S_\Gamma = B^{-1} S_0 B + T_1$, where T_1 is compact, $S_0 = S_{\Gamma_0}$, and B is the bounded and invertible operator defined by $(B\varphi)(\zeta) = \varphi(\alpha(\zeta))$. It can easily be checked that $B^* = |\beta'|\, B^{-1}$ with $\beta = \alpha^{-1}$ and that $(B^{-1})^* = |\alpha'|\, B$. By virtue of Theorem 4.2 the commutator $S_0\, |\alpha'| - |\alpha'|\, S_0$ is compact and due to Corollary 4.1 we have $S_0^* = S_0$. Thus

$$\begin{aligned} S_\Gamma^* - S_\Gamma &= |\beta'|\, B^{-1}\, S_0\, |\alpha'|\, B + T_1^* - B^{-1}\, S_0 B - T_1 \\ &= |\beta'|\, |\alpha' \circ \beta|\, B^{-1}\, S_0 B - B^{-1}\, S_0 B + T_2 = T_2, \end{aligned}$$

where $(\alpha' \circ \beta)(\tau) = \alpha'(\beta(\tau))$ and T_2 is a compact operator.

2. Now let Γ be an open Lyapunov curve. Choose any closed Lyapunov curve $\tilde{\Gamma}$ so that $\Gamma \subset \tilde{\Gamma}$ and denote by $\chi(t)$ $(t \in \tilde{\Gamma})$ the characteristic function of the set Γ. Then the space $\mathbf{L}_p(\Gamma, \varrho)$ can obviously be identified with that subspace of $\mathbf{L}_p(\tilde{\Gamma}, \varrho)$ which consists of all functions of the form $\chi\tilde{\varphi}$ $(\tilde{\varphi} \in \mathbf{L}_p(\tilde{\Gamma}, \varrho))$. This subspace is invariant under the operator $A = \chi S_{\tilde{\Gamma}} \chi I$ and we have $A \mid \mathbf{L}_p(\Gamma, \varrho) = S_\Gamma$. By what has been proved in step 1, the operator $T = S_{\tilde{\Gamma}}^* - S_{\tilde{\Gamma}}$ is compact. But $A^* = \chi(S_{\tilde{\Gamma}} + T)\, \chi I$ and therefore $S_\Gamma^* - S_\Gamma$ is also compact.

3. Finally, we consider the general case, where Γ is a Lyapunov curve system consisting of the open or closed Lyapunov curves $\Gamma_1, \ldots, \Gamma_n$. Denote by $\chi_j(t)$ $(t \in \Gamma)$ the characteristic function of the set Γ_j $(j = 1, \ldots, n)$ and let R_j be the operator defined on $\mathbf{L}_p(\Gamma, \varrho)$ as $R_j = \chi_j I$. Then $S_\Gamma = \sum_{j,k=1}^{n} R_j S_\Gamma R_k$. The operators $R_j S_\Gamma R_k$ $(j \neq k)$ have a continuous kernel and are therefore compact. The restriction of the operator $R_j S_\Gamma R_j$ to the subspace $\mathbf{L}_p(\Gamma_j, \varrho) = R_j \mathbf{L}_p(\Gamma, \varrho)$ coincides with S_Γ. By what has already been proved we have $(R_j S_\Gamma R_j)^* = R_j S_\Gamma R_j + T_j$ with a compact operator T_j. Thus $S_\Gamma^* - S_\Gamma$ is compact. ▮

Remarks. It can be shown that S_Γ is a self-adjoint operator on the space $\mathbf{L}_2(\Gamma)$ if and *only* if Γ is a curve of the type mentioned in Corollary 5.1. This, as well as Theorem 5.1, was established by ITSKOVICH [3]. Theorem 5.2 for $p = 2$ and $\varrho = 1$ was stated by GOHBERG [4]. The proof of this theorem given here is taken from GOHBERG and KRUPNIK [4], Chapter I, § 7 (there it is also shown that Theorem 5.2 fails to be true if Γ is only required to be a piecewise Lyapunov curve system).

5.2. The singular integral with Hilbert kernel

An integral closely related to the Cauchy singular integral (1.3) is the *Hilbert singular integral*:

$$(H\varphi)(s) = \frac{1}{2\pi} \int_0^{2\pi} \cot \frac{\sigma - s}{2}\, \varphi(\sigma)\, d\sigma, \qquad 0 \leq s \leq 2\pi. \tag{3}$$

This integral is of course to be understood in the sense of the Cauchy principal value. It can be easily checked in a direct way that the integral (3) exists for every s and for every function $\varphi \in \mathbf{H}^\mu[0, 2\pi]$ $(0 < \mu \leq 1)$ satisfying $\varphi(0) = \varphi(2\pi)$. In this case

$$\frac{1}{2\pi} \int_0^{2\pi} \cot \frac{\sigma - s}{2}\, \varphi(\sigma)\, d\sigma = \frac{1}{2\pi} \int_0^{2\pi} \cot \frac{\sigma - s}{2}\, [\varphi(\sigma) - \varphi(s)]\, d\sigma.$$

Let Γ_0 be the unit circle with the parametric representation $t(s) = e^{is}$ $(0 \leqq s \leqq 2\pi)$. Then the difference between the Cauchy kernel $\frac{d\tau}{\tau - t}$ and the *Hilbert kernel* $\frac{1}{2} \cot \frac{\sigma - s}{2}$ $d\sigma$ is

$$\frac{d\tau}{\tau - t} - \frac{1}{2} \cot \frac{\sigma - s}{2} d\sigma = \frac{1}{2} \frac{d\tau}{\tau}.$$

Finally, by setting $\varphi(s) := \varphi[t(s)] = \varphi(t)$ we arrive at the following relation between the Hilbert and Cauchy singular integrals:

$$(H\varphi)(s) = i(S_{\Gamma_0}\varphi)(t) - \frac{1}{2\pi} \int_{\Gamma_0} \frac{\varphi(\tau)}{\tau} d\tau. \tag{4}$$

This and the Theorems 3.1 and 4.7 give

Theorem 5.3. *The operator H is bounded on the spaces $\mathbf{L}_p(\Gamma_0, \varrho)$ $(1 < p < \infty)$ and $\mathbf{H}^\mu(\Gamma_0)$ $(0 < \mu < 1)$.*

Taking into account identity (4.3) we readily obtain from (4) that

$$(H^2\varphi)(s) = -\varphi(s) + \frac{1}{2\pi} \int_0^{2\pi} \varphi(\sigma) d\sigma.$$

Furthermore, from (4) and (1.7) and from the boundedness of the operators S_{Γ_0} and H on $\mathbf{L}_p(\Gamma_0)$ $(1 < p < \infty)$ it is easy to derive the following representations: if $\varphi \in \mathbf{L}_p(\Gamma_0)$ has the Fourier series expansion[1]) $\varphi(t) = \sum_{k=-\infty}^{\infty} c_k t^k$ $(|t| = 1)$, then

$$(S_{\Gamma_0}\varphi)(t) = \sum_{k=0}^{\infty} c_k t^k - \sum_{k=-\infty}^{-1} c_k t^k, \tag{5}$$

$$(H\varphi)(t) = i \sum_{k=1}^{\infty} c_k t^k - i \sum_{k=-\infty}^{-1} c_k t^k. \tag{6}$$

5.3. The projections generated by the singular integral

Throughout this subsection assume Γ is a closed piecewise Lyapunov curve system.

Then $S_\Gamma^2 = I$ (see Theorem 3.1) and, consequently, the operators P_Γ and Q_Γ defined on the space $\mathbf{L}_p(\Gamma, \varrho)$ as

$$P_\Gamma := \frac{1}{2}(I + S_\Gamma), \qquad Q_\Gamma := \frac{1}{2}(I - S_\Gamma) \tag{7}$$

are complementary projections. Obviously,

$$P_\Gamma - Q_\Gamma = S_\Gamma, \qquad P_\Gamma Q_\Gamma = Q_\Gamma P_\Gamma = 0,$$

and (2) gives

$$P_\Gamma^* = H_\Gamma Q_\Gamma H_\Gamma, \qquad Q_\Gamma^* = H_\Gamma P_\Gamma H_\Gamma. \tag{8}$$

We define

$$\mathbf{L}_p^+(\Gamma, \varrho) := \operatorname{Im} P_\Gamma, \qquad \mathring{\mathbf{L}}_p^-(\Gamma, \varrho) := \operatorname{Im} Q_\Gamma. \tag{9}$$

---

[1]) It is well known that this series converges in the norm of $\mathbf{L}_p(\Gamma_0)$ (see ZYGMUND [3]).

Let now $\varphi \in \mathbf{L}_p(\Gamma, \varrho)$ and consider the Cauchy type integral

$$\Phi_\varphi(z) = \frac{1}{2\pi i} \int_\Gamma \frac{\varphi(\tau)}{\tau - z} \, d\tau \qquad (z \in \mathbf{C} \setminus \Gamma). \tag{10}$$

Since $\varphi \in \mathbf{L}_1(\Gamma)$ (see Section 3.1), $\Phi_\varphi(z)$ is an analytic function in $D_\Gamma^\pm$, we have $\Phi_\varphi(\infty) = 0$, and the nontangential limits[1])

$$\Phi_\varphi^+(t) := \lim_{\substack{z \to t \\ z \in D_\Gamma^+}} \Phi_\varphi(z), \qquad \Phi_\varphi^-(t) := \lim_{\substack{z \to t \\ z \in D_\Gamma^-}} \Phi_\varphi(z)$$

exist almost everywhere on Γ and are equal to

$$\Phi_\varphi^+(t) = (P_\Gamma \varphi)(t), \qquad \Phi_\varphi^-(t) = -(Q_\Gamma \varphi)(t) \tag{11}$$

(see PRIVALOV [2], p. 136).

In view of (7) a function $\varphi \in \mathbf{L}_p(\Gamma, \varrho)$ belongs to the subspace $\mathbf{L}_p^+(\Gamma, \varrho)$ (resp. $\mathring{\mathbf{L}}_p^-(\Gamma, \varrho)$) if and only if $S_\Gamma \varphi = \varphi$ (resp. $S_\Gamma \varphi = -\varphi$).

Theorem 1.1 shows that

$$\mathbf{R}(\Gamma) \cap \mathbf{L}_p^+(\Gamma, \varrho) = \mathbf{R}^+(\Gamma), \qquad \mathbf{R}(\Gamma) \cap \mathring{\mathbf{L}}_p^-(\Gamma, \varrho) = \mathring{\mathbf{R}}^-(\Gamma) \tag{12}$$

with $\mathring{\mathbf{R}}^-(\Gamma)$ the collection of all functions $\varphi \in \mathbf{R}^-(\Gamma)$ vanishing at infinity, $\varphi(\infty) = 0$. The relations (12) in particular imply that the sets $\mathbf{R}^+(\Gamma)$ and $\mathring{\mathbf{R}}^-(\Gamma)$ are dense in the subspaces $\mathbf{L}_p^+(\Gamma, \varrho)$ and $\mathring{\mathbf{L}}_p^-(\Gamma, \varrho)$, repectively.

Henceforth we denote by $D_1^+, \ldots, D_m^+$ and $D_1^-, \ldots, D_k^-$ the (maximal) connected components of the sets D_Γ^+ and D_Γ^-, respectively. Then the subspaces $\mathbf{L}_p^+(\Gamma, \varrho)$ and $\mathring{\mathbf{L}}_p^-(\Gamma, \varrho)$ can be characterized as follows.

Theorem 5.4. *For a function $\varphi \in \mathbf{L}_p(\Gamma, \varrho)$ the following assertions are equivalent:*

(i) $\varphi \in \mathbf{L}_p^+(\Gamma, \varrho)$;

(ii) *there exist points $\alpha_j^- \in D_j^-$ $(j = 1, \ldots, k)$ such that*

$$\int_\Gamma \frac{\varphi(\tau)\, d\tau}{(\tau - \alpha_j^-)^n} = 0 \qquad (j = 1, \ldots, k;\ n = 1, 2, \ldots); \tag{13}$$

(iii) (13) *holds for an arbitrary choice of the points $\alpha_j^- \in D_j^-$ $(j = 1, \ldots, k)$.*

Proof. The function $\Phi_\varphi(z)$ is analytic in D_Γ^- and therefore it can be expanded into a Taylor series in a neighborhood of each point $\alpha_j^- \in D_j^-$:

$$\Phi_\varphi(z) = \sum_{l=0}^\infty \frac{(z - \alpha_j^-)^l}{2\pi i} \int_\Gamma \frac{\varphi(\tau)\, d\tau}{(\tau - \alpha_j^-)^{l+1}}. \tag{14}$$

(i) $\Rightarrow$ (iii). Let $\varphi \in \mathbf{L}_p^+(\Gamma, \varrho)$. Then $Q_\Gamma \varphi = 0$ and, consequently, the nontangential limits $\Phi_\varphi^-(t)$ vanish almost everywhere on Γ. The well-known Luzin-Privalov uniqueness theorem then implies that $\Phi_\varphi(z) \equiv 0$ in D_Γ^- (see PRIVALOV [2], pp. 212–213). Thus, if $\alpha_j^- \in D_j^-$ $(j = 1, \ldots, k)$ are arbitrarily chosen points, equalities (13) immediately result from the representation (14).

(iii) $\Rightarrow$ (ii). This is trivial.

(ii) $\Rightarrow$ (i). From (14) we deduce that $\Phi_\varphi(z)$ vanishes identically in a neighborhood of each point α_j^- and since $\Phi_\varphi(z)$ is analytic, we obtain that $\Phi_\varphi(z) \equiv 0$ in D_Γ^-. Hence $Q_\Gamma \varphi = 0$, that is, $\varphi = P_\Gamma \varphi \in \mathbf{L}_p^+(\Gamma, \varrho)$. ∎

[1]) That is, the point $z \in D_\Gamma^\pm$ approaches the point $t \in \Gamma$ along any nontangential path. The formulas (11) are usually referred to as the *Sokhotski-Plemelj formulas*.

The following theorem can be proved analogously.

Theorem 5.5. *For a function $\varphi \in \mathbf{L}_p(\Gamma, \varrho)$ the following statements are equivalent:*

(i) $\varphi \in \mathring{\mathbf{L}}_p^-(\Gamma, \varrho)$;

(ii) *there exist points $\alpha_j^+ \in D_j^+$ $(j = 1, \ldots, m)$ such that*

$$\int_\Gamma \frac{\varphi(\tau)\,d\tau}{(\tau - \alpha_j^+)^n} = 0 \qquad (j = 1, \ldots, m;\ n = 1, 2, \ldots); \tag{15}$$

(iii) (15) *is fulfilled for every choice of the points $\alpha_j^+ \in D_j^+$ $(j = 1, \ldots, m)$.*

By expanding the function $\Phi_\varphi(z)$ into a Taylor series in a neighborhood of the point at infinity, similarly as above the following theorem can be proved.

Theorem 5.6. *Let Γ be a closed Lyapunov curve. Then a function $\varphi \in \mathbf{L}_p(\Gamma, \varrho)$ belongs to $\mathbf{L}_p^+(\Gamma, \varrho)$ if and only if*

$$\int_\Gamma \varphi(t)\, t^n\, dt = 0 \qquad (n = 0, 1, 2, \ldots). \tag{16}$$

Henceforth we denote by $\mathbf{L}_p^-(\Gamma, \varrho)$ the subspace of all functions of the form $\varphi = \varphi_1 + c$, where $\varphi_1 \in \mathring{\mathbf{L}}_p^-(\Gamma, \varrho)$ and $c = \text{const}$. Furthermore, we let $\mathbf{L}_\infty^\pm(\Gamma)$ denote the set of all functions which are analytic and bounded in $D_\Gamma^\pm$. Notice that $\mathbf{L}_\infty^\pm(\Gamma)$ can be identified with the subspace of all functions $\varphi \in \mathbf{L}_\infty(\Gamma)$ which almost everywhere coincide with the nontangential limits of a function $\varphi(z)$ analytic and bounded in $D_\Gamma^\pm$.

Indeed, by a theorem of Fatou (see Privalov [2], p. 129) every function $\varphi \in \mathbf{L}_\infty^\pm(\Gamma)$ possesses a nontangential limit $\varphi^\pm(t)$ almost everywhere on Γ and the boundary function belongs to $\mathbf{L}_\infty(\Gamma)$. Vice versa, due to the Luzin-Privalov uniqueness theorem the function $\varphi \in \mathbf{L}_\infty^\pm(\Gamma)$ is uniquely determined by its boundary function $\varphi^\pm \in \mathbf{L}_\infty(\Gamma)$.

Theorem 5.7. *The following inclusions hold:*

$$\mathbf{L}_\infty^+(\Gamma) \subset \mathbf{L}_p^+(\Gamma, \varrho), \qquad \mathbf{L}_\infty^-(\Gamma) \subset \mathbf{L}_p^-(\Gamma, \varrho). \tag{17}$$

Proof. Let $\varphi \in \mathbf{L}_\infty^+(\Gamma)$. Then $\varphi \in \mathbf{L}_p(\Gamma, \varrho)$. Consider a sequence of closed Lyapunov curve systems $\Gamma_n \subset D_\Gamma^+$ $(n = 1, 2, \ldots)$ approximating the curve system Γ as $n \to \infty$. Let $\beta_n\colon D_\Gamma^+ \to D_{\Gamma_n}^+$ be conformal mappings such that $\beta_n(t) \to t$ $(t \in \Gamma)$ as $n \to \infty$. Then, for $\alpha \in D_\Gamma^-$ and for N a positive integer, define the functions $f_n(t) = \varphi(\beta_n(t))\,(t - \alpha)^{-N}$. Since f_n is obviously analytic in $\overline{D_\Gamma^+}$, we have

$$\int_\Gamma f_n(\tau)\,d\tau = 0. \tag{18}$$

Since the sequence $f_n(\tau)$ $(n = 1, 2, \ldots)$ is uniformly bounded and converges to the function $\varphi(\tau)\,(\tau - \alpha)^{-N}$ almost everywhere on Γ, it is by the Lebesgue theorem legetimated to pass to the limit $n \to \infty$ in the integral in (18). Thus, the function φ satisfies conditions (13) and from Theorem 5.4 we conclude that $\varphi \in \mathbf{L}_p^+(\Gamma, \varrho)$.

If $\varphi \in \mathbf{L}_\infty^-(\Gamma)$, then analogously the function $\varphi_1 = \varphi - \varphi(\infty)$ can be shown to satisfy (15), and this implies that $\varphi_1 \in \mathring{\mathbf{L}}_p^-(\Gamma, \varrho)$ and $\varphi \in \mathbf{L}_p^-(\Gamma, \varrho)$. ▮

Theorem 5.8. *If $a_+ \in \mathbf{L}_\infty^+(\Gamma)$ and $a_- \in \mathbf{L}_\infty^-(\Gamma)$ then*

$$P_\Gamma a_+ P_\Gamma = a_+ P_\Gamma, \qquad Q_\Gamma a_- Q_\Gamma = a_- Q_\Gamma. \tag{19}$$

Proof. Let $r \in \mathbf{R}(\Gamma)$. Then, obviously, $a_+ P_\Gamma r = a_+ r_+ \in \mathbf{L}_\infty^+(\Gamma)$, and from Theorem 5.7 we deduce that $P_\Gamma a_+ P_\Gamma r = a_+ P_\Gamma r$. This proves the first equality in (19).

Analogously, $a_- Q_\Gamma r = a_- r_- \in \mathbf{L}_\infty^-(\Gamma)$. Since $a_-(\infty)\, r_-(\infty) = 0$, we have actually $a_- Q_\Gamma r \in \mathring{\mathbf{L}}_\infty^-(\Gamma)$ and therefore $Q_\Gamma a_- Q_\Gamma r = a_- Q_\Gamma r$. ▮

Denote by $\mathbf{C}^{\pm}(\Gamma)$ the set of all functions $\varphi \in \mathbf{C}(\Gamma)$ that admit an extension which is analytic in $D_\Gamma^{\pm}$ and continuous on $\overline{D_\Gamma^{\pm}}$. The Theorems 5.4 and 5.5 immediately imply that

$$\mathbf{C}^{+}(\Gamma) \subset \mathbf{L}_p^{+}(\Gamma, \varrho), \qquad \mathbf{C}^{-}(\Gamma) \subset \mathbf{L}_p^{-}(\Gamma, \varrho). \tag{20}$$

Remark. The essential results of this subsection go back to PRIVALOV [2]. The representation of the material was partially patterned after GOHBERG and KRUPNIK [4], Chapter II, § 4.

§ 6. The singular integral operator on spaces of differentiable functions

6.1. We first fix some notation.

1. A Lyapunov curve Γ is said to belong to the class C^m (m a positive integer) if the angle $\theta_\Gamma(t)$ between the tangent to Γ and the positive direction of the x-axis has continuous derivatives up to the order $m - 1$. If, moreover, the derivative $\theta_\Gamma^{(m-1)}(t)$ satisfies a Hölder condition with the exponent λ $(0 < \lambda \leqq 1)$, we write $\Gamma \in C^{m,\lambda}$. If $\Gamma \in C^m$ for all m, then the curve Γ is called a *curve of the class* C^∞.

A Lyapunov curve system Γ is said to belong to the class C^m (resp. $C^{m,\lambda}$) if all curves of Γ belong to C^m (resp. $C^{m,\lambda}$).

2. Let Γ be a Lyapunov curve system. We denote by $\mathbf{C}^m(\Gamma)$ the collection of all m-times continuously differentiable (complex-valued) functons on Γ and by $\mathbf{C}^{m,\lambda}(\Gamma)$ the set of all functions $f \in \mathbf{C}^m(\Gamma)$ for which $f^{(m)} \in \mathbf{H}^\lambda(\Gamma)$.

Provided with the norms

$$\|f\|_{\mathbf{C}^m(\Gamma)} = \sum_{k=0}^{m} \max_{t \in \Gamma} |f^{(k)}(t)|$$

and

$$\|f\|_{\mathbf{C}^{m,\lambda}(\Gamma)} = \|f\|_{\mathbf{C}^{m-1}(\Gamma)} + \|f^{(m)}\|_{\mathbf{H}^\lambda(\Gamma)} \tag{1}$$

the linear sets $\mathbf{C}^m(\Gamma)$ and $\mathbf{C}^{m,\lambda}(\Gamma)$ become Banach spaces. In the case $\lambda = 0$ we write $\mathbf{C}^{m,0}(\Gamma) = \mathbf{C}^m(\Gamma)$ and $\mathbf{H}^0(\Gamma) = \mathbf{C}(\Gamma)$.

Let Γ be a curve of the class $C^{m,\lambda}$ and $t = t(s)$ $(0 \leqq s \leqq \gamma)$ its parametric representation (s the arc length, measured from a fixed point on Γ). Then $t(s) \in \mathbf{C}^{m,\lambda}[0, \gamma]$ and $\mathbf{C}^{m,\lambda}(\Gamma)$ coincides with the set of all functions f on Γ for which the function $\tilde{f}(s) := f[t(s)]$ belongs to $\mathbf{C}^{m,\lambda}[0, \gamma]$; note that $\tilde{f}$ is a function of a real variable. Clearly, the norm (1) and the norm $\|\tilde{f}\|_{\mathbf{C}^{m,\lambda}[0,\gamma]}$ are equivalent.

3. As usual, we denote by $\mathbf{W}_p^{(m)}(\Gamma)$ $(1 < p < \infty)$ the Sobolev space of all functions $f \in \mathbf{L}_p(\Gamma)$ which possess generalized derivatives $f^{(k)} \in \mathbf{L}_p(\Gamma)$, $k = 1, 2, \ldots, m$ (see SOBOLEV [3] or SMIRNOV [1]). The norm of a function $f \in \mathbf{W}_p^{(m)}(\Gamma)$ is defined by

$$\|f\|_{\mathbf{W}_p^{(m)}(\Gamma)} = \sum_{k=0}^{m} \|f^{(k)}\|_{\mathbf{L}_p(\Gamma)}. \tag{2}$$

$\mathbf{W}_p^{(m)}(\Gamma)$ is the completion of the space $\mathbf{R}(\Gamma)$ of all rational functions without poles on Γ under the norm (2). If the curve system Γ consists of the curves Γ_j $(j = 1, \ldots, n)$, the norm (2) is equivalent to the norm $\sum_{j=1}^{n} \|f\|_{\mathbf{W}_p^{(m)}(\Gamma_j)}$.

If $\Gamma \in C^m$ then $\mathbf{W}_p^{(m)}(\Gamma)$ coincides with the set of all functions f on Γ such that $\tilde{f}(s) := f[t(s)] \in \mathbf{W}_p^{(m)}(0, \gamma)$, and in this case the norm (2) is equivalent to the norm

$$\|\tilde{f}\|_{\mathbf{W}_p^{(m)}(0,\gamma)} = \sum_{k=0}^{m} \|\tilde{f}^{(k)}\|_{\mathbf{L}_p(0,\gamma)}.$$

4. Let Γ be a Lyapunov curve system, let $t_1, \ldots, t_r$ be pairwise distinct points on Γ, and let $\Gamma_{t_j} \subset \Gamma$ be an arc of the curve containing t_j in its interior.

We denote by $\mathbf{H}^\lambda(\Gamma; (\Gamma_{t_j}, m_j)_{j=1}^r)$ $(0 \leqq \lambda \leqq 1)$ the collection of all functions $f \in \mathbf{H}^\lambda(\Gamma)$ whose restrictions to Γ_{t_j} belong to the space $\mathbf{C}^{m_j,\lambda}(\Gamma_{t_j})$. Analogously, let $\mathbf{L}_p(\Gamma; (\Gamma_{t_j}, m_j)_{j=1}^r)$ denote the collection of all functions $f \in \mathbf{L}_p(\Gamma)$ the restrictions to Γ_{t_j} of which are in $\mathbf{W}_p^{(m_j)}(\Gamma_{t_j})$. Endowed with the norms

$$\|f\|_{\mathbf{H}^\lambda(\Gamma)} + \sum_{j=1}^{r} \|f\|_{\mathbf{C}^{m_j,\lambda}(\Gamma_{t_j})}$$

and

$$\|f\|_{\mathbf{L}_p(\Gamma)} + \sum_{j=1}^{r} \|f\|_{\mathbf{W}_p^{(m_j)}(\Gamma_{t_j})}$$

the linear sets $\mathbf{H}^\lambda(\Gamma; (\Gamma_{t_j}, m_j)_{j=1}^r)$ and $\mathbf{L}_p(\Gamma; (\Gamma_{t_j}, m_j)_{j=1}^r)$ become Banach spaces. In the following, whenever the concrete look of the arcs Γ_{t_j} is meaningless, we shall in the notation of the above spaces replace Γ_{t_j} by t_j. In particular, $\mathbf{C}(\Gamma; (t_j, m_j)_{j=1}^r)$ denotes the class of all functions in $\mathbf{C}(\Gamma)$ whose restrictions to certain Γ_{t_j} $(j = 1, \ldots, r)$ are in $\mathbf{C}^m(\Gamma_{t_j})$.

6.2. The objective of this subsection is to prove the boundedness of the singular integral operator S_Γ and the compactness of the commutator $aS_\Gamma - S_\Gamma a$ on the spaces introduced above.

Lemma 6.1. *Let Γ be an open Lyapunov curve with the end points a and b and with the positive direction chosen from a to b. If $\varphi \in \mathbf{W}_p^{(1)}(\Gamma)$ $(1 < p < \infty)$, then*

$$\frac{\mathrm{d}}{\mathrm{d}t} \int_\Gamma \frac{\varphi(\tau)}{\tau - t} \,\mathrm{d}\tau = \frac{\varphi(a)}{a - t} - \frac{\varphi(b)}{b - t} + \int_\Gamma \frac{\varphi'(\tau)}{\tau - t} \,\mathrm{d}\tau \tag{3}$$

for all inner points t of the curve Γ.

Proof. Assume first that $\varphi \in \mathbf{C}^{1,\lambda}(\Gamma)$ $(0 < \lambda < 1)$. Cut the complex plane along any curve joining the point $t \in \Gamma$ $(t \neq a, t \neq b)$ to the point at infinity and lying on the right of Γ. Then choose a continuous branch of $\ln (z - t)$ on the z-plane with this cut.

Let $\varepsilon > 0$ be an arbitrary sufficiently small number. Denote by t_1 and t_2 the two points on Γ for which the arc lengths t_1t and tt_2 are equal to ε and denote by Γ_ε the arc t_1t_2. Now consider the function

$$\psi_\varepsilon(t) = \int_{\Gamma \setminus \Gamma_\varepsilon} \varphi'(\tau) \ln (\tau - t) \,\mathrm{d}\tau.$$

Integration by parts gives

$$\psi_\varepsilon(t) = -\int_{\Gamma \setminus \Gamma_\varepsilon} \frac{\varphi(\tau)}{\tau - t} \,\mathrm{d}\tau + \varphi(b) \ln (b - t) - \varphi(a) \ln (a - t)$$
$$+ \varphi(t_1) \ln (t_1 - t) - \varphi(t_2) \ln (t_2 - t).$$

As $\varepsilon \to 0$, the integral on the right-hand side tends to the singular integral $-\int_\Gamma \frac{\varphi(\tau)}{\tau - t}\,d\tau$, while the following two terms do not depend on ε. The last two terms can be written as

$$\varphi(t_1) \ln (t_1 - t) - \varphi(t_2) \ln (t_2 - t) = \varphi(t) [\ln (t_1 - t) - \ln (t_2 - t)]$$
$$+ [\varphi(t_1) - \varphi(t)] \ln (t_1 - t) - [\varphi(t_2) - \varphi(t)] \ln (t_2 - t).$$

The first term on the right-hand side of this equality converges to $\pi i \varphi(t)$ as $\varepsilon \to 0$, and since $\lim_{x\to 0} x^\lambda \ln x = 0$, the last two terms converge to zero as $\varepsilon \to 0$. Thus,

$$\lim_{\varepsilon\to 0} \psi_\varepsilon(t) = \pi i \varphi(t) + \varphi(b) \ln (b - t) - \varphi(a) \ln (a - t) - \int_\Gamma \frac{\varphi(\tau)}{\tau - t}\,d\tau. \tag{4}$$

Now differentiate the function $\psi_\varepsilon(t)$. Taking into account the differentiation rule for parameter integrals with variable limits of integration we get

$$\psi_\varepsilon'(t) = -\int_{\Gamma\setminus\Gamma_\varepsilon} \frac{\varphi'(\tau)}{\tau - t}\,d\tau + [\varphi'(t_1) \ln (t_1 - t)\, t'(s - \varepsilon) - \varphi'(t_2) \ln (t_2 - t)\, t'(s + \varepsilon)]\, \overline{t'(s)}.$$

By what has already been proved, the right-hand side converges to

$$-\int_\Gamma \frac{\varphi'(\tau)}{\tau - t}\,d\tau + \pi i \varphi'(t)$$

as $\varepsilon \to 0$. Moreover, this convergence is uniform with respect to $t \in \Gamma_0$, where $\Gamma_0 \subset \Gamma$ is an arbitrary closed curve which does not contain the endpoints a and b (cf. Muskhelishvili [1], p. 40). Hence

$$\left[\lim_{\varepsilon\to 0} \psi_\varepsilon(t)\right]' = \lim_{\varepsilon\to 0} \psi_\varepsilon'(t) = \pi i \varphi'(t) - \int_\Gamma \frac{\varphi'(\tau)}{\tau - t}\,d\tau.$$

This and (4) prove (3) for $\varphi \in \mathbf{C}^{1,\lambda}(\Gamma)$.

Let now $\varphi \in \mathbf{W}_p^{(1)}(\Gamma)$. Then there is a sequence of functions $\varphi_n \in \mathbf{C}^{1,\lambda}(\Gamma)$ such that

$$\|\varphi - \varphi_n\|_{\mathbf{W}_p^{(1)}(\Gamma)} \to 0.$$

Put

$$\psi(t) = \int_\Gamma \frac{\varphi(\tau)}{\tau - t}\,d\tau, \qquad \psi_n(t) = \int_\Gamma \frac{\varphi_n(\tau)}{\tau - t}\,d\tau,$$

and denote the right-hand side of (3) by $\chi(t)$.

Since (3) has already been proved for the functions φ_n and since S_Γ is bounded on $\mathbf{L}_p(\Gamma)$, we obtain that

$$\|\psi - \psi_n\|_{\mathbf{L}_p(\Gamma)} \to 0, \qquad \|\chi - \psi_n'\|_{\mathbf{L}_p(\Gamma_0)} \to 0, \qquad n \to \infty.$$

But the operator of generalized differentiation is closed (see Smirnov [1]), and so we deduce that $\psi \in \mathbf{W}_p^{(1)}(\Gamma_0)$ and that $\psi'(t) = \chi(t)$, and this is what was wanted. ▮

Let us record an immediate consequence of the preceding proof.

Corollary 6.1. *Let $\varphi \in \mathbf{W}_p^{(1)}(\Gamma)$ and suppose t is an inner point of the curve Γ. Then one has the following rule for integration by parts:*

$$\int_\Gamma \varphi'(\tau) \ln (\tau - t)\,d\tau = \pi i \varphi(t) + \varphi(b) \ln (b - t) - \varphi(a) \ln (a - t) - \int_\Gamma \frac{\varphi(\tau)}{\tau - t}\,d\tau.$$

In what follows let Γ denote a Lyapunov curve system. Suppose $t_1, t_2, \ldots, t_r$ are pairwise distinct points on Γ and let $\Gamma_{t_j}, \Gamma'_{t_j}$ $(j = 1, 2, \ldots, r)$ be arcs of Γ such that both Γ_{t_j} and Γ'_{t_j} contain t_j as an inner point and that the closure $\overline{\Gamma'_{t_j}}$ (i.e., the arc Γ'_{t_j} together with its end points) entirely lies in the interior of Γ_{t_j}.

Lemma 6.1 immediately implies the following.

Theorem 6.1. *The singular integral operator S_Γ maps the space $\mathbf{L}_p(\Gamma; (\Gamma_{t_j}, m_j)_{j=1}^r)$ continuously into the space $\mathbf{L}_p(\Gamma; (\Gamma'_{t_j}, m_j)_{j=1}^r)$.*

An analogous result is true for $\mathbf{H}^\lambda(\Gamma; (\Gamma_{t_j}, m_j)_{j=1}^r)$ $(0 < \lambda < 1)$.

Theorem 6.2. *Let Γ be a closed Lyapunov curve system. Then S_Γ is a homeomorphism of the space $\mathbf{W}_p^{(m)}(\Gamma)$ $(1 < p < \infty)$ onto itself, and, for every $\varphi \in \mathbf{W}_p^{(m)}(\Gamma)$,*

$$\frac{\mathrm{d}^k}{\mathrm{d}t^k} \int_\Gamma \frac{\varphi(\tau)}{\tau - t} \mathrm{d}\tau = \int_\Gamma \frac{\varphi^{(k)}(\tau)}{\tau - t} \mathrm{d}\tau \qquad (k = 1, \ldots, m). \tag{5}$$

Again, the analogue of this theorem holds for $\mathbf{C}^{m,\lambda}(\Gamma)$ $(0 < \lambda < 1)$.

For the proof of Theorem 6.2 notice that in the case of a closed curve Γ $(a = b)$ the first two terms on the right-hand side of (3) vanish and that $S_\Gamma^2 = I$.

Theorem 6.3. *Let $a \in \mathbf{H}^\lambda(\Gamma; (t_j, m_j)_{j=1}^r)$ $(0 < \lambda < 1)$ and $\Gamma \in C^{m,\lambda}$, where $m = \max_j m_j$. Then the commutator $T = aS_\Gamma - S_\Gamma a$ is a compact operator from the space $\mathbf{L}_p(\Gamma)$ $(1 < p < \infty)$ into the space $\mathbf{L}_p(\Gamma; (t_j, m_j)_{j=1}^r)$.*

Proof. Let $\varphi \in \mathbf{L}_p(\Gamma)$. Our first objective is to show that $\psi = T\varphi$ belongs to $\mathbf{L}_p(\Gamma; (t_j, m_j)_{j=1}^r)$. Because $\psi \in \mathbf{L}_p(\Gamma)$, it suffices to consider only the restriction of the function ψ to Γ_{t_j} $(j = 1, \ldots, r)$. Furthermore, we may assume that Γ is a Lyapunov curve of the class $C^{m,\lambda}$.

Put

$$\psi(t) = \int_\Gamma T(t, \tau)\, \varphi(\tau)\, \mathrm{d}\tau, \qquad T(t, \tau) = \frac{1}{\pi \mathrm{i}} \frac{a(t) - a(\tau)}{\tau - t},$$

and, for fixed j $(j = 1, \ldots, r)$, consider only points $t \in \Gamma_{t_j}$. It is easy to see that the function $T(t, \tau)$ has partial derivatives with respect to t up to the order $m_j - 1$, which, moreover, belong to the class $\mathbf{H}^\lambda(\Gamma_{t_j} \times \Gamma)$. The m_j-th derivative can be represented as

$$T_t^{(m_j)}(t, \tau) = \frac{K(t, \tau) - K(t, t)}{\tau - t}$$

with $K(t, \tau) \in \mathbf{H}^\lambda(\Gamma_{t_j} \times \Gamma)$ (see Muskhelishvili [1], § 7). Thus, $T_t^{(m_j)}(t, \tau)$ is a weakly singular kernel. Furthermore, we have

$$\psi^{(k)}(t) = \int_\Gamma T_t^{(k)}(t, \tau)\, \varphi(\tau)\, \mathrm{d}\tau \qquad (t \in \Gamma_{t_j};\ k = 1, \ldots, m_j). \tag{6}$$

Indeed, if $k = 1, \ldots, m_j - 1$, then, by virtue of the continuity of the derivatives $T_t^{(k)}(t, \tau)$, this formula is a well-known result about parameter integrals (see, for instance, Fichtenholz [1], Bd. II). Thus we are left with the case $k = m_j$. If $\varphi \in \mathbf{C}(\Gamma)$, consider, in analogy to the proof of Lemma 6.1, the integral

$$\int_{\Gamma \setminus \Gamma_\varepsilon} T_t^{(m_j - 1)}(t, \tau)\, \varphi(\tau)\, \mathrm{d}\tau,$$

differentiate it, and then pass to the limit $\varepsilon \to 0$. This will give the assertion. Now let $\varphi \in \mathbf{L}_p(\Gamma)$. In this case choose a sequence of continuous functions φ_n so that

$\|\varphi - \varphi_n\|_{\mathbf{L}_p(\Gamma)} \to 0$ and, for $t \in \Gamma_{t_j}$, put

$$\Phi_n(t) := \int_\Gamma T_t^{(m_j-1)}(t, \tau)\, \varphi_n(\tau)\, d\tau,$$

$$\chi(t) := \int_\Gamma T_t^{(m_j)}(t, \tau)\, \varphi(\tau)\, d\tau.$$

By what has already been proved,

$$\Phi_n'(t) = \int_\Gamma T_t^{(m_j)}(t, \tau)\, \varphi_n(\tau)\, d\tau.$$

Since the integral operators defined by the right-hand side of (6) are compact from $\mathbf{L}_p(\Gamma)$ into $\mathbf{L}_p(\Gamma_{t_j})$, we get as $n \to \infty$

$$\|\psi^{(m_j-1)} - \Phi_n\|_{\mathbf{L}_p(\Gamma_{t_j})} \leqq C\, \|\varphi - \varphi_n\|_{\mathbf{L}_p(\Gamma)} \to 0,$$

$$\|\chi - \Phi_n'\|_{\mathbf{L}_p(\Gamma_{t_j})} \leqq C\, \|\varphi - \varphi_n\|_{\mathbf{L}_p(\Gamma)} \to 0.$$

Taking into consideration that the operator of generalized differentiation is closed, we therefore obtain that $\psi^{(m_j-1)} \in \mathbf{W}_p^{(1)}(\Gamma_{t_j})$ and $\psi^{(m_j)}(t) = \chi(t)$. This proves (6) for $k = m_j$.

From (6) and the above mentioned compactness of the operators defined by the right-hand sides of (6) we now get the assertion of the theorem immediately. ▮

Remark. Theorem 6.3 is also true for $\lambda = 0$ (see CALDERON [4]).

Theorem 6.4. *Let* $a \in \mathbf{H}^\lambda(\Gamma; (t_j, m_j)_{j=1}^r)$ $(0 < \lambda \leqq 1)$. *Suppose* $\Gamma \in C^{m,\lambda}$ *with* $m = \max_j m_j$, *and let* $0 < \mu < \lambda$. *Then the commutator* $T = aS_\Gamma - S_\Gamma a$ *is a compact operator from each of the spaces* $\mathbf{H}^\nu(\Gamma)$ $(0 < \nu \leqq 1)$ *into the space* $\mathbf{H}^\mu(\Gamma; (t_j, m_j)_{j=1}^r)$.

This is again a consequence of the formulas (6), since the operators defined by their right-hand sides are compact from $\mathbf{H}^\nu(\Gamma)$ into $\mathbf{H}^\mu(\Gamma_{t_j})$ (see PRÖSSDORF [17], Corollary 4.3, Chapter 3).

Remark. Corollary 6.1 is due to MIKHLIN [7]. The proof of Lemma 6.1 given here is a modification of MIKHLIN's proof [7] of Corollary 6.1. Formula (5) (for functions $\varphi \in \mathbf{C}^{m,\lambda}(\Gamma)$) can be found in GAKHOV [1] (p. 49). The proof of Theorem 6.3 follows PRÖSSDORF [10].

Chapter III
One-dimensional singular integral equations with continuous coefficients on closed curves

§ 1. Abstract singular operators

The purpose of the present section is to establish a sufficiently abstract foundation for our further investigations of one-dimensional singular integral equations.

1.1. Paired operators

Let X be a Banach space, $P \in \mathscr{L}(X)$ a projection and $Q = I - P$ the complementary projection.

An operator of the form $AP + BQ$ or $PA + QB$ with $A \in \mathscr{L}(X)$ and $B \in \mathscr{L}(X)$ is called a *paired operator*. In this context the operators A and B are referred to as the *coefficients* of the corresponding paired operator. The paired operators $AP + BQ$ and $PB + QA$ are called *transposed* to each other, while the operators $AP + BQ$ and $PA + QB$ are called *dual* to each other.

It turns out that paired operators are intimately related to operators of the form[1]) $PCP \mid \operatorname{Im} P$ and $QCQ \mid \operatorname{Im} Q$ $(C \in \mathscr{L}(X))$. The latter operators are called *operators of Wiener-Hopf type* or simply *Wiener-Hopf operators*.

Consider first the special case in which one of the coefficients is the identity operator. For example, let $B = I$. Then, obviously,

$$AP + Q = (PAP + Q)(I + QAP). \tag{1}$$

Since $(QAP)^2 = 0$, the operator $I + QAP$ is invertible and its inverse is

$$(I + QAP)^{-1} = I - QAP. \tag{2}$$

It is not difficult to see that the operator $PAP + Q$ is a $\Phi_{\pm}$-operator or a one-sided invertible operator on X if and only if the operator PAP restricted to the subspace $\operatorname{Im} P$ has the corresponding property (cf. Chapter I). Clearly, the preceding arguments also apply with the roles of P and Q interchanged. Furthermore, one has the following analogues of the relations (1) and (2):

$$PA + Q = (I + PAQ)(PAP + Q) \tag{3}$$

and

$$(I + PAQ)^{-1} = I - PAQ. \tag{4}$$

There is a one-to-one correspondence between the solutions of the equation

$$PAP\varphi = Pf \tag{5}$$

and those of

$$(AB + Q)\,\psi = f. \tag{6}$$

---

[1]) Recall that $PAP \mid \operatorname{Im} P$ denotes the restriction of the operator PAP to the subspace $\operatorname{Im} P$.

Namely, if ψ is a solution of equation (6), then $\varphi = P\psi$ is obviously a solution of (5). Conversely, if φ is a solution of (5), then $\psi = (I - QA)\, P\varphi + Qf$ is a solution of the equation (6).

If A and B are invertible operators, formula (1) gives the representations

$$AP + BQ = B(PB^{-1}AP + Q)\,(I + QB^{-1}AP) \tag{7}$$

and

$$AP + BQ = A(P + QA^{-1}BQ)\,(I + PA^{-1}BQ)\,. \tag{8}$$

This together with the above arguments implies the following result.

Theorem 1.1. *If the operators $A \in \mathscr{L}(X)$ and $B \in \mathscr{L}(X)$ are invertible then*

$$\begin{aligned}\dim \operatorname{Ker} (AP + BQ) &= \dim \operatorname{Ker} (PB^{-1}AP \mid \operatorname{Im} P)\\ &= \dim \operatorname{Ker} (QA^{-1}BQ \mid \operatorname{Im} Q)\,,\end{aligned} \tag{9}$$

$$\begin{aligned}\dim \operatorname{Coker} (AP + BQ) &= \dim \operatorname{Coker} (PB^{-1}AP \mid \operatorname{Im} P)\\ &= \dim \operatorname{Coker} (QA^{-1}BQ \mid \operatorname{Im} Q)\,.\end{aligned} \tag{10}$$

Analogous relations can be obtained for the paired operator $PA + QB$. Then one has to take into account the representations

$$PA + QB = (I + PAB^{-1}Q)\,(PAB^{-1}P + Q)\,B \tag{7'}$$

and

$$PA + QB = (I + QBA^{-1}P)\,(P + QBA^{-1}Q)\,A\,. \tag{8'}$$

Formulae (2), (4), (7), and (7′) immediately yield the following simple relationship between dual paired operators.

Theorem 1.2. *Let $A \in \mathscr{L}(X)$ and $B \in \mathscr{L}(X)$ be invertible operators and suppose $AB = BA$. Then*

$$C_1(AP + BQ)\,C_2 = PA + QB,$$

where

$$C_1 = (I + PAB^{-1}Q)\,B^{-1}, \qquad C_2 = (I - QAB^{-1}P)\,B$$

are invertible operators.

Corollary 1.1. *Let $A \in \mathscr{L}(X)$ and $B \in \mathscr{L}(X)$ be invertible operators and suppose $AB = BA$. Then for the operator $AP + BQ$ to be a $\Phi_+(\Phi_-)$-operator or to admit a left (right) regularization it is necessary and sufficient that the operator $PA + QB$ have the corresponding property. Furthermore, one has*

$$\dim \operatorname{Ker} (AP + BQ) = \dim \operatorname{Ker} (PA + QB),$$

$$\dim \operatorname{Coker} (AP + BQ) = \dim \operatorname{Coker} (PA + QB)\,.$$

Theorem 1.3. *Let $A \in \mathscr{L}(X)$, $B \in \mathscr{L}(X)$ and let $AB = BA$.*

(a) *If* $\operatorname{Ker} A \cap \operatorname{Ker} B = \{0\}$, *then*

$$\dim \operatorname{Ker} (AP + BQ) \geqq \dim \operatorname{Ker} (PA + QB)\,. \tag{11}$$

(b) *If there are two operators A', $B' \in \mathscr{L}(X)$ such that $AA' + BB' = I$ and if A' and B' commute with A and B, then*

$$\dim \operatorname{Ker} (AP + BQ) = \dim \operatorname{Ker} (PA + QB)\,.$$

Proof. (a) Let $(PA + QB)\varphi = 0$ ($\varphi \in X$). Then $PA\varphi = -QB\varphi$ and since $PQ = QP = 0$, we see that $PA\varphi = QB\varphi = 0$.

Put $\psi = B\varphi - A\varphi$. Then, obviously, $P\psi = B\varphi$ and $Q\psi = -A\varphi$. Hence $(AP + BQ)\psi = 0$. If $\psi = 0$ then $\varphi \in \operatorname{Ker} A \cap \operatorname{Ker} B$ and, consequently, $\varphi = 0$. This proves (a).

(b) If $A'A + B'B = I$, then $\operatorname{Ker} A \cap \operatorname{Ker} B = \{0\}$ and therefore the inequality (11) holds. Let us show the reverse inequality.

Let $AP\varphi + BQ\varphi = 0$. Put $\psi = B'P\varphi - A'Q\varphi$. Then it is easy to see that $A\psi = -Q\varphi$, $B\psi = P\varphi$ and hence $(PA + QB)\psi = 0$. But if $\psi = 0$, then obviously $\varphi = 0$ and so $\dim \operatorname{Ker}(AP + BQ) \leq \dim \operatorname{Ker}(PA + QB)$. ∎

1.2. Abstract singular operators

In what follows we assume that the coefficients of the paired operator belong to some subalgebra $\mathfrak{M} \subset \mathcal{L}(X)$ subject to the following condition:

(A) *The commutator* $[P, A] = PA - AP$ *is compact on* X *for every operator* $A \in \mathfrak{M}$.

First of all notice that if condition (A) is satisfied, then, for every $A \in \mathfrak{M}$, the operators $[Q, A]$, PAQ, QAP are compact, too. This is obvious from the identities $[Q, A] = [A, P]$, $PAQ = P[A, Q] = [P, A]\,Q$.

An operator of the form

$$C := AP + BQ + T, \tag{12}$$

where $A, B \in \mathfrak{M}$ and T is a linear compact operator on X, is called an *abstract singular operator* or simply a *singular operator*. The operators A and B are sometimes referred to as the *coefficients* of the operator (12).

The operator $C_0 := AP + BQ$ is called the *characteristic* (or *dominant*) *part* and T the *compact part* of the singular operator C.

Let $\tilde{\mathfrak{M}}$ denote the collection of all singular operators of the form (12). $\tilde{\mathfrak{M}}$ is easily seen to be an algebra.

Indeed, it is obvious that $\tilde{\mathfrak{M}}$ is a linear set. So it only remains to show that the product of two operators from $\tilde{\mathfrak{M}}$ again belongs to $\tilde{\mathfrak{M}}$.

Let $C_1 = A_1P + B_1Q + T_1$ and $C_2 = A_2P + B_2Q + T_2$ be two operators from $\tilde{\mathfrak{M}}$. Then

$$C_1C_2 = A_1A_2P + B_1B_2Q + T \tag{13}$$

with

$$T = A_1[P, A_2]\,P + B_1[Q, B_2]\,Q + A_1PB_2Q + B_1QA_2P + C_1T_2 + T_1C_2 - T_1T_2.$$

But in view of condition (A) the operator T is compact and therefore (13) shows that $C_1C_2 \in \tilde{\mathfrak{M}}$.

The abstract singular operator (12) is said to be an *operator of normal type* if its coefficients A and B are invertible in $\mathfrak{M}$, that is, if the inverses A^{-1} and B^{-1} of A and B exist and belong to $\mathfrak{M}$.

It is immediate from (13) that the product of two singular operators of normal type is again an operator of normal type.

Theorem 1.4. *Let the algebra* $\mathfrak{M}$ *satisfy the condition* (A). *Then if the singular operator* C *is of normal type, the operator* $R = A^{-1}P + B^{-1}Q$ *is a two-sided regularizer of the operator* C.

Proof. In (13), put $A_1 = A^{-1}$, $B_1 = B^{-1}$, $A_2 = A$, and $B_2 = B$. What results is

$$RC = P + Q + T' = I + T'$$

with some compact operator T'. Analogously, by interchanging the roles of A and A^{-1} (resp. B and B^{-1}), we get $CR = I + T''$ with some compact operator T''. ▮

Corollary 1.2. *Under the hypotheses of Theorem 1.4, C is a Φ-operator.*

This is an immediate consequence of Theorem 3.1, Chapter I.

Corollary 1.3. *Let $A_1, B_1 \in \mathscr{L}(X)$ and $A_2, B_2 \in \mathfrak{M}$. Then*

$$(A_1P + B_1Q)(A_2P + B_2Q) = A_1A_2P + B_1B_2Q + T, \tag{14}$$

where T is a compact operator.

Note that (14) immediately results from (13).

Henceforth the equation

$$C\varphi := AP\varphi + BQ\varphi + T\varphi = f \tag{15}$$

will be called an *abstract singular equation* or simply *singular equation*. If C is an operator of normal type, (15) is said to be an *equation of normal type*. Equation (15) with $T = 0$ is referred to as the *characteristic* (or *dominant*) *singular equation*.

Remark. Theorem 1.1 was stated by Gohberg and Feldman [1] (Chapter V, § 1) and Theorem 1.2 by Gohberg and Krupnik [4] (Chapter VII, § 6). Theorem 1.3 is due to Prössdorf and Silbermann [5]. The definitions and results contained in Section 1.2 can be found in the papers of Halilov [1] and Cherski [1], and also in the monograph of D. Przeworska-Rolewicz and S. Rolewicz [1] (Chapter II).

§ 2. Singular integral operators with rational coefficients

Throughout this and the following sections of the present chapter (with the exception of § 7) Γ always denotes a closed piecewise Lyapunov curve system and ϱ the weight

$$\varrho(t) = \prod_{j=1}^{m} |t - t_j|^{\beta_j},$$

where $t_j \in \Gamma$, $-1/p < \beta_j < 1 - 1/p$ $(j = 1, \ldots, m)$, and $1 < p < \infty$ (see Chapter II, § 3).

2.1. Important examples of paired operators on the space $\mathbf{L}_p(\Gamma, \varrho)$ are the operators $A = cI + dS_\Gamma$ and $B = cI + S_\Gamma d$, where $c, d \in \mathbf{L}_\infty(\Gamma)$.[1]) The operators A and B are called *singular integral operators on the curve system Γ*. They can also be written in the form

$$A = aP_\Gamma + bQ_\Gamma, \qquad B = P_\Gamma a + Q_\Gamma b,$$

where $a = c + d$, $b = c - d$, $P_\Gamma = \frac{1}{2}(I + S_\Gamma)$, $Q_\Gamma = \frac{1}{2}(I - S_\Gamma)$. Obviously,

$$\|A\|, \|B\| \leqq \|a\|_{\mathbf{L}_\infty(\Gamma)} \|P_\Gamma\| + \|b\|_{\mathbf{L}_\infty(\Gamma)} \|Q_\Gamma\|.$$

For the adjoint operators A^* and B^*, acting on the space $\mathbf{L}_q(\Gamma, \varrho^{-1})$ $(1/p + 1/q = 1)$, we have

$$A^* = P_\Gamma^* \bar{a} + Q_\Gamma^* \bar{b}, \qquad B^* = \bar{a}P_\Gamma^* + \bar{b}Q_\Gamma^*.$$

and consequently, taking into account relations (5.8), Chapter II,

$$A^* = H_\Gamma(P_\Gamma b + Q_\Gamma a)\, H_\Gamma, \qquad B^* = H_\Gamma(bP_\Gamma + aQ_\Gamma)\, H_\Gamma. \tag{1}$$

[1]) Recall the convention made in Chapter II, Section 4.2, according to which both a function and the operator of multiplication by this function are denoted by the same letter.

Here H_Γ is the operator defined by $(H_\Gamma\varphi)(t) = \overline{h_\Gamma(t)\,\varphi(t)}$, with $h_\Gamma(t) = \exp[\mathrm{i}\theta_\Gamma(t)]$ and with $\theta_\Gamma(t)$ the angle between the tangent to Γ at the point t and the positive direction of the x-axis.

For our further investigations the following identities are of fundamental importance:

$$(aP_\Gamma + bQ_\Gamma)(f_+P_\Gamma + f_-Q_\Gamma) = af_+P_\Gamma + bf_-Q_\Gamma \tag{2}$$

and

$$(P_\Gamma f_- + Q_\Gamma f_+)(P_\Gamma a + Q_\Gamma b) = P_\Gamma af_- + Q_\Gamma bf_+. \tag{3}$$

Here $a, b \in \mathbf{L}_\infty(\Gamma)$ and $f_\pm \in \mathbf{L}_\infty^\pm(\Gamma)$.

Equalities (2) and (3) are immediate consequences of the relations (5.19), Chapter II.

The present section is devoted to the study of the singular integral operators A and B for the case of rational coefficients $a, b \in \mathbf{R}(\Gamma)$.

2.2. Consider first the operators

$$C_{n;\lambda} = (t-\lambda)^n P_\Gamma + Q_\Gamma,$$

where $n = \pm 1, \pm 2, \ldots$ and $\lambda \in \mathbb{C}$. From (2) we obtain immediately that

$$C_{n;\lambda} = C_{1;\lambda}^n \qquad (n = 1, 2, \ldots)$$

and

$$C_{-n;\lambda}C_{n;\lambda} = I \qquad (n = 1, 2, \ldots;\ \lambda \notin \Gamma). \tag{4}$$

If $\lambda \in D_\Gamma^-$, the factors on the left-hand side of (4) commute.

However, if $\lambda \in D_\Gamma^+$, then the factors on the left-hand side of (4) do not commute, since the operator $C_{1;\lambda}$ is not right-invertible. Indeed, we have

$$\operatorname{Im} C_{1;\lambda} = \{f \in \mathbf{L}_p(\Gamma, \varrho) : (P_\Gamma f)(\lambda) = 0\}$$

and therefore $\operatorname{Im} C_{1;\lambda} \neq \mathbf{L}_p(\Gamma, \varrho)$.

Theorem 2.1. *The operator $C_{n;\lambda}$ $(n = 1, 2, \ldots)$ is invertible if $\lambda \in D_\Gamma^-$ and left-invertible if $\lambda \in D_\Gamma^+$. The inverse (resp. left-inverse) of $C_{n;\lambda}$ is $C_{-n;\lambda}$.*

For $n = 1, 2, \ldots$ and $\lambda \in D_\Gamma^+$ we have

$$\operatorname{Ker} C_{-n;\lambda} = \mathfrak{L}\left\{(t-\lambda)^{n-1} - \frac{1}{t-\lambda},\ (t-\lambda)^{n-2} - \frac{1}{(t-\lambda)^2}, \ldots, 1 - \frac{1}{(t-\lambda)^n}\right\}^{1)} \tag{5}$$

and

$$\dim \operatorname{Ker} C_{-n;\lambda} = n, \qquad \dim \operatorname{Coker} C_{n;\lambda} = n. \tag{6}$$

If $\lambda \in \Gamma$, then the operators $C_{n;\lambda}$ $(n = \pm 1, \pm 2, \ldots)$ are neither Φ_+- nor Φ_--operators and, in particular, are not one-sided invertible.

Proof. The first assertion of the theorem immediately follows from (4).

Let us prove (5). Suppose $C_{-n;\lambda}\varphi = 0$ $(\varphi \in \mathbf{L}_p(\Gamma, \varrho))$, that is,

$$P_\Gamma\varphi + (t-\lambda)^n Q_\Gamma\varphi = 0. \tag{7}$$

From Theorem 5.5, Chapter II, it is easy to deduce that there exists a polynomial p_{n-1} of degree at most $n-1$ such that $(t-\lambda)^n Q_\Gamma\varphi + p_{n-1} \in \mathring{\mathbf{L}}_p^-(\Gamma, \varrho)$. Now let the operator P_Γ act on both sides of equality (7) to see that

$$P_\Gamma\varphi = p_{n-1}, \quad Q_\Gamma\varphi = -\frac{p_{n-1}}{(t-\lambda)^n}, \quad \varphi(t) = p_{n-1}(t)\left[1 - \frac{1}{(t-\lambda)^n}\right].$$

[1] $\mathfrak{L}(M)$ denotes the linear hull of the set M.

Since the polynomial p_{n-1} can be written as $p_{n-1}(t) = a_0 + a_1(t - \lambda) + \dots + a_{n-1}(t - \lambda)^{n-1}$, the relation (5) follows.

(6) immediately results from (4) and (5).

Finally, let $\lambda_0 \in \Gamma$ and assume $C_{n;\lambda_0}$ is a $\Phi_\pm$-operator. Since in every neighborhood of the point λ_0 there are both points $\lambda \in D_\Gamma^-$ and points $\lambda \in D_\Gamma^+$, we conclude that every neighborhood of the operator $C_{n;\lambda_0}$ contains both invertible operators $C_{n;\lambda}$ (with $\lambda \in D_\Gamma^-$) and one-sided but not two-sided invertible operators $C_{n;\lambda}$ (with $\lambda \in D_\Gamma^+$). This, however, contradicts Theorem 3.9, Chapter I, according to which all operators in a sufficiently small neighborhood of a $\Phi_\pm$-operator are themselves $\Phi_\pm$-operators and have equal index. ∎

The following theorem can be proved analogously.

Theorem 2.2. *The operator* $D_{n;\lambda} = P_\Gamma + (1 - \lambda t^{-1})^n Q_\Gamma$ $(n = 1, 2, \dots)$ *is invertible if* $\lambda \in D_\Gamma^+$ *and left-invertible if* $\lambda \in D_\Gamma^-$. *The inverse (resp. left-inverse) of* $D_{n;\lambda}$ *is*

$$D_{-n;\lambda} = P_\Gamma + (1 - \lambda t^{-1})^{-n} Q_\Gamma.$$

For $n = 1, 2, \dots$ *and* $\lambda \in D_\Gamma^-$ *we have*

$$\operatorname{Ker} D_{-n;\lambda} = \mathfrak{L}\{gt^{-1}, gt^{-2}, \dots, gt^{-n}\} \qquad \left(g(t) = \frac{t^n}{(t-\lambda)^n} - 1\right) \tag{8}$$

and

$$\dim \operatorname{Ker} D_{-n;\lambda} = n, \qquad \dim \operatorname{Coker} D_{n;\lambda} = n.$$

If $\lambda \in \Gamma$, *then the operators* $D_{n;\lambda}$ $(n = \pm 1, \pm 2, \dots)$ *are neither* Φ_+- *nor* Φ_-*-operators and, in particular, are not one-sided invertible.*

Note that, obviously, $D_{n;\lambda} = D_{1;\lambda}^n$ $(n = 1, 2, \dots; \lambda \in \mathbb{C})$.

The operators $P_\Gamma(t - \lambda)^n + Q_\Gamma$ and $P_\Gamma + Q_\Gamma(1 - \lambda t^{-1})^n$, which are the dual operators of $C_{n;\lambda}$ and $D_{n;\lambda}$, respectively, can be studied analogously.

2.3. We now turn to the investigation of singular integral operators with arbitrary rational coefficients. To this end, we need a special representation (factorization) of rational functions.

Let $r \in \mathbf{R}(\Gamma)$, that is, r is a rational function without poles on Γ. Furthermore, assume r has no zeros on Γ. Let $r = q_1/q_2$, where q_1 and q_2 are polynomials. Denote by t_j^+ $(j = 1, \dots, k^+)$ and t_j^- $(j = 1, \dots, k^-)$ all zeros of the polynomial q_1 in D_Γ^+ and D_Γ^-, respectively, each zero counted up to its multiplicity. Analogously, let τ_j^+ $(j = 1, \dots, l^+)$ and τ_j^- $(j = 1, \dots, l^-)$ be all zeros of the polynomial q_2 in D_Γ^+ and D_Γ^-, respectively. Then the function r can obviously be represented in the form

$$r = r_- t^\varkappa r_+, \tag{9}$$

where

$$r_-(t) = \gamma \frac{\prod_{j=1}^{k^+} (1 - t^{-1} t_j^+)}{\prod_{j=1}^{l^+} (1 - t^{-1} \tau_j^+)}, \qquad r_+(t) = \frac{\prod_{j=1}^{k^-} (t - t_j^-)}{\prod_{j=1}^{l^-} (t - \tau_j^-)},$$

$\varkappa = k^+ - l^+$ and $\gamma \in \mathbb{C}$. The integer $\varkappa$ is called the *index of the function* r and denoted by $\operatorname{ind} r$. The representation (9) is called *factorization of the function* r with respect to the curve system Γ. Clearly, the functions $r_\pm$ and $1/r_\pm$ are analytic in $D_\Gamma^\pm$.

Theorem 2.3. *Let $a, b \in \mathbf{R}(\Gamma)$. Then for the operator $A = aP_\Gamma + bQ_\Gamma$ (resp. $B = P_\Gamma a + Q_\Gamma b$) to be a Φ_+- or Φ_--operator on the space $\mathbf{L}_p(\Gamma, \varrho)$ it is necessary and sufficient that*

$$a(t) \neq 0, \qquad b(t) \neq 0 \qquad (t \in \Gamma). \tag{10}$$

If condition (10) *is satisfied, then the operator A (resp. B) is invertible, only left-sided invertible, or only right-sided invertible, if the integer $\varkappa = \operatorname{ind} a/b$ is zero, positive, or negative, respectively.*

If condition (10) *is fulfilled and if*

$$c = c_- t^\varkappa c_+ \tag{11}$$

is the factorization of the function $c = a/b$ with respect to Γ, then the corresponding inverses or one-sided inverses are

$$A^{-1} = (c_+^{-1} P_\Gamma + c_- Q_\Gamma)(t^{-\varkappa} P_\Gamma + Q_\Gamma)\, c_-^{-1} b^{-1}, \tag{12}$$

$$B^{-1} = c_+^{-1} b^{-1} (P_\Gamma t^{-\varkappa} + Q_\Gamma)(P_\Gamma c_-^{-1} + Q_\Gamma c_+). \tag{13}$$

Proof. 1. Suppose first that condition (10) is satisfied. Then, using factorization (11) and identity (2), we obtain that

$$A = bc_-(t^\varkappa c_+ P_\Gamma + c_-^{-1} Q_\Gamma) = bc_-(t^\varkappa P_\Gamma + Q_\Gamma)(c_+ P_\Gamma + c_-^{-1} Q_\Gamma). \tag{14}$$

The operators $bc_- I$ and $c_+ P_\Gamma + c_-^{-1} Q_\Gamma$ are invertible and

$$(bc_- I)^{-1} = b^{-1} c_-^{-1} I, \qquad (c_+ P_\Gamma + c_-^{-1} Q_\Gamma)^{-1} = c_+^{-1} P_\Gamma + c_- Q_\Gamma.$$

By Theorem 2.1, the operator $t^\varkappa P_\Gamma + Q_\Gamma$ is only left-sided invertible if $\varkappa > 0$ and only right-sided invertible if $\varkappa < 0$. The corresponding inverse is $t^{-\varkappa} P_\Gamma + Q_\Gamma$. This and (14) imply (12). Applying the identity (3) we get

$$B = (P_\Gamma c_- + Q_\Gamma c_+^{-1})(P_\Gamma t^\varkappa + Q_\Gamma)\, c_+ b$$

and this gives (13).

2. It remains to show that condition (10) is necessary for the operators A and B to be Φ_+- or Φ_--operators. Thus let A be a $\Phi_\pm$-operator. We first consider the case $b = 1$ and, in addition, we assume that $a(t_0) = 0$ for some point $t_0 \in \Gamma$. Then, obviously, $a(t) = (t - t_0)\, u(t)$ and $a(t) = (t^{-1} - t_0^{-1})\, v(t)$ with $u, v \in \mathbf{R}(\Gamma)$. A simple calculation combined with the relations (5.19), Chapter II, yields the following representations of the operator A:

$$A = (uP_\Gamma + Q_\Gamma)((t - t_0) P_\Gamma + Q_\Gamma), \tag{15}$$

$$A = ((t^{-1} - t_0^{-1}) P_\Gamma + Q_\Gamma)(P_\Gamma v P_\Gamma + (t^{-1} - t_0^{-1}) Q_\Gamma v P_\Gamma + Q_\Gamma). \tag{16}$$

If A is a Φ_+-operator, then (15) implies that $(t - t_0) P_\Gamma + Q_\Gamma$ is also a Φ_+-operator. On the other hand, if A is a Φ_--operator, then (16) shows that $(t^{-1} - t_0^{-1}) P_\Gamma + Q_\Gamma$ is Φ_--operator, too (see Theorem 3.7, Chapter I). This, however, contradicts Theorems 2.1 and 2.2.

Analogously it can be shown that $P_\Gamma + bQ_\Gamma$ is a $\Phi_\pm$-operator only if $b(t) \neq 0$ $(t \in \Gamma)$.

We now consider the general case: $A = aP_\Gamma + bQ_\Gamma$. Choose functions $f, g \in \mathbf{R}(\Gamma)$ so that

$$f(t)\, g(t) \neq 0 \qquad (t \in \Gamma), \qquad fa \in \mathbf{R}^+(\Gamma), \qquad gb \in \mathbf{R}^-(\Gamma).$$

Then

$$A = f^{-1}(P_\Gamma + fbQ_\Gamma)(faP_\Gamma + Q_\Gamma)$$

and

$$A = g^{-1}(gaP_\Gamma + Q_\Gamma)(P_\Gamma + gbQ_\Gamma).$$

If A is a Φ_+- or Φ_--operator, then either $faP_\Gamma + Q_\Gamma$ and $P_\Gamma + gbQ_\Gamma$ are Φ_+-operators or $P_\Gamma + fbQ_\Gamma$ and $gaP_\Gamma + Q_\Gamma$ are Φ_--operators (Theorem 3.7, Chapter I). By what has been proved above, this implies (10).

Finally, assume B is a $\Phi_\pm$-operator. By Theorem 4.2, Chapter II, the operator $T = A - B$ is compact on $\mathbf{L}_p(\Gamma, \varrho)$ and, hence, A is also a $\Phi_\pm$-operator (Theorem 3.8, Chapter I). Again by what has just been proved, this gives (10). ▮

Theorem 2.4. *Let $a, b \in \mathbf{R}(\Gamma)$ satisfy* (10) *and suppose* (11) *is the factorization of the function $c = a/b$.*

Then if $\varkappa = \operatorname{ind} c < 0$,

$$\operatorname{Ker}(aP_\Gamma + bQ_\Gamma) = \mathfrak{L}\{g, gt, \dots, gt^{|\varkappa|-1}\}, \tag{17}$$

where $g = c_+^{-1} - c_- t^\varkappa$. In case $\varkappa > 0$,

$$\operatorname{Coker}(aP_\Gamma + bQ_\Gamma) = \mathfrak{L}\{bc_-, bc_-t, \dots, bc_-t^{\varkappa-1}\} \tag{18}$$

and the equation $aP_\Gamma\varphi + bQ_\Gamma\varphi = f$ has a solution if and only if

$$\int_\Gamma f(t)\, b^{-1}(t)\, c_-^{-1}(t)\, t^{-j}\, \mathrm{d}t = 0 \qquad (j = 1, \dots, \varkappa). \tag{19}$$

Proof. Put $A = aP_\Gamma + bQ_\Gamma$. First let $\varkappa < 0$. From the representation (14) we deduce that

$$\operatorname{Ker} A = (c_+^{-1}P_\Gamma + c_-Q_\Gamma)\operatorname{Ker}(t^\varkappa P_\Gamma + Q_\Gamma)$$

and Theorem 2.1 gives

$$\operatorname{Ker} A = \mathfrak{L}\{g_1, g_2, \dots, g_{|\varkappa|}\}$$

with $g_j = (c_+^{-1}P_\Gamma + c_-Q_\Gamma)(t^{|\varkappa|-j} - t^{-j})$ $(j = 1, \dots, |\varkappa|)$. But, obviously,

$$g_j = c_+^{-1}t^{|\varkappa|-j} - c_-t^{-j} \qquad (j = 1, \dots, |\varkappa|),$$

and this proves (17).

Now let $\varkappa > 0$. Then (14) implies that $\operatorname{Im} A = bc_- \operatorname{Im}(t^\varkappa P_\Gamma + Q_\Gamma)$ and this gives (18), since $\operatorname{Im}(t^\varkappa P_\Gamma + Q_\Gamma)$ consists of all those functions $\varphi \in \mathbf{L}_p(\Gamma, \varrho)$ for which $P_\Gamma\varphi$ has a zero of order at least $\varkappa$ at the point $t = 0$. Further, A as a left-sided invertible operator (Theorem 2.3) is a Φ-operator (see § 3, Chapter I), and so equation $A\varphi = f$ has a solution if and only if

$$\int_\Gamma f(t)\, \overline{y_j(t)}\, |\mathrm{d}t| = 0 \qquad (j = 1, \dots, m), \tag{20}$$

where $y_1, \dots, y_m$ are all linearly independent solutions of the adjoint homogeneous equation $A^*y = 0$ in $\mathbf{L}_q(\Gamma, \varrho^{-1})$ $(1/p + 1/q = 1)$.

Now put $z_j = H_\Gamma(c_-^{-1}b^{-1}t^{-j})$ $(j = 1, \dots, \varkappa)$. Taking into consideration (1) we see that

$$H_\Gamma A^* z_j = (P_\Gamma b + Q_\Gamma a)\, c_-^{-1}b^{-1}t^{-j} = P_\Gamma c_-^{-1}t^{-j} + Q_\Gamma c_+ t^{\varkappa-j} = 0,$$

and thus $z_j \in \operatorname{Ker} A^*$ $(j = 1, \dots, \varkappa)$. But the functions z_j are obviously linearly independent, and since $\dim \operatorname{Ker} A^* = \dim \operatorname{Coker} A = \varkappa$ (see (18)), we conclude that $m = \varkappa$ and $z_j = y_j$ for all j. Furthermore, $\overline{y_j(t)}\, |\mathrm{d}t| = h_\Gamma(t)\, c_-^{-1}(t)\, b^{-1}(t)\, t^{-j}\, |\mathrm{d}t| = c_-^{-1}(t)\, b^{-1}(t)\, t^{-j}\, \mathrm{d}t$, and, consequently, (19) and (20) coincide. This completes the proof. ▮

The following theorem can be proved analogously.

Theorem 2.5. *Let the conditions of Theorem 2.4 be satisfied. Then if $\varkappa < 0$,*

$$\operatorname{Ker}(P_\Gamma a + Q_\Gamma b) = \mathfrak{L}\{b^{-1}c_+^{-1}, b^{-1}c_+^{-1}t, \dots, b^{-1}c_+^{-1}t^{|\varkappa|-1}\}.$$

If $\varkappa > 0$, *then*

$$\operatorname{Coker}(P_\Gamma a + Q_\Gamma b) = \mathfrak{L}\{P_\Gamma c_-, P_\Gamma(c_- t), \dots, P_\Gamma(c_- t^{\varkappa-1})\}$$

and the equation $P_\Gamma a\varphi + Q_\Gamma b\varphi = f$ *has a solution if and only if*

$$\int_\Gamma f(t)\,(c_-^{-1}(t) - c_+(t)\,t)\,t^{-j}\,\mathrm{d}t = 0 \qquad (j = 1, \dots, \varkappa).$$

Remark. A comprehensive treatment of singular integral equations with Hölder-continuous coefficients on the spaces $\mathbf{H}^\mu(\Gamma)$ can be found in the well-known monographs by MUSKHELISHVILI [1] and GAKHOV [1]. There one can also find a detailed history of that theory dated back to the fundamental work of F. NOETHER [1]. The proofs given here are partially taken from the book of GOHBERG and KRUPNIK [4] (Chapter III, § 3).

§ 3. Singular integral operators with continuous coefficients

In this section we shall generalize the basic results of the preceding section to the case of continuous coefficients.

3.1. Our first concern is the definition of the *index* of a continuous function $a \in C(\Gamma)$. Suppose $a(t) \neq 0$ for $t \in \Gamma$. If Γ is a closed Jordan curve, the *index* of the function a is defined as the integer

$$\operatorname{ind} a = \frac{1}{2\pi}[\arg a(t)]_\Gamma.$$

Here $[\,.\,]_\Gamma$ denotes the increment of the expression in the brackets as the result of a circuit around Γ in the positive direction (i.e., counter-clockwise). If Γ consists of n closed Jordan curves Γ_j, the index is defined as

$$\operatorname{ind} a = \frac{1}{2\pi}\sum_{j=1}^{n}[\arg a(t)]_{\Gamma_j}.$$

It is easy to see that ind a has the following properties:

1. $\operatorname{ind} a = \frac{1}{2\pi \mathrm{i}}[\ln a(t)]_\Gamma = \frac{1}{2\pi \mathrm{i}}\int_\Gamma \mathrm{d}(\ln a(t))$.
2. If $a_1, a_2 \in \mathbf{C}(\Gamma)$ and $a_1(t)\,a_2(t) \neq 0$ for all $t \in \Gamma$, then $\operatorname{ind} a_1a_2 = \operatorname{ind} a_1 + \operatorname{ind} a_2$.
3. If $a, x \in \mathbf{C}(\Gamma)$ and $\max_{t\in\Gamma} |x(t)/a(t)| < 1$, then $\operatorname{ind}(a + x) = \operatorname{ind} a$. In other words, the mapping $\mathbf{GC}(\Gamma) \mapsto \mathbb{Z}$, $a \mapsto \operatorname{ind} a$ is continuous.

Finally, the index defined for rational functions in Section 2.3 coincides with the index defined here.

3.2. In what follows let Γ and ϱ have the same meaning as in the preceding section.

Theorem 3.1. *Let* $a, b \subset \mathbf{C}(\Gamma)$. *Then for the operator* $A = aP_\Gamma + bQ_\Gamma$ (*resp.* $A = P_\Gamma a + Q_\Gamma b$) *to be a* Φ_+- *or* Φ_--*operator on the space* $\mathbf{L}_p(\Gamma, \varrho)$ *it is necessary and sufficient that*

$$a(t) \neq 0, \qquad b(t) \neq 0 \qquad (t \in \Gamma). \tag{1}$$

If conditions (1) *are fulfilled, then the operator* A *is invertible, only left-sided invertible or only right-sided invertible if the integer* $\varkappa = \operatorname{ind} a/b$ *is zero, positive or negative, respectively. Furthermore,*

$$\dim \operatorname{Ker} A = \max(-\varkappa, 0), \qquad \dim \operatorname{Coker} A = \max(\varkappa, 0). \tag{2}$$

Proof. 1. First consider the operator $A = aP_\Gamma + bQ_\Gamma$. Suppose the conditions (1) are satisfied. Choose a rational function $r \in \mathbf{R}(\Gamma)$ which approximates the function $c = a/b$ sufficiently well, namely, let

$$\max_{t\in\Gamma} |d(t)| < 1/\|P_\Gamma\|,$$

where $d = (c/r) - 1$. Then $c = r(1 + d)$ and since $\|P_\Gamma\| \geqq 1$, from the properties 2 and 3 of the index we deduce that $\varkappa = \operatorname{ind} c = \operatorname{ind} r$. Let $r = r_- t^\varkappa r_+$ be the factorization of the rational function r with respect to Γ (see Section 2.3).

2. Suppose $\varkappa \geqq 0$. From (2.2) we easily obtain the representation

$$A = br_-(I + dP_\Gamma)(r_+P_\Gamma + r_-^{-1}Q_\Gamma)(t^\varkappa P_\Gamma + Q_\Gamma).$$

The first three factors on the right-hand side are invertible operators (note that $\|dP_\Gamma\| < 1$). The operator $t^\varkappa P_\Gamma + Q_\Gamma$ is invertible if $\varkappa = 0$ and only left-sided invertible if $\varkappa > 0$ (Theorem 2.1). The same is therefore also true for A.

3. Let $\varkappa < 0$. Then the operator $B = at^{-\varkappa}P_\Gamma + bQ_\Gamma$ is invertible, since $\operatorname{ind} ct^{-\varkappa} = 0$. But $B = A(t^{-\varkappa}P_\Gamma + Q_\Gamma)$, and hence $A^{-1} = (t^{-\varkappa}P_\Gamma + Q_\Gamma) B^{-1}$ is a right-inverse of A. Furthermore, combining the preceding arguments with Theorem 2.1 we get (2).

4. Thus it remains to show the necessity of conditions (1). Assume that A is a $\Phi_\pm$-operator, but let at least one of the functions a, b have a zero on Γ. Choose rational functions r_1 and r_2 from $\mathbf{R}(\Gamma)$ which approximate the functions a and b sufficiently close in the uniform norm. Moreover, the functions r_1, r_2 can be chosen in such a way that at least one of them has a zero on Γ. Since the norm $\|r_1P_\Gamma + r_2Q_\Gamma - A\|$ is sufficiently small, $r_1P_\Gamma + r_2Q_\Gamma$ is also a $\Phi_\pm$-operator (see Theorem 3.9, Chapter I). This, however, contradicts Theorem 2.3 and therefore proves the necessity of the conditions (1).

5. Finally, consider the operator $A' = P_\Gamma a + Q_\Gamma b$. Since $A - A'$ (with $A = aP_\Gamma + bQ_\Gamma$) is compact, by Theorem 4.2 of Chapter II the operators A and A' are either simultaneously $\Phi_\pm$-operators or not, and, moreover, $\operatorname{Ind} A = \operatorname{Ind} A'$ (see Theorem 3.8, Chapter I). This yields the necessity of the conditions (1) for the operator A'. On the other hand, if the conditions (1) are satisfied, then Theorem 1.2 shows that $\dim \operatorname{Ker} A = \dim \operatorname{Ker} A'$ and, consequently, $\dim \operatorname{Coker} A = \dim \operatorname{Coker} A'$. This implies all assertions of the theorem for the operator A'. ∎

Letting $a = b$ in Theorem 3.1, we obtain the following result.

Corollary 3.1. *The operator of multiplication by a continuous function $a \in \mathbf{C}(\Gamma)$, defined as*

$$(a\varphi)(t) = a(t)\,\varphi(t) \qquad (t \in \Gamma), \tag{3}$$

is invertible on the space $\mathbf{L}_p(\Gamma, \varrho)$ if and only if $a(t) \neq 0$ for all $t \in \Gamma$.

Corollary 3.2. *The norm of the operator defined by* (3) *on $\mathbf{L}_p(\Gamma, \varrho)$ is equal to*

$$\|a\| = \max_{t\in\Gamma} |a(t)|.$$

Proof. Corollary 3.1 implies that the spectral radius of the operator a is $r(a) = \max_{t\in\Gamma} |a(t)|$. Hence $\|a\| \geqq \max_{t\in\Gamma} |a(t)|$. The reverse inequality is obvious. ∎

3.3. We now consider operators of the form

$$A := aP + bQ + T. \tag{4}$$

Here $a, b \in \mathbf{C}(\Gamma)$ are continuous functions on Γ and T is any compact linear operator on the space $\mathbf{L}_p(\Gamma, \varrho)$. An operator A of the form (4) is also called a *singular integral operator on the curve system Γ*. Since the algebra $\mathfrak{M} = \mathbf{C}(\Gamma)$ satisfies the condition (A) in Section 1.2

(Theorem 4.2, Chapter II), the operator (4) is a concrete realization of the abstract singular integral operator.

In accordance with the definition given in Section 1.2 and in view of Corollary 3.1, the singular integral operator (4) is called an *operator of normal type* if conditions (1) are fulfilled.

From Section 1.2 we know that the collection of all operators of the form (4) is an algebra. The function

$$A(t, \theta) := a(t) \frac{1+\theta}{2} + b(t) \frac{1-\theta}{2} \quad (t \in \Gamma, \theta = \pm 1) \tag{5}$$

of the two independent variables $t \in \Gamma$ and $\theta = \pm 1$ is referred to as the *symbol* of the singular integral operator A. It is easily seen from (1.13) that function (5) meets the usual requirements imposed upon a symbol, that is, the symbol of the product (the sum) of two singular integral operators is equal to the product (the sum) of the symbols of these operators. Thus, the symbol produces a homomorphism of the algebra of all singular integral operators of the form (4) onto the algebra of all functions of the form (5) (see § 5, Chapter I).

It is obvious that conditions (1) are satisfied if and only if the symbol $A(t, \theta)$ *does not degenerate*, that is, if $A(t, \theta) \neq 0$ for all $t \in \Gamma$ and $\theta \in \{-1, 1\}$.

Finally, note that the singular integral operator (4) can be written in the form $A = cI + dS + T$, where $c = (a + b)/2$, $d = (a - b)/2$. Accordingly, the symbol can be rewritten as $A(t, \theta) = c(t) + \theta\, d(t)$.

Putting the Theorems 1.4 and 3.1 together with Theorem 3.8 of Chapter I we arrive at the following result.

Theorem 3.2. *For the operator A defined by* (4) *to be a Φ_+- or Φ_--operator on the space $\mathbf{L}_p(\Gamma, \varrho)$ it is necessary and sufficient that conditions* (1) *be fulfilled.*

If conditions (1) *are satisfied, then A is a Φ-operator on $\mathbf{L}_p(\Gamma, \varrho)$, its index is* $\operatorname{Ind} A = \operatorname{ind} b/a$, *and the operator $R = a^{-1} P_\Gamma + b^{-1} Q_\Gamma$ is a two-sided regularizer of the operator A.*

Furthermore, one has

Theorem 3.3. *Let A be the operator defined by* (4) *on the space $\mathbf{L}_p(\Gamma, \varrho)$. Then*

$$\max_{t \in \Gamma, \theta = \pm 1} |A(t, \theta)| \leqq \|A\|. \tag{6}$$

Proof. We shall prove that $|a(t)| \leqq \|A\|$ for all $t \in \Gamma$. Assume the contrary, that is, assume there is a point $t_0 \in \Gamma$ such that $|a(t_0)| > \|A\|$. Put $\lambda = 1/a(t_0)$ and consider the operator $B = I - \lambda A$. Then $\|\lambda A\| = |\lambda|\, \|A\| < 1$, hence B is invertible and thus a Φ-operator. By Theorem 3.4, Chapter I, the operator

$$B + \lambda T = (1 - \lambda a)\, P_\Gamma + (1 - \lambda b)\, Q_\Gamma$$

is a Φ-operator, too. But this contradicts the first assertion of Theorem 3.1, since the function $1 - \lambda a(t)$ vanishes at the point $t_0 \in \Gamma$. Thus, $|a(t)| \leqq \|A\|$ $(t \in \Gamma)$. Equally one can show that $|b(t)| \leqq \|A\|$ $(t \in \Gamma)$ and this gives the desired inequality (6). ▮

We conclude by stating the following immediate consequence of Theorem 3.3.

Corollary 3.3. *The set of all singular integral operators of the form* (4) *with continuous coefficients a and b is a closed subalgebra of the Banach algebra $\mathscr{L}(\mathbf{L}_p(\Gamma, \varrho))$.*

Remark 1. Theorem 3.1 was established by Gohberg [4–6], Khvedelidze [1–2], and Simonenko [5]. The second part of Theorem 3.2, for the space $\mathbf{L}_2(\Gamma)$, is due to Mikhlin [7, 29]. These results were generalized by Khvedelidze [1–2] to the spaces $\mathbf{L}_p(\Gamma, \varrho)$ and independently by Fichera [1] to the spaces $\mathbf{L}_p(\Gamma)$. The first part of Theorem 3.2 and Theorem 3.3 go back to Gohberg [4–5].

Remark 2. The necessity portions of Theorem 3.1 and 3.2 can be strengthened as follows: if the operator $A = aP_\Gamma + bQ_\Gamma$ is normally solvable on the space $\mathbf{L}_p(\Gamma)$, then, on each closed curve forming a part of the curve system Γ, each of the functions a and b vanishes either everywhere or nowhere (see Leiterer, Markus [1]). The just mentioned conditions for the functions a and b are necessary and sufficient for operator (4) with a certain compact operator T to be normally solvable on $\mathbf{L}_p(\Gamma)$ (see Leiterer [2]). Leiterer [1] also obtained necessary and sufficient conditions for the normal solvability of the operator $aP_\Gamma + bQ_\Gamma$.

§ 4. Singular integral operators on the space $\mathbf{H}^\mu(\Gamma)$

We now consider the following singular integral equation:

$$(A\varphi)(t) := a(t)(P_\Gamma\varphi)(t) + b(t)(Q_\Gamma\varphi)(t) + \int_\Gamma T(t,\tau)\,\varphi(\tau)\,d\tau = f(t) \qquad (t \in \Gamma). \tag{1}$$

The functions a, b, and f are supposed to be in $\mathbf{H}^\mu(\Gamma)$ $(0 < \mu < 1)$ and the kernel $T(t,\tau)$ is assumed to be of the form

$$T(t,\tau) = \frac{K(t,\tau) - K(t,t)}{\tau - t},$$

where $K(t,\tau) \in \mathbf{H}^\lambda(\Gamma\times\Gamma)$ with $\mu < \lambda \leqq 1$.

Then the integral operator T with the kernel $T(t,\tau)$ is compact on the space $\mathbf{H}^\mu(\Gamma)$ (see Prössdorf [17], Corollary 4.3, Chapter 3) and, consequently, the operator A defined by equation (1) is a bounded linear operator on $\mathbf{H}^\mu(\Gamma)$ (see Theorem 4.7, Chapter II).

First let us prove the following basic result.

Theorem 4.1. *Under hypothesis* (3.1), *every solution* $\varphi \in \mathbf{L}_p(\Gamma,\varrho)$ *of* (1) *belongs to* $\mathbf{H}^\mu(\Gamma)$.

Proof. Let $\varphi \in \mathbf{L}_p(\Gamma,\varrho)$ be a solution of (1). Then it is easy to see (cf. also (1.13)) that

$$\varphi - V\varphi = Rf, \tag{2}$$

where

$$V = (b^{-1} - a^{-1})([P_\Gamma, a]\,P_\Gamma + [P_\Gamma, b]\,Q_\Gamma) - RT, \qquad R = a^{-1}P_\Gamma + b^{-1}Q_\Gamma.$$

If $f \in \mathbf{H}^\mu(\Gamma)$ then $g = Rf \in \mathbf{H}^\mu(\Gamma)$ (see Theorem 4.7, Chapter II). Furthermore, making use of the Poincaré-Bertrand commutation formula it is readily seen that V is an integral operator with a weakly singular kernel $V(t,\tau)$.

Consequently, the n-th iterated kernel $V_n(t,\tau)$ is bounded for all sufficiently large n (see Mikhlin [1], p. 85).

From (2) we deduce that

$$\varphi - V^n\varphi = \sum_{k=0}^{n-1} V^k g,$$

with the right-hand side belonging to $\mathbf{H}^\mu(\Gamma)$. Application of Hölder's inequality gives

$$|V^n\varphi| = \left|\int_\Gamma V_n(t,\tau)\,\varphi(\tau)\,d\tau\right| \leqq \|\varphi\|_{\mathbf{L}_p(\Gamma,\varrho)} \left\{\int_\Gamma |V_n(t,\tau)|^q \varrho^{-q}(\tau)\,|d\tau|\right\}^{1/q}$$

$(p^{-1} + q^{-1} = 1)$, and thus φ is a bounded function. This in turn implies $V\varphi \in \mathbf{H}^\nu(\Gamma)$ for all ν such that $0 < \nu < \mu$ (see Prössdorf [17], Corollary 4.1, Chapter 3), and hence, taking into account (2), we conclude that $\varphi \in \mathbf{H}^\nu(\Gamma)$. It follows that $V\varphi \in \mathbf{H}^\mu(\Gamma)$ (see Prössdorf [17], Lemma 4.1, Chapter 3) and therefore, again by (2), $\varphi \in \mathbf{H}^\mu(\Gamma)$. ∎

Besides the operator A we still consider its *transposed operator* A' defined as

$$(A'\psi)(t) := (P_\Gamma b\psi)(t) + (Q_\Gamma a\psi)(t) + \int_\Gamma T(\tau,t)\,\psi(\tau)\,d\tau. \tag{3}$$

With the help of the Poincaré-Bertrand commutation formula it can be easily checked that

$$\int_\Gamma (A\varphi)(t)\,\psi(t)\,dt = \int_\Gamma \varphi(t)\,(A'\psi)(t)\,dt \tag{4}$$

for any functions φ and ψ in $\mathbf{H}^\mu(\Gamma)$. Taking into consideration that $dt = h_\Gamma(t)\,|dt|$ we deduce from (4) that

$$A^* = H_\Gamma A' H_\Gamma, \tag{5}$$

where A^* denotes the adjoint operator, acting on $\mathbf{L}_q(\Gamma, \varrho^{-1})$ (compare also (2.1)).

Theorems 3.2 and 4.1. immediately imply the following result.

Theorem 4.2. *Suppose conditions* (3.1) *are satisfied. Then:*

(a) *For* (1) *to have a solution* $\varphi \in \mathbf{H}^\mu(\Gamma)$ *it is necessary and sufficient that*

$$\int_\Gamma f(t)\,\psi(t)\,dt = 0$$

for every solution $\psi \in \mathbf{H}^\mu(\Gamma)$ *of the homogeneous transposed equation* $A'\psi = 0$.

(b) *The numbers* $\alpha(A)$ *and* $\alpha(A')$ *of linearly independent solutions in* $H^\mu(\Gamma)$ *of the homogeneous equations* $A\varphi = 0$ *and* $A'\psi = 0$ *are finite.*

(c) *The difference* $\alpha(A) - \alpha(A')$ *only depends on the characteristic part of the operator* A *and is equal to*

$$\alpha(A) - \alpha(A') = \operatorname{ind} b/a.$$

The assertions (a), (b), (c) are usually called the *Noether theorems* for the singular integral equation (1).

Remark. Theorem 4.2 goes back to Noether [1]. Theorem 4.1 was obtained by Khvedelidze [1] (however, with other methods) for the first time. The proof given here is due to Prössdorf [17]. Note that conditions (3.1) are also necessary for assertions (a) and (b) of Theorem 4.2 to be valid (see Gohberg and Krupnik [4], Chapter V, § 7).

§ 5. Factorization of continuous functions

In Section 2.3 the representation (2.9), i.e., the factorization of a rational function played a crucial role. It yielded explicit formulae for the one-sided inverses of a singular integral operator with rational coefficients and allowed us to describe its kernel and cokernel effectively. It turns out that analogous results can be obtained in much more general situations, too. Therefore, our first concern will be the extension of the concept of factorization to more general classes of functions.

5.1. Factorization in *R*-algebras

Let Γ be a closed piecewise Lyapunov curve system. Given a continuous function $a \in \mathbf{C}(\Gamma)$, a representation of the form

$$a = a_- t^\varkappa a_+ \tag{1}$$

is called a *factorization* of the function a with respect to Γ if $\varkappa$ is an integer and if $a_\pm$ possesses an extension analytic in $D_\Gamma^\pm$ and continuous on $\overline{D_\Gamma^\pm}$ such that $a_+(t) \neq 0$ $(t \in \overline{D_\Gamma^+})$ and $a_-(t) \neq 0$ $(t \in \overline{D_\Gamma^-})$. Since $\operatorname{ind} a_+ = \operatorname{ind} a_- = 0$, the integer $\varkappa$ in (1) is determined uniquely and is nothing else than $\operatorname{ind} a$.

In definition (1) the factor $t^\varkappa$ may be replaced by any factor of the form

$$[(t - t^-)/(t - t^+)]^\varkappa, \tag{1'}$$

where $t^\pm \in D_\Gamma^\pm$. This must be done if $0 \in \Gamma$ or $\infty \in \Gamma$. It is easily seen that a function a admits a factorization with an arbitrary middle factor (1') if and only if it admits a factorization of the form (1).

In what follows we denote by $\mathbf{R}^\pm(\Gamma)$ the set of all rational functions with poles outside $\overline{D_\Gamma^\mp}$ and by $\mathbf{C}^\pm(\Gamma)$ the closure of the set $\mathbf{R}^\pm(\Gamma)$ in the norm of $\mathbf{C}(\Gamma)$. It is obvious that $\mathbf{C}^\pm(\Gamma)$ forms a (closed) subalgebra of $\mathbf{C}(\Gamma)$ and that $\mathbf{C}^\pm(\Gamma)$ is just the set of all continuous functions on Γ possessing an extension analytic in $D_\Gamma^\pm$ and continuous on $\overline{D_\Gamma^\pm}$. Finally, let $\mathring{\mathbf{C}}^-(\Gamma)$ denote the subalgebra of all functions $a \in \mathbf{C}^-(\Gamma)$ with $a(\infty) = 0$.

Now let $\mathfrak{A}$ be a Banach algebra consisting of continuous functions and having the following properties:

a) $\mathfrak{A}$ contains the set $\mathbf{R}(\Gamma)$;
b) if $a \in \mathfrak{A}$ and $a(t) \neq 0$ $(t \in \Gamma)$, then $1/a \in \mathfrak{A}$.

Put

$$\mathfrak{A}^\pm := \mathfrak{A} \cap \mathbf{C}^\pm(\Gamma), \qquad \mathring{\mathfrak{A}}^- := \mathfrak{A} \cap \mathring{\mathbf{C}}^-(\Gamma).$$

We have

$$\|a\|_{\mathbf{C}(\Gamma)} \leqq \|a\|_{\mathfrak{A}} \qquad (a \in \mathfrak{A}) \tag{2}$$

(see Gelfand, Raikov, and Shilov [1], § 9), and this implies that $\mathfrak{A}^\pm$ and $\mathring{\mathfrak{A}}^-$ are closed subalgebras of the algebra $\mathfrak{A}$. Furthermore, it is easily seen that the inverse $1/a$ of a function $a \in \mathfrak{A}^\pm$ such that $a(t) \neq 0$ $(t \in \overline{D_\Gamma^\pm})$ also belongs to $\mathfrak{A}^\pm$.

The algebra $\mathfrak{A}$ is called a *decomposing algebra* if it splits into the direct sum of its subalgebras $\mathfrak{A}^+$ and $\mathring{\mathfrak{A}}^-$, that is, if $\mathfrak{A} = \mathfrak{A}^+ \dotplus \mathring{\mathfrak{A}}^-$. The algebra $\mathfrak{A}$ is decomposing if and only if $S_\Gamma \in \mathscr{L}(\mathfrak{A})$.

Indeed, if $\mathfrak{A}$ is a decomposing algebra, then $P_\Gamma = \frac{1}{2}(I + S_\Gamma)$ is obviously the projection of $\mathfrak{A}$ onto $\mathfrak{A}^+$ parallel to $\mathring{\mathfrak{A}}^-$. Conversely, if $S_\Gamma \in \mathscr{L}(\mathfrak{A})$, then $P_\Gamma(\mathfrak{A}) \subset \mathfrak{A}^+$ and $Q_\Gamma(\mathfrak{A}) \subset \mathring{\mathfrak{A}}^-$ (with $Q_\Gamma = I - P_\Gamma$). But since $P_\Gamma + Q_\Gamma = I$, we deduce that actually $P_\Gamma(\mathfrak{A}) = \mathfrak{A}^+$ and $Q_\Gamma(\mathfrak{A}) = \mathring{\mathfrak{A}}^-$.

Examples of decomposing algebras are $\mathbf{H}^\lambda(\Gamma)$ and $\mathbf{C}^{m,\lambda}(\Gamma)$ $(0 < \lambda < 1)$ (see § 6.1, Chapter II). The algebra $\mathbf{C}(\Gamma)$ is not decomposing (see, for instance, Gohberg and Feldman [1], p. 51 or Remark 1 in § 2, Chap. II).

An algebra $\mathfrak{A}$ of continuous functions on Γ is called an *R-algebra* if it has the properties a) and b) and if, moreover, the set $\mathbf{R}(\Gamma)$ is dense in $\mathfrak{A}$. It is not difficult to see that the property b) is actually a consequence of the density of the set $\mathbf{R}(\Gamma)$ in $\mathfrak{A}$.

The algebra $\mathbf{C}(\Gamma)$ is obviously an R-algebra. Another example of an R-algebra is the *Wiener algebra* $\mathbf{W}$. This is the algebra of all functions a on the unit circle Γ_0 which can be expanded into an absolutely convergent Fourier series,

$$a(t) = \sum_{j=-\infty}^{\infty} a_j t^j \quad \left(|t| = 1, \sum_{j=-\infty}^{\infty} |a_j| < \infty\right).$$

Under the norm

$$\|a\|_{\mathbf{W}} := \sum_{j=-\infty}^{\infty} |a_j| < \infty$$

W is a Banach algebra. Clearly,

$$(P_{\Gamma_0} a)(t) = \sum_{j=0}^{\infty} a_j t^j,$$

and this implies that **W** is decomposing.

The following theorem states an extremely important property of R-algebras.

Theorem 5.1. *Let $\mathfrak{A}$ be an R-algebra. Then every function $a \in \mathfrak{A}$ with $a(t) \neq 0$ $(t \in \Gamma)$ admits a factorization* (1) *with $a_\pm \in \mathfrak{A}^\pm$ if and only if $\mathfrak{A}$ is a decomposing algebra.*

The proof will be based on the following general fact, which is of interest in connection with some other problems, too.

Lemma 5.1. *Let $\mathfrak{R}$ be a Banach algebra with identity element and let $\mathfrak{R}^\pm$ be two subalgebras of $\mathfrak{R}$ such that $\mathfrak{R} = \mathfrak{R}^+ \dotplus \mathfrak{R}^-$. Denote by P the projection of $\mathfrak{R}$ onto $\mathfrak{R}^+$ parallel to $\mathfrak{R}^-$ and put $Q = I - P$.*

If $a \in \mathfrak{R}$ and if

$$\|a\| < \min(\|P\|^{-1}, \|Q\|^{-1}), \tag{3}$$

then the element $e + a$ admits a factorization

$$e + a = (e + a_-)(e + a_+) \tag{4}$$

with $a_\pm \in \mathfrak{R}^\pm$ and $(e + a_\pm)^{-1} - e \in \mathfrak{R}^\pm$.

Proof. By virtue of (3), the equation

$$x + Pax = e \tag{5}$$

has a unique solution $x \in \mathfrak{R}$. Obviously, $x = e + x_+$ with some $x_+ \in \mathfrak{R}^+$. Thus, (5) leads to the equality

$$(e + a)(e + x_+) = e + a_- \qquad (a_- \in \mathfrak{R}^-). \tag{6}$$

Similarly, considering the equation

$$x + Qxa = e$$

we obtain in the same fashion that

$$(e + x_-)(e + a) = e + a_+ \qquad (x_- \in \mathfrak{R}^-, a_+ \in \mathfrak{R}^+)$$

or, equivalently,

$$(e + a_+)(e + a)^{-1} = e + x_-. \tag{7}$$

On multiplying the left-hand and the right-hand sides of (6) and (7) we arrive at the equality

$$a_+ + x_+ + a_+x_+ = x_- + a_- + x_-a_-.$$

But since $\mathfrak{R}^+$ and $\mathfrak{R}^-$ have only the zero element in common, both sides of the last equality are zero. Hence

$$(e + a_+)(e + x_+) = (e + x_-)(e + a_-) = e. \tag{8}$$

In case the algebra $\mathfrak{R}$ is commutative this ends the proof, since then the assertion of our lemma immediately follows from (6) and (8).

If $\mathfrak{R}$ is a non-commutative algebra, then (8) implies the one-sided invertibility of the lements $e + a_\pm$. In order to prove that they are actually invertible, notice first that the above arguments also apply if the element a is replaced by λa $(0 \leqq \lambda \leqq 1)$. The elements $a_\pm$ must then be replaced by certain analytic functions $a_\pm(\lambda)$ $(0 \leqq \lambda \leqq 1)$, and the element $e + a_+(\lambda)$ (resp. $e + a_-(\lambda)$) is right (resp. left) invertible for every λ. But the element $e + a_\pm(0) = e$ is invertible and so Lemma 3.2, Chapter I, can be applied to deduce that the element $e + a_\pm = e + a_\pm$ (1) is invertible, too. ▮

Proof of Theorem 5.1. 1. Let $\mathfrak{A}$ be a decomposing R-algebra and $a \in \mathfrak{A}$ a function which vanishes nowhere on Γ. Choose a function $r \in \mathbf{R}(\Gamma)$ so that for $b = ar^{-1} - 1$ the inequality

$$\|b\|_{\mathfrak{A}} < \min(\|P\|^{-1}, \|Q\|^{-1}) \tag{9}$$

holds; here P denotes the projection of $\mathfrak{A}$ onto $\mathfrak{A}^+$ parallel to $\mathring{\mathfrak{A}}^-$ and $Q = I - P$.

Application of Lemma 5.1 to the algebras

$$\mathfrak{R} = \mathfrak{A}, \quad \mathfrak{R}^+ = \mathfrak{A}^+, \quad \mathfrak{R}^- = \mathring{\mathfrak{A}}^-$$

yields a factorization of the function $1 + b$, that is, a representation of the form

$$ar^{-1} = b_- b_+, \tag{10}$$

where

$$b_\pm \in \mathfrak{A}^\pm, \qquad b_\pm(t) \neq 0 \qquad (t \in \overline{D_\Gamma^\mp}).$$

In view of (2) we have

$$\max_{t \in \Gamma} |a(t)\, r^{-1}(t) - 1| \leqq \|b\|_{\mathfrak{A}} < 1,$$

whence

$$r(t) \neq 0 \qquad (t \in \Gamma), \qquad \operatorname{ind} r = \operatorname{ind} a = \varkappa.$$

Thus, according to (2.9) the function $r \in \mathbf{R}(\Gamma)$ can be factorized as

$$r = r_- t^\varkappa r_+, \tag{11}$$

where $r_\pm \in \mathbf{R}^\pm(\Gamma)$, $r_\pm(t) \neq 0$ $(t \in \overline{D_\Gamma^\mp})$. Putting (10) and (11) together we obtain the desired factorization of the function a immediately.

2. We now prove the necessity part of Theorem 5.1. Given an arbitrary function $a \in \mathfrak{A}$ put $b = e^a$. Then $b(t) \neq 0$ $(t \in \Gamma)$ and $\operatorname{ind} b = 0$. Therefore the function $b \in \mathfrak{A}$ admits a factorization as

$$b = b_-\, b_+ \qquad (b_\pm \in \mathfrak{A}^\pm, b_-(\infty) = 1).$$

Since $\operatorname{ind} b = 0$, a theorem of Shilov [1] implies that

$$a_\pm = \ln b_\pm \in \mathfrak{A}^\pm, \qquad a_- \in \mathring{\mathfrak{A}}^-.$$

But then

$$a = a_+ + a_-$$

and this shows that the algebra $\mathfrak{A}$ is decomposing. ▮

Corollary 5.1. *Every function $a \in \mathbf{W}$ which vanishes nowhere on the unit circle admits a factorization* (1) *with $a_\pm \in \mathbf{W}^\pm$.*

Finally, note that $\mathbf{H}^\mu(\Gamma)$ $(0 < \mu < 1)$ as a non-separable space cannot be an R-algebra.

Remark. Corollary 5.1 was stated by M. G. Krein [1]. Theorem 5.1 is due to Gohberg [5]. The proof presented here is taken from the book of Gohberg and Feldman [1] (Chapter I, § 5), but see also Budyanu and Gohberg [1, 2].

5.2. Factorization in algebras with two norms

The proof of Theorem 5.1 suggests the possibility to factorize functions from more general algebras than R-algebras, namely, from certain algebras with two norms.

Let $\mathfrak{A}$ and $\tilde{\mathfrak{A}}$ be two Banach algebras of continuous functions on Γ which, in addition to the properties a) and b) stated in Section 5.1, possess the following four properties:

α) $\mathfrak{A} \subset \tilde{\mathfrak{A}}$, with the embedding being continuous (i.e., the embedding operator E: $\mathfrak{A} \to \tilde{\mathfrak{A}}$ is bounded).

β) $\mathbf{R}(\Gamma)$ is dense in $\mathfrak{A}$ with respect to the norm of $\tilde{\mathfrak{A}}$.

γ) $\mathfrak{A}$ and $\tilde{\mathfrak{A}}$ are decomposing algebras.

δ) There exists a constant $\varrho > 0$ such that for all functions $a \in \mathfrak{A}$ and $\varphi \in \tilde{\mathfrak{A}}$ with $\|1 - a\|_{\tilde{\mathfrak{A}}} < \varrho$ the following implication is true: if $\tilde{P}a\tilde{P}\varphi \in \mathfrak{A}$ then $a\tilde{P}\varphi \in \mathfrak{A}$. Here $\tilde{P}$ denotes the projection of $\tilde{\mathfrak{A}}$ onto $\tilde{\mathfrak{A}}^+$ parallel to $\mathring{\tilde{\mathfrak{A}}}^-$.

The conditions γ) and δ) are obviously satisfied if only $\tilde{\mathfrak{A}}$ is required to be a decomposing algebra and if $(\tilde{P}a - a\tilde{P})\,\tilde{P}\varphi \in \mathfrak{A}$ for all functions $a \in \mathfrak{A}$ and $\varphi \in \tilde{\mathfrak{A}}$. This is immediate from the identity

$$a\tilde{P}\varphi = (a\tilde{P} - \tilde{P}a)\,\tilde{P}\varphi + \tilde{P}a\tilde{P}\varphi.$$

A Banach algebra $\mathfrak{A}$ is said to have the properties α)–δ) if there exists a Banach algebra $\tilde{\mathfrak{A}}$ such that α)–δ) hold.

Clearly, a decomposing R-algebra $\mathfrak{A}$ satisfies the conditions α)–δ) with $\tilde{\mathfrak{A}} = \mathfrak{A}$.

Lemma 5.2. *The algebras* $\mathfrak{A} = \mathbf{H}^\mu(\Gamma)$ $(0 < \mu < 1)$ *and* $\mathfrak{A} = \mathbf{C}^{m,\mu}(\Gamma)$ *have the properties* α)–δ).

Proof. In the case $\mathfrak{A} = \mathbf{H}^\mu(\Gamma)$ put $\tilde{\mathfrak{A}} = \mathbf{H}^\nu(\Gamma)$ with $0 < \nu < \mu$. The property α), i.e., the boundedness of the embedding operator $\mathbf{H}^\mu(\Gamma) \to \mathbf{H}^\nu(\Gamma)$ is obvious. The property β) can be verified as follows. Each function in $\mathbf{H}^\mu(\Gamma)$ can be approximated by piecewise linear functions in the norm of $\mathbf{H}^\nu(\Gamma)$, these, in turn, can be approximated by continuously differentiable functions. Finally, the latter functions can be approximated by rational functions. The property γ) is also obvious, and the property δ) is by the remark preceding Lemma 5.2 a consequence of Theorem 4.8, Chapter II.

Using the Theorems 6.2 and 6.3 of Chapter II it can be proved analogously that the algebras $\mathfrak{A} = \mathbf{C}^{m,\mu}(\Gamma)$ and $\tilde{\mathfrak{A}} = \mathbf{C}^{m,\nu}(\Gamma)$ $(0 < \nu < \mu)$ meet all the requirements α)–δ). ∎

Lemma 5.3. *Let the conditions* α), γ), *and* δ) *be fulfilled and put* $\tilde{Q} = I - \tilde{P}$. *Then every function* $a \in \mathfrak{A}$ *subject to*

$$\|a - 1\|_{\tilde{\mathfrak{A}}} < \min\left(\|\tilde{P}\|^{-1}, \|\tilde{Q}\|^{-1}, \varrho\right) \tag{12}$$

admits a factorization

$$a = a_- a_+ \tag{13}$$

with $a_\pm$ *and* $1/a_\pm$ *in* $\mathfrak{A}^\pm$.

Proof. Let $a \in \mathfrak{A}$ be any function satisfying (12). Then, by Lemma 5.1, a admits a factorization (13) with $a_\pm$ and $1/a_\pm$ in $\tilde{\mathfrak{A}}^\pm$. We shall prove that these functions actually belong to $\mathfrak{A}^\pm$.

Putting $\varphi = a_+^{-1} - a_-$ $(\in \tilde{\mathfrak{A}})$ and taking into account that the function $a_- - c$ (with $c = a_-(\infty)$) belongs to $\mathring{\tilde{\mathfrak{A}}}^-$, we see that

$$a_+^{-1} = \tilde{P}\varphi + c, \qquad a_- = -\tilde{Q}\varphi + c. \tag{14}$$

Substituting this into (13) we arrive at the equality

$$a\tilde{P}\varphi + \tilde{Q}\varphi = f,$$

where $f = (1 - a)\,c \in \mathfrak{A}$. Thus

$$\tilde{P}a\tilde{P}\varphi = \tilde{P}f \in \mathfrak{A},$$

whence, by virtue of property δ), the function $a\tilde{P}\varphi$ and therefore also $\tilde{P}\varphi$ is seen to be in $\mathfrak{A}$. Consequently, $a_+^{-1} \in \mathfrak{A}$ and so $a_\pm \in \mathfrak{A}^\pm$ and $a_\pm^{-1} \in \mathfrak{A}^\pm$. ∎

Theorem 5.2. *Let the Banach algebra $\mathfrak{A}$ have the properties α)–δ). Then every function $a \in \mathfrak{A}$ which vanishes nowhere on Γ admits a factorization of the form* (1) *with $a_\pm$ and $1/a_\pm$ in $\mathfrak{A}^\pm$.*

Proof. Due to property β) there is a function $r \in \mathbf{R}(\Gamma)$ such that the function $b = ar^{-1} - 1 \in \mathfrak{A}$ satisfies the inequality

$$\|b\|_{\mathfrak{A}} < \min\,(\|\tilde{P}\|^{-1}, \|\tilde{Q}\|^{-1}, \varrho).$$

Thus, by Lemma 5.3, the function ar^{-1} admits a factorization of the form (10). To complete the proof it remains to repeat the arguments of the proof of Theorem 5.1. ∎

Corollary 5.2. *Every function $a \in \mathbf{H}^\mu(\Gamma)$ $(0 < \mu < 1)$ which vanishes nowhere on Γ admits a factorization of the form* (1) *with $a_\pm \in [\mathbf{H}^\mu(\Gamma)]^\pm$.*

Corollary 5.3. *Every function $a \in \mathbf{C}^{m,\mu}(\Gamma)$ $(0 < \mu < 1)$ which vanishes nowhere on Γ admits a factorization of the form* (1) *with $a_\pm \in [\mathbf{C}^{m,\mu}(\Gamma)]^\pm$.*

Remark 1. It is easily seen that all results of § 2 remain valid for singular integral operators of the form $aP_\Gamma + bQ_\Gamma$ and $P_\Gamma a + Q_\Gamma b$ if their coefficients a and b belong to an algebra $\mathfrak{A}$ which possesses the properties α)–δ). Furthermore, if a and b are in $\mathbf{H}^\mu(\Gamma)$, then these results remain true when the operator is considered on the space $\mathbf{H}^\mu(\Gamma)$ $(0 < \mu < 1)$.

Remark 2. Corollary 5.2 was established by Plemelj [1, 2] and Corollary 5.3 by Budyanu and Gohberg [2]. In this section we essentially followed Prössdorf [17] (Section 2.3.2). Theorem 5.2 generalizes some results of Budyanu and Gohberg [2].

5.3. Generalized factorization of continuous functions

We know from Theorem 5.1 that not every continuous function admits a factorization with respect to a closed piecewise Lyapunov curve system Γ. However, one has the following result.

Theorem 5.3. *Every continuous function $a \in \mathbf{C}(\Gamma)$ which vanishes nowhere on Γ admits a generalized factorization of the form*

$$a = a_- t^\varkappa a_+, \tag{15}$$

where $\varkappa = \operatorname{ind} a$ and a_+, a_- are functions having the following properties:

1. *$a_+, a_+^{-1} \in \mathbf{L}_p^+(\Gamma)$ and $a_-, a_-^{-1} \in \mathbf{L}_p^-(\Gamma)$ for all p with $1 < p < \infty$.*
2. *The operator $a_+^{-1} P_\Gamma a_-^{-1}$ is bounded on the space $\mathbf{L}_p(\Gamma)$ for all p with $1 < p < \infty$.*

Theorem 5.3 is a special version of a more general theorem of Simonenko [6, 7] on matrix functions, which will be proved in Chapter V.

Note that if a function $a \in \mathbf{C}(\Gamma)$ admits two generalized factorizations $a = a_- t^\varkappa a_+$ and $a = b_- t^\varkappa b_+$, then obviously $b_- = ca_-$ and $b_+ = c^{-1}a_+$ with $c \in \mathbb{C}$, since the function $b_- a_-^{-1} = a_+ b_+^{-1}$ lies in the intersection $\mathbf{L}_1^-(\Gamma) \cap \mathbf{L}_1^+(\Gamma)$ and must therefore be a constant.

We also remark that by virtue of (15) the operators

$$a_+^{-1}P_\Gamma a_-^{-1}, \qquad a_-P_\Gamma a_+, \qquad a_-P_\Gamma a_-^{-1}, \qquad a_+^{-1}P_\Gamma a_+$$

are either simultaneously bounded on $\mathbf{L}_p(\Gamma)$ $(1 < p < \infty)$ or not.

Clearly, a factorization of the form (1) is always a generalized factorization in the sense of Theorem 5.3. A factorization of the form (15) is usually referred to as a *generalized factorization of the function a with respect to Γ.*

§ 6. Effective solution of singular integral equations with continuous coefficients

Using the generalized factorization of continuous functions, in this section we extend the results of § 2 to singular integral operators with arbitrary continuous coefficients.

Theorem 6.1. *Let a and b be continuous functions on Γ and suppose*

$$a(t) \neq 0, \qquad b(t) \neq 0 \qquad (t \in \Gamma).$$

Put $\varkappa = \operatorname{ind} a/b$ *and let*

$$a/b = c_- t^\varkappa c_+ \tag{1}$$

be the generalized factorization of the function a/b.

Then the one-sided resp. two-sided inverse of the operator $A = aP_\Gamma + bQ_\Gamma$ on the space $\mathbf{L}_p(\Gamma, \varrho)$ takes the form

$$A^{-1} = (t^{-\varkappa}P_\Gamma + Q_\Gamma)\,(c_+^{-1}P_\Gamma + c_-Q_\Gamma)\,c_-^{-1}b^{-1}. \tag{2}$$

If $\varkappa < 0$,

$$\operatorname{Ker} A = \mathfrak{L}\{g, gt, \ldots, gt^{|\varkappa|-1}\} \tag{3}$$

with $g = c_+^{-1} - c_-t^\varkappa$.

If $\varkappa > 0$,

$$\operatorname{Coker} A = \mathfrak{L}\{bc_-, bc_-t, \ldots, bc_-t^{\varkappa-1}\} \tag{4}$$

and the equation $A\varphi = f$ has a solution if and only if

$$\int_\Gamma f(t)\, b^{-1}(t)\, c_-^{-1}(t)\, t^{-j}\, dt = 0 \qquad (j = 1, \ldots, \varkappa). \tag{5}$$

Proof. First let $\varkappa = 0$. Then, by Theorem 3.1, A is invertible. We shall prove that

$$A^{-1} = (c_+^{-1}P_\Gamma + c_-Q_\Gamma)\,c_-^{-1}b^{-1}. \tag{6}$$

The remark made at the end of § 5.3 implies that the operator (6) is bounded on $\mathbf{L}_p(\Gamma, \varrho)$.

Let $r \in \mathbf{R}(\Gamma)$ be any rational function. Then

$$(c_+^{-1}P_\Gamma + c_-Q_\Gamma)\,c_-^{-1}b^{-1}Ar = (c_+^{-1}P_\Gamma + c_-Q_\Gamma)\,(c_+r_+ + c_-^{-1}r_-), \tag{7}$$

where $r_+ = P_\Gamma r$, $r_- = Q_\Gamma r$. Since $c_+r_+ \in \mathbf{L}_p^+(\Gamma, \varrho)$ and $c^{-1}r_- \in \mathring{\mathbf{L}}_p^-(\Gamma, \varrho)$, the right-hand side of (7) is equal to r. Thus $A^{-1}Ar = r$ and, consequently, the operator defined by (6) is the inverse of A.

If $\varkappa > 0$, then the operator A can be represented in the form

$$A = b(ab^{-1}t^{-\varkappa}P_\Gamma + Q_\Gamma)\,(t^\varkappa P_\Gamma + Q_\Gamma),$$

and by what has already been proved it follows that the operator defined by (2) is a left inverse of A.

Now let $\varkappa < 0$. Then the operator

$$B = b(ab^{-1} t^{-\varkappa} P_\Gamma + Q_\Gamma) = A(t^{-\varkappa} P_\Gamma + Q_\Gamma)$$

is invertible and the inverse B^{-1} can be obtained using (6). This shows that the operator (2) is a right inverse of A.

The description (3) of the kernel and the solvability conditions (5) can be proved in a completely analogous manner as in the proof of Theorem 2.4.

Thus we are left with the proof of (4) for $\varkappa > 0$. Put $f_j = bc_- t^{j-1}$ $(j = 1, \ldots, \varkappa)$. Then

$$\int_\Gamma f_j(t)\, b^{-1}(t)\, c_-^{-1}(t)\, t^{-j}\, \mathrm{d}t = \int_\Gamma \frac{\mathrm{d}t}{t} \neq 0 \qquad (j = 1, \ldots, \varkappa)$$

and therefore $f_j \notin \operatorname{Im} A$. This combined with the equality $\dim \operatorname{Coker} A = \varkappa$ implies (4). ∎

An analogous theorem can also be proved for the operator $P_\Gamma a + Q_\Gamma b$ (see Theorem 2.3 and Theorem 2.5).

§ 7. The case of a composite curve system

A collection of finitely many piecewise Lyapunov curves which have no common points is called a *composite curve system*. A composite curve system is said to be *closed* if it consists of closed curves only.

Note that a composite closed curve system is not necessarily a closed curve system in the sense of the Definition 1.4° of § 1, Chapter II. To see this, consider, for instance, two concentric circles which are both positively (i.e., counter-clockwise) oriented. Although in the preceding paragraphs we made essential use of the hypothesis that Γ is a closed curve system, it turns out that the main results of the present chapter can be carried over to the case of a composite closed curve system.

Theorem 7.1. *All assertions of Theorem 3.1 remain true if Γ is a composite closed curve system and $a, b \in \mathrm{C}(\Gamma)$.*

Proof. Let the curve system Γ consist of the closed Lyapunov curves $\Gamma_1, \Gamma_2, \ldots, \Gamma_n$ and denote by $D^+_{\Gamma_k}$ $(k = 1, \ldots, n)$ the region enclosed by Γ_k and lying on the left of Γ_k.

For each curve Γ_k construct a closed Lyapunov curve Γ'_k lying in the intersection of $D^+_{\Gamma_k}$ and some sufficiently small neighborhood of the curve Γ_k. Observe that all curves Γ_k, Γ'_k $(k = 1, \ldots, n)$ can be assumed to be pairwise disjoint. Provide the curve Γ'_k with the orientation opposite to the one of Γ_k (see Fig. 4). Put $\Gamma' = \bigcup_{k=1}^{n} \Gamma'_k$ and $\tilde{\Gamma} = \Gamma \cup \Gamma'$. Then the curve $\tilde{\Gamma}$ is the boundary of a set consisting of n annular regions, and therefore it is a closed Lyapunov curve system in the sense of Definition 1.4° of § 1, Chapter II. Finally, continue the functions a and b to the curve $\tilde{\Gamma}$ by defning $a(t) = b(t) = 1$ for $t \in \Gamma'$.

Now Theorem 3.1 applies to the operator $\tilde{A} = aP_{\tilde{\Gamma}} + bQ_{\tilde{\Gamma}}$ considered as acting on the space $\mathbf{L}_p(\tilde{\Gamma}, \varrho)$. Note that the space $\mathbf{L}_p(\tilde{\Gamma}, \varrho)$ is obviously isomorphic to the direct sum $\mathbf{L}_p(\Gamma, \varrho) \dotplus \mathbf{L}_p(\Gamma')$. It is easily seen that

$$C\tilde{A}C^{-1} = \begin{pmatrix} A & B \\ 0 & I \end{pmatrix}. \tag{1}$$

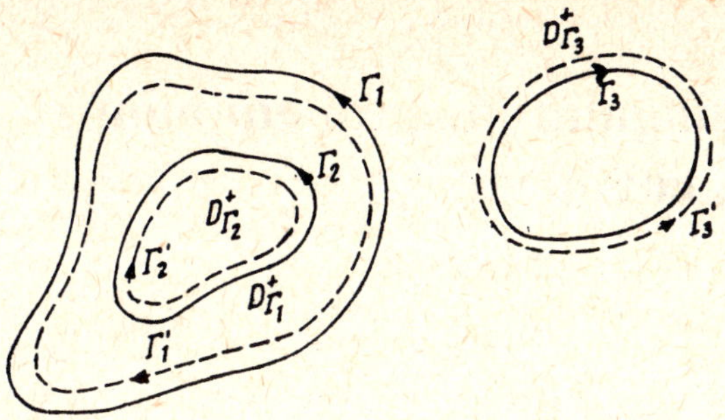

Fig. 4

Here $B: \mathbf{L}_p(\Gamma') \to \mathbf{L}_p(\Gamma, \varrho)$ is the bounded linear operator given by the formula

$$(B\varphi)(t) = \frac{1}{\pi i} \int_{\Gamma'} \frac{\varphi(\tau)}{\tau - t} d\tau \qquad (t \in \Gamma, \varphi \in \mathbf{L}_p(\Gamma'))$$

and $C: \mathbf{L}_p(\tilde{\Gamma}, \varrho) \to \mathbf{L}_p(\Gamma, \varrho) \dotplus \mathbf{L}_p(\Gamma')$ is the operator defined as $C\tilde{\psi} = \{\psi, \psi_1\}$ with $\psi = \tilde{\psi} \mid \Gamma$, $\psi_1 = \tilde{\psi} \mid \Gamma_1$. But (1) and the obvious identities

$$\begin{pmatrix} A & B \\ 0 & I \end{pmatrix} = \begin{pmatrix} I & B \\ 0 & I \end{pmatrix} \begin{pmatrix} A & 0 \\ 0 & I \end{pmatrix}, \begin{pmatrix} I & B \\ 0 & I \end{pmatrix}^{-1} = \begin{pmatrix} I & -B \\ 0 & I \end{pmatrix} \tag{2}$$

imply the assertion of the theorem for the operator $A = aP_\Gamma + bQ_\Gamma$.

The proof for the operator $P_\Gamma a + Q_\Gamma b$ is analogous. ∎

Finally, note that, with $A = aP_\Gamma + bQ_\Gamma$, equalities (1) and (2) immediately give that

$$A^{-1} = C^{-1} R \tilde{A}^{-1} C \mid \mathbf{L}_p(\Gamma, \varrho), \qquad \operatorname{Ker} A = \operatorname{Ker} \tilde{A} C^{-1},$$

where $R\{\psi, \psi_1\} = \{\psi, 0\}$.

Remark 1. In this paragraph we followed Gohberg and Krupnik [4] (Chapter III, § 11). The idea to extend the composite closed curve system Γ in order to obtain a closed Lyapunov curve system $\tilde{\Gamma}$ is due to Gakhov [1].

Remark 2. We conclude with noting that all results of the present chapter remain valid for the spaces $\mathbf{L}_p(\Gamma, \varrho)$, $1 < p < \infty$, with Γ and ϱ as in Theorem 3.3, Chap. II.

This can be seen combining Theorem 3.3 and the techniques employed above.

Chapter IV
One-dimensional singular integral equations with discontinuous coefficients

§ 1. Preliminaries

1.1. Alteration of the integration curve

Let $\tilde{\Gamma}$ be a piecewise continuous curve system and $\Gamma \subset \tilde{\Gamma}$ a part of $\tilde{\Gamma}$. The space $\mathbf{L}_p(\tilde{\Gamma}, \varrho)$[1]) can obviously be viewed as the direct sum of the spaces $\mathbf{L}_p(\Gamma, \varrho)$ and $\mathbf{L}_p(\tilde{\Gamma} \setminus \Gamma, \varrho)$. Denote by $\tilde{P}$ the projection of $\mathbf{L}_p(\tilde{\Gamma}, \varrho)$ onto $\mathbf{L}_p(\Gamma, \varrho)$ parallel to $\mathbf{L}_p(\tilde{\Gamma} \setminus \Gamma, \varrho)$ and let $\tilde{Q}$ denote the complementary projection.

Theorem 1.1. *Let $\tilde{a}$ and $\tilde{b}$ be functions in $\mathbf{L}_\infty(\tilde{\Gamma})$ and suppose*

$$\tilde{a}(t) = \tilde{b}(t) = 1 \qquad (t \in \tilde{\Gamma} \setminus \Gamma). \tag{1}$$

Denote by $a = \tilde{a} \mid \Gamma$ and $b = \tilde{b} \mid \Gamma$ the restrictions of the functions a and b to Γ.

Then the operator $A = aP_\Gamma + bQ_\Gamma$ is the restriction of the operator $\tilde{A} = \tilde{a}P_{\tilde{\Gamma}} + \tilde{b}Q_{\tilde{\Gamma}}$ to the space $\mathbf{L}_p(\Gamma, \varrho)$:

$$A = \tilde{A} \mid \mathbf{L}_p(\Gamma, \varrho). \tag{2}$$

Furthermore, one has

$$\tilde{A} = (I + \tilde{P}\tilde{A}\tilde{Q})(A\tilde{P} + \tilde{Q}), \tag{3}$$

and the operator $I + \tilde{P}\tilde{A}\tilde{Q}$ is invertible.

Proof. Because of (1),

$$\tilde{Q}\tilde{A} = \tilde{Q}\tilde{a}P_{\tilde{\Gamma}} + \tilde{Q}\tilde{b}Q_{\tilde{\Gamma}} = \tilde{Q}P_{\tilde{\Gamma}} + \tilde{Q}Q_{\tilde{\Gamma}} = \tilde{Q}$$

and hence

$$\tilde{A} = (\tilde{P} + \tilde{Q})\,\tilde{A} = \tilde{P}\tilde{A} + \tilde{Q}. \tag{4}$$

Consequently, $\tilde{P}\tilde{A}\tilde{P} = \tilde{A}\tilde{P} = A\tilde{P}$, that is, the space $\mathbf{L}_p(\Gamma, \varrho)$ is invariant under the operator $\tilde{A}$ and (2) holds. From (4) we also obtain that

$$\tilde{A} = \tilde{P}\tilde{A}\tilde{P} + \tilde{P}\tilde{A}\tilde{Q} + \tilde{Q} = (I + \tilde{P}\tilde{A}\tilde{Q})(A\tilde{P} + \tilde{Q}).$$

Finally, $I - \tilde{P}\tilde{A}\tilde{Q}$ is obviously the inverse of $I + \tilde{P}\tilde{A}\tilde{Q}$. ∎

We now provide some important consequences of Theorem 1.1.

---

[1]) Throughout the whole Chapter IV, ϱ has the same meaning as in Chapter III.

Corollary 1.1. *Suppose* (1) *holds. Then the operator A is normally solvable on the space* $\mathbf{L}_p(\Gamma, \varrho)$ *if and only if the operator $\tilde{A}$ is so on* $\mathbf{L}_p(\tilde{\Gamma}, \varrho)$. *Furthermore,*

$$\operatorname{Ker} A = \tilde{P} \operatorname{Ker} \tilde{A}, \qquad \operatorname{Coker} A = \tilde{P} \operatorname{Coker} \tilde{A}.$$

This is immediate from (2) and (3).

Corollary 1.2. *Suppose* (1) *holds. Then the operator A admits a left* (*resp. right*) *regularization if and only if the operator $\tilde{A}$ admits a left* (*resp. right*) *regularization.*

If $\tilde{R}$ is a left (*resp. right*) *regularizer of $\tilde{A}$, then* $R = \tilde{P}\tilde{R} \mid \mathbf{L}_p(\Gamma, \varrho)$ *is a left* (*resp. right*) *regularizer of A.*

Proof. The representation (3) immediately implies that $\tilde{A}$ is regularizable (of course, from the same side) if A is regularizable and that R is a left regularizer of A if $\tilde{R}$ is a left regularizer of $\tilde{A}$.

Now assume $\tilde{R}$ is a right regularizer of $\tilde{A}$. Then the operator $\tilde{A}\tilde{R} - I$ is compact and therefore so is the operator $\tilde{Q}\tilde{A}\tilde{R}\tilde{P} = \tilde{Q}\tilde{R}\tilde{P}$. Thus, the operator

$$\tilde{R}_1 = \tilde{P}\tilde{R}\tilde{P} + \tilde{P}\tilde{R}\tilde{Q} + \tilde{Q}\tilde{R}\tilde{Q}$$

is a right regularizer of $\tilde{A}$ and now it is clear that R is a right regularizer of A. ∎

Corollary 1.3. *Suppose* (1) *holds. Then the operator A is left* (*resp. right*) *invertible if and only if the operator $\tilde{A}$ is so.*

If $\tilde{A}^{-1}$ is a left (*resp. right*) *inverse of $\tilde{A}$, then the operator* $A^{-1} = \tilde{P}\tilde{A}^{-1} \mid \mathbf{L}_p(\Gamma, \varrho)$ *is a left* (*resp. right*) *inverse of A.*

The proof is analogous to the one of the preceding corollary.

Theorem 1.2. *Let $\Gamma_1, \ldots, \Gamma_n$ be piecewise Lyapunov curve systems which have no points in common, let* $\Gamma = \bigcup_{j=1}^{n} \Gamma_j$, *and suppose* $a, b \in \mathbf{L}_\infty(\Gamma)$. *Put* $a_j = a \mid \Gamma_j$ *and* $b_j = b \mid \Gamma_j$ $(j = 1, \ldots, n)$.

Then the operator $A = aP_\Gamma + bQ_\Gamma$ *is a* $\Phi_+(\Phi_-)$-*operator on* $\mathbf{L}_p(\Gamma, \varrho)$ *if and only if each operator* $A_j = a_jP_{\Gamma_j} + b_jQ_{\Gamma_j}$ *is a* $\Phi_+(\Phi_-)$-*operator on* $\mathbf{L}_p(\Gamma_j, \varrho)$. *In that case* $\operatorname{Ind} A = \sum_{j=1}^{n} \operatorname{Ind} A_j$.

Proof. Obviously, the space $\mathbf{L}_p(\Gamma, \varrho)$ can be identified with the direct sum $\mathbf{L}_p(\Gamma_1, \varrho) \dot{+} \mathbf{L}_p(\Gamma_2, \varrho) \dot{+} \cdots \dot{+} \mathbf{L}_p(\Gamma_n, \varrho)$. Denote by P_j $(j = 1, \ldots, n)$ the projection of $\mathbf{L}_p(\Gamma, \varrho)$ onto $\mathbf{L}_p(\Gamma_j, \varrho)$ parallel to the direct sum of the remaining spaces $\mathbf{L}_p(\Gamma_k, \varrho)$ $(k = 1, \ldots, n;$ $k \neq j)$. Then $P_1 + P_2 + \cdots + P_n = I$ and hence $A = \sum_{j,k=1}^{n} P_jAP_k$. It is easily seen that

$$P_jAP_j = A_jP_j, \qquad P_jAP_k = \tfrac{1}{2}(a_j - b_j)\, S_{\Gamma_k} \qquad (j \neq k).$$

Since the curves Γ_j $(j = 1, \ldots, n)$ are pairwise disjoint, the operators P_jAP_k $(j, k = 1, \ldots, n;\, j \neq k)$ are compact. Thus

$$A = \sum_{j=1}^{n} A_jP_j + T \tag{5}$$

with some compact operator T. Since the spaces $\mathbf{L}_p(\Gamma_j, \varrho)$ $(j = 1, \ldots, n)$ are invariant under the the operators A_j, by taking into consideration the results of Chapter I all assertions of the theorem can be readily obtained from (5). ∎

Another consequence of (5) is:

Corollary 1.4. *The operator* $A = aP_\Gamma + bQ_\Gamma$ *admits a left (resp. right) regularization if and only if each operator* $A_j = a_jP_{\Gamma j} + b_jQ_{\Gamma j}$ *admits a left (resp. right) regularization.*

Note that the results of this section can be carried over to singular operators of the form $P_\Gamma a + Q_\Gamma b$ as well as to the case of unbounded curves.

1.2. Separation of the singularities

In the present subsection we assume that Γ is a closed piecewise Lyapunov curve system. Thus, as we know, $P_\Gamma^2 = P_\Gamma$.

Let $a \in \mathbf{L}_\infty(\Gamma)$. The complement of the largest open subset of Γ in which the function a is continuous is called the *singular support of the function* a and is denoted by sing a. Furthermore, given a Banach algebra $\mathfrak{X}$ with identity element denote by $\mathbf{G}\mathfrak{X}$ the group of all invertible elements of $\mathfrak{X}$. In particular, $\mathbf{GL}_\infty(\Gamma)$ is the set of all functions $b \in \mathbf{L}_\infty(\Gamma)$ for which $\inf\limits_{t\in\Gamma} \text{ess}\, |b(t)| > 0$.

Theorem 1.3. *Let the singular support of a function* $a \in \mathbf{GL}_\infty(\Gamma)$ *be contained in the union of* n *pairwise disjoint open Jordan curves* $\gamma_1, \ldots, \gamma_n$ *and suppose we are given functions* $a_j \in \mathbf{GL}_\infty(\Gamma)$ $(j = 1, \ldots, n)$ *which have the following properties:*

1. *the functions* a_j *are continuous on* $\Gamma \setminus \gamma_j$ $(j = 1, \ldots, n)$;
2. $a_j(t) = a(t)$ *for* $t \in \gamma_j$ $(j = 1, \ldots, n)$.

Then the operator $aP_\Gamma + Q_\Gamma$ *is a* $\Phi_\pm(\Phi)$*-operator on* $\mathbf{L}_p(\Gamma, \varrho)$ *if and only if each operator* $a_jP_\Gamma + Q_\Gamma$ $(j = 1, \ldots, n)$ *is a* $\Phi_\pm(\Phi)$*-operator on* $\mathbf{L}_p(\Gamma, \varrho)$. *In that case*

$$\text{Ind}\,(aP_\Gamma + Q_\Gamma) = \sum_{j=1}^{n} \text{Ind}\,(a_jP_\Gamma + Q_\Gamma) - \text{ind}\, a_0,$$

where $a_0 = aa_1^{-1} \ldots a_n^{-1}$.

The proof will be based on two lemmas.

Lemma 1.1. *Let* $a_1, \ldots, a_n \in \mathbf{L}_\infty(\Gamma)$ *be functions whose singular supports are pairwise disjoint and put* $\tilde{a} = a_1 \ldots a_n$. *Then the operator* $P_\Gamma\tilde{a}P_\Gamma - P_\Gamma a_1P_\Gamma \ldots P_\Gamma a_nP_\Gamma$ *is compact on the space* $\mathbf{L}_p(\Gamma, \varrho)$.

Proof. In view of the identity

$$P_\Gamma a_1P_\Gamma a_2P_\Gamma a_3P_\Gamma = P_\Gamma a_1a_2P_\Gamma a_3P_\Gamma - (P_\Gamma a_1a_2P_\Gamma - P_\Gamma a_1P_\Gamma a_2P_\Gamma)\, a_3P_\Gamma$$

it suffices to prove the assertion for $n = 2$, i.e., to show that $P_\Gamma abP_\Gamma - P_\Gamma aP_\Gamma bP_\Gamma$ is compact for any $a, b \in L_\infty(\Gamma)$ such that sing $a \cap$ sing $b = \emptyset$. Note that this immediately results from Theorem 4.2, Chapter II, if at least one of the functions a, b is continuous on Γ.

Let us consider the general case. By our hypothesis, each point $t \in \Gamma$ has an open neighborhood $\gamma_t \in \Gamma$ such that $a \mid \gamma_t$ or $b \mid \gamma_t$ is continuous. Since Γ is compact, it can be covered by a finite collection $\{\gamma_{t_i}\}_{i=1}^r$ of such neighborhoods. Let $\sum\limits_{i=1}^{r} g_i = 1$ be a subordinate continuous partition of unity. Obviously, each g_i can be represented as $g_i = f_i^2$, where f_i is also continuous. Thus

$$
\begin{aligned}
&P_\Gamma abP_\Gamma - P_\Gamma aP_\Gamma bP_\Gamma \\
&= \sum_{i=1}^{r} [P_\Gamma af_i^2bP_\Gamma - P_\Gamma aP_\Gamma f_i^2P_\Gamma bP_\Gamma] \\
&= \sum_{i=1}^{r} [P_\Gamma af_iP_\Gamma f_ibP_\Gamma + P_\Gamma af_iQ_\Gamma f_ibP_\Gamma \\
&\quad - P_\Gamma aP_\Gamma f_iP_\Gamma f_iP_\Gamma bP_\Gamma - P_\Gamma aP_\Gamma f_iQ_\Gamma f_iP_\Gamma bP_\Gamma] \\
&= \sum_{i=1}^{r} [P_\Gamma aP_\Gamma f_iP_\Gamma f_iQ_\Gamma bP_\Gamma + P_\Gamma aQ_\Gamma f_iP_\Gamma f_iP_\Gamma bP_\Gamma \\
&\quad + P_\Gamma aQ_\Gamma f_iP_\Gamma f_iQ_\Gamma bP_\Gamma + P_\Gamma af_iQ_\Gamma f_ibP_\Gamma - P_\Gamma aP_\Gamma f_iQ_\Gamma f_iP_\Gamma bP_\Gamma]
\end{aligned}
$$

and since each item in the last sum only involves bounded operators and at least one operator of the form $P_\Gamma\varphi Q_\Gamma$ or $Q_\Gamma\varphi P_\Gamma$ with a continuous function φ, the assertion follows from Theorem 4.2, Chapter II. ∎

The following result is an immediate consequence of the just proved lemma.

Lemma 1.2. *If the singular supports of the functions $a_1, a_2 \in \mathbf{L}_\infty(\Gamma)$ are disjoint, then the operator $P_\Gamma a_1P_\Gamma a_2P_\Gamma - P_\Gamma a_2P_\Gamma a_1P_\Gamma$ is compact on the space $\mathbf{L}_p(\Gamma, \varrho)$.*

Proof of Theorem 1.3. Obviously, $a_0 \in \mathbf{GC}(\Gamma)$. Put

$$A = P_\Gamma aP_\Gamma + Q_\Gamma, \qquad A_j = P_\Gamma a_jP_\Gamma + Q_\Gamma \qquad (j = 0, 1, \ldots, n).$$

Lemma 1.1 implies that

$$A = A_0A_1 \ldots A_n + T \tag{6}$$

with some compact operator T. By the results of § 3, Chapter I, and Lemma 1.2, this implies that Theorem 1.3 is true if in its formulation the operator $aP_\Gamma + Q_\Gamma$ is replaced by A and the operators $a_jP_\Gamma + Q_\Gamma$ by A_j. But by virtue of the results of Section 1.1, Chapter III, the operators $aP_\Gamma + Q_\Gamma$ and $P_\Gamma aP_\Gamma + Q_\Gamma$ are either simultaneously $\Phi_\pm(\Phi)$-operators or not, and this completes the proof. ∎

Corollary 1.5. *Under the hypotheses of Theorem 1.3, the operator $aP_\Gamma + Q_\Gamma$ admits a left (resp. right) regularization if and only if each operator $a_jP_\Gamma + Q_\Gamma$ $(j = 1, \ldots, n)$ admits a left (resp. right) regularization.*

This can be easily deduced from Lemma 1.2, the representation (6), and (1.1) of Chapter III.

A direct consequence of Theorem 1.3 is the following result.

Theorem 1.4. *Let the singular supports of the functions $a, b \in \mathbf{GL}_\infty(\Gamma)$ be contained in the union of n pairwise disjoint open Jordan curves $\gamma_1, \ldots, \gamma_n$ and suppose the functions a_j, $b_j \in \mathbf{GL}_\infty(\Gamma)$ have the following properties: a_j and b_j are continuous on $\Gamma \setminus \gamma_j$ $(j = 1, \ldots, n)$ and $a_j(t) = a(t)$, $b_j(t) = b(t)$ for all $t \in \gamma_j$.*

Then the operator $A = aP_\Gamma + bQ_\Gamma$ is a $\Phi_\pm(\Phi)$-operator on the space $\mathbf{L}_p(\Gamma, \varrho)$ if and only if each operator $A_j = a_jP_\Gamma + b_jQ_\Gamma$ $(j = 1, \ldots, n)$ is a $\Phi_\pm(\Phi)$-operator on this space. If A is a Φ-operator, then

$$\operatorname{Ind} A = \sum_{j=1}^{n} \operatorname{Ind} A_j - \operatorname{ind} c,$$

where $c = ab^{-1}(a_1 \ldots a_n)^{-1}\, b_1 \ldots b_n$.

Remark. In principle, Theorem 1.1 was stated by GAKHOV [1] (see § 42) and Theorem 1.2 by GOHBERG and KRUPNIK [4] (Chapter VII, § 1). The results of Section 1.2 are due to GOHBERG and SEMENTSUL [1], the proof of Lemma 1.1 given here is taken from BÖTTCHER and SILBERMANN [2] (p. 36).

§ 2. Singular equations with bounded measurable coefficients

2.1. Necessary conditions for the Fredholm property

In this section we prove the following important result.

Theorem 2.1. *Suppose Γ is a composite curve system*[1]) *and let $a, b \in \mathbf{L}_\infty(\Gamma)$. If the operator $aP_\Gamma + bQ_\Gamma$ $(P_\Gamma a + Q_\Gamma b)$ is a Φ_+- or Φ_--operator on the space $\mathbf{L}_p(\Gamma, \varrho)$, then*

$$\operatorname*{inf\,ess}_{t\in\Gamma} |a(t)| > 0, \qquad \operatorname*{inf\,ess}_{t\in\Gamma} |b(t)| > 0. \tag{1}$$

The proof of this theorem is based on the following lemma.

Lemma 2.1. *Let Γ be a closed piecewise Lyapunov curve and $c \in \mathbf{L}_\infty(\Gamma)$. Suppose there exist two sets γ_1 and γ_2 $(\gamma_1, \gamma_2 \subset \Gamma)$ with positive Lebesgue measure such that $c(t) = 0$ for $t \in \gamma_1$ and $c(t) \neq 0$ for $t \in \gamma_2$. Then if A is any operator of the form $cP_\Gamma + Q_\Gamma$, $P_\Gamma + cQ_\Gamma$, $P_\Gamma c + Q_\Gamma$, or $P_\Gamma + Q_\Gamma c$, one has*

$$\dim \operatorname{Ker} A = \dim \operatorname{Ker} A^* = 0.$$

Proof. Let $\varphi \in \operatorname{Ker}(cP_\Gamma + Q_\Gamma)$. Then $cP_\Gamma\varphi = -Q_\Gamma\varphi$. There is a function $(Q_\Gamma\varphi)(z)$ holomorphic in D_Γ^- whose nontangential limits are $(Q_\Gamma\varphi)(t)$ a.e. on Γ. Since $(Q_\Gamma\varphi)(\infty) = 0$ and $(Q_\Gamma\varphi)(t) = 0$ for $t \in \gamma_1$, the Luzin-Privalov uniqueness teorem (see PRIVALOV [2], p. 212) implies that $(Q_\Gamma\varphi)(t) \equiv 0$. Consequently, $(P_\Gamma\varphi)(t) = 0$ for $t \in \gamma_2$ and again with the aid of the Luzin-Privalov uniqueness theorem we deduce that $(P_\Gamma\varphi)(t) \equiv 0$. Thus, $\varphi = P_\Gamma\varphi + Q_\Gamma\varphi = 0$.

Analogously it can be proved that $\dim \operatorname{Ker}(P_\Gamma + cQ_\Gamma) = 0$, and then Theorem 1.3, Chapter III, yields that

$$\dim \operatorname{Ker}(P_\Gamma c + Q_\Gamma) = \dim \operatorname{Ker}(P_\Gamma + Q_\Gamma c) = 0.$$

Finally, let A be any of the operators mentioned in the theorem. Denote by A' its transposed operator. Then, by (2.1) of Chapter III, $A^* = H_\Gamma A' H_\Gamma$ and $H_\Gamma^2 = I$. But this implies that $\dim \operatorname{Ker} A^* = \dim \operatorname{Ker} A' = 0$. ∎

Proof of Theorem 2.1. We first consider the case where Γ is a closed piecewise Lyapunov curve. Assume the contrary, that is, assume $A = aP_\Gamma + bQ_\Gamma$ is a Φ_+- or Φ_--operator, but let one of the conditions (1) be violated. For definiteness, let $\operatorname{inf\,ess} |a(t)| = 0$. Consider the functions

$$a_1(t) = \begin{cases} a(t), & \text{if } |a(t)| \geq \varepsilon, \\ 0, & \text{if } |a(t)| < \varepsilon, \end{cases} \qquad b_1(t) = \begin{cases} b(t), & \text{if } |b(t)| \geq \varepsilon, \\ \varepsilon, & \text{if } |b(t)| < \varepsilon, \end{cases}$$

where ε is a sufficiently small positive number. Obviously, $|a(t) - a_1(t)| < \varepsilon$ and $|b(t) - b_1(t)| < 2\varepsilon$ for all $t \in \Gamma$.

Because $\|A - A_1\| < 2\varepsilon(\|P_\Gamma\| + \|Q_\Gamma\|)$, the operator $A_1 = a_1P_\Gamma + b_1Q_\Gamma$ is also a Φ_+- or Φ_--operator. Put $c = a_1/b_1$ and $B = cP_\Gamma + Q_\Gamma$. Our assumption that $\operatorname{inf\,ess} |a(t)| = 0$ implies that the set $\gamma_1 = \{t \in \Gamma: c(t) = 0\}$ has positive measure. Together with A_1

---

[1]) For the definition of a composite curve system see § 7 of Chapter III.

the operator B is also a Φ_+- or Φ_--operator, and therefore the set $\gamma_2 = \{t \in \Gamma: c(t) \neq 0\}$ has positive measure, too. Now Lemma 2.1 gives that $\dim \operatorname{Ker} B = \dim \operatorname{Ker} B^* = 0$, and we conclude that B is invertible. Thus, there exists a φ_0 such that $B\varphi_0 = 1$. Since the function $1 - Q_\Gamma\varphi_0 = cP_\Gamma\varphi_0$ vanishes on the set γ_1, we obtain from the Luzin-Privalov uniqueness theorem that $1 - (Q_\Gamma\varphi_0)(z) \equiv 0$ in D_Γ^-. But this contradicts $(Q_\Gamma\varphi_0)(\infty) = 0$.

Let now Γ be a composite curve system. Construct a curve system $\tilde{\Gamma}$ containing Γ as a part and consisting only of closed piecewise Lyapunov curves $\tilde{\Gamma}_1, \ldots, \tilde{\Gamma}_n$ which have no common points. Denote by $\tilde{a}$ and $\tilde{b}$ the functions on $\tilde{\Gamma}$ that coincide with a and b, respectively, on Γ and that are 1 on $\tilde{\Gamma} \setminus \Gamma$. Put $\tilde{a}_j = \tilde{a} \mid \tilde{\Gamma}_j$ and $\tilde{b}_j = \tilde{b} \mid \tilde{\Gamma}_j$ $(j = 1, \ldots, n)$. From Theorem 1.1 we deduce that $\tilde{a}P_{\tilde{\Gamma}} + \tilde{b}Q_{\tilde{\Gamma}}$ is a $\Phi_\pm$-operator on the space $\mathbf{L}_p(\tilde{\Gamma}, \varrho)$, and Theorem 1.2 shows that each of the operators $\tilde{a}_jP_{\tilde{\Gamma}_j} + \tilde{b}_jQ_{\tilde{\Gamma}_j}$ is a $\Phi_\pm$-operator on the space $\mathbf{L}_p(\tilde{\Gamma}_j, \varrho)$ $(j = 1, \ldots, n)$. By what has already been proved,

$$\inf_{t\in\tilde{\Gamma}_j} \operatorname{ess} |\tilde{a}(t)| > 0, \qquad \inf_{t\in\tilde{\Gamma}_j} \operatorname{ess} |\tilde{b}_j(t)| > 0,$$

and this gives conditions (1) immediately. The proof is analogous for the operator $P_\Gamma a + Q_\Gamma b$. ∎

Remark. Heunemann [1] proved that for the operator $aP_\Gamma + bQ_\Gamma$ $(P_\Gamma a + Q_\Gamma b)$ to be normally solvable on the space $\mathbf{L}_p(\Gamma)$ $(1 < p < \infty)$ the following conditions are necessary:

(a) $\inf_{t \in \operatorname{supp} a} \operatorname{ess} |a(t)| > 0$, $\inf_{t \in \operatorname{supp} b} \operatorname{ess} |b(t)| > 0$.

(b) Either the sets supp a and supp b only differ by a set of measure zero or one of the coefficients a, b is identically zero and the other one vanishes on a set of measure zero only.

Here, for a function f on Γ, supp f denotes the set of all $t \in \Gamma$ such that $f(t) \neq 0$.

2.2. Theorems on the kernel and cokernel

In this section we derive several sufficient conditions which ensure that for the singular operator $A = aP_\Gamma + bQ_\Gamma$ $(A = P_\Gamma a + Q_\Gamma b)$ with coefficients $a, b \in \mathbf{L}_\infty(\Gamma)$ at least one of the integers $\dim \operatorname{Ker} A$ and $\dim \operatorname{Ker} A^*$ is zero.

Theorem 2.2. *Suppose Γ is a closed piecewise Lyapunov curve system, a and b are functions in $\mathbf{L}_\infty(\Gamma)$, and let $A = aP_\Gamma + bQ_\Gamma$ $(A = P_\Gamma a + Q_\Gamma b)$.*

If none of the functions a, b vanishes identically and if

$$\operatorname{mes} \{t \in \Gamma : a(t) = b(t) = 0\} = 0 \tag{2}$$

then at least one of the integers $\dim \operatorname{Ker} A$ *and* $\dim \operatorname{Ker} A^*$ *is equal to zero.*

Proof. Let $A = aP_\Gamma + bQ_\Gamma$, $\varphi \in \operatorname{Ker} A$, and $\psi \in \operatorname{Ker} A^*$. By Theorem 5.1, Chapter II, $A^* = H_\Gamma(P_\Gamma b + Q_\Gamma a) H_\Gamma$ and $H_\Gamma^2 = I$. Thus $(P_\Gamma b + Q_\Gamma a)\chi = 0$, where $\chi = H_\Gamma\psi$, and on multiplying this from the left by the projections P_Γ and Q_Γ we get $P_\Gamma b\chi = 0$ and $Q_\Gamma a\chi = 0$. Hence

$$h_+ := a\chi \in \mathbf{L}_q^+(\Gamma, \varrho^{-1}), \qquad h_- := b\chi \in \mathbf{L}_q^-(\Gamma, \varrho^{-1})$$

where $1/p + 1/q = 1$.

Put $\varphi_+ = P_\Gamma \varphi$ and $\varphi_- = Q_\Gamma \varphi$. After multiplying the equality $a\varphi_+ = -b\varphi_-$ by χ we have $h_+\varphi_+ = -h_-\varphi_-$. Since $h_+\varphi_+ \in \mathbf{L}_1^+(\Gamma)$, $h_-\varphi_- \in \mathring{\mathbf{L}}_1^-(\Gamma)$, and $\mathbf{L}_1^+(\Gamma) \cap \mathring{\mathbf{L}}_1^-(\Gamma) = \{0\}$, we obtain that $h_+\varphi_+ = h_-\varphi_- = 0$.

If the set on which at least one of the functions φ_+ and φ_- vanishes has measure zero, then $h_+ = h_- = 0$, whence $a\chi = b\chi = 0$. In view of (2) this implies that $\chi = 0$ and, consequently, $\psi = 0$ and thus $\dim \operatorname{Ker} A^* = 0$.

If at least one of the functions φ_+ and φ_- vanishes on a set of positive measure, then, by the Luzin-Privalov uniqueness theorem, this function vanishes identically on Γ. The equality $b\varphi_- = -a\varphi_+$ implies that the other function also vanishes on a set with positive measure. Again by the Luzin-Privalov uniqueness theorem, that function is identically zero on Γ. Thus $\varphi = \varphi_+ + \varphi_- = 0$ and therefore $\dim \operatorname{Ker} A = 0$. ∎

Remark 1. The conditions given in Theorem 2.2 are essential for the validity of this theorem. So one has, for instance, $\dim \operatorname{Ker} A = \dim \operatorname{Ker} A^* = \infty$ if $A = P_\Gamma$ $(b = 0)$, if $A = Q_\Gamma$ $(a = 0)$, or if A is the operator of multiplication by the characteristic function of an arbitrary subarc $\gamma \subset \Gamma$.

Remark 2. If at least one of the conditions required in Theorem 2.2 is violated, then it is easily seen that $\dim \operatorname{Ker} A = \infty$ or $\dim \operatorname{Ker} A^* = \infty$ for $A = aP_\Gamma + bQ_\Gamma$.

Theorem 2.3. *Suppose Γ is a closed piecewise Lyapunov curve system, a and b are in $\mathbf{L}_\infty(\Gamma)$, and let $A = aP_\Gamma + bQ_\Gamma$. Then if both the integers* $\dim \operatorname{Ker} A$ *and* $\dim \operatorname{Ker} A^*$ *are finite, at least one of them is zero.*

This is an immediate consequence of Theorem 2.2 and Remark 2.

Theorem 2.4. *Let Γ be an arbitrary composite curve system, let $a, b \in \mathbf{GL}_\infty(\Gamma)$, and let $A = aP_\Gamma + bQ_\Gamma$. Then at least one of the integers* $\dim \operatorname{Ker} A$ *and* $\dim \operatorname{Ker} A^*$ *is equal to zero.*

Proof. We first consider the case of a closed curve system Γ. Let $\varphi \in \operatorname{Ker} A$ and $\psi \in \operatorname{Ker} A^*$. Repeating the arguments from the first part of the proof of Theorem 2.2 we get $h_+\varphi_+ = 0$ with $h_+ = a\chi$, $\chi = H_\Gamma\psi$, $\varphi_+ = P_\Gamma\varphi$. If $h_+ = 0$, then $\psi = 0$ and hence $\dim \operatorname{Ker} A^* = 0$. In case $h_+ \neq 0$ the function φ_+ vanishes on a subset of Γ having positive measure. Since D_Γ^+ consists of finitely many connected components, some simple arguments based on the equality $a\varphi_+ = -b\varphi_-$ and on the Luzin-Privalov uniqueness theorem give $\varphi_+ = \varphi_- = 0$. Thus $\varphi = \varphi_+ + \varphi_- = 0$ and therefore $\dim \operatorname{Ker} A = 0$.

Now let Γ be an arbitrary composite curve system. Construct a closed curve system $\tilde{\Gamma}$ containing Γ as a part and denote by $\tilde{a}$ and $\tilde{b}$ the functions on $\tilde{\Gamma}$ coinciding with a and b, respectively, on Γ and taking the value 1 on $\tilde{\Gamma} \setminus \Gamma$. By what has already been proved, at least one of the integers $\dim \operatorname{Ker} \tilde{A}$ and $\dim \operatorname{Ker} \tilde{A}^*$, where $\tilde{A} = \tilde{a}P_{\tilde{\Gamma}} + \tilde{b}Q_{\tilde{\Gamma}}$, is equal to zero. This and Corollary 1.1 imply the assertion of the theorem. ∎

Theorem 2.5. *Suppose Γ is a composite curve system, a and b are functions in $\mathbf{L}_\infty(\Gamma)$ and let $A = aP_\Gamma + bQ_\Gamma$ $(A = P_\Gamma a + Q_\Gamma b)$.*

Then, if A is a Φ_+- or Φ_--operator, it is at least one-sided invertible.

For the operator $A = aP_\Gamma + bQ_\Gamma$ this is an immediate consequence of the Theorems 2.1 and 2.4. For the operator $A = P_\Gamma a + Q_\Gamma b$ the assertion then follows from Corollary 1.1, Chapter III.

Putting the Theorems 2.1 and 2.2 together we obtain the following.

Corollary 2.1. *Suppose Γ is a closed piecewise Lyapunov curve system and let $a \neq 0$ be a function in $\mathbf{L}_\infty(\Gamma)$. If the operator $aP_\Gamma + Q_\Gamma$ $(P_\Gamma a + Q_\Gamma)$ is normally solvable on the space $\mathbf{L}_p(\Gamma, \varrho)$, then* $\operatorname{inf\,ess} |a(t)| > 0$.

2.3. Reduction to the case of an invertible operator

Let Γ be a composite curve system and $a, b \in \mathbf{L}_\infty(\Gamma)$. Assume $A = aP_\Gamma + bQ_\Gamma$ is a Φ-operator and let $\varkappa = -\operatorname{Ind} A$. Then, by Theorem 2.5, the operator A is left-sided invertible, right-sided invertible, or two-sided invertible if $\varkappa > 0, \varkappa < 0$, or $\varkappa = 0$, respectively. The operator $A_\varkappa = at^{-\varkappa} P_\Gamma + bQ_\Gamma$ is invertible.

Indeed, Theorem 4.2, Chapter II, implies that the operator $A_\varkappa - A(t^{-\varkappa} P_\Gamma + Q_\Gamma)$ is compact and therefore $\operatorname{Ind} A_\varkappa = 0$. From Theorem 2.1 we deduce that $a, b \in \mathbf{GL}_\infty(\Gamma)$, and Theorem 2.4 gives the invertibility of $A_\varkappa$.

In case $\varkappa > 0$ the operator A can be written as

$$A = A_\varkappa(t^\varkappa P_\Gamma + Q_\Gamma)$$

and therefore a left-inverse of A can be given by

$$A^{-1} = (t^{-\varkappa} P_\Gamma + Q_\Gamma)\, A_\varkappa^{-1}. \tag{3}$$

It is esay to check by a straightforward calculation that the operator defined by (3) is a right-inverse of A if $\varkappa < 0$.

Furthermore, in the case $\varkappa < 0$ we have

$$\operatorname{Ker}(aP_\Gamma + bQ_\Gamma) = \mathfrak{L}\{g, gt, \ldots, gt^{-\varkappa-1}\} \tag{4}$$

with $g = P_\Gamma\varphi - t^\varkappa(1 - Q_\Gamma\varphi)$, where φ is the solution of the equation $A_\varkappa\varphi = b$.

Indeed, if $A_\varkappa\varphi = b$ then

$$(aP_\Gamma + bQ_\Gamma)(gt^j) = t^{j+\varkappa}(A_\varkappa\varphi - b) = 0 \qquad (j = 0, 1, \ldots, -\varkappa - 1).$$

Consequently, $\mathfrak{L}\{g, gt, \ldots, gt^{-\varkappa-1}\} \subset \operatorname{Ker} A$, and since the funtions $g, gt, \ldots, gt^{-\varkappa-1}$ are linearly independent and since $\dim \operatorname{Ker} A = -\varkappa$, we arrive at (4).

Analogous results can also be obtained for the dual singular operator $P_\Gamma a + Q_\Gamma b$.

Remark. Theorem 2.1 is due to Simonenko [7]. Theorem 2.2 goes back to Coburn [1]. Here we essentially followed Gohberg and Krupnik [4] (Chap. VII, §§ 4, 5, and 7).

§ 3. Generalized factorization of bounded measurable functions and the effective solution of singular equations

In § 5, Chapter III, we saw that every continuous function $a \in \mathbf{GC}(\Gamma)$ on a closed curve system Γ admits a generalized factorization with respect to Γ in the sense of Theorem 5.3, Chapter III. This, however, is not true for an arbitrary function $a \in \mathbf{GL}_\infty(\Gamma)$. In this section we introduce the concept of the generalized factorization of a function in the space $\mathbf{L}_p(\Gamma, \varrho)$, with the help of which the results of § 6, Chapter III, can be extended to singular equations with coefficients from $\mathbf{GL}_\infty(\Gamma)$.

3.1. Generalized factorization in the space $\mathbf{L}_p(\Gamma, \varrho)$

Let us begin with the following simple example.

Let Γ_0 be the unit circle and $a(t) = t^{1/2}$ $(t \in \Gamma_0)$. Then the function a $(\in \mathbf{GL}_\infty(\Gamma))$ does not admit a generalized factorization with respect to Γ_0 in the sense of Theorem 5.3, Chapter III.

To see this, assume the contrary and let $a = a_- t^\varkappa a_+$ be such a generalized factorization with respect to Γ_0 (see equation (5.15), Chapter III). The function a can also be written in the form

$$a(t) = (t - 1)^{1/2} (1 - 1/t)^{-1/2}. \tag{1}$$

Thus, $a_- t^{\varkappa} a_+ = b_- b_+$ with $b_+(t) = (t - 1)^{1/2}$ and $b_-(t) = (1 - 1/t)^{-1/2}$. Obviously, $b_\mp^{\pm 1} \in \mathbf{L}_p^+(\Gamma_0)$ and $b_-^{\pm 1} \in \mathbf{L}_p^-(\Gamma_0)$ for $1 < p < 2$. Hence, if $\varkappa < 0$ then $t^{\varkappa} a_- b_-^{-1} = b_+ a_+^{-1}$ with the left-hand side in $\mathring{\mathbf{L}}_1^-(\Gamma_0)$ and the right-hand side in $\mathbf{L}_1^+(\Gamma_0)$. Therefore $b_+ a_+^{-1} \equiv 0$, which is clearly impossible. On the other hand, in case $\varkappa > 0$ from the equality $t^{-1} b_- a_-^{-1} = t^{\varkappa - 1} a_+ b_+^{-1}$ we obtain analogously that $b_- a_-^{-1} \equiv 0$, which is again impossible. In the case $\varkappa = 0$ the equality $b_- a_-^{-1} = a_+ b_+^{-1}$ implies that $a_- = \lambda b_-$ with a certain complex number λ. But this is also impossible, since $b_- \notin \mathbf{L}_q(\Gamma_0)$ for $q > 2$.

However, representation (1) is a generalized factorization of the function $a(t) = t^{1/2}$ in the space $\mathbf{L}_p(\Gamma_0)$ $(1 < p < 2)$ in the sense of the following definition.

Definition. Let Γ be a closed piecewise Lyapunov curve system. A representation

$$a = a_- t^{\varkappa} a_+ \tag{2}$$

is called a *generalized factorization of the function* $a \in \mathbf{L}_\infty(\Gamma)$ *in the space* $\mathbf{L}_p(\Gamma, \varrho)$ if $\varkappa$ is an integer and if the functions a_+, a_- possess the following properties:

1. $a_- \in \mathbf{L}_p^-(\Gamma, \varrho)$, $a_+ \in \mathbf{L}_q^+(\Gamma, \varrho^{-1})$, $a_-^{-1} \in \mathbf{L}_q^-(\Gamma, \varrho^{-1})$, $a_+^{-1} \in \mathbf{L}_p^+(\Gamma, \varrho)$ $(1/p + 1/q = 1)$;

2. The operator $a_+^{-1} P_\Gamma a_-^{-1}$ is bounded on the space $\mathbf{L}_p(\Gamma, \varrho)$.

In connection with this definition recall that $\mathbf{L}_p^*(\Gamma, \varrho) = \mathbf{L}_q(\Gamma, \varrho^{-1})$. Furthermore, it is easy to see that every function which admits a generalized factorization with respect to Γ (cf. Section 5.3, Chapter III) admits a generalized factorization in every space

$$\mathbf{L}_p(\Gamma, \varrho) \left(1 < p < \infty,\ \varrho(t) = \prod_{k=1}^{m} |t - t_k|^{\beta_k},\ -1/p < \beta_k < 1 - 1/p\right).$$

Taking into account that $\mathbf{L}_1^+(\Gamma) \cap \mathbf{L}_1^-(\Gamma) = \mathbb{C}$, it is readily seen that the integer $\varkappa$ in (2) is determined uniquely. It is referred to as the *index of the function a in the space* $\mathbf{L}_p(\Gamma, \varrho)$ and is denoted by $\varkappa = \operatorname{ind} a \mid \mathbf{L}_p(\Gamma, \varrho)$. The factors a_+ and a_- in (2) are determined uniquely up to a constant factor. They are uniquely determined by the normalization $a_-(\infty) = 1$.

Theorem 3.1. *For a function* $a \in \mathbf{L}_\infty(\Gamma)$ *to admit a generalized factorization in the space* $\mathbf{L}_p(\Gamma, \varrho)$ *it is necessary and sufficient that the operator* $A = aP_\Gamma + Q_\Gamma$ *be a Φ-operator on the space* $\mathbf{L}_p(\Gamma, \varrho)$. *In that case* $\operatorname{ind} a \mid \mathbf{L}_p(\Gamma, \varrho) = -\operatorname{Ind} A$.

Proof. *Necessity.* Suppose the function $a \in \mathbf{L}_\infty(\Gamma)$ admits a generalized factorization (2) in the space $\mathbf{L}_p(\Gamma, \varrho)$. First consider the case where $\varkappa = \operatorname{ind} a \mid \mathbf{L}_p(\Gamma, \varrho) = 0$ and put $B = (a_+^{-1} P_\Gamma + a_- Q_\Gamma)\, a_-^{-1}$.

Let $r \in \mathbf{R}(\Gamma)$ be any rational function. Since $a_-^{-1} r \in \mathbf{L}_q(\Gamma, \varrho^{-1})$, $a_+^{-1} \in \mathbf{L}_p^+(\Gamma, \varrho)$, $a_- \in \mathbf{L}_p^-(\Gamma, \varrho)$, it follows that $a_+^{-1} P_\Gamma a_-^{-1} r \in \mathbf{L}_1^+(\Gamma)$ and $a_- Q_\Gamma a_-^{-1} r \in \mathring{\mathbf{L}}_1^-(\Gamma)$. Since, moreover, the operators $a_+^{-1} P_\Gamma a_-^{-1}$ and $a_- Q_\Gamma a_-^{-1} = I - a a_+^{-1} P_\Gamma a_-^{-1}$ are bounded on the space $\mathbf{L}_p(\Gamma, \varrho)$, we see that actually $a_+^{-1} P_\Gamma a_-^{-1} r \in \mathbf{L}_p^+(\Gamma, \varrho)$ and $a_- Q_\Gamma a_-^{-1} r \in \mathring{\mathbf{L}}_p^-(\Gamma, \varrho)$. Hence

$$ABr = a a_+^{-1} P_\Gamma a_-^{-1} r + a_- Q_\Gamma a_-^{-1} r = r.$$

Because $a_+ P_\Gamma r \in \mathbf{L}_q^+(\Gamma, \varrho^{-1})$ and $a^{-1} Q_\Gamma r \in \mathring{\mathbf{L}}_q^-(\Gamma, \varrho^{-1})$, we obtain analogously that $BAr = r$. Thus, the operator A is invertible and its inverse is B.

Now let $\varkappa \neq 0$. Then the function $at^{-\varkappa}$ admits the factorization $at^{-\varkappa} = a_- a_+$, and by what has already been proved, the operator $at^{-\varkappa} P_\Gamma + Q_\Gamma$ is invertible. If $\varkappa > 0$, then A can be represented in the form

$$A = (at^{-\varkappa}P_\Gamma + Q_\Gamma)\,(t^\varkappa P_\Gamma + Q_\Gamma)$$

and since $t^\varkappa P_\Gamma + Q_\Gamma$ is a Φ-operator with index $-\varkappa$ (see Thorem 2.3, Chapter III), the operator A is also a Φ-operator with index $-\varkappa$. In the case $\varkappa < 0$ we have

$$at^{-\varkappa}P_\Gamma + Q_\Gamma = A(t^{-\varkappa}P_\Gamma + Q_\Gamma)$$

and since $t^{-\varkappa}P_\Gamma + Q_\Gamma$ is a Φ-operator with index $\varkappa$, the operator A is a Φ-operator with index $-\varkappa$ (see § 3, Chapter I).

Sufficiency. Suppose $A = aP_\Gamma + Q_\Gamma$ is a Φ-operator with index $-\varkappa$. In Section 2.3 we showed that then the operator $A_\varkappa = at^{-\varkappa}P_\Gamma + Q_\Gamma$ is invertible on the space $\mathbf{L}_p(\Gamma, \varrho)$. Hence, $a \in \mathbf{GL}_\infty(\Gamma)$, by Theorem 2.1. Put $c = at^{-\varkappa}$. By virtue of (2.1), Chapter III, the operator $P_\Gamma + Q_\Gamma c$ is invertible on the space $\mathbf{L}_q(\Gamma, \varrho^{-1})$, and due to Theorem 1.2, Chapter III, the operator $c^{-1}P_\Gamma + Q_\Gamma = c^{-1}(P_\Gamma + cQ_\Gamma)$ is also invertible on $\mathbf{L}_q(\Gamma, \varrho^{-1})$.

Let $\varphi \in \mathbf{L}_p(\Gamma, \varrho)$ and $\psi \in \mathbf{L}_q(\Gamma, \varrho^{-1})$ be the solutions of the equations $(cP_\Gamma + Q_\Gamma)\,\varphi = 1$ and $(c^{-1}P_\Gamma + Q_\Gamma)\,\psi = 1$. Then

$$(P_\Gamma\varphi)\,(P_\Gamma\psi) = (1 - Q_\Gamma\varphi)\,(1 - Q_\Gamma\psi).$$

But the function $(1 - Q_\Gamma\varphi)\,(1 - Q_\Gamma\psi) - 1$ lies in $\mathring{\mathbf{L}}_1^-(\Gamma)$ and the function $(P_\Gamma\varphi)\,(P_\Gamma\psi) - 1$ belongs to $\mathbf{L}_1^+(\Gamma)$, and since $\mathbf{L}_1^+(\Gamma) \cap \mathring{\mathbf{L}}_1^-(\Gamma) = \{0\}$, we conclude that

$$(P_\Gamma\varphi)\,(P_\Gamma\psi) = (1 - Q_\Gamma\varphi)\,(1 - Q_\Gamma\psi) = 1.$$

Putting $a_+ = P_\Gamma\psi$ and $a_- = 1 - Q_\Gamma\varphi$ we obtain $c = a_+a_-$, whence $a = a_+t^\varkappa a_-$. It is obvious that the functions $a_+ = P_\Gamma\psi$, $a_- = 1 - Q_\Gamma\varphi$, $a_+^{-1} = P_\Gamma\varphi$, and $a_-^{-1} = 1 - Q_\Gamma\psi$ satisfy condition 1 in the definition of the generalized factorization in the space $\mathbf{L}_p(\Gamma, \varrho)$. Thus it remains to prove that the operator $a_+^{-1}P_\Gamma a_-^{-1}$ is bounded on the space $\mathbf{L}_p(\Gamma, \varrho)$. To this end we may without loss of generality assume that $|c(t)| \leqq m < 1$.

Again let $B = (a_+^{-1}P_\Gamma + a_-Q_\Gamma)\,a_-^{-1}$. Exactly as in the proof of the necessity part it can easily be verified that $BA_\varkappa\varphi = \varphi$ for all $\varphi \in \mathbf{L}_p(\Gamma, \varrho)$. Thus, it follows that $B = A_\varkappa^{-1}$ is bounded on $\mathbf{L}_p(\Gamma, \varrho)$. The representation $B = I + (1 - c)\,a_+^{-1}P_\Gamma a_-^{-1}$ then immediately implies the boundedness of the operator $a_+^{-1}P_\Gamma a_-^{-1}$ on $\mathbf{L}_p(\Gamma, \varrho)$. ∎

With the help of Theorem 3.1 it is easy to prove the following factorization criterion.

Theorem 3.2. *Let $a \in \mathbf{L}_\infty(\Gamma)$ and $b \in \mathbf{GC}(\Gamma)$. Then for the function a to admit a generalized factorization in the space $\mathbf{L}_p(\Gamma, \varrho)$ it is necessary and sufficient that the function ab admits a generalized factorization in the space $\mathbf{L}_p(\Gamma, \varrho)$.*

Proof. Since b is continuous, the operator $abP_\Gamma + Q_\Gamma - (aP_\Gamma + Q_\Gamma)\,(bP_\Gamma + Q_\Gamma)$ is compact on the space $\mathbf{L}_p(\Gamma, \varrho)$ (see Theorem 4.2, Chapter II). But $bP_\Gamma + Q_\Gamma$ is a Φ-operator, and therefore $abP_\Gamma + Q_\Gamma$ is a Φ-operator if and only if $aP_\Gamma + Q_\Gamma$ is a Φ-operator (cf. § 3, Chapter I). This and Theorem 3.1 imply the assertion of Theorem 3.2. ∎

3.2. Effective solution of singular equations with bounded measurable coefficients

The notion of the generalized factorization in the space $\mathbf{L}_p(\Gamma, \varrho)$ enables us to extend the results of § 6, Chapter III, to singular integral operators with arbitrary bounded measurable coefficients.

Theorem 3.3. *Let Γ be a closed piecewise Lyapunov curve system and $a, b \in \mathbf{L}_\infty(\Gamma)$. Then for the operator $A := aP_\Gamma + bQ_\Gamma$ to be a Φ_+- or Φ_--operator on the space $\mathbf{L}_p(\Gamma, \varrho)$ it is* *necessary that*

$$a \in \mathbf{GL}_\infty(\Gamma), \qquad b \in \mathbf{GL}_\infty(\Gamma). \tag{3}$$

Let conditions (3) *be fulfilled. Then A is a Φ-operator on $\mathbf{L}_p(\Gamma, \varrho)$ if and only if the function ab^{-1} admits a generalized factorization in the sapce $\mathbf{L}_p(\Gamma, \varrho)$.*

Let A be a Φ-operator and $\varkappa = \operatorname{ind} ab^{-1} \mid \mathbf{L}_p(\Gamma, \varrho)$. Then $\operatorname{Ind} A = -\varkappa$ and the operator A is left-sided invertible, right-sided invertible, or two-sided invertible if $\varkappa > 0$, $\varkappa < 0$, or $\varkappa = 0$, respectively. The corresponding (one- or two-sided) inverse is of the form

$$A^{-1} = (t^{-\varkappa} P_\Gamma + Q_\Gamma)\,(c_+^{-1} P_\Gamma + c_- Q_\Gamma)\, c_-^{-1} b^{-1}, \tag{4}$$

where $ab^{-1} = c_- t^\varkappa c_+$ is the generalized factorization of the function ab^{-1} in the space $\mathbf{L}_p(\Gamma, \varrho)$.

This theorem is a consequence of the Theorems 2.1, 2.4, 3.1, and the results of Section 2.3.

Theorem 3.4. *Let $a, b \in \mathbf{GL}_\infty(\Gamma)$ and suppose the function ab^{-1} admits a generalized factorization $ab^{-1} = c_- t^\varkappa c_+$ in the space $\mathbf{L}_p(\Gamma, \varrho)$.*

If $\varkappa < 0$, then

$$\operatorname{Ker} A = \mathfrak{L}\{g, gt, \ldots, gt^{|\varkappa|-1}\} \tag{5}$$

with $g = c_+^{-1} - c_- t^\varkappa$.

In the case $\varkappa > 0$ we have

$$\operatorname{Coker} A = \mathfrak{L}\{bc_-, bc_- t, \ldots, bc_- t^{\varkappa-1}\} \tag{6}$$

and the equation $A\varphi = f$ has a solution if and only if

$$\int_\Gamma f(t)\, b^{-1}(t)\, c_-^{-1}(t)\, t^{-j}\, dt = 0 \qquad (j = 1, \ldots, \varkappa). \tag{7}$$

The proof of this theorem is the proof of Theorem 6.1, Chapter III.

Analogous results hold for the dual singular operator $P_\Gamma a + Q_\Gamma b$. In that case the Theorems 3.3 and 3.4 remain valid if (4)–(7) are replaced by the following:

$$(P_\Gamma a + Q_\Gamma b)^{-1} = b^{-1} c_-^{-1} (P_\Gamma c_+^{-1} + Q_\Gamma c_-),$$
$$\operatorname{Ker} (P_\Gamma a + Q_\Gamma b) = \mathfrak{L}\{b^{-1} c_+^{-1}, b^{-1} c_+^{-1} t, \ldots, b^{-1} c_+^{-1} t^{-\varkappa-1}\},$$
$$\operatorname{Coker} (P_\Gamma a + Q_\Gamma b) = \mathfrak{L}\{g, gt, \ldots, gt^{\varkappa-1}\}$$

and

$$\int_\Gamma f(t)\, g(t)\, t^{-j}\, dt = 0 \qquad (j = 1, \ldots, \varkappa)$$

with $g = c_-^{-1} - c_+ t^\varkappa$.

Remark. The results of this section are due to Simonenko [7]. Note that these results also hold "locally" (see Simonenko [7], Douglas [1], Clancey and Gohberg [1]). Here we followed Gohberg and Krupnik [4] (Chap. VIII, §§ 3 and 4).

§ 4. Singular equations with piecewise continuous coefficients on closed curves

The purpose of this and the following two sections of the present chapter is to generalize the results of Chapter III to singular integral operators with piecewise continuous coefficients on the spaces $\mathbf{L}_p(\Gamma, \varrho)$. In particular, we shall derive necessary and sufficient conditions for such operators to be Fredholm on $\mathbf{L}_p(\Gamma, \varrho)$. The point is that, contrary to the case of continuous coefficients, these conditions as well as the index of the operator essentially depend on the value p and the weight ϱ. In the Sections 4, 5, and 6 we essentially follow Chapter IX of GOHBERG and KRUPNIK [4].

In the present section we restrict ourselves to the case of a closed piecewise Lyapunov curve system Γ.

4.1. Let z_1 and z_2 be two points in the complex plane and let δ be any real number such that $0 < \delta < \pi$. We denote by $K_\delta(z_1, z_2)$ the circular arc joining the points z_1 and z_2 and having the following two properties: (α) from the points $z \in K_\delta(z_1, z_2)$ ($z \neq z_1$, $z \neq z_2$) the straight line segment z_1z_2 is seen under the angle δ; (β) $K_\delta(z_1, z_2)$ is directed from z_1 to z_2 counter-clockwise. For $\pi < \delta < 2\pi$ we define $K_\delta(z_1, z_2) = K_{2\pi-\delta}(z_2, z_1)$ and for $\delta = \pi$ we let $K_\pi(z_1, z_2)$ denote the straight line segment z_1z_2. Then, for all values $\delta \in (0, 2\pi)$, the curve $K_\delta(z_1, z_2)$ can be given by the parametric representation

$$z = z_2 f_\delta(\mu) + z_1(1 - f_\delta(\mu)) \qquad (0 \leqq \mu \leqq 1),$$

where

$$f_\delta(\mu) = \begin{cases} \dfrac{\sin \theta\mu \; e^{i\theta\mu}}{\sin \theta \; e^{i\theta}} \quad (\theta = \pi - \delta) & \text{for} \quad 0 < \delta < 2\pi, \ \delta \neq \pi, \\ \mu & \text{for} \quad \delta = \pi. \end{cases} \tag{1}$$

4.2. In the following we assume that Γ is a closed piecewise Lyapunov curve system. Let $\mathbf{PC}(\Gamma)$ denote the algebra of all piecewise continuous[1]) and left-sided continuous (complex-valued) functions on Γ.

Let $p, \beta_1, \ldots, \beta_m$ be real numbers satisfying $1 < p < \infty$, $-1/p < \beta_k < 1 - 1/p$ ($k = 1, \ldots, m$) and let $\varrho(t) = \prod\limits_{k=1}^{m} |t - t_k|^{\beta_k}$, where $t_1, \ldots, t_m$ are pairwise distinct points on Γ.

With every function $a \in \mathbf{PC}(\Gamma)$ we associate a function $a_{p,\varrho} : \Gamma \times [0, 1] \to \mathbb{C}$ by the formula

$$a_{p,\varrho}(t, \mu) = a(t + 0) f(t, \mu) + a(t)(1 - f(t, \mu)) \qquad (t \in \Gamma, \ 0 \leqq \mu \leqq 1), \tag{2}$$

where $f(t, \mu) = f_{\delta(t)}(\mu)$ and

$$\delta(t) = \begin{cases} \dfrac{2\pi}{p}, & \text{if} \quad t \in \Gamma \setminus \{t_1, \ldots, t_m\}, \\ 2\pi\left(\dfrac{1}{p} + \beta_k\right), & \text{if} \quad t = t_k \ (k = 1, \ldots, m). \end{cases} \tag{3}$$

The range (image) of the function $a_{p,\varrho}$ is a continuous closed curve in the complex plane. It will be denoted by $V_{p,\varrho}(a)$. This curve is obtained from the range of a (the values $a(t)$, $t \in \Gamma$) by filling in the arcs $K_{\delta(\tau_j)}(a(\tau_j), a(\tau_j + 0))$ ($j = 1, \ldots, r$) at the points of discontinuity $\tau_1, \ldots, \tau_r$ of the function a. The curve $V_{p,\varrho}(a)$ is oriented in a natural way: on the intervals of continuity of the function a the motion of the point t along Γ in the positive direction induces a direction on $V_{p,\varrho}(a)$ and the arcs $K_{\delta(\tau_j)}(a(\tau_j), a(\tau_j + 0))$ ($j = 1, \ldots, r$) are oriented from $a(\tau_j)$ to $a(\tau_j + 0)$ counter-clockwise.

---

[1]) A function a given on Γ is said to be *piecewise continuous* if at each point $t \in \Gamma$ the one-sided finite limits $a(t \pm 0)$ (with respect to the orientation of Γ) exist and if the number of its discontinuities is finite.

A function $a \in \mathbf{PC}(\Gamma)$ is said to be *p, ϱ-regular* if the curve $V_{p,\varrho}$ does not pass through the origin. For a p, ϱ-regular function a the winding number of the curve $V_{p,\varrho}$ with respect to the origin is referred to as the *p, ϱ-index* of the function a and is denoted by $\operatorname{ind}_{p,\varrho} a$. In other words (see also § 3, Chapter III),

$$\operatorname{ind}_{p,\varrho} a := \frac{1}{2\pi}[\arg z]_{V_{p,\varrho}(a)} = \frac{1}{2\pi}[\arg a_{p,\varrho}(t, \mu)]_{t\in\Gamma, 0\leqq\mu\leqq1} = \operatorname{ind} a_{p,\varrho}.$$

Clearly, if a is continuous and $a(t) \neq 0$ for all $t \in \Gamma$, then $\operatorname{ind}_{p,\varrho} a = \operatorname{ind} a$. However, if a is a discontinuous function, the index $\operatorname{ind}_{p,\varrho} a$ depends on p and ϱ essentially. In the case $\varrho(t) \equiv 1$ one speaks of *p-regular functions* and the *p-index* of such functions.

A little thought shows that there exist p, ϱ-regular functions a and b such that $\operatorname{ind}_{p,\varrho} ab \neq \operatorname{ind}_{p,\varrho} a + \operatorname{ind}_{p,\varrho} b$. However, one has the following

Lemma 4.1. *If the p, ϱ-regular functions a and b have no common discontinuities, then their product $c = ab$ is also p, ϱ-regular and*

$$\operatorname{ind}_{p,\varrho} c = \operatorname{ind}_{p,\varrho} a + \operatorname{ind}_{p,\varrho} b.$$

Proof. It is easy to verify the identity

$$c_{p,\varrho}(t, \mu) - a_{p,\varrho}(t, \mu)\, b_{p,\varrho}(t, \mu) = [a(t + 0) - a(t)]\,[b(t + 0) - b(t)]\, f(t, \mu)\, (1 - f(t, \mu)).$$

Thus, under the hypotheses of the lemma we have $c_{p,\varrho}(t, \mu) = a_{p,\varrho}(t, \mu)\, b_{p,\varrho}(t, \mu)$ and this implies the assertion at once. ∎

Lemma 4.2. *A function $a \in \mathbf{PC}(\Gamma)$ is p, ϱ-regular if and only if it satisfies the following two conditions:*

1. $a(t \pm 0) \neq 0$ *for all* $t \in \Gamma$;

2. $a(t_0)/a(t_0 + 0) = e^{i\alpha_0}$ *with* $\delta(t_0) - 2\pi < \operatorname{Re} \alpha_0 < \delta(t_0)$ *at each point of discontinuity* t_0 *of the function* a.

Proof. The necessity of the first condition is obvious. On the other hand, if this condition is satisfied, then $a(t_0)/a(t_0 + 0)$ can always be written in the form $e^{i\alpha_0}$ with $\delta(t_0) - 2\pi < \operatorname{Re} \alpha_0 \leqq \delta(t_0)$. $\operatorname{Re} \alpha_0$ is just the angle under which the line segment $a(t_0), a(t_0 + 0)$ is seen from the origin $z = 0$. Thus, by construction, $0 \in K_{\delta(t_0)}(a(t_0), a(t_0 + 0))$ if and only if $\operatorname{Re} \alpha_0 \neq \delta(t_0)$. This gives the assertion of the lemma. ∎

4.3. We now turn to the investigation of the singular integral operator $A = aP_\Gamma + bQ_\Gamma$ ($A = P_\Gamma a + Q_\Gamma b$) with coefficients $a, b \in \mathbf{PC}(\Gamma)$ on the spaces $\mathbf{L}_p(\Gamma, \varrho)$. Our first concern is the case $b(t) \equiv 1$.

Theorem 4.1. *If $c \in \mathbf{PC}(\Gamma)$ then for the operator $A_c := cP_\Gamma + Q_\Gamma$ to be a Φ_+- or Φ_--operator on the space $\mathbf{L}_p(\Gamma, \varrho)$ it is necessary and sufficient that the function c be p, ϱ-regular.*

If c is p, ϱ-regular, then A_c is invertible, only left-sided invertible, or only right-sided invertible on the space $\mathbf{L}_p(\Gamma, \varrho)$, if the integer $\varkappa = \operatorname{ind}_{p,\varrho} c$ is zero, positive, or negative, respectively. In that case $\operatorname{Ind} A_c = -\varkappa$.

In order to prove this theorem we need an auxiliary result. Let $\gamma_1, \ldots, \gamma_r$ ($r \leqq m$) be complex numbers subject to

$$\frac{1}{p} + \beta_k - 1 < \operatorname{Re} \gamma_k < \frac{1}{p} + \beta_k \qquad (k = 1, \ldots, r).$$

Furthermore, let Γ_k ($\subset \Gamma$) be the closed curve containing the point t_k and denote by ψ_k the function defined on Γ as

$$\psi_k(t) = \begin{cases} (t - z_k)^{\varepsilon\gamma_k}, & \text{if} \quad t \in \Gamma_k, \\ 1, & \text{if} \quad t \in \Gamma \setminus \Gamma_k, \end{cases}$$

where $\varepsilon = 1$, $z_k \in D_\Gamma^+$ if Γ_k is oriented counter-clockwise and $\varepsilon = -1$, $z_k \in D_\Gamma^-$ if Γ_k is oriented clockwise. We choose the point z_k so that the straight line segment joining t_k and z_k has only the point t_k in common with Γ. Note that the function ψ_k is continuous at all points $t \neq t_k$.

With the help of the Lemmas 4.1 and 4.2 it is easy to check that the function $\psi = \psi_1 \dots \psi_r$ is p, ϱ-regular with p, ϱ-index zero.

Lemma 4.3. *The operator $A_\psi = \psi P_\Gamma + Q_\Gamma$ is invertible on $\mathbf{L}_p(\Gamma, \varrho)$.*

Proof. Let D_k ($k = 1, \dots, r$) be the bounded region in the complex plane whose boundary is Γ_k. Put $\Gamma_k' = \Gamma \cap D_k$, $\Gamma_k'' = \Gamma_k \cup \Gamma_k'$ if Γ_k is oriented clockwise and $\Gamma_k'' = \Gamma \cap D_k$, $\Gamma_k' = \Gamma_k \cup \Gamma_k''$ if Γ_k is oriented counter-clockwise. Then define

$$\psi_k^+(t) = \begin{cases} (t - t_k)^{\gamma_k}, & \text{if} \quad t \in \Gamma_k', \\ \left(\dfrac{t - t_k}{t - z_k}\right)^{\gamma_k}, & \text{if} \quad t \in \Gamma \setminus \Gamma_k', \end{cases}$$

$$\psi_k^-(t) = \begin{cases} (t - t_k)^{-\gamma_k}, & \text{if} \quad t \in \Gamma_k'', \\ \left(\dfrac{t - t_k}{t - z_k}\right)^{-\gamma_k}, & \text{if} \quad t \in \Gamma \setminus \Gamma_k''. \end{cases}$$

Here $\left(\dfrac{t - t_k}{t - z_k}\right)^{\gamma}$ denotes that branch of the function which is analytic in the complex plane cut along the straight line segment $z_k t_k$ and takes the value 1 at the point at infinity. By $(t - t_k)^\gamma$ we denote that branch of the function which is analytic in the complex plane cut along any line joining the point t_k with the point at infinity and having no points in common with D_Γ^+ and which makes the equality $\psi_k = \psi_k^+ \psi_k^-$ hold. Then $\psi = \psi_- \psi_+$, where $\psi_\pm = \psi_1^\pm \dots \psi_r^\pm$.

Obviously, $\psi_- \in \mathbf{L}_p^-(\Gamma, \varrho)$, $\psi_+ \in \mathbf{L}_q^+(\Gamma, \varrho^{-1})$, $\psi_-^{-1} \in \mathbf{L}_q^-(\Gamma, \varrho^{-1})$, $\psi_+^{-1} \in \mathbf{L}_p^+(\Gamma, \varrho)$ ($1/p + 1/q = 1$). Theorem 3.1, Chapter II, implies that the operator $\psi_+^{-1} P_\Gamma \psi_-^{-1}$ is bounded on the space $\mathbf{L}_p(\Gamma, \varrho)$. Therefore the function ψ admits a generalized factorization in the space $\mathbf{L}_p(\Gamma, \varrho)$ and $\operatorname{ind} \psi \mid \mathbf{L}_p(\Gamma, \varrho) = 0$. The assertion of the lemma now follows from the first part of the proof of Theorem 3.1. ∎

Proof of Theorem 4.1. *Sufficiency.* Let $c \in \mathbf{PC}(\Gamma)$ be p, ϱ-regular. Without loss of generality assume that $t_1, \dots, t_r$ ($r \leq m$) are all points of discontinuity of the function c. Then, by Lemma 4.2, $c(t_k)/c(t_k + 0) = e^{2\pi i \gamma_k}$, where

$$\delta(t_k)/2\pi - 1 < \operatorname{Re} \gamma_k < \delta(t_k)/2\pi \qquad (k = 1, \dots, r).$$

With these points t_k and these numbers γ_k construct the function ψ as above. Since

$$\psi_k(t_k)/\psi_k(t_k + 0) = e^{2\pi i \gamma_k} = c(t_k)/c(t_k + 0),$$

the function c/ψ is continuous. Therefore the function c can be written as $c = \psi r(1 + d)$, where r is a rational function without poles or zeros on Γ, $\varkappa = \operatorname{ind}_{p,\varrho} c = \operatorname{ind} r$, and $\max |d(t)|$ is so small that together with A_ψ the operator $A_{\psi + d\psi} = \psi P_\Gamma + Q_\Gamma + d\psi P_\Gamma$

is also invertible on $\mathbf{L}_p(\Gamma, \varrho)$. The function r has then the factorization

$$r(t) = r_-(t)\, t^{\varkappa}\, r_+(t)$$

(see (2.9), Chapter III).

Now we proceed as in the proof of Theorem 3.1, Chapter III. First let $\varkappa \geqq 0$. Then it is easily seen that $cP_\Gamma + Q_\Gamma = r_- A_{\varphi + d\varphi}\, r_-^{-1} A_r$. The operator $r_- A_{\varphi+d\varphi}\, r_-^{-1}$ is invertible on $\mathbf{L}_p(\Gamma, \varrho)$. The operator A_r is left-sided invertible and dim Coker $A_r = \varkappa$ (see Theorem 3.1, Chapter III). Consequently, the operator $A_c = cP_\Gamma + Q_\Gamma$ is invertible if $\varkappa = 0$ and only left-sided invertible if $\varkappa > 0$. In the latter case dim Coker $A_c = \varkappa$.

Finally, let $\varkappa < 0$. Then the operator $B = ct^{-\varkappa} P_\Gamma + Q_\Gamma$ is invertible, since $\operatorname{ind}_{p,\varrho}(ct^{-\varkappa}) = 0$. The operator $t^{-\varkappa} P_\Gamma + Q_\Gamma$ is only left-sided invertible and dim Coker $(t^{-\varkappa} P_\Gamma + Q_\Gamma) = -\varkappa$. Thus, the representation $B = A_c(t^{-\varkappa} P_\Gamma + Q_\Gamma)$ implies that the operator A_c is only right-sided invertible and that dim Ker $A_c = -\varkappa$.

Necessity. We assume the contrary, i.e., we assume that A_c is a Φ_+- or Φ_--operator on $\mathbf{L}_p(\Gamma, \varrho)$, but we let $0 \in V_{p,\varrho}(c)$.

We first consider the case where the curve $V_{p,\varrho}(c)$ is a Jordan curve in some neighborhood U of the origin $z = 0$. This neighborhood U can be supposed to be sufficiently small, so that for each $\lambda \in U$ the operator $A_{c-\lambda}$ is a $\Phi_\pm$-operator and that Ind $A_{c-\lambda} =$ Ind A_c (see § 3, Chapter I). But if the points $\lambda_1, \lambda_2 \in U$ are located on different sides of the curve $V_{p,\varrho}(c)$ then $\operatorname{ind}_{p,\varrho}(c - \lambda_1) \neq \operatorname{ind}_{p,\varrho}(c - \lambda_2)$, and this is a contradiction to the index formula proved above.

In the general case one can find a function $d \in \mathbf{PC}(\Gamma)$ having the following three properties: 1. $V_{p,\varrho}(d)$ is a Jordan curve in a sufficiently small neighborhood of the origin $z = 0$; 2. $\sup_{t\in\Gamma} |c(t) - d(t)| < \varepsilon$, where $\varepsilon > 0$ is small enough, so that A_d is a $\Phi_\pm$-operator; 3. $0 \in V_{p,\varrho}(d)$. However, by what has just been proved, the last condition contradicts the first two conditions. ∎

Putting the Theorems 3.1 and 4.1 together we arrive at the following.

Corollary 4.1. *A function $a \in \mathbf{PC}(\Gamma)$ admits a generalized factorization in the space $\mathbf{L}_p(\Gamma, \varrho)$ if and only if it is p, ϱ-regular. If a is p, ϱ-regular, then* $\operatorname{ind} a \mid \mathbf{L}_p(\Gamma, \varrho) = \operatorname{ind}_{p,\varrho} a$.

4.4. For the more general singular integral operator $A = aP_\Gamma + bQ_\Gamma$ $(A = P_\Gamma a + Q_\Gamma b)$ we have the following theorem.

Theorem 4.2. *Let $a, b \in \mathbf{PC}(\Gamma)$. Then for the operator $A = aP_\Gamma + bQ_\Gamma$ $(A = P_\Gamma a + Q_\Gamma b)$ to be a Φ_+- or Φ_--operator on the space $\mathbf{L}_p(\Gamma, \varrho)$ it is necessary and sufficient that*

$$a(t+0)\, b(t)\, f(t, \mu) + a(t)\, b(t+0)\, (1 - f(t, \mu)) \neq 0 \qquad (t \in \Gamma, 0 \leqq \mu \leqq 1). \quad (4)$$

If condition (4) is satisfied, then A is invertible, only left-sided invertible, or only right-sided invertible on the space $\mathbf{L}_p(\Gamma, \varrho)$ if the integer $\varkappa = \operatorname{ind}_{p,\varrho}(a/b)$ is zero, positive or negative, respectively. Furthermore, Ind $A = -\varkappa$.

Proof. If A is a $\Phi_\pm$operator on $\mathbf{L}_p(\Gamma, \varrho)$, then by virtue of Theorem 2.1

$$a(t) \neq 0, \qquad a(t+0) \neq 0, \qquad b(t) \neq 0, \qquad b(t+0) \neq 0 \qquad (t \in \Gamma) \quad (5)$$

and due to Theorem 4.1,

$$\frac{a(t+0)}{b(t+0)} f(t, \mu) + \frac{a(t)}{b(t)} (1 - f(t, \mu)) \neq 0 \qquad (t \in \Gamma, 0 \leqq \mu \leqq 1). \quad (6)$$

The conditions (5), (6) are obviously equivalent to condition (4).

The last assertion of Theorem 4.2 is an immediate consequence of Theorem 4.1. ∎

Remark 1. Singular integral equations with piecewise Hölder-continuous coefficients on special spaces of such functions were studied in detail in the monographs by MUSHKELISHVILI [1] and GAKHOV [1]. (With the help of functional-analytical methods such equations were systematically investigated by DUDUCHAVA (see his paper [1] and the references cited there).) Singular integral equations with piecewise continuous coefficients on the spaces $\mathbf{L}_p(\Gamma, \varrho)$ were considered by KHVEDELIDZE [1] for the first time, who estabilished sufficient conditions for such operators to be Fredholm. Under some additional hypotheses, Theorem 4.2 was stated by SHAMIR [1] for the case $\Gamma = \mathbf{R}, \varrho \equiv 1$ (see Theorem 6.1) and by WIDOM [2] for the case where Γ is the complex unit circle, $\varrho \equiv 1$, and $p = 2$. The generalizations of these results contained in the Sections 4, 5, 6 of the present chapter are due to GOHBERG and KRUPNIK [1, 2] (see also GOHBERG and KRUPNIK [4], Chapter IX).

Remark 2. Note also the following result obtained by HEUNEMANN [1]. If Γ is a closed Lyapunov curve and $a, b \in \mathbf{PC}(\Gamma)$, then for the operator $aP_\Gamma + bQ_\Gamma$ $(P_\Gamma a + Q_\Gamma b)$ to be normally solvable on the space $\mathbf{L}_p(\Gamma)$ it is necessary and sufficient that in addition to the conditions (a) and (b) of the remark made in Section 2.1 the following condition (c) be fulfilled:

(c) $$a(t+0)\, b(t)\, f_{2\pi/p}(\mu) + a(t)\, b(t+0)\, (1 - f_{2\pi/p}(\mu)) \neq 0$$

for all t belonging to the interior of the set $\operatorname{supp} a \cap \operatorname{supp} b$ and for all μ, $0 \leq \mu \leq 1$.

§ 5. Singular equations with piecewise continuous coefficients on non-closed curves

Let Γ be any composite curve system, that is, Γ consists of finitely many piecewise Lyapunov curves without common points. Assume Γ contains exactly l open curves, denote the end points of the j-th curve by t_j and t_{j+l} $(j = 1, \ldots, l)$, and suppose the positive direction is from t_j to t_{j+l}. The functions in $\mathbf{PC}(\Gamma)$ are always supposed to be continuous at the points t_j $(j = 1, \ldots, 2l)$. Furthermore, let $t_{2l+1}, \ldots, t_m$ be certain fixed points on Γ which are distinct from the points $t_1, \ldots, t_{2l}$, let $p, \beta_1, \ldots, \beta_m$ be real numbers satisfying $1 < p < \infty$, $-1/p < \beta_k < 1 - 1/p$ $(k = 1, \ldots, m)$, and let

$$\varrho(t) = \prod_{k=1}^{m} |t - t_k|^{\beta_k}.$$

Construct a closed piecewise Lyapunov curve system $\tilde{\Gamma}$ containing Γ as a part (see also § 7, Chapter III). Assume $0 \notin \Gamma$, and let $\tilde{\Gamma}$ be chosen so that the point $z = 0$ lies in the region $D_{\tilde{\Gamma}}^+$ whose boundary is $\tilde{\Gamma}$.

We consider the singular integral operator $A = aP_\Gamma + bQ_\Gamma$ $(A = P_\Gamma a + Q_\Gamma b)$ with coefficients $a, b \in \mathbf{PC}(\Gamma)$ on the space $\mathbf{L}_p(\Gamma, \varrho)$. In order to carry over Theorem 4.2 to the situation at hand, continue the functions a and b from Γ to $\tilde{\Gamma}$ as follows:

$$\tilde{a}(t) = \begin{cases} a(t), & \text{if } t \in \Gamma, \\ 1, & \text{if } t \in \tilde{\Gamma} \setminus \Gamma, \end{cases} \qquad \tilde{b}(t) = \begin{cases} b(t), & \text{if } t \in \Gamma, \\ 1, & \text{if } t \in \tilde{\Gamma} \setminus \Gamma. \end{cases}$$

Then $\tilde{a}$ and $\tilde{b}$ are obviously in $\mathbf{PC}(\tilde{\Gamma})$. A function $a \in \mathbf{PC}(\Gamma)$ is said to be *p, ϱ-regular* if the function $\tilde{a}$ is so. The integer $\operatorname{ind}_{p,\varrho} a := \operatorname{ind}_{p,\varrho} \tilde{a}$ is then referred to as the *p, ϱ-index* of the p, ϱ-regular function a.

Theorem 5.1. *Suppose Γ is a composite curve system and let $a, b \in \mathbf{PC}(\Gamma)$. Then for the operator $A = aP_\Gamma + bQ_\Gamma$ $(A = P_\Gamma a + Q_\Gamma b)$ to be a Φ_+- or Φ_--operator on the space*

$\mathbf{L}_p(\Gamma, \varrho)$ *it is necessary and sufficient that the following conditions are satisfied:*

1) $a(t+0)\,b(t)\,f(t, \mu) + a(t)\,b(t+0)\,(1 - f(t, \mu)) \neq 0 \qquad (t \in \Gamma \setminus \{t_1, \ldots, t_{2l}\})$;

2) $a(t_k)\,f(t_k, \mu) + b(t_k)\,(1 - f(t_k, \mu)) \neq 0 \qquad (k = 1, \ldots, l)$;

3) $a(t_k)\,(1 - f(t_k, \mu)) + b(t_k)\,f(t_k, \mu) \neq 0 \qquad (k = l+1, \ldots, 2l); \qquad 0 \leq \mu \leq 1.$

If these conditions are fulfilled, then A is invertible, only left-sided invertible, or only right-sided invertible on the space $\mathbf{L}_p(\Gamma, \varrho)$, if the integer $\varkappa = \operatorname{ind}_{p,\varrho}(a/b)$ is zero, positive, or negative, respectively. In that case $\operatorname{Ind} A = -\varkappa$.

This theorem follows from Theorem 4.2 and Theorem 1.1 immediately, since the conditions 1)–3) of Theorem 5.1 are equivalent to the validity of (4.4) for the functions $\tilde{a}$ and $\tilde{b}$ on $\tilde{\Gamma}$.

Corollary 5.1. *Let $a, b \in \mathbf{PC}(\Gamma)$ and $a(t \pm 0)\,b(t \pm 0) \neq 0$ for all $t \in \Gamma$. Then the operator $A = aP_\Gamma + bQ_\Gamma$ $(A = P_\Gamma a + Q_\Gamma b)$ has a finite index on $\mathbf{L}_p(\Gamma, \varrho)$.*

Proof. A little thought shows that one can find numbers p_1, p_2 satisfying $1 < p_1 < p < p_2 < \infty$ such that the function a/b is both p_1, ϱ-regular and p_2, ϱ-regular. In view of Theorem 5.1 the operator A is then a Φ-operator on both $\mathbf{L}_{p_1}(\Gamma, \varrho)$ and $\mathbf{L}_{p_2}(\Gamma, \varrho)$. Thus, $\dim \operatorname{Ker} A < \infty$, since $\mathbf{L}_p(\Gamma, \varrho) \subset \mathbf{L}_{p_1}(\Gamma, \varrho)$, and $\dim \operatorname{Ker} A^* < \infty$, because $\mathbf{L}_{p_2}(\Gamma, \varrho) \subset \mathbf{L}_p(\Gamma, \varrho)$. ∎

Remark. If the conditions 1)–3) of Theorem 5.1 are satisfied, then the function a/b admits a generalized factorization in the space $\mathbf{L}_p(\tilde{\Gamma}, \varrho)$ (see Corollary 4.1). Hence, if this factorization is given explicitly, the results of the Sections 1.1 and 3.2 enable us to construct one-sided inverses and to describe the kernel and cokernel of the operator $A = aP_\Gamma + bQ_\Gamma$ $(A = P_\Gamma a + Q_\Gamma b)$.

§ 6. Singular equations with piecewise continuous coefficients on the real line

In this section we consider singular integral operators with piecewise continuous coefficients on the space $\mathbf{L}_p(\mathbb{R}, \varrho_\infty)$, where the weight is given by

$$\varrho_\infty(t) = (1 + t^2)^{\beta/2} \prod_{k=1}^{m} |t - t_k|^{\beta_k},$$

the numbers $p, \beta, \beta_1, \ldots, \beta_m$ are subject to the conditions

$$1 < p < \infty, \quad -\frac{1}{p} < \beta_k < 1 - \frac{1}{p} \quad (k = 1, \ldots, m),$$

$$-\frac{1}{p} < \beta + \sum_{k=1}^{m} \beta_k < 1 - \frac{1}{p},$$

and $t_1, \ldots, t_m$ are pairwise distinct points on the real line.

We denote by $\mathbf{PC}(\overline{\mathbb{R}})$ the collection of all piecewise continuous and left-sided continuous functions on the real line having finite limits $a(\infty)$ and $a(-\infty)$. To each function $a \in \mathbf{PC}(\overline{\mathbb{R}})$ we assign a funtion $a_{p,\varrho_\infty} : \overline{\mathbb{R}} \times [0, 1] \to \mathbb{C}$ defined by

$$a_{p,\varrho_\infty}(t, \mu) = a(t+0)\,f(t, \mu) + a(t)\,(1 - f(t, \mu)) \quad (-\infty < t < \infty, 0 \leq \mu \leq 1)$$

and

$$a_{p,\varrho_\infty}(\infty, \mu) = a_{p,\varrho_\infty}(-\infty, \mu) = a(-\infty)\,f_\delta(\mu) + a(\infty)\,(1 - f_\delta(\mu)) \quad (0 \leq \mu \leq 1),$$

where $\delta = 2\pi\left(1 - \frac{1}{p} - \beta - \sum_{k=1}^{m} \beta_k\right)$ and where the functions $f_\delta(\mu)$ and $f(t, \mu)$ have the same meaning as in § 4.

The range of the function a_{p,ϱ_∞} is then a continuous closed curve in the complex plane. If this curve does not pass through the point $z = 0$, the function a is said to be p, ϱ_∞-*regular*; the winding number of this curve with respect to the origin $z = 0$ is called the p, ϱ_∞-*index* of the function a and is denoted by $\operatorname{ind}_{p,\varrho_\infty} a := \operatorname{ind} a_{p,\varrho_\infty}$.

Now put $P_\infty = \frac{1}{2}(I + S_\infty)$ and $Q_\infty = \frac{1}{2}(I - S_\infty)$, where S_∞ is the singular integral operator defined by

$$(S_\infty \varphi)(t) = \frac{1}{\pi i} \int_{-\infty}^{\infty} \frac{\varphi(\tau)}{\tau - t} d\tau \qquad (-\infty < t < \infty)$$

(see also Section 3.3, Chapter II).

Theorem 6.1. *Let $a, b \in \mathbf{PC}(\overline{\mathbb{R}})$. Then for the operator $A = aP_\infty + bQ_\infty$ $(A = P_\infty a + Q_\infty b)$ to be a Φ_+- or Φ_--operator on the space $\mathbf{L}_p(\mathbb{R}, \varrho_\infty)$ it is necessary and sufficient that the following conditions be satisfied:*

$$a(t+0)\, b(t)\, f(t, \mu) + a(t)\, b(t+0)\, (1 - f(t, \mu)) \neq 0$$
$$(-\infty < t < \infty,\ 0 \leqq \mu \leqq 1),$$
$$a(-\infty)\, b(\infty)\, f_\delta(\mu) + a(\infty)\, b(-\infty)\, (1 - f_\delta(\mu)) \neq 0 \qquad (0 \leqq \mu \leqq 1). \qquad (1)$$

If conditions (1) *are satisfied, then A is invertible, only left-sided invertible, or only right-sided invertible on the space $\mathbf{L}_p(\mathbb{R}, \varrho_\infty)$, if the integer $\varkappa = \operatorname{ind}_{p,\varrho_\infty}(a/b)$ is zero, positive, or negative, respectively. In that case* $\operatorname{Ind} A = -\varkappa$.

Proof. Let Γ_0 denote the complex unit circle, put

$$\zeta_k = (t_k + i)(t_k - i)^{-1} \ (k = 1, \ldots, m), \ \zeta_0 = 1,$$

$$\beta_0 = 1 - (2/p) - \beta - \sum_{k=1}^{m} \beta_k, \text{ and } \varrho_0(\zeta) = \prod_{k=0}^{m} |\zeta - \zeta_k|^{\beta_k}.$$

In Section 3.3, Chapter II, we proved that $BS_\infty B^{-1} = S_{\Gamma_0}$, where $B\colon \mathbf{L}_p(\mathbb{R}, \varrho_\infty) \to \mathbf{L}_p(\Gamma_0, \varrho_0)$ is the invertible operator acting by the rule

$$(B\varphi)(\zeta) = \frac{1}{\zeta - 1} \varphi\left(i \frac{\zeta + 1}{\zeta - 1}\right).$$

It can be easily verified that for every function $c \in \mathbf{PC}(\overline{\mathbb{R}})$ the identity $BcB^{-1} = c\left(i\frac{\zeta + 1}{\zeta - 1}\right)$ holds. Put $c^0(\zeta) := c\left(i\frac{\zeta + 1}{\zeta - 1}\right)$. Then $c(t_k \pm 0) = c^0(\zeta_k \pm 0)$, $c(\pm\infty) = c^0(1 \mp 0)$ and, hence, the range of the function c_{p,ϱ_∞} coincides with the range of the function c^0_{p,ϱ_0}. To complete the proof it remains to apply Theorem 4.2 to the operator $BAB^{-1} = a^0 P_{\Gamma_0} + b^0 Q_{\Gamma_0}$. ∎

§ 7. Norm estimates for the singular integral operator

The results of the Sections 4 and 5 can be successfully applied to obtain lower bounds for the essential norm of the operators S_Γ, P_Γ, and Q_Γ on the space $\mathbf{L}_p(\Gamma)$.

Theorem 7.1. *Let Γ be a composite curve system and abbreviate $\|\cdot\|_{\mathbf{L}_p(\Gamma)}$ to $\|\cdot\|_p$ $(1 < p < \infty)$. Then, for every $p \geqq 2$,*

$$\inf_T \|P_\Gamma + T\|_p \geqq 1/\sin\frac{\pi}{p}, \quad \inf_T \|Q_\Gamma + T\|_p \geqq 1/\sin\frac{\pi}{p}, \tag{1}$$

$$\inf_T \|S_\Gamma + T\|_p \geqq \cot\frac{\pi}{2p}, \tag{2}$$

where the infimum is taken over all compact operators on $\mathbf{L}_p(\Gamma)$.

The estimates (1) *and* (2) *remain valid for $1 < p < 2$ if on their right-hand sides p is replaced by q $(1/p + 1/q = 1)$.*

Proof. It suffices to consider only the case $p \geqq 2$; the assertion for $1 < p < 2$ follows by passing to the adjoint operators.

Assume there exists a p such that

$$\inf \|P_\Gamma + T\|_p < 1/\sin\frac{\pi}{p}.$$

Consider the operator $aP_\Gamma + Q_\Gamma$, where $a \in \mathbf{PC}(\Gamma)$ is a function which assumes only the two values

$$\left(\cos\frac{\pi}{p}\right) \exp\left(\pm \mathrm{i}\frac{\pi}{p}\right).$$

Then $|a(t) - 1| = \sin(\pi/p)$, hence $\inf \|(a - 1) P_\Gamma + T\|_p < 1$, and therefore the operator $I + (a - 1) P_\Gamma = aP_\Gamma + Q_\Gamma$ is a Φ-operator on $\mathbf{L}_p(\Gamma)$ (cf. § 3, Chapter I). But this contradicts Theorem 4.1, since the function a is not p-regular.

For the proof of the second estimate in (1) take a function $a \in \mathbf{PC}(\Gamma)$ assuming only the two values

$$\left(\sec\frac{\pi}{p}\right) \exp\left(\pm \mathrm{i}\frac{\pi}{p}\right).$$

Now $|(1 - a)/a| = \sin(\pi/p)$ and we get a contradiction as above, since the function a is not p-regular and, consequently, the operator $aP_\Gamma + Q_\Gamma = a(I + (1 - a)/aQ_\Gamma)$ cannot be Fredholm on $\mathbf{L}_p(\Gamma)$.

Finally, using a function $a \in \mathbf{PC}(\Gamma)$ assuming only the two values $\exp(\pm \mathrm{i}(\pi/p))$ and taking into consideration the identity

$$aP_\Gamma + Q_\Gamma = \frac{a + 1}{2}\left(I + \frac{a - 1}{a + 1} S_\Gamma\right)$$

we obtain the estimate (2) analogously. ∎

Thus, combining the estimates (1), (2), and Lemma 2.1 of Chapter II, we arrive at the following.

Corollary 7.1. *Let Γ_0 be the complex unit circle and put $S_0 = S_{\Gamma_0}$. Then, for $n = 1, 2, \ldots$,*

$$\|S_0\|_p = \begin{cases} \cot\dfrac{\pi}{2p} & \text{for} \quad p = 2^n, \\[2ex] \tan\dfrac{\pi}{2p} & \text{for} \quad p = 2^n/(2^n - 1). \end{cases}$$

Remark. 1. The results of this section are due to GOHBERG and KRUPNIK [1, 5]. Actually much more can be proved (see GOHBERG and KRUPNIK [4], German transl., pp. 328–330):

Theorem 7.2. *Let Γ be an arbitrary composite curve system and let $1 < p < \infty$. Then*

$$\inf_T \|S_\Gamma + T\|_p = \|S_0\|_p = \begin{cases} \cot \dfrac{\pi}{2p} & \text{for } 2 \leqq p < \infty, \\[2ex] \tan \dfrac{\pi}{2p} & \text{for } 1 < p \leqq 2. \end{cases}$$

Moreover, if $\Gamma \neq \Gamma_0$, then $\|S_\Gamma\|_2 > \|S_0\|_2 = 1$. Thus, the norm $\|S_\Gamma\|_2$ depends on the form of the curve Γ (see GOHBERG and KRUPNIK [4], p. 331).

VERBITSKI and KRUPNIK [1, 2] generalized Theorem 7.2 to the spaces $\mathbf{L}_p(\Gamma, \varrho)$ (see also KRUPNIK [9], §§ 6–7):

Theorem 7.3. (i) *Let Γ be a closed Lyapunov curve and $\varrho(t) = \prod_{k=1}^{m} |t - t_k|^{\beta_k}$. Then*

$$\inf_T \|S_\Gamma + T\|_{\mathbf{L}_p(\Gamma,\varrho)} = \max_k \nu(p, \beta_k), \tag{3}$$

where

$$\nu(p, \beta) = \begin{cases} \cot \dfrac{\pi}{2}\left(\dfrac{1}{q} - \beta\right) & \text{for } 1 - \dfrac{2}{p} < \beta < 1 - \dfrac{1}{p}, \\[2ex] \cot \dfrac{\pi}{2p} & \text{for } 0 \leqq \beta \leqq 1 - \dfrac{2}{p}, \\[2ex] \cot \dfrac{\pi}{2}\left(\dfrac{1}{p} + \beta\right) & \text{for } -\dfrac{1}{p} < \beta < 0 \end{cases}$$

if $2 \leqq p < \infty$, and $\nu(p, \beta) = \nu(q, -\beta)$ if $1 < p < 2$ $(1/p + 1/q = 1)$. Furthermore, $\|S_0\|_{\mathbf{L}_p(\Gamma,\varrho)} = \nu(p, \beta_1)$ if $m = 1$, whereas $\|S_0\|_{\mathbf{L}_p(\Gamma_0,\varrho)}$ depends on the location of the points t_k if $m > 1$.

(ii) *Let Γ be a curve system consisting of a finite number of open Lyapunov curves Γ_k $(k = 1, \ldots, m)$ which have no points in common. Denote the initial and final points of Γ_k by t_k and t_{k+m}, respectively, and let*

$$\varrho(t) = \prod_{k=1}^{m} |t - t_k|^{\beta_k}\, |t - t_k|^{\beta_{k+m}}.$$

Then (3) *holds with*

$$\nu(p, \beta) = \begin{cases} \cot \pi\left(\dfrac{1}{q} - \beta\right) & \text{for } 1 - \dfrac{3}{2p} < \beta < 1 - \dfrac{1}{p}, \\[2ex] \cot \dfrac{\pi}{2p} & \text{for } -\dfrac{1}{2p} \leqq \beta \leqq 1 - \dfrac{3}{2p}, \\[2ex] \cot \pi\left(\dfrac{1}{p} + \beta\right) & \text{for } -\dfrac{1}{p} < \beta < -\dfrac{1}{2p} \end{cases}$$

if $2 \leqq p < \infty$ and $\nu(\beta, p) = \nu(q, -\beta)$ if $1 < p < 2$ $(1/p + 1/q = 1)$.

Note also the following result of A. B. ALEKSANDROV [1]: the norm of the operator S_0 considered on the subspace of $H^\mu(\Gamma_0)$ consisting of all functions f for which $\int_{\Gamma_0} \frac{f(z)}{z}\, dz = 0$ is equal to

$$\|S_0\|_{H^\mu(\Gamma_0)} = \frac{1}{\pi} B\left(\frac{\mu}{2}, \frac{1 - \mu}{2}\right) \qquad (0 < \mu < 1),$$

where B denotes the Euler beta function.

Remark 2. Soldatov [1, 2] and Krupnik [10] have recently studied singular integral operators of the form

$$(A\varphi)(t) := a(t)\,\varphi(t) + \int_{\Gamma} \frac{b(t, \tau)\,\varphi(\tau)}{\tau - t}\,d\tau,$$

where a and b belong to certain classes of piecewise continuous functions, on spaces $\mathbf{L}_p$ with weight.

Chapter V
Systems of one-dimensional singular equations

§ 1. Two theorems on operator matrices

1.1. We first fix the notation that will be used throughout the following.

Given any linear space X we denote by X^n (n a positive integer) the linear space of all n-dimensional vectors with components from X and by $X^{n \times n}$ the linear space of all square matrices of order n with entries from X.

If X is a Fréchet space (a normed or unitary space), then X^n can be provided with a countable system of seminorms (a norm resp. scalar product), e.g., by defining the p-th seminorm (norm resp. scalar product) of a vector $x = (x_1, \ldots, x_n) \in X^n$ as the sum of the p-th seminorms (the norms resp. scalar products) of its components:

$$\|x\|_p = \sum_{j=1}^{n} \|x_j\|_p \qquad (p = 1, 2, \ldots).$$

If X is a normed space, then the norm of a matrix $A = (a_{jk})_{j,k=1}^n \in X^{n \times n}$ can, for instance, be defined by

$$\|A\| = n \max_{j,k} \|a_{jk}\| .$$

If $\mathfrak{X}$ is a Banach algebra, then $\mathfrak{X}^{n \times n}$ equipped with this norm also becomes a Banach algebra.

If X is a Fréchet space and $\mathscr{L}(X)$, as usual, denotes the collection of all linear continuous operators on X, then $[\mathscr{L}(X)]^{n \times n}$ can be identified with $\mathscr{L}(X^n)$. In other words, every operator $A \in \mathscr{L}(X^n)$ can be written as a matrix $A = (A_{jk})_{j,k=1}^n$ with $A_{jk} \in \mathscr{L}(X)$. Note that an operator A is compact on X^n if and only if each of the operators A_{jk} is compact on X.

To the adjoint operator A^* the adjoint matrix $A^* = (A_{kj}^*)_{k,j=1}^n$ corresponds.

1.2. The collection of all compact linear operators on a Banach space X forms a two-sided ideal in the Banach algebra $\mathscr{L}(X)$. Henceforth we denote this ideal by $\mathfrak{T} = \mathfrak{T}(X)$. Let $\mathfrak{A} \subset \mathscr{L}(X)$ be a subalgebra of $\mathscr{L}(X)$ having the following two properties:

(i) $\mathfrak{T} \subset \mathfrak{A}$.

(ii) $AB - BA \in \mathfrak{T}$ for all operators $A, B \in \mathfrak{A}$.

Theorem 1.1. *Suppose X is a Banach space, $\mathfrak{A} \subset \mathscr{L}(X)$ is a subalgebra satisfying the conditions* (i) *and* (ii), *and let $A = (A_{jk})_{j,k=1}^n \in \mathfrak{A}^{n \times n}$. Then A is a Φ-operator on X^n if and only if* det A *is a Φ-operator on X.*[1])

[1]) The determinant det A is defined as in the case of a matrix formed by complex numbers. The arrangement of the (in general, not commuting) factors is not of significance, since all possible results will only differ by a compact operator (see Theorem 3.4, Chapter I).

The proof of this theorem is based on the following lemma.

Lemma 1.1. *Let $\mathfrak{R}$ be a Banach algebra with identity element and let $a = (a_{jk})_{j,k=1}^n \in \mathfrak{R}^{n\times n}$ be a matrix whose entries $a_{jk} \in \mathfrak{R}$ commute pairwise. Then a is invertible in $\mathfrak{R}^{n\times n}$ if and only if the element $\det a$ is invertible in $\mathfrak{R}$.*

Proof. 1. Let $\det a$ be invertible in $\mathfrak{R}$. Then $(\det a)^{-1} a_{jk} = (\det a)^{-1} a_{jk} (\det a) (\det a)^{-1} = a_{jk} (\det a)^{-1}$. This implies that the inverse of a is $(\det a)^{-1} (b_{jk})_{j,k=1}^n$, where b_{jk} denotes the algebraic complement of a_{kj}, that is, $(-1)^{j+k}$ times the determinant of the matrix obtained from the matrix a by cancelling the row and the column which contains a_{kj}.

2. Now suppose a is invertible in $\mathfrak{R}^{n\times n}$ and let $a^{-1} = (c_{jk})_{j,k=1}^n$ ($c_{jk} \in \mathfrak{R}$). It suffices to prove that the elements c_{jk} commute pairwise and that they commute with all elements a_{jk}, since then the identity $e = \det (a^{-1} a) = \det (a^{-1}) \det a$ implies the assertion.

Let $\mathfrak{Z}$ denote the set of all commutative subalgebras of the algebra $\mathfrak{R}$ containing all entries a_{jk} of the matrix a. The set $\mathfrak{Z}$ is partially ordered by inclusion. Thus, by Zorn's lemma, $\mathfrak{Z}$ contains at least one maximal element $\mathfrak{U}$. The commutative subalgebra $\mathfrak{U} \subset \mathfrak{R}$ then possesses the following property: if $a \in \mathfrak{R}$ and if $ax = xa$ for all $x \in \mathfrak{U}$ then $a \in \mathfrak{U}$. Since $a_{jk} \in \mathfrak{U}$, for every $x \in \mathfrak{U}$ the equalities $a^{-1} x = a^{-1} xaa^{-1} = a^{-1} axa^{-1} = xa^{-1}$ hold and, hence, $xc_{jk} = c_{jk}x$ ($j, k = 1, \ldots, n$) for every $x \in \mathfrak{U}$. This implies that $c_{jk} \in \mathfrak{U}$, and therefore the elements c_{jk} commute pairwise and with all elements a_{jk}. ▮

Proof of Theorem 1.1. Let $\mathfrak{R}$ denote the quotient algebra $\mathscr{L}(X)/\mathfrak{T}$. Then $\mathfrak{R}^{n\times n}$ can be identified with the quotient algebra $\mathscr{L}(X^n)/\mathfrak{T}^{n\times n}$. By virtue of Theorem 3.1, Chapter I, $A \in \mathscr{L}(X^n)$ is a Φ-operator if and only if the coset of $\mathfrak{R}^{n\times n}$ containing A is invertible in $\mathfrak{R}^{n\times n}$. Theorem 1.1 is therefore an immediate consequence of Lemma 1.1. ▮

Remark 1. Theorem 1.1 remains true for Φ_+- and Φ_--operators and also for the case where X is a Fréchet space (see Prössdorf [17], Section 7.1).

1.3. Under the hypotheses of Theorem 1.1, the index of the operator A is, in general, not equal to the index of the operator $\det A$. To see this, let, for example, $\mathfrak{A}$ be the algebra of all multidimensional singular integral operators with continuous symbol on the space $X = \mathbf{L}_p(\mathbb{R}^m)$ (these operators will be studied in detail in the Chapters XI–XV); then $\operatorname{Ind} \det A = 0$ for every Φ-operator $A \in \mathfrak{A}^{n\times n}$, whereas, in general, $\operatorname{Ind} A \neq 0$.

The following theorem states a sufficient condition for the validity of the equality $\operatorname{Ind} A = \operatorname{Ind} \det A$.

Theorem 1.2. *Let X be a Banach space and let $\mathfrak{A} \subset \mathscr{L}(X)$ be a subalgebra which in addition to* (i) *and* (ii) *possesses the following property:*

(iii) *The collection of all Φ-operators is dense in $\mathfrak{A}$.*

Then, if $A = (A_{jk})_{j,k=1}^n \in \mathfrak{A}^{n\times n}$ is a Φ-operator, the equality

$$\operatorname{Ind} A = \operatorname{Ind} \det A$$

holds.

Proof. 1. Let $\mathfrak{R} = \mathscr{L}(X)/\mathfrak{T}$, $\hat{\mathfrak{A}} = \mathfrak{A}/\mathfrak{T}$, and denote by $\mathfrak{Z}$ the set of all commutative subalgebras of $\mathfrak{R}$ containing the commutative subalgebra $\hat{\mathfrak{A}}$. $\mathfrak{Z}$ is partially ordered by inclusion. By Zorn's lemma, there exists at least one maximal element $\mathfrak{U}$ in $\mathfrak{Z}$. Then, if the element $c \in \mathfrak{U}$ has an inverse $c^{-1} \in \mathfrak{R}$, the element c^{-1} actually belongs to $\mathfrak{U}$. Indeed, for every $d \in \mathfrak{U}$ we have $c^{-1} d = c^{-1} dcc^{-1} = c^{-1} c\, dc^{-1} = dc^{-1}$, and hence $c^{-1} \in \mathfrak{U}$ because of the maximality of $\mathfrak{U}$.

2. We now show that the coset $a = (a_{jk})_{j,k=1}^n$ ($a_{jk} \in \hat{\mathfrak{A}}$) containing the Φ-operator A is in $\mathfrak{U}$ homotopic to a diagonal matrix $b = (b_{jk})_{j,k=1}^n$ with $b_{jk} = 0$, $j \neq k$, and $b_{jj} \in \mathfrak{U}$.

If a_{11} is invertible in $\mathfrak{R}$, then, by what has just been proved, $a_{11}^{-1} \in \mathfrak{U}$, and the matrices

$$a(t) = \begin{pmatrix} a_{11} & a_{12} - ta_{11}^{-1}a_{12}a_{11} \dots a_{1n} - ta_{11}^{-1}a_{1n}a_{11} \\ \cdots & \cdots \\ a_{n1} & a_{n2} - ta_{11}^{-1}a_{12}a_{n1} \dots a_{nn} - ta_{11}^{-1}a_{1n}a_{n1} \end{pmatrix}, \quad 0 \leqq t \leqq 1, \tag{2}$$

realize a homotopy between a and the matrix

$$\begin{pmatrix} a_{11} & 0 & \dots & 0 \\ a_{21} & d_{22} & \dots & d_{2n} \\ \cdots & & & \\ a_{n1} & d_{n2} & \dots & d_{nn} \end{pmatrix}; \; d_{jk} = a_{jk} - a_{11}^{-1}a_{1k}a_{j1} \qquad (j, k = 2, \dots, n).$$

The subsequent homotopy

$$a(t) = \begin{pmatrix} a_{11} & 0 & \dots & 0 \\ (1-t)\, a_{21} & d_{22} & \dots & d_{2n} \\ \cdots & & & \\ (1-t)\, a_{n1} & d_{n2} & \dots & d_{nn} \end{pmatrix} \tag{2'}$$

then gives the desired result. Note that $\det a(t) = \det a$ $(0 \leqq t \leqq 1)$ for both homotopies (2) and (2′).

If a_{11} is not invertible in $\mathfrak{R}$, then by virtue of (iii) for every sufficiently small positive number ε there exists an invertible element $\tilde{a}_{11} \in \hat{\mathfrak{A}}$ such that $\|a_{11} - \tilde{a}_{11}\| < \varepsilon$ and $\|\det a - \det \tilde{a}\| < \varepsilon$, where $\tilde{a} = (\tilde{a}_{jk})_{j,k=1}^n$ with $\tilde{a}_{jk} := a_{jk}$ for $(j, k) \neq (1, 1)$. A homotopy between a and $\tilde{a}$ is then given by $a(t) = a + t(\tilde{a} - a)$ $(\in \mathfrak{U})$.

Note that this approximation by invertible elements is also possible for the elements $d_{jk} \in \mathfrak{U}$, since $a_{11}\, d_{jk} = a_{11}a_{jk} - a_{1k}a_{j1}$ $(\in \hat{\mathfrak{A}})$. By applying the described operations a finite number of times, we obtain a diagonal matrix $b \in \mathfrak{U}^{n \times n}$ which is homotopic to a and, moreover, satisfies the inequality

$$\|\det a - \det b\| < \varepsilon. \tag{3}$$

3. Let $B_{jj} \in \mathscr{L}(X)$ $(j = 1, \dots, n)$ be operators belonging to the cosets b_{jj} (note that B_{jj} is up to a compact summand determined uniquely). Then the diagonal matrix B whose diagonal entries are B_{jj} is a Φ-operator, since, by virtue of (3), the invertibility of $\det a$ implies that of $\det b$ (see Theorem 1.1). Because a and b are homotopic, an application of Theorem 3.11, Chapter I, gives

$$\operatorname{Ind} A = \operatorname{Ind} B \tag{4}$$

and, analogously, since $\det a$ is homotopic to $\det b$, we deduce that

$$\operatorname{Ind} \det A = \operatorname{Ind} \det B. \tag{5}$$

Combining (4), (5), and the obvious equality $\operatorname{Ind} B = \operatorname{Ind} \det B$ we arrive at the asserted formula (1). ∎

Remark 2. Theorem 1.1 and Lemma 1.1 are due to Krupnik [3]. Köhler and Silbermann [1, 2] generalized Theorem 1.1 to the case of Φ_+- and Φ_--operators on locally convex linear topological spaces. About at the same time (1973), Silbermann established Theorem 1.2 and communicated its proof to the authors. This proof is published here for the first time. In its second part it makes use of some arguments of Bojarski [1]. Recently Theorem 1.2 has also been published by Krupnik [4], whose proof, however, is incomplete, since it rests on a false auxiliary result (Lemma 6.1). (See also Krupnik [9], Chap. I).

Remark 3. Recently MARKUS and FELDMAN [1] stated the following theorem: *Suppose X is a Hilbert space (or a Banach space having the approximation property) and let* $A = (A_{jk})_{j,k=1}^n \in \mathscr{L}(X^n)$. *Then, if the operators* A_{jk} *commute pairwise up to a nuclear operator, equality* (1) *holds.* (See also MARKUS and FELDMAN [2] and KRUPNIK [9], Chap. I).

§ 2. Systems of singular integral equations with continuous coefficients on closed curves

Let Γ be a closed piecewise Lyapunov curve system and ϱ the weight

$$\varrho(t) = \prod_{j=1}^{m} |t - t_j|^{\beta_j},$$

where $t_j \in \Gamma$, $-1/p < \beta_j < 1 - 1/p$ $(j = 1, \ldots, m)$, and $1 < p < \infty$.

In the space $\mathbf{L}_p^n(\Gamma, \varrho) := [\mathbf{L}_p(\Gamma, \varrho)]^n$ we consider the system of singular integral equations of the form

$$(A\varphi)(t) := C(t)\,\varphi(t) + \frac{D(t)}{\pi i} \int_\Gamma \frac{\varphi(\tau)}{\tau - t}\, d\tau + (T\varphi)(t) = f(t) \qquad (t \in \Gamma). \tag{1}$$

Here C and D are continuous square-matrix functions (of order n) on Γ, T is an arbitrary compact linear operator on $\mathbf{L}_p^n(\Gamma, \varrho)$ and φ, f are vector functions from $\mathbf{L}_p^n(\Gamma, \varrho)$.

The matrix function

$$A(t, \theta) = C(t) + \theta\, D(t) \qquad (t \in \Gamma;\ \theta = \pm 1)$$

is referred to as the *symbol* of the operator A defined on the space $\mathbf{L}_p^n(\Gamma, \varrho)$ by (1). Furthermore, put

$$A(t) = C(t) + D(t), \qquad B(t) = C(t) - D(t) \tag{2}$$

and

$$S = (S_1\, \delta_{jk})_1^n, \qquad P = \tfrac{1}{2}\,(I + S), \qquad Q = \tfrac{1}{2}\,(I - S) \tag{3}$$

with S_1 the scalar singular operator defined on the space $\mathbf{L}_p(\Gamma, \varrho)$ as

$$(S_1 g)(t) = \frac{1}{\pi i} \int_\Gamma \frac{g(\tau)}{\tau - t}\, d\tau \qquad (t \in \Gamma). \tag{4}$$

With (2) and (3), operator (1) can be written in the form $A = AP + BQ + T$, where A and B denote the operators of multiplication by the matrix functions $A(t)$ and $B(t)$ on $\mathbf{L}_p^n(\Gamma, \varrho)$, respectively.

Now it is easy to carry over all results of Section 3.3, Chapter III, concerned with the scalar case to operators of the form (1). In particular, we have the following.

Theorem 2.1. *For the operator A defined by* (1) *to be a Φ-operator on the space* $\mathbf{L}_p^n(\Gamma, \varrho)$ *it is necessary and sufficient that*

$$\det A(t) \neq 0, \qquad \det B(t) \neq 0 \qquad (t \in \Gamma). \tag{5}$$

If the conditions (5) *are satisfied, then*

$$\operatorname{Ind} A = \operatorname{ind} \left[\frac{\det B}{\det A}\right], \tag{6}$$

and the operator $R = A^{-1} P + B^{-1} Q$ *is a two-sided regularizer of the operator* $\mathbf{A}$.

Taking into account that the scalar operator det A on the space $\mathbf{L}_p(\Gamma, \varrho)$ can be written in the form

$$\det A = (\det A)\, P_1 + (\det B)\, Q_1 + T_1,$$

where $P_1 = \frac{1}{2}(I + S_1)$, $Q_1 = \frac{1}{2}(I - S_1)$, and T_1 is a compact operator (see Theorem 4.2, Chapter II), Theorem 2.1 is seen to result from the Theorems 1.1 and 1.2 of the present chapter and the Theorems 1.3 and 3.2 of Chapter III immediately.

Corollary 2.1. *The operator of multiplication by a continuous matrix function defined by*

$$(A\varphi)(t) := A(t)\, \varphi(t) \qquad (t \in \Gamma)$$

is invertible on the space $\mathbf{L}_p^n(\Gamma, \varrho)$ *if and only if*

$$\det A(t) \neq 0 \qquad (t \in \Gamma).$$

Theorem 2.2. *For the operator* $\mathbf{A}$ *defined on the space* $\mathbf{L}_p^n(\Gamma, \varrho)$ *by* (1) *the estimate*

$$\max_{t \in \Gamma, \theta = \pm 1} \det |A(t, \theta)| \leqq \|\mathbf{A}\|.$$

holds.

This can be proved in the same way as Theorem 3.3, Chapter III.

Remark 1. Note that the conditions (5) are even necessary for $\mathbf{A}$ to be a Φ_+- or Φ_--operator on $\mathbf{L}_p^n(\Gamma, \varrho)$ (see Remark 1, § 1, and Theorem 3.2, Chapter III). Also notice that the matrix analogue of Remark 2, § 3, Chapter III, holds (see Leiterer [1, 2]).

Remark 2. The results of § 4, Chapter III, can be immediately extended to the corresponding systems of singular integral equations on the spaces $[\mathbf{H}^\mu(\Gamma)]^n$ $(0 < \mu < 1)$.

Remark 3. The index formula (6) for the case where the operator is considered on the space $[\mathbf{H}^\mu(\Gamma)]^n$ was found in 1943 by Muskhelishvili and N. Vekua [1] (see also Muskhelishvili [1] and N. P. Vekua [1]). The regularizer of the operator $\mathbf{A}$ as well as conditions for the solvability of (1) in the space $[\mathbf{H}^\mu(\Gamma)]^n$ (cp. §§ 3, 4 of the present Chapter) had somewhat earlier already been given by Giraud [4]. Giraud's results were carried over to the space $\mathbf{L}_2^n(\Gamma)$ by Mikhlin [7]. The results of this section were proved for the spaces $\mathbf{L}_p^n(\Gamma, \varrho)$ by Gohberg [4, 5], Khvedelidze [1, 2], Mandzhavidze and Khvedelidze [1], and Simonenko [6]. By means of topological (especially, homotopical) methods the index formula (6) was independently obtained by Bojarski [1] and Mikhlin [30].

What was said in Remark 2, p. 91, also applies to the case at hand.

§ 3. Factorization of matrix functions

During the following two sections, we retain all notations introduced in § 5, Chapter III.

3.1. General factorization theorems

Throughout what follows let Γ denote an arbitrary closed piecewise Lyapunov curve system.

Let $A \in [\mathbf{L}_\infty(\Gamma)]^{n \times n} =: \mathbf{L}_\infty^{n \times n}(\Gamma)$ be a non-singular matrix function on Γ, that is, $A \in \mathbf{G}[\mathbf{L}_\infty(\Gamma)]^{n \times n} =: \mathbf{GL}_\infty^{n \times n}(\Gamma)$. A representation of the form

$$A(z) = A_-(z)\, D(z)\, A_+(z) \qquad (z \in \Gamma), \tag{1}$$

where D is a diagonal matrix function of the form

$$D(z) = (z^{\varkappa_j}\, \delta_{jk})_1^n$$

with certain integers $\varkappa_1 \geqq \varkappa_2 \geqq \dots \geqq \varkappa_n$ and where $A_-^{\pm 1} \in [\mathbf{L}_p^-(\Gamma)]^{n\times n}$ and $A_+^{\pm 1} \in [\mathbf{L}_p^+(\Gamma)]^{n\times n}$ for some $p \geqq 2$, is called a *right factorization* of the matrix function A.

A factorization arising from (1) by changing the roles of A_+ and A_- is called a *left factorization* of the matrix function A. In an obvious way every right (left) factorization of the matrix function A generates a left (right) factorization of the transposed matrix function A' as well as of the inverse matrix function A^{-1}.

Theorem 3.1. *If the matrix function $A \in \mathbf{L}_\infty^{n\times n}(\Gamma)$ admits a right (left) factorization, then the integers $\varkappa_j = \varkappa_j(A)$ $(j = 1, \dots, n)$ are determined by the matrix function A uniquely.*

Proof. Assume that A beside the factorization (1) has another right factorization

$$A(z) = \tilde{A}_-(z)\, \tilde{D}(z)\, \tilde{A}_+(z), \tag{2}$$

where $\tilde{D}(z) = (z^{\tilde{\varkappa}_j}\, \delta_{jk})_{j,k=1}^n$.

From (1) and (2) we obtain

$$B_-(z)\, D(z) = \tilde{D}(z)\, B_+(z) \tag{3}$$

with

$$B_-(z) = \tilde{A}_-^{-1}(z)\, A_-(z), \qquad B_+(z) = \tilde{A}_+(z)\, A_+^{-1}(z). \tag{4}$$

After denoting the entries of the matrix functions $B_\pm$ by $b_{jk}^\pm$, equality (3) can be written in the form

$$b_{jk}^-(z)\, z^{\varkappa_k} = z^{\tilde{\varkappa}_j} b_{jk}^+(z) \qquad (j, k = 1, \dots, n).$$

This implies that

$$b_{jk}^-(z) = b_{jk}^+(z) = 0 \qquad (z \in \Gamma) \tag{5}$$

for all j, k such that $\varkappa_k < \tilde{\varkappa}_j$. Indeed, since the intersection $\mathbf{L}_1^+(\Gamma) \cap \mathbf{L}_1^-(\Gamma)$ contains only the constants, (5) is a simple consequence of the equalities

$$b_{jk}^-(z) = z^{\tilde{\varkappa}_j - \varkappa_k} b_{jk}^+(z), \qquad \varkappa_k < \tilde{\varkappa}_j.$$

Now assume that contrary to the assertion of the theorem $\varkappa_r \neq \tilde{\varkappa}_r$ for some r between 1 and n. Without loss of generality let $\varkappa_r < \tilde{\varkappa}_r$. Then obviously $\varkappa_k < \tilde{\varkappa}_j$ $(j = 1, \dots, r;$ $k = r, \dots, n)$ and, consequently, $b_{jk}^+(z) = 0$ $(j = 1, \dots, r;\ k = r, \dots, n)$.

The last equalities imply that each minor of order r formed by the first r rows of the matrix function B_+ vanishes identically. The Laplace expansion theorem for determinants then gives that $\det B_+(z) \equiv 0$. This, however, is a contradiction, which proves the theorem for the case of the right factorization. The proof for the case of the left factorization is analogous. ∎

According to the type of the factorization, the integers $\varkappa_j$ $(j = 1, \dots, n)$ are called *right (left) indices* or also *partial indices* of the matrix function A. The sum $\varkappa = \sum_{j=1}^n \varkappa_j$ is referred to as the *total index* or *sum index* of the matrix function A. Note that the right and left indices of a matrix function do in general not coincide.

The general form of the factorization of a matrix function is described by the following theorem.

Theorem 3.2. *If a matrix function $A \in \mathbf{L}_\infty^{n\times n}(\Gamma)$ admits a right factorization* (1), *then every right factorization of this matrix function is of the form* (2) *with $\tilde{D}(z) = D(z)$ and*

$$\tilde{A}_+(z) = B_+(z)\, A_+(z), \qquad \tilde{A}_-(z) = A_-(z)\, D(z)\, B_+^{-1}(z)\, D^{-1}(z). \tag{6}$$

Here B_+ is an arbitrary non-singular matrix function whose entries satisfy the following conditions:

1. $b_{jk}^+(z) = 0$ *if* $\varkappa_k < \varkappa_j$;
2. $b_{jk}^+(z)$ *is a constant if* $\varkappa_k = \varkappa_j$;
3. $b_{jk}^+(z)$ *is a polynomial in* z *of degree at most* $\varkappa_k - \varkappa_j$ *if* $\varkappa_k > \varkappa_j$.

An analogous theorem holds for the case of the left factorization, too.

Proof. Assume we are given two right factorizations (1) and (2) of the matrix function A. Then, first of all, $\tilde{D}(z) = D(z)$ by Theorem 3.1. Define the matrix functions $B_\pm$ by (4) and denote their entries by $b_{jk}^\pm$. Then (3) implies that

$$b_{jk}^-(z) = z^{\varkappa_j - \varkappa_k} b_{jk}^+(z) \qquad (j, k = 1, \ldots, n). \tag{7}$$

Since the intersection $\mathbf{L}_1^+(\Gamma) \cap \mathbf{L}_1^-(\Gamma)$ contains only the constants, we deduce from (7) that the functions b_{jk}^+ satisfy the conditions 1–3 of our theorem. Thus, the matrix function B_+ is of the form $B_+ = (M_{jk})_1^m$, where $M_{jk} = 0$ for $k > j$, M_{jj} $(j = 1, \ldots, m)$ are non-singular constant matrices, and the entries of the matrix functions M_{jk} $(k < j)$ are polynomials in z. As a consequence of this, we firstly see that (3) and (4) imply (6) and we secondly get that the determinant of the matrix function B_+ is the non-zero constant

$$\det B_+ = \prod_{j=1}^{m} \det M_{jj}.$$

Lastly, it is easily seen that through the formulae (6) any non-singular matrix function B_+ with the properties 1–3 transforms a given right factorization (1) into another right factorization of the matrix function A. ▮

Finally, notice the following consequence of Theorem 3.1: If the matrix function $A \in \mathbf{L}_\infty^{n\times n}(\Gamma)$ admits a factorization (1) and if $z^\pm \in D_\Gamma^\pm$ are arbitrarily chosen points, then A possesses a factorization of the form

$$A(z) = \tilde{A}_-(z) \left(\left(\frac{z - z^+}{z - z^-}\right)^{\varkappa_j} \delta_{jk}\right)_1^n \tilde{A}_+(z). \tag{1'}$$

Vice versa, any factorization of the type (1′) generates a factorization (1) without changing the indices $\varkappa_j$.

3.2. Canonical factorization of matrix functions

In what follows let $\mathfrak{A}(\Gamma)$ denote a Banach algebra of continuous functions on Γ which satisfies the conditions a) and b) of Section 5.1, Chapter III, that is,

a) $\mathfrak{A}(\Gamma)$ contains the set $\mathbf{R}(\Gamma)$ of all rational functions without poles on Γ;

b) if $a \in \mathfrak{A}(\Gamma)$ and $a(z) \neq 0$ for all $z \in \Gamma$, then $1/a \in \mathfrak{A}(\Gamma)$.

Such algebras have obviously the following property: if $A \in \mathfrak{A}^{n\times n}(\Gamma)$ $([\mathfrak{A}^\pm(\Gamma)]^{n\times n})$ and $\det A(z) \neq 0$ for all $z \in \Gamma (z \in \overline{D_\Gamma^\mp})$, then $A^{-1} \in \mathfrak{A}^{n\times n}(\Gamma)$ $([\mathfrak{A}^\pm(\Gamma)]^{n\times n})$.

A right factorization (1) of a non-singular matrix function $A \in \mathfrak{A}^{n\times n}(\Gamma)$ is said to be *canonical* if $A_+^{\pm 1} \in [\mathfrak{A}^+(\Gamma)]^{n\times n}$ and $A_-^{\pm 1} \in [\mathfrak{A}^-(\Gamma)]^{n\times n}$. Analogously the left canonical factorization is defined.

Thus, the point in the definition of the canonical factorization is that the factors $A_\pm^{\pm 1}$ are required to belong to the same algebra as A.

If a matrix function $A \in \mathfrak{A}^{n\times n}(\Gamma)$ admits a right or left canonical factorization, then

$$\det A(z) = \det A_-(z) \det D(z) \det A_+(z)$$

and, hence,

$$\sum_{j=1}^{n} \varkappa_j = \text{ind} \det A .$$

Thus, in that case the total index does not depend on the type of the factorization.

The following theorem ensures the existence of a canonical factorization in certain algebras with two norms.

Theorem 3.3. *Let $\mathfrak{A}(\Gamma)$ be a Banach algebra of continuous functions on Γ which in addition to* a) *and* b) *satisfies the conditions* α)–δ) *of Section 5.2, Chapter III.*

Then every non-singular matrix function $A \in \mathfrak{A}^{n\times n}(\Gamma)$ admits a right (left) canonical factorization.

For the proof of this theorem we need the following auxiliary fact.

Lemma 3.1. *Let $\mathfrak{A}(\Gamma)$ be an arbitrary algebra of continuous functions on Γ satisfying the conditions* a) *and* b).

If the function $a \in \mathfrak{A}^{\pm}(\Gamma)$ takes the value zero at a point $z_0 \in D_\Gamma^{\pm}$, then $(z - z_0)^{-1} a(z) \in \mathfrak{A}^{\pm}(\Gamma)$.

Proof. For the sake of definiteness, let $a \in \mathfrak{A}^{+}(\Gamma)$ and $a(z_0) = 0$ for some $z_0 \in D_\Gamma^{+}$. Then, by virtue of the conditions a) and b), $(z - z_0)^{-1} a(z) \in \mathfrak{A}(\Gamma)$. Let p_n $(n = 1, 2, \ldots)$ be a sequence of functions from $\mathbf{R}^{+}(\Gamma)$ converging to a uniformly on Γ. Obviously,

$$r_n(z) = (z - z_0)^{-1} [p_n(z) - p_n(z_0)] \in \mathbf{R}^{+}(\Gamma)$$

and since $p_n(z_0) \to 0$, the sequence r_n converges to $(z - z_0)^{-1} a(z)$ uniformly on Γ. Therefore, the function $(z - z_0)^{-1} a(z)$ belongs to $\mathbf{C}^{+}(\Gamma)$ and thus also to $\mathfrak{A}^{+}(\Gamma)$. ∎

Proof of Theorem 3.3. We restrict ourselves to the proof of the existence of a right canonical factorization. The proof for the left canonical factorization is analogous.

First let $A \in [\mathfrak{A}^{+}(\Gamma)]^{n\times n}$. Denote by $z_1, z_2, \ldots, z_q$ the zeros of the function $\det A$ in the region D_Γ^{+} and let $m_1, m_2, \ldots, m_q$ denote their multiplicities. For what follows it will be convenient to assume that $z_q = 0$. If that point is not a zero of the function $\det A$, we put $m_q = 0$.

Let f_j be the j-th row of the matrix function $A = (a_{jk})_{j,k=1}^{n}$, i.e.,

$$f_j = (a_{j1}, a_{j2}, \ldots, a_{jn}) \qquad (j = 1, 2, \ldots, n)$$

and denote by p_j $(j = 1, 2, \ldots, n)$ the order of the zero $z = z_1$ as a zero of the vector function f_j. Obviously, $\sum_{j=1}^{n} p_j \leq m_1$.

Without loss of generality assume $p_1 \geq p_2 \geq \cdots \geq p_n$. Let $\sum_{j=1}^{n} p_j < m_1$. Then there are complex numbers $c_1, c_2, \ldots, c_l$ $(l \leq n,\ c_l = 1)$ such that the vector function

$$f(z) = \sum_{j=1}^{l} \frac{c_j f_j(z)}{(z - z_1)^{p_j}} \qquad (z \in \Gamma)$$

vanishes at the point $z = z_1$. By virtue of Lemma 3.1, the components of this vector function belong to the algebra $\mathfrak{A}^{+}(\Gamma)$. The vector function

$$\hat{f}(z) = f(z)\,(z - z_1)^{p_l} = \sum_{j=1}^{l} \frac{c_j f_j(z)}{(z - z_1)^{p_j - p_l}}$$

has then obviously a zero of order $\hat{p}_l > p_l$ at the point $z = z_1$.

Now form the matrix function

$$\hat{B}_1(z) = \begin{pmatrix} 1 & 0 & \dots & 0 & \dots & 0 \\ 0 & 1 & \dots & 0 & \dots & 0 \\ \dots & \dots & \dots & \dots & \dots & \dots \\ \frac{c_1}{(z-z_1)^{p_1-p_l}} & \frac{c_2}{(z-z_1)^{p_2-p_l}} & \dots & 1 & \dots & 0 \\ \dots & \dots & \dots & \dots & \dots & \dots \\ 0 & 0 & \dots & 0 & \dots & 1 \end{pmatrix} \quad l\text{-th row.}$$

Clearly, $\det \hat{B}_1(z) = 1$ and $\hat{B}_1^{\pm 1} \in [\mathfrak{A}^-(\Gamma)]^{n\times n}$.

Since the matrix function $\hat{B}_1 A$ can obviously be obtained from the matrix function A by replacing the l-th row of A by the row vector $\hat{f}$, it follows that $\hat{B}_1 A \in [\mathfrak{A}^+(\Gamma)]^{n\times n}$. If thereby $\sum_{j\neq l} p_j + \hat{p}_l < m_1$, then the same construction can be repeated with the matrix function $\hat{B}_1 A$ in place of A. After a finite number of steps we obtain a matrix function $B_1 A$ whose row vectors have zeros of orders $\hat{p}_1 \geqq \hat{p}_2 \geqq \dots \geqq \hat{p}_n$ with $\sum_{j=1}^{n} \hat{p}_j = m_1$ at the point $z = z_1$. Thereby $\det B_1(z) \equiv 1$ and $B_1^{\pm 1} \in [\mathfrak{A}^-(\Gamma)]^{n\times n}$.

Next, form the diagonal matrix function $D_1 \in [\mathfrak{A}^-(\Gamma)]^{n\times n}$ as follows:

$$D_1(z) = \left(\left(\frac{z}{z-z_1}\right)^{\hat{p}_j} \delta_{jk}\right)_1^n \qquad (z \in \Gamma).$$

It is readily seen that the determinant of the matrix function $A_1 = D_1 B_1 A$ does not vanish at the point $z = z_1$. Therefore, in the region D_Γ^+ the function $\det A_1$ has only the zeros $z_2, z_3, \dots, z_q$ whose orders are $m_2, m_3, \dots, m_q + m_1$, respectively.

Similar operations can now be carried out for the matrix function A_1 and the zero $z = z_2$. What results is a matrix function $A_2 = D_2 B_2 A_1$, the determinant of which has only the zeros $z_3, z_4, \dots, z_q$ with the multiplicities $m_3, m_4, \dots, m_q + m_1 + m_2$, respectively, in the region D_Γ^+. Continuing this procedure we arrive at a matrix function

$$A_{q-1} = D_{q-1} B_{q-1} A_{q-2} \in [\mathfrak{A}^+(\Gamma)]^{n\times n}$$

with $B_{q-1}^{\pm 1}$ and $D_{q-1}^{\pm 1}$ in $[\mathfrak{A}^-(\Gamma)]^{n\times n}$ and with $\det A_{q-1}$ a function having only a single zero in the region D_Γ^+, namely, a zero of order $\sum_j m_j$ at the point $z = 0$.

Finally, by means of the above described procedure we can construct a matrix function B having the following properties:

$$BA_{q-1} \in [\mathfrak{A}^+(\Gamma)]^{n\times n}, \qquad B^{\pm 1} \in [\mathfrak{A}^-(\Gamma)]^{n\times n},$$

the only zero of the function $\det BA_{q-1}$ is a zero of order $\sum_j m_j$ at the point $z = 0$, and the sum of the orders $\varkappa_1 \geqq \varkappa_2 \geqq \dots \geqq \varkappa_n$ of the zero $z = 0$ as a zero of the corresponding row vectors of the matrix function BA_{q-1} is equal to $\sum_j m_j$.

Now form the diagonal matrix function $D(z) = (z^{\varkappa_j} \delta_{jk})_{j,k=1}^n$. Then, by Lemma 3.1,

$$D^{-1} BA_{q-1} = A_+ \in [\mathfrak{A}^+(\Gamma)]^{n\times n}, \tag{8}$$

where $\det A_+(z) \neq 0$ for $z \in \overline{D_\Gamma^+}$. From (8) we obtain that

$$B(z)\, D_{q-1}(z)\, B_{q-1}(z) \dots D_1(z)\, B_1(z)\, A(z) = D(z)\, A_+(z)$$

and, consequently,

$$A(z) = A_-(z)\, D(z)\, A_+(z) \qquad (z \in \Gamma)$$

with $A_- = [BD_{q-1}\, B_{q-1} \ldots D_1 B_1]^{-1}$. Thereby $A_-^{\pm 1} \in [\mathfrak{A}^-(\Gamma)]^{n\times n}$ and $A_+^{\pm 1} \in [\mathfrak{A}^+(\Gamma)]^{n\times n}$. This proves the theorem for the case where $A \in [\mathfrak{A}^+(\Gamma)]^{n\times n}$.

Now take a general non-singular matrix function $A \in \mathfrak{A}^{n\times n}(\Gamma)$. Let $\tilde{P}$ denote the continuous projection of the algebra $\tilde{\mathfrak{A}}^{n\times n}(\Gamma)$ onto $[\tilde{\mathfrak{A}}^+(\Gamma)]^{n\times n}$ parallel to $[\mathring{\tilde{\mathfrak{A}}}^-(\Gamma)]^{n\times n}$ and put $\tilde{Q} = I - \tilde{P}$. By virtue of the condition β), there exists a matrix function $R = (r_{jk})_{j,k=1}^n \in \mathbf{R}^{n\times n}(\Gamma)$ such that $R^{-1} \in \mathbf{R}^{n\times n}(\Gamma)$ and

$$\|I - AR^{-1}\|_{\tilde{\mathfrak{A}}^{n\times n}} < \min\,(\|\tilde{P}\|^{-1}, \|\tilde{Q}\|^{-1}, \varrho).^{1)}$$

By repeating the arguments of the proof of Lemma 5.3, Chapter III, we arrive at a factorization

$$AR^{-1} = B_- B_+$$

of the matrix function AR^{-1} with $B_-^{\pm 1} \in [\mathfrak{A}^-(\Gamma)]^{n\times n}$ and $B_+^{\pm 1} \in [\mathfrak{A}^+(\Gamma)]^{n\times n}$. Consequently,

$$B_+ R = B_-^{-1} A. \tag{9}$$

Now let $\lambda_1^+, \lambda_2^+, \ldots, \lambda_s^+$ denote all poles of the entries r_{jk} in the region D_Γ^+ (each pole counted up to its multiplicity) and put

$$r(z) = \prod_{i=1}^{s} (z - \lambda_i^+).$$

Then, obviously,

$$\tilde{A} = rB_+ R \in [\mathfrak{A}^+(\Gamma)]^{n\times n}$$

and $\det \tilde{A}(z) \neq 0$ for all $z \in \Gamma$. By what has already been proved, the matrix function A admits a factorization

$$\tilde{A} = \tilde{A}_-\, \tilde{D} \tilde{A}_+.$$

This combined with (9) gives

$$A = B_- \tilde{A}_-\, \hat{D} \tilde{A}_+ \tag{10}$$

with $\hat{D} = r^{-1}\, \tilde{D}$.

The matrix D is of diagonal form and the entries on its principal diagonal, d_j ($j = 1, \ldots, n$), belong to $\mathbf{R}(\Gamma)$. Thus (see (2.9), Chapter III), each function d_j admits a factorization

$$d_j(z) = d_j^-(z)\, z^{\varkappa_j}\, d_j^+(z) \qquad (z \in \Gamma),$$

where $d_j^\pm \in \mathbf{R}^\pm(\Gamma)$ ($j = 1, \ldots, n$) and, without loss of generality, we may assume that $\varkappa_1 \geqq \varkappa_2 \geqq \ldots \geqq \varkappa_n$. This yields a factorization for $\hat{D}$ of the form

$$\hat{D} = D_- D D_+, \tag{11}$$

where $D(z) = (z^{\varkappa_j}\, \delta_{jk})_{j,k=1}^n$, $D_+^{\pm 1} \in [\mathbf{R}^+]^{n\times n}$, and $D_-^{\pm 1} \in [\mathbf{R}^-]^{n\times n}$.

---

1) The number ϱ was defined in condition δ) of Section 5.2, Chapter III.

From (10) and (11) we obtain the desired factorization of the matrix function A:

$$A = A_- DA_+,$$

where $A_- = B_-\tilde{A}_- D_-$ and $A_+ = D_+\tilde{A}_+$. The proof of Theorem 3.3 is complete. ∎

Remark. If the algebra $\mathfrak{A}(\Gamma)$ satisfies the conditions α), β), and γ) of Section 5.2, Chapter III, then in order that every non-singular matrix function $A \in \mathfrak{A}^{n\times n}(\Gamma)$ admit a right (left) canonical factorization it is necessary that the condition δ) is also fulfilled (see Prössdorf and Unger [1]).

Since every decomposing R-algebra satisfies the conditions of Theorem 3.3 we have the following

Theorem 3.4. *Let $\mathfrak{A}(\Gamma)$ be a decomposing R-algebra. Then every non-singular matrix function $A \in \mathfrak{A}^{n\times n}(\Gamma)$ admits a right (left) canonical factorization.*

Corollary 3.1. *Every non-singular matrix function $A \in W^{n\times n}$ admits a right (left) canonical factorization.*

Here is an immediate consequence of Theorem 3.3 and Lemma 5.2, Chapter III:

Theorem 3.5. *Every non-singular matrix function*

$$A \in [\mathbf{H}^\mu(\Gamma)]^{n\times n} \qquad (A \in [\mathbf{C}^{m,\mu}(\Gamma)]^{n\times n}), \qquad 0 < \mu < 1,$$

admits a right (left) canonical factorization.

The hypotheses of Theorem 3.3 can be relaxed essentially. Namely, one has the following

Theorem 3.6. *Let $\tilde{\mathfrak{A}}(\Gamma)$, $\mathfrak{A}(\Gamma)$, and $\mathfrak{A}^0(\Gamma)$ be (not necessarily normed) algebras of continuous function on the curve Γ satisfying in addition to* a) *and* b) *the following conditions:*

1. $\mathfrak{A}(\Gamma) \subset \mathfrak{A}^0(\Gamma) \subset \tilde{\mathfrak{A}}(\Gamma)$.
2. *$\tilde{\mathfrak{A}}(\Gamma)$ is a decomposing algebra.*
3. *$\mathbf{R}(\Gamma)$ is dense in $\mathfrak{A}(\Gamma)$ with respect to the norm of $\tilde{\mathfrak{A}}(\Gamma)$.*
4. *For each function $a \in \mathfrak{A}(\Gamma)$ the operator T_a defined as $T_a := S_\Gamma a - aS_\Gamma$ (S_Γ the singular integral operator) maps $\tilde{\mathfrak{A}}(\Gamma)$ into $\mathfrak{A}^0(\Gamma)$ and $\mathfrak{A}^0(\Gamma)$ into $\mathfrak{A}(\Gamma)$.*

Then every non-singular matrix function $A \in \mathfrak{A}^{n\times n}(\Gamma)$ admits a right (left) canonical factorization.

Proof. Let $\tilde{P}$ denote the continuous projection of $\tilde{\mathfrak{A}}^{n\times n}(\Gamma)$ onto $[\tilde{\mathfrak{A}}^+(\Gamma)]^{n\times n}$ parallel to $[\mathring{\tilde{\mathfrak{A}}}^-(\Gamma)]^{n\times n}$ and put $\tilde{Q} = I - \tilde{P}$. Further, let $B \in \mathfrak{A}^{n\times n}(\Gamma)$ be any matrix function such that

$$\|I - B\|_{\tilde{\mathfrak{A}}^{n\times n}} < \min(\|\tilde{P}\|^{-1}, \|\tilde{Q}\|^{-1}).$$

Then, by Lemma 5.1, Chapter III, B admits a factorization $B = B_-B_+$ with $B_-^{\pm 1} \in [\tilde{\mathfrak{A}}^-(\Gamma)]^{n\times n}$ and $B_+^{\pm 1} \in [\tilde{\mathfrak{A}}^+(\Gamma)]^{n\times n}$. Repeating the arguments of the proof of Lemma 5.2, Chapter III, and taking into account condition 4 of our theorem we deduce that actually $B_-^{\pm 1} \in [\mathfrak{A}^{0-}(\Gamma)]^{n\times n}$ and $B_+^{\pm 1} \in [\mathfrak{A}^{0+}(\Gamma)]^{n\times n}$. Once again using this reasoning we finally conclude that $B_-^{\pm 1} \in [\mathfrak{A}^-(\Gamma)]^{n\times n}$ and $B_+^{\pm 1} \in [\mathfrak{A}^+(\Gamma)]^{n\times n}$. The proof can now be completed by proceeding in the same way as in the proof of Theorem 3.3. ∎

The following theorem is an example of a concrete situation covered by the theorem just proved.

Theorem 3.7.[1]) *Let* $\mathfrak{A}(\Gamma) = \mathbf{H}^\mu(\Gamma; (t_j, m_j)_{j=1}^r)$, *where* $0 < \mu < 1$, $t_1, \dots, t_r$ *are pairwise distinct points on* Γ, $m_1, \dots, m_r$ *are positive integers, and suppose* $\Gamma \in C^{m,\mu}$ *with* $m = \max_j m_j$.

Then every non-singular matrix function $A \in \mathfrak{A}^{n\times n}(\Gamma)$ *admits a right (left) canonical factorization.*

Proof. Put $\mathfrak{A}(\Gamma) = \mathbf{H}^\mu(\Gamma; (t_j, m_j)_{j=1}^r)$, $\mathfrak{A}^0(\Gamma) = \mathbf{H}^\lambda(\Gamma; (t_j, m_j)_{j=1}^r)$, and $\tilde{\mathfrak{A}}(\Gamma) = \mathbf{H}^\lambda(\Gamma)$, where $0 < \lambda < \mu$. Then the conditions of Theorem 3.6 are satisfied.

Indeed, Theorem 6.4, Chapter II, implies that T_a maps $\tilde{\mathfrak{A}}(\Gamma)$ into $\mathfrak{A}^0(\Gamma)$. That $T_a\varphi$ is in $\mathfrak{A}(\Gamma)$ whenever $a \in \mathfrak{A}(\Gamma)$ and $\varphi \in \mathfrak{A}^0(\Gamma)$ can be seen as follows.

Let $\Gamma_0 \subset \Gamma$ be any neighborhood of the point t_j the endpoints of which are α and β, and define

$$(S_0\varphi)(t) = \int_{\Gamma_0} \frac{\varphi(\tau)}{\tau - t}\, d\tau \qquad (t \in \Gamma_0), \qquad \psi_0 = (S_0 a - a S_0)\,\varphi.$$

Then (see Lemma 6.1, Chapter II)

$$\psi_0'(t) = [(S_0 a' - a' S_0)\,\varphi](t) + [(S_0 a - a S_0)\,\varphi'](t)$$
$$+ [a(\alpha) - a(t)]\frac{\varphi(\alpha)}{\alpha - t} + [a(t) - a(\beta)]\frac{\varphi(\beta)}{\beta - t}.$$

It remains to apply Theorem 4.8, Chapter II. ∎

Remark. Both the results of Section 3.1 and Corollary 3.1 are due to GOHBERG and M. G. KREIN [2]. Theorem 3.5 was found by MUSKHELISHVILI and N. P. VEKUA [1] (see also MUSKHELISHVILI [1] or N. P. VEKUA [1]). Theorem 3.4 is GOHBERG's [5]. The Theorems 3.3 and 3.6 (as well as Theorem 3.7) were established by PRÖSSDORF [17]; they are a generalization of some results of BUDYANU and GOHBERG [1, 2], whose proof, with minor modifications, is repeated here (see also PRÖSSDORF and UNGER [1]).

Recently GOHBERG, LERER, and RODMAN [1] have obtained an explicit formula for the partial indices of a polynomial matrix function.

For a detailed treatment and a survey of the current state of the theory of factorization of both continuous and certain discontinuous matrix functions, various techniques, different results and applications we refer the reader to the book CLANCEY, GOHBERG [2] and to the recent work HEINIG, SILBERMANN [1].

§ 4. Generalized factorization of continuous matrix functions and its application

4.1. Let Γ be a closed piecewise Lyapunov curve system. We know from Theorem 5.1, Chapter III, that not every continuous (matrix) function admits a canonical factorization. In this connection note, however, the following theorem.

Theorem 4.1. *Every non-singular continuous matrix function* $A \in \mathbf{C}^{n\times n}(\Gamma)$ *admits a right factorization* (3.1), *where*

1. $A_+^{\pm 1} \in [\mathbf{L}_p^+(\Gamma)]^{n\times n}$ *and* $A_-^{\pm 1} \in [\mathbf{L}_p^-(\Gamma)]^{n\times n}$ *for all* p *with* $1 < p < \infty$.
2. *The operator* $A_- P A_-^{-1}$ *is bounded on* $\mathbf{L}_p^n(\Gamma)$ *for all* p *with* $1 < p < \infty$.[2])

[1]) Recall Section 6.1, Chapter II, for the notation used here. Thus, what Theorem 3.7 says is that if in some neighborhood of the points t_j ($j = 1, \dots, r$) the matrix function $A \in [\mathbf{H}^\mu(\Gamma)]^{n\times n}$ has derivatives up to the order m_j satisfying a Hölder condition with the exponent μ, then the factors $A_\pm$ occuring in (1) have the same property (possibly, in a smaller neighborhood of the points t_j).

[2]) Here and in the following sections we maintain the notation introduced in § 2.

3. *The equality* $\sum_{j=1}^{n} \varkappa_j = \operatorname{ind} \det A$ *holds, where* $\varkappa_j$ $(j = 1, \ldots, n)$ *are the right indices of the matrix function* A.

Notice that an analogous theorem holds for the left factorization. Property 2 has then to be replaced by the following one: the operator $A_+QA_+^{-1}$ is bounded on $\mathbf{L}_p^n(\Gamma)$ for all p with $1 < p < \infty$.

Proof. 1. Let B be a continuous matrix function. Put $Q = I - P$. For $F \in \mathbf{L}_p^{n\times n}(\Gamma)$ let $(U_BF)(z) = P(B(z) F(z))$ and $(V_BF)(z) = Q(F(z) B(z))$. Then the operators U_B and V_B defined in this way are bounded on $\mathbf{L}_p^{n\times n}(\Gamma)$ for all p with $1 < p < \infty$. Denote their norms by $\| U_B\|_p$ and $\| V_B\|_p$, respectively.

We claim the following: if the norms of the entries of the matrix function B as elements of $\mathbf{C}(\Gamma)$ are sufficiently small, so that for certain p_1 and p_2 satisfying $1 < p_1 \leqq 2 \leqq p_2 < \infty$ the inequality

$$\max_{p_1\leqq r\leqq p_2} \{\| U_B\|_r, \| V_B\|_r\} < 1 \tag{1}$$

holds, then the matrix function $I + B$ admits a factorization

$$I + B(z) = B_-(z)\, B_+(z) \tag{2}$$

with $B_+^{\pm 1} \in [\mathbf{L}_r^+(\Gamma)]^{n\times n}$ and $B_-^{\pm 1} \in [\mathbf{L}_r^-(\Gamma)]^{n\times n}$ for $p_1 \leqq r \leqq p_2$.

Indeed, this follows almost at once by applying the arguments of the first part of the proof of Lemma 5.1, Chapter III, to the equations $X + PBX = I$ and $X + QXB = I$ considered on the space $\mathbf{L}_r^{n\times n}(\Gamma)$.

Our next objective is to show that the operator $K = B_+^{-1}PB_-^{-1}$ is bounded on all spaces $\mathbf{L}_r^n(\Gamma)$ $(p_1 \leqq r \leqq p_2)$. Since for all vectors $F \in [\mathbf{L}_r^-(\Gamma)]^n$ the equality $(KF)(z) = B_+^{-1}(z)\, B_-^{-1}(\infty)\, F(\infty)$ holds, it remains to show that K is bounded on $[\mathbf{L}_r^+(\Gamma)]^n$.

Because of (1), for every $w \in [\mathbf{L}_r^+(\Gamma)]^n$ the equation $F + PBF = G$ has a unique solution $F \in [\mathbf{L}_r^+(\Gamma)]^n$ $(p_1 \leqq r \leqq p_2)$ and there is an $H \in [\mathbf{L}_r^-(\Gamma)]^n$ such that

$$(I + B)\, F = G + H.$$

Taking into account (2), we conclude that $F = B_+^{-1}PB_-^{-1}G$. Consequently, the operators $(I + U_B)^{-1}$ and K take the same values on the subspace $[\mathbf{L}_r^+(\Gamma)]^n$, and this implies that K is bounded on $[\mathbf{L}_r^+(\Gamma)]^n$.

Finally, since $B_-PB_-^{-1} = (I + B)\, B_+^{-1}PB_-^{-1}$, the operator $B_-PB_-^{-1}$ is bounded on $\mathbf{L}_r^n(\Gamma)$ for all r such that $p_1 \leqq r \leqq p_2$.

2. Let now A be an arbitrary non-singular continuous matrix function, let p_1 $(1 < p_1 \leqq 2)$ be an arbitrary fixed number, and put $1/p_2 = 1 - 1/p_1$. Then there exists a matrix function R whose entries are (rational) polynomials in z such that the matrix function $B = AR^{-1} - I$ satisfies the inequality (1). By what has already been proved, then (2) holds and therefore we have the representation $A = B_-B_+R$.

It is not difficult to see that the construction used in the proof of Theorem 3.3 for non-singular matrix functions $A \in [\mathfrak{A}^+(\Gamma)]^{n\times n}$ also applies to matrix functions $A \in [\mathbf{L}_r^+(\Gamma)]^{n\times n}$ with $A^{-1} \in [\mathbf{L}_r^+(\Gamma)]^{n\times n}$. Hence, B_+R admits the factorization $B_+R = C_-\, DA_+$, where D is the corresponding diagonal matrix, $A_+^{\pm 1}$ are in $[\mathbf{L}_r^+(\Gamma)]^{n\times n}$ and $C_-^{\pm 1}$ $(\in [\mathbf{L}_r^-(\Gamma)]^{n\times n})$ are matrix functions with rational entries. Thus,

$$A = A_-\, DA_+ \tag{3}$$

with $A_- = B_-C_-$, so that $A_-^{\pm} \in [\mathbf{L}_r^-(\Gamma)]^{n\times n}$ for $p_1 \leqq r \leqq p_2$.

We now show that the operator

$$A_-PA_-^{-1} = B_-PB_-^{-1} + B_-C_-(PC_-^{-1} - C_-^{-1}P)\, B_-^{-1}$$

is bounded on the spaces $\mathbf{L}_r^n(\Gamma)$. The boundedness of the first item has already been proved above. The boundedness of the second one follows from the representation

$$[(PC_-^{-1} - C_-^{-1}P)\, B_-^{-1}\varphi]\,(t) = \frac{1}{\pi i} \int_\Gamma \frac{C_-^{-1}(\tau) - C_-^{-1}(t)}{\tau - t}\, B_-^{-1}(\tau)\, \varphi(\tau)\, d\tau,$$

(see (2.3) and (2.4)), the right-hand side of which defines a bounded operator from $\mathbf{L}_r^n(\Gamma)$ to $\mathbf{L}_\infty^n(\Gamma)$.

3. Now we prove that the factors $A_\pm$ in (3) satisfy the conditions 1 to 3 of the theorem for all p with $1 < p < \infty$. Let q_1 be any fixed number such that $1 < q_1 < p$ and put $q_2 = q_1/(1 - q_1)$. The results proved above imply the existence of a right factorization $A = \tilde{A}_- \tilde{D}\tilde{A}_+$ whose factors possess the properties 1 and 2 for all p with $q_1 \leqq p \leqq q_2$. By virtue of Theorem 3.1, $\tilde{D}(z) = D(z)$ for all $z \in \Gamma$, and in view of Theorem 3.2, the matrix functions $\tilde{A}_\pm$ are up to polynomial factors equal to $A_\pm$. From this it is easy to deduce that the matrix functions $A_\pm$ occuring in (3) satisfy the conditions 1 and 2 of Theorem 4.1 for all $q_1 \leqq p \leqq q_2$, too.

Finally, the property 3 is a consequence of property 1 and the equality

$$\det A = \det A_- \det D \det A_+. \blacksquare$$

Corollary 4.1. *Let $A \in \mathbf{C}^{n\times n}(\Gamma)$ be any non-singular continuous matrix function and* (3) *its right factorization. Then the operator $B = A_+^{-1}P\, D^{-1}PA_-^{-1}$ is bounded on $\mathbf{L}_p^n(\Gamma)$ for all p such that $1 < p < \infty$.*

Proof. The factorization (3) and the property 2 give the boundedness of the operator $A_+^{-1}D^{-1}PA_-^{-1} = A^{-1}A_-PA_-^{-1}$. But it is easy to see that the operator $A_+^{-1}D^{-1}PA_-^{-1} - B$ has finite rank, and therefore the operator B is also bounded. $\blacksquare$

4.2. By means of the factorization founded in the preceding section the results of § 6, Chapter III, can be generalized to systems of singular integral equations with continuous coefficients.

Theorem 4.2. *Let A and B be arbitrary continuous matrix functions on Γ such that*

$$\det A(t) \neq 0, \qquad \det B(t) \neq 0 \qquad (t \in \Gamma)$$

and let

$$C(t) = C_-(t)\, D(t)\, C_+(t), \qquad D(t) = (t^{\varkappa_j}\,\delta_{jk})_1^n \tag{4}$$

be a right factorization of the matrix function $C = B^{-1}A$.

Then the operator $\mathbf{A} = AP + BQ$ is generalized invertible on the space $\mathbf{L}_p^n(\Gamma)$ $(1 < p < \infty)$ and one has

$$\dim \operatorname{Ker} \mathbf{A} = -\sum_{\varkappa_j<0} \varkappa_j, \qquad \dim \operatorname{Coker} \mathbf{A} = \sum_{\varkappa_j>0} \varkappa_j. \tag{5}$$

A generalized inverse of $\mathbf{A}$ is

$$\mathbf{A}^{(-1)} = (C_+^{-1}P + C_-Q)\,(D^{-1}P + Q)\, C_-^{-1}B^{-1}. \tag{6}$$

The equation

$$\mathbf{A}\varphi = f \qquad (f \in \mathbf{L}_p^n(\Gamma)) \tag{7}$$

has a solution $\varphi \in \mathbf{L}_p^n(\Gamma)$ if and only if the vector function

$$g(t) = C_-^{-1}(t)\, B^{-1}(t)\, f(t) =: (g_1(t), \ldots, g_n(t))$$

satisfies

$$\int_\Gamma t^k g_j(t)\,dt = 0 \qquad (k = 0, 1, \ldots, \varkappa_j - 1) \tag{8}$$

for all $j \in \{1, \ldots, n\}$ *with* $\varkappa_j > 0$. *If* (8) *holds, then* $\varphi = A^{(-1)} f$ *is a solution of* (7).

An analogous result can also be stated for the operator $A = PA + QB$; then C must be defined as AB^{-1}.

Proof. The factorization (4) yields a representation of the operator $AP + BQ$ in the form

$$AP + BQ = BC_-(DP + Q)(C_+P + C_-^{-1}Q). \tag{9}$$

Taking into account that the outer factors on the right of (9) are invertible, (5) is easily seen to result from Theorem 3.1, Chapter III.

By virtue of Theorem 4.1 and Corollary 4.1 the operator defined by (6) is bounded on $\mathbf{L}_p(\Gamma)$ $(1 < p < \infty)$. The equality

$$AA^{(-1)}A = A \tag{10}$$

is readily checked, and therefore $A^{(-1)}$ is a generalized inverse of A. From (10) we also deduce immediately that $A^{(-1)}f$ is a solution of (7) if (7) is solvable.

Finally, the representation (9) implies that (7) is solvable if and only if the equation $(DP + Q)\,\psi = 0$ is so. But due to Theorem 6.1, Chapter III, the latter equation has a solution if and only if conditions (8) are fulfilled. ▮

Corollary 4.2. *The general solution of the homogeneous equation* $(AP + BQ)\,\varphi = 0$ *is given by the formula*

$$\varphi(t) = [C_+^{-1}(t) - C_-(t)\,D(t)]\,P(t), \qquad P = (P_{-\varkappa_1-1}, \ldots, P_{-\varkappa_n-1}),$$

where $P_{-\varkappa_j-1}(t)$ *denotes an arbitrary polynomial of degree at most* $-\varkappa_j - 1$ $(P_{-\varkappa_j-1}(t) \equiv 0$ *for* $\varkappa_j \geqq 0)$.

This easily follows from the representation (9) and Theorem 6.1, Chapter III.

Corollary 4.3. *Under the hypotheses of Theorem 4.1, the operator* $AP + BQ$ *is invertible* (*left-sided invertible, right-sided invertible*) *if and only if all right indices of the matrix function* $B^{-1}A$ *are zero* (*non-negative, non-positive*). *The inverse* (*a left inverse, a right inverse*) *is given by* (6).

Remark. Theorem 4.1 was established by Mandzhavidze and Khvedelidze [1] and by Simonenko [4, 6]. The proof given here follows Gohberg and Feldman [1].

For the space $[\mathrm{H}^\mu(\Gamma)]^n$ and Hölder-continuous matrix functions A, B the results of Section 4.2 were obtained by Muskhelishvili and N. P. Vekua [1]. Their generalization to the case considered here is due to Khvedelidze [1, 2], Mandzhavidze and Khvedelidze [1], Simonenko [4, 6], and Gohberg [5].

§ 5. Systems of singular integral equations with bounded measurable coefficients

The results of the Sections 2 and 3 of Chapter IV can be carried over to systems of singular equations. Here we restrict ourselves to the formulation of the most important results.

5.1. Theorem 5.1. *Let Γ be a piecewise Lyapunov curve system and $A, B \in \mathbf{L}_\infty^{n\times n}(\Gamma)$. If the operator $AP + BQ$ $(PA + QB)$ is a Φ_+- or Φ_--operator on the space $\mathbf{L}_p^n(\Gamma, \varrho)$, then*

$$\inf_{t\in\Gamma} \operatorname{ess} |\det A(t)| > 0, \qquad \inf_{t\in\Gamma} \operatorname{ess} |\det B(t)| > 0.$$

This theorem was established by SIMONENKO [4, 7] and KRUPNIK [4].

Remark. Recently KRUPNIK [4] has also generalized the results of Section 1.2, Chapter IV, to the system case.

5.2. The concept of the generalized factorization of a function in the space $\mathbf{L}_p(\Gamma, \varrho)$ can be extended to the case of matrix functions as follows.

Suppose Γ is a closed piecewise Lyapunov curve system. A representation of the form $A = A_- D A_+$ is called a *generalized factorization of the matrix function* $A \in \mathbf{GL}_\infty^{n\times n}(\Gamma)$ *in the space* $\mathbf{L}_p^n(\Gamma, \varrho)$ if $D = (t^{\varkappa_j} \delta_{jk})_{j,k=1}^n$ with certain integers $\varkappa_1 \geqq \varkappa_2 \geqq \dots \geqq \varkappa_n$ and if the factors A_- and A_+ satisfy the following conditions:

1) $A_- \in [\mathbf{L}_p^-(\Gamma, \varrho)]^{n\times n}, \qquad A_+ \in [\mathbf{L}_q^+(\Gamma, \varrho^{-1})]^{n\times n},$

$A_-^{-1} \in [\mathbf{L}_q^-(\Gamma, \varrho^{-1})]^{n\times n}, \qquad A_+^{-1} \in [\mathbf{L}_p^+(\Gamma, \varrho)]^{n\times n} \qquad (p^{-1} + q^{-1} = 1).$

2. The operator $A_- P A_-^{-1}$ is bounded on the space $\mathbf{L}_p^n(\Gamma, \varrho)$.

This definition as well as the following two theorems are due to SIMONENKO [7].

Theorem 5.2. *For a matrix function $A \in \mathbf{GL}_\infty^{n\times n}(\Gamma)$ to admit a generalized factorization in the space $\mathbf{L}_p^n(\Gamma, \varrho)$ it is necessary and sufficient that the operator $AP + Q$ is a Φ-operator on the space $\mathbf{L}_p^n(\Gamma, \varrho)$.*

Theorem 5.3. *Let the matrix function $A \in \mathbf{GL}_\infty^{n\times n}(\Gamma)$ admit a generalized factorization in the space $\mathbf{L}_p^n(\Gamma, \varrho)$. Then for the operators $\mathbf{A} = AP + Q$ and $\mathbf{B} = PA + Q$ the following equalities hold*:

$$\dim \operatorname{Ker} \mathbf{A} = \dim \operatorname{Ker} \mathbf{B} = -\sum_{\varkappa_j<0} \varkappa_j,$$

$$\dim \operatorname{Coker} \mathbf{A} = \dim \operatorname{Coker} \mathbf{B} = \sum_{\varkappa_j>0} \varkappa_j.$$

§ 6. Systems of singular integral equation with piecewise continuous coefficients

The purpose of this section is to extend the results of the Sections 4 and 5 of Chapter IV to the case of singular operators with piecewise continuous matrix coefficients. Furthermore, for such operators we shall define a symbol whose properties sufficiently well reflect those of the operator. Throughout the following we use the notations introduced in § 4, Chapter IV, and in § 2 of the present chapter.

6.1. The case of closed curves

Suppose Γ is a closed piecewise Lyapunov curve system, $t_1, \dots, t_m$ are pairwise distinct points on Γ, and $p, \beta_1, \dots, \beta_m$ are real number satisfying

$$1 < p < \infty, \qquad -1/p < \beta_j < 1 - 1/p \qquad (j = 1, \dots, m), \tag{1}$$

and put

$$\varrho(t) = \prod_{j=1}^m |t - t_j|^{\beta_j}.$$

Let $a \in \mathbf{PC}^{n\times n}(\Gamma)$ be a matrix function. Thus, the entries of a are piecewise continuous functions on Γ which are continuous from the left. In analogy to the case $n = 1$ (see Section 4.2, Chapter IV), we assign to the matrix function a a continuous matrix function $a_{p,\varrho}: \Gamma \times [0, 1] \to \mathbf{C}^{n\times n}$ by

$$a_{p,\varrho}(t, \mu) = f(t, \mu)\, a(t + 0) + (1 - f(t, \mu))\, a(t) \qquad (t \in \Gamma,\ 0 \leqq \mu \leqq 1),$$

where the function $f(t, \mu)$ has the same meaning as in Section 4.2, Chapter IV. Recall that $f(t, \mu) = \mu$ for $p = 2$ and $\varrho(t) \equiv 1$.

The matrix function a is said to be *p, ϱ-regular* if

$$\det a_{p,\varrho}(t, \mu) \neq 0 \qquad (t \in \Gamma,\ 0 \leqq \mu \leqq 1).$$

The following theorem is the generalization of Theorem 4.2, Chapter IV, to the matrix case.

Theorem 6.1. *Let a and b be matrix functions from $\mathbf{PC}^{n\times n}(\Gamma)$. Then the operator $A = aP + bQ$ is a Φ-operator on the space $\mathbf{L}_p^n(\Gamma, \varrho)$ if and only if the following two conditions are satisfied:*

(i) *$\det b\,(t \pm 0) \neq 0$ for all $t \in \Gamma$.*

(ii) *The matrix function $g = b^{-1} a$ is p, ϱ-regular.*

If the conditions (i) *and* (ii) *are fulfilled then* $\operatorname{Ind} A = -\operatorname{ind} \det g_{p,\varrho}$.

An analogous theorem is also valid for the operator $A = Pa + Qb$; then one has to put $g = ab^{-1}$.

For the proof we need two lemmas.

Lemma 6.1. *Let c, d be non-singular continuous matrix functions and h a bounded measurable matrix function on Γ. Then the operator $chdP + Q$ $(Pchd + Q)$ is a Φ-operator on the space $\mathbf{L}_p^n(\Gamma, \varrho)$ if and only if $hP + Q$ $(Ph + Q)$ is a Φ-operator on $\mathbf{L}_p^n(\Gamma, \varrho)$.*

Proof. The commutators $aP - Pa$ and $aQ - Qa$ are compact on $\mathbf{L}_p^n(\Gamma, \varrho)$ for any continuous matrix function a, and the hypotheses of the lemma imply that the operators $Pd + Qc^{-1}$ and $cP + d^{-1}Q$ are Fredholm on $\mathbf{L}_p^n(\Gamma,\varrho)$. Therefore, the assertion follows from the obvious identities

$$\begin{aligned} chdP + Q &= c[(hP + Q)(Pd + Qc^{-1}) + h(dP - Pd) + c^{-1} Q - Qc^{-1}],\\ Pchd + Q &= [(cP + d^{-1} Q)(Ph + Q) + (Pc - cP)h + Q d^{-1} - d^{-1} Q]\, d \end{aligned} \tag{2}$$

and the results of § 3, Chapter I. ∎

Lemma 6.2. *Let $g \in \mathbf{PC}^{n\times n}(\Gamma)$ and suppose* $\det g(t_j \pm 0) \neq 0$ *at all its points of discontinuity t_j $(j = 1, \ldots, m)$. Then the matrix function g admits a representation in the form*

$$g = chd \tag{3}$$

with certain non-singular continuous matrix functions c and d (of order n) and a triangular matrix function $h \in \mathbf{PC}^{n\times n}(\Gamma)$.

Proof. First choose non-singular (constant) matrices d_j $(j = 1, \ldots, m)$ so that $d_j g^{-1}(t_j + 0)\, g(t_j)\, d_j^{-1}$ are upper triangular matrices. Then let $d(t)$ be a non-singular continuous matrix function on Γ such that $d(t_j) = d_j$ for $j = 1, \ldots, m$. Finally, choose a non-singular upper triangular matrix function $h \in \mathbf{PC}^{n\times n}(\Gamma)$ having the following properties: it is continuous at all points distinct from the points t_j; $h(t_j)$ $(= h(t_j - 0))$ is equal to I_n, the identity matrix; $h(t_j + 0) = d_j g^{-1}(t_j)\, g(t_j + 0)\, d_j^{-1}$ $(j = 1, \ldots, m)$. Then, by construction,

$$g(t_j + 0)\, d_j^{-1} h^{-1}(t_j + 0) = g(t_j)\, d_j^{-1} h^{-1}(t_j)$$

and thus the matrix function $c = g\, d^{-1}\, h^{-1}$ is continuous. ∎

Proof of Theorem 6.1. *Sufficiency*. Let the conditions (i) and (ii) be satisfied. Then, by Lemma 6.2, the matrix function $g = b^{-1}\, a$ admits the factorization (3), and since h is obviously a p, ϱ-regular matrix function, all diagonal entries of h are p, ϱ-regular functions, too. Now we deduce from Theorem 4.1, Chapter IV, that all diagonal entries of the operator matrix $hP + Q$ are Φ-operators on $\mathbf{L}_p(\Gamma, \varrho)$ and this implies that $hP + Q$ itself is a Φ-operator on $\mathbf{L}_p^n(\Gamma, \varrho)$. To conclude finally that $aP + bQ$ is a Φ-operator on $\mathbf{L}_p^n(\Gamma, \varrho)$, it remains to recall Lemma 6.1. The index formula results immediately from (2), (3), and Theorem 4.1, Chapter IV.

Necessity. Let $aP + bQ$ be a Φ-operator. We first show that then $\det a(t \pm 0) \neq 0$ and $\det b(t \pm 0) \neq 0$ for all $t \in \Gamma$.

Assume the contrary, that is, without loss of generality, assume $\det a$ is not invertible in $\mathbf{L}_\infty(\Gamma)$. Then it is easy to find a continuous matrix function c possessing the following properties:

1. The norm $\|cP\|$ is sufficiently small, so that $(a + c)\, P + bQ$ is also a Φ-operator.
2. There are an arc $\gamma \subset \Gamma$ and a point $t_0 \in \gamma$ such that, with t_1 and t_2 the endpoints of γ,

$$\det (a + c)\, (t_1 + 0) \neq 0, \qquad \det (a + c)\, (t_2) \neq 0, \qquad \det (a + c)\, (t_0) = 0.$$

Now let d be any continuous matrix function which coincides with $a + c$ on γ and which is non-singular on $\Gamma \setminus \gamma$. Then $a + c$ can be written as $a + c = h\, d$ with some $h \in \mathbf{PC}^{n \times n}(\Gamma)$. Since the operator

$$(a + c)\, P + bQ = (hP + bQ)\, (Pd + Q) + h(dP - Pd)$$

is Fredholm and since $dP - Pd$ is compact, $Pd + Q$ is a Φ_+-operator (see § 3, Chapter I). This, however, is a contradiction to $\det d(t_0) = 0$ (see Theorem 2.1 and the Remark 1 in § 2). Hence $\det a(t \pm 0) \neq 0$ for all $t \in \Gamma$. It can be shown equally that $\det b(t \pm 0) \neq 0$ for all $t \in \Gamma$.

Thus, it remains to show that the condition (ii) is satisfied. Again by Lemma 6.2, we can represent the matrix function $g = b^{-1}\, a$ in the form (3). The operator $gP + Q$ is Fredholm, hence, by Lemma 6.1, $hP + Q$ is also a Φ-operator. This and Theorem 4.1, Chapter IV, imply that all diagonal entries of the matrix function h are p, ϱ-regular. But since h is a triangular matrix, h itself and therefore also g is a p, ϱ-regular matrix function. ∎

6.2. The case of non-closed curves

Let Γ be a curve system in the complex plane consisting of a finite number of closed or open piecewise Lyapunov curves which have no points in common. Let Γ contain l open curves, denote the endpoints of the j-th open curve by t_j and t_{l+j} ($j = 1, \ldots, l$), and let the positive direction be the one from t_j to t_{j+1}. Furthermore, let $t_{2l+1}, \ldots, t_m$ be certain fixed points on Γ distinct from the points $t_1, \ldots, t_{2l}$, and let p and β_j ($j = 1, \ldots, m$) be real numbers again subject to (1). We denote by $\mathbf{PC}(\Gamma)$ the collection of all functions that are continuous on Γ except, possibly, at the points $t_{2l+1}, \ldots, t_m$; at these points the functions from $\mathbf{PC}(\Gamma)$ are supposed to possess finite limits from the right and to be continuous from the left.

Let a and b be matrix functions from $\mathbf{PC}^{n\times n}(\Gamma)$ and suppose $\det b(t \pm 0) \neq 0$ for all $t \in \Gamma$. With the operator $A = aP + bQ$ we associate the matrix function $W(A; t, \mu)$ $(t \in \Gamma, 0 \leqq \mu \leqq 1)$ defined as follows:

$$W(A; t, \mu) = \begin{cases} f_{\delta_k}(\mu)\, b^{-1}(t_k)\, a(t_k) + (1 - f_{\delta_k}(\mu))\, I_n & (k = 1, \dots, l); \\ (1 - f_{\delta_k}(\mu))\, b^{-1}(t_k)\, a(t_k) + f_{\delta_k}(\mu)\, I_n & (k = l + 1, \dots, 2l); \\ f_{\delta_k}(\mu)\, b^{-1}(t_k + 0)\, a(t_k + 0) + (1 - f_{\delta_k}(\mu))\, b^{-1}(t_k)\, a(t_k) & \\ \qquad\qquad (k = 2l + 1, \dots, m); & \\ b^{-1}(t)\, a(t) \qquad (t \neq t_1, \dots, t_m;\ t \in \Gamma). & \end{cases}$$

Here I_n denotes the $n\times n$ identity matrix, $\delta_k = 2\pi(1/p + \beta_k)$ $(k = 1, \dots, m)$, and $f_\delta(\mu)$ is defined by (4.1), Chapter IV, i.e.,

$$f_\delta(\mu) = \begin{cases} \dfrac{\sin \theta\mu\, e^{i\theta\mu}}{\sin \theta\, e^{i\theta}} & (\theta = \pi - \delta) \quad \text{for} \quad 0 < \delta < 2\pi,\ \delta \neq \pi, \\ \mu & \text{for} \quad \delta = \pi. \end{cases} \tag{4}$$

In particular, $f_{\delta_k}(\mu) = \mu$ for $p = 2$ and $\beta_k = 0$.

Let Γ_0 denote a curve obtained from Γ by adding certain loops and straight line segments in the following way: think of each point t_k $(k = 2l + 1, \dots, m)$ as splitted up into two points t_k^- and t_k^+ and join t_k^- to t_k^+ by a loop, and to each point t_k $(k = 1, \dots, 2l)$ attach a straight line segment; along each added loop or segment let the parameter μ run from 0 to 1.

Obviously, $W(A; t, \mu)$ is a continuous matrix function on Γ_0. It is easily seen that in the case where Γ is a closed curve system the conditions (i) and (ii) of Theorem 6.1 are equivalent to the following two conditions:

$$\det b(t \pm 0) \neq 0 \qquad (t \in \Gamma) \tag{5}$$

and

$$\det W(A; t, \mu) \neq 0 \qquad (t \in \Gamma, 0 \leqq \mu \leqq 1). \tag{6}$$

If these conditions are satisfied, then, by Theorem 6.1,

$$\operatorname{Ind} A = -\frac{1}{2\pi} [\arg \det W(A; t, \mu)]_{(t,\mu)\in\Gamma_0}, \tag{7}$$

where $[\dots]_{(t,\mu)\in\Gamma_0}$ denotes the increment of the expression in the brackets as the result of one circuit of the point (t, μ) around the oriented curve Γ_0.

The results for the scalar case (see § 5, Chapter IV) can now easily be carried over to the situation considered in the present section. The methods to prove these results are the same as in the scalar case. Thus, we have the following

Theorem 6.2. *Let $a, b \in \mathbf{PC}^{n\times n}(\Gamma)$. Then the operator $A = aP + bQ$ is a Φ-operator on the space $\mathbf{L}_p^n(\Gamma, \varrho)$ if and only if the conditions* (5) *and* (6) *are satisfied. In that case its index can be computed through formula* (7).

6.3. The definition of the symbol

Recall that the symbol of the singular integral operator $A = aP + bQ$ with continuous coefficients $a, b \in \mathbf{C}(\Gamma)$ on a closed curve Γ was defined as the function

$$A(t, \theta) = a(t)\,(1 + \theta)/2 + b(t)\,(1 - \theta)/2 \qquad (t \in \Gamma, \theta = \pm 1).$$

This function can obviously also be viewed as a pair of functions $\mathscr{A}(t) = (a(t), b(t))$, or, somewhat otherwise, as a matrix of second order,

$$\mathscr{A}(t) = \begin{pmatrix} a(t) & 0 \\ 0 & b(t) \end{pmatrix}.$$

The latter definition can be immediately extended to the case of matrix coefficients $a, b \in \mathbf{C}^{n\times n}(\Gamma)$.

Now let Γ be a general curve system of the kind considered in Section 6.2. We are going to define the symbol of the singular operator $A = aP + bQ$ with coefficients $a, b \in \mathbf{PC}^{n\times n}(\Gamma)$ on the space $\mathbf{L}_p^n(\Gamma, \varrho)$. It will turn out that in this situation the symbol essentially depends on p and ϱ as well as on the location of the point t on Γ.

To begin with, let first $n = 1$. Denote by $h_\delta(\mu)$ a fixed branch of the root $\sqrt{f_\delta(\mu)\,(1 - f_\delta(\mu))}$, where $f_\delta(\mu)$ $(0 \leqq \mu \leqq 1)$ is given by (4). Furthermore, let $\delta_k = 2\pi(1/p + \beta_k)$ $(k = 1, \ldots, m)$.

The *symbol* of the operator $A = aP + bQ$ $(a, b \in \mathbf{PC}(\Gamma))$ on the space $\mathbf{L}_p(\Gamma, \varrho)$ is the matrix function $\mathscr{A}(t, \mu)$ $(t \in \Gamma,\ 0 \leqq \mu \leqq 1)$ defined as follows:

At all points $t \in \Gamma$ distinct from the points $t_1, \ldots, t_m$ let

$$\mathscr{A}(t, \mu) = \begin{pmatrix} a(t) & 0 \\ 0 & b(t) \end{pmatrix}; \tag{8}$$

at the initial points t_k $(k = 1, \ldots, l)$ put

$$\mathscr{A}(t_k, \mu) = \begin{pmatrix} f_{\delta_k}(\mu)\, a(t_k) + (1 - f_{\delta_k}(\mu))\, b(t_k) & 0 \\ 0 & b(t_k) \end{pmatrix}; \tag{9}$$

at the final points t_k $(k = l + 1, \ldots, 2l)$ set

$$\mathscr{A}(t_k, \mu) = \begin{pmatrix} (1 - f_{\delta_k}(\mu))\, a(t_k) + f_{\delta_k}(\mu)\, b(t_k) & 0 \\ 0 & b(t_k) \end{pmatrix}; \tag{10}$$

and at the points t_k $(k = 2l + 1, \ldots, m)$ define

$$\mathscr{A}(t_k, \mu) = \begin{pmatrix} f_{\delta_k}(\mu)\, a(t_k + 0) + (1 - f_{\delta_k}(\mu))\, a(t_k) & h_{\delta_k}(\mu)\, (b(t_k + 0) - b(t_k)) \\ h_{\delta_k}(\mu)\, (a(t_k + 0) - a(t_k)) & (1 - f_{\delta_k}(\mu))\, b(t_k + 0) + f_{\delta_k}(\mu)\, b(t_k) \end{pmatrix}. \tag{11}$$

In what follows the symbol $\mathscr{A}(t, \mu)$ will be written in the form

$$\mathscr{A}(t, \mu) = \begin{pmatrix} a_{11}(t, \mu) & a_{12}(t, \mu) \\ a_{21}(t, \mu) & a_{22}(t, 1 - \mu) \end{pmatrix}. \tag{12}$$

It is easily seen that the functions $a_{jk}(t, \mu)$ $(j, k = 1, 2)$ are continuous on Γ_0. The function $\det \mathscr{A}(t, \mu)$ is, in general, not continuous on Γ_0. However, if $a_{22}(t, 1) \neq 0$ and $a_{22}(t, 0) \neq 0$ for all $t \in \Gamma$ (i.e., $b(t \pm 0) \neq 0$ for all $t \in \Gamma$), then the function $\det \mathscr{A}(t, \mu)/a_{22}(t, 0)\, a_{22}(t, 1)$ is continuous on Γ_0. This follows from the equality

$$\det \mathscr{A}(t, \mu) = W(A; t, \mu)\, a_{22}(t, 0)\, a_{22}(t, 1), \tag{13}$$

which in turn is a special case of a more general formula which will be proved in Lemma 6.3 below.

From Theorem 5.1, Chapter IV, (see also Theorem 6.2 for $n = 1$) and (13) we immediately obtain the following result.

Theorem 6.3. *For the operator* $A = aP + bQ$ $(a, b \in \mathbf{PC}(\Gamma))$ *to be a* Φ_+*- or* Φ_-*-operator on the space* $\mathbf{L}_p(\Gamma, \varrho)$ *it is necessary and sufficient that*

$$\det \mathscr{A}(t, \mu) \neq 0 \qquad (t \in \Gamma, 0 \leqq \mu \leqq 1). \tag{14}$$

If condition (14) *is fulfilled, then the operator* A *is at least one-sided invertible on* $\mathbf{L}_p(\Gamma, \varrho)$ *and its index is*

$$\operatorname{Ind} A = -\frac{1}{2\pi}\left[\arg \det \mathscr{A}(t, \mu)/a_{22}(t, 0)\, a_{22}(t, 1)\right]_{(t,\mu)\in\Gamma_0}.$$

The symbol of the singular operator with matrix coefficients can now be defined in a completely analogous fashion: the *symbol* of the operator $A = aP + bQ$ $(a, b \in \mathbf{PC}^{n\times n}(\Gamma))$ is the matrix function $\mathscr{A}(t, \mu)$ of order $2n$ defined by (8) to (11), but with $a(t)$ and $b(t)$ now being the corresponding matrix functions of order n. Exactly as in the case $n = 1$, both the conditions (5), (6) and formula (7) can be formulated in terms of the symbol.

Theorem 6.4. *For the operator* $A = aP + bQ$ $(a, b \in \mathbf{PC}^{n\times n}(\Gamma))$ *to be a* Φ*-operator on the space* $\mathbf{L}_p^n(\Gamma, \varrho)$ *it is necessary and sufficient that*

$$\det \mathscr{A}(t, \mu) \neq 0 \qquad (t \in \Gamma, 0 \leqq \mu \leqq 1). \tag{15}$$

In that case

$$\operatorname{Ind} A = -\frac{1}{2\pi}\left[\arg \det \mathscr{A}(t,\mu)/\det a_{22}(t, 0) \det a_{22}(t, 1)\right]_{(t,\mu)\in\Gamma_0}. \tag{16}$$

The proof of this theorem is based on the following lemma.

Lemma 6.3. *Let* $a, b \in \mathbf{PC}^{n\times n}(\Gamma)$ *and suppose* $\det b(t \pm 0) \neq 0$ *for all* $t \in \Gamma$. *Then*

$$\det \mathscr{A}(t, \mu) = \det W(A; t, \mu) \det a_{22}(t, 0) \det a_{22}(t, 1). \tag{17}$$

Proof. It suffices to check (17) for $t = t_k$ $(k = 2l + 1, \ldots, m)$ and $\mu \neq 0, 1$, since at the remaining points $\mathscr{A}(t, \mu)$ is a quasidiagonal matrix and therefore the equality (17) is then obvious.

Think of $\mathscr{A}(t, \mu)$ as a block matrix of second order (compare (11)), multiply its first row by $f_{\delta_k}(\mu)/h_{\delta_k}(\mu)$ and its first column by $h_{\delta_k}(\mu)/f_{\delta_k}(\mu)$ and then subtract the second row from the first. After that, multiply the second column from the right by $b^{-1}(t)\, a(t)$ and then add it to the first column. What results is a block triangular matrix. By taking its determinant we obtain the desired formula (17). ∎

Proof of Theorem 6.4. Let condition (15) be satisfied. On putting $\mu = 0$ and $\mu = 1$ we get $\det b(t \pm 0) \neq 0$ for all $t \in \Gamma$. (17) implies that $\det W(A; t, \mu) \neq 0$ for all $t \in \Gamma$ and $0 \leqq \mu \leqq 1$, and thus, by Theorem 6.2, A is a Φ-operator on $\mathbf{L}_p^n(\Gamma, \varrho)$. The index formula (16) immediately results from (7) and (17).

Now let A be a Φ-operator on $\mathbf{L}_p^n(\Gamma, \varrho)$. We then deduce from Theorem 6.2 that the conditions (5) and (6) are satisfied. But this combined with (17) gives (15). ∎

Corollary 6.1. *Let* $a, b \in \mathbf{PC}^{n\times n}(\Gamma)$ *and* $\det a(t \pm 0)\, b(t \pm 0) \neq 0$ *for all* $t \in \Gamma$. *Then the operator* $A = aP + bQ$ *has a finite index on* $\mathbf{L}_p^n(\Gamma, \varrho)$.

Making of use Theorem 6.2, this can be proved in the same way as in the case $n = 1$ (see Corollary 5.1, Chapter IV).

Remark. The results of Section 6.1 are due to Gohberg and Krupnik [3, 6]. Sufficient conditions for a singular operator to be Fredholm had earlier already been given by N. P. Vekua [1] and Khvedelidze [1, 2] (for the case of piecewise Hölder continuous coefficients) and by Simonenko [4] (for the case of arbitrary bounded measurable coefficients and the space $\mathbf{L}_p^n(\Gamma)$). Necessary and sufficient conditions for the Fredholm property of singular operators with piecewise differentiable coefficients on the space $\mathbf{L}_p$ over the real line were established by SHAMIR [1] for the first time.

In the Sections 6.2 and 6.3 we followed GOHBERG and KRUPNIK [3].

§ 7. Productsums of singular operators with piecewise continuous coefficients

This section is concerned with operators of the form

$$A = \sum_{j=1}^{k} A_{j1}A_{j2} \dots A_{jr}, \tag{1}$$

where $A_{jl} = a_{jl}P + b_{jl}Q$ and $a_{jl}, b_{jl} \in \mathbf{PC}^{n\times n}(\Gamma)$. The collection of all operators of the form (1) will be denoted by $\mathfrak{R}_n = \mathfrak{R}_n(\Gamma)$.

If Γ is a closed curve system and if all coefficients a_{jl}, b_{jl} are continuous, then every operator (1) can be written in the form

$$A = aP + bQ + T \tag{1'}$$

with certain continuous matrix functions a and b and a certain compact operator T on $\mathbf{L}_p^n(\Gamma, \varrho)$. This is an immediate consequence of Theorem 4.2, Chapter II, according to which the commutator $T_c = cP - Pc$ $(= \frac{1}{2}\, cS - \frac{1}{2}\, Sc)$ is compact on $\mathbf{L}_p^n(\Gamma, \varrho)$ for every continuous matrix function c. However, in the case of piecewise continuous coefficients an operator of the form (1) can, in general, not be written in the form (1').[1])

In what follows let Γ be the (in general, non-closed) curve system introduced in Section 6.2. With an operator A represented in the form (1) we associate the matrix function $\mathscr{A}(t, \mu)$ $(t \in \Gamma,\ 0 \leqq \mu \leqq 1)$ defined as

$$\mathscr{A}(t, \mu) = \sum_{j=1}^{k} \mathscr{A}_{j1}(t, \mu)\, \mathscr{A}_{j2}(t, \mu) \dots \mathscr{A}_{jr}(t, \mu), \tag{2}$$

where $\mathscr{A}_{jl}(t, \mu)$ denotes the symbol of the operator A_{jl}. This matrix function $\mathscr{A}(t, \mu)$ is referred to as the *symbol* of the operator (1). Below it will be shown that the symbol of an operator A only depends on the operator A itself and not on its specific representation in the form (1). In the meanwhile, however, when speaking about the symbol of an operator $A \in \mathfrak{R}_n$, we shall always keep in mind its specific representation (1).

Notice that if $A, B \in \mathfrak{R}_n$ and $\mathscr{A}(t, \mu)$, $\mathscr{B}(t, \mu)$ are the corresponding symbols, then the operators $C = A + B$ and $D = AB$ can be written in the form (1) in such a way that their symbols $\mathscr{C}(t, \mu)$ and $\mathscr{D}(t, \mu)$ are connected with $\mathscr{A}(t, \mu)$ and $\mathscr{B}(t, \mu)$ by the formulae

$$\mathscr{C}(t, \mu) = \mathscr{A}(t, \mu) + \mathscr{B}(t, \mu), \qquad \mathscr{D}(t, \mu) = \mathscr{A}(t, \mu)\, \mathscr{B}(t, \mu). \tag{3}$$

Theorem 7.1. *An operator $A \in \mathfrak{R}_n$ given by* (1) *is a Φ-operator on the space $\mathbf{L}_p^n(\Gamma, \varrho)$ if and only if its symbol $\mathscr{A}(t, \mu) = (a_{jk}(t, \mu))_{j,k=1}^2$ satisfies the condition*

$$\det \mathscr{A}(t, \mu) \neq 0 \qquad (t \in \Gamma,\ 0 \leqq \mu \leqq 1). \tag{4}$$

In that case

$$\operatorname{Ind} A = -\frac{1}{2\pi} [\arg \det \mathscr{A}(t, \mu)/\det a_{22}(t, 0) \det a_{22}(t, 1)]_{(t,\mu)\in\Gamma_0}. \tag{5}$$

The proof of this theorem is based on the following auxiliary result.

---

[1]) If c is a piecewise continuous function, then the commutator T_c is compact on $\mathbf{L}_p(\Gamma, \varrho)$ if and only if c is continuous (cf. Corollary 7.4).

Lemma 7.1. *Let A_{jl} ($j = 1, \ldots, k$; $l = 1, \ldots, r$) be linear bounded operators on a Banach space E and let A be the operator on E defined by* (1). *Then*

$$\begin{pmatrix} Z & X \\ Y & 0 \end{pmatrix} = \begin{pmatrix} I_{k(r+1)} & 0 \\ W & I \end{pmatrix} \begin{pmatrix} I_{k(r+1)} & 0 \\ 0 & A \end{pmatrix} \begin{pmatrix} Z & X \\ 0 & I \end{pmatrix}. \tag{6}$$

Here Z is the square matrix of order $k\,(r+1)$

$$Z = \begin{pmatrix} I_k & B_1 & 0 \ldots 0 \\ 0 & I_k & B_2 \ldots 0 \\ \multicolumn{3}{c}{\cdots\cdots\cdots\cdots\cdots} \\ 0 & 0 & 0 \ldots B_r \\ 0 & 0 & 0 \ldots I_k \end{pmatrix},$$

with I_k the identity operator on E^k ($I = I_1$) and

$$B_j = \begin{pmatrix} A_{1j} & 0 & \ldots & 0 \\ 0 & A_{2j} & \ldots & 0 \\ \cdots & \cdots & \cdots & \cdots \\ 0 & 0 & \ldots & A_{kj} \end{pmatrix},$$

while X is the one-column matrix and Y, W are the one-row matrices defined by

$$X = \begin{bmatrix} \left.\begin{matrix} 0 \\ \vdots \\ 0 \end{matrix}\right\} kr \\ \left.\begin{matrix} -I \\ \vdots \\ -I \end{matrix}\right\} k \end{bmatrix}, \quad Y = (\underbrace{I, \ldots, I}_{k}, \underbrace{0, \ldots, 0}_{kr}), \quad W = (M_0, M_1, \ldots, M_r)$$

with $M_0 = (\underbrace{I, \ldots, I}_{k})$ and

$$M_j = (A_{11}A_{12} \ldots A_{1j}, A_{21}A_{22} \ldots A_{2j}, \ldots, A_{k1}A_{k2} \ldots A_{kj}) \qquad (j = 1, \ldots, r).$$

The validity of (6) can be verified by a straightforward computation.

The matrix on the left-hand side of (6), which is generated by the operators A_{jl}, is sometimes called the *linear dilation of the operator* A and will be denoted by $\mathfrak{D}(A_{jl})$. Thus, $\mathfrak{D}(A_{jl})$ is a linear operator acting on the space E^t, where $t = k(r+1) + 1$.

The two outer factors on the right of (6) are invertible operators, and therefore A is a Φ or $\Phi_\pm$-operator if and only if $\mathfrak{D}(A_{jl})$ has the corresponding property. Moreover, we have dim Ker A = dim Ker $\mathfrak{D}(A_{jl})$ and dim Coker A = dim Coker $\mathfrak{D}(A_{jl})$.

Proof of Theorem 7.1. With the operator (1), considered as acting on the space $\mathbf{L}_p^n(\Gamma, \varrho)$, associate its linear dilation $B = \mathfrak{D}(A_{jl})$ thought of as acting on the space $\mathbf{L}_p^s(\Gamma, \varrho)$, where $s = n(kr + k + 1)$. It is easily seen that B is just the singular integral operator $B = aP + bQ$ whose matrix coefficients are $a(t) = \mathfrak{D}(a_{jl}(t))$ and $b(t) = \mathfrak{D}(b_{jl}(t))$.

As already mentioned above, A is a Φ-operator on $\mathbf{L}_p^n(\Gamma, \varrho)$ if and only if B is a Φ-operator on $\mathbf{L}_p^s(\Gamma, \varrho)$. The symbol $\mathscr{B}(t, \mu)$ of the operator B is a square matrix of order $2s$. We regard $\mathscr{B}(t, \mu)$ as a block matrix of order $2(kr + k + 1)$ and interchange its rows and columns by the following rule. Interchange the j-th row with the $(2j - 1)$-st row if $j \leqq kr + k + 1$ and with the $2(j - kr - k - 1)$-st row if $j > kr + r + 1$. Then do the same with the columns. As the result of this, the matrix function $\mathscr{B}(t, \mu)$ has been transformed into the matrix function $\mathfrak{D}(A_{jl}(t, \mu))$. Consequently, det $\mathscr{B}(t, \mu)$ = det $\mathfrak{D}(A_{jl}(t, \mu))$.

The matrix function $\mathfrak{D}(A_{jl}(t, \mu))$ can now be written in the form (6), whence without difficulty the equality

$$\det \mathfrak{D}(\mathscr{A}_{jl}(t, \mu)) = \det \sum_{j=1}^{k} \mathscr{A}_{j1}(t, \mu) \mathscr{A}_{j2}(t, \mu), \dots \mathscr{A}_{jr}(t, \mu)$$

follows. Thus, we have proved that

$$\det \mathscr{B}(t, \mu) = \det \mathscr{A}(t, \mu). \tag{7}$$

(7) and Theorem 6.3 now imply that A is a Φ-operator on $\mathbf{L}_p^n(\Gamma, \varrho)$ if and only if condition (4) is fulfilled. To complete the proof it remains to verify the index formula (5).

Let condition (4) be satisfied and write $\mathscr{A}_{jl}(t, \mu) = (a_{u,v}^{(j,l)})_{u,v=1}^{2}$. Theorem 6.4 gives

$$\operatorname{Ind} B = -\frac{1}{2\pi}[\arg \det \mathscr{B}(t, \mu)/\det \mathfrak{D}(a_{22}^{(j,l)}(t, 0)) \det \mathfrak{D}(a_{22}^{(j,l)}(t, 1))]_{(t,\mu)\in\Gamma}, \tag{8}$$

and since

$$\mathscr{A}(t, 0) = \sum_{j=1}^{k} \prod_{l=1}^{r} \begin{pmatrix} a_{11}^{(j,l)}(t, 0) & 0 \\ 0 & a_{22}^{(j,l)}(t, 1) \end{pmatrix},$$

it follows that

$$\det a_{22}(t, 0) = \det \sum_{j=1}^{k} \prod_{l=1}^{r} a_{22}^{(j,l)}(t, 0) = \det \mathfrak{D}(a_{22}^{(j,l)}(t, 0)). \tag{9}$$

Analogously one can show that

$$\det a_{22}(t, 1) = \det \mathfrak{D}(a_{22}^{(j,l)}(t, 1)). \tag{10}$$

Putting (7)–(10) together and taking into account that $\operatorname{Ind} A = \operatorname{Ind} B$ we get the desired formula (5). ∎

Corollary 7.1. *Let $A \in \mathfrak{R}_n$ be the operator defined on the space $\mathbf{L}_p^n(\Gamma, \varrho)$ by* (1) *and suppose* $\det \mathscr{A}(t, 0) \mathscr{A}(t, 1) \neq 0$ *for all $t \in \Gamma$. Then the index of A is finite.*

Proof. Write $\mathfrak{D}(A_{jl})$ as $aP + bQ$ with $a, b \in \mathbf{PC}^{s\times s}(\Gamma)$. Since $\det \mathscr{A}(t, 0) \mathscr{A}(t, 1) \neq 0$, it follows that $\det a(t \pm 0) b(t \pm 0) \neq 0$ for all $t \in \Gamma$. Thus, by virtue of Corollary 6.1, the index of $\mathfrak{D}(A_{jl})$ and therefore also that of A is finite. ∎

Theorem 7.2. *If the operator $A \in \mathfrak{R}_n$ is compact then the symbol $\mathscr{A}(t, \mu)$ vanishes identically.*

Proof. We restrict ourselves to the case $n = 1$.

Let the operator A ($\in \mathfrak{R}_1$) be of the form (1) and let $\mathscr{A}(t, \mu) = (a_{jk}(t, \mu))_{j,k=1}^{n}$ denote its symbol. We first show that $a_{11}(t, \mu) \equiv 0$. To that end assume the contrary, that is, assume $a_{11}(t_0, \mu_0) = \delta \neq 0$ for some point (t_0, μ_0). We introduce a new operator B, which is defined as follows: if t_0 is distinct from the points $t_{2l+1}, \dots, t_m$ then put $B = \delta I - A$, and let $B = \delta I - PAP$ otherwise. Since the operator A is compact, B is a Φ-operator. It can be easily checked that $\mathscr{B}(t_0, \mu_0) = 0$, where $\mathscr{B}(t, \mu)$ denotes the symbol of the operator

$$\delta I - \sum_{j=1}^{k} A_{j1} \dots A_{jr} \quad \text{or} \quad \delta I - \sum_{j=1}^{k} PA_{j1} \dots A_{jr}P,$$

respectively. This, however, contradicts Theorem 7.1. Therefore $a_{11}(t, \mu) \equiv 0$. Analogously it can be proved that $a_{22}(t, \mu) \equiv 0$.

We now show that $a_{12}(t, \mu) \equiv 0$. Again assume the contrary, i.e., assume there is a point (t_k, μ_0) $(k = 2l + 1, \dots, m;\ 0 < \mu_0 < 1)$ such that $a_{12}(t_k, \mu_0) = \delta \neq 0$. Choose a function $c \in \mathbf{PC}(\Gamma)$ so that $c(t_k + 0) - c(t_k) = 1$ and consider the operator $C = PcP + QcQ$. Let λ be any regular value of the operator C. Then

$$B = \delta(C - \lambda I) + [\lambda^2/h_{\delta_k}(\mu_0) - h_{\delta_k}(\mu_0)] \sum_{j=1}^{k} PA_{j1} \dots A_{jr}Q$$

is a Φ-operator, since the operator $B - \delta(C - \lambda I)$ is compact. But it can be easily verified that $\mathscr{B}(t_k, \mu_0) = 0$, which is a contradiction to Theorem 6.1. Thus, $a_{12}(t, \mu) \equiv 0$. It can be shown similarly that $a_{21}(t, \mu) \equiv 0$. ▮

Remark. A proof of Theorem 7.2 for the case $n > 1$ can be found in the paper of Gohberg and Krupnik [3] (§ 4). A major part of the material presented in this section is taken from § 3 of that paper.

The following results are almost obvious consequences of Theorem 7.2.

Corollary 7.2. *The symbol of an operator $A \in \mathfrak{R}_n$ does not depend on its concrete representation in the form* (1).

Corollary 7.3. *Let $A, B \in \mathfrak{R}_n$, $C = A + B$ and $D = AB$. Then for the corresponding symbols we have*

$$\mathscr{C}(t, \mu) = \mathscr{A}(t, \mu) + \mathscr{B}(t, \mu), \qquad \mathscr{D}(t, \mu) = \mathscr{A}(t, \mu)\, \mathscr{B}(t, \mu).$$

Corollary 7.4. *If Γ is a closed curve system, then each of the operators $aP - Pa$, PaQ, QaP $(a \in \mathbf{PC}^{n\times n}(\Gamma))$ is compact on the space $\mathbf{L}_p^n(\Gamma, \varrho)$ if and only of a is continuous.*

Corollary 7.5. *If Γ is a closed curve system, then the operator $PaPbP - PabP$ $(a, b \in \mathbf{PC}^{n\times n}(\Gamma))$ is compact on $\mathbf{L}_p^n(\Gamma, \varrho)$ if and only if a and b have no common points of discontinuity.*

As for Corollary 7.5, note that the symbol of $PaPbP - PabP$ is a matrix function $(a_{jl}(t, \mu))_{j,l=1}^2$ all entries of which, with the exception of $a_{11}(t, \mu)$, are identically zero, while

$$a_{11}(t_k, \mu) = f_{\delta_k}(\mu)\, (f_{\delta_k}(\mu) - 1)\, [a(t_k + 0) - a(t_k)]\, [b(t_k + 0) - b(t_k)]$$

at the points of discontinuity t_k of the functions a and b and $a_{11}(t, \mu) = 0$ at all other points. This and Lemma 1.2, Chapter IV, then yield Corollary 7.5. A similar reasoning gives Corollary 7.4 (see also Theorem 4.2, Chapter II).

Notice that the singular integral operator $B = Pa + Qb$ $(a, b \in \mathbf{PC}^{n\times n}(\Gamma))$ is an operator of the form (1). This is seen from writing it as

$$Pa + Qb = aP + bQ + PaQ - QaP + QbP - PbQ.$$

This representation also shows that the symbol $\mathscr{B}(t, \mu)$ $(t \in \Gamma,\ 0 \leq \mu \leq 1)$ of the operator B is just the matrix obtained from the symbol matrix $\mathscr{A}(t, \mu)$ associated with the operator $A = aP + bQ$ by blockwise transposition:

$$\mathscr{B}(t, \mu) = \begin{pmatrix} a_{11}(t, \mu) & a_{21}(t, \mu) \\ a_{12}(t, \mu) & a_{22}(t, 1 - \mu) \end{pmatrix}.$$

Here the matrix functions a_{jk} are the same ones as in (6.12). The Theorems 7.1 and 6.3 combined with Theorem 1.2, Chapter III, now imply the following

Theorem 7.3. *The operator $B = Pa + Qb$ with $a, b \in \mathbf{PC}^{n\times n}(\Gamma)$ is a Φ-operator on the space $\mathbf{L}_p^n(\Gamma, \varrho)$ if and only if* $\det \mathscr{B}(t, \mu) \neq 0$ *for all $t \in \Gamma$ and $0 \leq \mu \leq 1$.*

In that case Ind B = Ind $(aP + bQ)$, *and if* $n = 1$, *the operator* B *is at least one-sided invertible.*

The following theorem, which includes Theorem 7.2 as a special case, is of particular importance.

Theorem 7.4. *Let* A *be an operator in* $\mathfrak{K}_n$ *and let* $\mathscr{A}(t, \mu) = (a_{uv}(t, \mu))_{u,v=1}^{2n}$ *be its symbol. Then*

$$\max |a_{uv}(t, \mu)| \leqq K \inf_T \|A + T\|, \tag{11}$$

with the infimum on the right-hand side taken over all compact linear operators on $\mathbf{L}_p^n(\Gamma, \varrho)$ *and with* K *a constant depending only on the space* $\mathbf{L}_p^n(\Gamma, \varrho)$.

Theorem 7.4 was established by Gohberg and Krupnik [3, 7]. We confine ourselves to the proof of this theorem for the space $\mathbf{L}_2^n(\Gamma)$, with Γ a closed curve system. In that case, if the norm $\|\mathscr{A}\|$ of the symbol $\mathscr{A}(t, \mu)$ is defined by

$$\|\mathscr{A}\| = \max_{t \in \Gamma; 0 \leqq \mu \leqq 1} s_1(\mathscr{A}(t, \mu)),$$

where $s_1^2(\mathscr{A}(t, \mu))$ denotes the greatest eigenvalue of the matrix $\mathscr{A}(t, \mu)\,(\mathscr{A}(t, \mu))^*$, even the equality

$$\|\mathscr{A}\| = \inf_T \|A + T\| \tag{12}$$

holds.

Proof (Gohberg and Krupnik [7]). Let $\mathfrak{R}$ denote the algebra of all linear bounded operators on $\mathbf{L}_2^n(\Gamma)$ and $\mathfrak{S}_\infty$ the two-sided ideal of all compact operators in $\mathfrak{R}$. By a well-known Theorem (cf. Gohberg and Krein [3], p. 85, Corollary 7.1), for every $X \in \mathfrak{R}$ the equalities

$$\inf_{T \in \mathfrak{S}_\infty} \|X + T\|^2 = \inf_{T \in \mathfrak{S}_\infty} \|XX^* + T\| = \sup_{\lambda \in \sigma(XX^*)} \lambda$$

hold, where $\sigma(XX^*)$ is the spectrum of the coset $XX^* + \mathfrak{S}_\infty$ in the algebra $\mathfrak{R}/\mathfrak{S}_\infty$. The set $\sigma(XX^*)$ coincides with the set of all numbers λ for which $XX^* - \lambda I$ is not Fredholm (cf. § 3, Chapter I).

Since both $P - P^*$ and $Q - Q^*$ are in $\mathfrak{S}_\infty$ (see Section 5.1, Chapter II), the operator $A^* - A'$, where $A' = \sum_{j=1}^{k} \prod_{l=1}^{r} (P\bar{a}_{jl} + Q\bar{b}_{jl})$, also belongs to $\mathfrak{S}_\infty$.

The equalities $\sigma(AA^*) = \sigma(AA')$ and $\mathscr{A}'(t, \mu) = (\mathscr{A}(t, \mu))^*$ together with Theorem 7.1 show that the set $\sigma(AA^*)$ coincides with the spectrum of the matrix $\mathscr{A}(t, \mu)\,(\mathscr{A}(t, \mu))^*$. Hence $\sup_{\lambda \in \sigma(AA^*)} \lambda = \|\mathscr{A}\mathscr{A}^*\|$. The asserted equality (12) now follows from the relation $\|\mathscr{A}\mathscr{A}^*\| = \|\mathscr{A}\|^2$. ∎

Remark. Let $\Lambda(\Gamma)$ denote the collection of all piecewise continuous functions on Γ which are continuous at the points $t_1, \ldots, t_{2l}$ (the endpoints of the open curves belonging to Γ) and continuous from the left at all other points of Γ. Thus, unlike the class $\mathbf{PC}(\Gamma)$, the functions in $\Lambda(\Gamma)$ are allowed to have discontinuities also at points distinct from the points $t_{2l+1}, \ldots, t_m$, which occur in the weight $\varrho(t) = \prod_{k=1}^{m} |t - t_k|^{\beta_k}$. All results of the present section remain valid if the coefficients a_{jl} and b_{jl} are taken from $\Lambda^{n \times n}(\Gamma)$. If one of the matrix functions a_{jl} and b_{jl} has a discontinuity at a point t_k distinct from $t_1, \ldots, t_m$, then the symbol of the operator $a_{jl}P + b_{jl}Q$ at this point is defined by (6.11) with $\delta_k = 2\pi/p$ (i.e., $\beta_k = 0$).

§ 8. The algebra generated by singular operators with piecewise continuous coefficients

Let $\mathfrak{R}_n$ be the Banach algebra of all linear bounded operators on the space $\mathbf{L}_p^n(\Gamma, \varrho)$ and let $\mathfrak{A}_n = \mathfrak{A}_n(\Gamma)$ denote the closure of the set $\mathfrak{K}_n$ in $\mathfrak{R}_n$. Clearly, $\mathfrak{A}_n$ is also a Banach algebra. The results of the preceding chapter can be extended to the algebra $\mathfrak{A}_n$. Here we restrict ourselves to the formulation of the corresponding results.

Let A be any operator in $\mathfrak{A}_n$, A_ν a sequence of operators from $\mathfrak{K}_n$ converging to the operator A uniformly, and let $\mathscr{A}_\nu(t, \mu) = (a_{uv}^{(\nu)}(t, \mu))_{u,v=1}^2$ denote the symbols of the operators A_ν. Then, in view of the inequality (7.11), the sequence of the matrix functions $a_{uv}^{(\nu)}(t, \mu)$ converges uniformly with respect to (t, μ) as $\nu \to \infty$ to a matrix function $a_{uv}(t, \mu)$. The matrix function $\mathscr{A}(t, \mu) = (a_{uv}(t, \mu))_{u,v=1}^2$ will be called the *symbol* of the operator $A \in \mathfrak{A}_n$.

It is easily seen that the symbol of the operator A does not depend on the choice of the approximating sequence $\{A_\nu\}$. It only depends on the operator A itself, the number p and the weight ϱ. The symbol $\mathscr{A}(t, \mu)$ is a matrix function of order $2n$, $\mathscr{A}(t, \mu) = (g_{rs}(t, \mu))_{r,s=1}^{2n}$, and, again by the inequality (7.11),

$$\max |g_{rs}(t, \mu)| \leqq K \inf_{T \in \mathfrak{T}_n} \|A + T\|,$$

where $\mathfrak{T}_n$ denotes the two-sided ideal of all compact linear operators on $\mathbf{L}_p^n(\Gamma, \varrho)$ and K is a constant depending only on the space $\mathbf{L}_p^n(\Gamma, \varrho)$.

It can be shown that $\mathfrak{T}_n \subset \mathfrak{A}_n$. Theorem 7.4 implies that all operators A belonging to the same coset $\boldsymbol{A}$ of the quotient algebra $\mathfrak{A}_n/\mathfrak{T}_n$ have the same symbol. That symbol will be denoted by $\boldsymbol{A}(t, \mu)$. The mapping $\boldsymbol{A} \mapsto \boldsymbol{A}(t, \mu)$ is then a homomorphism of the quotient algebra $\mathfrak{A}_n/\mathfrak{T}_n$ onto the algebra of all symbols associated with the operators from $\mathfrak{A}_n$.

Let $A \in \mathfrak{K}_n$ and let $\mathscr{A}(t, \mu) = (a_{uv}(t, \mu))_{u,v=1}^2$ be its symbol. If $\det \mathscr{A}(t, \mu) \neq 0$, then $\det (a_{22}(t, 1)\, a_{22}(t, 0)) \neq 0$ and the function

$$\Psi(A; t, \mu) := \det \mathscr{A}(t, \mu)/\det (a_{22}(t, 1)\, a_{22}(t, 0))$$

is continuous on Γ_0 (cf. Section 6.2 of § 7). The integer

$$\operatorname{ind} \Psi(A; t, \mu) := \frac{1}{2\pi} [\arg \Psi(A; t, \mu)]_{(t,\mu)\in\Gamma_0}$$

will be referred to as the *index* of the function $\Psi(A; t, \mu)$.

Now let $A \in \mathfrak{A}_n$, $\Psi(A; t, \mu) \neq 0$, and $A_\nu \to A$ $(A_\nu \in \mathfrak{K}_n)$. Then the functions $\Psi(A_\nu; t, \mu)$ converge uniformly to the function $\Psi(A; t, \mu)$. Thus, for $\nu > \nu_0$ (ν_0 sufficiently large), the integers $\operatorname{ind} \Psi(A_\nu; t, \mu)$ do not depend on ν and this gives rise to the definition

$$\operatorname{ind} \Psi(A; t, \mu) := \operatorname{ind} \Psi(A\nu; t, \mu).$$

The index of the function $\Psi(A; t, \mu)$ is obviously independent of the choice of the sequence A_ν $(A_\nu \to A)$.

Now we have the following theorem.

Theorem 8.1. *For an operator $A \in \mathfrak{A}_n$ to be a Φ_+- or Φ_--operator on the space $\mathbf{L}_p^n(\Gamma, \varrho)$ it is necessary and sufficient that*

$$\det \mathscr{A}(t, \mu) \neq 0 \qquad (t \in \Gamma, 0 \leqq \mu \leqq 1). \tag{1}$$

In that case A is a Φ-operator on $\mathbf{L}_p^n(\Gamma, \varrho)$ and its index is given by the formula $\operatorname{Ind} A = -\operatorname{ind} \Psi(A; t, \mu)$.

Remark. Theorem 8.1 is due to GOHBERG and KRUPNIK. In their paper [7] they proved it for the space $\mathbf{L}_2^n(\Gamma)$ (Γ a closed curve) and in their paper [3] for the general case.

Recently the results of the Sections 7 and 8 have been extended to the case of piecewise Lyapunov curves Γ with a finite number of corner points by COSTABEL [1–3], DUDUCHAVA [2, 3], NYAGA [1, 2], and KRUPNIK and NYAGA [1]. The corresponding algebra $\mathfrak{A}(\Gamma)$ studied by COSTABEL contains, e.g., the operator of the double layer potential; this operator is not compact in the presence of corners.

KRUPNIK [5–7] gave a characterization of all operator algebras in which a matrix symbol can be introduced. He proved, e.g., that it is impossible to introduce a scalar symbol in the algebra $\mathfrak{A}_1(\Gamma)$.

Chapter VI
One-dimensional singular equations with degenerate symbol

The present chapter is concerned with singular integral operators whose symbol has zeros. Our emphasis is on the case of finitely many zeros of integral orders and on the case of $\mathbf{L}_p$ as the underlying space; however, all results can be extended to the spaces $\mathbf{L}_p$ with weight without undue effort and by means of the same methods. The last section deals with the case where the determinant of the symbol of a singular matrix operator vanishes identically.

§ 1. Reduction of an operator with finite index to a Fredholm operator

In this section we present a method with the help of which certain classes of not normally solvable equations can be reduced to normally solvable ones. This method is based upon a special factorization of the operator defined by the equation and consists in an appropriate restriction of the space in which the right-hand side of the equation is given and a corresponding extension of the space in which the solution is sought.

Let X be a Banach space and $A \in \mathscr{L}(X)$ a linear bounded operator on X. Assume the operator A is not normally solvable, but let it admit a factorization of the form

$$A = BCD, \tag{1}$$

where $C \in \mathscr{L}(X)$ is a normally solvable operator and the operators B and D (both in $\mathscr{L}(X)$) are subject to the following conditions:

(i) $\dim \operatorname{Ker} B = 0$.

(ii) *The operator D possesses a continuous extension $\widetilde{D}$ which maps a certain larger Banach space $\widetilde{X}$ ($\supset X$) one-to-one onto X.*

We introduce the space $\overline{X} = \operatorname{Im} B$ (the image space of the operator B) by defining the norm of an element $y = Bf \in \overline{X}$ as

$$\|y\|_{\overline{X}} = \|f\|_X. \tag{2}$$

Then, obviously, $\overline{X}$ becomes a Banach space and for every $y \in \overline{X}$ we have the estimate $\|y\|_X \leqq \|B\|_{\mathscr{L}(X)} \|y\|_{\overline{X}}$, that is, the space $\overline{X}$ is continuously embedded into the space X. From the open mapping theorem it is easy to deduce that the norms $\|\cdot\|_X$ and $\|\cdot\|_{\overline{X}}$ given on $\overline{X}$ are equivalent if and only if the operator $B \in \mathscr{L}(X)$ is normally solvable.

In what follows let $\overline{B}$ denote the operator B thought of as acting from X onto $\overline{X}$:

$$\overline{B}x = Bx \qquad (x \in X).$$

Because of (2) the operator $\overline{B}$ is an isometric isomorphism of X onto $\overline{X}$. Thus, $\overline{B}$ is in particular invertible. Due to condition (ii), $\tilde{D} \in \mathscr{L}(\tilde{X}, X)$ is invertible, too.

The representation (1) and the condition (ii) imposed on D allow us to extend the domain of the operator A to $\tilde{X}$. We regard this extension of the operator A as an operator from $\tilde{X}$ into X and denote it by $\tilde{A}$:

$$\tilde{A} = \overline{B}C\tilde{D}. \tag{3}$$

The operator $\tilde{A}$ now obtained is obviously normally solvable.

Furthermore, (3) also implies that C is a Φ-operator (a $\Phi_{\pm}$-operator or a one-sided invertible operator) if and only if $\tilde{A}$ has the corresponding property. Moreover, we have

$$\dim \operatorname{Ker} \tilde{A} = \dim \operatorname{Ker} C, \qquad \dim \operatorname{Coker} \tilde{A} = \dim \operatorname{Coker} C. \tag{4}$$

Thus, what we can say about the solvability of the equation

$$\tilde{A}x = y \tag{5}$$

is the following: for (5) to possess a solution $x \in \tilde{X}$ it is necessary that $y \in \overline{X}$; if $y \in \overline{X}$, then (5) has a solution $x \in \tilde{X}$ if and only if $(B^{-1}y, f) = 0$ for every solution $f \in X^*$ of the equation $C^*f = 0$; if the latter condition is fulfilled, then the general solution of (5) is of the form $x = \tilde{D}^{-1}z$, where z is the general solution of the equation $Cz = B^{-1}y$.

It is easily seen that the condition (ii) is satisfied if the following condition is fulfilled:

(iii) *There exist both a linear operator $D^{(-1)}$ defined on X with a range $\tilde{X} = D^{(-1)}(X) \supset X$ and a linear extension $\tilde{D}$ of D defined on $\tilde{X}$ such that*

$$D^{(-1)}Dx = x, \qquad \tilde{D}D^{(-1)}x = x \qquad \forall x \in X. \tag{6}$$

Note that the operator $D^{(-1)}$ appearing in condition (6) is nothing else than a certain "formal" inverse of D. Now we define a norm in $\tilde{X}$ by

$$\|f\|_{\tilde{X}} = \|\tilde{D}f\|_X \qquad (f \in \tilde{X}).$$

Taking into account the second equality in (6), it is readily seen that $\tilde{X}$ is a Banach space and that $\tilde{D}$ maps $\tilde{X}$ isometrically onto X. The operator $D^{(-1)}$ is the inverse of $\tilde{D}$. Finally, from the first equality in (6) we deduce that X is continuously embedded into $\tilde{X}$.

We shall see in the following sections that when working with concrete realizations of the abstract scheme given here the condition (iii) can be always checked in a relatively simple fashion. Also notice that it is necessarily satisfied if D is an operator in $\mathscr{L}(X)$ such that

$$\dim \operatorname{Ker} D = \dim \operatorname{Coker} D = 0. \tag{7}$$

Indeed, in that case $\tilde{X}$ can be chosen as the completion of the space X in the norm $|x| = \|Dx\|_X$. Then X is continuously and densely embedded into $\tilde{X}$. If $\tilde{x} = \{x_n\}$ $(x_n \in X)$ is any element of the completion $\tilde{X}$, then $\|Dx_n - Dx_m\|_X \to 0$, and consequently, there exists an element $x \in X$ such that $\|Dx_n - x\|_X \to 0$. Put

$$\tilde{D}\tilde{x} = x. \tag{8}$$

It is easily seen that the definition of $\tilde{D}$ through (8) is correct and that $\tilde{D}$ is a linear extension of the operator D. Furthermore, since $\tilde{D}$ is a one-to-one mapping of $\tilde{X}$ onto X, it has an inverse $\tilde{D}^{-1}$ and the operator $D^{(-1)} := \tilde{D}^{-1}$ then satisfies (6).

Remark 1. The above constructions are obviously performable in the case where each operator in (1) acts between two (possibly different) Banach spaces, too.

Remark 2. It can be shown (see Haikin [3]) that every operator A whose index is finite has a representation in the form (1), where B satisfies (i), D is subject to (7), and C is a Φ-operator. But note that the factors in (1) are not determined uniquely. In fact, one can always achieve the two limiting cases $B = I$ or $D = I$; the corresponding pairs of spaces are then $(\tilde{X}, X)$ or $(X, \bar{X})$, respectively. In these cases (5) admits the following interpretation:

1° $B = I$: The right-hand side of (5) is given in the space X and we are looking for a "generalized" solution $x \in \tilde{X}$.

2° $D = I$: The right-hand side y is taken from $\bar{X}$ and an "ordinary" solution $x \in X$ is sought.

The problem we shall be concerned with in the following sections, when applying the above construction to concrete situations, is to choose the factors in the representation (1) in such a way that the spaces $\tilde{X}$ and $\bar{X}$ can be analytically described as simply as possible.

Remark 3. The method described in this section is frequently also called the *method of normalization*. In various special forms and modifications it was applied by Chebotarev [1, 2], Haikin [1, 3], Dybin [1–3], Dybin and Karapetyants[1] in the setting of Wiener-Hopf integral equations and their discrete analogues, and by Prössdorf [6, 7, 10, 17] in the setting of singular integral equations.

In the general form presented here this method was worked out simultaneously by Sementsul [1] and Prössdorf [13].

§ 2. Factorization of abstract singular operators

In order to realize the method described in the preceding section, we need several factorizations of paired operators. These will be presented in this section.

Let X be a Banach space, $P \in \mathscr{L}(X)$ a projection, and $Q = I - P$ the complementary projection. Denote by $\mathscr{L}^+(X)$ (resp. $\mathscr{L}^-(X)$) the subalgebra of $\mathscr{L}(X)$ consisting of all operators $A \in \mathscr{L}(X)$ satisfying $AP = PAP$ (resp. $PA = PAP$). It is easy to see that an operator $A \in \mathscr{L}(X)$ belongs to $\mathscr{L}^+(X)$ (resp. $\mathscr{L}^-(X)$) if and only if $QA = QAQ$ (resp. $AQ = QAQ$).

Now let $R_+ \in \mathscr{L}^+(X)$ and $R_- \in \mathscr{L}^-(X)$ be given operators. If the coefficients $A, B \in \mathscr{L}(X)$ are of the form

(a) $A = A_1R_+, \qquad B = B_1R_-,$

where $A_1, B_1 \in \mathscr{L}(X)$, then the paired operator $AP + BQ$ can be represented in the following form:

$$AP + BQ = (A_1P + B_1Q)\,(R_+P + R_-Q). \tag{1}$$

If, in addition to (a),

(b) $A - A_2R_- \in \mathscr{L}^-(X), \quad B - B_2R_+ \in \mathscr{L}^+(X)$

with certain $A_2, B_2 \in \mathscr{L}(X)$, then we have a representation

$$PA + QB = (PA_1P + QB_1Q + PA_2Q + QB_2P)\,(R_+P + R_-Q). \tag{2}$$

If the condition (a) is satisfied and if

$$R_+R_- = R_-R_+, \qquad A_1 - A_2R_- \in \mathscr{L}^-(X), \qquad B_1 - B_2R_+ \in \mathscr{L}^+(X),$$

then

$$PA + QB = [PA_1 + QB_1 + (QB_2P - PA_2Q)\,(R_- - R_+)]\,(PR_+ + QR_-). \tag{3}$$

For coefficients of the form

(c) $A = R_-A_1, \qquad B = R_+B_1$

we have

$$PA + QB = (PR_- + QR_+)(PA_1 + QB_1). \tag{4}$$

If, in addition to (c), the condition

(d) $A - R_+A_2 \in \mathscr{L}^+(X), \qquad B - R_-B_2 \in \mathscr{L}^-(X)$

is fulfilled, then the representation

$$AP + BQ = (PR_- + QR_+)(PA_1P + QB_1Q + PB_2Q + QA_2P) \tag{5}$$

holds. Finally, if the condition (c) is satisfied and if

$$R_+R_- = R_-R_+, \qquad A_1 - R_+A_2 \in \mathscr{L}^+(X), \qquad B_1 - R_-B_2 \in \mathscr{L}^-(X),$$

then

$$AP + BQ = (R_-P + R_+Q)[A_1P + B_1Q + (R_+ - R_-)(PB_2Q - QA_2P)]. \tag{6}$$

The validity of the formulas (1)–(6) can be checked straightforwardly.

Remark. The factorizations (2), (3) and (5), (6) were used by Silbermann [8] to study singular integral operators with degenerate discontinuous coefficients (see §§ 6–9 of the present chapter).

§ 3. Some classes of differentiable functions

3.1. Let Γ be a piecewise Lyapunov curve system, f a (complex-valued) function defined on Γ, and α an arbitrary point on Γ. We denote by $f^{\{n\}}(\alpha)$ $(n = 0, 1, 2, \ldots)$ the n-th *Taylor derivative* of the function f at the point α, that is, the limit (if it exists)

$$f^{\{n\}}(\alpha) := n! \lim_{\substack{z \to \alpha \\ z \in \Gamma}} \frac{f(z) - \sum_{k=0}^{n-1} \frac{1}{k!} f^{\{k\}}(\alpha)(z - \alpha)^k}{(z - \alpha)^n}, \tag{1}$$

where $f^{\{0\}}(\alpha) = \lim_{z \to \alpha} f(z)$. Obviously, if the usual derivative $f^{(n)}(\alpha)$ exists, then the Taylor derivative (1) also exists and both derivatives coincide. The converse is in general not true.

3.2. Now let $\mathfrak{A} = \mathfrak{A}(\Gamma)$ $(\subset \mathbf{L}_\infty(\Gamma))$ be an algebra of measurable bounded functions on Γ which contains all rational functions without poles on Γ. For α an arbitrary point on Γ and m an arbitrary positive integer, we denote by $\mathfrak{A}(\alpha, m)$ the algebra of all functions $a \in \mathfrak{A}$ possessing a representation in the form

$$a(z) = \sum_{j=0}^{m-1} a_j(z - \alpha)^j + \tilde{a}(z)(z - \alpha)^m \tag{2}$$

with certain complex numbers a_j $(j = 0, 1, \ldots, m - 1)$ and a function $\tilde{a} \in \mathfrak{A}$. The representation (2) of a function $a \in \mathfrak{A}(\alpha, m)$ is obviously unique. If $\mathfrak{A}$ is a Banach algebra under the norm $\|\cdot\|_{\mathfrak{A}}$, then $\mathfrak{A}(\alpha, m)$ provided with the norm

$$\|a(z)\|_{\mathfrak{A}(\alpha,m)} = \sum_{j=0}^{m-1} |a_j| + \|\tilde{a}(z)\|_{\mathfrak{A}}$$

also becomes a Banach algebra.

It is obvious that $a \in \mathfrak{A}(\alpha, m)$ if and only if the Taylor derivatives $a^{\{k\}}(\alpha) = k!\, a_k$ $(k = 0, 1, \ldots, m - 1)$ exist and if the function

$$\tilde{a}(z) = \left[a(z) - \sum_{k=0}^{m-1} \frac{1}{k!} a^{\{k\}}(\alpha)(z - \alpha)^k\right](z - \alpha)^{-m}$$

belongs to $\mathfrak{A}$. It is also not difficult to see that for the algebra $\mathfrak{A} = \mathbf{H}^\lambda(\Gamma)$ $(0 \leqq \lambda \leqq 1)$ the inclusion $\mathbf{H}^\lambda(\Gamma; (\Gamma_\alpha, m)) \subset \mathbf{H}^\lambda(\alpha, m)$ holds (see Section 6.1 and PRÖSSDORF [17], p. 203).

3.3. Let now $\alpha = (\alpha_1, \alpha_2, \ldots, \alpha_r)$ be a system of pairwise distinct points $\alpha_j \in \Gamma$ $(j = 1, 2, \ldots, r)$ and $m = (m_1, m_2, \ldots, m_r)$ a system of positive integers. Put

$$\mathfrak{A}(\alpha, m) = \bigcap_{j=1}^{r} \mathfrak{A}(\alpha_j, m_j); \qquad \mathfrak{A}(\alpha, m; \beta, n) = \mathfrak{A}(\alpha, m) \cap \mathfrak{A}(\beta, n).$$

If $\mathfrak{A}$ is a Banach algebra, then $\mathfrak{A}(\alpha, m)$ equipped with the norm

$$\|a(z)\|_{\mathfrak{A}(\alpha,m)} = \sum_{j=1}^{r} \|a(z)\|_{\mathfrak{A}(\alpha_j,m_j)} \qquad (a(z) \in \mathfrak{A}(\alpha, m))$$

becomes a Banach algebra, too.

Here are some properties of $\mathfrak{A}(\alpha, m)$ which will be important in what follows:

1° *If* $a \in \mathfrak{A}(\alpha, m)$ *and* $a(\alpha_j) = 0$, *then* $a(z)\,(z - \alpha_j)^{-1} \in \mathfrak{A}(\alpha, m')$ *with* $m' = (m_1, \ldots, \ldots, m_j - 1, \ldots, m_r)$.

Proof. Clearly, it suffices to assume that $r = 2$ and $j = 1$. Then the function a can be written in the form

$$a(z) = f(z)\,(z - \alpha_1) = b_0 + g(z)\,(z - \alpha_2),$$

where

$$f(z) = \sum_{j=1}^{m_1-1} a_j(z - \alpha_1)^{j-1} + \tilde{a}(z)\,(z - \alpha_1)^{m_1-1},$$

$$g(z) = \sum_{j=1}^{m_2-1} b_j(z - \alpha_2)^{j-1} + \tilde{b}(z)\,(z - \alpha_2)^{m_2-1}$$

with $\tilde{a}, \tilde{b} \in \mathfrak{A}$. Obviously, $f \in \mathfrak{A}(\alpha_1, m_1 - 1)$ and $g \in \mathfrak{A}(\alpha_2, m_2 - 1)$. Furthermore, it is easy to see that

$$f(z) = g^{\{0\}}(\alpha_1) + \frac{f(z) - g(z)}{\alpha_1 - \alpha_2}\,(z - \alpha_2) \qquad (z \in \Gamma). \tag{3}$$

This immediately implies that $f \in \mathfrak{A}(\alpha_2, 1)$. Successive application of formula (3) gives that $f \in \mathfrak{A}(\alpha_2, m_2 - 1)$. Thence, again by (3), $f \in \mathfrak{A}(\alpha_2, m_2)$, and, consequently, $a(z)\,(z - \alpha_1)^{-1} = f(z) \in \mathfrak{A}(\alpha, m')$. ∎

Property 1° immediately yields an extremely important property of the algebra $\mathfrak{A}(\alpha, m)$.

2° *If* $a \in \mathfrak{A}(\alpha, m)$, *then there exists a polynomial* p *of degree* $\sum_{j=1}^{r} m_j - 1$ *such that*

$$[a(z) - p(z)] \prod_{j=1}^{r} (z - \alpha_j)^{-m_j} \in \mathfrak{A}.$$

This (obviously, uniquely determined) polynomial p is the *Hermitean interpolation polynomial* of the function a with the knots α_j.

The following three properties hold for algebras $\mathfrak{A}$ of continuous functions; they are of interest in connection with the canonical factorization of matrix function (see Section 3.2, Chapter V).

3° *If* $\mathfrak{A}$ *is an R-algebra, then* $\mathfrak{A}(\alpha, m)$ *is also an R-algebra.*

4° *If* $\mathfrak{A}$ *is a decomposing algebra, then the algebra* $\mathfrak{A}(\alpha, m)$ *is also decomposing.*

Property 3° can be checked straightforwardly. In order to prove the property 4°, it suffices to show that the projection P of $\mathfrak{A}$ onto $\mathfrak{A}^+$ is a continuous operator on $\mathfrak{A}(\alpha, m)$ $(\alpha \in \Gamma)$. First notice that for any polynomial q and any function $\varphi \in \mathfrak{A}$ the expression $f = (Pq - qP)\,\varphi$ is a polynomial.

Indeed, if k is any non-negative integer, then

$$g = (Pz^k - z^kP)\,\varphi = (P - z^kPz^{-k})\,z^k\varphi$$

belongs to $\mathfrak{A}^+$. It is easily seen that every function $g \in \mathfrak{A}^+$ can be written in the form

$$g(z) = \sum_{j=0}^{k-1} g_jz^j + z^kh(z)$$

with certain $h \in \mathfrak{A}^+$ and certain complex numbers g_j (cf. Lemma 3.1, Chapter V). The last two equalities give

$$(Pz^k - z^kP)\,\varphi = \sum_{j=0}^{k-1} g_jz^j.$$

Thus,

$$(Pq - qP)\,\varphi = \sum_{j=0}^{l} f_jz^j \tag{4}$$

and

$$\sum_{j=0}^{l} |f_j| \leqq c_1 \|f\|_{\mathfrak{A}} \leqq c_2 \|\varphi\|_{\mathfrak{A}}.$$

Let now $a \in \mathfrak{A}(\alpha, m)$. Taking into account the representation (4), we obtain

$$Pa = \sum_{j=0}^{m-1} a_j(z - \alpha)^j + [P(z - \alpha)^m - (z - \alpha)^m P]\,\tilde{a} + (z - \alpha)^m P\tilde{a}.$$

Hence $Pa \in \mathfrak{A}(\alpha, m)$ and

$$\|Pa\|_{\mathfrak{A}(\alpha,m)} \leqq c_3 \left(\sum_{j=0}^{m-1} |a_j| + \|\tilde{a}\|_{\mathfrak{A}}\right) = c_3 \|a\|_{\mathfrak{A}(\alpha,m)}.$$

This proves the assertion 4°.

From the properties 3° and 4° we easily deduce the following property.

5° *If the algebras* $\mathfrak{A} \subset \mathfrak{A}^0 \subset \tilde{\mathfrak{A}}$ *are subject to the conditions of Theorem* 3.6, *Chapter V, then the same holds for the algebras* $\mathfrak{A}(\alpha, m) \subset \mathfrak{A}^0(\alpha, m) \subset \tilde{\mathfrak{A}}(\alpha, m)$.

3.4. Let a be a function from an algebra $\mathfrak{A} \subset \mathbf{PC}(\Gamma)$ (see Section 4.2, Chapter IV). We call $\alpha = (\alpha_1, \alpha_2, \ldots, \alpha_r)$ $(\alpha_j \in \Gamma,\ j = 1, 2, \ldots, r)$ a *system of zeros* of the function a with the *multiplicity* $m = (m_1, m_1, \ldots, m_r)$ if the function a can be represented in the form

$$a(z) = \prod_{j=1}^{r} (z - \alpha_j)^{m_j}\, b(z),$$

where $b \in \mathfrak{A}$ and $b(\alpha_j \pm 0) \neq 0$ for $j = 1, \ldots, r$.

Theorem 3.1. *Let $A \in \mathfrak{A}^{n\times n}$ be a matrix function of order n, let $\alpha = (\alpha_1, \alpha_2, \ldots, \alpha_r)$ be the system of all zeros of the function* $\det A(z)$ $(z \in \Gamma)$, *and let $m = (m_1, m_2, \ldots, m_r)$ be the multiplicity of the system α.*

If the entries of the matrix function A belong to the algebra $\mathfrak{A}(\alpha, m')$, where $m' = (n_1, n_2, \ldots, n_r)$ with $n_j \geqq m_j$ $(j = 1, \ldots, r)$, then A admits the following representations:

$$A(z) = R_-(z)\, D_-(z)\, A_1(z) = A_2(z)\, D_-(z)\, S_-(z) \tag{5}$$

and

$$A(z) = R_+(z)\, D_+(z)\, A_3(z) = A_4(z)\, D_+(z)\, S_+(z). \tag{6}$$

Here $D_\pm$ *are diagonal matrix functions of the form*

$$D_\pm(z) = \left(\prod_{k=1}^{r} (z^{\pm 1} - \alpha_k^{\pm 1})^{\mu_j^{(k)}}\, \delta_{jl}\right)_{j,l=1}^{n}, \tag{7}$$

$\mu_1^{(k)} \geqq \mu_2^{(k)} \geqq \dots \geqq \mu_n^{(k)} \geqq 0$ $(k = 1, \dots, r)$ *are integers,* $R_\pm$ *and* $S_\pm$ *are polynomial matrices with constant and non-zero determinant, and* A_i $(i = 1, \dots, 4)$ *are matrix functions with entries in the algebra* $\mathfrak{A}(\alpha, m' - m)$ $(m' - m = (n_1 - m_1, \dots, n_r - m_r))$ *such that* $\det A_i(z \pm 0) \neq 0$ *for all* $z \in \Gamma$.

Proof. We shall construct the representation (5). The representation (6) can be obtained analogously.

We first show that A can be written in the form

$$A(z) = R_1(z)\, D_1(z)\, B_1(z), \tag{8}$$

where D_1 is a diagonal matrix function of the form

$$D_1(z) = ((z^{-1} - \alpha_1^{-1})^{\mu_j^{(1)}}\, \delta_{jl})_{j,l=1}^{n},$$

R_1 is a polynomial matrix in z^{-1} with constant and non-vanishing determinant, and B_1 is a matrix function whose entries belong to the algebra $\mathfrak{A}(\alpha, n')$ $(n' = (n_1 - m_1, n_2, \dots, n_r))$ and for which $\det B_1(\alpha_1 \pm 0) \neq 0$.

Let a_j $(j = 1, 2, \dots)$ denote the j-th row vector of the matrix function A. Then there are integers $\nu_1 \geqq \nu_2 \geqq \dots \geqq \nu_n \geqq 0$ such that

$$a_j(z) = (z^{-1} - \alpha_1^{-1})^{\nu_j}\, a_j^0(z), \qquad a_j^0 \in \mathfrak{A}(\alpha, n''),$$

where $n'' = (n_1 - \mu_1, n_2, \dots, n_r)$ and $\mu_1 = \sum_{j=1}^{r} \nu_j$.

If $\mu_1 = m_1$, then we have our representation (8). Thus assume $\mu_1 < m_1$. Then the vectors $a_j^0(\alpha_1)$ $(j = 1, 2, \dots, n)$ are linearly dependent, i.e., we have

$$\sum_{j=1}^{n} c_j a_j^0(\alpha_1) = 0, \tag{9}$$

with at least one non-zero c_j. Let ν_l denote the greatest number among $\nu_1, \nu_2, \dots, \nu_n$ for which $c_j \neq 0$. Then form the polynomial matrix $T^{(1)}$ whose entries T_{jk} $(j, k = 1, \dots, n)$ are defined by

$$T_{jk}(z) = \delta_{jk} (j \neq l), \qquad T_{lk}(z) = c_k(z^{-1} - \alpha_1^{-1})^{\nu_l - \nu_k}.$$

Obviously, $\det T^{(1)}(z) \equiv c_l \neq 0$. Taking into account the equality (9) and the above property 1°, it is easy to see that in the matrix function $A^{(1)}(z) = T^{(1)}(z)\, A(z)$ the place of the integer ν_l is occupied by $\nu_l + 1$. Repeating the preceding construction with the matrix function $A^{(1)}$ in place of A and continuing this procedure as long as necessary, after a finite number of steps we arrive at a representation (8) having all required properties.

By applying the same construction to the matrix function B_1 occuring in (8) and the points $\alpha_2, \dots, \alpha_r$ etc., we finally obtain a representation of A in the form

$$A(z) = R_1(z)\, D_1(z)\, R_2(z)\, D_2(z) \dots R_r(z)\, D_r(z)\, B'(z), \tag{10}$$

where D_k $(k = 1, 2, \dots, r)$ are diagonal matrix functions of the form

$$D_k(z) = ((z^{-1} - \alpha_k^{-1})^{\mu_j^{(k)}}\, \delta_{jl})_{j,l=1}^{n},$$

R_k are polynomial matrices in z^{-1} with constant and non-vanishing determinant, B' is a matrix function with entries belonging to the algebra $\mathfrak{A}(\alpha, m' - m)$ and $\det B'(z \pm 0) \neq 0$ for all $z \in \Gamma$.

By a well-known theorem on polynomial matrices (see GANTMACHER [1], Theorem 3, Chapter VI) we have

$$R_1(z)\, D_1(z) \ldots R_r(z)\, D_r(z) = R_-(z)\, D_1(z) \ldots D_r(z)\, R'(z), \tag{11}$$

where R_- and R' are polynomial matrices in z^{-1} with constant and non-zero determinant. The equalities (10) and (11) now immediately yield the first representation (5) with $A_1(z) = R'(z)\, B'(z)$.

If we change the roles of rows and columns in the preceding arguments, we analogously obtain a representation

$$A(z) = A_2(z)\, \overline{D}_-(z)\, S_-(z), \tag{12}$$

where $\overline{D}_-$ is a diagonal matrix function of the form (7) with certain integers $\overline{\mu}_1^{(k)} \geqq \overline{\mu}_2^{(k)} \geqq \ldots \geqq \overline{\mu}_n^{(k)} \geqq 0$ $(k = 1, 2, \ldots, r)$. It remains to show that $\overline{\mu}_j^{(k)} = \mu_j^{(k)}$ for all $j = 1, 2, \ldots, n$ and $k = 1, 2, \ldots, n$.

The representations (5) and (12) imply that

$$A'(z)\, D_-(z) = D_-(z)\, A''(z), \tag{13}$$

where $A' = A_2^{-1} R_-$ and $A'' = S_- A_1^{-1}$. Now assume that there are certain k ($k = 1, \ldots, r$) and s ($s = 1, \ldots, n$) such that, say, $\overline{\mu}_s^{(k)} > \mu_s^{(k)}$. Then $\overline{\mu}_j^{(k)} > \mu_l^{(k)}$ for all $j = 1, 2, \ldots, s$ and $l = s, s+1, \ldots, n$. Using (13), from this it can be readily deduced that the entries a'_{jl} of the matrix function A' satisfy the following conditions:

$$a'_{jl}(\alpha_k) = 0 \qquad (j = 1, 2, \ldots, s;\, l = s, s+1, \ldots, n).$$

The last equalities imply that each minor of order s formed by the first s rows of the matrix $A'(\alpha_k)$ is equal to zero. Thus, by the Laplace expansion theorem for determinants, $\det A'(\alpha_k) = 0$. This, however, is a contradiction, since $\det A'(z \pm 0) \neq 0$ for all $z \in \Gamma$. ∎

The arguments of the last part of the preceding proof show that the integers $\mu_j^{(k)}$ ($j = 1, \ldots, n$; $k = 1, \ldots, r$) which enter into the representations (5) and (6) through (7) coincide and that they are uniquely determined by the matrix function A. We shall call these numbers the *partial multiplicities of the system of zeros* α of the determinant $\det A(z)$ $(z \in \Gamma)$. Note that

$$m_k = \sum_{j=1}^{n} \mu_j^{(k)} \qquad (k = 1, \ldots, r). \tag{14}$$

3.5. There is a reasonable generalization of the notion of the Taylor derivative, which is fitted to integrable functions and is obtained by taking the mean of the difference quotient in (1). This generalization will play an important role in what follows.

Let $[a, b]$ be a finite interval and f a function measurable and Lebesgue integrable on this interval: $f \in \mathbf{L}(a, b)$. Put $f(x) = 0$ for $x \notin [a, b]$. Given any $h > 0$ we denote by f_h the *Steklov mean function* defined as

$$f_h(x) = \frac{1}{2h} \int_{x-h}^{x+h} f(t)\, dt \quad \text{for} \quad a < x < b$$

and

$$f_h(a) = \frac{1}{h} \int_{a}^{a+h} f(t)\, dt, \qquad f_h(b) = \frac{1}{h} \int_{b-h}^{b} f(t)\, dt.$$

Let $x_0 \in [a, b]$ be an arbitrarily given point. If the limit $\alpha := \lim\limits_{h\to 0} f_h(x_0)$ exists, then it is called the *mean limit* of the function f at the point x_0 and we write $\alpha = \overset{\frown}{\lim\limits_{x\to x_0}} f(x)$.

Obviously, $\overset{\frown}{\lim\limits_{x\to x_0}} f(x) = f(x_0)$, if x_0 is a Lebesgue point of the function f.

Lemma 3.1. *If f possesses an ordinary limit at the point x_0, then the mean limit exists at that point and both limits coincide.*

Proof. First let $\alpha = \lim\limits_{x\to x_0} f(x)$ be finite and suppose $a < x_0 < b$. Thus, for any $\varepsilon > 0$ there exists a $\delta = \delta(\varepsilon) > 0$ such that $|f(t) - \alpha| < \varepsilon$ whenever

$$|t - x_0| < \delta, \qquad a \leqq t \leqq b, \qquad t \neq x_0.$$

Hence, if $h < \delta$ then

$$|f_h(x_0) - \alpha| \leqq \frac{1}{2h} \int\limits_{x_0-h}^{x_0+h} |f(t) - \alpha| \, dt < \varepsilon,$$

that is, $\overset{\frown}{\lim\limits_{x\to x_0}} f(x) = \alpha$.

The proof is analogous for $x_0 = a$, $x_0 = b$, or $\alpha = \infty$. ▮

Notice that the function $f(x) = \sin 1/x$ $(-1 \leqq x \leqq 1)$ does not possess a limit at the point $x = 0$, but that $\overset{\frown}{\lim\limits_{x\to 0}} f(x) = 0$.

We now denote $\overset{\frown}{\lim\limits_{x\to x_0}} f(x)$ by $f^{\{0\}}(x_0)$. If $f^{\{n-1\}}(x_0)$ has already been defined and is finite, then we put

$$f^{\{n\}}(x_0) := n! \, \overset{\frown}{\lim\limits_{x\to x_0}} \frac{f(x) - \sum\limits_{k=0}^{n-1} \frac{1}{k!} f^{\{k\}}(x_0)\,(x - x_0)^k}{(x - x_0)^n}.$$

If the limit $f^{\{n\}}(x_0)$ exists, then it is called the *mean derivative of order n* of the function f at the point x_0.

Lemma 3.1 shows that the mean derivative $f^{\{n\}}(x_0)$ of a function f at a point x_0 exists if its Taylor derivative (see 3.1) at x_0 exists and that in this case both derivatives coincide. In particular, $f^{\{n\}}(x_0) = f^{(n)}(x_0)$ if the ordinary derivative $f^{(n)}(x_0)$ exists.

Notice that there are situations where the mean derivative exists, although the ordinary (or even Taylor) derivative does not exist. Examples are the functions

$$f(x) = x \sin \frac{1}{x} \; (f(0) = 0); \qquad f(x) = |x|, \qquad -1 \leqq x \leqq 1,$$

whose mean derivatives at the point $x_0 = 0$ are $f^{\{1\}}(0) = 0$.

Lemma 3.2. *Let $g \in L_p(a, b)$ $(1 \leqq p < \infty)$, m be a non-negative integer, $\mu \geqq 1/p$ be a real number, and $x_0 \in [a, b]$. Then the function $f(x) = (x - x_0)^{m+\mu} g(x)$ possesses mean derivatives up to the order m at the point x_0 and $f^{\{k\}}(x_0) = 0$ $(k = 0, 1, \ldots, m)$.*

Proof. Using Hölder's inequality it is readily seen that

$$|f_h(x_0)| \leqq \frac{1}{\sqrt[p]{2}} h^{m+\mu-1/p} \left[\int\limits_{x_0-h}^{x_0+h} |g(t)|^p \, dt \right]^{1/p},$$

whence by virtue of the absolute continuity of the integral the assertion for $k = 0$ follows. The assertion for $k = 1, \ldots, m$ can be proved by induction. ▮

Remark. The definition of the mean derivative can be easily extended to functions of several variables. Then the Steklov mean functions have to be replaced by the Sobolev mean functions.

The notion of the mean derivative can without difficulty be carried over to functions defined on a sufficiently smooth Lyapunov curve. Let Γ be such a curve and let $t = t(s)$ $(0 \leqq s \leqq \gamma)$ be its parametric representation, where s denotes the arc coordinate on Γ. A function $f \in \mathbf{L}_1(\Gamma)$ is said to possess the mean derivative $f^{\{1\}}(t_0)$ at the point $t_0 = t(s_0) \in \Gamma$ if the function $f(s) := f(t(s))$ (which is a function of the real variable s) possesses the mean derivative $f^{\{1\}}(s_0)$ at the point $s_0 \in [0, \gamma]$. In that case we put

$$f^{\{1\}}(t_0) = \overline{t'(s_0)}\, f^{\{1\}}(s_0) \qquad \left(t'(s) = \frac{\mathrm{d}t(s)}{\mathrm{d}s}\right).$$

Finally, by induction we define

$$f^{\{k+1\}}(t) = \overline{t'(s)}\, \frac{\mathrm{d}}{\mathrm{d}s} f^{\{k\}}(t);$$

here $\mathrm{d}f^{\{k\}}(t)/\mathrm{d}s$ on the right-hand side is to be understood as follows: first compute $\mathrm{d}f^{\{k\}}(t)/\mathrm{d}s$ formally by the product rule and then in the expression obtained replace $\mathrm{d}f^{\{l\}}(s)/\mathrm{d}s$ by $f^{\{l+1\}}(s)$. Note that Lemma 3.2 obviously remains true.

Remark. The results of this section are due to Prössdorf [8, 16, 17].

§ 4. Function spaces

4.1. Let Γ be a closed piecewise Lyapunov curve system. As usual, denote by S_Γ the corresponding singular integral operator with Cauchy kernel, put

$$P_\Gamma = \frac{1}{2}(I + S_\Gamma), \qquad Q_\Gamma = \frac{1}{2}(I - S_\Gamma),$$

and let $\mathbf{L}_p^+(\Gamma)$ and $\mathring{\mathbf{L}}_p^-(\Gamma)$ be the subspaces of $\mathbf{L}_p(\Gamma)$ $(1 < p < \infty)$ generated by the projections P_Γ and Q_Γ, respectively (see Section 5.3, Chapter II).

Furthermore, let $\alpha = (\alpha_1, \alpha_2, \ldots, \alpha_r)$ and $\beta = (\beta_1, \beta_2, \ldots, \beta_r)$ be systems of points on Γ, and let $m = (m_1, m_2, \ldots, m_r)$ and $n = (n_1, n_2, \ldots, n_r)$ be systems of non-negative integers. Such a system m will be called a *multi-index of dimension* r. The *length* of a multi-index m is defined as $|m| = \sum_{j=1}^{r} m_j$. We also introduce the abbreviations

$$(t - \alpha)^m = \prod_{j=1}^{r} (t - \alpha_j)^{m_j}, \qquad (t - \alpha)^{-m} = \prod_{j=1}^{r} (t - \alpha_j)^{-m_j},$$

and $(t^{-1} - \alpha^{-1})^m$, $(t^{-1} - \alpha^{-1})^{-m}$ are defined similarly. By Theorem 5.8, Chapter II, we have

$$(t - \alpha)^m P_\Gamma = P_\Gamma(t - \alpha)^m P_\Gamma, \qquad (t^{-1} - \alpha^{-1})^m Q_\Gamma = Q_\Gamma(t^{-1} - \alpha^{-1})^m Q_\Gamma. \quad (1)$$

Lemma 4.1. *Let $f_+ \in \mathbf{L}_p^+(\Gamma)$ and $f_- \in \mathring{\mathbf{L}}_p^-(\Gamma)$. If the functions $\varrho_+(t) = (t - \alpha)^m$ and $\varrho_-(t) = (t^{-1} - \beta^{-1})^n$ have no common zeros on Γ, then*

$$\varrho_+(t) f_+(t) = \varrho_-(t) f_-(t) \qquad (t \in \Gamma) \quad (2)$$

always implies that $f_+ = f_- = 0$.

Proof. The assertion is trivial for $|m| = |n| = 0$, since $\mathbf{L}_p^+(\Gamma) \cap \mathring{\mathbf{L}}_p^-(\Gamma) = \{0\}$. Thus let $|m| + |n| > 0$. We first show that f_+ and f_- are polynomials. To that end, we introduce the polynomials $g_-(t) = t^{-|m|}\varrho_+(t)$ and $g_+(t) = t^{|n|}\varrho_-(t)$, and we write (2) in the form

$$t^{|m|+|n|}\, g_-(t)\, f_-(t) = g_+(t)\, f_+(t). \tag{3}$$

Now let the projection P_Γ act on both sides of the last equality. This shows that $h_+ := g_+ f_+$ is a polynomial of degree at most $|m| + |n| - 1$. This polynomial is (in the ring of polynomials) divisible by the polynomial g_+, since otherwise the function $h_+/g_+ = f_+$ would not belong to $\mathbf{L}_p^+(\Gamma)$ (note that $\beta_k \in \Gamma$). Hence, f_+ is a polynomial. By similar arguments, it follows that the polynomial h_+ is divisible by the polynomial $t^{|m|+|n|}\, g_-(t) = t^{|n|}\, \varrho_+(t)$. Thus, in view of (3), $f_-(t) = h_+(t)/t^{|m|+|n|}\, g_-(t)$ is also a polynomial.

Since by our hypotheses the zeros of the polynomials ϱ_+ and ϱ_- are distinct, (2) implies that $f_+/\varrho_+ = f_-/\varrho_-$ is a polynomial and, consequently, $f_+ = f_- = 0$. ∎

Corollary 4.1. *Let $f_+ \in \mathbf{L}_p^+(\Gamma)$ (resp. $f_- \in \mathring{\mathbf{L}}_p^-(\Gamma)$). Then, if the function*

$$\varphi_1(t) = f_+(t)\, (t - \alpha)^{-m} \quad (\textit{resp. } \varphi_2(t) = f_-(t)\, (t^{-1} - \alpha^{-1})^{-m})$$

belongs to $\mathbf{L}_p(\Gamma)$, it is in $\mathbf{L}_p^+(\Gamma)$ (resp. $\mathring{\mathbf{L}}_p^-(\Gamma)$).

Proof. Let $\varphi_1 \in \mathbf{L}_p(\Gamma)$. Since $P_\Gamma \varphi_1 = \varphi_1 - Q_\Gamma \varphi_1$, we obtain

$$(t - \alpha)^m\, P_\Gamma \varphi_1 - f_+ = -(t - \alpha)^m\, Q_\Gamma \varphi_1,$$

that is, an equality of the form

$$\varphi_+(t) = (t - \alpha)^m\, \varphi_-(t) \qquad (t \in \Gamma)$$

with $\varphi_+ \in \mathbf{L}_p^+(\Gamma)$ and $\varphi_- \in \mathring{\mathbf{L}}_p^-(\Gamma)$. Thus, by Lemma 4.1, $\varphi_+ = \varphi_- = 0$, hence $Q_\Gamma \varphi_1 = 0$ and therefore $\varphi_1 = P_\Gamma \varphi_1 \in \mathbf{L}_p^+(\Gamma)$. The assertion for the function φ_2 can be proved analogously. ∎

4.2. We now consider the operator

$$D = D(\alpha, m; \beta, n) := (t - \alpha)^m\, P_\Gamma + (t^{-1} - \beta^{-1})^n\, Q_\Gamma, \tag{4}$$

where it is supposed that the polynomials $(t - \alpha)^m$ and $(t^{-1} - \beta^{-1})^n$ have no common zeros on Γ. Denote by $D^{(-1)}$ the operator defined for $f \in \mathbf{L}_p(\Gamma)$ by the formula

$$D^{(-1)} f := (t - \alpha)^{-m}\, P_\Gamma f + (t^{-1} - \beta^{-1})^{-n}\, Q_\Gamma f. \tag{5}$$

The collection of all functions of the form $D^{(-1)} f$ with $f \in \mathbf{L}_p(\Gamma)$ will be denoted by $\tilde{\mathbf{L}}_p(\alpha, m; \beta, n)$.

Due to (1) we have

$$D^{(-1)}\, Dg = g \tag{6}$$

for every $g \in \mathbf{L}_p(\Gamma)$. Consequently, $\mathbf{L}_p(\Gamma)$ is a subset of $\tilde{\mathbf{L}}_p(\alpha, m; \beta, n)$. The operator D can be extended to an operator $\tilde{D}$ defined on $\tilde{\mathbf{L}}_p(\alpha, m; \beta, n)$ by putting

$$\tilde{D}\varphi := (t - \alpha)^m\, \tilde{P}_\Gamma \varphi + (t^{-1} - \beta^{-1})^n\, \tilde{Q}_\Gamma \varphi$$

for $\varphi \in \tilde{\mathbf{L}}_p(\alpha, m; \beta, n)$. Here $\tilde{P}_\Gamma$ and $\tilde{Q}_\Gamma$ are the operators defined for a function $\varphi = D^{(-1)} f \in \tilde{\mathbf{L}}_p(\alpha, m; \beta, n)$ by

$$\tilde{P}_\Gamma \varphi := (t - \alpha)^{-m}\, P_\Gamma f, \qquad \tilde{Q}_\Gamma \varphi := (t^{-1} - \beta^{-1})^{-n}\, Q_\Gamma f.$$

Note that the operators $\tilde{P}_\Gamma$ and $\tilde{Q}_\Gamma$ are well defined, since, by Lemma 4.1, the function $f \in \mathbf{L}_p(\Gamma)$ is uniquely determined by the function $\varphi \in \tilde{\mathbf{L}}_p(\alpha, m; \beta, n)$. Furthermore, on account of (6) we have $\tilde{P}_\Gamma g = P_\Gamma g$, $\tilde{Q}_\Gamma g = Q_\Gamma g$, and, hence, $\tilde{D}g = Dg$ for every $g \in \mathbf{L}_p(\Gamma)$. It is also clear that $\tilde{D}D^{(-1)} f = f$ for all $f \in \mathbf{L}_p(\Gamma)$.

Thus, the operator D satisfies the condition (iii) of § 1. It follows that $\tilde{\mathbf{L}}_p(\alpha, m; \beta, n)$ endowed with the norm $\|\varphi\|_{\tilde{\mathbf{L}}_p} := \|\tilde{D}\varphi\|_{\mathbf{L}_p(\Gamma)} = \|f\|_{\mathbf{L}_p(\Gamma)}$ becomes a Banach space which is isometrically mapped onto $\mathbf{L}_p(\Gamma)$ by the operator $\tilde{D}$. Finally, notice that the embedding $\mathbf{L}_p(\Gamma) \subset \tilde{\mathbf{L}}_p(\alpha, m; \beta, n)$ is continuous.

Remark. If $(t - \alpha)^m$ and $(t^{-1} - \beta^{-1})^n$ have common zeros on Γ, then it is easily seen that Ker $D^{(-1)}$ has a positive finite dimension. In that case $\tilde{\mathbf{L}}_p(\alpha, m; \beta, n)$ is a Banach space under the norm $\|\varphi\|_{\tilde{\mathbf{L}}_p} = \inf_{D^{(-1)}f=\varphi} \|f\|_{\mathbf{L}_p}$.

4.3. We now introduce the operator

$$B = B(\alpha, m; \beta, n) := P_\Gamma(t^{-1} - \alpha^{-1})^m + Q_\Gamma(t - \beta)^n. \tag{7}$$

From (1) it is obvious that

$$B = \prod_{j=1}^{r} B(\alpha_j, m_j; \beta_j, n_j). \tag{8}$$

Lemma 4.2. dim Ker $B = 0$.

Proof. Let $B\varphi = 0$ ($\varphi \in \mathbf{L}_p(\Gamma)$). Action with the projection P_Γ on both sides of this equality and a simple application of (1) give $P_\Gamma(t^{-1} - \alpha^{-1})^m P_\Gamma\varphi = 0$, or,

$$(t^{-1} - \alpha^{-1})^m P_\Gamma\varphi = Q_\Gamma(t^{-1} - \alpha^{-1})^m P_\Gamma\varphi.$$

Hence, by Lemma 4.1, $P_\Gamma\varphi = 0$. Analogously we deduce that $Q_\Gamma(t - \beta)^n Q_\Gamma\varphi = 0$, i.e.,

$$(t - \beta)^n Q_\Gamma\varphi = P_\Gamma(t - \beta)^n Q_\Gamma\varphi.$$

Again using Lemma 4.1, we see that $Q_\Gamma\varphi = 0$ and thus $\varphi = 0$. ∎

In accordance with § 1, we denote by $\overline{\mathbf{L}}_p(\alpha, m; \beta, n)$ the image space Im B thought of as provided with the norm

$$\|f\|_{\overline{\mathbf{L}}_p} = \|B^{-1}f\|_{\mathbf{L}_p(\Gamma)} \qquad (f \in \overline{\mathbf{L}}_p(\alpha, m; \beta, n)).$$

Then $\overline{\mathbf{L}}_p(\alpha, m; \beta, n) \subset \mathbf{L}_p(\Gamma)$, the embedding being continuous, and the operator B maps $\mathbf{L}_p(\Gamma)$ onto the Banach space $\overline{\mathbf{L}}_p(\alpha, m; \beta, n)$ isometrically.

Our next concern is the analytical description of the space $\overline{\mathbf{L}}_p(\alpha, m; \beta, n)$.

Theorem 4.1. *For a function $f \in \mathbf{L}_p(\Gamma)$ to belong to the space $\overline{\mathbf{L}}_p(\alpha, m; \beta, n)$ it is necessary and sufficient that the following two conditions are satisfied:*

(i) *There exists a polynomial* $q_-(t) = \frac{a_1}{t} + \dots + \frac{a_{|m|}}{t^{|m|}}$ *such that* $\varphi_1(t) := [P_\Gamma f - q_-(t)]\,(t^{-1} - \alpha^{-1})^{-m} \in \mathbf{L}_p(\Gamma)$.

(ii) *There exists a polynomial* $q_+(t) = b_1 + \dots + b_{|n|}\, t^{|n|-1}$ *such that* $\varphi_2(t) := [Q_\Gamma f - q_+(t)]\,(t - \beta)^{-n} \in \mathbf{L}_p(\Gamma)$.

If the conditions (i) *and* (ii) *are fulfilled, then* $B^{-1} f = \varphi_1 + \varphi_2$ *and the norm in the space* $\overline{\mathbf{L}}_p(\alpha, m; \beta, n)$ *is equivalent to the norm* $\|f\|' = \|\varphi_1\|_{\mathbf{L}_p(\Gamma)} + \|\varphi_2\|_{\mathbf{L}_p(\Gamma)}$.

Proof. Abbreviate $(t - \beta)^n$ and $(t^{-1} - \alpha^{-1})^m$ to $\varrho_+(t)$ and $\varrho_-(t)$, respectively.

Necessity. Let $f \in \overline{\mathbf{L}}_p(\alpha, m; \beta, n)$, that is, $f = B\varphi$ with $\varphi \in \mathbf{L}_p(\Gamma)$. Let the operators P_Γ and Q_Γ act on both sides of the equality $f = B\varphi$ and take into consideration the identities (1). What results is

$$P_\Gamma \varrho_- P_\Gamma \varphi = P_\Gamma f, \qquad Q_\Gamma \varrho_+ Q_\Gamma \varphi = Q_\Gamma f.$$

From this, after some elementary computations, we obtain that

$$t^{|m|} f_- = t^{|m|} P_\Gamma f - t^{|m|} \varrho_- P_\Gamma \varphi \tag{9}$$

and

$$Q_\Gamma f - \varrho_+ Q_\Gamma \varphi = f_+ \tag{10}$$

with certain $f_+ \in \mathbf{L}_p^+(\Gamma)$ and $f_- \in \mathring{\mathbf{L}}_p^-(\Gamma)$. The right-hand sides of the last two equalities belong to the subspace $\mathbf{L}_p^+(\Gamma)$; we make their left-hand sides belong to the subspace $\mathring{\mathbf{L}}_p^-(\Gamma)$ by subtracting certain polynomials of the form $q(t) = a_{|m|} + a_{|m|-1} t + \dots + a_1 t^{|m|-1}$ for (9) and $q_+(t) = b_1 + b_2 t + \dots + b_{|n|} t^{|n|-1}$ for (10). Thus, acting with the operator Q_Γ on both sides of (9) and with the operator P_Γ on both sides of (10), we get

$$t^{|m|} f_-(t) = q(t), \qquad f_+(t) = q_+(t). \tag{11}$$

But (9)–(11) give (i) and (ii) immediately.

Sufficiency. Vice versa, suppose now that the conditions (i) and (ii) are satisfied. Then, by Corollary 4.1, $\varphi_1 \in \mathbf{L}_p^+(\Gamma)$ and $\varphi_2 \in \mathring{\mathbf{L}}_p^-(\Gamma)$. Hence, for the function $\varphi = \varphi_1 + \varphi_2$ we have $P_\Gamma \varphi = \varphi_1$ and $Q_\Gamma \varphi = \varphi_2$. Again using the identities (1) we therefore obtain that

$$B\varphi = P_\Gamma \varrho_- \varphi_1 + Q_\Gamma \varrho_+ \varphi_2 = P_\Gamma(P_\Gamma f - q_-) + Q_\Gamma(Q_\Gamma f - q_+) = P_\Gamma f + Q_\Gamma f = f.$$

Finally, the asserted equivalence of the two norms immediately follows from the estimates

$$\|\varphi_1\|_{\mathbf{L}_p(\Gamma)} \leqq \|P_\Gamma\| \, \|\varphi\|_{\mathbf{L}_p(\Gamma)}, \qquad \|\varphi_2\|_{\mathbf{L}_p(\Gamma)} \leqq \|Q_\Gamma\| \, \|\varphi\|_{\mathbf{L}_p(\Gamma)}$$

and the obvious inequality $\|\varphi\|_{\mathbf{L}_p(\Gamma)} \leqq \|f\|'$. ∎

The following theorem shows that it is sufficient to have an analytical description of the space $\overline{\mathbf{L}}_p(\alpha, m; \beta, n)$ for the case $r = 1$.

Theorem 4.2. *We have*

$$\overline{\mathbf{L}}_p(\alpha, m; \beta, n) = \bigcap_{j=1}^{r} \overline{\mathbf{L}}_p(\alpha_j, m_j; \beta_j, n_j), \tag{12}$$

and the norm in the space $\overline{\mathbf{L}}_p(\alpha, m; \beta, n)$ *is equivalent to the norm*

$$|f| = \sum_{j=1}^{r} \|f\|_{\overline{\mathbf{L}}_p(\alpha_j, m_j; \beta_j, n_j)}.$$

Proof. It is clear that it suffices to prove the assertion for $r = 2$. Thus, what we must prove is that

$$\operatorname{Im} B_1 B_2 = \operatorname{Im} B_1 \cap \operatorname{Im} B_2,$$

where $B_j = P_\Gamma(t^{-1} - \alpha_j^{-1})^{m_j} + Q_\Gamma(t - \beta_j)^{n_j}$ $(j = 1, 2)$.

The inclusion $\operatorname{Im} B_1 B_2 \subset \operatorname{Im} B_1 \cap \operatorname{Im} B_2$ is obvious, since B_1 and B_2 commute. So we are left with the reverse inclusion. Because $\alpha_1 \neq \alpha_2$ and $\beta_1 \neq \beta_2$, there are polynomials $R_j^{\pm}$ $(j = 1, 2)$ in $z^{\pm 1}$ such that

$$(z^{-1} - \alpha_1^{-1})^{m_1} R_1^-(z) + (z^{-1} - \alpha_2^{-1})^{m_2} R_2^-(z) \equiv 1,$$
$$(z - \beta_1)^{n_1} R_1^+(z) + (z - \beta_2)^{n_2} R_2^+(z) \equiv 1.$$

After putting $C_j = P_\Gamma R_j^- + Q_\Gamma R_j^+$ $(j = 1, 2)$ we have

$$B_1 C_1 + B_2 C_2 = I. \tag{13}$$

Now let $f = B_1 g_1 = B_2 g_2 \in \operatorname{Im} B_1 \cap \operatorname{Im} B_2$ and set $h = C_1 g_2 + C_2 g_1$. Then (13) implies that $f = B_1 B_2 h \in \operatorname{Im} B_1 B_2$ and this proves (12).

The assertion on the equivalence of the norms results from the estimates

$$\|h\| \leqq \|C_1\| \, \|g_2\| + \|C_2\| \, \|g_1\|$$

and

$$\|g_1\| \leqq \|B_2\| \, \|h\|, \qquad \|g_2\| \leqq \|B_1\| \, \|h\|;$$

here $\|\cdot\| := \|\cdot\|_{\mathbf{L}_p(\Gamma)}$. ∎

Corollary 4.2. *Let $\Gamma \in C^{\overline{m}}$, where $\overline{m} = \max\{m_1, \ldots, m_r, n_1, \ldots, n_r\}$. Then*

$$\mathbf{L}_p(\Gamma; (\Gamma_{\alpha_j}, m_j)_1^r, (\Gamma_{\beta_j}, n_j)_1^r) \subset \overline{\mathbf{L}}_p(\alpha, m; \beta, n) \tag{14}$$

with the embedding being continuous.[1])

Proof. By virtue of Theorem 4.2 it suffices to show the inclusion (14) for $r = 1$.

Given a function $f \in \mathbf{L}_p(\Gamma; (\Gamma_\alpha, m))$ $(\alpha \in \Gamma)$ we have

$$g(t) := \frac{f(t) - \sum_{k=0}^{m-1} \frac{1}{k!} f^{(k)}(\alpha) \, (t - \alpha)^k}{(t - \alpha)^m} \in \mathbf{L}_p(\Gamma)$$

and $\|g\|_{\mathbf{L}_p(\Gamma)} \leqq c \, \|f\|_{\mathbf{L}_p(\Gamma; (\Gamma_\alpha, m))}$ with some constant $c > 0$ (see, for example, Prössdorf [17], Lemma 1.2, Chapter 6). This combined with Theorem 6.1, Chapter II, and Theorem 4.1 readily gives the inclusion (14). ∎

Taking into consideration Lemma 3.2 it is easily seen that Theorem 4.1 can also be formulated as follows.

Theorem 4.3. *Let α, β be any points on Γ and let m, n be non-negative integers. Then for a function $f \in \mathbf{L}_p(\Gamma)$ to belong to the space $\overline{\mathbf{L}}_p(\alpha, m; \beta, n)$ it is necessary and sufficient that the following two conditions are satisfied:*

(i) *The mean derivatives $a_k = (P_\Gamma f)^{\{k\}}(\alpha)$ $(k = 0, 1, \ldots, m - 1)$ exist and*

$$\left[P_\Gamma f - \sum_{k=0}^{m-1} \frac{1}{k!} a_k (t - \alpha)^k\right] (t - \alpha)^{-m} \in \mathbf{L}_p(\Gamma).$$

(ii) *The mean derivatives $b_l = (Q_\Gamma f)^{\{l\}}(\beta)$ $(l = 0, 1, \ldots, n - 1)$ exist and*

$$\left[Q_\Gamma f - \sum_{l=0}^{n-1} \frac{1}{l!} b_l (t - \beta)^l\right] (t - \beta)^{-n} \in \mathbf{L}_p(\Gamma).$$

Corollary 4.3. *The inverse of the operator*

$$B = P_\Gamma (t^{-1} - \alpha^{-1})^m + Q_\Gamma (t - \beta)^n$$

[1]) For notation see Section 6.1, Chapter II.

is given by

$$B^{-1}f = (t^{-1} - \alpha^{-1})^{-m} \left[P_\Gamma f - \sum_{k=0}^{m-1} (t^m P_\Gamma f)^{\{k\}}(\alpha) \frac{(t-\alpha)^k}{k!\, t^m}\right]$$
$$+ (t-\beta)^{-n} \left[Q_\Gamma f - \sum_{l=0}^{n-1} (Q_\Gamma f)^{\{l\}}(\beta) \frac{(t-\beta)^l}{l!}\right]. \tag{15}$$

4.4. We finally consider the operator

$$\hat{B} = \hat{B}(\alpha, m; \beta, n) := (t^{-1} - \alpha^{-1})^m P_\Gamma + (t-\beta)^n Q_\Gamma.$$

If the functions $(t^{-1} - \alpha^{-1})^m$ and $(t-\beta)^n$ have no common zeros on Γ, then Theorem 1.3 of Chapter III and Lemma 4.2 imply that dim Ker $\hat{B} = 0$. Provided with the norm $\|\varphi\|_{\hat{\mathbf{L}}_p} = \|\hat{B}^{-1} f\|_{\mathbf{L}_p}$ the image Im $\hat{B}$ also becomes a Banach space. It will be denoted by $\hat{\mathbf{L}}_p(\alpha, m; \beta, n)$.

Theorem 4.4. *If $(t^{-1} - \alpha^{-1})^m$ and $(t-\beta)^n$ have no common zeros on Γ, then $\hat{\mathbf{L}}_p(\alpha, m; \beta, n) = \overline{\mathbf{L}}_p(\alpha, m; \beta, n)$ (more precisely: both spaces are topologically equivalent).*

Proof. Again abbreviate $(t^{-1} - \alpha^{-1})^m$ and $(t-\beta)^n$ to $\varrho_-(t)$ and $\varrho_+(t)$, respectively. These functions can be represented as

$$\varrho_- = \varrho_+ g_1 + h_+, \qquad \varrho_+ = \varrho_- g_2 + h_-$$

with certain $g_1, g_2 \in \mathbf{C}(\Gamma)$ and $h_\pm \in \mathbf{C}^\pm(\Gamma)$ (see § 3). By formula (2.5) we have

$$\hat{B} = B(I + T), \qquad T = Q_\Gamma g_1 P_\Gamma + P_\Gamma g_2 Q_\Gamma. \tag{16}$$

The operator T is compact on $\mathbf{L}_p(\Gamma)$, hence, by virtue of Atkinson's theorem, (16) implies that $\hat{B} \in \Phi(\mathbf{L}_p(\Gamma), \overline{\mathbf{L}}_p(\alpha, m; \beta, n))$ and Ind $\hat{B} = 0$. Since dim Ker $\hat{B} = 0$, we conclude that $\hat{B}$ is a continuous and invertible operator from $\mathbf{L}_p(\Gamma)$ onto $\overline{\mathbf{L}}_p(\alpha, m; \beta, n)$.

Because the operator $\hat{B} \in \mathscr{L}(\mathbf{L}_p(\Gamma), \hat{\mathbf{L}}_p(\alpha, m; \beta, n))$ has the same properties, there exists a continuous linear and one-to-one mapping of $\overline{\mathbf{L}}_p(\alpha, m; \beta, n)$ onto $\hat{\mathbf{L}}_p(\alpha, m; \beta, n)$, and therefore these spaces are topologically equivalent. ∎

Remark. (16) shows that always Im $\hat{B} \subset$ Im B. If the hypothesis of Theorem 4.4 is not satisfied, then Im $\hat{B} \neq$ Im B; indeed, in that case $1 \notin$ Im $\hat{B}$, whereas $1 \in$ Im B (see Corollary 4.2).

4.5. Theorem 4.5. *Let $a \in \mathbf{C}(\alpha, m; \beta, n)$. Then the commutator $T_a = aS_\Gamma - S_\Gamma a$ is both a compact operator from $\mathbf{L}_p(\Gamma)$ into $\overline{\mathbf{L}}_p(\alpha, m; \beta, n)$ and from $\tilde{\mathbf{L}}_p(\alpha, m; \beta, n)$ into $\mathbf{L}_p(\Gamma)$.*

Proof. We first prove that the commutator $[P_\Gamma, aI] = -\frac{1}{2} T_a$ is compact from $\mathbf{L}_p(\Gamma)$ into $\overline{\mathbf{L}}_p(\alpha, m; \beta, n)$. In view of Theorem 4.2 it is sufficient to consider the case $r = 1$.

Thus, let α, β be any points on Γ and m, n any non-negative integers. By means of induction with respect to k it can be easily deduced from the identities

$$[P_\Gamma, t^k I] = -\alpha P_\Gamma (t^{-1} - \alpha^{-1}) P_\Gamma t^k Q_\Gamma + \alpha P_\Gamma t^{k-1} Q_\Gamma \qquad (k = 1, 2, \ldots)$$

that the commutators $[P_\Gamma, t^k I]$ $(k = 1, 2, \ldots)$ map the space $\mathbf{L}_p(\Gamma)$ into the space Im $B = \overline{\mathbf{L}}_p(\alpha, m; \beta, n)$. Being finite-rank operators these commutators are compact.

On account of property 2°, § 3, the function $a \in \mathbf{C}(\alpha, m; \beta, n)$ admits the representations

$$a(t) = q_1(t) + g(t)\,(t^{-1} - \alpha^{-1})^m = q_2(t) + h(t)\,(t-\beta)^n \tag{17}$$

with $g, h \in \mathbf{C}(\Gamma)$ and certain polynomials (in t) q_1, q_2. Hence

$$[P_\Gamma, aI] = [P_\Gamma, q_1 I] + P_\Gamma (t^{-1} - \alpha^{-1})^m [P_\Gamma, gI] - Q_\Gamma (t^{-1} - \alpha^{-1})^m g P_\Gamma.$$

By what has already been proved, the commutator $[P_\Gamma, q_1 I]$ is compact from $\mathbf{L}_p(\Gamma)$ into $\overline{\mathbf{L}}_p(\alpha, m; \beta, n)$. The same is valid for the operator

$$P_\Gamma(t^{-1} - \alpha^{-1})^m [P_\Gamma, gI] = BP_\Gamma[P_\Gamma, gI],$$

since $[P_\Gamma, gI]$ is compact on $\mathbf{L}_p(\Gamma)$. Using (17) we finally obtain

$$Q_\Gamma(t^{-1} - \alpha^{-1})^m gP_\Gamma = Q_\Gamma(q_2 - q_1) P_\Gamma + BQ_\Gamma hP_\Gamma,$$

and consequently, this operator is also compact from $\mathbf{L}_p(\Gamma)$ into $\overline{\mathbf{L}}_p(\alpha, m; \beta, n)$.

Making use of the identities

$$[P_\Gamma, t^{-k}I] = \alpha^{-1} Q_\Gamma t^{-k} P_\Gamma(t - \alpha) P_\Gamma - \alpha^{-1} Q_\Gamma t^{-k+1} P_\Gamma \qquad (k = 1, 2, \ldots)$$

it can be shown with the help of analogous arguments that the commutator $[P_\Gamma, aI]$ is a compact operator from $\tilde{\mathbf{L}}_p(\alpha, m; \beta, n)$ into $\mathbf{L}_p(\Gamma)$. ∎

Corollary 4.4. *The operator of multiplication by a function $a \in \mathbf{C}(\alpha, m; \beta, n)$ is bounded both on $\overline{\mathbf{L}}_p(\alpha, m; \beta, n)$ and $\tilde{\mathbf{L}}_p(\alpha, m; \beta, n)$.*

This is an immediate consequence of Theorem 4.5 and the identities

$$aB = Ba + [P_\Gamma, aI] \{(t - \beta)^n - (t^{-1} - \alpha^{-1})^m\},$$

$$aD = Da + \{(t^{-1} - \beta^{-1})^n - (t - \alpha)^m\} [P_\Gamma, aI].$$

Remark. The spaces of the type $\overline{\mathbf{L}}_p$ and $\tilde{\mathbf{L}}_p$ were first and simultaneously introduced by Prössdorf [13] and Sementsul [1]. The proof of Lemma 4.1 is due to Sementsul [1]. Theorem 4.4 is Silbermann's [8]. All remaining results of the present section were established by Prössdorf [13–14, 16, 17].

§ 5. Singular operators with degenerate continuous coefficients

Let Γ be a closed piecewise Lyapunov curve system, let $\alpha = (\alpha_1, \alpha_2, \ldots, \alpha_r)$ and $\beta = (\beta_1, \beta_2, \ldots, \beta_r)$ be systems of points on Γ, and let m, n be multi-indices of dimension r.

5.1. For singular integral operators with coefficients from the algebra $\mathbf{C}(\alpha, m; \beta, n)$, Theorem 3.1 of Chapter III can be carried over to the spaces $\overline{\mathbf{L}}_p(\alpha, m; \beta, n)$ and $\tilde{\mathbf{L}}_p(\alpha, m; \beta, n)$.

Theorem 5.1. *Suppose a and b are functions from $\mathbf{C}(\alpha, m; \beta, n)$ and let $A = aP_\Gamma + bQ_\Gamma$ ($A = P_\Gamma a + Q_\Gamma b$). Then A is a linear bounded operator both on $\overline{\mathbf{L}}_p(\alpha, m; \beta, n)$ and $\tilde{\mathbf{L}}_p(\alpha, m; \beta, n)$.*

In order that A is a Φ_+- or Φ_--operator on these spaces it is necessary and sufficient that

$$a(t) \neq 0, \quad b(t) \neq 0 \quad (t \in \Gamma). \tag{1}$$

If conditions (1) are satisfied, then A has the same index on each of the spaces $\mathbf{L}_p(\Gamma)$, $\overline{\mathbf{L}}_p(\alpha, m; \beta, n)$, and $\tilde{\mathbf{L}}_p(\alpha, m; \beta, n)$.

Proof. From Theorem 4.5 and Corollary 4.4 we deduce that $AB = BA + T_1$ and $AD = DA + T_2$ with certain compact operators $T_1: \mathbf{L}_p(\Gamma) \to \overline{\mathbf{L}}_p(\alpha, m; \beta, n)$ and $T_2: \tilde{\mathbf{L}}_p(\alpha, m; \beta, n) \to \mathbf{L}_p(\Gamma)$. Consequently, A is a $\Phi_\pm$-operator on $\overline{\mathbf{L}}_p(\alpha, m; \beta, n)$ or $\tilde{\mathbf{L}}_p(\alpha, m; \beta, n)$, if and only if it is a $\Phi_\pm$-operator on $\mathbf{L}_p(\Gamma)$. Now all assertions of the theorem follow from Theorem 3.1, Chapter III. ∎

Our next concern is the case where the continuous coefficients a and b of the singular integral operator have finitely many zeros on Γ.

Theorem 5.2. *Suppose the coefficients a and b ($\in \mathbf{C}(\alpha, m; \beta, n)$) are of the form*

$$a(t) = (t^{-1} - \alpha^{-1})^m\, c(t), \qquad b(t) = (t - \beta)^n\, d(t) \tag{2}$$

with certain functions $c \in \mathbf{C}(\beta, n)$ and $d \in \mathbf{C}(\alpha, m)$ vanishing nowhere on Γ.

Then, for any multi-indices m_k and n_k ($k = 1, 2$) such that $m_1 + m_2 = m$ and $n_1 + n_2 = n$, the operator $A = aP_\Gamma + bQ_\Gamma$ ($A = P_\Gamma a + Q_\Gamma b$) is a continuous Φ-operator from $\tilde{\mathbf{L}}_p(\alpha, m_1; \beta, n_1)$ to $\overline{\mathbf{L}}_p(\alpha, m_2; \beta, n_2)$ with index

$$\operatorname{Ind} A = \operatorname{ind} d/c + |m_1| + |n_1|. \tag{3}$$

If $\alpha_j \neq \beta_k$ for $j, k = 1, \ldots, r$, then A is at least one-sided invertible on that pair of spaces.

Proof. Write the coefficients a and b in the form

$$a(t) = (t^{-1} - \alpha^{-1})^{m_2}\, (t - \alpha)^{m_1}\, c_0(t),$$
$$b(t) = (t - \beta)^{n_2}\, (t^{-1} - \beta^{-1})^{n_1}\, d_0(t)$$

where $c_0(t) = c(t)\,(-t\alpha)^{-m_1}$, $d_0(t) = d(t)\,(-t\beta)^{n_1}$. Furthermore, put

$$B = B(\alpha, m_2; \beta, n_2), \qquad D = D(\alpha, m_1; \beta, n_1)$$

(see (4.4) and (4.7)).

Using the factorization (2.1) we get

$$A = (c_1 P_\Gamma + d_1 Q_\Gamma)\, D \tag{4}$$

with $c_1(t) = (t^{-1} - \alpha^{-1})^{m_2}\, c_0(t)$, $d_1(t) = (t - \beta)^{n_2}\, d_0(t)$. Obviously, $c_1 \in \mathbf{C}(\alpha, m_2; \beta, n)$ and $d_1 \in \mathbf{C}(\alpha, m; \beta, n_2)$, and Theorem 4.5 gives

$$c_1 P_\Gamma + d_1 Q_\Gamma = P_\Gamma c_1 + Q_\Gamma d_1 + T \tag{5}$$

with a compact operator $T\colon \mathbf{L}_p(\Gamma) \to \overline{\mathbf{L}}_p(\alpha, m_2; \beta, n_2)$. From (4), (5), and (2.4) we obtain

$$A = [B(P_\Gamma c_0 + Q_\Gamma d_0) + T]\, D.$$

This combined with Theorem 3.1, Chapter III, shows that $A : \tilde{\mathbf{L}}_p(\alpha, m_1; \beta, n_1) \to \overline{\mathbf{L}}_p(\alpha, m_2; \beta, n_2)$ is a continuous Φ-operator whose index is

$$\operatorname{Ind} A = \operatorname{ind} d_0/c_0 = \operatorname{ind} d/c + |m_1| + |n_1|.$$

If $\alpha_j \neq \beta_k$ ($j, k = 1, \ldots, r$), then the functions c_1 and d_1 have no common zeros on Γ and hence, by Theorem 1.3 of Chapter III,

$$\dim \operatorname{Ker} (c_1 P_\Gamma + d_1 Q_\Gamma) = \dim \operatorname{Ker} (P_\Gamma c_1 + Q_\Gamma d_1) = \dim \operatorname{Ker} (P_\Gamma c_0 + Q_\Gamma d_0).$$

Again applying Theorem 3.1 of Chapter III we see that $\dim \operatorname{Ker} A = \max(-\varkappa, 0)$ and $\dim \operatorname{Coker} A = \max(\varkappa, 0)$ with $\varkappa = \operatorname{ind} c_0/d_0$.

This completes the proof for the operator $A = aP_\Gamma + bQ_\Gamma$. The proof for the operator $A = P_\Gamma a + Q_\Gamma b$ is analogous, since, for arbitrary coefficients $c, d \in \mathbf{C}(\alpha, m; \beta, n)$ which have no common zeros on Γ, the kernels of the operators $cP_\Gamma + dQ_\Gamma$ and $P_\Gamma c + Q_\Gamma d$, both considered as acting on the space $\tilde{\mathbf{L}}_p(\alpha, m; \beta, n)$ (see Theorem 5.1), have equal dimension (see the proof of Theorem 1.3, Chapter III). ∎

Remark 1. Another proof of Theorem 5.2, which also applies to the more general case of piecewise continuous coefficients, will be given in § 6. It can also be shown that for the singular integral operator A with coefficients $a, b \in \mathbf{C}(\alpha, m; \beta, n)$ to be a Φ-operator from $\tilde{\mathbf{L}}_p(\alpha, m_1; \beta, n_1)$ to $\overline{\mathbf{L}}_p(\alpha, m_2; \beta, n_2)$

it is necessary that the coefficients a, b admit the representation (2) (see PRÖSSDORF [17], Theorem 4.3, Chapter IV).

5.2. Now let a and b be functions which admit a representation in the form (2) and furthermore suppose that a and b belong to the class $\mathbf{C}^l(\Gamma)$, where $l = \max(m_1, \ldots, m_r, n_1, \ldots, n_r)$. Then the singular operator $A = aP_\Gamma + bQ_\Gamma$ $(A = P_\Gamma a + Q_\Gamma b)$ generates a closed operator $\mathbf{A} : \mathbf{L}_p(\Gamma) \to \mathbf{W}_p^{(l)}(\Gamma)$ defined by

$$\mathbf{A}\varphi = A\varphi, \qquad D(\mathbf{A}) = \{\varphi \in \mathbf{L}_p(\Gamma) : A\varphi \in \mathbf{W}_p^{(l)}(\Gamma)\}.$$

Taking into account that the embedding $\mathbf{W}_p^{(l)}(\Gamma) \subset \overline{\mathbf{L}}_p(\alpha, m; \beta, n)$ is dense (see Corollary 4.2), it is easy to derive the following theorem from Theorem 5.2.

Theorem 5.3. *The operator* $\mathbf{A} : \mathbf{L}_p(\Gamma) \to \mathbf{W}_p^{(l)}(\Gamma)$ *is a closed operator with index* $\operatorname{Ind} \mathbf{A} = \operatorname{ind} d/c$.

Remark 2. The results presented in this section are essentially due to PRÖSSDORF [10, 13–14] (see also [17]). These results were generalized to the case where the coefficients a and b have a finite number of zeros of non-integral orders by PRÖSSDORF and SILBERMANN [1, 2]. S. MEYER [1, 2] considered zeros of logarithmic type and also situations where the coefficients a, b vanish identically on whole arcs. For further results and for remarks concerning the history of the theory of singular integral operators with degenerate continuous coefficients we refer to PRÖSSDORF [17], Chapter 6.

Remark 3. In the case where the coefficients of the singular operator have common zeros it is advantageous to consider (in addition to the function spaces iutroduced in § 4) two further Banach spaces:

$$\overline{\mathbf{L}}_p(\alpha, m) := \{\varphi := (t - \alpha)^m f : f \in \mathbf{L}_p(\Gamma)\}$$

with the norm $\|\varphi\|_{\overline{\mathbf{L}}_p(\alpha,m)} = \|f\|_{\mathbf{L}_p(\Gamma)}$ and

$$\widetilde{\mathbf{L}}_p(\alpha, m) := \{\psi := (t - \alpha)^{-m} f : f \in \mathbf{L}_p(\Gamma)\}$$

with the norm $\|\psi\|_{\widetilde{\mathbf{L}}_p(\alpha,m)} = \|f\|_{\mathbf{L}_p(\Gamma)}$.

Let $\alpha_j = \beta_j$ $(j = 1, \ldots, r; 1 \leqq s \leqq r)$ be the common zeros of the functions a and b, let $l_j = \min(m_j, n_j)$, and let $l = (l_1, \ldots, l_s, 0, \ldots, 0)$ be the corresponding multi-index of dimension r. Then the functions $(t - \alpha)^{m-l}$ and $(t - \beta)^{n-l}$ have no common zeros on Γ. The functions a, b can be represented in the form

$$a(t) = (t - \alpha)^l (t - \alpha)^{m-l} c_0(t), \qquad b(t) = (t - \alpha)^l (t^{-1} - \beta^{-1})^{n-l} d_0(t)$$

with certain functions c_0 and d_0 vanishing nowhere on Γ. Accordingly, the operator $A = aP_\Gamma + bQ_\Gamma$ can be written in the form

$$A = (t - \alpha)^l (c_0 P_\Gamma + d_0 Q_\Gamma) D(\alpha, m - l; \beta, n - l)$$

and this implies immediately that the operator

$$aP_\Gamma + bQ_\Gamma : \widetilde{\mathbf{L}}_p(\alpha, m - l; \beta, n - l) \to \overline{\mathbf{L}}_p(\alpha, l)$$

is continuous, at least one-sided invertible, and has the index $\operatorname{Ind} A = \operatorname{ind} d_0/c_0$. It is easy to see that $\operatorname{ind} d_0/c_0 = \operatorname{ind} d/c + |m| + |n - l|$ (see (2) and (3)).

An analogous result can be stated for the operator

$$P_\Gamma a + Q_\Gamma b : \widetilde{\mathbf{L}}_p(\alpha, l) \to \overline{\mathbf{L}}_p(\alpha, m - l; \beta, n - l).$$

In connection with certain systems of singular integral equations, in § 3 of Chapter VII we shall return to these ideas once more.

Remark 4. Note that all results and arguments of the Sections 4 and 5 remain valid if everywhere $\mathbf{L}_p(\Gamma)$ and $\mathbf{C}(\Gamma)$ is formally replaced by $\mathbf{H}^\lambda(\Gamma)$ $(0 < \lambda < 1)$ and $\mathbf{W}_p^{(l)}(\Gamma)$ by $\mathbf{C}^{l,\lambda}(\Gamma)$.

§ 6. Singular operators with degenerate piecewise continuous coefficients

In the present section we shall generalize the results of § 5 to the case of piecewise continuous coefficients. We retain the notation introduced in the preceding sections of this chapter.

6.1. Let Γ be a closed piecewise Lyapunov curve system. On the space $\mathbf{L}_p(\Gamma)$ $(1 < p < \infty)$ we consider the singular integral operator $A = aP_\Gamma + bQ_\Gamma$ with coefficients $a, b \in \mathbf{PC}(\Gamma)$ of the following form

$$a(t) = (t^{-1} - \alpha^{-1})^m\, c(t), \qquad b(t) = (t - \beta)^n\, d(t). \tag{1}$$

Herein it is supposed that

$$c \in \mathbf{PC}(\beta, n), \qquad d \in \mathbf{PC}(\alpha, m) \tag{2}$$

and that the functions $\varrho_-(t) = (t^{-1} - \alpha^{-1})^m$ and $\varrho_+(t) = (t - \beta)^n$ have no common zeros on Γ:

$$\{t \in \Gamma : \varrho_+(t) = 0\} \cap \{t \in \Gamma : \varrho_-(t) = 0\} = \varnothing\,. \tag{3}$$

Theorem 6.1. *Let the conditions* (1)–(3) *be satisfied. Then* $A = aP_\Gamma + bQ_\Gamma \in \mathscr{L}(\mathbf{L}_p(\Gamma), \overline{\mathbf{L}}_p(\alpha, m; \beta, n))$ *and this operator is a* Φ_+-, Φ_--, *or* Φ-*operator, if and only if*

$$\det C(t, \mu) \neq 0 \qquad (t \in \Gamma, 0 \leq \mu \leq 1), \tag{4}$$

where $C(t, \mu)$ *denotes the symbol of the operator* $C = cP_\Gamma + dQ_\Gamma \in \mathscr{L}(\mathbf{L}_p(\Gamma))$.

If the condition (4) *is satisfied, then the operator* $A \in \mathscr{L}(\mathbf{L}_p(\Gamma), \overline{\mathbf{L}}_p(\alpha, m; \beta, n))$ *is at least one-sided invertible and* Ind A = Ind C.

The proof of this theorem is based on the following auxiliary proposition.

Lemma 6.1. *Let* $c = c_1c_2 + c_3$, $d = d_1d_2 + d_3$, *where* $c_1, d_1 \in \mathbf{PC}(\Gamma)$ *and* $c_j, d_j \in \mathbf{C}(\Gamma)$ $(j = 2, 3)$. *Then the operators*

$$C = cP_\Gamma + dQ_\Gamma \quad \text{and} \quad C_1 = C + (c_2 - d_2)(P_\Gamma d_1 Q_\Gamma - Q_\Gamma c_1 P_\Gamma)$$

are either simultaneously $\Phi_\pm$-*operators on* $\mathbf{L}_p(\Gamma)$ *or not. If* $C \in \Phi(\mathbf{L}_p(\Gamma))$, *then* Ind C = Ind C_1.

Proof. It is easy to see that C_1 is an operator of the form (7.1) of Chapter V. Thus, for the symbols of the operators C and C_1 we have (cf. § 7, Chapter V)

$$C(t, \mu) = \begin{pmatrix} c_{11}(t, \mu) & h_p'(\mu)\, c_2(t)\, (d_1(t+0) - d_1(t)) \\ h_p'(\mu)\, d_2(t)\, (c_1(t+0) - c_1(t)) & c_{22}(t, 1-\mu) \end{pmatrix},$$

$$C_1(t, \mu) = \begin{pmatrix} c_{11}(t, \mu) & h_p'(\mu)\, d_2(t)\, (d_1(t+0) - d_1(t)) \\ h_p'(\mu)\, c_2(t)\, (c_1(t+0) - c_1(t)) & c_{22}(t, 1-\mu) \end{pmatrix},$$

where $c_{11}(t, \mu) = f_p'(\mu)\, c(t+0) + (1 - f_p'(\mu))\, c(t)$, $c_{22}(t, 1-\mu) = (1 - f_p'(\mu))\, d(t+0) + f_p'(\mu)\, d(t)$, and where we defined $f_p'(\mu) = f_{2\pi/\delta}(\mu)$ and $h_p'(\mu) = h_{2\pi/\delta}(\mu)$ (see § 6, Chapter V). Obviously, det $C(t, \mu)$ = det $C_1(t, \mu)$. Taking into account that both matrices have the same principal diagonal, Lemma 6.1 is seen to be a direct consequence of the Theorems 6.3 and 7.1 of Chapter V. ∎

Proof of Theorem 6.1. The representations (2) and the results of § 3 yield the representations

$$c = \varrho_+ c_1 + h_+, \qquad d = \varrho_- d_1 + h_- \tag{5}$$

with certain functions $c_1, d_1 \in \mathbf{PC}(\Gamma)$ and $h_\pm \in \mathbf{C}^\pm(\Gamma)$. Formula (2.6) gives the factorization

$$A = (\varrho_- P_\Gamma + \varrho_+ Q_\Gamma)\,[cP_\Gamma + dQ_\Gamma + (\varrho_+ - \varrho_-)\,(P_\Gamma\, d_1 Q_\Gamma - Q_\Gamma c_1 P_\Gamma)]. \qquad (6)$$

As was shown in the proof of Theorem 4.4, the operator $\hat{B} := \varrho_- P_\Gamma + \varrho_+ Q_\Gamma \in \mathscr{L}(\mathbf{L}_p(\Gamma), \overline{\mathbf{L}}_p(\alpha, m; \beta, n))$ is invertible. Thus, we deduce from (6) that $A \in \mathscr{L}(\mathbf{L}_p(\Gamma), \overline{\mathbf{L}}_p(\alpha, m; \beta, n))$ and that this operator is a Φ_+-, Φ_--, or Φ-operator if and only if $C_1 = cP_\Gamma + dQ_\Gamma + (\varrho_+ - \varrho_-)\,(P_\Gamma\, d_1 Q_\Gamma - Q_\Gamma c_1 P_\Gamma)$ is such an operator on $\mathbf{L}_p(\Gamma)$. By Theorem 6.3 of Chapter V and Lemma 6.1, the operator C_1 is a Φ_+-, Φ_--, or Φ-operator on $\mathbf{L}_p(\Gamma)$ if and only if (4) holds; in that case $\operatorname{Ind} A = \operatorname{Ind} C$.

Now let condition (4) be satisfied. We shall show that then the operator $A \in \mathscr{L}(\mathbf{L}_p(\Gamma), \overline{\mathbf{L}}_p(\alpha, m; \beta, n))$ is at least one-sided invertible. In view of the decomposition (6) it remains to prove that the operator C_1 is at least one-sided invertible on $\mathbf{L}_p(\Gamma)$. Since $\operatorname{Ind} C = \operatorname{Ind} C_1$ (Lemma 6.1) and since C is one-sided invertible on $\mathbf{L}_p(\Gamma)$ (Theorem 6.3, Chapter V), this, on its hand, will follow as soon as we have proved that

$$\dim \operatorname{Ker} C = \dim \operatorname{Ker} C_1. \qquad (7)$$

The latter equality will be shown to follow from the identity

$$(\varrho_+ P_\Gamma + \varrho_- Q_\Gamma)\, C = C_1 (\varrho_+ P_\Gamma + \varrho_- Q_\Gamma), \qquad (8)$$

which, in turn, can be easily verified.

To see that (7) results from (8), note first that (8) and $\dim \operatorname{Ker} (\varrho_+ P_\Gamma + \varrho_- Q_\Gamma) = 0$ immediately imply that $\dim \operatorname{Ker} C \leqq \dim \operatorname{Ker} C_1$. Passing to the adjoint operators in (8) and taking into account that $\dim \operatorname{Ker} (\varrho_+ P_\Gamma + \varrho_- Q_\Gamma)^* = \dim \operatorname{Ker} (P_\Gamma \varrho_- + Q_\Gamma \varrho_+) = 0$ (see Theorem 5.1, Chapter II, and Lemma 4.2), we obtain $\dim \operatorname{Coker} C_1 \leqq \dim \operatorname{Coker} C$. This and the equality $\operatorname{Ind} C = \operatorname{Ind} C_1$ give (7). ∎

6.2. In this subsection we consider the singular integral operator $A = P_\Gamma a + Q_\Gamma b$. Its coefficients are supposed to be of the form

$$a(t) = (t - \alpha)^m\, c(t), \qquad b(t) = (t^{-1} - \beta^{-1})^n\, d(t). \qquad (9)$$

Theorem 6.2. *Let the conditions* (2), (3), *and* (9) *be fulfilled. Then* $A = P_\Gamma a + Q_\Gamma b \in \mathscr{L}(\tilde{\mathbf{L}}_p(\alpha, m; \beta, n), \mathbf{L}_p(\Gamma))$ *and this operator is a* Φ_+-, Φ_--, *or* Φ-*operator, if and only if*

$$\det C(t, \mu) \neq 0 \qquad (t \in \Gamma, 0 \leqq \mu \leqq 1), \qquad (10)$$

with $C(t, \mu)$ *the symbol of the operator* $C = P_\Gamma c + Q_\Gamma d \in \mathscr{L}(\mathbf{L}_p(\Gamma))$.

If condition (10) *is satisfied, then the operator* $A \in \mathscr{L}(\tilde{\mathbf{L}}_p(\alpha, m; \beta, n), \mathbf{L}_p(\Gamma))$ *is at least one-sided invertible and* $\operatorname{Ind} A = \operatorname{Ind} C$.

To prove this theorem, we need the following analogue of Lemma 6.1.

Lemma 6.2. *Under the hypotheses of Lemma 6.1, the operators*

$$C = P_\Gamma c + Q_\Gamma d \quad \text{and} \quad C_1 = C + (Q_\Gamma d_1 P_\Gamma - P_\Gamma c_1 Q_\Gamma)\,(d_2 - c_2)$$

are either both $\Phi_\pm$-*operators on the space* $\mathbf{L}_p(\Gamma)$ *or not. If* $C \in \Phi(\mathbf{L}_p(\Gamma))$, *then* $\operatorname{Ind} C = \operatorname{Ind} C_1$.

Proof. The proof is the same as that of Lemma 6.1, since the symbols of the operators C and C_1 are now of the following form (again see §§ 6, 7, Chapter V):

$$C(t, \mu) = \begin{pmatrix} c_{11}(t, \mu) & h_p'(\mu)\, d_2(t)\, (c_1(t+0) - c_1(t)) \\ h_p'(\mu)\, c_2(t)\, (d_1(t+0) - d_1(t)) & c_{22}(t, 1-\mu) \end{pmatrix},$$

$$C_1(t, \mu) = \begin{pmatrix} c_{11}(t, \mu) & h_p'(\mu)\, c_2(t)\, (c_1(t+0) - c_1(t)) \\ h_p'(\mu)\, d_2(t)\, (d_1(t+0) - d_1(t)) & c_{22}(t, 1-\mu) \end{pmatrix}$$

with $c_{11}(t, \mu) = f_p'(\mu)\, c(t + 0) + (1 - f_p'(\mu))\, c(t)$, $c_{22}(t, 1 - \mu) = (1 - f_p'(\mu))\, d(t + 0) + f_p'(\mu)\, d(t)$. ▮

Proof of Theorem 6.2. We abbreviate $(t - \alpha)^m$ and $(t^{-1} - \beta^{-1})^n$ to $\varrho_+(t)$ and $\varrho_-(t)$, respectively. In view of (2) we then have the representations (see § 3) $c = \varrho_- c_1 + h_-$, $d = \varrho_+ d_1 + h_+$ with certain functions $c_1, d_1 \in \mathbf{PC}(\Gamma)$ and $h_\pm \in \mathbf{C}^\pm(\Gamma)$. (2.3) gives the decomposition

$$A = [P_\Gamma c + Q_\Gamma d + (Q_\Gamma d_1 P_\Gamma - P_\Gamma c_1 Q_\Gamma)\,(\varrho_- - \varrho_+)]\,(P_\Gamma \varrho_+ + Q_\Gamma \varrho_-). \tag{11}$$

Now extend the operator $D_1 = P_\Gamma \varrho_+ + Q_\Gamma \varrho_-$ continuously to the operator $\tilde{D}_1 = P_\Gamma \varrho_+ \tilde{P}_\Gamma + Q_\Gamma \varrho_- \tilde{Q}_\Gamma$ (cf. Section 4.2). It is clear that $\tilde{D}_1$ maps the space $\tilde{\mathbf{L}}_p(\alpha, m; \beta, n)$ onto $\mathbf{L}_p(\Gamma)$ isometrically. Thus, all assertions of the lemma, excepted the one-sided invertibility of the operator A, follow from the representation (11) and Lemma 6.2. Using the equality $C(P_\Gamma \varrho_- + Q_\Gamma \varrho_+) = (P_\Gamma \varrho_- + Q_\Gamma \varrho_+)\, C_1$, it can be shown in the same fashion as in the proof of Theorem 6.1 that A is one-sided invertible. ▮

6.3. By writing the coefficients as in the proof of Theorem 5.2, taking into consideration the formulas (2.1), (2.4), and applying the Theorems 6.1 and 6.2 we obtain the following analogue of Theorem 5.2.

Theorem 6.3. *Suppose the conditions* (1)–(3) *are satisfied, let* $A = aP_\Gamma + bQ_\Gamma$ $(A = P_\Gamma a + Q_\Gamma b)$, *and let* m_k, n_k $(k = 1, 2)$ *be any multi-indices such that* $m_1 + m_2 = m$, $n_1 + n_2 = n$.

Then $A \in \mathcal{L}(\tilde{\mathbf{L}}_p(\alpha, m_1; \beta, n_1), \bar{\mathbf{L}}_p(\alpha, m_2; \beta, n_2))$ *and this operator is a* Φ_+-, Φ_--, *or* Φ-*operator if and only if the symbol of the operator* $C = cP_\Gamma + dQ_\Gamma$ $(C = P_\Gamma c + Q_\Gamma d) \in \mathcal{L}(\mathbf{L}_p(\Gamma))$ *satisfies the condition* (4).

If condition (4) *is fulfilled, then the operator* $A \in \mathcal{L}(\tilde{\mathbf{L}}_p(\alpha, m_1; \beta, n_1), \bar{\mathbf{L}}_p(\alpha, m_2; \beta, n_2))$ *is at least one-sided invertible and* $\operatorname{Ind} A = \operatorname{Ind} C + |m_1| + |n_1|$.

Remark 1. All results of this section are due to Silbermann [5–8], who also considered the case of zeros α, β of non-integral orders (see also Prössdorf [20]).

Remark 2. M. I. Haikin [4, 5] (see also [6]) studied singular operators of the form $A = aP_\Gamma + bQ_\Gamma$ $(a, b \in \mathbf{PC}(\Gamma))$ on the space $\mathbf{L}_p(\Gamma)$ with symbols degenerating in another sense: he assumed that

$$\inf_\Gamma |a(t)| > 0, \qquad \inf_\Gamma |b(t)| > 0,$$

but that the function a/b is not p-regular.

§ 7. Singular operators with degenerate coefficients on non-closed curves

The purpose of this section is to extend the results of § 6 to the case of non-closed curves. To do this, we apply the method of altering the integration curve described in § 1 of Chapter IV.

7.1. Let $\tilde{\Gamma}$ be a closed piecewise Lyapunov curve system and $\Gamma \subset \tilde{\Gamma}$ a part of $\tilde{\Gamma}$ containing certain open curves. Assume Γ is closed, that is, assume the end points of the open subcurves of Γ belong to Γ.

It is well-known that $\mathbf{L}_p(\tilde{\Gamma}) = \mathbf{L}_p(\Gamma) \dotplus \mathbf{L}_p(\tilde{\Gamma} \setminus \Gamma)$ $(1 < p < \infty)$. Let $\tilde{P}$ denote the projection of $\mathbf{L}_p(\tilde{\Gamma})$ onto $\mathbf{L}_p(\Gamma)$ parallel to $\mathbf{L}_p(\tilde{\Gamma} \setminus \Gamma)$ and denote by $\tilde{Q}$ the complementary projection:

$$(\tilde{P}\varphi)(t) = \begin{cases} \varphi(t), & t \in \Gamma, \\ 0, & t \in \tilde{\Gamma} \setminus \Gamma \end{cases}; \qquad (\tilde{Q}\varphi)(t) = \begin{cases} 0, & t \in \Gamma, \\ \varphi(t), & t \in \tilde{\Gamma} \setminus \Gamma \end{cases}$$

($\varphi \in \mathbf{L}_p(\tilde{\Gamma})$). Furthermore, let $\alpha = (\alpha_1, \alpha_2, \ldots, \alpha_r)$ and $\beta = (\beta_1, \beta_2, \ldots, \beta_r)$ be systems consisting of inner points of the curve Γ and let m, n be any multi-indices of dimension r. Finally suppose the condition (6.3) is fulfilled, that is, $(t - \alpha)^m$ and $(t - \beta)^n$ have no common zeros on Γ.

Consider the spaces introduced in § 4,[1])

$$\overline{\mathbf{L}}_p(\tilde{\Gamma}; \alpha, m; \beta, n) := \operatorname{Im}\left(P_{\tilde{\Gamma}}(t^{-1} - \alpha^{-1})^m + Q_{\tilde{\Gamma}}(t - \beta)^n\right),$$

$$\tilde{\mathbf{L}}_p(\tilde{\Gamma}; \alpha, m; \beta, n) := \operatorname{Im}\left((t - \alpha)^{-m} P_{\tilde{\Gamma}} + (t^{-1} - \beta^{-1})^{-n} Q_{\tilde{\Gamma}}\right)$$

and define

$$\overline{\mathbf{L}}_p(\Gamma; \alpha, m; \beta, n) := \tilde{P}\overline{\mathbf{L}}_p(\tilde{\Gamma}; \alpha, m; \beta, n), \tilde{\mathbf{L}}_p(\Gamma; \alpha, m; \beta, n) := \tilde{P}\tilde{\mathbf{L}}_p(\tilde{\Gamma}; \alpha, m; \beta, n).$$

For the sake of brevity, denote these spaces by $\overline{\mathbf{L}}_p(\tilde{\Gamma})$, $\tilde{\mathbf{L}}_p(\tilde{\Gamma})$, $\overline{\mathbf{L}}_p(\Gamma)$, and $\tilde{\mathbf{L}}_p(\Gamma)$, respectively.

Lemma 7.1. *We have*

$$\overline{\mathbf{L}}_p(\tilde{\Gamma}) = \overline{\mathbf{L}}_p(\Gamma) \dot{+} \mathbf{L}_p(\tilde{\Gamma} \setminus \Gamma), \qquad \tilde{\mathbf{L}}_p(\tilde{\Gamma}) = \tilde{\mathbf{L}}_p(\Gamma) \dot{+} \mathbf{L}_p(\tilde{\Gamma} \setminus \Gamma). \tag{1}$$

Proof. Let us prove the first of the relations (1). Obviously, $\tilde{Q}\varphi \in \overline{\mathbf{L}}_p(\tilde{\Gamma})$ for every $\varphi \in \mathbf{L}_p(\tilde{\Gamma})$ (see Corollary 4.2) and, hence, $\mathbf{L}_p(\tilde{\Gamma} \setminus \Gamma) \subset \overline{\mathbf{L}}_p(\tilde{\Gamma})$. Therefore $\tilde{P}\psi = \psi - \tilde{Q}\psi \in \overline{\mathbf{L}}_p(\tilde{\Gamma})$ for every $\psi \in \overline{\mathbf{L}}_p(\tilde{\Gamma})$ and thus $\overline{\mathbf{L}}_p(\Gamma) \subset \overline{\mathbf{L}}_p(\tilde{\Gamma})$.

Since $\overline{\mathbf{L}}_p(\tilde{\Gamma})$ is continuously embedded in $\mathbf{L}_p(\tilde{\Gamma})$, the operator $\tilde{Q}$ is closed on the space $\overline{\mathbf{L}}_p(\tilde{\Gamma})$ and thus we conclude from the closed graph theorem that $\tilde{Q}$ is continuous on $\overline{\mathbf{L}}_p(\tilde{\Gamma})$. This operator is obviously a projection, that is, $\mathbf{L}_p(\tilde{\Gamma} \setminus \Gamma)$ is closed in the norm of $\overline{\mathbf{L}}_p(\tilde{\Gamma})$. Banach's theorem on the inverse operator implies that this norm is equivalent to the usual norm in $\mathbf{L}_p(\tilde{\Gamma} \setminus \Gamma)$. Furthermore, from $\tilde{P} = I - \tilde{Q}$ we conclude that $\tilde{P} \in \mathscr{L}(\overline{\mathbf{L}}_p(\tilde{\Gamma}))$ and $\tilde{P}^2 = \tilde{P}$. This proves the first relation in (1). The second one can be proved similarly. ∎

Corollary 7.1. *We have*

$$\mathbf{L}_p(\Gamma; (\Gamma_{\alpha_j}, m_j)_1^r, (\Gamma_{\beta_j}, n_j)_1^r) \subset \overline{\mathbf{L}}_p(\Gamma; (\alpha, m); (\beta, n)),$$

the embedding being continuous.

7.2. We now consider the singular integral operator

$$A = aP_\Gamma + bQ_\Gamma \qquad (A = P_\Gamma a + Q_\Gamma b)$$

with coefficients $a, b \in \mathbf{PC}(\Gamma)$, which can be assumed to be continuous at the end points of the open subcurves of Γ. In accordance with the procedure described in Section 1.1, Chapter IV, with the operator A we associate the operator $\tilde{A} = \tilde{a}P_{\tilde{\Gamma}} + \tilde{b}Q_{\tilde{\Gamma}}$ ($\tilde{A} = P_{\tilde{\Gamma}}\tilde{a} + Q_{\tilde{\Gamma}}\tilde{b}$) whose coefficients are given by

$$\tilde{a}(t) = \begin{cases} a(t), & t \in \Gamma, \\ 1, & t \in \tilde{\Gamma} \setminus \Gamma, \end{cases} \qquad \tilde{b}(t) = \begin{cases} b(t), & t \in \Gamma, \\ 1, & t \in \tilde{\Gamma} \setminus \Gamma. \end{cases}$$

[1]) In order to avoid misunderstandings, we now add the integration curve to the collection of parameters characterizing the spaces $\overline{\mathbf{L}}_p$ and $\tilde{\mathbf{L}}_p$.

Again assume the coefficients a and b are subject to the conditions (6.1)–(6.3). Since α_j and β_j $(j = 1, \ldots, r)$ are supposed to be inner points of Γ, the coefficients a and b can be written in the form

$$\tilde{a}(t) = (t^{-1} - \alpha^{-1})^m \,\tilde{c}(t), \qquad \tilde{b}(t) = (t - \beta)^n \,\tilde{d}(t),$$

where

$$\tilde{c}(t) = \begin{cases} c(t), & t \in \Gamma, \\ (t^{-1} - \alpha^{-1})^{-m}, & t \in \tilde{\Gamma} \setminus \Gamma, \end{cases} \qquad \tilde{d}(t) = \begin{cases} d(t), & t \in \Gamma, \\ (t - \beta)^{-n}, & t \in \tilde{\Gamma} \setminus \Gamma \end{cases} \tag{2}$$

are piecewise continuous functions from $\mathbf{PC}(\Gamma)$.

By repeating the arguments of the proof of Theorem 1.1, Chapter IV, taking into account Lemma 7.1, and applying Theorem 6.3, we get the following theorem.

Theorem 7.1. *Let the conditions* (6.1)–(6.3) *be satisfied, let* $A = aP_\Gamma + bQ_\Gamma$ $(A = P_\Gamma a + Q_\Gamma b)$, *and let* m_k, n_k $(k = 1, 2)$ *be any multi-indices such that* $m_1 + m_2 = m$ *and* $n_1 + n_2 = n$.

Then $A \in \mathscr{L}(\tilde{\mathbf{L}}_p(\Gamma; \alpha, m_1; \beta, n_1), \overline{\mathbf{L}}_p(\Gamma; \alpha, m_2; \beta, n_2))$ *and this operator is a* Φ_+-, Φ_--, *or* Φ-*operator if and only if the symbol of the operator* $\tilde{C} = \tilde{c}P_{\tilde{\Gamma}} + \tilde{d}Q_{\tilde{\Gamma}}$ $(\tilde{C} = P_{\tilde{\Gamma}}\tilde{c} + Q_{\tilde{\Gamma}}\tilde{d}) \in \mathscr{L}(\mathbf{L}_p(\tilde{\Gamma}))$ *whose coefficients are given by* (2) *satisfies the condition*

$$\det \tilde{C}(t, \mu) \neq 0 \qquad (t \in \tilde{\Gamma}, 0 \leqq \mu \leqq 1). \tag{3}$$

If condition (3) *is fulfilled, then the operator* A *is at least one-sided invertible in that pair of spaces and* $\operatorname{Ind} A = \operatorname{Ind} \tilde{C} + |m_1| + |n_1|$.

Remark. The results of this section were established by Silbermann [8]; he also considered the case where α and β are systems of zeros of non-integral orders (see also Prössdorf [20]). Note that absolutely nothing is known in the case where one of the points α_j, β_j $(j = 1, \ldots, r)$ is an end point of Γ.

However, in this connection we refer to Schüppel [1], dealing with generalizations of the results of Prössdorf [1, 2, 8, 9, 10] on the non-bounded regularization of singular operators with degenerate continuous coefficients to the case of non-closed curves (see also Prössdorf and Teichmann [1]).

§ 8. Singular operators with degenerate measurable coefficients

The aim of this section is to give an idea how the methods developed in the present chapter can be applied to certain classes of singular integral equations with degenerate measurable bounded coefficients.

Let Γ be a closed piecewise Lyapunov curve system, $\alpha = (\alpha_1, \alpha_2, \ldots, \alpha_r)$ and $\beta = (\beta_1, \beta_2, \ldots, \beta_r)$ systems of points on Γ, and m, n any multi-indices of dimension r. We consider the singular integral operator $A = aP_\Gamma + bQ_\Gamma$ with coefficients $a, b \in \mathbf{L}_\infty(\Gamma)$ possessing the following representation:

$$a(t) = (t^{-1} - \alpha^{-1})^m \,c(t), \qquad b(t) = (t - \beta)^n \,d(t). \tag{1}$$

Here we suppose that

$$c^{\pm 1}, d^{\pm 1} \in \mathbf{L}_\infty(\Gamma); \qquad c, d \in \mathbf{L}_\infty(\beta, n); \qquad d \in \mathbf{L}_\infty(\alpha, m) \tag{2}$$

and that the functions $\varrho_-(t) = (t^{-1} - \alpha^{-1})^m$, $\varrho_+(t) = (t - \beta)^n$ have no common zeros on Γ:

$$\{t \in \Gamma : \varrho_+(t) = 0\} \cap \{t \in \Gamma : \varrho_-(t) = 0\} = \emptyset. \tag{3}$$

From (2) and the results of § 3 we obtain the following representations

$$\begin{aligned} &d = \varrho_+ d_1 + h_1, \quad d = \varrho_- d_2 + h_2, \quad d^{-1} = \varrho_+ d_3 + h_3, \\ &d^{-1} = \varrho_- d_4 + h_4; \quad g = d^{-1} c = \varrho_+ g_1 + q_1, \quad g^{-1} = \varrho_+ g_2 + q_2, \end{aligned} \tag{4}$$

where $d_j \in \mathbf{L}_\infty(\Gamma)$ $(j = 1, \ldots, 4)$, $g_i \in \mathbf{L}_\infty(\Gamma)$ $(i = 1, 2)$, $h_j \in \mathbf{L}_\infty^+(\Gamma)$ $(j = 1, 3)$, $h_j \in \mathbf{L}_\infty^-(\Gamma)$ $(j = 2, 4)$, $q_i \in \mathbf{L}_\infty^+(\Gamma)$ $(i = 1, 2)$.

The operator $A = aP_\Gamma + bQ_\Gamma$ can be written in the form

$$A = d(g\varrho_- P_\Gamma + \varrho_+ Q_\Gamma). \tag{5}$$

Lemma 8.1. *The operator of multiplication by the function d is defined, continuous, and invertible on the space* $\overline{\mathbf{L}}_p(\alpha, m; \beta, n)$.

Proof. In view of (4) and (2.6) we have

$$\begin{aligned} d(\varrho_- P_\Gamma + \varrho_+ Q_\Gamma) &= (\varrho_- P_\Gamma + \varrho_+ Q_\Gamma)\,[dI + (\varrho_+ - \varrho_-)\,(P_\Gamma d_2 Q_\Gamma - Q_\Gamma d_1 P_\Gamma)], \\ d^{-1}(\varrho_- P_\Gamma + \varrho_+ Q_\Gamma) &= (\varrho_- P_\Gamma + \varrho_+ Q_\Gamma)\,[d^{-1} I + (\varrho_+ - \varrho_-)\,(P_\Gamma d_3 Q_\Gamma - Q_\Gamma d_4 P_\Gamma)]. \end{aligned}$$

Since $\varrho_- P_\Gamma + \varrho_+ Q_\Gamma \in \mathscr{L}(\mathbf{L}_p(\Gamma), \overline{\mathbf{L}}_p(\alpha, m; \beta, n))$ is an invertible operator (see the proof of Theorem 4.4), we deduce from the last equalities that the operators dI and $d^{-1} I$ are defined and continuous on $\overline{\mathbf{L}}_p(\alpha, m; \beta, n)$. Finally, because $dd^{-1}\varphi = d^{-1} d\varphi = \varphi$ $(\varphi \in \overline{\mathbf{L}}_p(\alpha, m; \beta, n))$ we conclude that $d^{-1} I$ is the inverse of dI. ▮

(2.6) shows that the second factor in (5) can be written in the form

$$g\varrho_- P_\Gamma + \varrho_+ Q_\Gamma = (\varrho_- P_\Gamma + \varrho_+ Q_\Gamma)\,[gP_\Gamma + Q_\Gamma + (\varrho_- - \varrho_+)\,Q_\Gamma g_1 P_\Gamma]. \tag{6}$$

The operator $G := gP_\Gamma + Q_\Gamma + (\varrho_- - \varrho_+)\,Q_\Gamma g_1 P_\Gamma$ is normally solvable, a $\Phi_\pm$-operator, or one-sided invertible on $\mathbf{L}_p(\Gamma)$ if and only if the operator $gP_\Gamma + Q_\Gamma$ has the corresponding property. This is an immediate consequence of the following lemma.

Lemma 8.2. *We have*

$$G = G_1(gP_\Gamma + Q_\Gamma)\,G_2 \tag{7}$$

with certain operators G_1 *and* G_2 *invertible on* $\mathbf{L}_p(\Gamma)$.

Proof. It can be checked straightforwardly that (7) holds with

$$\begin{aligned} G_1 &= [gI + (\varrho_+ - I)\,(P_\Gamma g Q_\Gamma - Q_\Gamma g_1 P_\Gamma)]\,g^{-1}, \\ G_2 &= (I - P_\Gamma g^{-1} Q_\Gamma + \varrho_+ P_\Gamma g^{-1} Q_\Gamma)\,[I + (\varrho_- - I)\,Q_\Gamma g_1 P_\Gamma]. \end{aligned}$$

Since $\varrho_\pm \in \mathbf{C}^\pm(\Gamma)$, the first factor of the operator G_2 is seen to be of the form $I + PBQ$ and the second one of the form $I + QCP$. Therefore G_2 is invertible (see Section 1.1, Chapter III). It remains to prove that G_1 is also invertible.

The representations (4) and (2.6) give

$$\begin{aligned} g(P_\Gamma + \varrho_+ Q_\Gamma) &= (P_\Gamma + \varrho_+ Q_\Gamma)\,[gI + (\varrho_+ - I)\,(P_\Gamma g Q_\Gamma - Q_\Gamma g_1 P_\Gamma)], \\ g^{-1}(P_\Gamma + \varrho_+ Q_\Gamma) &= (P_\Gamma + \varrho_+ Q_\Gamma)\,[g^{-1} I + (\varrho_+ - I)\,(P_\Gamma g^{-1} Q_\Gamma - Q_\Gamma g_2 P_\Gamma). \end{aligned}$$

This shows that the operators gI and $g^{-1} I$ are continuous and invertible on the space $\hat{\mathbf{L}}_p(\alpha, 0; \beta, n) = \mathrm{Im}\,(P_\Gamma + \varrho_+ Q_\Gamma)$ (cf. Section 4.4). Hence the operator $gI + (\varrho_+ - I)$ $(P_\Gamma g Q_\Gamma - Q_\Gamma g_1 P_\Gamma)$ is invertible on $\mathbf{L}_p(\Gamma)$ and, consequently, so also is G_1. ▮

The equalities (5)–(7) and the Lemmas 8.1 and 8.2 yield the following theorem.

Theorem 8.1. *Let the conditions* (1)–(3) *be fulfilled. Then* $A = aP_\Gamma + bQ_\Gamma \in \mathscr{L}(\mathbf{L}_p(\Gamma), \overline{\mathbf{L}}_p(\alpha, m; \beta, n))$ *and this operator is a* Φ_+-, Φ_--, *or* Φ-*operator if and only if the operator* $C = cP_\Gamma + dQ_\Gamma \in \mathscr{L}(\mathbf{L}_p(\Gamma))$ *has the corresponding property.*

If $C \in \Phi(\mathbf{L}_p(\Gamma))$, then $\operatorname{Ind} A = \operatorname{Ind} C$ *and the operators A and C are either both one-sided invertible (on the quoted spaces) or not.*

An analogous theorem holds for the singular integral operator $A = P_\Gamma a + Q_\Gamma b$ as well as for the pair of spaces $\tilde{\mathbf{L}}_p(\alpha, m; \beta, n)$, $\mathbf{L}_p(\Gamma)$.

Remark 1. The results of this section are due to SILBERMANN [8], who also studied the case where the zeros α, β are of non-integral orders (see also PRÖSSDORF [20]).

A simpler (though somewhat less general) approach to the theory of singular operators with degenerate coefficients can be found in the book of PRÖSSDORF and SILBERMANN [11] (Chapter 5, § 1).

More general cases of degenerate coefficients (e.g. coefficients with a countable set of zeros) have been recently investigated by M. I. HAIKIN [7] and DYBIN [6, 7].

Remark 2. In particular, Theorem 8.1 applies to coefficients c and d (see (1)) which are continuous on Γ with the exception of finitely many points where they are allowed to have discontinuities of the second kind of almost periodic type. (An isolated point of discontinuity z_0 of a function f given on the unit circle $|z| = 1$ is called a *point of discontinuity of almost periodic type* if there exists an uniformly almost periodic function $p(\lambda)$, $-\infty < \lambda < \infty$, such that

$$\lim_{z \to z_0} \left[f(z) - p\left(-\mathrm{i} \frac{z + z_0}{z - z_0} \right) \right] = 0.)$$

For such coefficients the theory of the operators $cP_\Gamma + dQ_\Gamma$ and $P_\Gamma c + Q_\Gamma d$ was worked out by COBURN and DOUGLAS [1], GOHBERG and SEMENTSUL [1], SEMENTSUL [1] and then generalized and developed further by SARASON [1], SAGINASHVILI [1], ABRAHAMCE [1], POWER [1], DYBIN [4–5, 7], GRUDSKI and DYBIN [1], GRUDSKI [1], DYBIN and GAPONENKO [1].

Thus, the results of the present section yield the possibility to construct examples of singular integral operators with continuous and pointwise degenerate coefficients whose kernel resp. cokernel is of infinite dimension. (In this connection see also SEMENTSUL [1], where the method described in § 1 is applied to investigate such operators.)

§ 9. Singular matrix operators with degenerate coefficients

9.1. Let Γ be a closed piecewise Lyapunov curve system, $\alpha = (\alpha_1, \alpha_2, \ldots, \alpha_r)$ and $\beta = (\beta_1, \beta_2, \ldots, \beta_r)$ systems of points on Γ, and $m = (m_1, m_2, \ldots, m_r)$, $n = (n_1, n_2, \ldots, n_r)$ any multi-indices. Furthermore, let $R_\pm$ be two arbitrary polynomial matrix functions in $z^{\pm 1}$ of the order s with constant and non-vanishing determinant, and let $D_\pm$ be diagonal matrix functions of the form

$$D_-(z) = \left(\prod_{k=1}^{r} (z^{-1} - \alpha_k^{-1})^{\mu_j^{(k)}} \delta_{jl} \right)_1^s,$$

$$D_+(z) = \left(\prod_{k=1}^{r} (z - \beta_k)^{\nu_j^{(k)}} \delta_{jl} \right)_1^s. \tag{1}$$

Here we suppose that

$$\mu_1^{(k)} \geqq \mu_2^{(k)} \geqq \cdots \geqq \mu_s^{(k)} \geqq 0, \qquad \nu_1^{(k)} \geqq \nu_2^{(k)} \geqq \cdots \geqq \nu_s^{(k)} \geqq 0 \qquad (k = 1, \ldots, r) \tag{2}$$

are integers such that

$$\sum_{j=1}^{s} \mu_j^{(k)} = m_k, \qquad \sum_{j=1}^{s} \nu_j^{(k)} = n_k \qquad (k = 1, \ldots, r). \tag{3}$$

Throughout this section, if there is no fear of misunderstandings, we shall denote the systems of numbers (2) by μ and ν, respectively, and write $D_-(z) =: (z^{-1} - \alpha^{-1})^\mu$, $D_+(z) =: (z - \beta)^\nu$.

9.2. Put $P := (P_\Gamma\,\delta_{jl})_{j,l=1}^s$, $Q := (Q_\Gamma\,\delta_{jl})_{j,l=1}^s$ (see also § 2, Chapter V) and consider the singular matrix operator

$$\boldsymbol{B} := PR_-D_- + QR_+D_+ = (PR_- + QR_+)\,(PD_- + QD_+) \tag{4}$$

on the space $\mathbf{L}_p^s(\Gamma)$ $(1 < p < \infty)$ (compare (2.4)). The inverse of the operator $PR_- + QR_+$ on $\mathbf{L}_p^s(\Gamma)$ is $PR_-^{-1} + QR_+^{-1}$. The operator $PD_- + QD_+$ is of diagonal form and Lemma 4.2 implies that $\dim \operatorname{Ker} (PD_- + Q\,D_+) = 0$. Denote by $\overline{\mathbf{L}}_p^s = \overline{\mathbf{L}}_p^s(R_-, \alpha, \mu;\, R_+, \beta, \nu)$ the image space $\operatorname{Im} \boldsymbol{B}$ provided with the norm $\|f\|_{\overline{\mathbf{L}}_p^s} = \|\boldsymbol{B}^{-1} f\|_{\mathbf{L}_p^s}$. The results of Section 4.3 immediately yield an analytical description of the space $\overline{\mathbf{L}}_p^s$. Corollary 4.2 gives the inclusion

$$\mathbf{L}_p^s(\Gamma;\, (\Gamma_{\alpha_j}, \mu_1^{(j)})_1^r,\, (\Gamma_{\beta_j}, \nu_1^{(j)})_1^r) \subset \overline{\mathbf{L}}_p^s(R_-, \alpha, \mu;\, R_+, \beta, \nu). \tag{5}$$

9.3. For the sake of simplicity assume now that $\alpha_j \neq \beta_k$ $(j, k = 1, \ldots, r)$ and consider the operator

$$\boldsymbol{D} := (t - \alpha)^\mu\, S_+P + (t^{-1} - \beta^{-1})^\nu\, S_-Q, \tag{6}$$

where $S_\pm$ are polynomial matrix functions in $t^{\pm 1}$ with constant and non-vanishing determinant. Denote by $\boldsymbol{D}^{(-1)}$ the operator defined for $f \in \mathbf{L}_p^s(\Gamma)$ by the formula

$$\boldsymbol{D}^{(-1)} f := S_+^{-1}(t - \alpha)^{-\mu}\, Pf + S_-^{-1}(t^{-1} - \beta^{-1})^{-\nu}\, Qf. \tag{7}$$

The collection of all vectors of the form (7) with f ranging over $\mathbf{L}_p^s(\Gamma)$ will be denoted by $\tilde{\mathbf{L}}_p^s = \tilde{\mathbf{L}}_p^s(\alpha, \mu, S_+;\, \beta, \nu, S_-)$. Equipped with the norm

$$\|\varphi\|_{\tilde{\mathbf{L}}_p^s} = \|f\|_{\mathbf{L}_p^s} \qquad (\varphi = \boldsymbol{D}^{(-1)} f \in \tilde{\mathbf{L}}_p^s)$$

$\tilde{\mathbf{L}}_p^s$ becomes a Banach space.

Obviously, $\mathbf{L}_p^s(\Gamma) \subset \tilde{\mathbf{L}}_p^s$ and the operator $\tilde{\boldsymbol{D}}$ defined by $\tilde{\boldsymbol{D}}\varphi = f$ is a linear continuous operator mapping the space $\tilde{\mathbf{L}}_p^s$ isometrically onto $\mathbf{L}_p^s(\Gamma)$. It is clear that $\tilde{\boldsymbol{D}} \,|\, \mathbf{L}_p^s(\Gamma) = \boldsymbol{D}$ and $\tilde{\boldsymbol{D}}^{-1} = \boldsymbol{D}^{(-1)}$.

9.4. We consider the singular matrix operators $AP + BQ$ and $PA + QB$ with coefficients $A, B \in \mathbf{C}^{s\times s}(\alpha, m;\, \alpha, n)$. Assume α (resp. β) is the system of all zeros of the function $\det A$ (resp. $\det B$) on Γ and let its multiplicity be m (resp. n). Then, by virtue of Theorem 3.1, the matrix functions A and B can be represented in the form

$$A(t) = R_-(t)\,(t^{-1} - \alpha^{-1})^\mu\, A_1(t) = A_2(t)\,(t - \alpha)^\mu\, S_+(t), \tag{8}$$

$$B(t) = R_+(t)\,(t - \beta)^\nu\, B_1(t) = B_2(t)\,(t^{-1} - \beta^{-1})^\nu\, S_-(t) \tag{9}$$

with certain non-singular continuous matrix functions A_j and B_j $(j = 1, 2)$.

From (4), (6), (8), (9) and (2.1), (2.4) we obtain the representations

$$AP + BQ = (A_2P + B_2Q)\,\boldsymbol{D}, \qquad PA + QB = \boldsymbol{B}(PA_1 + QB_1). \tag{10}$$

Taking into account Theorem 4.5 and Corollary 4.4 it is easy to see that

$$AP + BQ = \boldsymbol{B}(PA_1 + QB_1 + T_1), \qquad PA + QB = (A_2P + B_2Q + T_2)\,\boldsymbol{D}, \tag{11}$$

where T_1 and T_2 are compact operators on $\mathbf{L}_p^s(\Gamma)$.

(10), (11) and Theorem 2.1, Chapter V, give the following theorem.

Theorem 9.1. *Suppose the assumptions of Section 9.4 are satisfied and let* $\boldsymbol{A} = AP + BQ$ $(\boldsymbol{A} = PA + QB)$. *Then:*

(i) $\mathbf{A} \in \Phi(\mathbf{L}_p^s(\Gamma), \overline{\mathbf{L}}_p^s(R_-, \alpha, \mu; R_+, \beta, \nu))$ *with*

$$\operatorname{Ind} \mathbf{A} = \operatorname{ind} \left[\frac{\det B(t)}{\det A(t)} \frac{(t^{-1} - \alpha^{-1})^{|m|}}{(t - \beta)^{|n|}}\right]. \tag{12}$$

(ii) $\mathbf{A} \in \Phi(\widetilde{\mathbf{L}}_p^s(\alpha, \mu, S_+; \beta, \nu, S_-), \mathbf{L}_p^s(\Gamma))$ *with*

$$\operatorname{Ind} \mathbf{A} = \operatorname{ind} \left[\frac{\det B(t)}{\det A(t)} \frac{(t - \alpha)^{|m|}}{(t^{-1} - \beta^{-1})^{|n|}}\right]. \tag{13}$$

Remark. Theorem 9.1 was established by Prössdorf [10, 14]. Obviously, from this theorem the corresponding analogue of Theorem 5.3 can be derived. Furthermore, Theorem 9.1 can be extended to the case of zeros α, β of non-integral orders (in this connection see Prössdorf and Silbermann [2] and Prössdorf [14]). For further results in this direction we refer to Prössdorf [20].

9.5. We now carry over the results of the preceding subsection to the case of piecewise continuous matrix functions $A, B \in \mathbf{PC}^{s\times s}(\alpha, m; \beta, n)$. Again assume that α (resp. β) is the system of all zeros of the determinant $\det A$ (resp. $\det B$) on Γ and that its multiplicity is m (resp. n). Then, by Theorem 3.1, we have the representations (8), (9) with certain non-singular matrix functions $A_j \in \mathbf{PC}^{s\times s}(\beta, n)$ and $B_j \in \mathbf{PC}^{s\times s}(\alpha, m)$ $(j = 1, 2)$. Once more applying Theorem 3.1, we deduce that there exist two polynomial matrix functions $T_\pm \in [\mathbf{C}^\pm(\Gamma)]^{s\times s}$ such that (I the identity matrix)

$$A(t) = (t - \beta)^{|n|} IC_1(t) + T_+(t) = R_+(t) (t - \beta)^\nu C(t) + T_+(t),$$
$$B(t) = (t^{-1} - \alpha^{-1})^{|m|} ID_1(t) + T_-(t) = R_-(t) (t^{-1} - \alpha^{-1})^\mu D(t) + T_-(t),$$

where $C, D \in \mathbf{PC}^{s\times s}(\Gamma)$. Using these representations and the formula (2.5) we get

$$\mathbf{A}P + \mathbf{B}Q = \mathbf{BC}, \qquad \mathbf{C} = PA_1P + QB_1Q + PDQ + QCP. \tag{14}$$

By § 7, Chapter V, the symbol $\mathcal{C}(t, \mu)$ $(t \in \Gamma,\ 0 \leqq \mu \leqq 1)$ of the operator $\mathbf{C} \in \mathscr{L}(\mathbf{L}_p^s(\Gamma))$ is

$$\mathcal{C}(t, \mu) = \begin{pmatrix} C_{11}(t, \mu) & D_{12}(t, \mu) \\ C_{21}(t, \mu) & D_{22}(t, \mu) \end{pmatrix},$$

where

$$C_{11}(t, \mu) = f_p'(\mu) A_1(t + 0) + (1 - f_p'(\mu)) A_1(t),$$
$$C_{21}(t, \mu) = h_p'(\mu) (C(t + 0) - C(t)),$$
$$D_{12}(t, \mu) = h_p'(\mu) (D(t + 0) - D(t)),$$
$$D_{22}(t, \mu) = (1 - f_p'(\mu)) B_1(t + 0) + f_p'(\mu) B_1(t).$$

Theorem 9.2. *Let* (8) *and* (9) *hold with* $A_1 \in \mathbf{PC}^{s\times s}(\beta, n)$ *and* $B_1 \in \mathbf{PC}^{s\times s}(\alpha, m)$ *and let* $\mathbf{A} = AP + BQ$.

(a) *Then* $\mathbf{A} \in \mathscr{L}(\mathbf{L}_p^s(\Gamma), \overline{\mathbf{L}}_p^s(R_-, \alpha, \mu; R_+, \beta, \nu))$ *and this operator is a* Φ*-operator if and only if*

$$\det \mathcal{C}(t, \lambda) \neq 0 \qquad (t \in \Gamma,\ 0 \leqq \lambda \leqq 1). \tag{15}$$

If condition (15) *is satisfied, then* $\operatorname{Ind} \mathbf{A} = \operatorname{Ind} \mathbf{C}$.

(b) *If the matrix functions* A_1 *and* B_1 *have no common points of discontinuity on* Γ, *then the operators* $\mathbf{A} \in \mathscr{L}(\mathbf{L}_p^s(\Gamma), \overline{\mathbf{L}}_p^s(R_-, \alpha, \mu; R_+, \beta, \nu))$ *and* $\mathbf{A}_1 = A_1P + B_1Q \in \mathscr{L}(\mathbf{L}_p^s(\Gamma))$ *are either simultaneously* Φ*-operators or not. If* $\mathbf{A}_1 \in \Phi(\mathbf{L}_p^s(\Gamma))$, *then* $\operatorname{Ind} \mathbf{A} = \operatorname{Ind} \mathbf{A}_1$.

Proof. (a) This is an immediate consequence of Theorem 7.1., Chapter V, and (14).

(b) According to § 6, Chapter V, the symbol of the operator $A_1 = A_1P + B_1Q \in \mathscr{L}(\mathbf{L}_p^s(\Gamma))$ is

$$A_1(t, \lambda) = \begin{pmatrix} C_{11}(t, \lambda) & h_p'(\lambda)\,(B_1(t+0) - B_1(t)) \\ h_p'(\lambda)\,(A_1(t+0) - A_1(t)) & D_{22}(t, \lambda) \end{pmatrix}.$$

It is easily seen that if A_1 and B_1 have no common discontinuities on Γ, then the matrix functions C and D have no common discontinuities on Γ, too. Consequently, $C(t, \lambda)$ and $A_1(t, \lambda)$ are quasi-diagonal matrices. But this implies that

$$\det C(t, \lambda) = \det C_{11}(t, \lambda)\, D_{22}(t, \lambda) = \det A_1(t, \lambda)\,.$$

The last equality, Theorem 6.4 of Chapter V, and the already proved assertion (a) now imply assertion (b). ∎

An analogus theorem holds for the singular matrix operator $A = PA + QB$ as well as for the pair of spaces

$$\tilde{\mathbf{L}}_p^s(\alpha, \mu, S_+; \beta, \nu, S_-), \qquad \mathbf{L}_p^s(\Gamma)\,.$$

Remark 1. If the matrix functions A_1 and B_1 have common discontinuities on Γ, then it can happen that one of the operators $\mathbf{C}$ and $A_1 = A_1P + B_1Q$ (see (14)) is a Φ-operator on $\mathbf{L}_p^s(\Gamma)$ while the other one is not even normally solvable (Theorem 6.1 shows that such an effect is impossible in the scalar case $s = 1$). Situations of that kind occur, for instance, when singular integral operators with degenerate matrix coefficients are considered (with the methods of the Sections 7 and 9) on nonclosed curves.

Remark 2. Theorem 9.2 and its generalization to the case of zeros α, β of non-integral orders go back to Silbermann [8] (see also Prössdorf [20] and M. I. Haikin [8]). Recently, M. I. Haikin [9] studied singular equations with degenerate matrix coefficients as improperly posed problems and constructed regularizers in the sense of A. N. Tikhonov.

§ 10. Singular operators on the spaces $\mathbf{C}^\infty(\Gamma)$ and $\mathbf{C}^{-\infty}(\Gamma)$

The theory of singular integral operators with degenerate coefficients takes an especially simple form in the countably normed space $\mathbf{C}^\infty(\Gamma)$ of all infinitely differentiable functions on a closed curve system Γ and in the distribution space $\mathbf{C}^{-\infty}(\Gamma) = [\mathbf{C}^\infty(\Gamma)]^*$. In these spaces one has the following remarkable analogue to the theory in the space $\mathbf{L}_p(\Gamma)$: The singular integral operator (with coefficients from $\mathbf{C}^\infty(\Gamma)$) is Fredholm if and only if its symbol has at most finitely many zeros of finite order.

10.1. Let Γ be a closed Lyapunov curve system of the class C^∞. We denote by $\mathbf{C}^\infty(\Gamma)$ the countably normed space of all infinitely differentiable (complex-valued) functions on Γ. The topology in $\mathbf{C}^\infty(\Gamma)$ is defined through the countable system of norms

$$\|f\|_n := \|f\|_{\mathbf{C}^n(\Gamma)} = \sum_{j=0}^{n} \max_{t\in\Gamma} |f^{(j)}(t)| \qquad (n = 0, 1, 2, \ldots).$$

$\mathbf{C}^\infty(\Gamma)$ is a perfect space, and hence it is reflexive.[1])

Let $[\mathbf{C}^\infty(\Gamma)]^s$ denote the countably normed space of all s-dimensional vectors $f = (f_k)_{k=1}^s$ with $f_k \in \mathbf{C}^\infty(\Gamma)$. The system of norms in $[\mathbf{C}^\infty(\Gamma)]^s$ is given by

$$\|f\|_n = \sum_{k=1}^{s} \|f_k\|_n \qquad (n = 0, 1, 2, \ldots).$$

[1]) A countably normed space is called *perfect* if all its bounded subsets are relatively compact. The perfectness of $\mathbf{C}^\infty(\Gamma)$ can be easily proved with the help of the Arzelá theorem (see Gelfand and Shilov [2], p. 47).

We denote by $[\mathbf{C}^{-\infty}(\Gamma)]^s$ the conjugate space of $[\mathbf{C}^{\infty}(\Gamma)]^s$.

We now consider the singular integral operator

$$\boldsymbol{A} = AP + BQ + T, \tag{1}$$

with P, Q the projections defined in 9.1 and with A, B matrix functions from $[\mathbf{C}^{\infty}(\Gamma)]^{s\times s}$. First let T be any compact linear operator on the space $[\mathbf{C}^{\infty}(\Gamma)]^s$.

Let $\varphi \in [\mathbf{C}^{\infty}(\Gamma)]^s$. From the equality

$$(P\varphi)(t) = \frac{1}{2\pi i} \int_\Gamma \frac{\varphi(\tau) - \varphi(t)}{\tau - t} \mathrm{d}\tau + \varphi(t)$$

we obtain immediately that

$$\|P\varphi\|_n \leqq M_n \|\varphi\|_{n+1} \qquad (n = 0, 1, 2, \ldots), \qquad M_n = \text{const},$$

and the conclusion is that the operator P is continuous on the space $[\mathbf{C}^{\infty}(\Gamma)]^s$. Thus, Q and $\boldsymbol{A}$ are also continuous operators on $[\mathbf{C}^{\infty}(\Gamma)]^s$.

Lemma 10.1. *Let* $T \in \mathscr{L}(\mathbf{L}_p^s(\Gamma), [\mathbf{C}^{\infty}(\Gamma)]^s)$ *and* $f \in [\mathbf{C}^{\infty}(\Gamma)]^s$. *If*

$$\det A(t) \neq 0, \qquad \det B(t) \neq 0 \qquad (t \in \Gamma),$$

then every solution $\varphi \in \mathbf{L}_p^s(\Gamma)$ $(1 < p < \infty)$ *of the equation*

$$\boldsymbol{A}\varphi = f \tag{2}$$

belongs to $[\mathbf{C}^{\infty}(\Gamma)]^s$.

Proof. Put $\boldsymbol{B} = A^{-1}P + B^{-1}Q$ and let $\boldsymbol{B}$ act on both sides of (2) to get $\varphi + T_1\varphi = \boldsymbol{B}f$. It is easily seen that $T_1 \in \mathscr{L}(\mathbf{L}_p^s(\Gamma), [\mathbf{C}^{\infty}(\Gamma)]^s)$, and therefore $\varphi = \boldsymbol{B}f - T_1\varphi \in [\mathbf{C}^{\infty}(\Gamma)]^s$. ∎

Lemma 10.2. *If* $k(t, \tau) \in \mathbf{C}^{\infty}(\Gamma \times \Gamma)$, *then the operator* K *defined by*

$$(K\varphi)(t) = \int_\Gamma k(t, \tau) \varphi(\tau) \mathrm{d}\tau$$

is compact on the space $\mathbf{C}^{\infty}(\Gamma)$.

This follows immediately from Lemma 4.1, Chapter I, since it is easy to check with the aid of Arzelá's theorem that $K: \mathbf{C}(\Gamma) \to \mathbf{C}^n(\Gamma)$ is compact for all $n = 0, 1, 2, \ldots$

10.2. Now we study the singular operator (1) on the space $[\mathbf{C}^{\infty}(\Gamma)]^s$. As in Section 9.4, assume the determinants

$$\det A(t), \det B(t) \tag{3}$$

have only finitely many zeros of integral orders on Γ. Denote by $\alpha = (\alpha_1, \alpha_2, \ldots, \alpha_r)$ and $\beta = (\beta_1, \beta_2, \ldots, \beta_r)$ the corresponding systems of all zeros of the functions (3) and by $m = (m_1, m_2, \ldots, m_r)$ and $n = (n_1, n_2, \ldots, n_r)$ their multiplicities.

Theorem 10.1. *The singular operator* (1) *is Fredholm on* $[\mathbf{C}^{\infty}(\Gamma)]^s$ *and its index is given by formula* (9.12).

Proof. In view of Theorem 3.4, Chapter I, and Lemma 10.2 it suffices to prove the assertion for the operator $PA + QB$. For this operator we have the second representation (9.10), where the operator $\boldsymbol{B}$ is invertible on $[\mathbf{C}^{\infty}(\Gamma)]^s$ (see Corollary 4.3) and where $A_1, B_1 \in [\mathbf{C}^{\infty}(\Gamma)]^{s\times s}$. From the Lemmas 10.1 and 10.2 it is easy to deduce that every solution $\varphi \in \mathbf{L}_p^s(\Gamma)$ of the equation $PA_1\varphi + QB_1\varphi = f$, where $f \in [\mathbf{C}^{\infty}(\Gamma)]^s$, is also in $[\mathbf{C}^{\infty}(\Gamma)]^s$. This implies that $PA_1 + QB_1$ is a Φ-operator on $[\mathbf{C}^{\infty}(\Gamma)]^s$ and that its indices on the spaces $[\mathbf{C}^{\infty}(\Gamma)]^s$ and $\mathbf{L}_p^s(\Gamma)$ coincide. The assertion now follows from (9.10) and Theorem 2.1, Chapter V. ∎

Theorem 10.2. *Let $A, B \in [\mathbf{C}^\infty(\Gamma)]^{s\times s}$. Then for $AP + BQ$ $(PA + QB)$ to be a Fredholm operator on the space $[\mathbf{C}^\infty(\Gamma)]^s$ it is necessary and sufficient that each of the two functions* (3) *have at most finitely many zeros of integral order on Γ.*

To prove this, we need the following lemma.

Lemma 10.3. *Let Γ be a closed Lyapunov curve of the class C^∞, let $a \in \mathbf{C}^\infty(\Gamma)$, let $t_0 \in \Gamma$, and suppose $a^{(k)}(t_0) = 0$ for all $k = 0, 1, 2, \ldots$ Then there exists a sequence of real-valued functions $f_n \in \mathbf{C}^\infty(\Gamma)$ $(n = 1, 2, \ldots)$ such that*

a) $0 \leqq f_n \leqq 1, f_n(t_0) = 1$;

b) $\operatorname{supp} f_n \to t_0$ *as* $n \to \infty$ *(that is, for any neighborhood $U \subset \Gamma$ of t_0 there is a positive integer $N(U)$ with* $\operatorname{supp} f_n \subset U$ *for all $n > N(U)$)*;[1])

c) af_n *converges to zero as $n \to \infty$ in the topology of the space* $\mathbf{C}^\infty(\Gamma)$.

Proof. Let γ be the length of the curve Γ and $t = t(s)$ $(0 \leqq s \leqq \gamma)$ its equation. Without loss of generality let the arc abscissa s_0 of the point $t_0 = t(s_0)$ satisfy $0 < s_0 < \gamma$. We denote by $\hat{\mathbf{C}}^\infty[0, \gamma]$ the following subspace of $\mathbf{C}^\infty[0, \gamma]$:

$$\hat{\mathbf{C}}^\infty[0, \gamma] := \{f \in \mathbf{C}^\infty[0, \gamma] : f^{(k)}(0) = f^{(k)}(\gamma), k = 0, 1, \ldots\}.$$

The mapping $\psi : \mathbf{C}^\infty(\Gamma) \to \hat{\mathbf{C}}^\infty[0, \gamma]$ defined by $(\psi f)(s) = f(t(s))$ is obviously a topological isomorphism of the algebra $\mathbf{C}^\infty(\Gamma)$ onto the algebra $\hat{\mathbf{C}}^\infty[0, \gamma]$. Therefore it suffices to prove the assertion for $\hat{\mathbf{C}}^\infty[0, \gamma]$ in place of $\mathbf{C}^\infty(\Gamma)$.

Thus, suppose $a \in \hat{\mathbf{C}}^\infty[0, \gamma]$ and $a^{(k)}(s_0) = 0$ for $k = 0, 1, \ldots$ Let $f \in \hat{\mathbf{C}}^\infty[0, \gamma]$ denote any real-valued function on the interval $[0, \gamma]$ satisfying $0 \leqq f \leqq 1$, $f(s_0) = 1$ and vanishing outside a sufficiently small neighborhood of s_0, say outside the interval $|s - s_0| \leqq c$.

Now let $p \in \{0, 1, 2, \ldots\}$ be an arbitrary but fixed number and write $A = \sup\limits_{\substack{0 \leqq s \leqq \gamma \\ 0 \leqq k \leqq p}} |f^{(k)}(s)|$. Also put $f_n(s) = f(s_0 + n(s - s_0))$. Then

$$|f_n^{(k)}(s)| \leqq n^k A = O(n^k) \qquad (k = 0, 1, \ldots, p).$$

On the other hand, using Taylor's formula (e.g. with the remainder in Peano's form) it is easy to obtain the following estimates for the derivatives of the function $a(s)$ on the interval $|s - s_0| \leqq d$:

$$|a^{(r)}(s)| = o(d^{p-r}) \qquad (r = 0, 1, \ldots, p). \tag{4}$$

We now estimate the norm $\|af_n\|_p$. Since the function $f_n(s)$ vanishes outside the interval $|s - s_0| \leqq c/n$, we may put $d = c/n$ in (4). Applying the Leibniz rule we then find

$$|(af_n)^{(m)}| \leqq \sum_{k=0}^{m} \binom{m}{k} |f_n^{(k)}| \, |a^{(m-k)}| = \sum_{k=0}^{m} \binom{m}{k} O(n^k) \, o\left(\frac{1}{n^{p-m+k}}\right) = o\left(\frac{1}{n^{p-m}}\right)$$

for every $m = 0, 1, \ldots, p$. Consequently, $\lim\limits_{n\to\infty} \|af_n\|_p = 0$ $(p = 0, 1, \ldots)$. It remains to recall how f_n was constructed to see that the sequence $\{f_n\}$ satisfies the required conditions a)–c). ∎

Remark. The sequence $\{f_n\}$ constructed above does not depend on the function a but only on the point $t_0 \in \Gamma$.

Proof of Theorem 10.2. The sufficiency part follows from Theorem 10.1. Due to Theorem 1.1, Chapter V (see also the Remark 1 following after it) and owing to Lemma 10.2, it is enough to prove the necessity portion for the case $s = 1$.

Thus, suppose the scalar operator $A_\Gamma := aP_\Gamma + bQ_\Gamma$ is a Φ-operator on the space

---

[1]) By $\operatorname{supp} f$ we denote the *support of the function f*.

$C^\infty(\Gamma)$. What must be proved is that the coefficients $a, b \in \mathbf{C}^\infty(\Gamma)$ have at most finitely many zeros of integral order on Γ.

Assume that the function a does not satisfy this condition. Then there is a point $t_0 \in \Gamma$ such that $a^{(k)}(t_0) = 0$ for all $k = 0, 1, \ldots$ Let $f_n \in \mathbf{C}^\infty(\Gamma)$ denote the functions constructed in Lemma 10.3. We have

$$aP_\Gamma f_n = P_\Gamma a f_n + (aP_\Gamma - P_\Gamma a) f_n .$$

The kernel of the integral operator $aP_\Gamma - P_\Gamma a$ is infinitely differentiable. This and the properties b) and c) of Lemma 10.3 imply that $aP_\Gamma f_n \to 0$ as $n \to \infty$, with the convergence in the topology of the space $\mathbf{C}^\infty(\Gamma)$. We claim that even $P_\Gamma f_n \to 0$, again in the topology of $\mathbf{C}^\infty(\Gamma)$.

Indeed, since A_Γ was supposed to be a Φ-operator on $\mathbf{C}^\infty(\Gamma)$, with the help of Lemma 2.1 of Chapter I it is easy to check that the multiplication operator

$$\hat{a} := a \mid \operatorname{Im} P_\Gamma : \operatorname{Im} P_\Gamma \to \mathbf{C}^\infty(\Gamma)$$

is a Φ-operator. The Luzin-Privalov uniqueness theorem then shows that $\operatorname{Ker} \hat{a} = 0$. Thus, $\hat{a}^{-1} : \mathbf{C}^\infty(\Gamma) \to \operatorname{Im} P_\Gamma$ is a continuous operator, and so $aP_\Gamma f_n \to 0$ implies that $P_\Gamma f_n \to 0$.

Now assume $\Gamma = K$ is the unit circle. Then we get a contradiction to the conditions a) and b) imposed upon the sequence $\{f_n\}$ as follows. For any function $f \in \mathbf{C}^\infty(\Gamma)$ one has

$$f(t) = \sum_{j=-\infty}^{\infty} \xi_j t^j, \qquad (P_K f)(t) = \sum_{j=0}^{\infty} \xi_j t^j, \qquad (Q_K f)(t) = \sum_{j=-\infty}^{-1} \xi_j t^j \qquad (|t| = 1)$$

and hence, if f is real-valued, then $\bar{\xi}_j = \xi_{-j}$ and $P_K f = \overline{Q_K f} + \xi_0$ with ξ_0 real. Applied to our sequence $\{f_n\}$ this gives

$$f_n - c_n = P_K f_n + \overline{P_K f_n} \qquad (c_n \text{ real};\ n = 1, 2, \ldots). \tag{5}$$

Since $P_K f_n \to 0$ and since $0 \leq f_n \leq 1$, we deduce from (5) the existence of a convergent subsequence $\{f_{n_k}\} \subset \mathbf{C}^\infty(K)$. This, however, is impossible by virtue of the properties of the sequence $\{f_n\}$ mentioned above. The contradiction obtained shows that in the case $\Gamma = K$ the function a has at most finitely many zeros of integral orders. Analogously the same can be proved for the function b.

Now consider the case of an arbitrary curve system Γ. Denote by $\Gamma_0 \subset \Gamma$ the closed curve containing the point t_0. Clearly, $A_0 = A_{\Gamma_0}$ is a Φ-operator on the space $\mathbf{C}^\infty(\Gamma_0)$. There is a one-to-one correspondence between the points $t \in \Gamma_0$ and the points $s \in K$:

$$t = t(s), \tag{6}$$

where $t(s) \in \mathbf{C}^\infty(K)$ and $t'(s) \neq 0$ for $s \in K$. The following operator on $\mathbf{C}^\infty(K)$ corresponds to the operator A_0 through the mapping (6):

$$(\tilde{A}_0\varphi)(s) := a(s)(P_K\varphi)(s) + b(s)(Q_K\varphi)(s) + \int_K k(s, \sigma)\,\varphi(\sigma)\,d\sigma \qquad (s \in K).$$

Here $a(s) = a(t(s))$, $b(s) = b(t(s))$ are in $\mathbf{C}^\infty(K)$ and $k(s, \sigma)$ is in $\mathbf{C}^\infty(K \times K)$. Since A_0 is a Φ-operator, $\tilde{A}_0$ is a Φ-operator on $\mathbf{C}^\infty(K)$. However, by what has already been proved, this contradicts the fact that $a^{(k)}(s_0) = 0$ for $k = 0, 1, 2, \ldots$ and with $t_0 := t(s_0)$. Thus, the function a has an at most finite number of zeros of integral orders on Γ. The same assertion for b can be proved equally. ∎

Remark. The preceding proof is SILBERMANN'S [3]. It actually shows that the conditions of Theorem 10.2 are even necessary for A to be a Φ_+-operator on $[\mathbf{C}^\infty(\Gamma)]^s$. The corresponding statement for

Φ_--operators is also true (see Köhler and Silbermann [2]). Theorem 10.1 was established by Prössdorf [4–5, 7].

10.3. We now consider the singular operator

$$\mathbf{A} = AP + BQ + T,$$

where

$$(T\varphi)(t) := \int_\Gamma T(t, \tau)\, \varphi(\tau)\, d\tau \quad \text{and} \quad T \in [\mathbf{C}^\infty(\Gamma\times\Gamma)]^{s\times s},$$

on the distribution space $[\mathbf{C}^{-\infty}(\Gamma)]^s = [\mathbf{C}^\infty(\Gamma)]^{*,s}$. The matrix functions A and B (belonging to $[\mathbf{C}^\infty(\Gamma)]^{s\times s}$) are subject to the same conditions as in the previous subsection.

Put $X = [\mathbf{C}^\infty(\Gamma)]^s$, $E = H = \mathbf{L}_2^s(\Gamma)$, and recall the Sections 4.6 and 4.7 of Chapter I. Assign to each vector function $f \in \mathbf{L}_2^s(\Gamma)$ a regular functional $f \in [\mathbf{C}^{-\infty}(\Gamma)]^s$ through the formula

$$(\varphi, f) = \int_\Gamma \varphi(t)\, f(t)\, dt \qquad (\varphi \in [\mathbf{C}^\infty(\Gamma)]^s). \tag{7}$$

In this way an embedding $E \subset [\mathbf{C}^{-\infty}(\Gamma)]^s$ is defined. The Poincaré-Bertrand commutation formula and the convention (7) give

$$(\varphi, \mathbf{A}f) = (\mathbf{A}'\varphi, f) \qquad (\varphi \in [\mathbf{C}^\infty(\Gamma)]^s) \tag{8}$$

for every $f \in \mathbf{L}_2^s(\Gamma)$ (see also formula (4.4), Chapter III). Herein $\mathbf{A}'$ denotes the transposed integral operator

$$\mathbf{A}' = PB' + QA' + T',$$

where A', B' are the matrix functions resulting from A, B by transposition and where

$$(T'\varphi)(t) = \int_\Gamma T'(\tau, t)\, \varphi(\tau)\, d\tau.$$

By (8) the operator $\mathbf{A}$ is defined on the whole distribution space $[\mathbf{C}^{-\infty}(\Gamma)]^s$.

The Theorems 10.1 and 10.2 together with Theorem 4.1, Chapter I, yield the following theorem.

Theorem 10.3. *For the singular operator* $\mathbf{A}$ *to be Fredholm on the distribution space* $[\mathbf{C}^{-\infty}(\Gamma)]^s$ *it is necessary and sufficient that each of the two functions* (3) *have at most finitely many zeros of integral orders on* Γ.

If this condition is fulfilled, then the index of the operator $\mathbf{A}$ *is given by*

$$\operatorname{Ind}_* \mathbf{A} = \operatorname{ind} \left[\frac{\det B(t)}{\det A(t)} \, \frac{(t - \alpha)^{|m|}}{(t^{-1} - \beta^{-1})^{|n|}}\right]. \tag{9}$$

Remark. The proof of Theorem 10.3 given here is due to Prössdorf [4, 7].

The scalar singular integral operator ($s = 1$) on the distribution space $\mathbf{C}^{-\infty}(\Gamma)$ was considered by Kosulin [1] for the first time (however, under very restrictive hypotheses on the zeros of the symbol). He was also the first to state the theorem on the normal solvability of the corresponding equation in generalized functions. Notice, however, that the proof of this assertion given in Kosulin [2] is incorrect. Kosulin [3] also considered the case of open curves. In certain other classes of distributions and by means of other methods (using integrals in the sense of Hadamard), singular integral equations with degenerate coefficients were studied by Chikin [1] (see also Rogozhin [1]).

10.4. Now consider the paired operator $\mathbf{A} = AP + BQ$ ($\mathbf{A} = PA + QB$) under the hypotheses of Section 10.2. It follows from the formulas (9) and (9.13) that $\operatorname{Ind}_* \mathbf{A}$ is equal to the index of the operator $A \in \mathscr{L}(\tilde{\mathbf{L}}_p^s, \mathbf{L}_p^s(\Gamma))$, where $\tilde{\mathbf{L}}_p^s = \tilde{\mathbf{L}}_p^s(\alpha, \mu, S_+; \beta, \nu, S_-)$ (cf. 9.3). Let $\boldsymbol{D}$ denote the singular operator defined by formula (9.6) and let $\boldsymbol{B} = \boldsymbol{D}'$ be the transposed operator. Given any $f \in \tilde{\mathbf{L}}_p^s$ put

$$f(\varphi) = \int_\Gamma (\boldsymbol{B}^{-1}\varphi)(t)\,(\boldsymbol{D}f)(t)\,dt \qquad (\varphi \in [\mathbf{C}^\infty(\Gamma)]^s).$$

So $\tilde{\mathbf{L}}_p^s \subset [\mathbf{C}^{-\infty}(\Gamma)]^s$ and Af (compare 9.4) coincides with Af as given by formula (8). Thus, formula (8) provides an extension of the operator $A \in \mathscr{L}(\tilde{\mathbf{L}}_p^s, \mathbf{L}_p^s(\Gamma))$ to the distribution space $[\mathbf{C}^{-\infty}(\Gamma)]^s$. Still taking into account that A is Fredholm on both $[\mathbf{C}^{-\infty}(\Gamma)]^s$ and the pair of spaces $\tilde{\mathbf{L}}_p^s, \mathbf{L}_p^s(\Gamma)$ it is easy to obtain the following theorem.

Theorem 10.4. *The generalized solutions $\varphi \in [\mathbf{C}^{-\infty}(\Gamma)]^s$ of the homogeneous equation $A\varphi = 0$ coincide with the solutions of this equation in the space $\tilde{\mathbf{L}}_p^s(\Gamma)$ $(1 < p < \infty)$.*

§ 11. Singular matrix operators with degenerate coefficients of constant rank

Let Γ be a closed Lyapunov curve systems of the class C^∞. We consider the following system of singular integral equations in the space $[\mathbf{C}^\infty(\Gamma)]^n$:

$$A\varphi := AP\varphi + BQ\varphi + T\varphi = f, \tag{1}$$

where

$$(T\varphi)(t) := \frac{1}{2\pi i}\int_\Gamma K_1(t,\tau)\ln\left(1 - \frac{t}{\tau}\right)\varphi(\tau)\,d\tau$$
$$+ \frac{1}{2\pi i}\int_\Gamma K_2(t,\tau)\ln\left(1 - \frac{\tau}{t}\right)\varphi(\tau)\,d\tau + \int_\Gamma K_3(t,\tau)\,\varphi(\tau)\,d\tau. \tag{1'}$$

Here we suppose that $A(t)$, $B(t)$, and $K_j(t,\tau)$ $(j = 1, 2, 3)$ are infinitely differentiable matrix functions of the order n. Given a fixed $t \in \Gamma$ the expression $\ln(1 - \tau/t)$ (resp. $\ln(1 - t/\tau)$) is defined as that branch of the function which is continuous for $\tau \in D_\Gamma^+ \cup \Gamma$, $\tau \neq t$ (resp. $\tau \in D_\Gamma^- \cup \Gamma$, $\tau \neq t$) and takes the value zero at $\tau = 0$ (resp. $\tau = \infty$). We still introduce the following notation:

$$K_1(t,t) =: \gamma(t), \qquad K_2(t,t) =: \delta(t). \tag{2}$$

In the sequel we assume that the condition

$$\det A(t) \neq 0, \qquad \det B(t) \neq 0 \qquad (t \in \Gamma) \tag{3}$$

is violated at *all* points on Γ. We also make the following additional assumptions:

The rank of the matrix functions A and B is constant on Γ (4)

and

$$\det G_1(t) \neq 0, \qquad \det G_2(t) \neq 0 \qquad (t \in \Gamma), \tag{5}$$

where the $n \times n$ matrix functions G_1 and G_2 are defined as follows. For arbitrarily fixed $t \in \Gamma$ denote by $\theta_1(t)$, $\theta_2(t)$ matrices with a maximal number of linearly independent columns such that $A(t)\,\theta_1(t) = 0$, $B(t)\,\theta_2(t) = 0$. Then let $G_1(t)$ be formed by the linearly independent columns of the matrix $A(t)$ and the columns of the matrix $\gamma(t)\, G_1(t)$; $G_2(t)$ is defined analogously.

Now put $r_1 = \operatorname{rank} A(t)$ and $r_2 = \operatorname{rank} B(t)$. Furthermore, let σ_1 and σ_2 be matrix functions in $[\mathbf{C}^\infty(\Gamma)]^{n\times n}$ satisfying the following conditions:

$$\det \sigma_1(t) \neq 0, \qquad \det \sigma_2(t) \neq 0 \qquad (t \in \Gamma), \tag{6}$$

$$A(t)\, \sigma_1^{(j)}(t) = 0, \qquad B(t)\, \sigma_2^{(p)}(t) = 0 \qquad (t \in \Gamma) \tag{7}$$

$$(j = r_1 + 1, \ldots, n;\ p = r_2 + 1, \ldots, n),$$

with $\sigma_1^{(l)}(t), \sigma_2^{(l)}(t)$ $(l = 1, \ldots, n)$ the columns of the matrices $\sigma_1(t), \sigma_2(t)$. The existence of such matrices follows from our hypothesis (4).

Finally, let $\Omega_1(t)$ denote the matrix whose columns are $A(t)\, \sigma_1^{(k)}(t)$, $-t\gamma(t)\, \sigma_1^{(j)}(t)$ $(k = 1, \ldots, r_1;\ j = r_1 + 1, \ldots, n)$ and $\Omega_2(t)$ the matrix whose columns are $B(t)\, \sigma_2^{(l)}(t)$, $t\, \delta(t)\, \sigma_2^{(p)}(t)$ $(l = 1, \ldots, r_2;\ p = r_2 + 1, \ldots, n)$. Under the hypotheses (4), the condition (5) is equivalent to the following one:

$$\det \Omega_1(t) \neq 0, \qquad \det \Omega_2(t) \neq 0 \qquad (t \in \Gamma). \tag{8}$$

Beside equation (1) we still consider the homogeneous transposed equation on $[\mathbf{C}^\infty(\Gamma)]^n$

$$A'\psi := BP'\psi + QA'\psi + T'\psi = 0, \tag{9}$$

where

$$(T'\psi)(t) := -\frac{1}{2\pi i} \int_\Gamma K_1'(\tau, t) \ln \left(1 - \frac{\tau}{t}\right) \psi(\tau)\, d\tau$$
$$+ \frac{1}{2\pi i} \int_\Gamma K_2'(\tau, t) \ln \left(1 - \frac{t}{\tau}\right) \psi(\tau)\, d\tau + \int_\Gamma K_3'(\tau, t)\, \psi(\tau)\, d\tau,$$

with K' the transposed matrix of K.

Theorem 11.1. *Let the conditions* (4) *and* (5) *be satisfied. Then both*

$$\alpha = \dim \operatorname{Ker} \mathbf{A}, \qquad \alpha' = \dim \operatorname{Ker} \mathbf{A}'$$

are finite, and the equation (1) *has a solution if and only if*

$$\int_\Gamma f(t)\, \psi^{(j)}(t)\, dt = 0 \qquad (j = 1, \ldots, \alpha'), \tag{10}$$

where $\psi^{(1)}, \ldots, \psi^{(\alpha')}$ are the linearly independent solutions of equation (9).

Theorem 11.2. *If the conditions* (4) *and* (5) *are fulfilled, then*

$$\alpha - \alpha' = \operatorname{ind} \left[\frac{\det (\Omega_2 \sigma_1)}{\det (\Omega_1 \sigma_2)}\right]. \tag{11}$$

Corollary 11.1. *If the conditions* (4) *and* (5) *are satisfied, then* $\mathbf{A}$ *is a Fredholm operator on* $[\mathbf{C}^\infty(\Gamma)]^n$ *and its index is given by* (11).

The proof of the Theorems 11.1 and 11.2 is based on the following lemma.

Lemma 11.1. *For any functions $\varphi, \psi \in \mathbf{C}^\infty(\Gamma)$ there exist uniquely determined functions $\mu, \nu \in \mathbf{C}^\infty(\Gamma)$ such that*

$$\varphi = \mu + P_\Gamma t\mu' =: T_1\mu, \tag{12}$$

$$\psi = P_\Gamma \nu + tQ_\Gamma \nu' =: T_2\nu, \tag{13}$$

where $\mu'(t) = d\mu/dt$.

Proof. Consider the Cauchy type integral

$$\Phi_\mu(z) = \frac{1}{2\pi i} \int_\Gamma \frac{\mu(\tau)\, d\tau}{\tau - z}.$$

A few application of the Sokhotski-Plemelj formulas (see (5.11), Chapter II) and the uniqueness theorem for analytic functions shows that the representations (12), (13) are equivalent to the following equations:

$$\begin{aligned} \Phi_\varphi(z) &= \frac{d}{dz} \Phi_{t\mu}(z) \quad \text{for} \quad z \in D_\Gamma^+, \\ \Phi_\varphi(z) &= \Phi_\mu(z) \quad \text{for} \quad z \in D_\Gamma^-; \end{aligned} \tag{14}$$

$$\begin{aligned} \Phi_\psi(z) &= \Phi_\nu(z) \quad \text{for} \quad z \in D_\Gamma^+, \\ \Phi_\psi(z) &= z \frac{d}{dz} \Phi_\nu(z) \quad \text{for} \quad z \in D_\Gamma^-. \end{aligned} \tag{15}$$

(14) and (15), on its hand, are equivalent to the following:

$$\begin{aligned} -\frac{1}{2\pi i} \int_\Gamma \varphi(\tau) \ln\left(1 - \frac{z}{\tau}\right) d\tau + c &= \Phi_{t\mu}(z) \quad \text{for} \quad z \in D_\Gamma^+, \\ \Phi_\varphi(z) &= \Phi_\mu(z) \quad \text{for} \quad z \in D_\Gamma^-; \end{aligned} \tag{16}$$

$$\begin{aligned} \Phi_\nu(z) &= \Phi_\psi(z) \quad \text{for} \quad z \in D_\Gamma^+, \\ \Phi_\nu(z) &= -\frac{1}{2\pi i} \int_\Gamma \psi(\tau)\, \tau^{-1} \ln\left(1 - \frac{\tau}{z}\right) d\tau \quad \text{for} \quad z \in D_\Gamma^-; \end{aligned} \tag{17}$$

here c denotes an arbitrary constant.

Again using the Sokhotski-Plemelj formulas we see that (16) and (17) are satisfied if and only if

$$\mu(t) = -\frac{t^{-1}}{2\pi i} \int_\Gamma \varphi(\tau) \ln\left(1 - \frac{t}{\tau}\right) d\tau + (Q_\Gamma \varphi)(t),$$

$$\nu(t) = \frac{1}{2\pi i} \int_\Gamma \psi(\tau)\, \tau^{-1} \ln\left(1 - \frac{\tau}{t}\right) d\tau + (P_\Gamma \psi)(t).$$

The assertion of the lemma now follows from the last two equations and Corollary 6.1, Chapter II. ∎

Proof of the Theorems 11.1 and 11.2. In view of Lemma 11.1, the solution of (1) may be sought in the form

$$\varphi = \sigma_1 V_1 \sigma_1^{-1} \sigma_2 V_2 \sigma_2^{-1} \mu =: V\mu \tag{18}$$

with $\mu = (\mu_1, \ldots, \mu_n) \in [C^\infty(\Gamma)]^n$. The operators V_1 and V_2 are defined as follows. If one puts $g = V_1 f$ and $h = V_2 f$, then

$$g_k = f_k, \quad k = 1, \ldots, r_1; \qquad g_k = T_1 f_k, \quad k = r_1 + 1, \ldots, n;$$

$$h_j = f_j, \quad j = 1, \ldots, r_2; \qquad h_j = T_2 f_j, \quad j = r_2 = 1, \ldots, n.$$

Lemma 11.1 implies that V is invertible on $[C^\infty(\Gamma)]^n$.

Now insert (18) into the equation (1) and apply the Poincaré-Bertrand formula and the following formulae for integration by parts (see also Corollary 6.1, Chapter II):

$$\int_\Gamma k(t,\tau)\, f'(\tau) \ln\left(1-\frac{\tau}{t}\right) d\tau = \pi i k(t,t)\, f(t) - \int_\Gamma \frac{k(t,\tau)\, f(\tau)\, d\tau}{\tau - t} - \int_\Gamma \frac{dk(t,\tau)}{d\tau} f(\tau) \ln\left(1-\frac{\tau}{t}\right) d\tau,$$

$$\int_\Gamma k(t,\tau)\, f'(\tau) \ln\left(1-\frac{t}{\tau}\right) d\tau = -\pi i k(t,t)\, f(t) - \int_\Gamma \frac{t k(t,\tau)\, f(\tau)\, d\tau}{\tau(\tau - t)} - \int_\Gamma \frac{dk(t,\tau)}{d\tau} f(\tau) \ln\left(1-\frac{t}{\tau}\right) d\tau.$$

What results is the following singular integral equation in μ, which is equivalent to (1):

$$A^{(1)}\mu := A^{(1)}P\mu + B^{(1)}Q\mu + T^{(1)}\mu = f, \tag{19}$$

where

$$A^{(1)}(t) = \Omega_1(t)\, \sigma_1^{-1}(t), \qquad B^{(1)}(t) = \Omega_2(t)\, \sigma_2^{-1}(t).$$

The operator $T^{(1)}$ is given by (1') with certain well-defined matrix functions $K_j^{(1)}(t,\tau) \in [\mathbf{C}^\infty(\Gamma\times\Gamma)]^{n\times n}$ ($j = 1, 2, 3$).

Since the coefficients $A^{(1)}$, $B^{(1)}$ are subject to condition (3), every solution $\mu \in \mathbf{L}_p^n(\Gamma)$ ($1 < p < \infty$) of (19) belongs to $[\mathbf{C}^\infty(\Gamma)]^n$ (cf. § 2, Chapter V, and § 4, Chapter III). Analogously, if $\psi \in \mathbf{L}_q^n(\Gamma)$ ($1/p + 1/q = 1$) is a solution of (9), then $\psi \in [\mathbf{C}^\infty(\Gamma)]^n$. Because

$$\int_\Gamma (A\varphi)(t)\, \psi(t)\, dt = \int_\Gamma \varphi(t)\, (A'\psi)(t)\, dt,$$

for arbitrary $\varphi \in \mathbf{L}_p^n(\Gamma)$ and $\psi \in \mathbf{L}_q^n(\Gamma)$, these results together with those of § 2, Chapter V, give the Theorems 11.1 and 11.2. ▮

Remark 1. To obtain (19) from (1) through the transformation (18) we made use of condition (4) only (and did not use condition (5)). Thus, the Theorems 11.1 and 11.2 can be generalized as follows.

Define the operators $A^{(j)}$ as AV^j ($j = 1, 2, \ldots$) and suppose there is a positive integer k with the following properties: the coefficients of the operators $A^{(j)}$ satisfy condition (4) for $j = 0, 1, \ldots, k-1$ and condition (3) for $j = k$. Then the statement of Theorem 11.1 remains valid and the index of the operator A is equal to the index of the operator $A^{(k)}$.

Remark 2. The results of this section are due to Tovmasyan [1].

Chapter VII
Some problems leading to singular integral equations

§ 1. Singular integro-differential equations

We consider the singular integro-differential equation

$$(D\varphi)(t) := \sum_{j=0}^{m} \left[a_j(t)\, \varphi^{(j)}(t) + \frac{1}{\pi i} \int_\Gamma \frac{K_j(t, \tau)\, \varphi^{(j)}(\tau)\, d\tau}{\tau - t} \right] = f(t). \tag{1}$$

Here Γ is a Lyapunov curve, a_j, K_j, f are given functions, and $\varphi^{(j)}$ denotes the j-th derivate of the unknown function φ ($\varphi^{(0)} := \varphi$).

In the special case $m = 1$, $a_1(t) \equiv 0$, equation (1) is the well-known Prandtl integro-differential equation, which plays an important role in the theory of aircraft wings. Equations of the more general form (1) were studied by Magnaradze [1, 2] for the first time. In the following two subsections we present a simpler method for their treatment, which goes back to N. P. Vekua [2] (see also Mushkelishvili [1], § 117).

1.1. First assume $\Gamma = ab$ is an open Lyapunov curve. By putting $\varphi^{(m)} = \psi$ we then obtain

$$\varphi^{(k)}(t) = \int_\Gamma \omega_{m-k-1}(t, t_1)\, \psi(t_1)\, dt_1 + \frac{c_1 t^{m-k-1}}{(m-k-1)!} + \dots + c_{m-k}, \tag{2}$$

where

$$\omega_0(t, t_1) = 1 \quad \text{for} \quad t_1 \in at, \qquad \omega_0(t, t_1) = 0 \quad \text{for} \quad t_1 \notin at,$$

$$\omega_{k-1}(t, t_1) = \int_\Gamma \omega_0(t, t_2)\, \omega_{k-2}(t_2, t_1)\, dt_2, \qquad k = 2, 3, \dots, m,$$

with arbitrary constants $c_1, c_2, \dots, c_m$.

After inserting the expressions (2) into (1) we arrive at a singular integral equation in ψ of the form

$$(A\psi)(t) := a_m(t)\, \psi(t) + \frac{1}{\pi i} \int_\Gamma \frac{K_m(t, \tau)\, \psi(\tau)\, d\tau}{\tau - t} + \int_\Gamma k(t, \tau)\, \psi(\tau)\, d\tau$$

$$= f(t) - \sum_{k=1}^{m} c_k g_k(t) \tag{3}$$

with

$$k(t, \tau) = \sum_{k=0}^{m-1} a_k \omega_{m-k-1}(t, \tau) + \sum_{j=0}^{m-1} \frac{1}{\pi i} \int_\Gamma \frac{K_j(t, t_1)\, \omega_{m-j-1}(t_1, \tau)}{t_1 - t}\, dt_1$$

and with g_k ($k = 1, \dots, m$) certain well-defined linearly independent functions which can be easily determined explicitly. Note that (3) is an equation of the form (3.9), Chapter I. It is equivalent to the original equation (1) in the following sense:

If ψ satisfies (3) with certain constants $c_1, \ldots, c_m$, then the function φ given by (2) is a solution of (1). Conversely, if φ is a solution of (1), then $\psi = \varphi^{(m)}$ satisfies equation (3) for appropriately chosen constants $c_1, \ldots, c_m$.

Theorem 3.12, Chapter I, shows that (3) is a Fredholm problem[1]) (in $\mathbf{H}^\lambda(\Gamma)$, $0 < \lambda < 1$, or in $\mathbf{L}_p(\Gamma)$, $1 < p < \infty$) if and only if the singular integral operator A is Fredholm and that the index of the problem (3) is then given by

$$\tilde{\varkappa}(A) := \tilde{\alpha}(A) - \tilde{\beta}(A) = \operatorname{Ind} A + m.$$

From the equivalence of the equations (1) and (3) we obtain

$$\tilde{\alpha}(A) = \dim \operatorname{Ker} D, \qquad \tilde{\beta}(A) = \dim \operatorname{Coker} D$$

and thus $\tilde{\varkappa}(A) = \operatorname{Ind} D$.

Conditions for A to be a Fredholm operator were given in Chapter IV.

1.2. Now assume Γ is a closed Lyapunov curve. In that case the Vekua method described above also applies, however, one has to take into account that the functions $\varphi^{(k)}$, $k = 0, 1, \ldots, m-1$, resulting from $\varphi^{(m)} = \psi$ by the formulae (2) may be multiple-valued. Therfeore, in the case of a closed curve Γ we choose a somewhat other way.

For the sake of simplicity, assume that the functions $a_j(t)$ and $b_j(t) = K_j(t, t)$ ($j = 0, 1, \ldots, m$) are continuous on Γ and that the kernels $K_j(t, \tau)$ are of the form

$$K_j(t, \tau) = b_j(t) + (\tau - t)\, T_j(t, \tau) \qquad (j = 0, 1, \ldots, m),$$

with $T_j(t, \tau)$ kernels generating compact integral operators on $\mathbf{L}_p(\Gamma)$ $(1 < p < \infty)$. Then the operator D defined by (1) is a linear bounded operator from $\mathbf{W}_p^{(m)}(\Gamma)$ into $\mathbf{L}_p(\Gamma)$. Consider (1) with $f \in \mathbf{L}_p(\Gamma)$ and $\varphi \in \mathbf{W}_p^{(m)}(\Gamma)$. We have $D = D_0 + T$, where

$$(D_0\varphi)(t) := a_m(t)\, \varphi^{(m)}(t) + \frac{1}{\pi \mathrm{i}} \int_\Gamma \frac{K_m(t, \tau)\, \varphi^{(m)}(\tau)\, \mathrm{d}\tau}{\tau - t} \tag{4}$$

and

$$(T\varphi)(t) := \sum_{j=0}^{m-1} \left[a_j(t)\, \varphi^{(j)}(t) + \frac{1}{\pi \mathrm{i}} \int_\Gamma \frac{K_j(t, \tau)\, \varphi^{(j)}(\tau)\, \mathrm{d}\tau}{\tau - t} \right].$$

By virtue of Sobolev's embedding theorem, the operator T is compact from $\mathbf{W}_p^{(m)}(\Gamma)$ into $\mathbf{L}_p(\Gamma)$.

The operator (4) can be written as product $D_0 = KC$, with K the singular integral operator

$$(K\varphi)(t) := a_m(t)\, \varphi(t) + \frac{1}{\pi \mathrm{i}} \int_\Gamma \frac{K_m(t, \tau)\, \varphi(\tau)\, \mathrm{d}\tau}{\tau - t}$$

and C the simple differential operator $(C\varphi)(t) := \varphi^{(m)}(t)$. For the operator $C \in \mathscr{L}(\mathbf{W}_p^{(m)}(\Gamma), \mathbf{L}_p(\Gamma))$ we have $\dim \operatorname{Ker} C = m$ and $\dim \operatorname{Coker} C = m$, since the equation $C\varphi = 0$ has the m linearly independent solutions $\varphi_j(t) = t^{j-1}$ ($j = 1, \ldots, m$) and since the equation $C\varphi = \psi$ is solvable if and only if ψ satisfies the following conditions:

$$\int_\Gamma \psi(\tau_1)\, \mathrm{d}\tau_1 = 0, \qquad \int_\Gamma \int_{t_0}^{\tau_2} \psi(\tau_1)\, d\tau_1\, d\tau_2 = 0, \ldots,$$

$$\int_\Gamma \int_{t_0}^{\tau_m} \cdots \int_{t_0}^{\tau_2} \psi(\tau_1)\, \mathrm{d}\tau_1 \ldots \mathrm{d}\tau_{m-1}\, \mathrm{d}\tau_m = 0.$$

[1]) For the definition of a Fredholm problem recall Section 3.5, Chapter I.

Here t_0 is a fixed point on Γ. Due to Theorem 3.1, Chapter III, the operator K is a Φ-operator on $\mathbf{L}_p(\Gamma)$ if and only if

$$a_m(t) + b_m(t) \neq 0, \qquad a_m(t) - b_m(t) \neq 0 \qquad (t \in \Gamma); \tag{5}$$

in that case

$$\operatorname{Ind} K = \operatorname{ind}\left[\frac{a_m(t) - b_m(t)}{a_m(t) + b_m(t)}\right].$$

Putting these things together and taking into consideration the results of § 3, Chapter I, we arrive at the following:

The operator $D \in \mathscr{L}(\mathbf{W}_p^{(m)}(\Gamma), \mathbf{L}_p(\Gamma))$ is a Φ-operator if and only if the condition (5) is fulfilled. If it is satisfied, then

$$\operatorname{Ind} D = \operatorname{ind}\left[\frac{a_m(t) - b_m(t)}{a_m(t) + b_m(t)}\right].$$

In the case of Hölder-continuous datas in equation (1), the same statement holds for the operator $D \in \mathscr{L}(\mathbf{C}^{m,\lambda}(\Gamma), \mathbf{H}^{\lambda}(\Gamma))$.

Finally, notice that all arguments of the present section apply to the system case.

Remark. In Section 1.2 we followed Prössdorf and v. Wolfersdorf [1].

1.3. Elschner [1–4] studied the degenerate singular integro-differential equation (for $m \geqq 1$), that is, the case where the condition (5) is violated. In this section we present some of his results. For lack of place, the proofs will be omitted.

Let the coefficients a_j and b_j be subject to the following conditions:

$$c_j(t) := a_j(t) + b_j(t) = [\varrho_1(t)]^j\, \tilde{c}_j(t),$$
$$d_j(t) := a_j(t) - b_j(t) = [\varrho_2(t)]^j\, \tilde{d}_j(t), \qquad j = 0, \ldots, m, \tag{6}$$

where

$$\varrho_1(t) = \prod_{i=1}^{r} (t - \alpha_i), \quad \alpha_i \in \Gamma, \quad \alpha_i \neq \alpha_h, \quad i \neq h, \quad i, h = 1, \ldots, r;$$

$$\varrho_2(t) = \prod_{l=1}^{s} (t - \beta_l), \quad \beta_l \in \Gamma, \quad \beta_l \neq \beta_n, \quad l \neq n, \quad l, n = 1, \ldots, s;$$

$$\tilde{c}_j, \tilde{d}_j \in \mathbf{C}^{k+1}(\Gamma), \qquad j = 0, \ldots, m; \tag{7}$$

$$\tilde{c}_m(t) \neq 0, \qquad \tilde{d}_m(t) \neq 0 \qquad (t \in \Gamma). \tag{8}$$

Here Γ denotes a sufficiently smooth closed Lyapunov curve (say, of the class C^∞).

Consider the singular integro-differential operator

$$(D\varphi)(t) := \sum_{j=0}^{m} \left[a_j(t)\, \varphi^{(j)}(t) + \frac{b_j(t)}{\pi \mathrm{i}} \int_\Gamma \frac{\varphi^{(j)}(\tau)\, \mathrm{d}\tau}{\tau - t}\right],$$

which can also be written in the form[1])

$$D = \sum_{j=0}^{m} (c_j\, D_t^j P_\Gamma + d_j D_t^j Q_\Gamma), \qquad D := \frac{\mathrm{d}}{\mathrm{d}t}. \tag{9}$$

[1]) Note that $D_t S_\Gamma \varphi = S_\Gamma D_t \varphi$ for all $\varphi \in \mathbf{W}_p^{(1)}(\Gamma)$ (see Theorem 6.2, Chapter II).

Under the hypotheses (6), the operator $\sum_{j=0}^{m} c_j D_t^j$ (resp. $\sum_{j=0}^{m} d_j D_t^j$) is just a *Fuchsian differential operator*[1]) with the singularities located at the points α_i, $i = 1, \ldots, r$ (resp. β_l, $l = 1, \ldots, s$). In the case of only one singularity and $\Gamma = [a, b] \subset \mathbb{R}^1$ such operators were studied by GLUSHKO [2] (see also PRÖSSDORF and SILBERMANN [11], § 3, Chapter III, ELSCHNER [4], Chapter 1, ELSCHNER and SILBERMANN [1]).

If the domain is defined as

$$X_p^k := \{\varphi \in \mathbf{W}_p^{(k)}(\Gamma) : \varrho_1^j D_t^j P_\Gamma \varphi, \varrho_2^j D_t^j Q_\Gamma \varphi \in \mathbf{W}_p^{(k)}(\Gamma), j = 1, \ldots, m\},$$

the operator D is a closed operator on the space $\mathbf{W}_p^{(k)}(\Gamma)$, $k \geqq 0$. Further, under the norm

$$\|\varphi\| = \sum_{j=0}^{m} (\|\varrho_1^j D_t^j P_\Gamma \varphi\|_{\mathbf{W}_p^{(k)}(\Gamma)} + \|\varrho_2^j D_t^j Q_\Gamma \varphi\|_{\mathbf{W}_p^{(k)}(\Gamma)})$$

X_p^k is a Banach space and D is in $\mathcal{L}(X_p^k, \mathbf{W}_p^{(k)}(\Gamma))$.

With the operator (9) associate the characteristic equations:

$$\begin{aligned} a_m^{(i)}\mu(\mu - 1) \ldots (\mu - m + 1) + a_{m-1}^{(i)}\mu(\mu - 1) \ldots (\mu - m + 2) + \ldots \\ + a_1^{(i)}\mu + a_0^{(i)} = 0, \qquad i = 1, \ldots, r \end{aligned} \tag{10}$$

and

$$\begin{aligned} b_m^{(l)}\nu(\nu - 1) \ldots (\nu - m + 1) + b_{m-1}^{(l)}\nu(\nu - 1) \ldots (\nu - m + 2) + \ldots \\ + b_1^{(l)}\nu + b_0^{(l)} = 0, \qquad l = 1, \ldots, s, \end{aligned} \tag{11}$$

where

$$a_j^{(i)} = \frac{c_j(t)}{(t - \alpha_i)^j}\bigg|_{t=\alpha_i}, \qquad b_j^{(l)} = \frac{d_j(t)}{(t - \beta_l)^j}\bigg|_{t=\beta_l}$$

($i = 1, \ldots, r$; $l = 1, \ldots, s$; $j = 0, \ldots, m$). Let $\mu_j^{(i)}$ and $\nu_j^{(l)}$ ($j = 1, \ldots, m$) be the roots of the equations (10) and (11), respectively. Assume that

$$\operatorname{Re} \mu_j^{(i)} \neq k - \frac{1}{p}, \qquad \operatorname{Re} \nu_j^{(l)} \neq k - \frac{1}{p} \tag{12}$$

($i = 1, \ldots, r$; $l = 1, \ldots, s$; $j = 1, \ldots, m$). Denote by $\chi_i(k, p)$ (resp. $\tilde{\chi}_l(k, p)$) the number of roots of equation (10) (resp. (11)) satisfying the inequality $\operatorname{Re} \mu_j^{(i)} < k - 1/p$ (resp. $\operatorname{Re} \nu_j^{(l)} > k - 1/p$). Finally, put

$$\chi(k, p) = \sum_{i=1}^{r} \chi_i(k, p), \quad \tilde{\chi}(k, p) = \sum_{l=1}^{s} \tilde{\chi}_l(k, p).$$

Then we have the following

Theorem 1.1. *Under the hypotheses* (6), (7), (8), *and* (12), *the operator* $D \in \mathcal{L}(X_p^k, \mathbf{W}_p^{(k)}(\Gamma))$ *is a Φ-operator and its index is*

$$\operatorname{Ind} D = \frac{1}{2\pi} [\arg \tilde{d}_m(t)/\tilde{c}_m(t)]_\Gamma - \chi(k, p) + \tilde{\chi}(k, p).$$

If the condition (12) *is violated, then D is not Fredholm on that pair of spaces.*

An analogous theorem was established by ELSCHNER [1] for the space $\mathbf{C}^{k,\lambda}(\Gamma)$ ($0 < \lambda < 1$); the only difference is that then in condition (12) the number $-1/p$ has to be replaced by the number λ.

[1]) Cf. CODDINGTON and LEVINSON [1].

If (6) holds and if

$$\tilde{c}_j, \tilde{d}_j \in \mathbf{C}^\infty(\Gamma), \qquad j = 0, \ldots, m, \tag{7'}$$

then the operator D is a linear bounded operator on the space $\mathbf{C}^\infty(\Gamma)$. By passing to the projective limit of the spaces X_p^k and $\mathbf{W}_p^{(k)}(\Gamma)$ as $k \to \infty$, we immediately obtain from Theorem 1.1 the following result.

Theorem 1.2. *Under the hypotheses* (6), (7′), *and* (8), *the operator* $D \in \mathscr{L}(\mathbf{C}^\infty(\Gamma))$ *is a* Φ-*operator and its index is equal to*

$$\operatorname{Ind} D = \frac{1}{2\pi} [\arg \tilde{d}_m(t)/\tilde{c}_m(t)]_\Gamma - mr.$$

Thus, it turns out that under the hypotheses of Theorem 1.2 the index of the singular integro-differential operator D on the space $\mathbf{C}^\infty(\Gamma)$ depends on the coefficients c_m and d_m only. But notice that this is, in general, no longer true if all what we assume is merely that each of the functions c_m and d_m have finitely many zeros of integral order on Γ. The operator $D \in \mathscr{L}(\mathbf{C}^\infty(\Gamma))$ is nevertheless a Φ-operator in that case; but the dependence of its index on the coefficients c_j, d_j $(0 \leq j \leq m)$ is, in general, of rather intricate nature (in this connection see Elschner [3], and [4], Chapter 5). However, the above statement on the index again holds in the space $\mathbf{H}(\Omega)$ of all functions analytic in an open circular annulus Ω ($\Gamma \subset \Omega$ the unit circle) (see also Elschner [2]).

1.4. Elschner [1] also applied his results presented in the foregoing subsection to certain degenerate singular integral equations of the form (11.1)–(11.1′), Chapter VI. Here we introduce Elschner's main results in this direction.

Thus, consider the singular integral operator

$$(A\varphi)(t) := a(t)\,\varphi(t) + \frac{b(t)}{\pi i} \int_\Gamma \frac{\varphi(\tau)\,d\tau}{\tau - t} + (T\varphi)(t) \tag{13}$$

with

$$\begin{aligned}(T\varphi)(t) &= \frac{1}{2\pi i} \int_\Gamma K_1(t,\tau) \ln\left(1 - \frac{t}{\tau}\right) \varphi(\tau)\,d\tau \\ &\quad + \frac{1}{2\pi i} \int_\Gamma K_2(t,\tau) \ln\left(1 - \frac{\tau}{t}\right) \varphi(\tau)\,d\tau + \int_\Gamma K_3(t,\tau)\,\varphi(\tau)\,d\tau,\end{aligned} \tag{14}$$

suppose

$$c(t) := a(t) + b(t) = \varrho_1(t)\,\tilde{c}(t), \qquad d(t) := a(t) - b(t) = \varrho_2(t)\,\tilde{d}(t), \tag{15}$$

where $\varrho_1(t)$ and $\varrho_2(t)$ are of the same form as in 1.3; also suppose

$$\tilde{c}, \tilde{d} \in \mathbf{C}^\infty(\Gamma), \qquad K_i \subset \mathbf{C}^\infty(\Gamma \times \Gamma) \qquad (i = 1, 2, 3); \tag{16}$$

$$\tilde{c}(t) \neq 0, \qquad \tilde{d}(t) \neq 0 \qquad (t \in \Gamma). \tag{17}$$

Let the curve Γ be a closed Lyapunov one of the class C^∞.

By differentiating the integrals in (14) similarly as in the proof of Lemma 6.1 of Chapter II, it is easy to see that T is a compact operator on the space $\mathbf{W}_p^{(k)}(\Gamma)$ $(k \geq 0)$. Consequently, the operator (13) is a linear continuous operator on the space $\mathbf{W}_p^{(k)}(\Gamma)$.

With $D_p^{k+1}(A) := \{\varphi \in \mathbf{W}_p^{(k)}(\Gamma) : A\varphi \in \mathbf{W}_p^{(k+1)}(\Gamma)\}$ as domain, $A : \mathbf{W}_p^{(k)}(\Gamma) \to \mathbf{W}_p^{(k+1)}(\Gamma)$

is a closed operator. So $D_p^{k+1}(A)$ provided with the norm

$$\|\varphi\| = \|\varphi\|_{\mathbf{W}_p^{(k)}(\Gamma)} + \|A\varphi\|_{\mathbf{W}_p^{(k+1)}(\Gamma)}$$

becomes a Banach space and one has $A \in \mathscr{L}(D_p^{k+1}(A), \mathbf{W}_p^{(k+1)}(\Gamma))$.

By means of the substitution $\varphi = P_\Gamma \psi + t\, D_t \psi$ (see § 11, Chapter VI) Elschner transforms the equation (13) into an equivalent singular integro-differential equation the characteristic equations (10) and (11) of which have the following roots:

$$\mu^{(i)} = -c_0(t)\, c_1^{-1}(t)\, (t - \alpha_i)\big|_{t=\alpha_i} \qquad (i = 1, \ldots, r),$$

$$\nu^{(l)} = -d_0(t)\, d_1^{-1}(t)\, (t - \beta_l)\big|_{t=\beta_l} \qquad (l = 1, \ldots, s);$$

here $c_1(t) = tc(t)$, $d_1(t) = t\, d(t)$, $c_0(t) = c(t) - tK_1(t, t)$, $d_0(t) = tK_2(t, t)$. Suppose that

$$\operatorname{Re} \mu^{(i)} \neq k + 1 - \frac{1}{p}, \quad \operatorname{Re} \nu^{(l)} \neq k + 1 - \frac{1}{p} \tag{18}$$

and denote by $\chi(k + 1, p)$ (resp. $\tilde{\chi}(k + 1, p)$) the number of roots $\mu^{(i)}$ (resp. $\nu^{(l)}$) satisfying the inequality $\operatorname{Re} \mu^{(i)} < k + 1 - 1/p$ (resp. $\operatorname{Re} \nu^{(l)} > k + 1 - 1/p$). Then one has the following

Theorem 1.3. *Under the hypotheses* (15), (16), (17), *and* (18), *the singular integral operator* $A \in \mathscr{L}(D_p^{k+1}(A), \mathbf{W}_p^{(k+1)}(\Gamma))$ *is a Φ-operator with index*

$$\operatorname{Ind} A = \frac{1}{2\pi} [\arg \tilde{d}(t)/\tilde{c}(t)]_\Gamma - \chi(k + 1, p) + \tilde{\chi}(k + 1, p).$$

If the condition (18) *is violated, then A is not a Φ-operator on that pair of spaces.*

Elschner [1] also established an analogue of Theorem 1.3 for the pair of spaces $\mathbf{C}^{k,\lambda}(\Gamma)$, $\mathbf{C}^{k+1,\lambda}(\Gamma)$ $(0 < \lambda < 1)$; then in (18) the number $-1/p$ must be replaced by λ.

By passing to the projective limit of the spaces $D_p^{k+1}(A)$ and $\mathbf{W}_p^{(k+1)}(\Gamma)$ as $k \to \infty$, we immediately get from Theorem 1.2 the following theorem for the space $\mathbf{C}^\infty(\Gamma)$.

Theorem 1.4. *Under the hypotheses* (15), (16), *and* (17), *the singular integral operator* $A \in \mathscr{L}(\mathbf{C}^\infty(\Gamma))$ *is a Φ-operator with the index*

$$\operatorname{Ind} A = \frac{1}{2\pi} [\arg d(t)/c(t)]_\Gamma - r. \tag{19}$$

A remarkable feature of this result is the fact that the operators A and $A - T$ have the same index on the space $\mathbf{C}^\infty(\Gamma)$ (see the formula (19) as well as formula (9.12) and Theorem 10.1, Chapter VI), although, in general, the operator T is not compact on $\mathbf{C}^\infty(\Gamma)$.

§ 2. The generalized Riemann-Hilbert-Poincaré problem

2.1. Let G be a bounded and simply connected region in the complex plane $\mathbb{C}$ which contains the origin and whose boundary Γ is a closed Lyapunov curve.

Under the problem referred to in the title (also called *Problem V*) one understands the following.

To find functions $\Phi(z)$ which are holomorphic in G and satisfy the boundary condition

$$\operatorname{Re} (L\Phi)(t) = f(t) \qquad (t \in \Gamma), \tag{1}$$

where L is an integro-differential operator of the form

$$(L\Phi)(t) = \sum_{j=0}^{m} \{a_j(t)\,\Phi^{(j)}(t) + \int_\Gamma h_j(t,\tau)\,\Phi^{(j)}(\tau)\,d\sigma\} \tag{2}$$

with $\sigma = \sigma(\tau)$ the arc coordinate on Γ. The complex-valued functions $a_j(t)$ $(j = 0, 1, \ldots, m)$ and the real-valued function $f(t)$ are given on Γ and are supposed to be in $\mathbf{H}^\lambda(\Gamma)$ $(0 < \lambda < 1)$; the functions $h_j(t,\tau)$ $(j = 0, 1, \ldots, m)$ are also given and it is assumed that they are of the form

$$h_j(t,\tau) = \frac{h_j^0(t,\tau)}{|t-\tau|^\alpha} \qquad (0 \leqq \alpha < 1)$$

with certain functions $h_j^0(t,\tau)$ Hölder-continuous with respect to t and τ, and it is finally required that the corresponding integral operators

$$\int_\Gamma h_j(t,\tau)\,\varphi(\tau)\,d\sigma$$

map the space $\mathbf{H}^\lambda(\Gamma)$ into itself.

By $\Phi^{(j)}(t)$ will be understood the boundary values $[\Phi^{(j)}(t)]^+$ $(j = 0, 1, \ldots, m)$ of the j-th derivative of the function $\Phi(z)$; the existence of these boundary values is thus part of the hypothesis. Furthermore, we require that $\Phi^{(m)}(t) \in \mathbf{H}^\lambda(\Gamma)$.

In the special case $m = 0$ and $h_j(t,\tau) = 0$ the Problem V is known as the *Riemann-Hilbert-problem*; in that case the boundary condition (1) assumes the simpler form

$$\operatorname{Re}\,(a_0(t)\,\Phi(t)) = f(t) \qquad (t \in \Gamma).$$

The problem V was first raised and completely solved by I. N. Vekua [1]. Following I. N. Vekua we shall first transform the Problem V into an equivalent singular integral equation, which will then be studied with the methods of Section 3.5, Chapter I.

The well-known integral representation of I. N. Vekua (see also Muskhelishvili [1], § 69) gives

$$\Phi(z) = \int_\Gamma \mu(t)\left(1-\frac{z}{t}\right)^{m-1} \ln\left(1-\frac{z}{t}\right) ds + \int_\Gamma \mu(t)\,ds + iC \tag{3}$$

for $m \geqq 1$ and

$$\Phi(z) = \int_\Gamma \frac{\mu(t)\,ds}{1-z/t} + iC \quad \text{for} \quad m = 0 \tag{3'}$$

with a real-valued density function $\mu \in \mathbf{H}^\lambda(\Gamma)$ and certain real constant C. Note that μ and C are uniquely determined by Φ. Here $\ln(1 - z/t)$ is the branch vanishing (for fixed t) at $z = 0$; $s = s(t)$ denotes the arc coordinate on Γ.

Substitute (3) (resp. (3′)) into the boundary condition (1) and use the Sokhotski-Plemelj formulas. After a simple computation you get the following real singular integral equation in μ being equivalent to the Problem V (see also Muskhelishvili [1], §§ 71–72):

$$(A\mu)(t) := c(t)\,\mu(t) + \frac{d(t)}{\pi i}\int_\Gamma \frac{\mu(\tau)\,d\tau}{\tau - t} + \int_\Gamma K(t,\tau)\,\mu(\tau)\,d\tau = f(t) - Cg(t) \tag{4}$$

with

$$\begin{aligned} a(t) &:= c(t) + d(t) = (-1)^m\,(m-1)!\,\pi i t^{1-m}\overline{t'}a_m(t),\\ b(t) &:= c(t) - d(t) = (-1)^{m+1}\,(m-1)!\,\pi i \overline{t}^{1-m} t'\overline{a_m(t)},\\ g(t) &:= \operatorname{Re}\{i a_0(t) + i\int_\Gamma h_0(t,\tau)\,d\sigma\} \end{aligned} \tag{5}$$

and a weakly singular kernel of the form

$$K(t, \tau) = \frac{T(t, \tau) - T(t, t)}{t - \tau},$$

where $T(t, \tau)$ is a function Hölder-continuous with respect to t and τ; if $m = 0$, one has to set $(m-1)! = 1$ in the formulae (5).

In what follows we suppose that $a_m(t) \neq 0$ for $t \in \Gamma$. Then, by Chapter III, the operator A given by (4) is a Fredholm operator on the (complex or real[1]) space $\mathbf{H}^\lambda(\Gamma)$ and its index is equal to

$$\operatorname{Ind} A = \alpha(A) - \beta(A) = \operatorname{ind}\left[\frac{t^{m-1} t' \overline{a_m(t)}}{\bar{t}^{m-1}\, \bar{t}' a_m(t)}\right] = 2(m+n),$$

where $n = \operatorname{ind} \overline{a_m(t)} = (1/2\pi)\,[\arg \overline{a_m(t)}]_\Gamma$. Here $\alpha(A)$ is the number of linearly independent solutions of the homogeneous equation $A\mu = 0$ and $\beta(A)$ denotes the number of linearly independent solutions of the transposed homogeneous equation

$$(A'\nu)\,(t) := c(t)\,\nu(t) - \frac{1}{\pi i} \int_\Gamma \frac{d(\tau)\,\nu(\tau)}{\tau - t}\,d\tau + \int_\Gamma K(\tau, t)\,\nu(\tau)\,d\tau = 0. \tag{6}$$

Unless $g(t) \equiv 0$, equation (4) is a problem of the form (3.9), (3.10), Chapter I, with $p = 1$ and $q = 0$. By virtue of Theorem 3.12 of Chapter I, (4) and hence also Problem V is a Fredholm problem whose index is given by

$$\tilde{\varkappa}(A) = \tilde{\alpha}(A) - \tilde{\beta}(A) = 2(m+n) + 1. \tag{7}$$

The integers $\tilde{\alpha}(A)$ and $\tilde{\beta}(A)$ are, respectively, equal to the number of linearly independent solutions $\Phi_0(z)$ of the homogeneous Problem V[2] and the number of linearly independent solvability conditions for f in the inhomogeneous Problem V, with linear independence understood in the sense of linear combinations with real coefficients. Theorem 3.12, Chapter I, now gives that

$$\tilde{\alpha}(A) = \alpha(A) \quad \text{resp.} \quad \alpha(A) + 1, \qquad \tilde{\beta}(A) = \beta(A) - 1 \quad \text{resp.} \quad \beta(A). \tag{8}$$

Because of the arbitrary constant C in (3), if in the case $g(t) \equiv 0$ the integers $\tilde{\alpha}(A)$ and $\tilde{\beta}(A)$ are defined as above, (8) and (9) also hold in that case. The latter then takes the form

$$\tilde{\alpha}(A) = \alpha(A) + 1, \qquad \beta(A) = \beta(A). \tag{8'}$$

2.2. The question we shall ask now and which has already been asked by I. N. Vekua is under which conditions one has $\tilde{\beta}(A) = 0$, that is, which conditions ensure that the Problem V has a solution for every given function f. By virtue of (8'), if $g(t) \equiv 0$, then $\tilde{\beta}(A) = 0$ if and only if $\beta(A) = 0$. If $g(t) \not\equiv 0$, then, by Section 3.5 of Chapter I, for $\tilde{\beta}(A)$ to be 0 it is necessary and sufficient that $\beta(A) = 1 - r$, where $r = \dim Z_0$, $Z_0 = Z \cap \operatorname{Im} A$, Z consists of the multiples of the function g, and $\operatorname{Im} A$ consists of all functions $\varphi \in \mathbf{H}^\lambda(\Gamma)$ such that $\int_\Gamma \varphi(t)\,\nu(t)\,dt = 0$ for all solutions ν of equation (9).

---

[1]) See Theorem 3.13, Chapter I.

[2]) Look at equation (4) to see that $C = 0$ if $\mu(t) \equiv 0$, $f(t) \equiv 0$, and $g(t) \not\equiv 0$. So in the case of the homogeneous Problem V, to the trivial solution $\mu(t) \equiv 0$ of (4) there corresponds through (3) the trivial solution $\Phi(z) \equiv 0$ of the Problem V. Vice versa, if $\Phi(z) \equiv 0$, then, again through (3), $C = \operatorname{Im} \Phi(0) = 0$ and $\mu(t) \equiv 0$.

Therefore, owing to (8), in case $g(t) \not\equiv 0$ we have $\tilde{\beta}(A) = 0$ if and only if either $\beta(A) = 0$ (then $\operatorname{Im} A = \mathbf{H}^\lambda(\Gamma)$, $Z_0 = Z$ and, hence, $r = 1$) or $\beta(A) = 1$, where in the last case r must be equal to 0 (i.e. $Z \cap \operatorname{Im} A = \{0\}$), or, equivalently,

$$\int_\Gamma g(t)\, v(t)\, dt \neq 0 \tag{9}$$

with v the (up to a constant factor uniquely determined) solution of equation (6).

2.3. We finally look for conditions ensuring that $\tilde{\alpha}(A) = 0$, i.e., ensuring that the homogeneous Problem V has the trivial solution only. In view of (8′) this is impossible in case $g(t) \equiv 0$. If $g(t) \not\equiv 0$, then, due to Section 3.5 of Chapter I, $\tilde{\alpha}(A) = 0$ if and only if $\alpha(A) = 0$ and $g \notin \operatorname{Im} A$, that is, (9) must hold for at least one solution v of equation (6); in particular, one has necessarily $\beta(A) > 0$.

Because of (7), $\tilde{\varkappa}(A)$ is always an odd integer. Hence the equality $\tilde{\alpha}(A) = \tilde{\beta}(A)$ is impossible. On the other hand, we have $\tilde{\alpha}(A) = 1$ and $\tilde{\beta}(A) = 0$ in the case $g(t) \equiv 0$ if $\beta(A) = \alpha(A) = 0$ and in the case where $g(t) \not\equiv 0$ if $\beta(A) = \alpha(A) = 0$ or if $\beta(A) = \alpha(A) = 1$, where, however, (9) must be satisfied; in either case $\operatorname{Ind} A = 0$, that is, $n = -m$.

Remark. The argument of this section goes back to Prössdorf and v. Wolfersdorf [1]. It can be carried over to the case of the space $\mathbf{L}_p(\Gamma; \varrho)$ (in place of $\mathbf{H}^\lambda(\Gamma)$) and continuous datas, since Vekua's integral representation admits a corresponding generalization (see Khvedelidze [1]). Furthermore, it is not difficult to extend the results to the system case (see N. P. Vekua [1]).

2.4. Many important boundary value problems for partial differential equations of elliptic type can be transformed into Problem V (see I. N. Vekua [2, 3] and Khvedelidze [1]). As an example, we here consider the *Poincaré problem*, which arose from the mathematical theory of tides. The problem reads as follows:

Find a function u which is harmonic in some region G and satisfies the condition

$$A(s) \frac{du}{dn} + B(s) \frac{du}{ds} + c(s)\, u = f(s) \tag{10}$$

on the boundary Γ of G. Here $A(s), B(s), c(s), f(s) \in \mathbf{H}^\lambda(\Gamma)$ are given real-valued functions, s is the arc coordinate on Γ and n is the outer normal to Γ. The boundary condition (10) can also be written in the form

$$a(t) \frac{\partial u}{\partial x} + b(t) \frac{\partial u}{\partial y} + c(t)\, u = f(t) \qquad (t \in \Gamma), \tag{11}$$

where $a(t) = A(s) \sin \theta + B(s) \cos \theta$, $b(t) = -A(s) \cos \theta + B(s) \sin \theta$, $c(t) = c(s)$, $f(t) = f(s)$ with $t = x + iy$ and θ is the angle between the positive tangent to Γ and the x-axis.

The Poincaré problem involves the following well-known special cases:
1. The *Dirichlet problem*: $A(s) = B(s) \equiv 0$, $c(s) \equiv 1$.
2. The *Neumann problem*: $A(s) \equiv 1$, $B(s) = c(s) \equiv 0$.
3. The *problem of the oblique derivative*: $c(s) \equiv 0$.

In what follows we shall require that the partial derivatives $\partial u/\partial x$ and $\partial u/\partial y$ of the unknown function u have boundary values on Γ belonging to the class $\mathbf{H}^\lambda(\Gamma)$ $(0 < \lambda < 1)$. We also assume that the functions a and b do not vanish simultaneously on Γ: $a(t) + ib(t) \neq 0$ $(t \in \Gamma)$.

Let $\Phi(z) = u(x, y) + iv(x, y)$ denote a function holomorphic in G and such that $\operatorname{Re} \Phi(z) = u(x, y)$. Then (11) can be written in the form

$$\operatorname{Re} \{(a + ib)(t)\, \Phi'(t) + c(t)\, \Phi(t)\} = f(t) \qquad (t \in \Gamma). \tag{12}$$

But (12) is nothing else than the particular case of condition (1) for the Problem V obtained by letting

$$m = 1, \qquad h_0 = h_1 \equiv 0, \qquad a_1(t) = a(t) + ib(t), \qquad a_0(t) = c(t).$$

In accordance with the method presented in Section 2.1 we seek the unknown function $\Phi(z)$ in (12) in the form (3) with $m = 1$:

$$\Phi(z) = \int_\Gamma \mu(t) \ln\left(1 - \frac{z}{t}\right) \mathrm{d}s + \int_\Gamma \mu(t)\, \mathrm{d}s + \mathrm{i}C.$$

In the case at hand we have $g(t) \equiv 0$ and a little computation shows that the real singular integral equation (4) now assumes the simple form (cf. also MUSKHELISHVILI [1], § 74)

$$(A\mu)(t) := \operatorname{Re}\{-\pi \mathrm{i}\, \bar{t}'[a(t) + ib(t)]\}\, \mu(t) + \int_\Gamma \mu(\tau) \operatorname{Re}\left\{c(t) \ln \mathrm{e}\left(1 - \frac{t}{\tau}\right) - \frac{a(t) + ib(t)}{\tau - t}\right\} \mathrm{d}\sigma = f(t). \tag{13}$$

The homogeneous transposed equation of (13) is

$$(A'\nu)(t) := \operatorname{Re}\{-\pi \mathrm{i}\, \bar{t}'[a(t) + ib(t)]\}\, \nu(t) + \int_\Gamma \nu(\tau) \operatorname{Re}\left\{c(\tau) \ln \mathrm{e}\left(1 - \frac{\tau}{t}\right) + \frac{a(\tau) + ib(\tau)}{\tau - t}\right\} \mathrm{d}\sigma = 0. \tag{14}$$

Now the following *Theorem of I. N. Vekua* is seen to be an immediate consequence of the results of the Sections 2.1 and 2.2.

For the Poincaré problem to have a solution for every right-hand side f it is necessary and sufficient that the equation (14) *has no solutions distinct from zero. If this condition is fulfilled, then the homogeneous Poincaré problem* ($f = 0$) *has exactly* $2(n + 1)$ *linearly independent solutions*[1]), *where*

$$n = \frac{1}{2\pi} [\arg (a - ib)]_\Gamma.$$

Here is an example of what else results from the formulae (7) and (8):

If the homogeneous Poincaré problem does not possess solutions distinct from zero and if $n = -1$ (that is, $\operatorname{Ind} A = 0$), then the Poincaré problem is uniquely solvable for every right-hand side f.

Remark. It is easy to see that the operator (13) is of the form (1.13)–(1.14). Consequently, in the case of the degenerate Poincaré problem, more specifically, if the functions a and b have finitely many common simple zeros, the results of Section 1.4 apply. In the matrix case, that is, if $a, b, c \in [\mathrm{C}^\infty(\Gamma)]^{s \times s}$ and $f \in [\mathrm{C}^\infty(\Gamma)]^s$, under the condition that

$$\det (a(t) + ib(t)) = 0 \qquad \forall t \in \Gamma$$

the results of § 11, Chapter VI, can be applied to the equations (13), (14).

For the case of certain common multiple zeros of the functions a and b we refer to ELSCHNER [4], Chapter 5.

---

[1]) Taking into account the original problem (11) we here make the following convention: two solutions Φ_1 and Φ_2 of the problem (11) are regarded as different ones only if their real parts u_1 and u_2 are different; on the other hand, two solutions are identified if they merely differ by an additive constant of the form $\mathrm{i}C$ (C real).

§ 3. A boundary value problem for elliptic systems of first order in the plane

3.1. Let G be a bounded and simply connected region in the complex plane $\mathbb{C}$. Suppose the positively oriented boundary $\partial G = \Gamma$ is a Lyapunov curve and let $0 \in G$.

We consider the following elliptic system of first order in $2n$ unknown real functions $U = (u_1, u_2, \ldots, u_n)$ and $V = (v_1, v_2, \ldots, v_n)$ in G:

$$U_x - V_y = A_1U + B_1V + F_1, \qquad U_y + V_x = A_2U + B_2V + F_2. \tag{1'}$$

Here A_j, B_j $(j = 1, 2)$ are $n \times n$ matrices whose entries are given real-valued functions in G and F_1, F_2 are vectors of length n whose components are given real-valued functions in G.

It is simpler to write the system (1′) in the complex form as follows:

$$W_{\bar z} = AW + B\overline{W} + F \quad \text{in } G \tag{1}$$

with

$$W = U + iV, \; W_{\bar z} := \frac{\partial}{\partial \bar z} W = \frac{1}{2}\left(\frac{\partial W}{\partial x} + i\frac{\partial W}{\partial y}\right),$$

$$A = \frac{1}{4}(A_1 + B_2 + iA_2 - iB_1),$$

$$B = \frac{1}{4}(A_1 - B_2 + iA_2 + iB_1), \quad F = \frac{1}{2}(F_1 + iF_2).$$

Thus, in (1) A and B are given complex $n \times n$ matrix functions, F is a given complex vector function of length n, and W is a sought complex vector function of length n. In what follows $\partial/\partial\bar z$ will always denote the generalized derivative in the sense of Sobolev. $W_{\bar z} = 0$ is the *Cauchy-Riemann system.*

The solutions of the system (1) are required to satisfy the following Riemann-Hilbert boundary conditions:

$$\operatorname{Re}\{CW\} = \Phi \quad \text{on } \Gamma. \tag{2}$$

Here C is a given complex $n \times n$ matrix function on Γ and Φ a given real vector function on Γ with n components.

The boundary value problem (1), (2) is called the *generalized Riemann-Hilbert problem.* In the case of the Cauchy-Riemann system ($A \equiv B \equiv F \equiv 0$), this is the Riemann-Hilbert problem (see § 2), which was posed by Riemann in his famous Inauguraldissertation (1851) for the first time. At the third International Congress of Mathematicians (Heidelberg 1904) Hilbert showed how the Riemann-Hilbert problem can be reduced to a singular integral equation, in the study of which, however, he admitted some inaccuracies, which then led to incorrect results (see also Hilbert [1], page 219, Satz 43).

A complete solution of the generalized Riemann-Hilbert problem for the case $n = 1$ and $|C(z)| \equiv 1$ was given by I. N. Vekua ([3], Chapter 4) for the first time, and his method was based on the theory of singular integral equations. Vekua calls the homogeneous equation

$$W_{\bar z} = AW + B\overline{W} \text{ in } G \tag{1$_0$}$$

the *generalized Cauchy-Riemann system* and its (generalized) solutions *generalized analytic functions.* I. N. Vekua ([3], Chapters V and VI) applied these boundary value problems to several questions of the theory of infinitely small deformations of surfaces and of the theory of moment-free shells. Meister [1] gave various applications to Hydro- and Aerodynamics and to the theory of vibrations. Systems of the form (1′), $n > 1$, were studied by Bojarski [2] and Pascali [1, 2]. Under the aspect of the theory of Fredholm operators, boundary value problems of the form (1), (2) with the *Lopatinski condition* being satisfied, that is, with

$$\det C \neq 0 \text{ on } \Gamma \tag{3}$$

were investigated in the monograph of WENDLAND [1]. Assuming that the condition (3) is violated at finitely many points, that the entries of A and B have compact support in G, and that A, B, C are sufficiently smooth, WENDLAND [1, 2] showed that the operator corresponding to the boundary value problem (1), (2) is a densely defined and closed Fredholm operator on appropriate function spaces and computed its index. To obtain these results, Wendland made use of certain results on singular integral equations with degenerate coefficients going back to PRÖSSDORF [10] (see also Section 5.2, Chapter VI). Recently, in KOPP [1], the boundary value problem (1), (2) for the case of pointwise violated Lopatinski condition has been systematically studied with the methods of the Sections 1–5 of Chapter VI.

In this section we present some of these results on the generalized Riemann-Hilbert problem as well as some of their generalizations. Apart from certain simplifications in the proofs, in the Sections 3.2–3.5 we essentially follow KOPP [1]. The results of Section 3.6 are new; there we essentially follow PRÖSSDORF [23].

Remarks. 1. The system (1′) is elliptic in the sense of DOUGLIS and NIRENBERG (see, e.g., MIRANDA [1]).

2. Consider a general elliptic system of first order in G of the form

$$U_x + \tilde{A} U_y + \tilde{B} U + \tilde{C} = 0, \tag{4}$$

where $\tilde{A}$, $\tilde{B}$ are real $2n \times 2n$ matrix functions and $\tilde{C}$ is a real vector function with $2n$ components. If the coefficients are subject to suitable regularity conditions and if $n = 1$, then (4) can always be transformed into the form (1′). In the case $n > 1$, however, (1′) is only a canonical form of a certain subclass of the general elliptic system (4) (see WENDLAND [1], § 3.1).

3. The strongly elliptic system of second order

$$\sum_{i=1}^{2} (U_{x_i x_i} + B_i U_{x_i}) + CU = F$$

(B_1, B_2, and C real $m \times m$ matrix functions) with boundary conditions of Poincaré type (2.11) (such systems were considered by BITSADZE [1], Chapter VI) can be also transformed into a boundary value problem (1), (2) (see PASCALI [1]).

3.2. In what follows let $\mathbf{C}^{m,\alpha}(\overline{G})$ ($m \geqq 0$ an integer, $0 < \alpha < 1$), $\mathbf{C}^{\alpha}(\overline{G}) = \mathbf{C}^{0,\alpha}(\overline{G})$, denote the Banach space of (in general, complex valued) vector functions of length n in $\overline{G} = G \cup \Gamma$ whose partial derivatives of m-th order satisfy a Hölder condition with exponent α in G; the norm is defined in analogy to (6.1), Chapter II. Furthermore, put $\mathbf{C}^{\alpha}_{\bar{z}}(\overline{G}) := \{W : W_{\bar{z}} \in \mathbf{C}^{\alpha}(\overline{G})\}$. Provided with the norm $\| W \|_{\mathbf{C}^{\alpha}_{\bar{z}}(\overline{G})} := \| W_{\bar{z}} \|_{\mathbf{C}^{\alpha}(\overline{G})}$ the linear set $\mathbf{C}^{\alpha}_{\bar{z}}(\overline{G})$ becomes a Banach space. So $\mathbf{C}^{\alpha}(\overline{G}) \cap \mathbf{C}^{\alpha}_{\bar{z}}(\overline{G})$ endowed with the usual norm in the intersection of two Banach spaces,

$$\| W \|_{\cap} := \| W \|_{\mathbf{C}^{\alpha}(\overline{G})} + \| W_{\bar{z}} \|_{\mathbf{C}^{\alpha}(\overline{G})},$$

also becomes a Banach space.

The solution of the boundary value problem (1), (2) will be sought in the space $\mathbf{C}^{\alpha}(\overline{G}) \cap \mathbf{C}^{\alpha}_{\bar{z}}(\overline{G})$. Thus, in the sequel we consider the boundary value problem

$$RW := \begin{cases} W_{\bar{z}} - AW - B\overline{W} = F & \text{in } G, \\ \operatorname{Re} CW = \Phi & \text{on } \Gamma, \end{cases} \tag{5}$$

with the datas required to satisfy the following regularity conditions

$$A, B, F \in \mathbf{C}^{\alpha}(\overline{G}); \qquad C \in \mathbf{C}^{\alpha}(\Gamma); \qquad \Phi \in \mathbf{C}^{\alpha}(\Gamma). \tag{6}$$

First, for obvious reasons, the boundary value problem operator R defined by (5) will be considered on the following pair of spaces:

$$R : \mathrm{C}^\alpha(\overline{G}) \cap \mathrm{C}^\alpha_{\bar z}(\overline{G}) \to \mathrm{C}^\alpha(\overline{G}) \times \mathrm{C}^\alpha_{\mathrm{R}}(\Gamma)^{1)}.$$

It is easy to check that the operator R is $\mathbb{R}$-linear and continuous between these spaces.

3.3. We now transform the boundary value problem (5) into an equivalent system of singular integral equations. To do this, we use the following integral representation for the solutions of (5), which is due to I. N. VEKUA [3]:

Lemma 3.1. *Let $W \in \mathrm{C}^\alpha(\overline{G}) \cap \mathrm{C}^\alpha_{\bar z}(\overline{G})$ be a solution of* (1). *Then W admits the following representation in $\overline{G}$:*

$$W - T_G(AW + B\overline{W}) = \Psi M + \mathrm{i}K + TF. \tag{7}$$

Herein both the vector function $M \in \mathrm{C}^\alpha_{\mathrm{R}}(\Gamma)$ and the vector $K \in \mathbb{R}^n$ are uniquely determined by W; T_G and Ψ denote the integral operators defined as follows[2]:

$$(T_G W)(z) = -\frac{1}{\pi} \iint_G \frac{W(\zeta)\,\mathrm{d}\xi\,\mathrm{d}\eta}{\zeta - z} \qquad (\zeta = \xi + \mathrm{i}\eta),$$

$$(\Psi M)(z) = \frac{1}{2\pi \mathrm{i}} \int_\Gamma \frac{M(\tau)\,\mathrm{d}\tau}{\tau - z} \qquad (z \in \mathbb{C} \setminus \Gamma).$$

Proof. Put $\tilde F = W - T(AW + B\overline{W}) - TF$. Since $(Tf)_{\bar z} = f$ for every $f \in \mathrm{L}_1(\overline{G})$ we have $\tilde F_{\bar z} = 0$. Consequently, $\tilde F$ satisfies the Cauchy-Riemann differential equations and $\tilde F(z) = (\Psi \tilde F)(z)$ for $z \in G$ (see I. N. VEKUA [3], Chapter I, §§ 4 and 7). Since $T \in \mathscr{L}(\mathrm{C}^\alpha(\overline{G}))$, we have also $\tilde F \in \mathrm{C}^\alpha(\overline{G})$. Now Vekua's integral representation (2.3′) immediately yields the representation $\tilde F = \Psi M + \mathrm{i}K$. The uniqueness of M and K follows from the generalized Harnack theorem (see MUSKHELISHVILI [1], § 30). ∎

Remark 1. One has $T \in \mathscr{L}(\mathrm{C}^\alpha(\overline{G}), \mathrm{C}^\beta(\overline{G}))$ for any $0 < \alpha, \beta < 1$ (see I. N. VEKUA [3], Chapter I). It is well known that the embedding operator $\mathrm{C}^\gamma(\overline{G}) \to \mathrm{C}^\alpha(\overline{G})$ $(0 < \alpha < \gamma < 1)$ is compact (see, for instance, PRÖSSDORF [17], Corollary 4.3, Chapter 3). Therefore, T is a compact operator on the space $\mathrm{C}^\alpha(\overline{G})$ $(0 < \alpha < 1)$.

For what follows we need a further property of the integral operator $UW := W - T(AW + B\overline{W})$ defined by the left-hand side of (7). This operator is obviously in $\mathscr{L}(\mathrm{C}^\alpha(\overline{G}))$ (the linearity with respect to $\mathbb{R}$). We require:

$$\textit{U is continuously invertible on } \mathrm{C}^\alpha(\overline{G}). \tag{8}$$

Remark 2. The similarity principle for generalized analytic functions (see I. N. VEKUA [3]) implies that condition (8) is always satisfied in the case $n = 1$. If $n > 1$, then one can construct an $\mathbb{R}$-linear finite rank operator $\tilde T$ having the following two properties: $(\tilde T W)_{\bar z} = 0$ for $W \in \mathrm{C}^\alpha(\overline{G})$ and the opera-

---

[1]) In what follows the subscript $\mathbb{R}$ in the notation of a function space indicates that the corresponding space of real valued functions over $\mathbb{R}$ as scalar field is meant.

[2]) The operator T_G, introduced by D. Pompeiu, plays a fundamental role in the theory of generalized analytic functions (see I. N. VEKUA [3]; there, in Chapter I, the most important properties of the operator T_G can be found). Henceforth we omit the subscript G. Ψ is the integral operator of Cauchy type. ΨM is defined as the corresponding nontangential limit $(\Psi M)^+(t) = (P_\Gamma M)(t)$ (see formula (5.11), Chapter II) at the points t on Γ.

tor $\tilde{U} := U + \tilde{T}$ is continuously invertible on $C^\alpha(\overline{G})$ (see WENDLAND [1], pp. 157–159). It is easy to see that all the following arguments remain true when U is replaced by $\tilde{U}$.

Here is a converse of Lemma 3.1.

Lemma 3.2. *Given $M \in C^\alpha_{\mathbb{R}}(\Gamma)$ and $K \in \mathbb{R}^n$, the system of integral equations*

$$UW = \Psi M + iK + TF \tag{9}$$

has exactly one solution $W \in C^\alpha(\overline{G})$. Moreover, W belongs to $C^\alpha_{\bar z}(\overline{G})$ and is a solution of the system (1).

Proof. By the Sokhotski-Plemelj formulae (5.11) in Chapter II, the right-hand side of (9) is in $C^\alpha(\overline{G})$. Because of the requirement (8), the system (9) has therefore exactly one solution $W \in C^\alpha(\overline{G})$. After differentiating (9) with respect to $\bar z$ (in G) we get $W_{\bar z} - (AW + B\overline{W}) = F$, since ΨM is holomorphic in G and hence $(\Psi M)_{\bar z} = 0$. Thus, W is a solution of (1). Furthermore, $W_{\bar z} \in C^\alpha(\overline{G})$, that is, $W \in C^\alpha_{\bar z}(\overline{G})$. ∎

Theorem 3.1. (a) *Let $W \in C^\alpha(\overline{G}) \cap C^\alpha_{\bar z}(\overline{G})$ be a solution of the boundary value problem* (5). *Then the pair $(M, K) \in C^\alpha_{\mathbb{R}}(\Gamma) \times \mathbb{R}^n$ uniquely determined by* (7) *satisfies the following system of singular integral equations on Γ*[1])

$$\begin{aligned} \Lambda(M, K) &:= CPM + \overline{C}QM + \overline{C}VM + C(U^{-1} - I)\,\Psi M \\ &\quad + \overline{C(U^{-1} - I)\,\Psi M} + CU^{-1}\,iK + \overline{CU^{-1}\,iK} = 2\Phi + 2\,\mathrm{Re}\,\{CU^{-1}\,TF\}. \end{aligned} \tag{10}$$

Here P and Q are the projections introduced in § 2 of Chapter V[2]) *and V is the following potential of a double layer:*

$$(VM)(t) = -\frac{1}{\pi}\int_\Gamma \frac{\cos(r, \hat n)}{r} M(\tau)\,d\sigma \qquad (t \in \Gamma),$$

where $r = |t - \tau|$, $\tau = \tau(\sigma)$, and $(r, \hat n)$ denotes the angle between the vector $\overrightarrow{\tau t}$ and the outer normal $\hat n$ to Γ at the point $\tau \in \Gamma$.

(b) *Conversely, if $(M, K) \in C^\alpha_{\mathbb{R}}(\Gamma) \times \mathbb{R}^n$ is a solution of the system* (10) *of integral equations on Γ, then the function $W \in C^\alpha(\overline{G}) \cap C^\alpha_{\bar z}(\overline{G})$ uniquely determined by* (9) *satisfies the boundary value problem* (5).

Proof. (a) First of all, by Lemma 3.1, W satisfies (7) on $\overline{G}$ with certain uniquely determined $(M, K) \in C^\alpha_{\mathbb{R}}(\Gamma) \times \mathbb{R}^n$; in particular, W satisfies (7) on Γ. Applying the Sokhotski-Plemelj formulas (5.11) of Chapter II to ΨM we get

$$W = T(AW + B\overline{W}) + \frac{1}{2}M + \frac{1}{2}SM + iK + TF \quad \text{on } \Gamma.$$

Taking into account the boundary conditions (3) and the identity $2\,\mathrm{Re}\,CW = CW + \overline{CW}$ we obtain

$$\begin{aligned} &\frac{1}{2}(C + \overline{C})\,M + \frac{1}{2}(CSM + \overline{CSM}) + CT(AW + B\overline{W}) \\ &+ \overline{CT(AW + B\overline{W})} + i(C - \overline{C})\,K = 2\Phi - 2\,\mathrm{Re}\{CTF\}. \end{aligned} \tag{11}$$

[1]) In KOPP [1], the system (10) formally looks somewhat more complicated. Writing it in the form (10) as above will be more advantageous for what follows.

[2]) That is, $P = \frac{1}{2}(I + S)$ and $Q = \frac{1}{2}(I - S)$, with S the Cauchy singular integral operator.

Further, we have[1])

$$\overline{SM} = -SM + 2VM. \tag{12}$$

Due to (9) the equality

$$T(AW + B\overline{W}) = (U^{-1} - I)\,(\Psi M + \mathrm{i}K + TF). \tag{13}$$

holds on $\overline{G}$. After substituting (12), (13) into (11) and using the equality

$$\frac{1}{2}(C + \overline{C})\,M + \frac{1}{2}(C - \overline{C})\,SM = CPM + \overline{C}QM$$

we arrive at (10).

(b) Owing to Lemma 3.2, $W \in \mathbf{C}^\alpha(\overline{G}) \cap \mathbf{C}^\alpha_{\bar z}(\overline{G})$ is a solution of the system (1). Since W satisfies the system of integral equations (9), which is equivalent to (7), the values of $\operatorname{Re} CW$ on Γ can be computed as in the proof of assertion (a). The integral equation (10) then shows that the boundary condition (2) is fulfilled. ∎

Corollary 3.1. *The correspondence between the solutions $W \in \mathbf{C}^\alpha(\overline{G}) \cap \mathbf{C}^\alpha_{\bar z}(\overline{G})$ of the boundary value problem* (5) *and the solutions $(M, K) \in \mathbf{C}^\alpha_{\mathbb{R}}(\Gamma) \times \mathbb{R}^n$ of the system of integral equations* (10) *established by Theorem 3.1 is linear and bijective.*

An immediate consequence of this is the following

Corollary 3.2. *The boundary value problem operator $R: \mathbf{C}^\alpha(\overline{G}) \cap \mathbf{C}^\alpha_{\bar z}(\overline{G}) \to \mathbf{C}^\alpha(\overline{G}) \times \mathbf{C}^\alpha_{\mathbb{R}}(\Gamma)$ is a Φ-operator if and only if the operator $\Lambda: \mathbf{C}^\alpha_{\mathbb{R}}(\Gamma) \times \mathbb{R}^n \to \mathbf{C}^\alpha_{\mathbb{R}}(\Gamma)$ defined by the left-hand side of* (10) *is a Φ-operator. In that case,* $\operatorname{Ind} R = \operatorname{Ind} \Lambda$.

3.4. We now require that $\Gamma \in \mathbf{C}^{2,\alpha}$. This hypothesis enables us to prove the following theorem.

Theorem 3.2. *The operator $\Lambda: \mathbf{C}^\alpha_{\mathbb{R}}(\Gamma) \times \mathbb{R}^n \to \mathbf{C}^\alpha_{\mathbb{R}}(\Gamma)$ is a Φ-operator if and only if the Lopatinski condition* (3) *is satisfied. In that case*

$$\operatorname{Ind} \Lambda = n - 2 \operatorname{ind} \det C. \tag{14}$$

Proof. Put

$$A = CP + \overline{C}Q, \qquad A_1 = A + \overline{C}V,$$

$$A_2M = A_1M + C(U^{-1} - I)\,\Psi M + \overline{C(U^{-1} - I)\,\Psi M}.$$

From the maximum principle for analytic functions and the Plemelj-Privalov theorem it is easy to see that $\Psi \in \mathscr{L}(\mathbf{C}^\alpha(\Gamma), \mathbf{C}^\alpha(\overline{G}))$. The operator

$$U^{-1} - I = T(AU^{-1} + B\overline{U^{-1}}): \mathbf{C}^\alpha(\overline{G}) \to \mathbf{C}^\alpha(\overline{G})$$

is compact (see Remark 1 of 3.3). Therefore A_2 is in $\mathscr{L}(\mathbf{C}^\alpha_{\mathbb{R}}(\Gamma))$ and $A_2 - A_1 \in \mathscr{L}(\mathbf{C}^\alpha_{\mathbb{R}}(\Gamma))$ is compact. By virtue of Corollary 3.2, Chapter I, the operators $\Lambda \in \mathscr{L}(\mathbf{C}^\alpha_{\mathbb{R}}(\Gamma) \times \mathbb{R}^n, \mathbf{C}^\alpha_{\mathbb{R}}(\Gamma))$ and $A_2 \in \mathscr{L}(\mathbf{C}^\alpha_{\mathbb{R}}(\Gamma))$ are simultaneously Φ-operators and

$$\operatorname{Ind} \Lambda = n + \operatorname{Ind} A_2. \tag{15}$$

In view of Theorem 3.4, Chapter I, the operators A_1 and A_2 are simultaneously Φ-operators on the space $\mathbf{C}^\alpha_{\mathbb{R}}(\Gamma)$ and their indices coincide. Since $\Gamma \in \mathbf{C}^{2,\alpha}$, the kernel of

---

[1]) This can be seen by a reasoning similar to that of Muskhelishvili [1], § 12 (see also Section 5.1, Chapter II).

the integral operator V belongs to the class $\mathbf{C}_{\mathbf{R}}^{\alpha}(\Gamma\times\Gamma)$ (see MUSKHELISHVILI [1], § 7) and consequently, V is compact (both on $\mathbf{C}^{\alpha}(\Gamma)$ and on $\mathbf{C}_{\mathbf{R}}^{\alpha}(\Gamma)$). From (12) it follows easily that A_1M is real-valued for each $M \in \mathbf{C}_{\mathbf{R}}^{\alpha}(\Gamma)$. Hence $A_1 \in \mathscr{L}(\mathbf{C}^{\alpha}(\Gamma)) \cap \mathscr{L}(\mathbf{C}_{\mathbf{R}}^{\alpha}(\Gamma))$ and Theorem 3.13 of Chapter I shows that $A_1 \in \Phi(\mathbf{C}_{\mathbf{R}}^{\alpha}(\Gamma))$ if and only if $A_1 \in \Phi(\mathbf{C}^{\alpha}(\Gamma))$ and that then the indices of the operators coincide. Again using Theorem 3.4, Chapter I, we see that the operators A_1 and A_2 are simultaneously Φ-operators on the space $\mathbf{C}^{\alpha}(\Gamma)$ and that they have the same index. Finally, from the results of § 2, Chapter V, we conclude that $A \in \Phi(\mathbf{C}^{\alpha}(\Gamma))$ if and only if the condition (3) is satisfied and that then

$$\operatorname{Ind} A = \operatorname{ind}[\det \overline{C}/\det C] = -2 \operatorname{ind} \det C. \tag{16}$$

Formula (14) now results from (15) and (16). ∎

Putting Corollary 3.2 and Theorem 3.2 together we obtain the following

Corollary 3.3. *The boundary value problem operator* $R: \mathbf{C}^{\alpha}(\overline{G}) \cap \mathbf{C}_{\bar{z}}^{\alpha}(\overline{G}) \to \mathbf{C}^{\alpha}(\overline{G}) \times \mathbf{C}_{\mathbf{R}}^{\alpha}(\Gamma)$ *is a Φ-operator if and only if the Lopatinski condition* (3) *is fulfilled. In that case*

$$\operatorname{Ind} R = n - 2 \operatorname{ind} \det C.$$

Remark 1. If all what we suppose is that $\Gamma \in \mathbf{C}^{1,\alpha}$ (i.e. that Γ is a Lyapunov curve), then the integral operator V is still compact on $\mathbf{C}^{\beta}(\Gamma)$ for $0 < \beta < \alpha/2$ (see, for instance, PRÖSSDPRF [17], Corollary 4.5, Chapter 3). Consequently, in this case the results of the present section remain valid in spaces with sufficiently small Hölder exponent ($0 < \beta < \alpha/2$).

Remark 2. In the case of an $(m+1)$-fold connected region the above argument combined with the generalized Harnack theorem gives the index formula

$$\operatorname{Ind} R = n(1-m) - 2 \operatorname{ind} \det C.$$

3.5. We finally investigate the case where the Lopatinski condition (3) is violated at finitely many points.

Let $\tilde{\alpha} = (\alpha_1, \alpha_2, \ldots, \alpha_r)$ ($\alpha_r \in \Gamma$; $j = 1, \ldots, r$) be the system of all zeros of the function $\det C(z)$ ($z \in \Gamma$), let $\tilde{m} = (m_1, m_2, \ldots, m_r)$ be its multiplicity, and let $C \in [\mathbf{H}^{\alpha}(\tilde{\alpha}, \tilde{m})]^{n\times n}$ ($0 < \alpha < 1$). Then, by virtue of Theorem 3.1 of Chapter VI, the matrix function C admits the representation

$$C = R_+ D_+ C_0. \tag{17}$$

Here R_+ is a polynomial matrix with constant and non-vanishing determinant; C_0 belongs to $[\mathbf{H}^{\alpha}(\Gamma)]^{n\times n}$ and $\det C_0(z) \neq 0$ for $z \in \Gamma$; D_+ is a diagonal matrix function of the form

$$D_+(z) = \left(\prod_{k=1}^{r} (z-\alpha_k)^{\mu_j^{(k)}} \delta_{jl}\right)_{j,l=1}^{n},$$

the non-negative integers $\mu_1^{(k)} \geqq \mu_2^{(k)} \geqq \ldots \geqq \mu_n^{(k)} \geqq 0$ ($k = 1, \ldots, r$) being the (uniquely determined by C) partial multiplicaties of the zero α_k of $\det C$. Note that

$$m_k = \sum_{j=1}^{n} \mu_j^{(k)} \quad (k = 1, \ldots, r).$$

In the present subsection we make the following restrictive assumption:

$$\textit{The polynomial matrix } R_+ \textit{ in } (17) \textit{ is real.} \tag{18}$$

Remark 1. In the case $n = 1$ the hypothesis (18) is no restriction, since we may assume that $\det R_+(z) \equiv 1$.

2. The assumption (18) implies that R_+ is necessarily constant.

Indeed, assume R_+ is not constant. Then at least one entry of R_+ is a polynomial p on Γ with a degree greater than or equal to 1. Clearly, p can be extended analytically into the whole plane $\mathbb{C}$.

Thus, $\operatorname{Im} p$ is harmonic in G, is continuous on $\overline{G}$, and we have $\operatorname{Im} p \equiv 0$ on Γ. Since the Dirichlet problem is solvable uniquely, we deduce that $\operatorname{Im} p = 0$ on $\overline{G}$. It follows that $\operatorname{Re} p$ is constant on $\overline{G}$ and in particular on Γ. This, however, is a contradiction.

We have $f_k(z) = \dfrac{\bar z - \bar\alpha_k}{z - \alpha_k} \in \mathbf{H}^\alpha(\Gamma)$ and $f_k(z) \neq 0$ for $z \in \Gamma$ (see, for instance, Prössdorf [17], Section 6.1.2). So (17) and (18) yield the following factorization of $\overline{C}$:

$$\overline{C} = R_+ D_+ D_0 \tag{19}$$

with

$$D_0(z) = \left(\prod_{k=1}^{r} [f_k(z)]^{\mu_j^{(k)}} \delta_{jl} \right)_{j,l=1}^{n} \overline{C_0(z)} \in [H^\alpha(\Gamma)]^{n\times n},$$

$$\det D_0(z) = \prod_{k=1}^{r} [f_k(z)]^{m_k} \overline{\det C_0(z)} \neq 0 \qquad (z \in \Gamma).$$

Obviously, $\det C_0(z) = \det C(z) \Big/ \prod_{k=1}^{r} (z - \alpha_k)^{m_k}$. It is easily seen that $\operatorname{ind} f_k = -1$ $(k = 1, \ldots, r)$ and consequently

$$\operatorname{ind} \det D_0(z) = \operatorname{ind} \overline{\det C_0(z)} - \sum_{k=1}^{r} m_k. \tag{20}$$

Indeed, if Γ is the unit circle, then $f_k(z) = -(\alpha_k z)^{-1}$ and hence $\operatorname{ind} f_k = -1$. As for the general case, note that the index of the function $f_k \in \mathbf{H}^\alpha(\Gamma)$ remains untouched when the curve Γ is mapped homeomorphically onto the unit circle.

We now introduce the following spaces (compare Remark 3, Section 5.2, Chapter VI):

$$\overline{\mathbf{C}}^\alpha(\Gamma) := \{\Phi := R_+ D_+ F : F \in \mathbf{C}^\alpha(\Gamma)\}$$

with the norm given by $\|\Phi\|_{\overline{\mathbf{C}}^\alpha(\Gamma)} = \|F\|_{\mathbf{C}^\alpha(\Gamma)}$; and

$$\overline{\mathbf{C}}^\alpha_{\mathrm{R}}(\Gamma) := \{\Phi \in \overline{\mathbf{C}}^\alpha(\Gamma) : \Phi \text{ is real valued}\}.$$

$\overline{\mathbf{C}}^\alpha(\Gamma)$ and $\overline{\mathbf{C}}^\alpha_{\mathrm{R}}(\Gamma)$ are obviously Banach spaces and we have $\overline{\mathbf{C}}^\alpha(\Gamma) = \overline{\mathbf{C}}^\alpha_{\mathrm{R}}(\Gamma) \dotplus i\overline{\mathbf{C}}^\alpha_{\mathrm{R}}(\Gamma)$.

By virtue of the factorizations (17) and (19), each item in the representation (10) of the operator Λ involves the factor $R_+ D_+$. Therefore, Λ (and thus also the boundary value problem operator R) can be viewed as an operator mapping into the space $\overline{\mathbf{C}}^\alpha_{\mathrm{R}}(\Gamma)$. So Corollary 3.1 immediately gives the following

Corollary 3.2′. *The boundary value problem operator* $R : \mathbf{C}^\alpha(\overline{G}) \cap \mathbf{C}^\alpha_{\bar z}(\overline{G}) \to \mathbf{C}^\alpha(\overline{G}) \times \overline{\mathbf{C}}^\alpha_{\mathrm{R}}(\Gamma)$ *is a Φ-operator if and only if the operator* $\Lambda : \mathbf{C}^\alpha_{\mathrm{R}} \times \mathbb{R}^n \to \overline{\mathbf{C}}^\alpha_{\mathrm{R}}(\Gamma)$ *is a Φ-operator. In that case* $\operatorname{Ind} R = \operatorname{Ind} \Lambda$.

By applying the arguments of the proof of Theorem 3.2 to the operators

$$\Lambda \in \mathscr{L}(\mathbf{C}^\alpha_{\mathrm{R}}(\Gamma) \times \mathbb{R}^n, \overline{\mathbf{C}}^\alpha_{\mathrm{R}}(\Gamma)), \qquad A_2 \in \mathscr{L}(\mathbf{C}^\alpha_{\mathrm{R}}(\Gamma), \overline{\mathbf{C}}^\alpha_{\mathrm{R}}(\Gamma)),$$

$$A_1, A \in \mathscr{L}(\mathbf{C}^\alpha_{\mathrm{R}}(\Gamma), \overline{\mathbf{C}}^\alpha_{\mathrm{R}}(\Gamma)) \cap \mathscr{L}(\mathbf{C}^\alpha(\Gamma), \overline{\mathbf{C}}^\alpha(\Gamma))$$

we arrive at the following analogue of that theorem.

Theorem 3.2′. *The operator* $\Lambda : \mathbf{C}^\alpha_{\mathrm{R}}(\Gamma) \times \mathbb{R}^n \to \overline{\mathbf{C}}^\alpha_{\mathrm{R}}(\Gamma)$ *is a Φ-operator and its index is*

$$\operatorname{Ind} \Lambda = n + \operatorname{ind} [\det D_0 / \det C_0] = n - 2 \operatorname{ind} \det C_0 - \sum_{k=1}^{r} m_k. \tag{21}$$

Note that the last equality is a consequence of (20). Finally, combining Corollary 3.2′ and Theorem 3.2′ we get the following.

Corollary 3.3′. *Under the hypotheses of Section 3.5, the boundary value problem operator* $R : \mathbf{C}^\alpha(\overline{G}) \cap \mathbf{C}^\alpha_{\bar{z}}(\overline{G}) \to \mathbf{C}^\alpha(\overline{G}) \times \mathbf{C}^\alpha_{\mathbf{R}}(\Gamma)$ *is a Φ-operator with index*

$$\operatorname{Ind} R = n - 2 \operatorname{ind} \det C_0 - \sum_{k=1}^{r} m_k. \tag{22}$$

3.6. We now consider the case where the Lopatinski condition (3) is violated at finitely many points but the condition (18) is not fulfilled.

Again let $\tilde{\alpha} = (\alpha_1, \alpha_2, \ldots, \alpha_r)$ $(\alpha_j \in \Gamma;\ j = 1, \ldots, r)$ be the system of all zeros of the function $\det C(z)$ $(z \in \Gamma)$ and $\tilde{m} = (m_1, m_2, \ldots, m_r)$ its multiplicity. Suppose that in some neighborhood of the points α_j all entries of the matrix function $C(z)$ $(z \in \Gamma)$ have derivatives up to the order m_j inclusively $(j = 1, \ldots, r)$ and that all these derivatives belong to the class $\mathbf{H}^\beta(\Gamma)$ $(0 < \beta < 1)$. Then Theorem 3.1, Chapter VI, implies that C besides (17) also admits the representation

$$C = R_- D_- C_1, \tag{23}$$

where R_- is a polynomial matrix in z^{-1} with constant and non-zero determinant; C_1 belongs to $[\mathbf{H}^\beta(\Gamma)]^{n \times n}$ and $\det C_1(z) \neq 0$ for $z \in \Gamma$; D_- is a diagonal matrix of the form

$$D_-(z) = \left(\prod_{k=1}^{r} (z^{-1} - \alpha_k^{-1})^{\mu_j^{(k)}} \delta_{jl} \right)_{j,l=1}^{n}.$$

By virtue of the same Theorem 3.1, Chapter VI, the matrix function $\overline{C(z)}$ can be represented in the form

$$\overline{C} = S_+ D_+ D_1, \tag{23′}$$

where S_+ is a polynomial matrix in z with constant and nonvanishing determinant and where $D_1 \in [\mathbf{H}^\beta(\Gamma)]^{n \times n}$ and $\det D_1(z) \neq 0$ for $z \in \Gamma$.

Obviously,

$$\det C_1(z) = \gamma_1 \frac{\det C(z)}{\prod_{k=1}^{r} (z^{-1} - \alpha_k^{-1})^{m_k}} = \gamma_2 \det C_0(z) \prod_{k=1}^{r} (-\alpha_k z)^{m_k},$$

$$\det D_1(z) = \gamma_3 \frac{\det \overline{C(z)}}{\prod_{k=1}^{r} (z - \alpha_k)^{m_k}} = \gamma_4 \det \overline{C_0(z)} \prod_{k=1}^{r} [f_k(z)]^{m_k}$$

with $\gamma_i = \text{const}$ $(i = 1, \ldots, 4)$ and consequently,

$$\operatorname{ind} \det C_1(z) = \operatorname{ind} \det C_0(z) + \sum_{k=1}^{r} m_k,$$

$$\operatorname{ind} \det D_1(z) = -\operatorname{ind} \det C_0(z) - \sum_{k=1}^{r} m_k.$$

Now suppose $\Gamma \in \mathbf{C}^{m+2,\alpha}$, where $m = \max_k m_k$ and $0 < \alpha < \beta$, and consider the following Banach spaces (see Section 9.2, Chapter VI):

$$\mathfrak{C}^\alpha(\Gamma) := \{\Phi := (PR_- D_- + QS_+ D_+)\varphi : \varphi \in \mathbf{C}^\alpha(\Gamma)\}$$

with the norm defined by $\|\Phi\|_{\mathfrak{C}^\alpha(\Gamma)} = \|\varphi\|_{\mathbf{C}^\alpha(\Gamma)}$; and

$$\mathfrak{C}^\alpha_{\mathbb{R}}(\Gamma) := \{\Phi \in \mathfrak{C}^\alpha(\Gamma) : \Phi \text{ real valued}\}.$$

From the results of § 4, Chapter VI, we conclude that

$$\mathbf{C}^\alpha(\Gamma; (\alpha_k, \mu_1^{(k)})_1^r) \subset \mathfrak{C}^\alpha(\Gamma). \tag{24}$$

Taking into account Theorem 6.4 of Chapter II it can be shown as in Section 9.4 of Chapter VI that the operator $A = CP + \overline{C}Q : \mathbf{C}^\alpha(\Gamma) \to \mathfrak{C}^\alpha(\Gamma)$ is a linear continuous Φ-operator whose index is $\operatorname{Ind} A = \operatorname{ind} [\det D_1/\det C_1]$.

Because $\Gamma \in C^{m+2,\alpha}$, the kernel of the integral operator V belongs to the class $\mathbf{C}^{m,\alpha}_{\mathbb{R}}(\Gamma \times \Gamma)$ (see Muskhelishvili [1], §§ 7 and 12). It follows immediately from the inclusion (24) that the operators V and $\overline{C}V$ are compact from $\mathbf{C}^\alpha(\Gamma)$ into $\mathfrak{C}^\alpha(\Gamma)$.

We now replace condition (18) by the following one:

The entries of the matrix functions $A(z)$ and $B(z)$ $(z \in G)$ vanish in some neighborhood of the points α_j $(j = 1, \ldots, r)$. (25)

Due to this condition the operator $A_2 - A_1$ is obviously compact from $\mathbf{C}^\alpha(\Gamma)$ into the space $\mathbf{C}^\alpha(\Gamma; (\alpha_k, m_k)_1^r)$ and thus also into the space $\mathfrak{C}^\alpha(\Gamma)$ (see (24)).

By repeating the arguments of the preceding subsections of the present section we obtain the following analogues of the results of Section 3.5.

Corollary 3.2*. *The boundary value problem operator* R: $\mathbf{C}^\alpha(\overline{G}) \cap \mathbf{C}^\alpha_{\bar z}(\overline{G}) \to \mathbf{C}^\alpha(\overline{G}) \times \mathfrak{C}^\alpha_{\mathbb{R}}(\Gamma)$ *is a Φ-operator if and only if the operator* $\Lambda : \mathbf{C}^\alpha_{\mathbb{R}}(\Gamma) \times \mathbb{R}^n \to \mathfrak{C}^\alpha_{\mathbb{R}}(\Gamma)$ *is a Φ-operator. In that case* $\operatorname{Ind} R = \operatorname{Ind} \Lambda$.

Theorem 3.2*. *Under the hypotheses of this subsection the singular operator* $\Lambda : \mathbf{C}^\alpha_{\mathbb{R}}(\Gamma) \times \mathbb{R}^n \to \mathfrak{C}^\alpha_{\mathbb{R}}(\Gamma)$ *is a Φ-operator with index*

$$\operatorname{Ind} \Lambda = n + \operatorname{ind} [\det D_1/\det C_1] = n - 2 \operatorname{ind} \det C_0 - 2 \sum_{k=1}^r m_k. \tag{26}$$

Corollary 3.3*. *Under the hypotheses of the present subsection the boundary value problem operator* $R : \mathbf{C}^\alpha(\overline{G}) \cap \mathbf{C}^\alpha_{\bar z}(\overline{G}) \to \mathbf{C}^\alpha(\overline{G}) \times \mathfrak{C}^\alpha_{\mathbb{R}}(\Gamma)$ *is a Φ-operator and its index is*

$$\operatorname{Ind} R = n - 2 \operatorname{ind} \det C_0 - 2 \sum_{k=1}^r m_k. \tag{27}$$

Remarks. 1. If the condition (18) is satisfied, then $\overline{\mathbf{C}}^\alpha(\Gamma) \subset \mathfrak{C}^\alpha(\Gamma)$ and $\overline{\mathbf{C}}^\alpha_{\mathbb{R}}(\Gamma) \subset \mathfrak{C}^\alpha_{\mathbb{R}}(\Gamma)$.

Indeed, (17) and (23) imply that $R_+ D_+ = R_- D_- C_2$ with $C_2 = C_1 C_0^{-1}$. Thus, for every $F \in \mathbf{C}^\alpha(\Gamma)$,

$$R_+ D_+ F = (PR_- D_- + QR_+ D_+)\, \varphi, \qquad \varphi = PC_2 F + QF,$$

and this gives the assertion at once.

The reason for the different index formulae (22) and (27) is that $\overline{\mathbf{C}}^\alpha_{\mathbb{R}}(\Gamma)$ has the co-dimension $\sum_{k=1}^r m_k$ in $\mathfrak{C}^\alpha_{\mathbb{R}}(\Gamma)$.

2. The results of Section 3.6 remain valid if condition (25) is replaced by the following weaker condition: for any points $z \in \Gamma$ and $\zeta \in G$ in some sufficiently small neighborhood of the point α_k $(k = 1, \ldots, r)$ both $A(\zeta)/(\zeta - z)^{m_k}$ and $B(\zeta)/(\zeta - z)^{m_k}$ regarded as functions of ζ belong to the class C^α.

Indeed, it is not difficult to see that then the functions TAg, TBg $(g \in \mathbf{C}^\alpha(\overline{G}))$ have derivatives up to the order m_k inclusively in some neighborhood of the points α_k and that these derivatives can be obtained upon differentiating under the integral sign. Hence TA, TB, and therefore also $A_2 - A_1$ are compact operators from $\mathbf{C}^\alpha(\Gamma)$ into the space $\mathfrak{C}^\alpha(\Gamma)$.

3. The following result, which is due to WENDLAND ([1], §§ 3.5, 5.1, and [2]), can be easily derived from Corollary 3.3*:

Let the hypotheses of the present subsection be fulfilled, suppose supp $A \subset G$, supp $B \subset G$, and put

$$X := \{W \in \mathbf{C}^{\alpha}(\overline{G}) : W_{\bar{z}} \in \mathbf{C}^{m-1,\alpha}(\overline{G})\},$$

$$Y := \mathbf{C}_{\mathrm{R}}^{m,\alpha}(\Gamma), \qquad m = \max_k m_k .$$

Then the boundary value problem operator $R: X \to \mathbf{C}^{m-1,\alpha}(\overline{G}) \times Y$ is a densely defined and closed operator whose index is given by formula (27).

4. The results of this section remain valid if the corresponding operators are considered on the spaces $\mathbf{L}_p$ $(1 < p < \infty)$ instead on C^{α}.

It is still an open question whether or not the extra assumptions required in 2. can be removed.

§ 4. On algebraic operators

In this section, we present a new general method which provides us with a relatively universal formalism to reduce the investigation of several classes of operator equations (e.g. singular integral equations with Carleman shift or Tricomi integral equations, see §§ 5 and 6) to the study of singular integral equations. This method is due to Ch. Meyer. The material of the following paragraphs is essentially taken from the works of CH. MEYER [1, 2].

Definition (PRZEWORSKA-ROLEWICZ [1]). Let X be a real or complex Banach space. An operator $J \in \mathscr{L}(X)$ is called an *algebraic operator* if there exists an algebraic polynomial $P(\lambda)$ $(\lambda \in \mathbf{C})$ such that $P(J) \equiv 0$. The polynomial $P(\lambda)$ is referred to as the *characteristic polynomial* of the operator J and the solutions λ_j $(j = 1, \ldots, n;\ n = \text{degree } P \geqq 2)$ of the equation $P(\lambda) = 0$ are referred to as the *characteristic roots* of the operator J.

In what follows we shall always suppose that J is a given algebraic operator whose characteristic roots are non-zero and pairwise distinct (i.e., all characteristic roots are of multiplicity 1), and also that they are real if X is a real space.

Lemma 4.1. (PRZEWORSKA-ROLEWICZ [1]). *The space X can be decomposed into the direct sum $X = X_1 \dotplus \cdots \dotplus X_n$ of the eigenspaces $X_j = \{x \in X : Jx = \lambda_j x\}$ associated with the characteristic roots of the algebraic operator J. The corresponding decomposition of the elements $x \in X$ is*

$$x = \sum_{j=1}^{n} x_j \quad \textit{with} \quad x_j = \left(\prod_{\substack{k=1\\k \neq j}}^{n} \frac{J - \lambda_k I}{\lambda_j - \lambda_k} \right) x \in X_j . \tag{1}$$

Proof. We have

$$\prod_{k=1}^{n} (J - \lambda_k I) \equiv 0 . \tag{2}$$

Put

$$P_j(\lambda) = \prod_{\substack{k=1\\k \neq j}}^{n} \frac{\lambda - \lambda_k}{\lambda_j - \lambda_k}, \qquad P_j = P_j(J) \qquad (j = 1, \ldots, n).$$

Then obviously $P_j \in \mathscr{L}(X)$. The Lagrange interpolation formula $\sum_{j=1}^{n} P_j(\lambda) \equiv 1$ $(\lambda \in \mathbf{C})$ gives

$$\sum_{j=1}^{n} P_j = I . \tag{3}$$

(2) implies that $x_j = P_j x$ is in X_j for every $x \in X$, and from (3) we get (1). From (2) and (3) it follows immediately that

$$P_j P_k = 0 \qquad (j \neq k), \qquad P_j^2 = P_j$$

and this proves the uniqueness of the representation $x = \sum_{j=1}^{n} x_j$, $x \in X$. ∎

Remark. The following converses of Lemma 4.1 can be easily obtained from the preceding proof.

(i) Suppose the space X is the direct sum $X = X_1 \dotplus \dots \dotplus X_n$ of its closed subspaces $X_1, \dots, X_n$ and let $\lambda_1, \dots, \lambda_n$ be any non-zero and pairwise distinct numbers (real if X is a real space and complex if X is a complex space). Then there exists an algebraic operator J whose characteristic roots are λ_j and the corresponding eigenspaces $\{x \in X : Jx = \lambda_j x\}$ of which coincide with X_j $(j = 1, \dots, n)$. The characteristic polynomial of J is

$$P(\lambda) = \prod_{j=1}^{n} (\lambda - \lambda_j). \tag{4}$$

(ii) Let $J \in \mathscr{L}(X)$ and let $\lambda_1, \dots, \lambda_n$ be any numbers subject to the same conditions as in (i). If X is the direct sum of the n closed subspaces $X_j = \{x \in X : Jx = \lambda_j x\}$, then J is an algebraic operator and (4) is its characteristic polynomial.

Let $Y \subset X$ be a Banach space which is continuously embedded in X. Denote by Y_j the closed subspace of Y defined as $Y_j = \{y \in Y : Jy = \lambda_j y\}$.

Lemma 4.2. *If $J(Y) \subset Y$ then $J \in \mathscr{L}(Y)$ and $Y = Y_1 \dotplus \dots \dotplus Y_n$.*

Proof. We show that J is a closed operator on Y. Let $\{y_k\}_1^\infty$ be a sequence of elements in Y such that $\|y_k - y\|_Y \to 0$ and $\|Jy_k - z\|_Y \to 0$ as $k \to \infty$. Since Y is continuously embedded in X, we conclude that $\|y_k - y\|_X \to 0$ and $\|Jy_k - z\|_X \to 0$. Because J is continuous on X, we have $Jy = z$, hence J is closed on Y and thus ,by the closed graph theorem, it is continuous on Y. Finally, since the restriction of J to Y satisfies the hypotheses of Lemma 4.1, the second assertion of Lemma 4.2 follows. ∎

Remark. In general, the decomposition $Y = Y_1 \dotplus \dots \dotplus Y_n$ cannot be guaranteed without the hypothesis that $J(Y) \subset Y$. This is seen from the following simple example:

Let $X = \mathbb{R}^2$, $Y = \{(x_1, x_2) \in X : x_1 = x_2\}$, and $J(x_1, x_2) = (x_1, -x_2)$. Then $J^2 - I \equiv 0$ and we have $X_1 = \{(x, 0) \in X : x \in \mathbb{R}\}$, $X_2 = \{(0, x) \in X : x \in \mathbb{R}\}$, $Y_1 = Y_2 = \{(0, 0)\}$. Thus $Y_1 \dotplus Y_2 \neq Y$.

Now consider the following situation: let $Y \subset X$ and suppose we are given an operator $W \in \mathscr{L}(X, Y)$ such that

$$JW = WJ. \tag{5}$$

From (5) we get

$$JWx = WJx = W(\lambda_j x) = \lambda_j Wx \qquad \forall x \in X_j, \qquad j = 1, \dots, n,$$

and, hence, $W(X_j) \subset Y_j$. Thus, the restrictions $W_j := W \mid X_j$ are in $\mathscr{L}(X_j, Y_j)$ $(j = 1, \dots, n)$.

Theorem 4.1. *For the operator W to be a Φ-operator (resp. $\Phi_\pm$-operator, normally solvable, left-sided invertible, right-sided invertible) it is necessary and in case $J(Y) \subset Y$ also sufficient that all operators W_j $(j = 1, \dots, n)$ are Φ-operators (resp. $\Phi_\pm$-operators, normally solvable, left-sided invertible, right-sided invertible).*

Proof. Let $Y' = Y_1 \times \dots \times Y_n$ be the cartesian product of the spaces $Y_1, \dots, Y_n$. By Lemma 4.1, every element $x \in X$ has a unique representation in the form $x = \sum_{j=1}^{n} x_j$ with $x_j \in X_j$. Define the operator $W' \in \mathscr{L}(X, Y')$ by $W'x = (W_1 x_1, \dots, W_n x_n)$.

Since the space $X = X_1 \dotplus \dots \dotplus X_n$ is naturally isomorphic to the space $X' = X_1 \times \dots \times X_n$, we can think of W' as an operator belonging to $\mathscr{L}(X', Y')$. So W' can be regarded as the diagonal operator $W' = (W_j \, \delta_{jk})_1^n$ and this implies at once that W' is a Φ-operator (resp. $\Phi_\pm$-operator, normally solvable, left-sided invertible, right-sided invertible) if and only if all operators W_j $(j = 1, \dots, n)$ have the corresponding property. Some elementary arguments also show that if W possesses one of those properties, then the operator W' possesses that property, too. Furthermore, one has

$$\alpha(W') = \sum_{j=1}^{n} \alpha(W_j) = \alpha(W), \tag{6}$$

$$\beta(W') = \sum_{j=1}^{n} \beta(W_j) \leqq \beta(W). \tag{7}$$

If $J(Y) \subset Y$, then, by virtue of Lemma 4.2, the spaces Y and Y' are isomorphic and therefore the operators W and W' have the same properties. ▮

Theorem 4.2. *Let $J(Y) \subset Y$. If W is a Φ-operator, then* $\operatorname{Ind} W = \sum_{j=1}^{n} \operatorname{Ind} W_j$.

Proof. Due to Lemma 4.2 both the spaces Y and Y' and the spaces $\operatorname{Im} W$ and $\operatorname{Im} W'$ are isomorphic. This implies that the quotient spaces $Y/\operatorname{Im} W$ and $Y'/\operatorname{Im} W'$ are also isomorphic. Hence $\beta(W) = \beta(W')$ and the assertion follows from (6) and (7). ▮

Lemma 4.3. *If there is a Banach space $\tilde{Y}$ such that $J(\tilde{Y}) \subset \tilde{Y}$ and $Y = \operatorname{Im} W + \tilde{Y}$, then $J(Y) \subset Y$.*

Proof. Lemma 4.2 shows that $\tilde{Y} = \tilde{Y}_1 \dotplus \dots \dotplus \tilde{Y}_n$, where $\tilde{Y}_j = Y_j \cap \tilde{Y}$, and also that $\operatorname{Im} W = \operatorname{Im} W_1 \dotplus \dots \dotplus \operatorname{Im} W_n$. Thus, $Y = \operatorname{Im} W + \tilde{Y} = (\operatorname{Im} W_1 + \tilde{Y}_1) \dotplus \dots \dotplus (\operatorname{Im} W_n + \tilde{Y}_n)$. Since the sets $\operatorname{Im} W_j + \tilde{Y}_j$ are linear and since they are contained in Y_j $(j = 1, \dots, n)$, we deduce that $J(Y) \subset Y$. ▮

Remark. If W is a Φ-operator and if, in addition, $\tilde{Y}$ is dense in Y, then $Y = \operatorname{Im} W + \tilde{Y}$ (see Chapter I).

§ 5. Singular integral equations with Carleman shift

5.1. Let Γ be a sufficiently smooth Lyapunov curve and let $v = v(t)$ be a continuous automorphism of the curve Γ preserving the orientation of Γ and possessing the following properties:

a) $v(t) \not\equiv t$.

b) There is an integer $n \geqq 2$ such that $v_n(t) \equiv t$ and $v_j(t) \not\equiv t$ for $j = 1, \dots, n - 1$, where $v_i(t) = v_{i-1}(v(t))$ and $v_0(t) \equiv t$.[1])

c) The derivative v' exists everywhere on Γ and $v' \in \mathbf{H}^\lambda(\Gamma)$ $(0 < \lambda < 1)$.

We consider the integral operator

$$K := \sum_{i=0}^{n-1} D_v^i H_i = \sum_{i=0}^{n-1} A_i(v_i(t)) \, D_v^i + B_i(v_i(t)) \, D_v^i S \tag{1}$$

[1]) The properties a) and b) imply that $v_i(t) \neq t$ for all $t \in \Gamma$ and $i = 0, 1, \dots, n - 1$ (see Kravchenko and Litvinchuk [1] or Litvinchuk [2]).

on the space $\mathbf{L}_p^k(\Gamma)$ $(1 < p < \infty)$ of all vector functions $f = (f_1, \ldots, f_k)$ with $f_j \in \mathbf{L}_p(\Gamma)$ $(j = 1, \ldots, k)$. Here D_v^i denotes the linear, continuous, and invertible operator defined by $(D_v^i f)(t) = f(v_i(t))$, H_i denotes the singular integral operator $H_i := A_i I + B_i S$ with coefficients $A_i, B_i \in \mathbf{L}_\infty^{k \times k}(\Gamma)$ $(i = 0, 1, \ldots, n - 1)$, and S is the Cauchy singular integral operator

$$(Sf)(t) = \frac{1}{\pi i} \int_\Gamma \frac{f(\tau)}{\tau - t} d\tau .$$

The operator K is called a *singular integral operator with Carleman shift.* With the operator K associate the $kn \times kn$ matrix functions

$$A(t) = \begin{pmatrix} A_0(t) & A_1(v_1(t)) & \ldots & A_{n-1}(v_{n-1}(t)) \\ A_{n-1}(t) & A_0(v_1(t)) & \ldots & A_{n-2}(v_{n-1}(t)) \\ \cdots & \cdots & \cdots & \cdots \\ A_1(t) & A_2(v_1(t)) & \ldots & A_0(v_{n-1}(t)) \end{pmatrix}$$

and

$$B(t) = \begin{pmatrix} B_0(t) & B_1(v_1(t)) & \ldots & B_{n-1}(v_{n-1}(t)) \\ B_{n-1}(t) & B_0(v_1(t)) & \ldots & B_{n-2}(v_{n-1}(t)) \\ \cdots & \cdots & \cdots & \cdots \\ B_1(t) & B_2(v_1(t)) & \ldots & B_0(v_{n-1}(t)) \end{pmatrix}.$$

Theorem 5.1 (LITVINCHUK [1, 2]). *Let A_i and B_i $(i = 0, 1, \ldots, n - 1)$ be continuous matrix functions. Then the operator K is a Φ-operator on the space $\mathbf{L}_p^k(\Gamma)$ if and only if*

$$\det (A + B)(t) \neq 0, \qquad \det (A - B)(t) \neq 0 \qquad (t \in \Gamma). \tag{2}$$

If (2) *is satisfied, then*

$$\operatorname{Ind} K = \frac{1}{n} \operatorname{ind} [\det (A - B)/\det (A + B)]. \tag{3}$$

In this section we shall prove an essentially generalized version of Theorem 5.1. In particular, we allow the condition (2) to be violated at a finite number of points on Γ. This will be done with the help of the results of the § 4 and of the preceding chapters.

5.2. For what follows we need two auxiliary propositions.

Lemma 5.1. *Let $C = A \pm B$, then* $\det C(t) = \det C(v_i(t))$ *for* $i = 0, 1, \ldots, n - 1$.

Proof. Put $C_i = A_i \pm B_i$ and let $n = 2$. Then

$$\det C(t) = \det \begin{pmatrix} C_0(t) & C_1(v(t)) \\ C_1(t) & C_0(v(t)) \end{pmatrix} = (-1)^k \det \begin{pmatrix} C_1(t) & C_0(v(t)) \\ C_0(t) & C_1(v(t)) \end{pmatrix}$$

$$= (-1)^{2k} \det \begin{pmatrix} C_0(v(t)) & C_1(t) \\ C_1(v(t)) & C_0(t) \end{pmatrix} = \det C(v(t)),$$

and this proves the assertion for $n = 2$. The proof for $n > 2$ is analogous. ∎

Thus, if $t_0 \in \Gamma$ is a zero of order m_0 of the function $\det (A \pm B)$, then each of the points $v(t_0), v_2(t_0), \ldots, v_{n-1}(t_0)$ $(\in \Gamma)$ is also a zero of order m_0 of that function.

Lemma 5.2.[1]) *Let $v' \in \mathbf{H}^\lambda(\Gamma; (t_j, m_j)_{j=1}^r)$ $(0 < \lambda \leq 1, t_j \in \Gamma)$. Then the commutator $T_v = D_v S - S D_v$ is compact from $\mathbf{L}_p(\Gamma)$ $(1 < p < \infty)$ into the space $\mathbf{L}_p(\Gamma; (t_j, m_j)_{j=1}^r)$.*

[1]) For notation see Section 6.1, Chapter II.

Proof. Since $D_v^n = I$, we have $T_v = (D_v S D_v^{n-1} - S) D_v$. Let $\varphi \in \mathbf{L}_p(\Gamma)$ and put $T_1 = D_v S D_v^{n-1} - S$. The substitution $u = v_{n-1}(\tau)$, $\tau = v(u)$ in the integral $D_v S D_v^{n-1} \varphi$ gives

$$(T_1\varphi)(t) = \frac{1}{\pi i} \int_\Gamma T(t, u)\, \varphi(u)\, du$$

with

$$T(t, u) = \frac{K(t, u) - K(u, u)}{u - t} v'(u), \quad K(t, u) = \frac{t - u}{v(t) - v(u)}.$$

Now it can be shown as in the proof of Theorem 6.3, Chapter II, that T_1 has the property stated in the lemma. Then it follows that the same is also true for the operator T. ∎

5.3. For the sake of simplicity, in what follows we assume that each of the conditions (2) is violated at n points on Γ. Let $\alpha = (\alpha_1, \alpha_2, \dots, \alpha_n)$, $\alpha_i = v_{i-1}(\alpha_1) \in \Gamma$, be the zeros of the function $\det(A + B)$ and let $\beta = (\beta_1, \beta_2, \dots, \beta_n)$, $\beta_i = v_{i-1}(\beta_1) \in \Gamma$, be the zeros of the function $\det(A - B)$. Let $m_1 \geqq 0$ be the multiplicity of the zeros α_i and $n_1 \geqq 0$ the multiplicity of the zeros β_i (recall what was said after Lemma 5.1). Suppose the functions A, B, and v are sufficiently smooth in some neighborhood of the points α_i and β_i. Consider the space

$$\overline{\mathbf{L}}_p^{kn}(\Gamma) := \overline{\mathbf{L}}_p^{kn}(R_-, \alpha, \mu; R_+, \beta, \nu)$$

introduced in Section 9.2 of Chapter VI. Herein we denote by

$$\mu_1^{(i)} \geqq \dots \geqq \mu_{kn}^{(i)} \geqq 0, \qquad \nu_1^{(i)} \geqq \dots \geqq \nu_{kn}^{(i)} \geqq 0 \qquad (i = 1, \dots, n)$$

the partial multiplicities of the zeros α_i and β_i of the functions $\det(A + B)$ and $\det(A - B)$, respectively. Thus

$$\sum_{j=1}^{kn} \mu_j^{(i)} = m_1, \quad \sum_{j=1}^{kn} \nu_j^{(i)} = n_1 \qquad (i = 1, \dots, n).$$

Due to Section 9.4 of Chapter VI, the singular integral operator $V := AI + BS$ is a linear continuous operator from $\mathbf{L}_p^{kn}(\Gamma)$ to the space $\overline{\mathbf{L}}_p^{kn}(\Gamma)$. In § 9 of Chapter VI we also stated sufficient conditions for V to be in $\Phi(\mathbf{L}_p^{kn}(\Gamma), \overline{\mathbf{L}}_p^{kn}(\Gamma))$ and we computed the index of the operator V.

We still introduce the following space:

$$\mathfrak{L}_1 := \{f \in \mathbf{L}_p^k(\Gamma) : F = (f(t), f(v_1(t)), \dots, f(v_{n-1}(t)) \in \overline{\mathbf{L}}_p^{kn}(\Gamma)\}.$$

Endowed with the norm $\|f\|_{\mathfrak{L}_1} = \|F\|_{\overline{\mathbf{L}}_p^{kn}(\Gamma)}$, $\mathfrak{L}_1$ is a Banach space which is continuously embededd in $\mathbf{L}_p^k(\Gamma)$.

Denote by J the linear continuous operator on $\mathbf{L}_p^{kn}(\Gamma)$ defined by

$$J = \begin{bmatrix} 0 & 0 & \dots & 0 & D_v \\ D_v & 0 & \dots & 0 & 0 \\ 0 & D_v & \dots & 0 & 0 \\ \dots & \dots & \dots & \dots & \dots \\ 0 & 0 & \dots & D_v & 0 \end{bmatrix}. \tag{4}$$

Obviously, $J^n - J \equiv 0$, and consequently, J is an algebraic operator.

Our next objective is to prove the following theorem from which in the special case $m_1 = n_1 = 0$ (that is, $\overline{\mathbf{L}}_p^{kn}(\Gamma) = \mathbf{L}_p^{kn}(\Gamma)$) Theorem 5.1 follows.

Theorem 5.2. (i) Im K *is a subset of* $\mathfrak{L}_1$ *and the operator* $K : \mathbf{L}_p^k(\Gamma) \to \mathfrak{L}_1$ *is continuous.*

(ii) *If* $V \in \Phi(\mathbf{L}_p^{kn}(\Gamma), \overline{\mathbf{L}}_p^{kn}(\Gamma))$, *then* $K \in \Phi(\mathbf{L}_p^k(\Gamma), \mathfrak{L}_1)$.

(iii) *If there exists an operator* $\tilde{V} \in \Phi_-(\mathbf{L}_p^{kn}(\Gamma), \overline{\mathbf{L}}_p^{kn}(\Gamma))$ *such that* $\tilde{V}J = J\tilde{V}$, *then* $K \in \Phi(\mathbf{L}_p^k(\Gamma), \mathfrak{L}_1)$ *implies that also* $\tilde{V} \in \Phi(\mathbf{L}_p^{kn}(\Gamma), \overline{\mathbf{L}}_p^{kn}(\Gamma))$ *and, furthermore, that* Ind $K = (1/n)$ Ind V.

Corollary 5.1. *If* $V \in \Phi(\mathbf{L}_p^{kn}(\Gamma), \overline{\mathbf{L}}_p^{kn}(\Gamma))$ *then* $K \in \Phi(\mathbf{L}_p^k(\Gamma), \mathfrak{L}_1)$ *and* Ind $K = (1/n)$ Ind V.

In order to prove Theorem 5.2, we use the results of § 4 in the following setting: $X = \mathbf{L}_p^{kn}(\Gamma)$, $Y = \overline{\mathbf{L}}_p^{kn}(\Gamma)$, and

$$W = \begin{pmatrix} H_0 & D_vH_1D_v^{n-1} & \dots & D_v^{n-1}H_{n-1}D_v \\ H_{n-1} & D_vH_0D_v^{n-1} & \dots & D_v^{n-1}H_{n-2}D_v \\ \dots & \dots & \dots & \dots \\ H_1 & D_vH_2D_v^{n-1} & \dots & D_v^{n-1}H_0D_v \end{pmatrix}.$$

It is easy to see that $JW = WJ$ (J the operator given by (4)) and also that $W = AI + BS + T$, where

$$T = \begin{pmatrix} 0 & D_vB_1(SD_v^{n-1} - D_v^{n-1}S) & \dots & D_v^{n-1}B_{n-1}(SD_v - D_vS) \\ 0 & D_vB_0(SD_v^{n-1} - D_v^{n-1}S) & \dots & D_v^{n-1}B_{n-2}(SD_v - D_vS) \\ \dots & \dots & \dots & \dots \\ 0 & D_vB_2(SD_v^{n-1} - D_v^{n-1}S) & \dots & D_v^{n-1}B_0(SD_v - D_vS). \end{pmatrix}.$$

Lemma 5.2 together with the inclusion (9.5) of Chapter VI shows that the operator $T : \mathbf{L}_p^{kn}(\Gamma) \to \overline{\mathbf{L}}_p^{kn}(\Gamma)$ is compact. Thus, $W \in \mathscr{L}(X, Y)$ and the operators W and $V = AI + BS$ $(\in \mathscr{L}(X, Y))$ are simultaneously Φ-operators and have the same index (see Theorem 3.4, Chapter I).

Now denote by $\lambda_1 = 1$ and λ_j $(j = 2, \dots, n)$ the solutions of the equation $\lambda^n - 1 = 0$ and let K_j $(j = 1, 2, \dots, n)$ denote the operator

$$K_j = \sum_{i=0}^{n-1} \lambda_j^{n-i} D_v^i H_i.$$

Then $K_1 = K$. Furthermore, let $f_j \in \mathbf{L}_p^{kn}(\Gamma)$ $(j = 1, \dots, n)$ be the vector function defined by

$$f_j(t) = (f(t), \lambda_j^{n-1}f(v_1(t)), \dots, \lambda_j f(v_{n-1}(t))) \quad (f \in \mathbf{L}_p^k(\Gamma)) \tag{5}$$

and let P_j be the operator defined as $P_jf_j = f$. The set $\mathfrak{L}_j := \{f \in \mathbf{L}_p^k(\Gamma) : f_j \in \overline{\mathbf{L}}_p^{kn}(\Gamma)\}$ $(j = 1, \dots, n)$ is a Banach space under the norm $\|f\|_{\mathfrak{L}_j} = \|f_j\|_{\overline{\mathbf{L}}_p^{kn}}$. It is easy to check straightforwardly that every element $x_j \in X_j$ $(X_j = \{x \in X : Jx = \lambda_j x\})$ is of the form (5) and that $K_jf = P_jW_jP_j^{-1}f$ for every $f \in \mathbf{L}_p^k(\Gamma)$, where $W_j = W \mid X_j$. This implies at once that $K_j \in \mathscr{L}(\mathbf{L}_p^k(\Gamma), \mathfrak{L}_j)$. Finally, Theorem 4.1 shows that $K_j \in \Phi(\mathbf{L}_p^k(\Gamma), \mathfrak{L}_j)$ $(j = 1, \dots, n)$ in case $V \in \Phi(X, Y)$ (or, equivalently, in case $W \in \Phi(X, Y)$). So the proof of the assertions (i) and (ii) of Theorem 5.2 is complete.

To prove the third assertion of Theorem 5.2, we need some lemmas. The first what we shall show is that under the hypotheses of (iii) the inclusion $J(Y) \subset Y$ holds.

Recall the notation fixed at the beginning of this subsection and put $m_0 = \max_i \mu_1^{(i)}$, $n_0 = \max_i \nu_1^{(i)}$, $\tilde{Y} = [\overline{\mathbf{L}}_p(\alpha, m_0; \beta, n_0)]^{kn}$ (see Section 4.3, Chapter VI).

Lemma 5.3. *The inclusion* $J(\tilde{Y}) \subset \tilde{Y}$ *holds.*

Proof. Clearly, it suffices to show that D_v maps the space $\overline{\mathbf{L}}_p(\alpha, m_0; \beta, n_0) = \operatorname{Im}(P\varrho_- + Q\varrho_+)$ into itself. Here

$$\varrho_-(t) = \prod_{j=1}^{n} (t - \alpha_j)^{m_0}, \quad \varrho_+(t) = \prod_{j=1}^{n} (t - \beta_j)^{n_0}.$$

If $f = (P\varrho_- + Q\varrho_+)\, g$ with $g \in \mathbf{L}_p(\Gamma)$, then

$$D_v f = (PD_v\varrho_- + QD_v\varrho_+)\, g + \frac{1}{2}(D_vS - SD_v)(\varrho_- g - \varrho_+ g)$$

$$= [P\varrho_-(v(t)) + Q\varrho_+(v(t))]\, g + T_1 g.$$

Lemma 5.2 implies that $T_1 g \in \overline{\mathbf{L}}_p(\alpha, m_0; \beta, n_0)$.

Denote by s the arc abscissa and let $t = t(s)$ be the parametric equation of the curve Γ. Then

$$v(t) - v_i(\alpha_1) = [t - v_{i-1}(\alpha_1)]\frac{v'(\xi_i)}{t'(\theta_i)} \quad (i = 0, \dots, n-1;\ v_{-1}(t) = v_{n-1}(t)),$$

where $v'(\xi_i)$ and $t'(\theta_i)$ are the derivatives with respect to s taken at certain points ξ_i and θ_i on Γ resulting from the mean value theorem. Hence

$$\varrho_-(v(t))\, g(t) = \varrho_-(t)\, g_1(t), \qquad \varrho_+(v(t))\, g(t) = \varrho_+(t)\, g_2(t)$$

with $g_1, g_2 \in \mathbf{L}_p(\Gamma)$. Therefore

$$[P\varrho_-(v(t)) + Q(\varrho_+(v(t)))]\, g = (P\varrho_- + Q\varrho_+)(Pg_1 + Qg_2)$$

and thus $D_v f \in \operatorname{Im}(P\varrho_- + Q\varrho_+)$. ∎

Lemma 5.4. *The set $\tilde{Y}$ is dense in the space Y.*

Proof. Put

$$R_1 = \left(\prod_{i=1}^{n} (t^{-1} - \alpha_i^{-1})^{m_0 - \mu_j^{(i)}}\, \delta_{ij}\right)_1^{kn}, \quad R_2 = \left(\prod_{i=1}^{n} (t - \beta_i)^{n_0 - \nu_j^{(i)}}\, \delta_{ij}\right)_1^{kn}.$$

Then, obviously,

$$\varrho_- I = (t^{-1} - \alpha^{-1})^{\mu} R_1, \qquad \varrho_+ I = (t - \beta)^{\nu} R_2, \tag{6}$$

where I is the identity matrix of order kn and $(t^{-1} - \alpha^{-1})^{\mu}$, $(t - \beta)^{\nu}$ are the diagonal matrix functions defined by the formulae (9.1), Chapter VI. Now set $V_1 = PR_-(t^{-1} - \alpha^{-1})^{\mu} + QR_+(t - \beta)^{\nu}$. Then $Y = \operatorname{Im} V_1$, and from (6) we get

$$P\varrho_- I + Q\varrho_+ I = V_1(PR_1 + QR_2)(PR_-^{-1} + QR_+^{-1}). \tag{7}$$

Since, by virtue of the inclusion (9.5) in Chapter VI, every sufficiently smooth vector function belongs to $\operatorname{Im}(PR_1 + QR_2)$, we deduce from (7) that each vector function $f \in Y$ can be approximated by elements from $\tilde{Y} = \operatorname{Im}(P\varrho_- I + Q\varrho_+ I)$ as closely as desired. ∎

We now turn to the proof of assertion (iii). Thus, assume there is an operator $\tilde{V} \in \Phi_-(X, Y)$ such that $\tilde{V}J = J\tilde{V}$. If all operators K_j $(j = 1, \dots, n)$ (and, consequently, also all operators W_j) are Φ-operators, then Theorem 4.1 implies that W is a Φ-operator, too. Further, from Lemma 4.3, the remark following it, and from the Lemmas 5.3 and 5.4 we deduce that $J(Y) \subset Y$ and then Theorem 4.2 gives that $\operatorname{Ind} V = \sum_{j=1}^{n} \operatorname{Ind} K_j$. Therefore it remains to show that all operators K_j $(j = 1, \dots, n)$ are either simultaneously Φ-operators or not and that they have the same index. To do this, we proceed as follows.

Let t_0 be a point on Γ distinct from the points α_j and β_j ($j = 1, \ldots, n$). Then the curve Γ can be divided into the n pairwise disjoint pieces $\Gamma_1 = (t_0, v_1(t_0)), \ldots, \Gamma_n = (v_{n-1}(t_0), t_0)$. Let u_j ($j = 2, \ldots, n$) be any functions on the closure $\overline{\Gamma}_1$ of Γ_1 which are sufficiently smooth in some neighborhood of the points α_1 and β_1 (without loss of generality assume $\alpha_1, \beta_1 \in \Gamma_1$). Extend u_j to the whole curve in the following way:

$$\tilde{u}_j(t) = \begin{cases} u_j(t), & t \in \overline{\Gamma}_1, \\ x_{1,j}(t)\, u_j(v_{-1}(t)), & t \in \overline{\Gamma}_2, \\ \cdots\cdots\cdots\cdots\cdots\cdots \\ x_{n-1,j}(t)\, u_j(v_{-n+1}(t)), & t \in \overline{\Gamma}_n, \end{cases}$$

where $v_{-i}(t) = v_{n-i}(t)$ and $x_{i,j}$ ($i = 1, \ldots, n-1$; $j = 2, \ldots, n$) are certain functions which will be specified below. For the function $z_j(t) = \tilde{u}_j(t) + \lambda_j \tilde{u}_j(v_1(t)) + \ldots + \lambda_j^{n-1} \tilde{u}_j(v_{n-1}(t))$ we have

$$z_j(v_i(t)) = \lambda_j^{n-i} z_j(t) \qquad (i = 1, \ldots, n-1;\ j = 2, \ldots, n). \tag{8}$$

Now choose the functions $x_{i,j}$ so that they are sufficiently smooth in some neighborhoods of the points α_k and β_k ($k = 1, \ldots, n$), that $\tilde{u}_j$ is continuous at the points $v_i(t_0)$ ($i = 0, \ldots, n-1$), and that $z_j(t) \neq 0$ on $\overline{\Gamma}_1$.

Lemma 5.5. *If $f \in Y$ then $z_j f \in Y$ ($j = 2, \ldots, n$).*

Proof. Let $f \in Y$, that is, $f = V_1 g$ with $g \in \mathbf{L}_p^{kn}(\Gamma)$. Then

$$\begin{aligned} V_1 z_j g &= z_j V_1 g + (P z_j I - z_j P)\, R_-(t^{-1} - \alpha^{-1})^\mu\, g + (Q z_j I - z_j Q)\, R_+(t - \beta)^\nu\, g \\ &= z_j f + \tilde{g}. \end{aligned}$$

Due to the local smoothness of the function z_j the operators $P z_j I - z_j P$ and $Q z_j I - z_j Q$ map into the space $Y = \operatorname{Im} V_1$ (see Theorem 4.5, Chapter VI). Hence $\tilde{g} \in Y$ and thus $z_j f \in Y$. ∎

By combining Lemma 5.5 with the closed graph theorem we conclude that the operator of multipliciaton by the function z_j is continuous on the space Y.

If $f \in \mathfrak{L}_j$, then, by (8), $z_j f \in \mathfrak{L}_1$ ($j = 2, \ldots, n$). Therefore the operator of multiplication by z_j maps the space $\mathfrak{L}_j$ continuously onto the space $\mathfrak{L}_1$. This mapping is one-to-one, since $z_j(t) \neq 0$. Now consider the operators

$$\begin{aligned} K z_j(t) - z_j(t) K &= \sum_{i=0}^{n-1} D_v^i (A_i + B_i S)\, z_j(t) - z_j(t) \sum_{i=0}^{n-1} \lambda_j^{n-i} D_v^i (A_i + B_i S) \\ &= \sum_{i=0}^{n-1} z_j(v_i(t))\, D_v^i (A_i + B_i S) - \sum_{i=0}^{n-1} \lambda_j^{n-i} z_j(t)\, D_v^i (A_i + B_i S) \\ + \sum_{i=0}^{n-1} D_v^i B_i [S z_j(t) - z_j(t) S] &- \sum_{i=0}^{n-1} [z_j(v_i(t)) - \lambda_j^{n-i} z_j(t)]\, D_v^i H_i + T_2 = T_2. \end{aligned} \tag{9}$$

Herein the last equality results from (8).

Because of the local smoothness of the functions z_j, the operator T_2 is compact from $\mathbf{L}_p^k(\Gamma)$ into $\mathfrak{L}_1$. Since the operators of multiplication by z_j are invertible on $\mathbf{L}_p^k(\Gamma)$ and from $\mathfrak{L}_j$ onto $\mathfrak{L}_1$, it follows from (9) that the operators K_j ($j = 1, \ldots, n$) are either simultaneously Φ-operators or not and that they have the same index. This completes the proof of Theorem 5.2. ∎

Remarks. 1. The parts (i) and (ii) of Theorem 5.2 were established by CH. MEYER [1, 2], the part (iii) is due to MEYER and SILBERMANN [1]. When proving that the operators K_j have the same index we used a construction going back to KRAVCHENKO and LITVINCHUK [1].

2. It is not difficult to derive analogues of the results of this section for the case where the shift $v(t)$ alters the orientation of the curve Γ (in this connection see GOHBERG and KRUPNIK [8] and CH. MEYER [1]).

§ 6. The Tricomi equation

Consider the linear continuous integral operators given on the space $\mathbf{L}_2(0, 1)$ by

$$(K_1\varphi)(x) := \tilde{c}(x)\,\varphi(x) + \frac{\tilde{d}(x)}{\pi i} \int_0^1 \left(\frac{1}{y-x} - \frac{1}{x+y-2xy}\right) \varphi(y)\,dy,$$

$$(K_2\varphi)(x) := \tilde{c}(x)\,\varphi(x) + \frac{\tilde{d}(x)}{\pi i} \int_0^1 \left(\frac{1}{y-x} + \frac{1}{x+y-2xy}\right) \varphi(y)\,dy,$$

where $\tilde{c}$ and $\tilde{d}$ are piecewise continuous functions on the interval $[0, 1]$, which are without loss of generality supposed to be continuous from the left. The equation $K_1\varphi = f$ is usually called the *Tricomi equation*.

With the operators K_1 and K_2 we associate the following singular integral operator on the space $\mathbf{L}_2(\Gamma)$ over the unit circle Γ:

$$(W\psi)(t) := c(t)\,\psi(t) + \frac{d(t)}{\pi i} \int_\Gamma \frac{\psi(\tau)}{\tau - t}\,d\tau.$$

The piecewise continuous functions $c, d \in \mathbf{PC}(\Gamma)$ are defined as follows:

$$c(t) = \begin{cases} \tilde{c}\left(i\dfrac{1-t}{1+t}\right), & t \in (\widehat{1, i}], \\[2ex] \tilde{c}\left(\dfrac{i(1-t)}{2i(1-t)-(1+t)}\right), & t \notin (\widehat{1, i}], \end{cases}$$

$$d(t) = \begin{cases} \tilde{d}\left(i\dfrac{1-t}{1+t}\right), & t \in (\widehat{1, i}], \\[2ex] -\tilde{d}\left(\dfrac{i(1-t)}{2i(1-t)-(1+t)}\right), & t \notin (\widehat{1, i}]. \end{cases}$$

Here $(\widehat{1, i}]$ denotes the arc of the unit circle which joins the points 1 and i counterclockwise and which does not contain the point $t = 1$. Put $u(t) = \dfrac{i(1-t)-1}{i(1-t)-t}$. It is easily seen that u maps the unit circle Γ onto itself, that u alters the orientation of Γ, and that u transforms $(\widehat{i, 1})$ onto $(\widehat{1, i})$. The points $t = 1$ and $t = i$ are the only fixpoints of u and one has $u(u(t)) \equiv t$.

In the sequel the functions

$$a(t) = c(t) + d(t), \qquad b(t) = c(t) - d(t)$$

are allowed to degenerate. For the sake of simplicity we assume that a has a single zero t_0 of integral order on Γ,

$$a(t) = (t^{-1} - t_0^{-1})^m\, a_1(t),$$

where $t_0 \neq 1$, $t_0 \neq \mathrm{i}$, and $m \geqq 0$ is an integer. Because $b(t) = a(u(t+0))$, the function b has a zero of order m at the point $u(t_0)$:

$$b(t) = (t - u(t_0))^m\, b_1(t).$$

Suppose $a_1 \in \mathbf{PC}(u(t_0), m)$ and, hence, $b_1 \in \mathbf{PC}(t_0, m)$. Then Theorem 6.1, Chapter VI, immediately gives the following.

Theorem 6.1. *The operator* $W \in \mathcal{L}(\mathbf{L}_2(\Gamma), \overline{\mathbf{L}}_2(t_0, m; u(t_0), m))$ *is a* Φ*-operator if and only if*

$$\mu a_1(t+0)\, b_1(t) + (1-\mu)\, a_1(t)\, b_1(t+0) \neq 0 \qquad (|t| = 1, 0 \leqq \mu \leqq 1). \tag{1}$$

If the condition (1) *is satisfied, then W is at least one-sided invertible and* Ind $W = \mathrm{ind}_2(b_1/a_1)$

Now define

$$\mathfrak{L}_j := \{f(x) \in \mathbf{L}_2(0, 1) : f_j(t) \in \overline{\mathbf{L}}_2(t_0, m; u(t_0), m)\},$$

where $j = 1, 2$ and

$$f_j(t) = \begin{cases} \dfrac{1}{1+t} f\left(\mathrm{i}\dfrac{1-t}{1+t}\right) & \text{for } |t| = 1,\ t \in \widehat{(1, \mathrm{i}]}, \\[2ex] \dfrac{(-1)^{j+1}}{2\mathrm{i}(1-t) - (1+t)} f\left(\dfrac{\mathrm{i}(1-t)}{2\mathrm{i}(1-t) - (1+t)}\right) & \text{for } t \notin \widehat{(1, \mathrm{i}]}. \end{cases} \tag{2}$$

When provided with the norm $\|f\|_{\mathfrak{L}_j} = \|f_j\|_{\mathbf{L}_2}$ the linear set $\mathfrak{L}_j$ $(j = 1, 2)$ becomes a Banach space. It can be shown that the operators $K_j \colon \mathbf{L}_2(0, 1) \to \mathfrak{L}_j$ $(j = 1, 2)$ are continuous.

Theorem 6.2. *The operators* $K_j : \mathbf{L}_2(0, 1) \to \mathfrak{L}_j$ $(j = 1, 2)$ *are* Φ*-operators if and only if condition* (1) *is fulfilled.*

If the condition (1) *is satisfied, then the operators* K_j *are left-sided invertible, right-sided invertible, or invertible if the integer* $\mathrm{ind}_2\,(b_1/a_1)$ *is negative, positive, or zero, respectively. In each case,* Ind K_1 + Ind $K_2 = \mathrm{ind}_2\,(b_1/a_1)$.

Proof. Consider the linear continuous operator defined on $\mathbf{L}_2(\Gamma)$ by

$$(Jf)\,(t) = \frac{1}{\mathrm{i}(1-t) - t} f(u(t)).$$

It can be easily checked that $J^2 = I$ and $JW = WJ$. Now apply the results of § 4 in the following setting: $X = \mathbf{L}_2(\Gamma)$, $Y = \overline{\mathbf{L}}_2(t_0, m; u(t_0), m)$, $\lambda_1 = 1$, $\lambda_2 = -1$. Let f_j be the function given by (2) and define the operator $P_j : \mathbf{L}_2(0, 1) \to \mathbf{L}_2(\Gamma)$ $(j = 1, 2)$ by the formula $(P_j f)\,(t) = f_j(t)$ $(t \in \Gamma)$. It can be shown that $J(Y) \subset Y$ and that $K_j = P_j^{-1} W_j P_j$. Now Theorem 6.2 is seen to be a direct consequence of the Theorems 6.1, 4.1, and 4.2. ∎

Finally, notice the following theorem (see Ch. Meyer [1] for a proof).

Theorem 6.3. *Let the condition* (1) *be satisfied and suppose* Ind $W = 2k$, *where k is any integer. Then* Ind K_1 = Ind $K_2 = k$.

Remark 1. All results of this section are due to Ch. Meyer [1, 2].

Remark 2. Ch. Meyer [1, 2] also applied the results of § 4 to singular integral equations with conjugate complex unknowns and degenerate coefficients. However, in this situation it is simpler to proceed as follows.

Suppose, for instance, we are given an equation of the form

$$(Au)(t) := a(t)\,u(t) + b(t)\,\overline{u(t)} + \frac{c(t)}{\pi \mathrm{i}} \int\limits_{\Gamma} \frac{u(\tau)}{\tau - t}\,\mathrm{d}\tau + \frac{d(t)}{\pi \mathrm{i}} \int\limits_{\Gamma} \frac{\overline{u(\tau)}}{\tau - t}\,\mathrm{d}\tau + (T_1 u)(t) + (T_2 \bar{u})(t) = f(t)$$

with certain compact operators T_1 and T_2. Then add the equation $\overline{(Au)(t)} = \overline{f(t)}$. What results is a system of two singular integral equations with the two unknown functions u and $\bar{u}$, which is equivalent to the original equation $Au = f$. Now it remains to apply the results of § 9, Chapter VI, to the system obtained in this way.

Chapter VIII
Some further subsidiaries

§ 1. Stereographic projection

Stereographic projection is the mapping of the m-dimensional Euclidean space $\mathbb{R}^m$ onto the unit sphere Σ of the $(m + 1)$-dimensional space $\mathbb{R}^{m+1}$ given by the formulae

$$\xi_k = \frac{2x_k}{|x|^2 + 1}, \qquad k = 1, 2, \dots, m; \qquad \xi_{m+1} = \frac{|x|^2 - 1}{|x|^2 + 1}. \tag{1}$$

The stereographic projection can be interpreted geometrically as follows. We identify the space $\mathbb{R}^m$ with the plane $\xi_{m+1} = 0$ of the space $\mathbb{R}^{m+1}$. Let a point x with coordinates $x_1, x_2, \dots, x_m$ be given in that plane. We connect that point by a straight line with the point $N = (0, 0, \dots, 0, 1)$ on the sphere Σ (Fig. 5). Then (1) gives the (second) intersection point of that line with the sphere Σ which corresponds to the original point x.

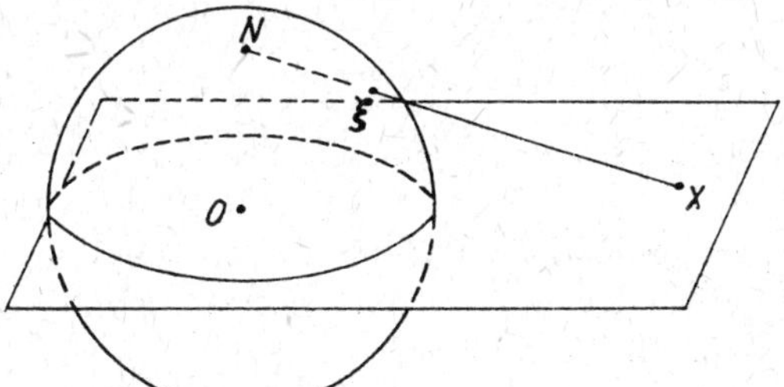

Fig. 5

Let r denote the distance of the points x and y of the space $\mathbb{R}^m$ and likewise ϱ the distance of their images by stereographic projection, ξ and η, on the sphere Σ. Then

$$\varrho^2 = \sum_{j=1}^{m+1} (\xi_j - \eta_j)^2 = 4 \sum_{j=1}^{m} \left(\frac{x_j}{|x|^2 + 1} - \frac{y_j}{|y|^2 + 1}\right)^2 + \left(\frac{|x|^2 - 1}{|x|^2 + 1} - \frac{|y|^2 - 1}{|y|^2 + 1}\right)^2.$$

Simple transformations reduce that expression to $4r^2(|x|^2 + 1)^{-1}(|y|^2 + 1)^{-1}$, and we obtain the relation

$$r^2 = \frac{\varrho^2}{4} (|x|^2 + 1)(|y|^2 + 1). \tag{2}$$

By $\mathrm{d}y$ we denote the differential volume element in the space $\mathbb{R}^m$ and by $\mathrm{d}\sigma$ the surface element on the sphere Σ. We shall now prove the relation

$$\mathrm{d}y = \left(\frac{|y|^2 + 1}{2}\right)^m \mathrm{d}\sigma. \tag{3}$$

In the $(m + 1)$-dimensional space with coordinates $\eta_1, \eta_2, \ldots, \eta_{m+1}$ we take a 2-dimensional plane Ξ passing through the origin and the points y, η (Fig. 6). The plane Ξ intersects the sphere Σ in some circle and the space $\mathbb{R}^m$ in the straight line PP which passes the origin and is perpendicular to the line ON. The elements $\mathrm{d}\sigma$ and $\mathrm{d}y$ intersect Ξ in the circular arc $\eta\eta'$ and in the segment yy', resp. Since stereographic projection is a conformal mapping, we have up to an infinitely small term

$$\frac{\mathrm{d}y}{\mathrm{d}\sigma} = \frac{|y' - y|^m}{|\eta' - \eta|^m}.$$

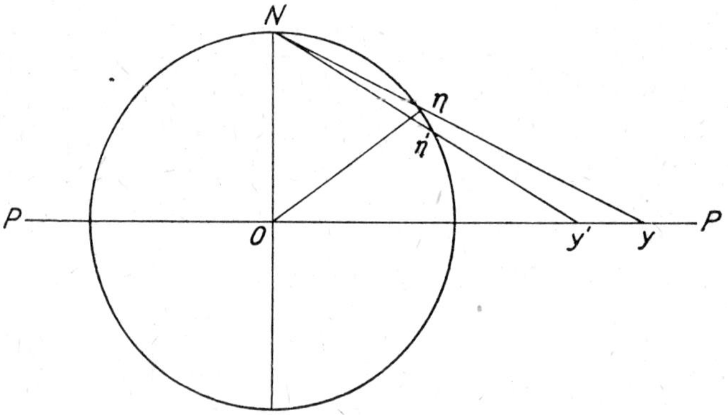

Fig. 6

Equation (2) shows that up to the terms mentioned $\dfrac{|y' - y|}{|\eta' - \eta|} = \dfrac{|y|^2 + 1}{2}$ and (3) is proved.

In case $m = 2$ the stereographic projection is nothing else than the well-known mapping of the complex plane onto the Riemann sphere. This is the reason why we shall call Σ the *Riemann sphere* in the sequel.

§ 2. Some function spaces

We introduce some function spaces which we shall use in the following.

2.1. Dini spaces. Let M be a compact set endowed with a distance, and let $u(x)$ be a continuous real- or complex-valued function defined on M. The *modulus of continuity* of the function u is the non-negative function $\omega(u, t)$, $t \geqq 0$, defined by the formula

$$\omega(u, t) = \sup_{\substack{x,y \in M \\ \varrho(x,y) \leqq t}} |u(y) - u(x)| \tag{1}$$

where $\varrho(x, y)$ denotes the distance between the points $x, y \in M$. The Banach space $\mathbf{Di}(M)$ (Dini space) consists of all functions u defined on M such that

$$\int_0^\delta \frac{\omega(u, t)}{t}\,\mathrm{d}t < \infty, \qquad \delta = \text{const} > 0. \tag{2}$$

The norm in $\mathbf{Di}(M)$ is given by the formula

$$\|u\| = \max_{x\in M} |u(x)| + \int_0^{\delta} \frac{\omega(u, t)}{t} \, dt \tag{3}$$

or any other norm which is equivalent to the norm (3).

The condition (2) is called *Dini condition.*

2.2. An important special case of the Dini spaces is provided by the *Lipschitz spaces* $\mathbf{Lip}_\alpha(M)$, $0 < \alpha \leqq 1$. Those spaces consist of functions defined on the set M which have the properties as indicated in Section 2.1 and satisfy on M the inequality

$$|u(x) - u(y)| \leqq C\varrho^\alpha(x, y) \tag{4}$$

where the constant C may depend on the function u. The norm in $\mathbf{Lip}_\alpha(M)$ is usually given by

$$\|u\|_{\mathbf{Lip}_\alpha(M)} = \max_{x\in M} |u(x)| + \sup_{x,y\in M} \varrho^{-\alpha}(x, y) \, |u(x) - u(y)| . \tag{5}$$

In case $\alpha < 1$ the Lipschitz spaces are frequently also called *Hölder spaces* and are denoted by $\mathbf{H}^\alpha(M)$.

2.3. We suppose that the reader is familiar with the notion of the spaces $\mathbf{L}_p(M)$ and with the notion of the *Sobolev-Slobodetski spaces* $\mathbf{W}_p^{(k)}(\Omega)$, where M is an arbitrary measurable set and Ω a domain in the Euclidean space $\mathbb{R}^m$. In what follows we need the spaces $\mathbf{W}_p^{(k)}(S)$ with S being the unit sphere. They can be introduced by means of two equivalent constructions.

a) We take a ball layer $B = \{x: \varrho_1 < |x| < \varrho_2\}$ in $\mathbb{R}^m$ where $0 < \varrho_1 < \varrho_2 < \infty$. Any function defined on S can be extended to the ball layer B by setting it constant on every ray starting from the origin. The space $\mathbf{W}_p^{(k)}(S)$ is then defined as the space of functions defined on S with the extension constructed that way belonging to the Sobolev-Slobodetski space $\mathbf{W}_p^{(k)}(B)$. As the norm of a function $u \in \mathbf{W}_p^{(k)}(S)$ we take the norm of its extension in $\mathbf{W}_p^{(k)}(B)$.

b) Let S_j, $j = 1, 2, \ldots, j_0 < \infty$, be a finite covering of the sphere S and denote by $\varphi_j(\theta)$, $j = 1, 2, \ldots, j_0$, the corresponding partition of unity. We suppose that each of the sets S_j can be mapped 1-to-1 and sufficiently smooth onto a certain domain D_j of the Euclidean space $\mathbb{R}^{m-1}$; the mapping will be denoted by $x = \tau_j(\theta)$ where θ is a varying point on S. Let $u(\theta)$ be a function defined on S. We shall say that u belongs to $\mathbf{W}_p^{(k)}(S)$ iff $u_j(x) := u(\tau_j^{-1}(x)) \in \mathbf{W}_p^{(k)}(D_j)$, $j = 1, 2, \ldots, j_0$. The norm in $\mathbf{W}_p^{(k)}(S)$ is defined by

$$\|u\|_{\mathbf{W}_p^{(k)}(S)} = \sum_{j=1}^{j_0} \|u_j\|_{\mathbf{W}_p^{(k)}(D_j)} . \tag{6}$$

§ 3. Weakly singular integral operators

3.1. In the present section we shall give some theorems on the compactness of weakly singular integral operators in function spaces which are frequently used; some of these theorems will be given without proof.

Remember that a weakly singular integral operator in the Euclidean space is of the form

$$(tu)(x) = \int_D \frac{A(x, y) \, u(y)}{|x - y|^\lambda} \, dy \tag{1}$$

where D is a bounded measurable subset in $\mathbb{R}^m$, x and y are points belonging to D, and $A(x,y)$ is a bounded measurable function. The constant λ satisfies the restriction $0 < \lambda < m$, and $\mathrm{d}y$ stands for the volume element with respect to the Lebesgue measure in $\mathbb{R}^m$.

In the general case a weakly singular integral operator is defined by

$$(tu)(x) = \int_M A(x,y)\,\varrho^{-\lambda}(x,y)\,u(y)\,\mathrm{d}_y M. \tag{1_1}$$

Here M denotes a bounded m-dimensional manifold imbedded in the Euclidean space $\mathbb{R}^{m'}$ $(m' \geqq m)$, $x, y \in M$, and $\varrho(x, y)$ stands for the distance between the points x and y in the metric of $\mathbb{R}^{m'}$. The constant λ is the constant above $(0 < \lambda < m)$, and $\mathrm{d}M$ is the element of the measure on M which is induced by the imbedding of M in $\mathbb{R}^{m'}$. The manifold M is supposed to be sufficiently smooth.

3.2. Theorem 3.1. *The operator* (1) *considered as a mapping from* $\mathbf{L}_p(D)$ *into* $\mathbf{L}_q(D)$, $1 \leqq p \leqq \infty$, *is bounded provided that*

$$p \leqq q \leqq q^*, \qquad q^* := \begin{cases} \dfrac{mp}{m - p(m-\lambda)}, & p < \dfrac{m}{m-\lambda}; \\ \infty, & p \geqq \dfrac{m}{m-\lambda}. \end{cases}$$

The operator is completely continuous, if $p \leqq q < q^*$ *or (in case* $q^* = \infty$*)* $p < q$.

Theorem 3.2. *If the set* M *is a closed one and* $A \in \mathbf{C}(M \times M)$, *then the operator* (1_1) *is completely continuous in* $\mathbf{C}(M)$.

We shall not prove the Theorems 3.1 and 3.2. The latter is well-known; it is an immediate consequence of Arzelà's theorem on compactness in $\mathbf{C}(M)$. The proof of Theorem 3.1 can be found in the book by Sobolev [3]. We remark only that Theorem 3.1 is also true for the operator (1_1).

3.3. Theorem 3.3. *Let the point* $x = 0$ *belong to the set* $\bar{D}$. *The operator* (1) *is completely continuous in* $\mathbf{L}_p(D, |x|^\beta)$, $1 \leqq p \leqq \infty$, *if the constant* β *satisfies the inequality*

$$-\frac{m}{p} < \beta < \frac{m}{p'}; \quad \frac{1}{p} + \frac{1}{p'} = 1. \tag{2}$$

Here $\mathbf{L}_p(D, |x|^\beta)$ denotes the space of functions which are summable over the set D in the p-th power with respect to the weight function $|x|^{p\beta}$ [1]).

For $\beta = 0$ Theorem 3.3 is a special case of Theorem 3.1, therefore we may assume $\beta \neq 0$ in the following.

To begin with, we show that under the assumptions of the Theorem the operator (1) is bounded. Let $\beta > 0$, put $|x - y| = r$, $|x| = R$, $|y| = \varrho$. With the notations $v(x) = (tu)(x)$, $A_0 = \sup |A(x, y)|$ we have

$$|R^\beta v(x)| \leqq A_0 R^\beta \int_D \frac{1}{r^\lambda} |u(y)|\,\mathrm{d}y = A_0 R^\beta \int_D \frac{1}{r^{\lambda/p'} \varrho^\beta} \frac{|u(y)|\varrho^\beta}{r^{\lambda/p}}\,\mathrm{d}y \leqq$$

$$\leqq A_0 R^\beta \left\{ \int_D \frac{\mathrm{d}y}{r^\lambda \varrho^{\beta p'}} \right\}^{1/p'} \left\{ \int_D \frac{|u(y)|^p \varrho^{\beta p}}{r^\lambda}\,\mathrm{d}y \right\}^{1/p}. \tag{3}$$

[1]) The boundedness of the operator (1) under the assumptions of Theorem 3.3 was proved by Glushko [1] in a different way.

We estimate the first integral of the right-hand side in (3). By a theorem on the product of weakly singular integral operators (SOBOLEV [1], cf. also MIKHLIN [1]),

$$\int_D \frac{dy}{r^\lambda \varrho^{\beta p'}} \leqq \begin{cases} C, & \lambda + \beta p' < m, \\ C \ln \frac{c_1}{R}, & \lambda + \beta p' = m_1;\ c_1 = \text{const}, \\ CR^{-(\lambda+\beta p'-m)}, & \lambda + \beta p' > m. \end{cases} \tag{4}$$

Here C is a constant depending on the set D; it remains bounded if D degenerates into a point. Denote the right-hand side in the inequality (4) by $C\chi(R)$, then we have by (3)

$$|v(x)|^p R^{\beta p} \leqq A_0^p C^p R^{\beta p} [\chi(R)]^{p/p'} \int_D \frac{|u(y)|^p \varrho^{\beta p}}{r^\lambda} dy$$

and hence ($\|\cdot\| = \|\cdot\|_{\mathbf{L}_p(D,|x|^\beta)}$)

$$\|v\|^p \leqq A_0^p C^p \int_D |u(y)|^p \varrho^{\beta p}\, dy \int_D \frac{R^{\beta p}[\chi(R)]^{p/p'}}{r^\lambda} dy. \tag{5}$$

If $\lambda + \beta p' > m$ then

$$R^{\beta p}[\chi(R)]^{p/p'} = R^{(m-\lambda)p/p'}.$$

The exponent at R is positive, the inner integral in (5) is bounded by some constant B^p, therefore we obtain

$$\|v\| \leqq A_0 BC \|u\| \tag{6}$$

and that was to be shown. In case $\lambda + \beta p' = m$ we have $\beta = (m - \lambda)/p' > 0$, the inner integral in (5) is bounded once again, and we obtain again the inequality (6). Finally, let $\lambda + \beta p' < m$, then $\chi(R) = 1$. If $\beta > 0$ the inner integral in (5) remains to be bounded, i.e. the inequality (6) is true.

In case $\beta < 0$ we consider the adjoint operator

$$(t^* u)(x) = \int_D \frac{\overline{A(y, x)}}{r^\lambda} u(y)\, dy. \tag{7}$$

As one can easily see the relation $t \in \mathscr{L}(\mathbf{L}_p(D, R^\beta), \mathbf{L}_p(D, R^\beta))$ implies $t^* \in \mathscr{L}(\mathbf{L}_{p'}(D, R^{-\beta}), \mathbf{L}_{p'}(D, R^{-\beta}))$ and vice versa. Therefore, it is sufficient to show that the conjugate space of $X = \mathbf{L}_p(D, g)$ is $X^* = \mathbf{L}_{p'}(D, g^{-1})$. Let $u \in \mathbf{L}_p(D, g)$ and put $\tilde{u} = gu$. That transformation is an isometric mapping of $\mathbf{L}_p(D, g)$ to $\mathbf{L}_p(D)$. Let f be a bounded linear functional over $\mathbf{L}_p(D, g)$. We define a functional $\tilde{f}$ by the relation $\tilde{f}(\tilde{u}) := f(g^{-1}\tilde{u})$ which is bounded on $\mathbf{L}_p(D)$:

$$|\tilde{f}(\tilde{u})| \leqq \|f\|_{\mathbf{L}_p(D,g)} \|g^{-1}\tilde{u}\|_{\mathbf{L}_p(D,g)} = \|f\|_{\mathbf{L}_p(D,g)} \|\tilde{u}\|_{\mathbf{L}_p(D)}.$$

But then $\tilde{f}(\tilde{u}) = \int_D \overline{\tilde{F}(x)}\, \tilde{u}(x)\, dx$ with a certain function $\tilde{F} \in \mathbf{L}_{p'}(D)$. Consequently, $f(u) = \tilde{f}(gu) = \int_D \overline{F(x)}\, u(x)\, dx$ where $F(x) = g(x)\, \tilde{F}(x)$, and hence $F \in \mathbf{L}_{p'}(D, g^{-1})$.

Now let us consider the operator t^* again. Under the assumptions made above we have $-m < \beta < 0$, i.e., $0 < -\beta < m$, and that is the same as the inequality (2). We proved that the operator t^* is bounded in $\mathbf{L}_{p'}(D, R^{-\beta})$, hence the operator t is bounded in $\mathbf{L}_p(D, R^\beta)$.

3.4. Now we shall prove that the operator (1) is completely continuous. We shall assume that β is positive; for $\beta < 0$ it is sufficient to consider the adjoint operator. We rewrite (1) as

$$(tu)(x) = \int\limits_{D\cap(r<\eta)} \frac{A(x, y)}{r^\lambda} u(y)\, dy + \int\limits_{D\cap(r>\eta)} \frac{A(x, y)}{r^\lambda} u(y)\, dy = (t'u)(x) + (t''u)(x)$$

where η is an arbitrarily small positive number. We put

$$A'(x, y) = \begin{cases} A(x,y), & r \leqq \eta, \\ 0, & r > \eta \end{cases}; \quad A''(x, y) = \begin{cases} 0, & r < \eta, \\ A(x, y), & r \geqq \eta \end{cases}$$

then

$$(t'u)(x) = \int\limits_D \frac{A'(x, y)}{r^\lambda} u(y)\, dy, \qquad (t''u)(x) = \int\limits_D \frac{A''(x, y)}{r^\lambda} u(y)\, dy .$$

The kernel of the operator t'' is bounded, say $|r^{-\lambda} A''(x, y)| \leqq C_1 = \text{const}$. We put $\tilde{u} = uR^\beta$ (then $\tilde{u} \in \mathbf{L}_p(D)$) and show that the operator $\tilde{t}$: $\tilde{t}\tilde{u} = t''(\tilde{u}R^{-\beta})$ is completely continuous in $\mathbf{L}_p(D)$. We have

$$(\tilde{t}\tilde{u})(x) = \int\limits_D \frac{A''(x, y)\, \varrho^{-\beta}}{r^\lambda} u(y)\, dy$$

and hence

$$(\tilde{t}^*\tilde{v})(x) = R^{-\beta} \int\limits_D \frac{\overline{A''(y, x)}}{r^\lambda} \tilde{v}(y)\, dy = R^{-\beta}(T\tilde{v})(x). \tag{8}$$

The kernel of the operator T is bounded, and T can be considered as a weakly singular operator with the exponent $\lambda = 0$. Due to Theorem 3.1 the operator T is completely continuous from $\mathbf{L}_p(D)$ into $\mathbf{L}_\infty(D)$.

Let $\tilde{M}$ be a set which is bounded in $\mathbf{L}_p(D)$. By the operator T it is mapped into the set $T\tilde{M}$ which is compact in $\mathbf{L}_\infty(D)$, i.e. there is a sequence $\{\tilde{u}_n\}$, $\tilde{u}_n \in \tilde{M}$, such that

$$\| T(\tilde{u}_n - \tilde{u}_k) \|_{\mathbf{L}_\infty(D)} \xrightarrow[n,k\to\infty]{} 0 .$$

We have

$$\| \tilde{t}^*(\tilde{u}_n - \tilde{u}_k) \|^{p'}_{\mathbf{L}_{p'}(D)} = \int\limits_D |\tilde{t}^*(\tilde{u}_n - \tilde{u}_k)|^{p'}\, dx =$$

$$= \int\limits_D R^{-\beta p'} |T(\tilde{u}_n - \tilde{u}_k)|^{p'}\, dx \leqq \| T(\tilde{u}_n - \tilde{u}_k) \|^{p'}_{\mathbf{L}_\infty(D)} \int\limits_D R^{-\beta p'}\, dx .$$

Now $\beta p' < m$ because of $\beta < m/p'$, and the last integral in the estimate is convergent. Thus the operator $\tilde{t}^*$ is completely continuous in $\mathbf{L}_{p'}(D)$ and so is the operator $\tilde{t}$ in $\mathbf{L}_p(D)$.

Next we are going to show that the operator t'' in $\mathbf{L}_p(D, |x|^\beta)$ is completely continuous. Let M be a set of functions which is bounded in the space mentioned. The transformation $\tilde{u}(x) = R^\beta u(x)$ is an isometric mapping of the space $\mathbf{L}_p(D, |x|^\beta)$ to the space $\mathbf{L}_p(D)$,

$$\| \tilde{u} \|^p_{\mathbf{L}_p(D)} = \int\limits_D |x|^{\beta p} |u(x)|^p\, dx = \| u \|^p_{\mathbf{L}_p(D, |x|^\beta)} .$$

Therefore the set $\tilde{M}$ the image of M under that transformation is bounded in $\mathbf{L}_p(D)$, and the corresponding set $t\tilde{M}$ is compact in $\mathbf{L}_p(D)$. Hence, there is a sequence $\{\tilde{u}_n\}$, $\tilde{u}_n \in \tilde{M}$, such that

$$\|\tilde{t}(\tilde{u}_n - \tilde{u}_k)\|^p_{\mathbf{L}_p(D)} = \int_D |\tilde{t}(\tilde{u}_n - \tilde{u}_k)|^p \, dx \xrightarrow[n,k\to\infty]{} 0 .$$

Put $\tilde{u}_n = R^\beta u_n(x)$, then $u_n \in M$ and $\tilde{t}\tilde{u}_n = t'' u_n$. Now

$$\|t''(u_n - u_k)\|^p_{\mathbf{L}_p(D,|x|^\beta)} = \int_D |x|^{\beta p} |t''(u_n - u_k)|^p \, dx$$

$$\leqq H^{\beta p} \int_D |t''(u_n - u_k)|^p \, dx = H^{\beta p} \int_D |\tilde{t}(\tilde{u}_n - \tilde{u}_k)|^p \, dx \xrightarrow[n,k\to\infty]{} 0 ,$$

where H denotes the diameter of the set D. This relation proves that t'' is a completely continuous operator in $\mathbf{L}_p(D, |x|^\beta)$.

Finally we shall show that $\lim\limits_{\eta\to 0} \|t'\|_{\mathbf{L}_p(D,|x|^\beta)} = 0$. We denote $v(x) = (t'u)(x)$, $x \in D$, then analogously to inequality (3) we find

$$R^\beta |v(x)| \leqq A R^\beta \left\{ \int\limits_{r<\eta} \frac{dy}{r^\lambda \varrho^{\beta p'}} \right\}^{1/p'} \left\{ \int\limits_{r<\eta} \frac{|u(y)|^p \varrho^{\beta p}}{r^\lambda} dy \right\}^{1/p}$$

and, furthermore, by (4)

$$|v(x)|^p \, R^{\beta p} \leqq A^p C^p R^{\beta p} [\chi(R)]^{p/p'} \int\limits_{r<\eta} \frac{|u(y)|^p \, \varrho^{\beta p}}{r^\lambda} dy .$$

Integrating over D and changing the order of integration with respect to x and y we obtain

$$\|v\|^p_{\mathbf{L}_p(D,|x|^\beta)} \leqq A^p C^p \int_D |u(y)|^p \varrho^{\beta p} dy \int\limits_{r<\eta} \frac{R^{\beta p}[\chi(R)]^{p/p'}}{r^\lambda} dy . \tag{9}$$

As we stated above, for $\beta > 0$ and $\lambda + \beta p' \neq m$ the numerator in the inner integral is bounded by some constant, say C_1^p. The double integral can be estimated from above by the expression

$$C_1^p \, \|u\|^p_{\mathbf{L}_p(D,|x|^\beta)} \int\limits_{r<\eta} \frac{dx}{r^\lambda} = C_1^p \, \|u\|^p_{\mathbf{L}_p(D,|x|^\beta)} |S| \, \eta^{m-\lambda}/(m - \lambda)$$

where $|S| = 2\pi^{m/2}/\Gamma(m/2)$ is the measure of the surface of the unit ball in $\mathbf{R}^m$. Now the relation (9) shows that

$$\|t'\|_{\mathbf{L}_p(D,|x|^\beta)} \leqq ACC_1 \left(\frac{|S|}{m - \lambda} \right)^{1/p} \eta^{(m-\lambda)/p} \xrightarrow[\eta\to 0]{} 0 .$$

The case $\lambda + \beta p' = m$ can be eliminated in the following way: It suffices to represent the kernel of the operator (1) in the form $A(x, y) \, r^\alpha / r^{\alpha+\lambda}$ where $0 < \alpha < m - \lambda$, then $\lambda + \alpha + \beta p' \neq m$.

Thus we obtained $\|t - t''\|_{\mathbf{L}_p(D,|x|^\beta)} \xrightarrow[\eta\to 0]{} 0$, i.e. the operator t is the limit (in the sense of norm convergence) of the completely continuous in $\mathbf{L}_p(D, |x|^\beta)$ operators t'' and consequently it is itself completely continuous in that space. Thus, Theorem 3.3 is proved.

3.5. Theorem 3.3 remains to hold true, obviously, if the weight function is not equal but equivalent to R^β, i.e. its product with $R^{-\beta}$ is bounded from below and from above by positive constants. Then it is easy to see that the theorem is valid also if D is replaced by the Riemann sphere Σ (see § 1) and the weight function by the function $(1-\xi_{m+1})^{\beta/2}$, because the weights $(1-\xi_{m+1})^{\beta/2}$ and $\left(\sum_{k=1}^{m}\xi_k^2\right)^{\beta/2}$ are equivalent on Σ for ξ_{m+1} close to 1.

Theorem 3.4. *The operator*

$$\int_{\mathbb{R}^m}\frac{\varphi(y,\theta)-\varphi(x,\theta)}{r^m}f(x,y)\,u(y)\,dy, \qquad r=|y-x|, \qquad \theta=(y-x)/r \tag{10}$$

is completely continuous in $\mathbf{L}_p(\mathbb{R}^m)$ *provided that the function* $f(x,y)$ *is bounded, and* $\varphi(x,\theta)$ *satisfies the inequality*

$$|\varphi(y,\theta)-\varphi(x,\theta)|\leqq A\varrho^\lambda \tag{11}$$

where A *and* λ *are positive constants and* ϱ *denotes the distance of those points on the Riemann sphere which correspond to the points* x *and* y *in the stereographic projection.*

We denote the integral (10) by $v(x)$ and map the space $\mathbb{R}^m$ stereographically onto the Riemann sphere Σ. Then the integral (10) becomes

$$v(x)=\int_\Sigma\frac{\varphi(y,\theta)-\varphi(x,\theta)}{\varrho^m}f(x,y)\left(\frac{|y|^2+1}{|x|^2+1}\right)^{m/2}u(y)\,d\sigma.$$

Put

$$\left(\frac{|x|^2+1}{2}\right)^{m/2}u(x)=u'(\xi), \qquad \left(\frac{|x|^2+1}{2}\right)^{m/2}v(x)=v'(\xi) \tag{12}$$

then

$$v'(\xi)=\int_\Sigma\frac{\varphi(y,\theta)-\varphi(x,\theta)}{\varrho^m}f(x,y)\,u'(\eta)\,d\sigma. \tag{13}$$

Eq. (12) realizes an isometric mapping of the space $\mathbf{L}_p(\mathbb{R}^m)$ onto the space $\mathbf{L}_p(\Sigma,q)$ where

$$q(\xi)=\left(\frac{|x|^2+1}{2}\right)^{-m\left(\frac{p}{2}-1\right)}=(1-\xi_{m+1})^{-m\left(\frac{p}{2}-1\right)}.$$

The exponent $\beta=m(p-2)/p$ satisfies the inequality (2), hence, by Theorem 3.3 the operator (13) is completely continuous in $\mathbf{L}_p(\Sigma,q)$, but then the operator (10) is completely continuous in $\mathbf{L}_p(\mathbb{R}^m)$.

§ 4. On the powers of the Beltrami operator

Fundamental information on the Beltrami operator and on spherical functions in the space $\mathbb{R}^m$ of arbitrary dimension you can find, for instance, in the books by Mikhlin [12] and Sobolev [1].

4.1. The Laplace operator for spherical coordinates $(\varrho,\theta)=(\varrho,\vartheta_1,\vartheta_2,\ldots,\vartheta_{m-1})$ has the form

$$\Delta=\frac{\partial^2}{\partial\varrho^2}+\frac{m-1}{\varrho}\frac{\partial}{\partial\varrho}-\frac{1}{\varrho^2}\delta, \tag{1}$$

where the so-called *Beltrami operator* δ is defined by the relations

$$\delta = -\sum_{j=1}^{m-1} \frac{1}{q_j \sin^{m-j-1} \vartheta_j} \frac{\partial}{\partial \vartheta_j} \left(\sin^{m-j-1} \vartheta_j \frac{\partial}{\partial \vartheta_j} \right), \tag{2}$$

$$q_1 = 1; \qquad q_j = (\sin \vartheta_1 \sin \vartheta_2 \dots \sin \vartheta_{j-1})^2, \qquad j > 1. \tag{3}$$

The Beltrami operator is symmetric in the space $\mathbf{L}_2(S)$, the corresponding quadratic form is

$$(\delta f, f) = \int_S \sum_{j=1}^{m-1} \frac{1}{q_j} \left| \frac{\partial f}{\partial \vartheta_j} \right|^2 \mathrm{d}S. \tag{4}$$

Equation (4) shows that the Beltrami operator is non-negative in $\mathbf{L}_2(S)$, and therefore it admits of an extension to a self-adjoint operator (cf. FRIEDRICHS [1] and also MIKHLIN [14]). In the sequel δ will denote exactly that extension.

The spectrum of the Beltrami operator consists of the eigenvalues

$$\lambda_n = n(n + m - 2), \qquad n = 0, 1, 2, \dots, \tag{5}$$

and to each eigenvalue there are $k_{n,m}$ linearly independent eigenfunctions $Y_{n,m}^{(k)}(\theta)$ which are the *spherical functions of order* n in the m-dimensional space. The upper index k numbers the linearly independent spherical functions of one and the same order n, and it varies within the bounds

$$1 \leq k \leq k_{n,m} := (2n + m - 2) \frac{(n + m - 3)!}{(m - 2)!\, n!}.$$

The functions $Y_{n,m}^{(k)}(\theta)$ are supposed to be orthonormal in the metric of the space $\mathbf{L}_2(S)$.

Every function $f \in \mathbf{L}_2(S)$ admits of an expansion into a series with respect to the spherical functions,

$$f(\theta) = \sum_{n=0}^{\infty} \sum_{k=1}^{k_{n,m}} a_n^{(k)} Y_{n,m}^{(k)}(\theta), \tag{6}$$

where, obviously, the series

$$\sum_{n=0}^{\infty} \sum_{k=1}^{k_{n,m}} |a_n^{(k)}|^2 \tag{7}$$

converges. If $f \in D(\delta^q)$, where q is an arbitrary positive number, then it is easy to see that

$$\gamma_n^{(k)} = n^q (n + m - 2)^q\, a_n^{(k)} \tag{8}$$

where $a_n^{(k)}$ and $\gamma_n^{(k)}$ denote the coefficients in the expansion of the functions f and $\delta^q f$, resp., with respect to the spherical functions.

4.2. Theorem 4.1. *On the set of functions*

$$\mathbf{W}_2^{(2q)}(S) \cap D(\delta^q) \cap \bar{\mathbf{L}}_2(S)$$

we have the relation

$$\|f\|_{\mathbf{W}_2^{(2q)}(S)} \sim \|\delta^q f\|_{\mathbf{L}_2(S)}, \qquad \forall q > 0. \tag{9}$$

Here $\bar{\mathbf{L}}_2(S)$ *denotes the subspace of* $\mathbf{L}_2(S)$ *which is orthogonal to* 1, *and the symbol* $\sim$ *is an abbreviation for the existence of a two-sided estimate with positive constants that do not depend on* f.

To begin with, we indicate a formula which is implied by (8): Let $f \in D(\delta^q)$, and let the expansion of $f(\theta)$ with respect to the spherical functions have the form (6) such that

$$\delta^q f(\theta) = \sum_{n=1}^{\infty} \sum_{k=1}^{k_{n,m}} n^q (n + m - 2)^q a_n^{(k)} Y_{n,m}^{(k)}(\theta) \tag{10}$$

and

$$\| \delta^q f \|_{\mathbf{L}_2(S)}^2 = \sum_{n=1}^{\infty} \sum_{k=1}^{k_{n,m}} n^{2q} (n + m - 2)^{2q} | a_n^{(k)} |^2 . \tag{11}$$

From that relation we obtain obviously the following assertion: The domain $D(\delta^q)$ of the operator δ^q consists of those and only those functions $f(\theta)$ for which

$$\sum_{n=1}^{\infty} \sum_{k=1}^{k_{n,m}} n^{4q} | a_n^{(k)} |^2 < \infty \tag{12}$$

where $a_n^{(k)}$ are the coefficients in the series (6). The range of the operator δ^q coincides with the subspace $\overline{\mathbf{L}}_2(S)$; by the way, it follows that negative powers of the operator δ are defined on $\overline{\mathbf{L}}_2(S)$.

The further proof consists of several steps.

4.3. Let $\eta(\varrho)$ be a finite and infinitely differentiable function on the interval $(0, \infty)$ such that $\eta(\varrho) = 1$ in some neighborhood of the point $\varrho = 1$. Put $\psi(x) = \eta(\varrho) f(\theta)$ where $f \in \mathbf{W}_2^{(l)}(S)$, then obviously

$$\| f \|_{\mathbf{W}_2^{(l)}(S)} \sim \| \psi \|_{\mathbf{W}_2^{(l)}(\mathbf{R}^m)} \qquad \forall l \geqq 0 . \tag{13}$$

4.4. Denote by $x = (x_1, \ldots, x_m) = (\varrho, \theta)$ and $\xi = (\xi_1, \ldots, \xi_m)$ points of the space $\mathbf{R}^m$, and by F the Fourier transformation,

$$(F\varphi)(x) = \int_{\mathbf{R}^m} \mathrm{e}^{-2\pi \mathrm{i}(x,\xi)} \varphi(\xi) \, \mathrm{d}\xi = \int_{\mathbf{R}^m} \mathrm{e}^{-2\pi \mathrm{i} R\varrho \cos\gamma} \varphi(\xi) \, \mathrm{d}\xi ,$$

where γ is the angle between the vectors x and ξ. From the representation (1) of the Laplace operator we obtain

$$(\delta F\varphi)(x) = 4\pi^2 R^2 (F(\varrho^2(\varphi(\xi)))(x) - 4\pi^2 \sum_{j,k=1}^{m} x_j x_k (F(\xi_j \xi_k \varphi(\xi)))(x)$$

$$- 2\pi \mathrm{i}(m - 1) \sum_{j=1}^{m} x_j (F(\xi_j \varphi(\xi)))(x) . \tag{14}$$

4.5. Now we shall show the relation

$$\| \psi \|_{\mathbf{W}_2^{(2q)}(\mathbf{R}^m)} \sim \| \delta^q \psi \|_{\mathbf{L}_2(\mathbf{R}^m)} \tag{15}$$

if $0 < q < 1$.

By the Parseval equation,

$$I := \int_{\mathbf{R}^m} \{ | \delta^q \psi |^2 + | \psi |^2 \} \, \mathrm{d}x = \int_{\mathbf{R}^m} \{ | F \, \delta^q \psi |^2 + | F\psi |^2 \} \, \mathrm{d}\xi .$$

The operator δ is symmetric, hence ($x = R\theta$, $\xi = \varrho\theta'$)

$$(F \, \delta\psi)(\xi) = \int_{\mathbf{R}^m} \mathrm{e}^{-2\pi \mathrm{i} R\varrho \cos\gamma} \, \delta\psi(x) \, \mathrm{d}x = \int_0^{\infty} R^{m-1} \, \mathrm{d}R \int_S \mathrm{e}^{-2\pi \mathrm{i} R\varrho \cos\gamma} \delta\psi \, \mathrm{d}S$$

$$= \int_0^{\infty} R^{m-1} \, \mathrm{d}R \int_S \psi \delta_\theta \, \mathrm{e}^{-2\pi \mathrm{i} R\varrho \cos\gamma} \mathrm{d}S = \int_0^{\infty} R^{m-1} \, \mathrm{d}R \int_S \psi \delta_{\theta'} \, \mathrm{e}^{-2\pi \mathrm{i} R\varrho \cos\gamma} \, \mathrm{d}S = (\delta F\psi)(\xi) ,$$

i.e. the operators F and δ commute. But then the operator F and any function of the operator δ do so, and, in particular, $F\delta^q = \delta^q F$ such that

$$I = \int_0^\infty \varrho^{m-1} \mathrm{d}\varrho \int_S \{|\delta^q F\psi|^2 + |F\psi|^2\} \,\mathrm{d}_{\theta'}S.$$

Eq. (11) holds for any function $g \in D(\delta^q)$, and we may write it in the form

$$\int_S |\delta^q g|^2 \,\mathrm{d}S = \sum_{n=1}^{\infty} \left[n^2(n+m-2)^2 \sum_{k=1}^{k_{n,m}} |b_n^{(k)}|^2 \right]^q \left[\sum_{k=1}^{k_{n,m}} |b_n^{(k)}|^2 \right]^{1-q};$$

here $b_n^{(k)}$ are the coefficients of the expansion of the function $g(\theta)$ into a series with respect to the spherical functions.

Applying Hölder's inequality with exponents q^{-1} and $(1-q)^{-1}$ we obtain for the $\mathbf{L}_2(S)$ norm the inequality

$$\|\delta^q g\| \leqq \|\delta g\|^q \, \|g\|^{1-q}. \tag{16}$$

Put $g = F\psi$ and find

$$I \leqq C \int_0^\infty R^{m-1} \mathrm{d}R \left(\int_S |F\psi|^2 \,\mathrm{d}S \right)^{1-q} \left(\int_S [|\delta F\psi|^2 + |F\psi|^2] \,\mathrm{d}S \right)^q .$$

From (14) you can easily see that

$$|\delta F\psi|^2 + |F\psi|^2 \leqq (1+R^2)^2 \sum_{j,k=1}^{m} |F\beta_{jk}\psi|^2$$

with certain polynomials β_{jk}. Hence,

$$I \leqq C \int_0^\infty (1+R^2)^{2q} R^{m-1} \mathrm{d}R \left(\int_S \sum_{j,k=1}^{m} |F\beta_{jk}\psi|^2 \,\mathrm{d}S \right)^q \left(\int_S |F\psi|^2 \mathrm{d}S \right)^{1-q},$$

and by Hölder's inequality

$$I \leqq C \left\{ \|\psi\|^2_{\mathbf{W}_2^{(2q)}(\mathbb{R}^m)} + \sum_{j,k=1}^{m} \|\beta_{jk}\psi\|^2_{\mathbf{W}_2^{(2q)}(\mathbb{R}^m)} \right\}.$$

Taking into account that the function ψ is finite, we can estimate here the second term in the right-hand side by the first one such that

$$I \leqq C \|\psi\|^2_{\mathbf{W}_2^{(2q)}(\mathbb{R}^m)}. \tag{17}$$

Thus, we proved one side of the relation (15),

$$\|\delta^q \psi\|_{\mathbf{L}_2(\mathbb{R}^m)} \leqq \sqrt{I} \leqq C \, \|\psi\|_{\mathbf{W}_2^{(2q)}(\mathbb{R}^m)}.$$

4.6. Now we shall derive the opposite estimate. A norm in the space $\mathbf{W}_2^{(2q)}(\mathbb{R}^m)$ can be defined by the formula

$$\|u\|^2_{\mathbf{W}_2^{(2q)}(\mathbb{R}^m)} = \int_{\mathbb{R}_m} |Fu|^2 (1+\varrho^2)^{2q} \,\mathrm{d}\xi. \tag{18}$$

At first a remark. If some function $g(\theta)$ is orthogonal to 1 in the space $\mathbf{L}_2(S)$ then so is the function $(F\omega)(\xi)$ where $\omega(x) = g(\theta)\,\eta(\varrho)$. Indeed, one has

$$\int_S (F\omega)(\xi) \,\mathrm{d}_{\theta'}S = \int_S g(\theta) \mathrm{d}_\theta S \int_0^1 \eta(\varrho)\, \varrho^{m-1} \,\mathrm{d}\varrho \int_S \mathrm{e}^{-\mathrm{i}\varrho R\cos\gamma} \,\mathrm{d}_{\theta'}S.$$

The integral $\int\limits_S e^{-i\varrho R\cos\gamma}\, d_{\theta'}S$ does not depend on θ. To see this draw the axis $0\xi_1$ into the direction of the vector x. But then the right-hand side of the last equation vanishes, and the assertion is proved.

The function $\delta^q\psi$ belongs to $\bar{\mathbf{L}}_2(S)$, and then also $F\,\delta^q\psi \in \bar{\mathbf{L}}_2(S)$, and the equality $F\psi = \delta^{-q}\,F\,\delta^q\psi$ holds. By (18) we obtain

$$\|\psi\|^2_{W_2^{(2q)}(R^m)} := J = \int\limits_{R^m} (1+\varrho^2)^{2q}\,|F\psi|^2 d\xi = \int\limits_0^\infty (1+\varrho^2)^{2q}\,\varrho^{m-1} d\varrho \int\limits_S |\delta^{-q}F\,\delta^q\psi|^2\, d_{\theta'}S.$$

The same arguments which led us to the inequality (16) give the relation

$$\int\limits_S |\delta^{-q}F\,\delta^q\psi|^2 d_{\theta'}S \leqq \Big(\int\limits_S |\delta^{-1}\,F\,\delta^q\psi|^2 d_{\theta'}S\Big)^q \Big(\int\limits_S |F\,\delta^q\psi|^2 d_{\theta'}S\Big)^{1-q}. \tag{19}$$

Inserting this estimate into the last but one formula and applying Hölder's inequality with exponents q^{-1} and $(1-q)^{-1}$, we obtain

$$J \leqq \Bigg[\int\limits_0^\infty (1+\varrho^2)^2\,\varrho^{m-1}\, d\varrho \int\limits_S |\delta^{-1}\,F\,\delta^q\psi|^2\, d_{\theta'}S\Bigg]^q \Bigg[\int\limits_0^\infty \varrho^{m-1}\, d\varrho \int\limits_S |F\,\delta^q\psi|^2\, d_{\theta'}S\Bigg]^{1-q}$$

which becomes

$$J \leqq \Big(\int\limits_{R^m} (1+\varrho^2)^2\,|F\,\delta^{q-1}\psi|^2 d\xi\Big)^q \Big(\int\limits_{R^m} |\delta^q\psi|^2\, d\psi\Big)^{1-q} \tag{20}$$

because of the equality $\delta^{-1}\,F\,\delta^q = F\,\delta^{q-1}$.

Let us estimate the first factor in (20). Note that

$$(1+\varrho^2)\,(F\,\delta^{q-1}\psi)\,(\xi) = (F(E-\Delta)\,\delta^{q-1}\psi)\,(\xi)$$

such that

$$\int\limits_{R^m} (1+\varrho^2)^2\,|F(\delta^{q-1}\psi)|^2\, d\xi = \int\limits_{R^m} |F(E-\Delta)\,\delta^{q-1}\psi|^2\, d\xi$$
$$= \int\limits_{R^m} (E-\Delta)\,\delta^{q-1}\psi|^2\, dx \leqq 2\int\limits_{R^m} |\delta^{q-1}\psi|^2\, dx + 2\int\limits_{R^m} |\Delta\delta^{q-1}\psi|^2\, dx. \tag{21}$$

Since $q < 1$, the operator δ^{q-1} is bounded in the $\mathbf{L}_2(S)$ metric, hence

$$\int\limits_{R^m} |\delta^{q-1}\psi|^2\, dx = \int\limits_0^\infty R^{m-1} dR \int\limits_S |\delta^{q-1}\psi|^2\, dS \leqq C\int\limits_0^\infty R^{m-1} dR \int\limits_S |\psi(x)|^2\, dx$$
$$= C\int\limits_{R^m} |\psi(x)|^2 dx.$$

On the other hand, (1) yields

$$|\Delta\delta^{q-1}\psi(x)|^2 \leqq 2\left|R^{1-m}\frac{\partial}{\partial R}\Big(R^{m-1}\frac{\partial}{\partial R}\,\delta^{q-1}\psi(x)\Big)\right|^2 + \frac{2}{R^4}\,|\delta^q\psi(x)|^2$$

such that

$$\int\limits_{R^m} |\Delta\delta^{q-1}\psi|^2\, dx \le 2\int\limits_0^\infty \left[R^{1-m}\frac{d}{dR}\Big(R^{m-1}\frac{d\eta}{dR}\Big)\right]^2 R^{m-1} dR$$
$$\times \int\limits_S |\delta^{q-1}f(\theta)|^2 dS + 2\int\limits_0^\infty \eta^2(R)\,R^{m-5}\, dR \int\limits_S |\delta^q f(\theta)|^2\, dS$$
$$\leqq C\int\limits_S \{|\delta^{q-1}f(\theta)|^2 + |\delta^q f(\theta)|^2\}\, dS \leqq C\int\limits_S \{|f(\theta)|^2 + |\delta^q f(\theta)|^2\}\, dS$$
$$\leqq C\int\limits_{R^m} \{|\psi(x)|^2 + |\delta^q\psi(x)|^2\}\, dx.$$

Thus, we have

$$J \leqq C\Big(\int_{R^m}\{|\psi|^2 + |\delta^q\psi|^2\}\,dx\Big)^q\Big(\int_{R^m}|\delta^q\psi|^2\,dx\Big)^{1-q} \leqq C\int_{R^m}\{|\psi|^2+|\delta^q\psi|^2\}\,dx.$$

To estimate this inequality further, we note that the function f belongs to $\overline{L}_2(S)$, and it is easy to see that for fixed ϱ the function $\psi(x)$ has the same property. But on $\overline{L}_2(S)$ the operator δ^{-q} is bounded such that

$$\int_{R^m}|\psi(x)|^2\,dx \leqq C\int_{R^m}|\delta^q\psi(x)|^2\,dx.$$

Thus, we arrived at

$$J = \|\psi\|_{W_2^{(2q)}(R^m)} \leqq C\,\|\delta^q\psi\|_{L_2(R^m)}, \tag{22}$$

and the relation (15) is proved completely.

4.7. We show that for some $\gamma > 0$ the relation $f \in L_2(S) \cap W_2^{(2\gamma+2)}(S)$ implies

$$\|\psi\|_{W_2^{(2\gamma+2)}(R^m)} \sim \|\delta\psi\|_{W_2^{(2\gamma)}(R^m)}. \tag{23}$$

By (1) we find $\delta\psi = \eta_1(R)\,f(\theta) - R^2\,\Delta\psi$ where $\eta_1(R)$ is a finite and infinitely differentiable function. Therefore,

$$\|\delta\psi\|_{W_2^{(2\gamma)}(R^m)} - \|R^2\Delta\psi\|_{W_2^{(2\gamma)}(R^m)} \leqq \|\eta_1(R)\,f(\theta)\|_{W_2^{(2\gamma)}(R^m)}. \tag{24}$$

Since the function $\eta(R)$ is finite and infinitely differentiable, too, we have $\|R^2\Delta\psi\|_{W_2^{(2\gamma)}(R^m)} \sim \|\Delta\psi\|_{W_2^{(2\gamma)}(R^m)}$ and, by (8), $\|\Delta\psi\|_{W_2^{(2\gamma)}(R^m)} \sim \|\psi\|_{W_2^{(2\gamma+2)}(R^m)}$. Now (24) yields the estimates

$$\|\delta\psi\|_{W_2^{(2\gamma)}(R^m)} \leqq C\,\{\|\psi\|_{W_2^{(2\gamma+2)}(R^m)} + \|\eta_1(R)\,f(\theta)\|_{W_2^{(2\gamma)}(R^m)}\}, \tag{26}$$

$$\|\psi\|_{W_2^{(2\gamma+2)}(R^m)} \leqq C\,\{\|\delta\psi\|_{W_2^{(2\gamma)}(R^m)} + \|\eta_1(R)\,f(\theta)\|_{W_2^{(2\gamma)}(R^m)}. \tag{27}$$

Using (13), we obtain

$$\|\eta_1(R)\,f(\theta)\|_{W_2^{(2\gamma)}(R^m)} \sim \|f\|_{W_2^{(2\gamma)}(S)} \sim \|\psi\|_{W_2^{(2\gamma)}(R^m)} \tag{28}$$

and that leads together with (26) to the estimate

$$\|\delta\psi\|_{W_2^{(2\gamma)}(R^m)} \leqq C\,\|\psi\|_{W_2^{(2\gamma+2)}(R^m)}. \tag{29}$$

Now, we show that an opposite inequality is also true. Applying Hölder's inequality with the exponents $(\gamma+1)/\gamma$ and $\gamma+1$, we obtain

$$\begin{aligned}\|\psi\|^2_{W_2^{(2\gamma)}(R^m)} &= \int_{R^m}(1+\varrho^2)^{2\gamma}\,|F\psi|^2\,d\xi \\ &\leqq \Big[\int_{R^m}(1+\varrho^2)^{2\gamma+2}\,|F\psi|^2\,d\xi\Big]^{\gamma/(\gamma+1)}\Big[\int_{R^m}|F\psi|^2\,d\xi\Big]^{1/(\gamma+1)} \\ &= \|\psi\|^{2\gamma/(\gamma+1)}_{W_2^{(2\gamma+2)}(R^m)}\,\|\psi\|^{2/(\gamma+1)}_{L_2(R^m)} \leqq C\,\|\psi\|^{2\gamma/(\gamma+1)}_{W_2^{(2\gamma+2)}(R^m)}\,\|\delta\psi\|^{2/(\gamma+1)}_{L_2(R^m)};\end{aligned}$$

the last estimate is a consequence of the boundedness of the operator δ^{-1} in $\overline{L}_2(S)$. The L_2 norm is weaker than the $W_2^{(2\gamma)}$ norm such that

$$\|\psi\|_{W_2^{(2\gamma)}(R^m)} \leqq C\,\|\psi\|^{\gamma/(\gamma+1)}_{W_2^{(2\gamma+2)}(R^m)}\,\|\delta\psi\|^{1/(\gamma+1)}_{W_2^{(2\gamma)}(R^m)}.$$

Due to the relations (27) and (28) we have

$$\|\psi\|_{W_2^{(2\gamma+2)}(R^m)} \leqq C\,[\|\delta\psi\|_{W_2^{(2\gamma)}(R^m)} + \|\psi\|^{\gamma/(\gamma+1)}_{W_2^{(2\gamma+2)}(R^m)}\,\|\delta\psi\|^{1/(\gamma+1)}_{W_2^{(2\gamma)}(R^m)}].$$

Further, by the inequality (29),

$$\| \delta\psi \|_{W_2^{(2\gamma)}(R^m)} \leqq C \| \psi \|^{\gamma/(\gamma+1)}_{W_2^{(2\gamma+2)}(R^m)} \| \delta\psi \|^{1/(\gamma+1)}_{W_2^{(2\gamma)}(R^m)}$$

such that we arrive at the estimate

$$\| \psi \|_{W_2^{(2\gamma+2)}(R^m)} \leqq C \| \delta\psi \|_{W_2^{(2\gamma)}(R^m)} .$$

Hence, the relation (23) is proved.

4.8. For any $q \geqq 0$ the relation

$$\| \psi \|_{W_2^{(2q)}(R^m)} \sim \| \delta^q \psi \|_{L_2(R^m)} \tag{30}$$

holds. Let $q = [q] + \alpha$, $0 \leqq \alpha < 1$, then by (15)

$$\| \delta^{[q]} \psi \|_{W_2^{(2\alpha)}(R^m)} \sim \| \delta^q \psi \|_{L_2(R^m)} .$$

Successively applying the relation (23) we obtain

$$\| \delta^{[q]} \psi \|_{W_2^{(2\alpha)}(R^m)} \sim \| \psi \|_{W_2^{(2q)}(R^m)} .$$

The relation (30) is a consequence of the last two formulas. Now it follows from (13) that the relations (30) and (9) are equivalent, and Theorem 4.1 is proved.

4.6. For integer q Theorem 4.1 was proved by Mikhlin [13] (see also Chap. X, § 6). The case of non-integer q was considered by Mikhailova-Gubenko [1, 3], and her proof is presented here with some unimportant changes.

Chapter IX
Singular integrals of higher dimensions in spaces with a uniform metric

§ 1. Basic notions

1.1. Let $w(x)$ be a function defined a.e. on a certain manifold Γ which is endowed with a metric. Let x_0 be a certain point on Γ; draw a ball with radius ε around x_0 and remove from Γ the intersection of that ball with Γ. Suppose that the function $w(x)$ is summable over the remaining part Γ_ε of the manifold Γ, and that for any $\varepsilon > 0$. If the limit

$$\lim_{\varepsilon \to 0} \int_{\Gamma_\varepsilon} w(x) \, d\Gamma$$

exists it is called *singular integral* of the function $w(x)$ over the manifold Γ. In particular, the manifold Γ can be the space $\mathbb{R}^m$ or a domain in $\mathbb{R}^m$.

The definition of a singular integral given here coincides with that given in § 1, Chap. II, provided that Γ is a one-dimensional manifold.

1.2. Let us consider the singular integral

$$v(x) = \int_{\mathbb{R}^m} r^{-m} f(x, \theta) \, u(y) \, dy, \tag{1}$$

where x and y are points in the space $\mathbb{R}^m$, and $r = |y - x|$, $\theta = (y - x)/r$. The point x is called *pole* and the function $f(x, \theta)$ *characteristic* of the singular integral (1). For the time being, we shall assume that the density and the characteristic have the following properties: 1. In every ball $\Omega_R = \{y\colon |y - x| \leq R\}$ around the pole the modulus of continuity $\omega(u, t)$ of the density $u(x)$ satisfies the Dini condition

$$\int_0^t \tau^{-1} \, \omega(u, \tau) \, d\tau < \infty.$$

Remember that

$$\omega(u, t) = \sup_{\substack{y, y_0 \in \Omega_R \\ |y - y_0| \leq t}} |u(y) - u(y_0)|.$$

2. For large $|x|$ the relation $u(x) = O(|x|^{-k})$, $k > 0$, holds.
3. The characteristic is bounded and, for fixed x, continuous with respect to θ.

Theorem 1.1. *Under the assumptions* 1.–3. *the singular integral* (1) *exists if and only if*

$$\int_S f(x, \theta) \, dS = 0, \tag{2}$$

where S denotes the unit sphere and θ varies in S.

Proof. We have

$$\int_{\mathbb{R}^m} r^{-m} f(x, \theta) \, u(y) \, dy = \int_{r>1} r^{-m} f(x, \theta) \, u(y) \, dy$$
$$+ \int_{r<1} r^{-m} f(x, \theta) \, [u(y) - u(x)] \, dy + u(x) \int_{r<1} r^{-m} f(x, \theta) \, dy.$$

The first two integrals right converge absolutely. For the third integral we substitute spherical coordinates with center at the point x. Then $\mathrm{d}y = r^{m-1}\,\mathrm{d}r\,\mathrm{d}S$, and

$$\int_{r<1} r^{-m} f(x, \theta)\,\mathrm{d}y = \lim_{\varepsilon\to 0} \int_{\varepsilon<r<1} r^{-m} f(x, \theta)\,\mathrm{d}y = \lim_{\varepsilon\to 0} \ln \varepsilon^{-1} \int_S f(x, \theta)\,\mathrm{d}S;$$

the limit exists if and only if condition (2) is fulfilled. ∎

In what follows we always shall assume the condition (2).

Because of (2) the integral (1) can be expressed by absolutely convergent integrals,

$$\int_{\mathrm{R}^m} r^{-m} f(x, \theta)\, u(y)\,\mathrm{d}y = \int_{r>1} r^{-m} f(x, \theta)\, u(y)\,\mathrm{d}y + \int_{r<1} r^{-m} f(x, \theta)\,[u(y) - u(x)]\,\mathrm{d}y. \tag{3}$$

Remark 1. Due to the definition of a singular integral one has

$$\int_{\mathrm{R}^m} r^{-m} f(x, \theta)\, u(y)\,\mathrm{d}y = \lim_{\varepsilon\to 0} \int_{r>\varepsilon} r^{-m} f(x, \theta)\, u(y)\,\mathrm{d}y.$$

If the conditions 1.–3. and (2) are satisfied then the integral on the right-hand side of the last equation converges in any finite domain uniformly with respect to x.

Indeed, under those assumptions,

$$\int_{r>\varepsilon} r^{-m} f(x, \theta)\, u(y)\,\mathrm{d}y = \int_{r>1} r^{-m} f(x, \theta)\, u(y)\,\mathrm{d}y + \int_{\varepsilon<r<1} r^{-m} f(x, \theta)\,[u(y) - u(x)]\,\mathrm{d}y.$$

The first integral right does not depend on ε whereas the second one converges uniformly to the limit

$$\int_{r<1} r^{-m} f(x, \theta)\,[u(y) - u(x)]\,\mathrm{d}y.$$

Therefore, if the assumptions mentioned above are satisfied and, moreover, the characteristic is continuous on $\mathrm{R}^m \times S$, then the integral (1) is continuous in R^m.

Remark 2. In case $m = 2$ and if the characteristic does not depend on the pole, Theorem 1.1 was proved by Tricomi [1, 2]. The general case was considered by Mikhlin ([4, 10], see also [7]).

1.3. We surround the point x by a domain of arbitrary shape, say σ_ε, the diameter of which tends to zero together with ε, and consider the limit

$$\lim_{\varepsilon\to 0} \int_{\mathrm{R}^m \setminus \sigma_\varepsilon} r^{-m} f(x, \theta)\, u(y)\,\mathrm{d}y. \tag{4}$$

Let $r = \alpha(\varepsilon, x, \theta)$ be the equation for the boundary of the domain σ_ε. We assume that the limit

$$\lim_{\varepsilon\to 0} \frac{\alpha(\varepsilon, x, \theta)}{\varepsilon} = \beta(x, \theta) > 0 \tag{5}$$

exists and that the convergence is uniform with respect to θ. Then

$$\lim_{\varepsilon\to 0} \int_{\mathrm{R}^m \setminus \sigma_\varepsilon} r^{-m} f(x, \theta)\, u(y)\,\mathrm{d}y = \int_{r>1} r^{-m} f(x, \theta)\, u(y)\,\mathrm{d}y$$

$$+ \lim_{\varepsilon\to 0} \int_{\alpha<r<1} r^{-m} f(x, \theta)\,[u(y) - u(x)]\,\mathrm{d}y - u(x) \lim_{\varepsilon\to 0} \int_S f(x, \theta) \ln \alpha(\varepsilon, x, \theta)\,\mathrm{d}S.$$

Due to condition (2) we may write the last integral as follows,

$$\int_S f(x, \theta) \ln \alpha(\varepsilon, x, \theta)\,\mathrm{d}S = \int_S f(x, \theta) \ln \frac{\alpha(\varepsilon, x, \theta)}{\varepsilon}\,\mathrm{d}S$$

and using (3) we obtain finally

$$\lim_{\varepsilon\to 0} \int_{\mathrm{R}^m \setminus \sigma_\varepsilon} r^{-m} f(x, \theta)\, u(y)\,\mathrm{d}y = \int_{\mathrm{R}^m} r^{-m} f(x, \theta)\, u(y)\,\mathrm{d}y - u(x) \int_S f(x, \theta) \ln \beta(x, \theta)\,\mathrm{d}S. \tag{6}$$

That relation remains to hold true if the integral is taken not over the whole space $\mathbb{R}^m$ but over some domain only.

For the existence of the limit (4) the condition (2) is not only sufficient but also necessary. Indeed, as we realized, the matter comes to the existence of the limit

$$\lim_{\varepsilon \to 0} \int_S f(x, \theta) \ln \alpha(\varepsilon, x, \theta) \, dS.$$

But we have

$$\int_S f(x, \theta) \ln \alpha(\varepsilon, x, \theta) \, dS = \int_S f(x, \theta) \ln \frac{\alpha(\varepsilon, x, \theta)}{\varepsilon} \, dS + \ln \varepsilon \int_S f(x, \theta) \, dS.$$

The limit of the first summand right exists and is equal to $\int_S f(x, \theta) \ln \beta(x, \theta) \, dS$ as $\varepsilon \to 0$, but the second summand has a limit only if condition (2) is fulfilled.

1.4. Let D be a domain in the space $\mathbb{R}^m$, and let the functions $f(x, \theta)$ and $u(x)$ meet the assumptions 1.–3. made at the beginning of the present section. Certainly, condition (2) is assumed, too.

Consider the singular integral

$$\int_D K(x, y) \, u(y) \, dy; \qquad K(x, y) = r^{-m} f(x, \theta).$$

By $x_1, x_2, \ldots, x_m$ we denote the Cartesian coordinates of a point x, and we assume that the functions

$$\xi_k = \xi_k(x_1, x_2, \ldots, x_m), \qquad k = 1, 2, \ldots, m, \tag{7}$$

give a 1-to-1 and continuous mapping of the domain D onto a certain domain $D' \subset \mathbb{R}^m$. The functions (7) are assumed to be twice continuously differentiable, and the Jacobian be non-singular.

By ξ and η we denote the images of the points x and y, resp., under the transformation (7), by σ_ε the image of the ball $r < \varepsilon$, and by $J(\xi)$ the Jacobian of the mapping (7). Furthermore, we put

$$\varrho = |\eta - \xi|, \qquad \theta' = (\eta - \xi)/\varrho, \qquad u(x) = u'(\xi), \qquad K(x, y) = K'(\xi, \eta).$$

Then the formulae

$$r^2 = \varrho^2 F^2(\xi, \theta') + O(\varrho^3),$$

$$K'(\xi, \eta) = \varrho^{-m} f'(\xi, \theta') + O(\varrho^{-\mu}), \qquad \mu < m,$$

are easily obtained. The boundary of the domain σ_ε is defined by the equation $\varrho^2 F^2(\xi, \theta') + O(\varrho^3) = \varepsilon^3$ such that

$$\lim_{\varepsilon \to 0} \frac{\varrho}{\varepsilon} = [F(\xi, \theta')]^{-1}.$$

Now (6) yields the following formula for a transformation of coordinates in a singular integral,

$$\int_D K(x, y) \, u(y) \, dy = \int_{D'} K'(\xi, \eta) \, J(\eta) \, u'(\eta) \, d\eta$$

$$+ u'(\xi) J(\xi) \int_S f'(\xi, \theta') \ln F(\xi, \theta') \, d_{\theta'} S. \tag{8}$$

1.5. The characteristics of the singular integrals as considered in the present section are functions which depend on the points x and θ, where the first one varies in the space $\mathbb{R}^m$

and the second one in the unit sphere. Equivalently, we may consider (and we shall frequently do so in the sequel) the point θ varying in the whole space $\mathbf{R}^m$ except the origin and infinity, and the characteristic (or any other function under consideration that depends on x and θ) as a positive-homogeneous function of degree 0 of θ such that it preserves a constant value on every radius from the origin in the θ-space.

1.6. The singular kernel $K(x - y) = r^{-m} f(\theta)$ generates a distribution of the class $\mathscr{D}'$ (cf. L. Schwartz [1]) which acts on an arbitrary function $u(x - y) \in \mathscr{D}$ (x fixed) according to the formula

$$(K(y), u(x - y)) = \int_{\mathbf{R}^m} K(y)\, u(x - y)\, dy = \int_{\mathbf{R}^m} K(x - y)\, u(y)\, dy$$
$$= \lim_{\varepsilon \to 0, N \to \infty} \int_{\varepsilon < r < N} K(x - y)\, u(y)\, dy;$$

that definition makes sense due to Theorem 1.1. The singular integral $\int_{\mathbf{R}^m} r^{-m} f(\theta)\, u(y)\, dy$ can also be considered as a convolution $K(x) * u(x) = \int_{\mathbf{R}^m} K(x - y)\, u(y)\, dy$.

For a characteristic depending on the pole we consider the distribution $K(z, y) = |y|^{-m} f(z, y/|y|)$ which depends on a parameter $z \in \mathbf{R}^m$. The singular integral (1) can be taken as the value of the convolution $K(z, x) * u(x)$ at $z = x$.

§ 2. Singular integrals over an arbitrary integration manifold

2.1. Singular integrals of the form

$$\int_{\Gamma} K'(\xi, \eta)\, u'(\eta)\, d_\eta \Gamma \tag{1}$$

where Γ is a Lyapunov manifold, i.e. a manifold of the class $\mathbf{C}^{1,\alpha}$, $\alpha > 0$, can be reduced under suitable assumptions to singular integrals over a certain domain of the Euclidean space. Let ξ and η be points on Γ and denote the distance between those points with respect to the intrinsic metric of the manifold by ϱ. Assume that the product $\varrho^m \cdot K'(\xi, \eta)$, $m = \dim \Gamma$, is bounded as $\eta \to \xi$. The singular integral (1) is defined as

$$\lim_{\varepsilon \to 0} \int_{\Gamma \setminus (\varrho < \varepsilon)} K'(\xi, \eta)\, u'(\eta)\, d_\eta \Gamma \tag{2}$$

provided that the limit exists. We divide Γ into two parts Γ_1 and Γ_2 such that Γ_1 admits of a 1-to-1 transformation onto a certain domain D in the Euclidean space $\mathbf{R}^m$; we may require that the transformation has first derivatives belonging to the class $\mathbf{Lip}_\alpha$. Furthermore, we assume the transformation to be conformal at the point ξ at which the limit (2) is computed (note that this assumption is not necessary). The point ξ is assumed to vary strictly in the interior of Γ_1. Then

$$\int_{\Gamma} K'(\xi, \eta)\, u'(\eta)\, d_\eta \Gamma = \int_{\Gamma_2} K'(\xi, \eta)\, u'(\eta)\, d_\eta \Gamma + \lim_{\varepsilon \to 0} \int_{\Gamma_1 \setminus (\varrho < \varepsilon)} K'(\xi, \eta)\, u'(\eta)\, d_\eta \Gamma. \tag{3}$$

Let x and y be the images of ξ and η, resp., under the transformation mentioned above, and put $r = |y - x|$, $\theta = (y - x)/r$. Since the transformation is conformal at the point ξ the limit $\lim_{\varrho \to 0} \varrho/r$ does not depend on θ. The product $K'(\xi, \eta)\, d\eta$ is transformed to a product of the form $K(x, y)\, dy$. We assume that the kernel $K(x, y)$ can be represented as

$$K(x, y) = K(x, x + r\theta) = r^{-m} f(x, \theta) + r^{-\mu} f_1(x, r, \theta) \tag{4}$$

where $\mu < m$, and f, f_1 are continuous functions. Using the relations (1.4) and (1.9)–(1.11) we obtain easily

$$\int_{\Gamma} K'(\xi, \eta)\, u'(\eta)\, \mathrm{d}_\eta\Gamma = \int_{\Gamma_2} K'(\xi, \eta)\, u'(\eta)\, \mathrm{d}_\eta\Gamma + \int_D r^{-\mu} f_1(x, r, \theta)\, u(y)\, \mathrm{d}y$$
$$+ \int_D r^{-m} f(x, \theta)\, u(y)\, \mathrm{d}y \qquad (5)$$

where $u(y) = u'(\eta)$; clearly, the characteristic has to meet condition (1.2).

A similar definition of a singular integral over a manifold is given in a paper by SEELEY [2] (see also Chap. XIII).

2.2. Let us consider in more detail the case where the singular integral is taken over the unit sphere Σ of the space $\mathbb{R}^{m+1}$. We take two points ξ and η on the unit sphere Σ and put $\varrho^2 = \sum_{j=1}^{m+1} (\xi_j - \eta_j)^2$ with ξ_j and η_j being the Cartesian coordinates of the points ξ and η. Then we draw the diameter of the sphere Σ which passes through the point ξ, and the plane P normal to that diameter passing through the point η (see Fig. 7). The intersection of the diameter mentioned with the plane P will be denoted by a. On the radius from the point a to η we mark the point Λ which has distance 1 to the point a. The number ϱ and the point Λ uniquely determine the point η. Note that Λ varies on the unit sphere S of the space $\mathbb{R}^m$ and ϱ within the segment $0 \leqq \varrho \leqq 2$.

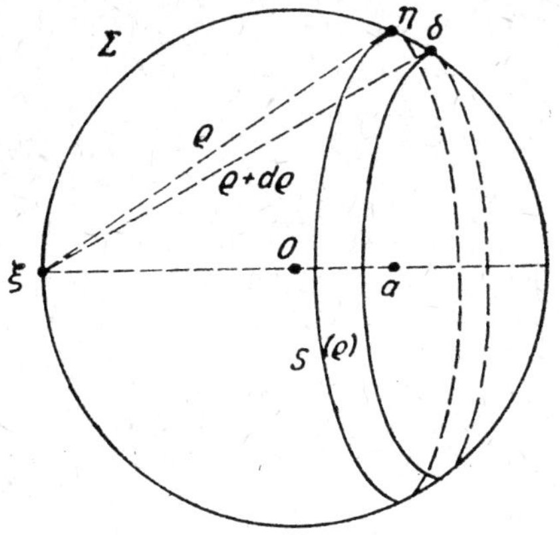

Fig. 7

By $\mathrm{d}\sigma$ and $\mathrm{d}S$ we denote the surface element on Σ and S, resp. We shall prove the relation

$$\mathrm{d}\sigma = \varrho^{m-1} \left(1 - \frac{\varrho^2}{4}\right)^{(m-2)/2} \mathrm{d}\varrho\, \mathrm{d}S. \qquad (6)$$

To that aim we consider in the space $\mathbb{R}^{m+1}$ a sphere of radius ϱ, $0 \leqq \varrho \leqq 2$, centered at the point ξ. The intersection of that sphere with Σ forms an $(m-1)$-dimensional sphere $S^{(\varrho)}$ with a radius equal to $\varrho \sqrt{1 - \varrho^2/4}$ as you can easily see from Fig. 7. The same figure shows that

$$\mathrm{d}\sigma = \delta\, \mathrm{d}S^{(\varrho)} = \delta \varrho^{m-1} (1 - \varrho^2/4)^{(m-1)/2}\, \mathrm{d}S.$$

Further, δ is the third side of a triangle which is inscribed in a circle of radius 1 and has two other sides ϱ and $\varrho + \mathrm{d}\varrho$. Therefore, we have $\delta = (1 - \varrho^2/4)^{-1/2}\, \mathrm{d}\varrho$ up to infinitesimal terms of higher order. Thus, (6) is proved.

Let us consider now the singular integral

$$v(\xi) = \int_{\Sigma} \varrho^{-m} f(\xi, \Lambda)\, u(\eta)\, d\sigma, \tag{7}$$

where the density $u(\eta)$ satisfies the Dini condition, and the characteristic is a measurable and bounded function. By definition,

$$\begin{aligned} v(\xi) &= \lim_{\varepsilon\to 0} \int_{S} dS \int_{\varepsilon}^{2} \varrho^{-1} f(\xi, \Lambda)\,(1-\varrho^2/4)^{(m-2)/2}\, u(\eta)\, d\varrho \\ &= \int_{S} dS \int_{0}^{2} \varrho^{-1} f(\xi, \Lambda)\,(1-\varrho^2/4)^{(m-2)/2}\, [u(\eta) - u(\xi)]\, d\varrho \\ &\quad - u(\xi) \int_{S} f(\xi,\Lambda)\, dS \int_{0}^{2} \varrho^{-1}[1-(1-\varrho^2/4)^{(m-2)/2}]\, d\varrho \\ &\quad + u(\xi) \lim_{\varepsilon\to 0} \ln\frac{2}{\varepsilon} \int_{S} f(\xi, \Lambda)\, dS. \end{aligned} \tag{8}$$

Thus, we obtain a necessary and sufficient condition for the existence of the integral (7),

$$\int_{S} f(\xi, \Lambda)\, dS = 0. \tag{9}$$

Under that condition the second integral in (8) vanishes, too, and we obtain

$$v(\xi) = \int_{\Sigma} \varrho^{-m} f(\xi, \Lambda)\, [u(\eta) - u(\xi)]\, d\sigma. \tag{10}$$

If the point ξ is surrounded not by a sphere but by some domain ω_ε with a boundary given by the equation $\varrho = \alpha(\varepsilon, \xi, \Lambda)$ and if the convergence

$$\frac{\alpha(\varepsilon, \xi, \Lambda)}{\varepsilon} \underset{\varepsilon\to 0}{\rightrightarrows} \beta(\xi, \eta) > 0$$

is uniform with respect to Λ then a formula similar to (1.4) is true,

$$\lim_{\varepsilon\to 0} \int_{\Sigma\setminus\omega_\varepsilon} \varrho^{-m} f(\xi, \Lambda)\, u(\eta)\, d\sigma = \int_{\Sigma} \varrho^{-m} f(\xi, \Lambda)\, u(\eta)\, d\sigma - u(\xi) \int_{S} f(\xi, \Lambda) \ln \beta(\xi, \Lambda)\, dS. \tag{11}$$

§ 3. The Zygmund inequality

3.1. Let Γ be a circle in the complex plane, by z and ζ we denote points of that circle and also their complex coordinates. Let $u(\zeta)$ be a function on Γ satisfying the Dini condition, and consider

$$v(z) = \int_{\Gamma} \frac{u(\zeta)}{\zeta - z}\, d\zeta. \tag{1}$$

It was shown by A. Zygmund [3] that the moduli of continuity of the density u and of the singular integral v are connected by the inequality

$$\omega(v, t) \le C\left[\int_{0}^{t} \frac{\omega(u,\tau)}{\tau}\, d\tau + t \int_{t}^{a} \frac{\omega(u, \tau)}{\tau^2}\, d\tau\right] \tag{2}$$

where a is the diameter of the circle, and the constant C does not depend neither on t nor on the function u. The inequality remains to hold true as one can easily see if Γ is replaced by an arbitrary sufficiently smooth curve.

As we shall see later on the inequality (2) is no longer true for singular integrals of higher dimensions; for such integrals a certain weaker inequality can be proved. In the following we shall indicate a necessary and sufficient condition for the Zygmund inequality in case of singular integrals of higher dimensions.

3.2. Let us consider the singular integral

$$v(x) = \int_G K(x, y)\, u(y)\, \mathrm{d}y, \qquad K(x, y) = r^{-m} f(x, \theta), \tag{3}$$

where G is a finite domain in the space $\mathbb{R}^m$. The characteristic $f(x, \theta)$ is assumed to be continuously differentiable with respect to the Cartesian coordinates of the points $x \in \overline{G}$ and $\theta \in S$. Then you can easily verify the relation

$$\operatorname{grad}_x K(x, y) = O(r^{-m-1}), \qquad r \to 0. \tag{4}$$

Under these assumptions we have

Theorem 3.1. *Let the density $u(y)$ of the singular integral* (3) *satisfy the Dini condition on $\overline{G}$ and let G' be an arbitrary interior subdomain of the domain G. Then, for sufficiently small t, in $\overline{G}$ the so-called weak Zygmund inequality*

$$\omega(v, t) \leqq C\left[t\,\|u\|_{C(\overline{G})} + \int_0^t \tau^{-1}\,\omega(u, \tau)\,\mathrm{d}\tau + t\int_t^a \tau^{-2}\,\omega(u, \tau)\,\mathrm{d}\tau\right] \tag{5}$$

is true where a denotes the diameter of the domain G, and the constant C does not depend neither on t nor on the function u.

Let δ be a positive number not greater than the smallest distance between points of the boundaries of G and G'. For $x \in \overline{G'}$ one can easily obtain a formula similar to (1.3),

$$v(x) = \int_{G \cap (r > \delta)} r^{-m} f(x, \theta)\, u(y)\, \mathrm{d}y + \int_{r < \delta} r^{-m} f(x, \theta)\,[u(y) - u(x)]\, \mathrm{d}y, \tag{6}$$

i.e. $v = v_1 + v_2 + v_3$ and

$$v_1(x) = u(x) \int_{G \cap (r > \delta)} r^{-m} f(x, \theta)\, \mathrm{d}y =: u(x)\, \psi(x),$$

$$v_2(x) = \int_{G \cap (r > \delta)} r^{-m} f(x, \theta)\,[u(y) - u(x)]\, \mathrm{d}y,$$

$$v_3(x) = \int_{r < \delta} r^{-m} f(x, \theta)\,[u(y) - u(x)]\, \mathrm{d}y.$$

Consider an arbitrary point x belonging to $\overline{G'}$ and a vector $h \in \mathbb{R}^m$ such that $(x + h) \in \overline{G'}$, put $|h| = t$. Then, obviously, $t \leqq a - 2\delta$. Now we shall estimate the moduli of continuity $\omega(v_j, t)$, $j = 1, 2, 3$, in $\overline{G'}$. We have

$$|v_1(x + h) - v_1(x)| \leqq |u(x)|\, |\psi(x + h) - \psi(x)| + |\psi(x)|\, |u(x + h) - u(x)|.$$

The function $\psi(x)$ is continuously differentiable, hence $|\psi(x + h) - \psi(x)| \leqq Ct$ and $|\psi(x)| \leqq C$ such that

$$\omega(v_1, t) \leqq Ct\,\|u\|_{C(\overline{G})} + C\omega(u, t). \tag{7}$$

The modulus of continuity is a monotone non-decreasing function of t such that

$$t\int_t^a \tau^{-2}\,\omega(u,\tau)\,\mathrm{d}\tau \geqq t\omega(u,t)\left(\frac{1}{t}-a\right) = \omega(u,t)\left(1-\frac{t}{a}\right) \geqq \frac{2\delta}{a}\,\omega(u,t)$$

and, hence,

$$\omega(u,t) \leqq \frac{a}{2\delta}\,t\int_t^a \tau^{-2}\,\omega(u,\tau)\,\mathrm{d}\tau\,. \tag{8}$$

Combining this estimate with (7) we see that the function $v_1(x)$ satisfies the inequality (5).

Now let us consider the function $v_2(x)$. The difference $v_2(x+h)-v_2(x)$ can be represented as

$$\begin{aligned} v_2(x+h)-v_2(x) &= \int_{G'} \{K(x+h,y)\,[u(y)-u(x+h)] - K(x,y)\,[u(y)-u(x)]\}\,\mathrm{d}y \\ &\quad + \int_{G''} K(x+h,y)\,[u(y)-u(x+h)]\,\mathrm{d}y \\ &\quad - \int_{G'''} K(x,y)\,[u(y)-u(x)]\,\mathrm{d}y =: I_1+I_2+I_3. \end{aligned} \tag{9}$$

Here we put $G' = G \cap (|x-y|>\delta) \cap (|x+h-y|>\delta)$, $G'' = (|x-y|<\delta) \cap (|x+h-y|>\delta)$, $G''' = (|x-y|>\delta) \cap (|x+h-y|<\delta)$, see Fig. 8.

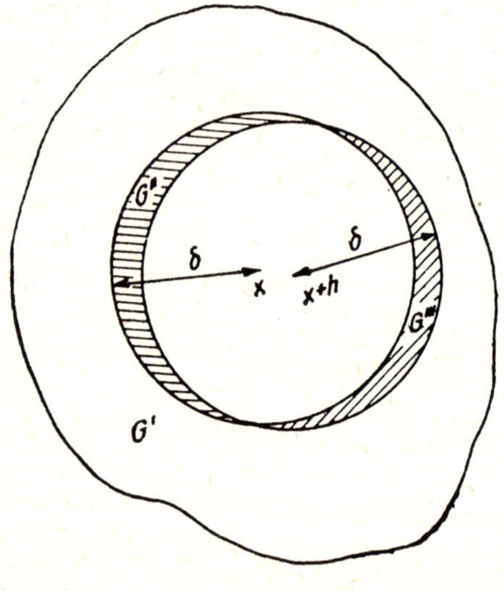

Fig. 8

Let us begin with estimating the integral I_3. Transforming it to spherical coordinates centered at x, we find

$$|I_3| \leqq C \int_{\delta<r<\delta+t} r^{-1}\omega(u,r)\,\mathrm{d}r\,\mathrm{d}S = C\int_\delta^{\delta+t} \tau^{-1}\omega(u,\tau)\,\mathrm{d}\tau \leqq Ct\,\delta^{-1}\,\omega(u,t+\delta)\,.$$

By (8) we have

$$\omega(u,t+\delta) \leqq C(t+\delta)\int_{t+\delta}^a \tau^{-2}\omega(u,\tau)\,\mathrm{d}\tau \leqq C\int_t^a \tau^{-2}\omega(u,\tau)\,\mathrm{d}\tau$$

such that finally

$$|I_3| \leq Ct \int_t^a \tau^{-2}\omega(u, \tau)\, d\tau. \tag{10}$$

Similarly we find, introducing spherical coordinates centered at the point $x + h$,

$$|I_2| \leq Ct \int_t^a \tau^{-2}\omega(u, \tau)\, d\tau. \tag{11}$$

Now we consider the integral I_1. The integrand of it can be written as

$$\begin{aligned} &K(x + h, y)\, [u(y) - u(x + h)] - K(x, y)\, [u(y) - u(x)] \\ &= [K(x + h, y) - K(x, y)\, [u(y) - u(x + h)] - K(x, y)\, [u(x + h) - u(x)] \end{aligned} \tag{12}$$

such that

$$\begin{aligned} I_1 = &\int_{G'} [K(x + h, y) - K(x, y)]\, [u(y) - u(x + h)]\, dy \\ &- [u(x + h) - u(x)] \int_{G'} K(x, y)\, dy =: I_{11} + I_{12}. \end{aligned}$$

In the following we assume $t < \delta/2$. Since the function $K(x, y)$ is continuous in G' we have

$$\begin{aligned} |I_{11}| &\leq Ct \int_{G'} |u(y) - u(x + h)|\, dy \leq Ct \int_{\delta<\tau<a} \omega(u, \tau)\, \tau^{m-1} d\tau\, dS \\ &\leq Ct \int_\delta^a \tau^{-2}\omega(u, \tau)\, d\tau \leq Ct \int_t^a \tau^{-2}\omega(u, \tau)\, d\tau. \end{aligned} \tag{13}$$

Further, $|I_{12}| \leq C\omega(u, t)$, and the relation (8) yields

$$|I_{12}| \leq Ct \int_t^a \tau^{-2}\omega(u, \tau)\, d\tau. \tag{14}$$

The inequalities (10), (11), (13), and (14) imply

$$\omega(v_2, t) \leq Ct \int_t^a \tau^{-2}\omega(t, \tau)\, d\tau \tag{15}$$

such that the function $v_2(x)$ satisfies the inequality (5), too.

3.3. Now it remains to show that the function $v_3(x)$ satisfies the weak Zygmund inequality. Using (12) and (1.2) we obtain

$$\begin{aligned} v_3(x + h) - v_3(x) = &\int_{|x-y|<\delta-t} K(x + h, y)\, [u(y) - u(x + h)]\, dy \\ &- \int_{|x-y|<\delta-t} K(x, y)\, [u(y) - u(x)]\, dy \\ &+ \int_{(|x+h-y|<\delta)\cap(|x-y|>\delta-t)} K(x + h, y)\, [u(y) - u(x + h)]\, dy \\ &- \int_{\delta-t<|x-y|<\delta} K(x, y)\, [u(y) - u(x)]\, dy \\ =: &J_1 - J_2 + J_3 - J_4. \end{aligned} \tag{16}$$

Introducing spherical coordinates centered at the point x we obtain

$$|J_4| = \left| \int_{\delta-t<r<\delta} f(x, \theta) \frac{u(y) - u(x)}{r}\, dr\, dS \right| \leq Ct\omega(u, \delta)$$

and, by the estimate (8),

$$|J_4| \leqq Ct \int_\delta^a \tau^{-2}\omega(u,\tau)\,d\tau \leqq Ct \int_t^a \tau^{-2}\omega(u,\tau)\,d\tau. \tag{17}$$

For J_3 we find

$$|J_3| \leqq C \int_{(|x+h-y|<\delta)\cap(|x-y|>\delta-t)} |u(x+h)-u(y)|\,dy;$$

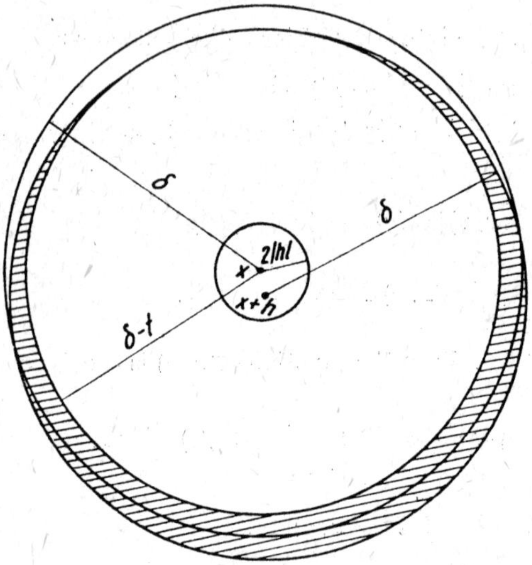

Fig. 9

here the domain of integration (shaded in Fig. 9) is a part of the ball layer $\delta - t < |x+h-y| < \delta$ such that

$$|J_3| \leqq C \int_{\delta-t<|x+h-y|<\delta} |u(x+h)-u(y)|\,dy.$$

Now we introduce spherical coordinates centered at $x+h$ and find

$$|J_3| \leqq C \int_{\delta-t}^{\delta} \omega(u,\tau)\,d\tau \leqq Ct\omega(u,\delta);$$

thence, in the same way as done above, we obtain

$$|J_3| \leqq Ct \int_t^a \tau^{-2}\omega(u,\tau)\,d\tau. \tag{18}$$

For J_1 and J_2 we divide the domain of integration into two parts, the ball $|x-y| < 2t$ and the ball layer $2t < |x-y| < \delta - t$. In the ball $|x-y| < 2t$ we have

$$|u(y)-u(x)| \leqq \left|u(y) - u\left(\frac{x+y}{2}\right)\right| + \left|u\left(\frac{x+y}{2}\right) - u(x)\right| \leqq 2\omega(u,t)$$

and, therefore,

$$\left|\int_{|x-y|<2t} K(x,y)\,[u(y)-u(x)]\,dy\right| \leqq C \int_0^{2t} r^{-1}\omega(u,r)\,dr$$

$$= C \int_0^t \tau^{-1}\omega(u,2\tau)\,d\tau \leqq 2C \int_0^t \tau^{-1}\omega(u,\tau)\,d\tau. \tag{19}$$

The ball $|x - y| < 2t$ is contained in the ball $|x + h - y| < 3t$. We obtain, denoting $r_1 = |x + h - y|$

$$\Big| \int_{|x-y|<2t} K(x+h, y)\,[u(y) - u(x+h)]\,dy \Big|$$
$$\leq C \int_{|x-y|<2t} r_1^{-1}\omega(u, r_1)\,dr_1\,dS \leq C \int_{r_1<3t} r_1^{-1}\omega(u, r_1)\,dr_1\,dS$$
$$= C\int_0^{3t} r_1^{-1}\omega(u, r_1)\,dr_1 = C\int_0^t \tau^{-1}\omega(u, 3\tau)\,d\tau \leq 3C\int_0^t \tau^{-1}\omega(u, \tau)\,d\tau. \tag{20}$$

By means of (12) the difference of the two integrals taken over the ball layer $2t < |x - y| < \delta - t$ is transformed to the integral

$$J = \int_{2t<|x-y|<\delta-t} [K(x+h, y) - K(x, y)]\,[u(y) - u(x+h)]\,dy.$$

By the Taylor formula, $K(x + h, y) - K(x, y) = (\mathrm{grad}_x K(\xi, y), h)$ where ξ is a certain point of the segment $(x, x + h)$. The inequality (2) yields

$$|K(x+h, y) - K(x, y)| \leq Ct\,|\xi - y|^{-m-1}.$$

Obviously, $|\xi - y| \geq r - t$, $r = |x - y|$, and since $r > 2t$ in the integration domain it is $|\xi - y| > r/2$. Thus, we arrive at

$$|J| \leq Ct\int_{2t}^{\delta-t} r^{-2}\omega(u, r)\,dr = Ct\int_t^{(\delta-t)/2} \tau^{-2}\,\omega(u, 2\tau)\,d\tau$$
$$\leq 2C\int_t^{(\delta-t)/2} \tau^{-2}\,\omega(u, \tau)\,d\tau \leq Ct\int_t^a \tau^{-2}\,\omega(u, \tau)\,d\tau. \tag{21}$$

The estimates (17)–(21) together show that the function $v_3(x)$ also satisfies the inequality (5). As we saw above so do the functions $v_1(x)$ and $v_2(x)$, and it thence follows that the sum $v(x) = v_1(x) + v_2(x) + v_3(x)$ satisfies that inequality. Theorem 3.1 is proved. ▮

3.4. Theorem 3.1 easily carries over to integrals over manifolds belonging to te class $C^{1,\alpha}$. In particular, we shall consider the case of a closed manifold Γ[1]).

Let $\Gamma_1, \Gamma_2, \ldots, \Gamma_k$ be a finite covering of Γ. From each set Γ_j we select an interior subset Γ_j' such that $\Gamma_1', \Gamma_2', \ldots, \Gamma_k'$ also forms a covering of Γ. Assume that the density of the singular integral satisfies on Γ the Dini condition, and let the kernel $K(\xi, \eta)$ be continuously differentiable if $\xi \neq \eta$ and satisfy the inequality

$$\left|\frac{\partial K}{\partial x_j}\right| \leq \frac{C}{\varrho^{m+1}}; \tag{22}$$

where x_j are the local coordinates of the point ξ with respect to the coordinate system with origin at the point η. Due to Theorem 3.1 the singular integral satisfies the weak Zygmund inequality on each set Γ_j' and, therefore, on the whole manifold Γ, too. Thus we arrive at

Theorem 3.2. *Let the kernel $K(\xi, \eta)$ of the singular integral taken over a closed $C^{1,\alpha}$-manifold Γ be continuously differentiable if $\xi \neq \eta$ and satisfy the inequality (22). Let the density of the integral satisfy the Dini condition on Γ. Then the weak Zygmund inequality holds on Γ for the singular integral.*

[1]) This is a compact manifold without boundary.

3.5. a) Tracing the arguments in the proof for Theorem 3.1 you can easily see that the term $\|u\|_{C(\bar{G})}$ in the inequality (5) is caused by the presence of a summand $v_1(x)$, more precisely, by the fact that the function $\psi(x)$ in that summand is not a constant one. For that reason, it is clear that the integral

$$w(x) = \int_G r^{-m} f(x, \theta)\,[u(y) - u(x)]\,\mathrm{d}y$$

satisfies the (original) Zygmund inequality (2). The same is, of course, true for the integral

$$w(\xi) = \int_\Gamma K(\xi, \eta)\,[u(\eta) - w(\xi)]\,\mathrm{d}_\eta\Gamma.$$

For two-dimensional manifolds Γ the Zygmund inequality for such integrals was proved by Gegelia [1].

b) We shall indicate two cases where

$$\psi(\xi) = \int_\Gamma K(\xi, \eta)\,\mathrm{d}_\eta\Gamma \equiv \text{const} \tag{23}$$

such that the singular integral satisfies the (original) Zygmund inequality.

1. Let Γ be a closed smooth curve in the complex plane. Then

$$\psi(\xi) = \int_\Gamma \frac{\mathrm{d}\eta}{\eta - \xi} = \pi \mathrm{i},$$

and this is the case which was considered by A. Zygmund.

2. Let Γ be a sphere and the singular integral be of the form (2.2). In that case we have $\psi(x) \equiv 0$ such that we can give the integral (2.2) the form (2.5). This case was treated in a paper by Ruzmetov [1].

c) Condition (23) is not only sufficient but also necessary for the singular integral to satisfy the original Zygmund inequality (2). Indeed, suppose $\psi(\xi) \not\equiv$ const. Put $u(\xi) \equiv 1$, then $v(\xi) \not\equiv$ const and, therefore, $\omega(v, t) \not\equiv 0$. The Zygmund inequality assumes the form $\omega(v, t) \leqq 0$, and this is impossible in the case considered.

3.6. Theorem 3.3. *Under the assumptions of Theorem 3.1 the weak Zygmund inequality holds for any $t \leqq a - 2\delta$ whereas under the assumptions of Theorem 3.2 it is true for any $t \leqq a$.*

It is sufficient to consider the case of Theorem 3.1. Let us estimate $|v(x)|$. Obviously, $\max_{x \in \bar{G}'} |v_k(x)| = C \max_{x \in \bar{G}} |u(x)|$, $k = 1, 2$. Furthermore, if $x \in \bar{G}'$ then

$$|v_3(x)| = \left|\int_{r<\delta} [u(y) - u(x)]\,r^{-m} f(x, \theta)\,\mathrm{d}y\right| \leqq C \int_0^\delta \tau^{-1}\omega(u, \tau)\,\mathrm{d}\tau \leqq C \int_0^a \tau^{-1}\omega(u, \tau)\,\mathrm{d}\tau.$$

Therefore,

$$\max_{x \in \bar{G}'} |v(x)| \leqq C\left\{\|u\|_{C(\bar{G})} + \int_0^a \tau^{-1}\,\omega(u, \tau)\,\mathrm{d}\tau\right\}. \tag{24}$$

For $x + h \in \bar{G}'$ the max $|v(x + h)|$ admits of the same estimate. In that case we have

$$|v(x + h) - v(x)| \leqq C\left\{\|u\|_{C(\bar{G})} + \int_0^t \tau^{-1}\omega(u, \tau)\,\mathrm{d}\tau + t\int_t^a t^{-1}\tau^{-1}\omega(u, \tau)\,\mathrm{d}\tau\right\}, \qquad t = |h|.$$

Let $t \geqq \delta/2$ then $\|u\|_{C(\bar{G})} \leqq (2t/\delta)\,\|u\|_{C(\bar{G})}$. Furthermore, if $t \leqq \tau \leqq a$ we obtain the relation $(t\tau)^{-1} = \tau^{-2}\,\tau/t < 2a/\delta\tau^2$, and the inequality (5) is proved.

§ 4. Consequences of the Zygmund inequality

4.1. Let us introduce the spaces $\mathbf{Di}(\overline{G})$ and $\mathbf{DiL}(\overline{G})$. The first one consists of all functions which satisfy on $\overline{G}$ the Dini condition (see § 2, Chap. VIII), and the second one is the space of functions which satisfy on $\overline{G}$ the somewhat more restrictive condition

$$\max_{x \in \overline{G}} |u(x)| + \int_0^a \frac{\omega(u, \tau)}{\tau} \ln \frac{2a}{t} \, dt < \infty; \tag{1}$$

the quantity (1) is a norm in $\mathbf{DiL}(\overline{G})$.

Theorem 4.1. *Let G be a bounded domain in the space $\mathbf{R}^m$ and G' an interior subdomain of G. If the characteristic $f(x, \theta)$ is continuously differentiable with respect to the Cartesian coordinates of the points x and θ then the singular integral operator* (3.3) *is bounded from* $\mathbf{DiL}(\overline{G})$ *into* $\mathbf{Di}(\overline{G}')$.

Proof. Inequality (3.5) implies

$$\int_0^{a'} t^{-1} \omega(v, t) \, dt \le C \left\{ a \|u\|_{C(\overline{G})} + \int_0^a \frac{dt}{t} \int_0^t \frac{\omega(u, \tau)}{\tau} d\tau + \int_0^{a'} dt \int_t^a \frac{\omega(u, \tau)}{\tau^2} d\tau \right\};$$

where a denotes the diameter of the domain G. Changing the order of integration we find

$$\int_0^a \frac{dt}{t} \int_0^t \frac{\omega(u, \tau)}{\tau} d\tau = \int_0^a \frac{\ln a/\tau}{\tau} \omega(u, \tau) \, d\tau \le \int_0^a \frac{\ln 2a/\tau}{\tau} \omega(u, \tau) \, d\tau;$$

$$\int_0^{a'} dt \int_t^a \frac{\omega(u, \tau)}{\tau^2} d\tau = \int_0^{a'} \frac{\omega(u, \tau)}{\tau} d\tau + (a - a') \int_{a'}^a \frac{\omega(u, \tau)}{\tau^2} d\tau$$

$$\le \int_0^{a'} \frac{\omega(u, \tau)}{\tau} d\tau + \frac{a - a'}{a'} \int_{a'}^a \frac{\omega(u, \tau)}{\tau} d\tau$$

such that

$$\int_0^{a'} \frac{\omega(v, t)}{t} dt \le C \left[\|u\|_{C(\overline{G})} + \int_0^a \frac{\ln 2a/t}{t} \omega(u, t) \, dt \right]. \tag{2}$$

The inequalities (2) and (3.24) together yield

$$\|v\|_{\mathbf{Di}(\overline{G}')} \le C \|u\|_{\mathbf{DiL}(\overline{G})}, \tag{3}$$

which was to be demonstrated. ∎

Remark. By means of Theorem 3.2 it is not difficult to show that singular integrals taken over a sufficiently smooth closed manifold Γ under the assumption that its characteristic is continuously differentiable satisfy a stronger inequality (3), namely

$$\|v\|_{\mathbf{Di}(\Gamma)} \le C \|u\|_{\mathbf{DiL}(\Gamma)}. \tag{4}$$

4.2. Let $u \in \mathbf{Lip}_\alpha(\overline{G})$ or $u \in \mathbf{Lip}_\alpha(\Gamma)$ with G and Γ as introduced above, and assume that the characteristic of the singular integral is continuously differentiable. In that case we have $\omega(u, t) \le Ct^\alpha$, and the inequality (3.5) yields the following estimate for the modulus of continuity of the singular iutegral:

$$\omega(v, t) \leqq \begin{cases} Ct^\alpha, & 0 < \alpha < 1, \\ Ct \ln \dfrac{2a}{t}, & \alpha = 1. \end{cases} \tag{5}$$

That estimate and the theorems of §§ 3, 4 imply

Theorem 4.2. *Let* $0 < \alpha < 1$. *Under the assumptions mentioned above in* 4.2 *the singular integral operator is bounded from* $\mathbf{Lip}_\alpha(\overline{G})$ *into* $\mathbf{Lip}_\alpha(\overline{G'})$. *If the integral is taken over a closed manifold* Γ *the corresponding singular operator is bounded in* $\mathbf{Lip}_\alpha(\Gamma)$.

Theorem 4.2 was proved by Privalov [1, 2] for the dimension $m = 1$; for an arbitrary dimension m it was proved by Giraud [2].

4.3. Gegelia [1] considered the case of a two-dimensional closed manifold. He showed that the class of functions u such that

$$\omega(u, t) = O(t^\alpha |\ln t|^p) \tag{6}$$

where $0 < \alpha < 1$ and p is an arbitrary real number, is invariant under the singular integral operator. Khvoles [3] indicated more general classes of that behavior: The class of functions u such that $\omega(u, t) \leqq C_u \varphi(t)$ is invariant under the singular integral operator, here φ satisfies the inequality

$$\int_0^t \tau^{-1} \varphi(\tau)\, d\tau + t \int_t^a \tau^{-2} \varphi(\tau)\, d\tau \leqq C_0 \varphi(t), \qquad C_0 = \text{const}. \tag{7}$$

§ 5. The order of a singular integral at infinity

5.1. We introduce the class $\mathbf{A}_{\alpha,k}$ of functions subject to the following conditions. A function u belongs to $\mathbf{A}_{\alpha,k}$ iff $u(x)$ is defined everywhere in $\mathbb{R}^m$ and satisfies the inequalities

$$|u(x) - u(y)| \leqq A r^\alpha (|x|^2 + 1)^{-k/2}; \qquad r \leqq 1, A = \text{const} \tag{1}$$

and

$$|u(x)| \leqq B(|x|^2 + 1)^{-k/2}, \qquad B = \text{const} \tag{2}$$

where $0 < \alpha < 1$ and $k > 0$. By $\mathbf{A}'_{\alpha,k}$ we denote the class of functions which are defined everywhere in $\mathbb{R}^m$, satisfy the inequality (1) and, for sufficiently large $|x|$ the inequality

$$|u(x)| \leqq B(|x|^2 + 1)^{-k/2} \ln (|x|^2 + 1), \qquad B = \text{const}. \tag{3}$$

Theorem 5.1. *Assume that the singular kernel* $K(x, y)$ *meets the conditions* of § 3. *For* $k \leqq m$ *the singular integral operator*

$$v(x) := \int_{\mathbb{R}^m} K(x, y)\, u(y)\, dy = \int_{\mathbb{R}^m} r^{-m} f(x, \theta)\, u(y)\, dy \tag{4}$$

maps the class $\mathbf{A}_{\alpha,k}$ *into the class* $\mathbf{A}'_{\alpha,k}$. *If* $k > m$ *the operator* (4) *maps the class* $\mathbf{A}_{\alpha,k}$ *into the class* $\mathbf{A}_{\alpha,m}$.

Proof. In order to show that the function $v(x)$ satisfies the inequality (1) it is sufficient to repeat almost word by word the arguments used in the proof to Theorem 3.1, one has to replace $\omega(u, t)$ only by the right-hand side of the inequality (1).

Thus it remains to show that $v(x)$ satisfies the inequality (3) for sufficiently large $|x|$. First, we note that the characteristic $f(x, \theta)$ is bounded under the assumptions of § 3. Next, due to (1.3) we have

$$v(x) = \int_{r>1} K(x, y)\, u(y)\, \mathrm{d}y + \int_{r<1} K(x, y)\, [u(y) - u(x)]\, \mathrm{d}y =: v_1(x) + v_2(x). \quad (5)$$

Using the inequality (1) we obtain

$$|v_2(x)| \leqq C(|x|^2 + 1)^{-k/2} \int_{r<1} \frac{\mathrm{d}y}{r^{m-\alpha}} = C(|x|^2 + 1)^{-k/2}, \quad (6)$$

and it remains now to estimate $v_1(x)$. To begin with, let $k < m$. According to the inequality (2) we have

$$|v_1(x)| \leqq C \int_{r>1} r^{-m}(|y|^2 + 1)^{-k/2}\, \mathrm{d}y.$$

Introducing spherical coordinates with the origin at x we obtain

$$|v_1(x)| \leqq C \int_S \mathrm{d}S \int_1^\infty \frac{\mathrm{d}r}{r(r^2 + |x|^2 + 1 - 2r\,|x|\cos\gamma)^{k/2}},$$

where γ denotes the angle between the vectors x and $y - x$. The x_1-axis we fix along the direction of the vector x. Then $\mathrm{d}S = \sin^{m-2}\gamma\, \mathrm{d}\gamma\, \mathrm{d}S'$ where S' stands for the unit sphere in the $(m-1)$-dimensional space. The angle γ varies in the segment $0 \leqq \gamma \leqq \pi$ if $m > 2$ or in the segment $-\pi \leqq \gamma \leqq \pi$ if $m = 2$. For the present, consider the case $m > 2$. Then

$$|v_1(x)| \leqq C \int_1^\infty \frac{\mathrm{d}r}{r} \int_0^\pi \frac{\sin^{m-2}\gamma\, \mathrm{d}\gamma}{(r^2 + |x|^2 + 1 - 2r\,|x|\cos\gamma)^{k/2}}.$$

We introduce the abbreviation $\sigma = 2r\,|x|/(r^2 + |x|^2 + 1)$ where $0 \leqq \sigma < 1$. Now,

$$|v_1(x)| \leqq C \int_1^\infty \frac{\mathrm{d}r}{r(r^2 + |x|^2 + 1)^{k/2}} \int_0^\pi \frac{\sin^{m-2}\gamma\, \mathrm{d}\gamma}{(1 - \sigma\cos\gamma)^{k/2}}. \quad (7)$$

The inner integral is divided into two integrals taken over the intervals $(0, \pi/2)$ and $(\pi/2, \pi)$. Since $0 \leqq \sigma < 1$ the second integral is bounded, and the corresponding summand in the estimate for $|v_1(x)|$ is not greater than

$$C \int_1^\infty \frac{\mathrm{d}r}{r(r^2 + |x|^2 + 1)^{k/2}}$$

$$= C(|x|^2 + 1)^{-k/2} \int_{(|x|^2+1)^{-1/2}}^\infty t^{-1}(t^2 + 1)^{-k/2}\, \mathrm{d}t \leqq C(|x|^2 + 1)^{-k/2} \ln(|x|^2 + 1). \quad (8)$$

The inequality (8) is true for sufficiently large $|x|$.

The first inner integral in (7) is equal to

$$\int_0^{\pi/2} \sin^{m-2}\gamma (1 - \sigma\cos\gamma)^{-k/2}\, \mathrm{d}\gamma. \quad (9)$$

If $k < m - 1$ the integral (9) is bounded. Indeed, in that case we have

$$\int_0^{\pi/2} \sin^{m-2}\gamma (1 - \sigma\cos\gamma)^{-k/2}\, \mathrm{d}\gamma \leqq \int_0^{\pi/2} \sin^{m-2}\gamma (1 - \cos\gamma)^{-k/2}\, \mathrm{d}\gamma < 2^{k/2} \int_0^{\pi/2} \sin^{m-k-2}\gamma\, \mathrm{d}\gamma,$$

and the last integral converges because of $m - k - 2 > -1$. Once the integral (9) is bounded the estimate (8) implies that $v_1(x)$ satisfies an inequality of the form (3), and the theorem is proved.

Now let us consider integral (9) in the case $m - 1 \leqq k < m$. Substituting $\cos\gamma = t$ we obtain

$$\int_0^{\pi/2} \frac{\sin^{m-2}\gamma \, d\gamma}{(1 - \sigma\cos\gamma)^{k/2}} = \int_0^1 \frac{(1 - t^2)^{(m-3)/2} \, dt}{(1 - \sigma t)^{k/2}} < 2^{(m-3)/2} \int_0^1 \left(\frac{1 - t}{1 - \sigma t}\right)^{(m-3)/2} \frac{dt}{(1 - \sigma t)^{\varepsilon+1}}$$

where we put $\varepsilon = (k - m + 1)/2$ with $\varepsilon \geqq 0$. Furthermore, $1 - t < 1 - \sigma t$ such that

$$\int_0^{\pi/2} \frac{\sin^{m-2}\gamma \, d\gamma}{(1 - \sigma\cos\gamma)^{k/2}} < 2^{(m-3)/2} \int_0^1 \frac{dt}{(1 - \sigma t)^{\varepsilon+1}} = \begin{cases} 2^{(m-3)/2} \dfrac{1}{\varepsilon\sigma}\left[\dfrac{1}{(1-\sigma)^{\varepsilon}} - 1\right], & \varepsilon > 0, \\ 2^{(m-3)/2} \dfrac{1}{\sigma} \ln(1 - \sigma), & \varepsilon = 0. \end{cases} \tag{10}$$

Let, firstly, $\varepsilon > 0$. The right-hand side in (10) is bounded if σ is close to $\sigma = 0$, and we obtain

$$\int_0^{\pi/2} \frac{\sin^{m-2}\gamma \, d\gamma}{(1 - \sigma\cos\gamma)^{k/2}} < \frac{C}{(1 - \sigma)^{\varepsilon}}. \tag{11}$$

Further, we have

$$1 - \sigma = \frac{1 + (r - |x|)^2}{1 + r^2 + |x|^2},$$

and inserting this into (11) we find that the corresponding summand in the inequality (7) is not greater than

$$C \int_1^{\infty} \frac{dr}{r(1 + r^2 + |x|^2)^{(k-2\varepsilon)/2} [1 + (r - |x|)^2]^{\varepsilon}}. \tag{12}$$

To that integral we apply the Hölder inequality with an exponent p which will be specified later,

$$\int_1^{\infty} \frac{dr}{r(1 + r^2 + |x|^2)^{(k-2\varepsilon)/2} [1 + (r - |x|)^2]^{\varepsilon}}$$
$$\leqq \left\{\int_1^{\infty} \frac{dr}{r(1 + r^2 + |x|^2)^{p'(k-2\varepsilon)/2}}\right\}^{1/p'} \left\{\int_1^{\infty} \frac{dr}{r[1 + (r - |x|)^2]^{p\varepsilon}}\right\}^{1/p} =: J_1^{1/p'} J_2^{1/p}. \tag{13}$$

The integral J_1 can be estimated by means of the inequality (8), and we obtain

$$J_1^{1/p'} \leqq C(|x|^2 + 1)^{-(k-2\varepsilon)/2} [\ln(|x|^2 + 1)]^{1/p'}. \tag{14}$$

The integral J_2 is divided as follows,

$$J_2 = \int_1^{|x|} \frac{dr}{r[1 + (|x| - r)^2]^{p\varepsilon}} + \int_{|x|}^{\infty} \frac{dr}{r[1 + (r - |x|)^2]^{p\varepsilon}} =: J_2' + J_2''. \tag{15}$$

We choose the number p close to 1 such that the inequality $2p\varepsilon < 1$ holds. Substituting $r = |x| (1 + t)$ we obtain

$$J_2'' = \int_0^1 \frac{dt}{(1+t)(1+|x|^2 t^2)^{p\varepsilon}} + \int_1^\infty \frac{dt}{(1+t)(1+|x|^2 t^2)^{p\varepsilon}}$$

$$< \int_0^1 \frac{dt}{(|x|\,t)^{2p\varepsilon}} + \int_1^\infty \frac{dt}{t^{1+2p\varepsilon}\,|x|^{2p\varepsilon}} = O(|x|^{-2p\varepsilon}) = O((|x|^2+1)^{-p\varepsilon}). \quad (16)$$

In the integral J_2' we substitute $r = (1-t)\,|x|$,

$$J_2' = \int_0^{1/2} \frac{dt}{(1-t)(1+|x|^2 t^2)^{p\varepsilon}} + \int_{1/2}^{1-1/|x|} \frac{dt}{(1-t)(1+|x|^2 t^2)^{p\varepsilon}}.$$

The first integral is estimated as follows,

$$\int_0^{1/2} \frac{dt}{(1-t)(1+|x|^2 t^2)^{p\varepsilon}} < 2\int_0^{1/2} \frac{dt}{(1+|x|^2 t^2)^{p\varepsilon}}$$

$$= \frac{2}{|x|}\left[\int_0^1 \frac{dy}{(1+y^2)^{p\varepsilon}} + \int_1^{|x|/2} \frac{dy}{(1+y^2)^{p\varepsilon}}\right]$$

$$< \frac{2}{|x|}\left[1 + \int_1^{|x|/2} y^{-2p\varepsilon}\,dy\right] = O(|x|^{-2p\varepsilon}) = O((1+|x|^2)^{-p\varepsilon}). \quad (17)$$

The second integral is estimated more easily,

$$\int_{1/2}^{1-1/|x|} \frac{dt}{(1-t)(1+|x|^2 t^2)^{p\varepsilon}} < \left(1 + \frac{|x|^2}{4}\right)^{-p\varepsilon} \int_{1/2}^{1-1/|x|} \frac{dt}{1-t} = O\left(\frac{\ln(|x|^2+1)}{(|x|^2+1)^{p\varepsilon}}\right). \quad (18)$$

The formulae (15)–(18) imply

$$J_2 = O[(|x|^2+1)^{-p\varepsilon} \ln(|x|^2+1)].$$

Now the relations (12)–(14) yield the estimate

$$v_1(x) = O[(|x|^2+1)^{-k/2} \ln(|x|^2+1)]$$

such that the theorem is proved in the case $m > 2$, $0 < \varepsilon < 1/2$.

Next, let $k = m - 1$ (i.e. $\varepsilon = 0$). For any $\eta > 0$ and $0 \leqq \sigma < 1$ we have $\sigma^{-1} \ln(1-\sigma) < C_\eta (1-\sigma)^\eta$, and further the arguments are the same as in the general case.

In order to complete the proof for $k < m$ we have to consider still the case $m = 2$ only. In that case we find instead of (7) the inequality

$$|v_1(x)| < C\int_0^\infty \frac{dr}{r(1+|x|^2+r^2)^{k/2}} \int_{-\pi}^{\pi} \frac{d\gamma}{(1-\sigma\cos\gamma)^{k/2}}$$

$$= 2C\int_0^\infty \frac{dr}{r(1+|x|^2+r^2)^{k/2}} \int_0^{\pi} \frac{d\gamma}{(1-\sigma\cos\gamma)^{k/2}},$$

such that the proof is reduced to estimating the integral

$$\int_0^\pi \frac{d\gamma}{(1-\sigma\cos\gamma)^{k/2}}.$$

If $k < 1$ the integral is bounded, and the proof is ready.

Let $1 < k < 2$. Put $t = \tan \gamma/2$, $a^2 = (1-\sigma)/(1+\sigma)$ then as you will easily see

$$\int_0^{\pi/2} \frac{d\gamma}{(1-\sigma\cos\gamma)^{k/2}} < C \int_0^1 \frac{dt}{(a^2+t^2)^{k/2}} < \frac{C}{a^{k-1}} \int_0^\infty \frac{dz}{(1+z^2)^{k/2}}$$

$$< \frac{C}{(1-\sigma)^{(k-1)/2}} = \frac{C}{(1-\sigma)^\varepsilon}.$$

Further the demonstration goes as in the general case. In the same manner the case $k = 1$ can be treated which corresponds to $\varepsilon = 0$.

Now let us consider the case $k \geqq m$. The integral (12) will be estimated in another way. First, we note that

$$\int_1^\infty \frac{dr}{r(1+r^2+|x|^2)^{(k-2\varepsilon)/2}\,[1+(r-|x|)^2]^\varepsilon} < \frac{1}{|x|^{k-2\varepsilon}} \int_1^\infty \frac{dr}{r[1+(r-|x|)^2]^\varepsilon}.$$

Next, we divide the domain of integration into four intervals: $(1, |x|/2)$, $(|x|/2, |x|)$, $(|x|, 2|x|)$, and $(2|x|, \infty)$. Taking into account that in the case considered $\varepsilon \geqq 1/2$ we obtain

$$\int_1^{|x|/2} \frac{dr}{r[1+(r-|x|)^2]^\varepsilon} < \frac{C}{|x|^{2\varepsilon}} \int_1^{|x|/2} \frac{dr}{r} < \frac{C\ln|x|}{|x|^{2\varepsilon}};$$

$$\int_{|x|/2}^{|x|} \frac{dr}{r[1+(r-|x|)^2]^\varepsilon} < \frac{2}{|x|} \int_{|x|/2}^{|x|} \frac{dr}{[1+(r-|x|)^2]^\varepsilon} = \frac{2}{|x|} \int_0^{|x|} \frac{dt}{(1+t^2)^\varepsilon}. \tag{19$_1$}$$

For $\varepsilon > 1/2$ (or $k > m$) the last integral is bounded, and for $\varepsilon = 1/2$ (or $k = m$) it is equal to $\ln(|x|/2 + \sqrt{1+|x|^2/4}) < C\ln|x|$ such that

$$\int_{|x|/2}^{|x|} \frac{dr}{r[1+(r-|x|)^2]^\varepsilon} \leqq \begin{cases} C|x|^{-1}, & k > m, \\ C|x|^{-1}\ln|x|, & k = m. \end{cases} \tag{19$_2$}$$

Similarly,

$$\int_{|x|}^{2|x|} \frac{dr}{r[1+(r-|x|)^2]^\varepsilon} \leqq \begin{cases} C|x|^{-1}, & k > m, \\ C|x|^{-1}\ln|x|, & k = m. \end{cases} \tag{19$_3$}$$

Finally, we obtain

$$\int_{2|x|}^\infty \frac{dr}{r[1+(r-|x|)^2]^\varepsilon} = \int_{|x|}^\infty \frac{dt}{(1+t)(1+t^2)^\varepsilon} < \int_{|x|}^\infty t^{-1-2\varepsilon}\,dt = C|x|^{-2\varepsilon}. \tag{19$_4$}$$

The formulae (19) show that the integral can be bounded by $O((|x|^2+1)^{m/2})$ if $k > m$, and $O((|x|^2+1)\ln(|x|^2+1))$ if $k = m$. Thus, the proof is complete in the case $m > 2$. The case $m = 2$ can be treated in the same way as done above. ▮

Remark. The case $k < m$ was considered in MIKHLIN's book [11], and the case $k \geqq m$ in a paper by T. A. TIMAN [1]. In that paper TIMAN constructed an example in order to show that the estimate of the singular integral at infinity given in Theorem 5.1 cannot be improved. We have to admit, however, that in Timan's example the characteristic of the singular integral is, no doubt, continuous but not continuously differentiable. Therefore, the question of how to improve Theorem 5.1 as posed in the book [11] by MIKHLIN seems to be still not completely solved.

5.2. By $\mathscr{A}$ we shall denote the class of singular kernels $K(x, x - y) = r^{-m} f(x, \theta)$ with the property that under stereographic transformation the product $[(1 + |x|^2)(1 + |y|^2)]^{m/2} K(x, x - y)$ corresponds to some kernel $\tilde{K}(\xi, \eta)$ which satisfies on the Riemann sphere Σ the conditions of Theorem 3.2 such that it is continuously differentiable if $\xi \neq \eta$ and satisfies the inequality $|\text{grad}_\xi \tilde{K}(\xi, \eta)| \leq C\varrho^{-m-1}$, $\varrho = |\xi - \eta|$. Furthermore, we denote by $\mathbf{A}_\alpha$ the class of functions $u(x)$ defined on $\mathbb{R}^m$ such that the product $(1 + |x|^2)^{m/2} u(x)$ belongs to $\mathbf{Lip}_\alpha(\Sigma)$, $0 < \alpha < 1$.

Theorem 5.2. *If $K \in \mathscr{A}$ the the singular operator* (4) *transforms every function of the class $\mathbf{A}_\alpha$ into a function belonging to the same class.*

Proof. Due to the definition of a singular integral, the formula (4) means

$$v(x) = \lim_{\varepsilon \to 0} \int_{r > \varepsilon} K(x, x - y)\, u(y)\, dy. \tag{20}$$

In that integral we perform the stereographic transformation. Then the sphere $r = \varepsilon$ corresponds to a $(m - 1)$-dimensionl sphere ω_ε lying on the sphere Σ and which is given by the equation

$$\varrho = \frac{2\varepsilon}{\sqrt{(1 + |x|^2)(1 + |y|^2)}};$$

if $\varepsilon \to 0$ the quotient ϱ/ε has the limit $\beta = 2(1 + |x|^2)^{-1}$. We put

$$(|x|^2 + 1)^{m/2} u(x) = \tilde{u}(\xi), \qquad (|x|^2 + 1)^{m/2} v(x) = \tilde{v}(\xi),$$

then the relation (20) corresponds to the following one:

$$\tilde{v}(\xi) = \lim_{\varepsilon \to 0} \int_{\Sigma \setminus \sigma_\varepsilon} \tilde{K}(\xi, \eta)\, \tilde{u}(\eta)\, d\eta; \tag{21}$$

here σ_ε denotes that part of the sphere Σ which is bounded by the surface ω_ε. To the integral (20) we apply formula (2.11). Then the second integral of the right-hand side vanishes since β does not depend on Λ, and we obtain

$$\tilde{v}(\xi) = \int_{\Sigma} \tilde{K}(\xi, \eta)\, \tilde{u}(\eta)\, d\sigma.$$

The function $\tilde{u}(\xi)$ satisfies a Lipschitz condition on Σ with a positive exponent α. Due to Theorem 4.2, so does the function $\tilde{v}(\xi)$ but this means exactly $v \in \mathbf{A}_\alpha$. ∎

Theorem 5.2 was obtained by Mikhlin [7].

§ 6. Singular integrals in some other spaces with a uniform metric

The theorems of §§ 3–5 have been generalized in various directions. In the present section we shall quote the results of some of those contributions without proving them.

6.1. Rode and Simonenko [1] introduce a space $\mathbf{H}_\varphi$ which is defined as follows. Let $\varphi(t)$ be a positive non-decreasing function defined for $t > 0$ such that $\lim_{t \to 0} \varphi(t) = 0$. Furthermore, let ν be a natural and k be a positive real number. A function $u(x)$ which is continuous on $\mathbb{R}^m$ belongs to the class $\mathbf{H}_{\varphi,k}^{(\nu)}$ iff

$$|\Delta_h^\nu u(x)| \leq A\varphi(|h|)\, \varkappa(x), \qquad |u(x)| \leq B\varkappa(x); \qquad \varkappa(x) = (1 + |x|^2)^{-k/2}. \tag{1}$$

Here Δ_h^ν denotes the difference of order ν and step length h,

$$\Delta_h^\nu u(x) = \sum_{\mu=0}^{\nu} (-1)^\mu \binom{\nu}{\mu} u(x + (\nu - \mu) h);$$

the constants A and B do not depend on x and h but may depend, in general, on u. The formula

$$\|u\|_{\mathbf{H}_{\varphi,k}^{(\nu)}} = \sup_{x\in\mathbb{R}^m, |h|\leq 1} \frac{|\Delta_h^\nu u(x)|}{\varphi(|h|)\varkappa(x)} + \sup_{x\in\mathbb{R}^m} \frac{|u(x)|}{\varkappa(x)} \tag{2}$$

defines a norm in the class $\mathbf{H}_{\varphi,k}^{(\nu)}$ which makes it a Banach space.

We shall require the function $\varphi(t)$ to satisfy the following inequalities:

$$t^\nu \int_t^2 \tau^{-\nu-1} \varphi(\tau)\, d\tau \leq C\varphi(t), \qquad t \leq 1; \tag{3}$$

$$t^l \int_t^2 \tau^{-l-1} \varphi_\alpha(\tau)\, d\tau \leq C[\varphi_\alpha(t)]^{l/\nu}; \qquad t \leq 1, \qquad \varphi_\alpha(t) = t^{-\alpha}\varphi(t), \qquad \alpha = \text{const},$$

$$0 < \alpha < 1. \tag{4}$$

The characteristic $f(x, \theta)$ is subject to the conditions

$$|f(x, \theta)| \leq C\psi(x), \qquad |\Delta_h^\nu f(x, \theta)| \leq C\varphi(|h|)\, \psi(x);$$

$$|h| \leq 1, \qquad \psi(x) = (1 + \ln^+ |x|)^{-1} \tag{5}$$

where $\Delta_h^\nu f(x, \theta)$ denotes the corresponding difference of the function $f(x, \theta)$ with respect to x and fixed θ.

Further we assume that there exists a positive ε such that

$$|f(x, \theta_1) - f(x, \theta)| \leq C\, |\theta_1 - \theta|^{\alpha+\varepsilon}\, \psi(x),$$

$$|\Delta_h^\nu[f(x, \theta_1) - f(x, \theta)]| \leq C\varphi_\alpha(|h|)\, |\theta_1 - \theta|^{\alpha+\varepsilon}\, \psi(x), \qquad |h| \leq 1. \tag{6}$$

Theorem 6.1. *Under the assumptions mentioned above the singular operator* (1.1) *is bounded in the space* $\mathbf{H}_{\varphi,k}^{(\nu)}$.

6.2. Under somewhat different restrictions for the characteristic SALAYEV [1] indicated another class of spaces which are invariant with respect to the singular operator (1.1). Namely, let the characteristic be continuous and bounded on $\overline{\mathbb{R}}^m \times S$. Put

$$\omega_f^{\mathrm{I}}(\delta, \xi) = \sup |f(x_1, \theta) - f(x_2, \theta)|, \tag{7}$$

$$\omega_f^{\mathrm{II}}(\delta) = \sup |f(x, \theta_1) - f(x, \theta_2)| \tag{8}$$

where the supremum in (7) is taken over the set

$$\{x_1, x_2 \in \mathbb{R}^m : \theta \in S;\ |x_1 - x_2| \leq \delta;\ (1 + |x_j|)^{-1} \leq \xi, j = 1, 2\}$$

and in (8) over the set $\{x \in \mathbb{R}^m\colon \theta_1, \theta_2 \in S;\ |\theta_1 - \theta_2| \leq \delta\}$. Additionally, we assign to every continuous and bounded function $w(x)$ on $\overline{\mathbb{R}}^m$ the two following quantities:

$$\Omega_w(\xi) = \sup |w(x)|, \qquad (1 + |x|)^{-1} \leq \xi; \tag{9}$$

$$\omega_w(\delta, \xi) = \sup |w(x_1) - w(x_2)|, \qquad |x_1 - x_2| \leq \delta,$$

$$(1 + |x_j|)^{-1} \leq \xi, \qquad j = 1, 2. \tag{10}$$

We define the class Φ consisting of pairs of functions $(\varphi(\xi), \psi(\delta, \xi))$, $0 < \xi \leqq 1$, $\delta > 0$, and its subclass $\Phi\mathbf{H}_m$. The pairs of functions belonging to the class Φ are subject to the following conditions: $\varphi(\xi) > 0$, $\psi(\delta, \xi) > 0$; these functions are non-decreasing in each of the arguments ξ and δ; the quotient $\psi(\delta, \xi)/\delta$ almost decreases with respect to δ uniformly in ξ, i.e. there is a constant C such that $h_2 > h_1$ implies the inequality $\psi(h_2, \xi)/h_2 \leqq C\psi(h_1, \xi)/h_1$; $\psi(\delta, \xi) = O(\varphi(\xi))$.

Let $(\varphi, \psi) \in \Phi$, then by $\mathbf{H}_{\varphi,\psi}$ we denote the subspace of functions continuous on $\overline{\mathbb{R}}^m$ which have finite norm

$$\|u\|_{\varphi,\psi} = \sup_{0<\xi\leq 1} \frac{\Omega_u(\xi)}{\varphi(\xi)} + \sup_{\substack{0<\xi\leq 1\\ \delta>0}} \frac{\omega_u(\delta, \xi)}{\psi(\delta, \xi)}. \tag{11}$$

It is easy to see that $\mathbf{H}_{\varphi,\psi}$ is a Banach space.

The following additional conditions define the subclass $\Phi\mathbf{H}_m$:

$$\int_0^\xi t^{-1}\varphi(t)\,dt + \xi^m \int_\xi^1 t^{-m-1}\varphi(t)\,dt = O(\varphi(\xi)),$$

$$\int_0^\delta t^{-1}\,\psi(t, \xi)\,dt + \delta \int_\delta^\xi t^{-2}\psi(t, \xi)\,dt = O(\psi(\xi, \delta)), \qquad 0 < \delta \leqq \xi^{-1},$$

$$\delta\xi\,\varphi(\xi) = O(\psi(\delta, \xi)). \tag{12}$$

Theorem 6.2. *If* $(\varphi, \psi) \in \Phi\mathbf{H}_m$ *and*

$$\omega_f^{\mathrm{I}}(\delta, \xi) = O\left(\frac{\psi(\delta, \xi)}{\varphi(\xi)}\right), \qquad \int_\delta^{1/\xi} t^{-1}\,\omega_f^{\mathrm{II}}(\delta/t)\,\psi(t, \xi)\,dt = O(\psi(\delta, \xi)) \tag{13}$$

then the singular operator (1.1) *is bounded in* $\mathbf{H}_{\varphi,\psi}$.

In the paper [1] Salayev proved also that under the assumptions $\varphi(\xi) \sim \xi^k$ and $\psi(\delta, \xi) \sim \delta^\alpha\xi^k(1 + \delta)^{-\alpha}$ (the symbol $\sim$ means that the quotient of both sides of the corresponding relation is bounded from below and from above by two positive constants) the norm in the class $\mathbf{A}_{\alpha,k}$,

$$\|u\|_{\mathbf{A}_{\alpha,k}} = \sup_{x\in\mathbb{R}^m} |u(x)|\,(1 + |x|^2)^{k/2} + \sup_{\substack{x,y\in\mathbb{R}^m\\ |x-y|\leq 1}} \frac{|u(x) - u(y)|}{|x - y|^\alpha}(1 + |x|^2)^{k/2}, \tag{14}$$

is equivalent to the norm in $\mathbf{H}_{\varphi,\psi}$ and that these two spaces consist of one and the same elements. At the same time, the pair of functions under consideration belong, no doubt, to the class Φ but not to the subclass $\Phi\mathbf{H}_m$ provided that $0 < \alpha \leqq 1$, $0 < k \leqq m$.

6.3. In the paper [3] Khvoles considered spaces $\mathbf{C}_\varphi$ defined as follows. Let $\varphi(t)$ be a function of the real variable $t \in [0, a]$ which satisfies the condition $\int_0^a t^{-1}\varphi(t)\,dt < \infty$. A function $u(x)$ defined on a measurable set M belongs to the space $\mathbf{C}_\varphi(M)$ iff the norm

$$\|u\|_{\mathbf{C}_\varphi(M)} = \sup_{x\in M} |u(x)| + \sup_{\substack{x\in M\\ x+h\in M\\ |h|\leq 1}} \frac{|u(x + h) - u(x)|}{\varphi(h)} \tag{15}$$

is finite. If, in addition, the function $\varphi(t)$ satisfies the condition

$$\int_0^t \tau^{-1}\varphi(\tau)\,d\tau + t\int_\tau^a \tau^{-2}\varphi(\tau)\,d\tau \leqq C\varphi(t) \tag{16}$$

then the singular operator (1.1) or (2.1) is bounded in $\mathbf{C}_\varphi(\mathbb{R}^m)$ or $\mathbf{C}_\varphi(\Gamma)$, resp. The last assertion was proved under rather weak assumptions on the characteristic of the singular operator.

§ 7. Differentiation of weakly singular integrals

As an application of singular integrals we shall give a differentiation rule for certain weakly singular integrals. Let G be a domain in the space $\mathbb{R}^m$ which may coincide with the whole space, and consider

$$w(x) = \int_G u(y) \frac{\varphi(x, \theta)}{r^{m-1}} \, dy, \qquad x \in G. \tag{1}$$

We assume that the function $\varphi(x, \theta)$ together with its first derivatives with respect to the Cartesian coordinates of the points x and θ is continuous and bounded on the set $G \times S$. The function $u(x)$ is assumed to satisfy a Dini condition on every bounded closed subset of the domain G, and at infinity (if the domain G is not bounded) be $u(x) = O(|x|^{-l})$, $l > 1$.

We shall show that under these assumptions the first derivatives of the integral (1) exist and are given by the formula

$$\frac{\partial}{\partial x_k} \int_G u(y) \frac{\varphi(x, \theta)}{r^{m-1}} \, dy = \int_G u(y) \frac{\partial}{\partial x_k} \left[\frac{\varphi(x, \theta)}{r^{m-1}}\right] dy - u(x) \int_S \varphi(x, \theta) \cos(r, x_k) \, dS \tag{2}$$

($k = 1, 2, \ldots, m$), where the x_k denote the Cartesian coordinates of the point x.

We surround the point x with a ball $r < \varepsilon$, put $G_\varepsilon := G \setminus (r < \varepsilon)$, and compute the derivative

$$\frac{\partial}{\partial x_k} \int_{G_\varepsilon} u(y) \frac{\varphi(x, \theta)}{r^{m-1}} \, dy = \int_{G_\varepsilon} u(y) \frac{\partial}{\partial x_k} \left[\frac{\varphi(x, \theta)}{r^{m-1}}\right] dy - \int_{r=\varepsilon} u(y) \frac{\varphi(x, \theta)}{\varepsilon^{m-1}} \cos(r, x_k) \, dS_\varepsilon. \tag{3}$$

Here $dS_\varepsilon = \varepsilon^{m-1} \, dS$ denotes the surface element on the sphere $r = \varepsilon$, and r is directed from x to y.

Let us introduce the following notation. By D_k' we denote the derivative with respect to x_k which is computed under the assumption that r and θ do not depend on x_k, whereas D_k'' denotes the derivative with respect to x_k under the assumption that only r and θ depend on x_k. With these notations we can rewrite (3) as follows:

$$\frac{\partial}{\partial x_k} \int_{G_\varepsilon} u(y) \frac{\varphi(x, \theta)}{r^{m-1}} \, dy = \int_{G_\varepsilon} u(y) \, D_k'' \left[\frac{\varphi(x, \theta)}{r^{m-1}}\right] dy$$

$$+ \int_{G_\varepsilon} u(y) \frac{1}{r^{m-1}} D_k'[\varphi(x, \theta)] \, dy - \int_S u(x + \varepsilon\theta) \, \varphi(x, \theta) \cos(r, x_k) \, dS. \tag{4}$$

If $\varepsilon \to 0$ the sum of the second and the third integral in the right-hand side converges uniformly to a limit which is equal to

$$\int_G u(y) \frac{1}{r^{m-1}} D_k'[\varphi(x, \theta)] \, dy - u(x) \int_S \varphi(x, \theta) \cos(r, x_k) \, dS.$$

Consider the first integral. The kernel can, obviously, be represented as

$$D_k'' \left[\frac{\varphi(x, \theta)}{r^{m-1}}\right] = \frac{f(x, \theta)}{r^m}.$$

We show that $f(x, \theta)$ satisfies the condition (1.2). We number the coordinate axes for the moment such that $k = 1$, and introduce spherical coordinates

$$\left.\begin{aligned} y_1 &= x_1 + r\cos\vartheta_1, \\ y_2 &= x_2 + r\sin\vartheta_1\cos\vartheta_2, \\ &\cdots\cdots\cdots\cdots\cdots\cdots \\ y_{m-1} &= x_{m-1} + r\sin\vartheta_1 \ldots \sin\vartheta_{m-2}\cos\vartheta_{m-1}, \\ y_m &= x_m + r\sin\vartheta_1 \ldots \sin\vartheta_{m-2}\sin\vartheta_{m-1}. \end{aligned}\right\} \tag{5}$$

Then

$$\mathrm{d}S = \sin^{m-2}\vartheta_1 \sin^{m-3}\vartheta_2 \ldots \sin\vartheta_{m-2}\, \mathrm{d}\vartheta_1\, \mathrm{d}\vartheta_2 \ldots \mathrm{d}\vartheta_{m-2}\, \mathrm{d}\vartheta_{m-1}.$$

Only the variables r and ϑ_1 depend on $x_k = x_1$, hence,

$$D_1''\left[\frac{\varphi(x, \theta)}{r^{m-1}}\right] = \frac{1}{r^m}\left[r\frac{\partial\varphi}{\partial\vartheta_1}\frac{\partial\vartheta_1}{\partial x_1} - (m-1)\,\varphi\frac{\partial r}{\partial x_1}\right].$$

Furthermore, $\partial r/\partial x_1 = (x_1 - y_1)/r = -\cos\vartheta_1$; after another differentiation we find $\partial\vartheta_1/\partial x_1 = \sin\vartheta_1/r$. Now,

$$D_1''\left[\frac{\varphi(x, \theta)}{r^{m-1}}\right] = \frac{1}{r^m}\left[(m-1)\,\varphi\cos\vartheta_1 + \frac{\partial\varphi}{\partial\vartheta_1}\sin\vartheta_1\right]$$

and, consequently,

$$f(x, \theta) = (m-1)\,\varphi\cos\vartheta_1 + \frac{\partial\varphi}{\partial\vartheta_1}\sin\vartheta_1.$$

We form the integral

$$\int_S f(y, \theta)\,\mathrm{d}S = \int_{-\pi}^{\pi}\mathrm{d}\vartheta_1 \int_0^{\pi}\sin\vartheta_{m-2}\,\mathrm{d}\vartheta_{m-2} \ldots \int_0^{\pi}\sin^{m-3}\vartheta_2\,\mathrm{d}\vartheta_2$$
$$\times\int_0^{\pi}\left[(m-1)\,\varphi\cos\vartheta_1 + \frac{\partial\varphi}{\partial\vartheta_1}\sin\vartheta_1\right]\sin^{m-2}\vartheta_1\,\mathrm{d}\vartheta_1.$$

The inner integral equals

$$\int_0^{\pi}\frac{\partial}{\partial\vartheta_1}[\sin^{m-1}\vartheta_1\,\varphi(x, \theta)]\,\mathrm{d}\vartheta_1 = 0,$$

and the condition (1.2) is fulfilled.

As shown in § 1, the first integral in the right-hand side of (4) converges uniformly in x to the singular integral

$$\int_G D_k''\left[\frac{\varphi(x, \theta)}{r^{m-1}}\right]u(y)\,\mathrm{d}y$$

as $\varepsilon \to 0$. But then the derivative

$$\frac{\partial}{\partial x_k}\int_{G_\varepsilon} u(y)\frac{\varphi(x, \theta)}{r^{m-1}}\,\mathrm{d}y$$

converges also uniformly to a certain limit, the integral (1) has a derivative which is equal to that limit, and (2) is proved.

Later on (see Chap. X) we shall extend that formula (2) to wider classes of functions. Equation (2) was obtained by TRICOMI [1, 2] in the case $m = 2$ and φ not depending on x. The more general case was considered by MIKHLIN [7, 14].

Differentiability properties of functions are studied by means of the concept of singular integrals at some length in the book of STEIN [1].

Chapter X
The symbol of multidimensional singular integral operators

§ 1. The Fourier transform of a singular kernel. The symbol

1.1. Let us consider the singular integral

$$(A_0 u)(x) = v(x) := \int_{\mathrm{R}^m} K(x-y)\, u(y)\, \mathrm{d}y; \qquad K(x-y) = r^{-m} f(\theta), \tag{1}$$

the characteristic of which does not depend on the pole. As mentioned in § 1, Chap. IX, the integral (1) can be considered as the convolution of the distribution $K(x)$ and the density $u(x)$. We shall show that the Fourier transform $\hat{K}(\xi) = F_{x\to\xi} K$ of the kernel $K(x)$ is a locally summable function. Thus we may introduce the notion of a symbol for operators (1) and as we shall see later on also for a certain wider class of operators. Therefore, it is necessary to study the Fourier transform of singular kernels. This is the topic of CALDERON and ZYGMUND [1]; in subsection 1.2 their results will be presented with some slight modifications.

Deviating somewhat from the usual notations in the theory of distributions we define the Fourier transforms of a singular kernel by the formula

$$\hat{w}(\xi) = F_{x\to\xi}\, w := \lim_{\varepsilon\to 0, N\to\infty} \int_{\varepsilon<|x|<N} \mathrm{e}^{-2\pi i x\xi}\, w(x)\, \mathrm{d}x. \tag{2}$$

We shall show that under sufficiently general assumptions the limit (2) does exist and that our definition leads to the well-known formula

$$\hat{v}(\xi) = \hat{K}(\xi)\, \hat{u}(\xi) \tag{3}$$

for the convolution.

1.2. Assume that the characteristic is bounded and as usually satisfies the condition (1.2), Chap. IX. We introduce the kernel

$$K_{\varepsilon,N}(x) = \begin{cases} K(x), & \varepsilon \leqq |x| \leqq N, \\ 0 & \text{otherwise,} \end{cases} \tag{4}$$

and consider the integral

$$v_{\varepsilon,N}(x) = \int_{\mathrm{R}^m} K_{\varepsilon,N}(x-y)\, u(y)\, \mathrm{d}y. \tag{5}$$

Let the function $u(y)$ be finite (i.e. non-zero only on a certain compact set) and infinitely differentiable. By the convolution theorem which can be applied here without any difficulty,

$$F v_{\varepsilon,N} = F K_{\varepsilon,N} F u. \tag{6}$$

Further, we have

$$FK_{\varepsilon,N} = \int\limits_{R^m} K_{\varepsilon,N}(x)\, e^{-2\pi i x\xi}\, dx = \int\limits_{\varepsilon}^{N} \frac{dR}{R} \int\limits_{S} f(\theta)\, e^{2\pi i \varrho R \cos\gamma}\, dS$$

$$= \int\limits_{\varepsilon\varrho}^{N\varrho} \frac{dt}{t} \int\limits_{S} f(\theta)\, e^{-2\pi i t\cos\gamma}\, dS,$$

where we made use of the notations $\varrho = |\xi|$, $R = |x|$, $\theta = -x/R$; the angle between the vectors x and ξ is denoted by γ. We assume $\varrho \neq 0$, otherwise the last transformation above is not allowed.

Due to (1.2), Chap. IX, the last equation can be represented as

$$FK_{\varepsilon,N} = \int\limits_{\varepsilon\varrho}^{N\varrho} \frac{dt}{t} \int\limits_{S} f(\theta)\,(e^{-2\pi i t\cos\gamma} - e^{-2\pi t})\, dS$$

$$= \int\limits_{S} f(\theta)\, dS \int\limits_{\varepsilon\varrho}^{N\varrho} \frac{e^{-2\pi i t\cos\gamma} - e^{-2\pi t}}{t}\, dt. \tag{7}$$

Put $2\pi\varepsilon\varrho = \varepsilon'$, $2\pi N\varrho = N'$. The inner integral is transformed to the form

$$\int\limits_{\varepsilon'}^{N'} \frac{e^{-it\cos\gamma} - e^{-t}}{t}\, dt = \int\limits_{\varepsilon'}^{1} \frac{e^{-it\cos\gamma} - e^{-t}}{t}\, dt + \int\limits_{1}^{N'} \frac{e^{-it\cos\gamma} - e^{-t}}{t}\, dt.$$

The first integral right tends to a similar integral with the bounds $t = 0$ and $t = 1$ as $\varepsilon' \to 0$. As you can easily see the assumption $|\cos\gamma| \geqq \delta$, $\delta = \text{const} > 0$, implies that the second integral converges uniformly to the integral with the bounds $t = 1$ and $t = \infty$. To see that it is sufficient to perform the integration by parts,

$$\int\limits_{1}^{N'} \frac{e^{-it\cos\gamma}}{t}\, dt = -\left.\frac{e^{-it\cos\gamma}}{it\cos\gamma}\right|_{1}^{N'} - \frac{1}{i t\cos\gamma}\int\limits_{1}^{N'} \frac{e^{-it\cos\gamma}}{t^2}\, dt$$

and to take into account that both terms in the right-hand side tend uniformly to their limit as $N' \to \infty$ provided that $|\cos\gamma| \geqq \delta$.

The arguments, in particular, imply that the limit of the inner integral in (7) for $\gamma = \pi/2$ exists as $\varepsilon' \to 0$ and $N' \to \infty$.

Now, we prove the inequality

$$\left|\int\limits_{\varepsilon'}^{N'} \frac{e^{-it\cos\gamma} - e^{-t}}{t}\, dt\right| \leqq \ln \frac{a}{|\cos\gamma|}, \qquad a = \text{const}.$$

In order to do so it is sufficient to show the relation

$$\left|\int\limits_{1}^{N'} t^{-1} e^{-it\alpha}\, dt\right| \leqq \ln \frac{1}{|\alpha|} + M, \qquad M = \text{const}.$$

Let, for example, $\alpha > 0$. Then

$$\int\limits_{1}^{N'} t^{-1} e^{-it\alpha}\, dt = \int\limits_{\alpha}^{\alpha N'} t^{-1} e^{-it}\, dt.$$

For $\alpha N' \leqq 1$ we have

$$\left|\int\limits_{\alpha}^{\alpha N'} t^{-1} e^{-it}\, dt\right| \leqq \ln N' \leqq \ln \frac{1}{\alpha},$$

and for $\alpha N' > 1$

$$\int_{\alpha}^{\alpha N'} t^{-1} e^{-it} dt = \int_{\alpha}^{1} t^{-1} e^{-it} dt + \int_{1}^{\alpha N'} t^{-1} e^{-it} dt.$$

The first integral right is in modulus not greater than $\ln 1/\alpha$, and the second one is bounded by some constant due to the convergence of the integral $\int_1^\infty t^{-1} e^{-it} dt$. Thus, we proved the inequality mentioned above. But then, since the function $f(\theta)$ is bounded the integral (7) is bounded if $\varrho \neq 0$ and tends to a certain limit uniformly as $\varepsilon \to 0$, $N \to \infty$ which is also bounded. This limit is considered as the Fourier transform of the kernel K, and thus we obtained a substantiation of the formula (3).

It is worth noting that the integral (7) does not depend on ϱ if $\varrho \neq 0$ – this is immediately clear from (7).

In order that eq. (3) makes sense it is not necessary to have the function $f(\theta)$ bounded; it suffices to require the condition

$$\int_S |f(\theta)| \ln^+ |f(\theta)| dS < \infty.$$

more details you will find in CALDERON and ZYGMUND [1].

Equation (6) now yields

$$\lim_{\varepsilon \to 0, N \to \infty} F v_{\varepsilon,N} = FKFu. \tag{8}$$

The finite and infinitely differentiable function $u(x)$ belongs in any case to the space $L_2(\mathbb{R}^m)$ to which also belongs Fu. Since the function $FK_{\varepsilon,N}$ is bounded and converges uniformly to FK the function $Fv_{\varepsilon,N}$ belongs to $L_2(\mathbb{R}^m)$, too, and the convergence in (8) can be understood as the convergence in $L_2(\mathbb{R}^m)$.

Put $Fv_{\varepsilon,N} = w_{\varepsilon,N}$, $\lim_{\varepsilon \to 0, N \to \infty} w_{\varepsilon,N} = w$; then $v_{\varepsilon,N} = F^{-1} w_{\varepsilon,N} \to F^{-1} w$. At the same time we have also $v_{\varepsilon,N}(x) \to v(x)$. Hence, $v(x) = F^{-1} w$ and $w = Fv$. Equation (2) is proved, at least for functions which are finite and infinitely differentiable. But then it is not difficult to extend it to functions which belong to larger function classes.

1.3. Now let us consider a singular operator of the form

$$(A_0 u)(x) := au(x) + \int_{\mathbb{R}^m} K(x-y)\, u(y)\, dy; \qquad K(x-y) = r^{-m} f(\theta). \tag{9}$$

It can be considered as the convolution of the functions u and $a\,\delta + K$. In accordance with the results of § 6, Chap. I, we may define the *symbol* of the operator A_0 by the formula

$$\operatorname{Smb} A_0 = a + \hat{K}(\xi). \tag{10}$$

Sum and product of operators (9) correspond to sum and product of symbols (10). It is worth noting that (8) leads to a simpler representation of the singular operator (9) by its symbol: if we denote $\operatorname{Smb} A_0$ briefly by $\Phi(\xi)$ we obtain

$$v = A_0 u = F^{-1}_{\xi \to x} \Phi(\xi)\, F_{x \to \xi} u. \tag{11}$$

For singular operators of a more general form,

$$(Au)(x) := a(x)\, u(x) + \int_{\mathbb{R}^m} K(x, x-y)\, u(y)\, dy; \qquad K(x, x-y) = r^{-m} f(x, \theta) \tag{12}$$

we define the *symbol*, for the present, formally by the relation

$$\operatorname{Smb} A = a(x) + \hat{K}(x, \xi), \tag{13}$$

where $\hat{K}(x, \xi)$ denotes the Fourier transform of the kernel $K(x, z)$ with respect to its argument z. With that definition, obviously, the sum of operators (12) corresponds to the sum of their symbols. It remains to show that there also the formula $\operatorname{Smb} A_1 A_2 = \operatorname{Smb} A_1 \operatorname{Smb} A_2$ holds true. Under some additional assumptions this will be done in § 7, Chap. XI.

1.4. It turns out that we may give a rather simple representation of the symbol of a singular operator by means of its characteristic. As we saw the symbol $(\operatorname{Smb} A_0)(\xi)$ does not depend on $|\xi|$. That means the symbol is a positive-homogeneous function of degree zero, and if $\operatorname{Smb} A_0 = \Phi_0$ we may, therefore, write $\Phi_0 = \Phi_0(\Lambda)$, $\Lambda = \xi/|\xi|$. Equivalently, we may speak of the symbol as a function defined on the unit sphere. By definition, we have

$$\Phi_0(\Lambda) = a + \int_{\mathbf{R}^m} \mathrm{e}^{-2\pi i x\xi} f(-\theta_0)\, R^{-m}\, \mathrm{d}x = a + \lim_{\varepsilon\to 0, N\to\infty} \int_{\varepsilon<R<N} \mathrm{e}^{-2\pi i x\xi} f(-\theta_0)\, R^{-m}\, \mathrm{d}x$$

where $R = |x|$ and $\theta_0 = x/R$. Note that we may write in these integrals $\xi = \Lambda$ or assume $|\xi| = 1$. Substituting $x = -x'$ and after that omitting the prime we obtain

$$\Phi_0(\Lambda) = a + \lim_{\varepsilon\to 0, N\to\infty} \int_{\varepsilon<R<N} \mathrm{e}^{2\pi i x\xi} f(\theta_0)\, R^{-m}\, \mathrm{d}x$$

$$= \int_S f(\theta)\, \mathrm{d}S \int_0^{1/2\pi} \frac{\mathrm{e}^{2\pi i x\xi} - 1}{R}\, \mathrm{d}R + \int_S f(\theta)\, \mathrm{d}\theta \int_{1/2\pi}^{\infty} \frac{e^{2\pi i x\xi}}{R}\, \mathrm{d}R.$$

Let γ be the angle between the vectors x and ξ; taking into account $|\xi| = 1$ we find

$$\int_0^{1/2\pi} \frac{\mathrm{e}^{2\pi i x\xi} - 1}{R}\, \mathrm{d}R = \int_0^{1/2\pi} \frac{\mathrm{e}^{2\pi i R\cos\gamma} - 1}{R}\, \mathrm{d}R; \qquad \int_{1/2\pi}^{\infty} \frac{\mathrm{e}^{2\pi i x\xi}}{R}\, \mathrm{d}R = \int_1^{\infty} \frac{\mathrm{e}^{2\pi i R\cos\gamma}}{R}\, \mathrm{d}R.$$

The substitution $2\pi R\,|\cos\gamma| = t$ yields

$$\int_0^{1/2\pi} \frac{\mathrm{e}^{2\pi i x\xi} - 1}{R}\, \mathrm{d}R + \int_{1/2\pi}^{\infty} \frac{\mathrm{e}^{2\pi i R\cos\gamma}}{R}\, \mathrm{d}R = \ln \frac{1}{|\cos\gamma|} + Q(\cos\gamma), \tag{14}$$

where Q is a certain bounded function. We shall now compute it.

Let $\cos\gamma > 0$. Then the substitution applied above leads to

$$\int_{1/2\pi}^{\infty} R^{-1}\, \mathrm{e}^{2\pi i R\cos\gamma}\, \mathrm{d}R = \int_{\cos\gamma}^{\infty} t^{-1}\, \mathrm{e}^{it}\, \mathrm{d}t = \int_{\cos\gamma}^{1} t^{-1} (\mathrm{e}^{it} - 1)\, \mathrm{d}t + \ln \frac{1}{|\cos\gamma|} + \int_1^{\infty} t^{-1}\, \mathrm{e}^{-it}.$$

On the other hand,

$$\int_0^{1/2\pi} R^{-1} (\mathrm{e}^{2\pi i R\cos\gamma} - 1)\, \mathrm{d}R = \int_0^{\cos\gamma} t^{-1} (\mathrm{e}^{it} - 1)\, \mathrm{d}t.$$

Thence

$$Q(\cos\gamma) = \int_0^1 t^{-1} (\mathrm{e}^{it} - 1)\, \mathrm{d}t + \int_1^{\infty} t^{-1}\, \mathrm{e}^{it}\, \mathrm{d}t$$

$$= -\int_0^1 \frac{1 - \cos t}{t}\, \mathrm{d}t + \int_1^{\infty} \frac{\cos t}{t}\, \mathrm{d}t + \mathrm{i} \int_0^{\infty} \frac{\sin t}{t}\, \mathrm{d}t = \alpha + \frac{\mathrm{i}\pi}{2}.$$

Now let $\cos\gamma < 0$. Put $-2\pi R\cos\gamma = t$ and repeat the foregoing calculations, then

$$Q(\cos\gamma) = -\int_0^1 \frac{1-\cos t}{t}\,dt + \int_1^\infty \frac{\cos t}{t}\,dt - i\int_0^\infty \frac{\sin t}{t}\,dt = \alpha - \frac{i\pi}{2}.$$

Due to condition (1.2), Chap. IX, we may omit the constant α in the kernel (14) and, therefore, assume that the kernel of the integral operator expressing the symbol of the singular integral by its characteristic has the form

$$\ln\frac{1}{|\cos\gamma|} + \frac{i\pi}{2}\operatorname{sign}\cos\gamma. \tag{15}$$

This formula was obtained by CALDERON and ZYGMUND [3].

Finally, we arrived at the following integral representation for the symbol $\Phi(\theta)$ of a singular integral by its characteristic $f(\theta)$,

$$\Phi(\theta) = \int_S f(\theta')\left[\ln\frac{1}{|\cos\gamma|} + \frac{i\pi}{2}\operatorname{sign}\cos\gamma\right] d_{\theta'}S. \tag{16}$$

If the characteristic depends on the pole, we have similarly

$$\Phi(x,\theta) = \int_S f(x,\theta')\left[\ln\frac{1}{|\cos\gamma|} + \frac{i\pi}{2}\operatorname{sign}\cos\gamma\right] d_{\theta'}S. \tag{17}$$

An important property of the operator (16) is the following,

$$\int_S \left|\ln\frac{1}{|\cos\gamma|} + \frac{i\pi}{2}\operatorname{sign}\cos\gamma\right|^p d_{\theta'}S = C_p = \text{const} < \infty, \qquad \forall p,\ 1 \leqq p < \infty. \tag{18}$$

In order to prove the relation (18) we introduce Cartesian coordinates such that the first of the axes passes through the point θ. Then $\gamma = \vartheta_1$ and

$$\int_S \left|\ln\frac{1}{\cos\gamma} + \frac{i\pi}{2}\operatorname{sign}\cos\gamma\right|^p d_{\theta'}S = \int_0^\pi \left|\ln\frac{1}{\cos\gamma} + \frac{i\pi}{2}\operatorname{sign}\cos\gamma\right|^p$$

$$\times \sin^{m-2}\gamma\,d\gamma \int_0^\pi \cdots \int_0^\pi \int_0^{2\pi} \sin^{m-3}\vartheta_2 \ldots \sin\vartheta_{m-2}\,d\gamma\,d\vartheta_1 \ldots d\vartheta_{m-2}\,d\vartheta_{m-1},$$

but this is a (finite) constant.

It is easy to show that the operator (16) is completely continuous in $\mathbf{L}_p(S)$ for any p, $1 < p < \infty$, simply use e.g. (18) and the results by ITSKOVICH [2].

1.5. The symbol for a matrix singular operator, too, is constructed easily. Let $a(x)$ and $K(x,z)$ be square matrices of order N in (12), then the symbol of this operator is also defined by (14). It is a matrix of the same order the elements of which are the symbols of the correspondig operators referring to elements of the matrix K. The matrix (13) will be called the *symbol matrix* and its determinant the *symbol determinant* of the singular matrix operator.

§ 2. Expansion of the symbol in a series with respect to spherical functions

2.1. We seek the Fourier transform of a specific singular kernel, namely

$$K(x-y) = \frac{Y_{n,m}^{(k)}(\theta)}{r^m}, \qquad n \geqq 1. \tag{1}$$

As we did before we put $R = |x|$, $\theta_0 = x/R$, and obtain $K(x) = R^{-m}\, Y_{n,m}^{(k)}(-\theta_0)$,

$$\hat{K}(\xi) = \lim_{\varepsilon\to 0, N\to\infty} \int_{\varepsilon<R<N} R^{-m}\, Y_{n,m}^{(k)}(-\theta_0)\, e^{-2\pi i(x,\xi)}\, d\xi .$$

We have $(x, \xi) = R\varrho \cos\gamma$ with $\varrho = |\xi|$ and γ denoting the angle between the vectors x and ξ. In the integral herein we replace x by $-x$, θ_0 by θ, and subtitute $\Lambda = \xi/\varrho$ such that

$$\begin{aligned}\hat{K}(\xi) &= \lim_{\varepsilon\to 0, N\to\infty} \int_{\varepsilon<R<N} R^{-m}\, Y_{n,m}^{(k)}(\theta)\; e^{2\pi i\varrho R\cos\gamma}\, dx \\ &= \lim_{\varepsilon\to 0, N\to\infty} \int_{\varepsilon}^{N} \frac{dR}{R} \int_S e^{2\pi i R\varrho\cos\gamma}\, Y_{n,m}^{(k)}(\theta)\, d_\theta S \\ &= \lim_{\varepsilon'\to 0, N'\to\infty} \int_{\varepsilon'}^{N'} \frac{dt}{t} \int_S e^{it\cos\gamma}\, Y_{n,m}^{(k)}(\theta)\, d_\theta S ;\end{aligned} \tag{2}$$

here we put $\varepsilon' = 2\pi\varrho\varepsilon$ and $N' = 2\pi\varrho N$.

Now we make use of the following theorem in a book by SOBOLEV [2].

In the integral equation

$$\varphi(\Lambda) = \lambda \int_S P(\cos\gamma)\, \varphi(\theta)\, d_\theta S \tag{3}$$

we assume that the kernel $P(\cos\gamma)$ is squared summable and admits of an expansion

$$P(\cos\gamma) = \sum_{n=0}^{\infty} a_n C_n^{(m/2-1)}(\cos\gamma)$$

with respect to the Gegenbauer polynomials $C_n^{(m/2-1)}$. Then the characteristic values of (3) are given by the formula

$$\lambda_n = a_n^{-1} (4\pi)^{-m/2}\, \Gamma\left(\frac{m}{2} - 1\right) (m + 2n - 2); \tag{4}$$

eigenfunctions to a characteristic value λ_n are all spherical functions of order n.

For the function $\exp(it\cos\gamma)$ the corresponding coefficients are known (cf. WATSON [1], Section 11.5, eq. (2)),

$$a_n = t^{1-m/2} 2^{m/2-1} \Gamma\left(\frac{m}{2} - 1\right) i^n \left(\frac{m}{2} + n - 1\right) J_{\frac{m}{2}+n-1}(t),$$

and the theorem cited above yields

$$\int_S e^{it\cos\gamma}\, Y_{n,m}^{(k)}(\theta)\, d_\theta S = t^{1-m/2} (2\pi)^{m/2}\, i^n J_{\frac{m}{2}+n-1}(t)\; Y_{n,m}^{(k)}(\Lambda).$$

Now, it follows from (2)

$$\begin{aligned}\hat{K}(\xi) &= (2\pi)^{m/2}\, i^n \lim_{\varepsilon'\to 0, N'\to\infty} \int_{\varepsilon'}^{N'} J_{\frac{m}{2}+n-1}(t)\, t^{-m/2}\, dt \\ &= (2\pi)^{m/2}\, i^n\, Y_{n,m}^{(k)}(\Lambda) \int_0^\infty t^{-m/2} J_{\frac{m}{2}+n-1}(t)\, dt .\end{aligned} \tag{5}$$

Due to a well-known formula in the theory of Bessel functions (cf. WATSON [1], Section 13.24),

$$\int_0^\infty t^{-m/2} J_{\frac{m}{2}+n-1}(t)\, dt = \frac{\Gamma(n/2)}{2^{m/2}\Gamma((m+n)/2)}$$

and, therefore,

$$\hat{K}(\xi) = \frac{\pi^{m/2} \mathrm{i}^n \Gamma(n/2)}{\Gamma((m+n)/2)} Y_{n,m}^{(k)}(\Lambda). \tag{6}$$

2.2. Now let us consider the operator (1.12). We expand its characteristic in a series with respect to spherical functions,

$$f(x, \theta) = \sum_{n=1}^{\infty} \sum_{k=1}^{k_{n,m}} a_n^{(k)}(x)\, Y_{n,m}^{(k)}(\theta), \tag{7}$$

where

$$k_{n,m} = \frac{(2n+m-2)\,(n+m-3)!}{(m-2)!\, n!}$$

denotes the number of linearly independent spherical functions of order n; the zero term vanishes due to condition (1.2), Chap. IX. The formulae (6) and (1.13) enable us to expand the symbol of the operator (1.12) into a series with respect to spherical functions,

$$\Phi_k(x, \Lambda) = \sum_{n=0}^{\infty} \sum_{k=1}^{k_{n,m}} \gamma_{m,n} a_n^{(k)}(x)\, Y_{n,m}^{(k)}(\Lambda); \tag{8}$$

where $a_0^{(1)}(x) = a(x)$, $\gamma_{m,0} = 1$, and

$$\gamma_{m,n} = \frac{\pi^{m/2} \mathrm{i}^n \Gamma(n/2)}{\Gamma((m+n)/2)}. \tag{9}$$

The formulae (8) and (9) were obtained by Mikhlin [4, 5] in case $m = 2$ and generalized by Giraud [1] for an arbitrary dimension m. Giraud did not publish a proof of those formulae, it was given by Mikhlin [8].

Remark. Bochner [1] obtained a formula for the Fourier transform of functions $P(x)\,\varphi(|x|)$ where $P(x)$ is a homogeneous harmonic polynomial, under the assumption that the integral

$$\int_{\mathbb{R}^m} P(x)\,\varphi(|x|)\, \mathrm{e}^{-2\mathrm{i}\pi(x,\xi)}\, \mathrm{d}x$$

is absolutely convergent. The function (1) does not satisfy that condition, nevertheless, (6) can be obtained formally as a particular case of Bochner's formula.

§ 3. Transformation of the symbol under a substitution of variables

3.1. Let us assume that the singular operator (1.12) undergoes a transformation of y- and likewise of x-coordinates. The transformation is assumed to be 1-to-1, to have infinity as a fixed point, to be continuously differentiable with the first derivatives satisfying a Hölder condition, and to have a Jacobian which is bounded from below and from above by positive constants. The question is how does the symbol change under such a transformation.

The problem posed needs some further specifications. For the sake of definiteness we assume that the characteristic $f(x, \theta)$ of the singular integral in (1.12) satisfies a Lipschitz condition with respect to the variable θ with some fixed exponent and a constant not dependiug on x. Furthermore, let the characteristic be bounded on the set $\mathbb{R}^m \times S$. Suppose that the transformation of coordinates has the form $x = \varphi(x')$, $y = \varphi(y')$, and denote by $A = A_{x'}$ its Jacobian computed at the point x. Then, in a certain vicinity of that point, $y = x + A_{x'}(y' - x') + \psi(x', y')$ where $\psi(x', y') = O(|x' - y'|^\alpha)$, $\alpha > 1$. Thence, as it is easy to see, the singular kernel

$$K(x, x - y) = |x - y|^{-m} f(x, \theta), \qquad \theta = (y - x)/r,$$

can be written as $K(x, x - y) = K_1(x', x' - y') + K_2(x', x' - y')$ where

$$K_1(x', x' - y') = |A(x' - y')|^{-m} f\left(\varphi(x'), \frac{A(y' - x')}{|A(y' - x')|}\right),$$

$$K_2(x', x' - y') = \begin{cases} O(|x' - y'|^{-m+\alpha}), & |y' - x'| \to 0, \\ O(|x' - y'|^{-m}), & |y' - x'| \to \infty. \end{cases}$$

For $z \in \mathrm{R}^m$ with $z = \varphi(z')$ we put $g(z') = |z|^2 = (Az', Az') + o(|z'|^2)$. The transformation of coordinates takes the singular integral

$$\int_{\mathrm{R}^m} K(x, x - y)\, u(y)\, dy = \lim_{\varepsilon \to 0} \int_{|x-y| > \varepsilon} K(x, x - y)\, u(y)\, dy$$

to the sum

$$\lim_{\varepsilon \to 0} \int_{g(x'-y') > \varepsilon^2} K_1(x', x' - y')\, J(y')\, u(\varphi(y'))\, dy' + \int_{\mathrm{R}^m} K_2(x', x' - y')\, J(y')\, u(\varphi(y'))\, dy',$$

where $J(y')$ denotes the determinant of the Jacobian of the mapping $y = \varphi(y')$. The limit necessarily exists. Representing the kernel $K_1(x, x - y)$ as

$$K_1(x', x' - y') = |x' - y'|^{-m} f'(x', \theta'), \qquad \theta' = \frac{y' - x'}{|y' - x'|}, \tag{1}$$

(what, obviously, is possible) the new characteristic satisfies condition (1.2), Chap. IX, as shown there.

Under the transformation of coordinates the operator (1.12) becomes

$$b(x')\, u(\varphi(x')) + \int_{\mathrm{R}^m} K_1(x', x' - y')\, u(\varphi(y'))\, J(y')\, dy + \int_{\mathrm{R}^m} K_2(x', x' - y')\, u(\varphi(y'))\, J(y')\, dy' \tag{2}$$

where, due to (1.6), Chap. IX,

$$b(x') = a(\varphi(x')) + J(x') \int_{|z|=1} K_1(x', -z) \ln \frac{|z|}{|Az|}\, dS. \tag{3}$$

Multiplying the operator (2) by $J(x')$ and putting $u'(x') := J(x')\, u(\varphi(x'))$ we give the operator considered the form

$$b(x')\, u'(x') + \int_{\mathrm{R}^m} J(x')\, K_1(x', x' - y')\, u'(y')\, dy' + \int_{\mathrm{R}^m} J(x')\, K_2(x', x' - y')\, u'(y')\, dy'. \tag{4}$$

3.2. Later on we shall see that we may assign a symbol to the non-singular integral operator with the kernel $K_2(x', x' - y')$ under rather general conditions, and that symbol is identically equal to zero. Therefore, the problem of transforming the symbol of the operator (1.12) under a substituting of variables can be considered as the problem of relating the symbol of the operator (1.12) to that of the operator K' where

$$(K'u')(x') = b'(x')\, u'(x') + \int_{\mathrm{R}^m} J(x')\, K_1(x', x' - y')\, u'(y')\, dy'. \tag{5}$$

Due to (1.13) we have for the symbol of the operator (5)

$$(\mathrm{Smb}\, K')(x', \xi) := \Phi(x, \xi) := b(x') + J(x') \int_{\mathrm{R}^m} K_1(x', z)\, \mathrm{e}^{-2\pi \mathrm{i}(z,\xi)}\, dz. \tag{6}$$

The symbol of the operator (1.12) denoted by $\Phi(x, \xi)$ is defined as

$$\Phi(x, \xi) = a(x) + \lim_{\varepsilon\to 0, N\to\infty} \int_{\varepsilon<|z|<N} K(x, z)\, e^{-2\pi i(z,\xi)}\, dz.$$

We introduce new variables by means of the formulas $z = Az_0$, $\xi = A\xi_0$. Additionally, we put $A^*A\xi_0 = \zeta$ where A^* means the adjoint matrix to A. Furthermore, let $g_0(z_0) = (Az_0, Az_0)$. Then

$$\begin{aligned}\Phi(x, \xi) &= a(x) + \lim_{\varepsilon\to 0, N\to\infty} J(x') \int_{\varepsilon^2<g_0(z_0)<N^2} K_1(x', z_0)\, e^{-2\pi i(Az_0, A\xi_0)}\, dz_0 \\ &= a(x) + J(x') \lim_{\varepsilon\to 0, N\to\infty} \int_{\varepsilon^2<g_0(z)<N^2} K_1(x', z)\, e^{-2\pi i(z,\zeta)}\, dz \\ &= a(x) + J(x') \lim_{\varepsilon\to 0} \int_{(g_0(z)>\varepsilon^2)\cap(|z|<1)} K_1(x', z)\, e^{-2\pi i(z,\zeta)}\, dz \\ &\quad + J(x') \lim_{N\to\infty} \int_{(g_0(z)<N^2)\cap(|z|>1)} K_1(x', z)\, e^{-2\pi i(z,\zeta)}\, dz. \qquad (7)\end{aligned}$$

In the first integral we introduce spherical coordinates $R = |z|$, $\theta' = z/R$ such that $K_1(x', z) = R^{-m} f'(x', \theta')$. Further, let the surface $g_0(z) = t^2$ in spherical coordinates have the equation $R = w(\theta', t)$; note that

$$w(\theta', t) = \frac{t}{\sqrt{(A\theta', A\theta')}} = \frac{t}{\sqrt{g_0(\theta')}}. \qquad (8)$$

The first summand in (7) becomes equal to

$$J(x') \lim_{\varepsilon\to 0} \int_S dS \int_{w(\theta',\varepsilon)}^{1} f'(x', \theta')\, e^{-2\pi i(z,\zeta)} \frac{dR}{R}$$

$$= J(x') \int_S f'(x', \theta')\, dS \int_0^1 [e^{-2\pi i(z,\zeta)} - 1] \frac{dR}{R} + \lim_{\varepsilon\to 0} J(x') \int_S f'(x', \theta') \ln \frac{1}{w(\theta', \varepsilon)}\, dS. \qquad (9)$$

As we saw before, the characteristic $f'(x', \theta')$ satisfies condition (1.2), Chap. IX. Thence, the first integral right in (9) equals the singular integral

$$J(x') \int_{|z|<1} K_1(x', z)\, e^{-2\pi i(z,\zeta)}\, d\zeta \qquad (10)$$

(the non-integrable singularity of the integrand is in the origin). The second integral can be represented due to the same condition (1.2), Chap. IX, as

$$\lim_{\varepsilon\to 0} J(x') \int_S f'(x', \theta') \ln \frac{\varepsilon}{w(\theta', \varepsilon)}\, dS.$$

The quotient $\varepsilon/w(\theta', \varepsilon) = \sqrt{g_0(\theta')} = |z|/|Az|$ does not depend on ε, and the lim-symbol can be omitted. The last integral is equal to

$$J(x') \int_{|z|=1} f'(x', \theta') \ln \frac{|z|}{|Az|}\, dS = b(x') - a(\varphi(x')). \qquad (11)$$

Equations (10) and (11) together imply that the first limit on the right-hand side in (7) is equal to

$$b(x') - a(\varphi(x')) + J(x') \int_{|z|=1} K_1(x', z)\, e^{-2\pi i(z,\zeta)}\, dz. \qquad (12)$$

Now let us consider the second limit in (7) which can be represented as

$$J(x') \lim_{N\to\infty} \int_{1<|z|<N} K_1(x', z)\, e^{-2\pi i(z,\zeta)}\, dz.$$

Let $\delta > 0$ be an arbitrary number, choose a number N_1 large enough such that

$$\left| J(x') \int_{|z|>N_1} K_1(x', z)\, \mathrm{e}^{-2\pi i(z,\zeta)}\, \mathrm{d}z \right| < \frac{\delta}{2}. \tag{13}$$

By λ_1 we denote the smallest eigenvalue of the quadratic form $g_0(z)$, then (8) implies $w(\theta, N) \leq \lambda_1^{-1/2} N$. For N_1 we require additionally $\lambda_1^{-1/2} N \leq N_1 \leq 2\lambda_1^{-1/2} N$ (obviously, this is possible for sufficiently large N) and consider the integral

$$J(x') \int_{(|z|<N_1)\cap(g_0(z)>N^2)} K_1(x', z)\, \mathrm{e}^{-2\pi i(z,\zeta)}\, \mathrm{d}z = J(x') \int_S f'(x', \theta')\, \mathrm{d}S \int_{w(\theta',N)}^{N_1} \mathrm{e}^{-2\pi i(z,\zeta)} \frac{\mathrm{d}R}{R}.$$

Substituting z by $N_1 z$ we bring that integral to the form

$$J(x') \int_S f(x', \theta')\, \mathrm{d}S \int_{w(\theta',N)/N_1}^{1} \mathrm{e}^{-2\pi i(z,\zeta)} \frac{\mathrm{d}R}{R}.$$

We remove from the domain of integration all points lying in a cone around the axis which is orthogonal to the vector ζ and with a small apex angle and apply to the remaining part of the bounded domain $w(\theta', N)/N_1 < R < 1$ the Riemann theorem. Then we find that the last integral tends to zero as $N \to \infty$ and, therefore, the inequality

$$\left| J(x') \int_{(|z|<N_1)\cap(g_0(z)>N^2)} K_1(x', z)\, \mathrm{e}^{-2\pi i(z,\zeta)}\, \mathrm{d}z \right| < \frac{\delta}{2} \tag{14}$$

holds for sufficiently large N. Equations (13) and (14) together yield

$$\lim_{N\to\infty} J(x') \int_{g_0(z)>N^2} K_1(x', z)\, \mathrm{e}^{-2\pi i(z,\zeta)}\, \mathrm{d}z = 0.$$

Thus, it is clear that the second limit on the right-hand side of (7) is equal to

$$J(x') \int_{|z|>1} K_1(x', z)\, \mathrm{e}^{-2\pi i(z,\zeta)}\, \mathrm{d}z. \tag{15}$$

Now, the formulae (4), (6), (12), and (15) imply $\Phi'(x', \zeta) = \Phi(x, \xi)$, and because of $\zeta = A^*A\xi' = A^*\xi$ we obtain finally

$$\Phi'(x', \zeta) = \Phi(\varphi(x'), (A^*)^{-1}\zeta). \tag{16}$$

Equation (16) gives the complete answer to the problem of transforming the symbol of a singular operator under a substitution of variables with infinity as a fixed point. If the matrix A is an orthogonal one, i.e. $(A^*)^{-1} = A$, we have $\Phi'(x', \zeta) = \Phi(\varphi(x'), A\zeta)$. And that means: If in the singular integral an orthogonal coordinate transformation is performed then the independent variables in the symbol undergo the same transformation.

Equation (16) implies, in particular, that the range of the symbol remains invariant under a coordinate transformation. In case $m = 2$ that assertion was proved earlier and in another way by ITSKOVICH [1].

§ 4. Transformation of the symbol under inversion

4.1. The previous section was devoted to the problem how a coordinate transformation with infinity as a fixed point affects on the symbol. On the other hand, it may be of interest the case where infinity is taken e.g. to the origin. Let us consider the simplest example of such a transformation, namely the inversion.

To that aim we consider a singular operator of the form (1.12) with a characteristic which is continuously differentiable with respect to the Cartesian coordinates of the point $\theta \in S$. First, we will find out how the integral operator denoted by K transforms under the inversion

$$x = x'/|x'|^2, \qquad y = y'/|y'|^2. \tag{1}$$

The similarity of the triangles $0xy$ and $0y'x'$ (see Fig. 10) yields $|x|/|y'| = |y - x|/|y' - x'|$ or with the notations $r = |y - x|$, $r' = |y' - x'|$,

$$r = \frac{r'\,|x|}{|y'|} = \frac{r'}{|x'|\,|y'|}.$$

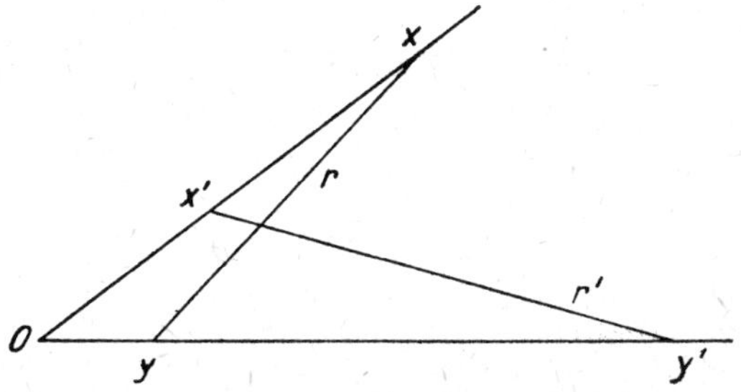

Fig. 10

Let $\theta = (y - x)/r$, $\theta' = (y' - x')/r'$. Simple calculations give

$$\theta = \theta' - 2x_0(x_0, \theta') + O(r'), \qquad x_0 = x'/|x'|. \tag{2}$$

and thence

$$f\left(\frac{x'}{|x'|^2}, \theta\right) = f\left(\frac{x'}{|x'|^2}, \theta' - 2x_0(x_0, \theta')\right) + r' f_0(x', r', \theta') \tag{3}$$

where the function $f_0(x', r', \theta')$ is bounded for every r' and of order $O(1/r')$ as $r' \to \infty$.

The expression Ku can be rewritten as follows,

$$(Ku)\,(x) = a\left(\frac{x'}{|x'|^2}\right) u\left(\frac{x'}{|x'|^2}\right) + \lim_{\varepsilon \to 0} \int_{r' > \varepsilon |x'|\,|y'|} r^{-m} f\left(\frac{x'}{|x'|^2}, \theta'\right) u\left(\frac{y'}{|y'|^2}\right) \frac{dy'}{|y'|^{2m}}.$$

The domain of integration herein is bounded by the sphere $r' = \varepsilon\,|x'|\,|y'|$, and from this equation we find $r' = \varepsilon\,|x'|^2 + O(\varepsilon^2)$ or $\lim_{\varepsilon \to 0} r'/\varepsilon = |x'|^2$. That limit does not depend on θ, and by (1.6), Chap. IX,

$$(Ku)\,(x) = a\left(\frac{x'}{|x'|^2}\right) u\left(\frac{x'}{|x'|^2}\right) + \int_{\mathrm{R}^m} r^{-m} f\left(\frac{x'}{|x'|^2}, \theta\right) u\left(\frac{y'}{|y'|^2}\right) \frac{dy'}{|y'|^{2m}}. \tag{4}$$

Put

$$a\left(\frac{x'}{|x'|^2}\right) = a'(x'), \qquad |x'|^{-m}\, u\left(\frac{x'}{|x'|^2}\right) = u'(x'),$$

$$f\left(\frac{x'}{|x'|^2}, \theta' - 2x_0(x_0, \theta')\right) = f'(x', \theta'),$$

then

$$|x'|^{-m}(Ku)(x) = a'(x')\,u'(x') + \int_{\mathrm{R}^m} \frac{f'(x',\theta')}{r'^m}\,u'(y')\,dy' + \int_{\mathrm{R}^m} \frac{f_0(x',r',\theta')}{r'^{m-1}}\,u'(y')\,dy'. \quad (5)$$

4.2. For $u \in \mathbf{A}_\alpha$ (cf. § 5, Chap. IX), $\alpha > 0$, the second integral in (5) is absolutely convergent. We show that the characteristic $f'(x', \theta')$ satisfies the condition (1.2), Chap. IX, and, therefore, the first integral in (5) does exist. To that aim we consider the transformation in R^m defined by the formula

$$\zeta = \eta - 2x_0(x_0, \eta) \quad (6)$$

where x_0 is some fixed unit vector. Squaring both sides of (6) we obtain

$$|\zeta|^2 = (\eta - 2x_0(x_0,\eta), \eta - 2x_0(x_0,\eta)) = |\eta|^2 - 4(x_0,\eta)^2 + 4(x_0,\eta)^2\,|x_0|^2 = |\eta|^2.$$

This means that the transformation (6) is an isometry. Multiplying (6) by x_0 we find $(\zeta, x_0) = -(\eta, x_0)$ and thence $\eta = \zeta - 2x_0(x_0, \zeta)$. Inserting this into (6) we obtain an identity. Therefore, the isometry (6) is invertible for any $\zeta \in \mathrm{R}^m$, it is, thence, unitary in R^m and leaves invariant the unit sphere S and also its measure element. Put $\theta' - 2x_0(x_0, \theta') = \theta_0$ then we obtain

$$\int_S f'(x', \theta')\,\mathrm{d}_{\theta'}S = \int_S f(x, \theta_0)\,\mathrm{d}_{\theta_0}S = 0,$$

which was to be demonstrated.

Let us introduce the notations

$$K'(x', x' - y') = \frac{f'(x', \theta')}{r'^m}, \quad (7)$$

$$(K'u')(x') = a'(x')\,u'(x') + \int_{\mathrm{R}^m} K'(x', x' - y')\,u'(y')\,dy'. \quad (8)$$

4.3. Speaking about the problem of transforming the symbol of the operator (1.12) under inversion we mean the relation between the symbols of the operators (8) and (1.12). The symbol of the operator (8) is given by

$$(\mathrm{Smb}\,K')(x', \xi) = a'(x') + \int_{\mathrm{R}^m} \mathrm{e}^{-2\pi\mathrm{i}(z,\xi)}\,K'(x', z)\,dz. \quad (9)$$

We denote the integral in (9) by $\Phi'(x', \xi)$ and shall compute it now. It can be represented as

$$\Phi'(x', \xi) = \int_{\mathrm{R}^m} \mathrm{e}^{-2\pi\mathrm{i}(z,\xi)}\,f\left(\frac{x'}{|x'|^2}, -\theta' + \frac{2x'}{|x'|^2}(x', \theta')\right)\frac{dz}{|z|^m}, \qquad \theta' = z/|z|.$$

We introduce spherical coordinates centered at the origin and put $z = R\theta'$, $R = |z|$. The the last integral becomes

$$\Phi'(x', \xi) = \int_S \mathrm{d}S \int_0^\infty \mathrm{e}^{-2\pi\mathrm{i}(z,\xi)}\,f\left(x, -\frac{1}{R}(z - 2x(x', z))\right)\frac{\mathrm{d}R}{R}.$$

Now, we introduce the new integration variable $\zeta = z - 2x(x', z) = z - 2x_0(x_0, z)$ such that $z = \zeta - 2x_0(x_0, \zeta)$, $|\zeta| = |z| = R$, and with $\zeta = R\theta$ we have $\theta = R^{-1}[z - 2x_0(x_0, z)]$. Hence,

$$\Phi'(x', \xi) = \int_S \mathrm{d}S \int_0^\infty \mathrm{e}^{-2\pi\mathrm{i}(\zeta - 2x_0(x_0,\zeta),\xi)}\,f(x, -\theta)\frac{\mathrm{d}R}{R} = \int_{\mathrm{R}^m} \mathrm{e}^{-2\pi\mathrm{i}(\zeta, \xi - 2x_0(x_0,\xi))}\,f(x, -\theta)\frac{\mathrm{d}\zeta}{|\zeta|^m}.$$

Denoting the symbol of the integral in (1.12) by $\Phi(x, \xi)$ we obtain $\Phi'(x', \xi) = \Phi(x, \xi - 2x_0(x_0, \xi))$ and, finally,

$$(\text{Smb } K')(x', \xi) = (\text{Smb } K)(x, \xi - 2x_0(x_0, \xi)). \tag{10}$$

§ 5. A theorem on the boundedness of the singular operator

5.1. Theorem 5.1. *If the symbol of a singular integral operator does not depend on the pole then that operator is bounded in* $\mathbf{L}_2(\mathbf{R}^m)$ *if and only if its symbol is bounded a.e. on* S. *In that case the norm of the singular operator is equal to the essential supremum of the symbol modulus.*

Remember that a function is called bounded a.e. on a measurable set iff there is a constant c_0 such that this function does not exceed in modulus a.e. that constant. The smallest of all those constants is called *essential supremum* of the function and is denoted by sup ess.

The proof of Theorem 6.1 is almost obvious. If the symbol does not depend on the pole the operator is of the form (1.9), and its symbol can be written as (1.10). Theorem 5.1 is an immediate consequence of the following facts: The operator of Fourier transformation in $\mathbf{L}_2(\mathbf{R}^m)$ is unitary, the operator of multiplication with a bounded a.e. function in that space is bounded, and this operator has a norm which is equal to the essential supremum of the modulus of that function. ▮

Theorem 5.1 implies that the singular operator admits of a norm preserving extension to the whole space $\mathbf{L}_2(\mathbf{R}^m)$ provided that its symbol does not depend on the pole. This extension can be given by (1.11).

Theorem 5.1 was proved first by MIKHLIN in [15] for a dimension $m = 2$ and in [16] for an arbitrary dimension.

5.2. Theorem 5.1 is easily extended to singular operators the integral of which is taken not over the whole space $\mathbf{R}^m$ but over some measurable subset Ω. Indeed, let

$$v(x) = (A_\Omega u)(x) := au(x) + \int_\Omega \frac{f(\theta)}{r^m} u(y)\, dy, \qquad x \in \Omega, \tag{1}$$

and $u \in \mathbf{L}_2(\Omega)$. We extend the functions $u(x)$ and $v(x)$ to the whole space $\mathbf{R}^m$ setting $u(x) = 0$, $x \in \mathbf{R}^m \setminus \Omega$, and

$$v(x) = au(x) + \int_{\mathbf{R}^m} \frac{f(\theta)}{r^m} u(y)\, dy. \tag{2}$$

Let the symbol of the operator (2) (which we shall call also though not quite correctly *symbol* of the operator (1)) be bounded on S a.e. Due to Theorem 5.1,

$$\int_{\mathbf{R}^m} |v(x)|^2\, dx \leq \sup\text{ess}\, |\Phi(\theta)|^2 \int_{\mathbf{R}^m} |u(x)|^2\, dx.$$

The integrals herein are replaced by integrals taken over Ω. The value of the integral right does not change and the integral left becomes not greater such that the last inequality remains to hold true,

$$\int_\Omega |v(x)|^2\, dx \leq \sup\text{ess}\, |\Phi(\theta)|^2 \int_\Omega |u(x)|^2\, dx.$$

But that means that the operator (1) is bounded in $\mathbf{L}_2(\Omega)$ and its norm does not exceed the essential supremum of its symbol modulus.

5.3. Now let us consider an example which will be used in the following section. The polyharmonic equation $(-1)^q \Delta_x^q u = \delta(x-y)$ with δ being the Dirac distribution has the fundamental solution

$$\Gamma(x,y) = \begin{cases} \dfrac{(-1)^{(m/2)-1}\, r^{2q-m} \ln r}{\Gamma(q)\,\Gamma\left(q-\dfrac{m}{2}+1\right) 2^{2q-1} \pi^{m/2}}, & m \text{ even and } 2q \geqq m, \\[2ex] \dfrac{(-1)^q\, \Gamma(m/2-q)\, r^{2q-m}}{\Gamma(q)\, 2^{2q} \pi^{m/2}}, & \text{otherwise}. \end{cases} \tag{3}$$

Let Ω be a finite domain in the space $\mathbb{R}^m$. Consider the integral

$$\psi(x) = \int_\Omega \Gamma(x,y)\, f(y)\, dy, \tag{4}$$

which may be called *polyharmonic potential*; the function $f(y)$ is called density of that potential. It is well-known that the density satisfies the equation

$$(-\Delta)^q\, \psi = f(x). \tag{5}$$

provided that $f \in \mathbf{Lip}_\alpha(\Omega)$, $\alpha > 0$.

We shall prove the following fact: If $f \in \mathbf{L}_2(\Omega)$ then $\psi \in \mathbf{W}_2^{(2q)}(\Omega)$, and the function ψ satisfies (5) a.e. in Ω.

Form the Sobolev mean function $f_h(x)$ (cf. Sobolev [3]) of the function $f(x)$ such that $\|f - f_h\|_{\mathbf{L}_2(\Omega)} \to 0$ as $h \to 0$. Consider the potential

$$\psi(x,h) = \int_\Omega \Gamma(x,y)\, f_h(y)\, dy. \tag{6}$$

The function $\Gamma(x,y)$ and its derivatives of order $< 2q$ are weakly singular kernels (in some cases these kernels are simply bounded), and the integral operators with these kernels are completely continuous and, the more, bounded in $\mathbf{L}_2(\Omega)$ and $\mathbf{C}(\overline{\Omega})$. Consequently, the derivatives

$$D^\alpha \psi(x,h) = \int_\Omega D_x^\alpha \Gamma(x,y)\, f_h(y)\, dy, \qquad |\alpha| < 2q, \tag{7}$$

exist and are continuous in Ω where $\|D^\alpha \psi(\cdot, h) - D^\alpha \psi\|_{\mathbf{L}_2(\Omega)} \to 0$ as $h \to 0$. Since the operator of generalized differentiation is closed we have $\psi \in \mathbf{W}_2^{(2q-1)}(\Omega)$.

The derivatives of Γ of order $2q$ are singular kernels the characteristics of which are infinitely differentiable with respect to the Cartesian coordinates of the point θ and do not depend on the pole. The results of § 7, Chap. IX, imply that all derivatives $D^\beta \psi(x,h)$, $|\beta| = 2q$, exist, and these derivatives can be expressed as singular operators in the form (2) with the density f_h. Now, the results of subsection 5.2 imply that the limits of $D^\beta \psi(x,h)$, $|\beta| = 2q$, as $h \to 0$ in the $\mathbf{L}_2(\Omega)$ metric do exist. That means, since the operator of generalized differentiation is closed, the potential (4) has all generalized derivatives of order $2q$ which are squared summable, and these derivatives coincide with the limits mentioned above, i.e. $\psi \in \mathbf{W}_2^{(2q)}(\Omega)$. The potential $\psi(x,h)$ satisfies (5), $(-\Delta)^q\, \psi(x,h) = f_h(x)$. Let $h \to 0$ here, then we obtain that the potential (4) satisfies (5). ∎

§ 6. On series with respect to spherical functions

6.1. Theorem 6.1. *Let l be a natural number. Then the sets $\mathbf{W}_2^{(l)}(S)$ and $D(\delta^{l/2})$ consist of one and the same elements (δ denotes here the Beltrami operator).*

Equations (4.10) and (4.12), Chap. VIII, allow to formulate Theorem 6.1 equivalently as follows:

Assume that a function $f(\theta)$ admits of an expansion

$$f(\theta) = \sum_{n=0}^{\infty} \sum_{k=1}^{k_{n,m}} a_n^{(k)} Y_{n,m}^{(k)}(\theta), \tag{1}$$

where $Y_{n,m}^{(k)}(\theta)$ are spherical functions orthonormal in the space $\mathbf{L}_2(S)$ and let l be a natural number. Then the function $f(\theta)$ belongs to the space $\mathbf{W}_2^{(l)}(S)$ if and only if

$$a_n^{(k)} = n^{-l} \beta_n^{(k)}, \qquad n \geqq 1 \tag{2}$$

where

$$\sum_{n=1}^{\infty} \sum_{k=1}^{k_{n,m}} |\beta_n^{(k)}|^2 < \infty. \tag{3}$$

Proof. *Necessity.* If $f(\theta) \in \mathbf{W}_2^{(l)}(S)$ and $q = l/2$ then $\delta^q f \in \mathbf{L}_2(S)$, and there is an expansion

$$\delta^q f = \sum_{n=1}^{\infty} \sum_{k=1}^{k_{n,m}} \gamma_n^{(k)} Y_{n,m}^{(k)}(\theta), \qquad \sum_{n=1}^{\infty} \sum_{k=1}^{k_{n,m}} |\gamma_n^{(k)}|^2 < \infty, \tag{4}$$

which is convergent in the $\mathbf{L}_2(S)$ metric. According to equations (4.6) and (4.8), Chap. VIII, it is now sufficient to put

$$\beta_n^{(k)} = \frac{n^q}{(n + m - 2)^q} \gamma_n^{(k)}. \tag{5}$$

Sufficiency. Put

$$\alpha_n^{(k)} = \frac{(n + m - 2)^q}{n^q} \beta_n^{(k)}, \qquad q = l/2, \tag{6}$$

$$\varphi(\theta) = \sum_{n=1}^{\infty} \sum_{k=1}^{k_{n,m}} \alpha_n^{(k)} Y_{n,m}^{(k)}(\theta); \tag{7}$$

obviously, $\varphi \in \mathbf{L}_2(S)$.

In the subspace $\tilde{\mathbf{L}}_2(S)$ invariant under the Beltrami operator the spectrum of this operator consists of the numbers $n(n + m - 2)$, $n = 1, 2, \ldots$ Hence, the Beltrami operator is positive definite in $\tilde{\mathbf{L}}_2(S)$, and the equation

$$\delta^q F = \varphi \tag{8}$$

has a unique solution in $\tilde{\mathbf{L}}_2(S)$. It is easy to see that $f(\theta) = F(\theta) + \text{const}$, and it remains to show the relation $F \in \mathbf{W}_2^{(l)}(S)$. The demonstration will be performed for l even and odd separately. Let $l = 2s$. We want to compute $(-\Delta)^s F$. Since $F(\theta)$ does not depend on ϱ we have $-\Delta F = \varrho^{-2}\, \delta F$. Consequently,

$$(-\Delta)^s F = \sum_{j=1}^{s} p_j(\varrho)\, \delta^j F,$$

where $p_j(\varrho)$ are certain polynomials in ϱ^{-1}. Now, equations (4.6), (4.8), Chap. VIII, and (7), (8) of the present section easily imply the relation

$$\delta^j F = \sum_{n=1}^{\infty} \sum_{k=1}^{k_{n,m}} \frac{\alpha_n^{(k)}}{[n(n + m - 2)]^{s-j}} Y_{n,m}^{(k)}(\theta). \tag{9}$$

Due to the conditions (2) and (3) the series

$$\sum_{n=1}^{\infty} \sum_{k=1}^{k_{n,m}} \frac{|\alpha_n^{(k)}|^2}{[n(n+m-2)]^{2(s-j)}}, \qquad j = 1, 2, \ldots, s,$$

converge and, therefore, $\delta^j F \in \mathbf{L}_2(S)$, $j = 1, 2, \ldots, s$. Put

$$\sum_{j=1}^{s} p_j(\varrho)\, \delta^j F =: \Phi(x); \tag{10}$$

obviously, $\Phi \in \mathbf{L}_2(\Sigma)$ where Σ denotes here the ball layer $\varrho_1 \leqq \varrho \leqq \varrho_2$. Let $\Gamma(x, y)$ be the fundamental solution to the equation $(-\Delta)^s u = 0$. Then $F(\theta)$ admits of a representation

$$F(\theta) = \int_{\Sigma} \Gamma(x, y)\, \Phi(y)\, dy + F_0(x) \tag{11}$$

with a function $F_0(x)$ which is polyharmonic in Σ. As it is well-known, the function $F_0(x)$ has derivatives of arbitrary order in every interior subdomain of Σ and, in particular, in every smaller ball layer $\Sigma' \subset \Sigma$. Concerning the first summand in (11), it belongs due to the results of subsection 5.3 to the space $\mathbf{W}_2^{(2s)}(\Sigma)$. But then $F \in \mathbf{W}_2^{(2s)}(\Sigma)$ or what is the same, $F \in \mathbf{W}_2^{(l)}(\Sigma)$.

Now, let $l = 2s + 1$. We seek a function $\varphi_1 \in \tilde{\mathbf{L}}_2(S)$ satisfying the equation $\delta^{1/2}\varphi_1(\theta) = \varphi(\theta)$. Such a function can be represented as a series

$$\varphi_1(\theta) = \sum_{n=1}^{\infty} \sum_{k=1}^{k_{n,m}} \frac{\alpha_n^{(k)}}{\sqrt{n(n+m-2)}}\, Y_{n,m}^{(k)}(\theta).$$

Due to (4.9), Chap. VIII, that function φ_1 belongs to $\mathbf{W}_2^{(1)}(S)$. If the function $F(\theta)$ satisfies the eq. (8) then

$$(-\Delta)^s F = \sum_{j=1}^{s} p_j\,(\varrho)\, \delta^j F = \Phi_1(x) \tag{12}$$

holds true and you can easily see that $\Phi_1(x) \in \mathbf{W}_2^{(1)}(S)$. As we did before, we have

$$F(\theta) = \int_{\Sigma} \Gamma(x, y)\, \Phi_1(y)\, dy + F_0(x). \tag{13}$$

Similarly, the polyharmonic function $F_0(x)$ has continuous derivatives of arbitrary order in any smaller ball layer Σ'. Concerning the integral in (13), it belongs due to the results of subsection 5.3 in any case to the space $W_2^{(2s)}(\Sigma)$. We show that it also belongs to the space $W_2^{(2s+1)}(\Sigma)$. In order to do so we compute the first derivatives of the integral mentioned which will be denoted by $F_1(x)$. We obtain

$$\frac{\partial F}{\partial x_j} = \int_{\Sigma} \frac{\partial \Gamma(x, y)}{\partial x_j}\, \Phi_1(y)\, dy = -\int_{\Sigma} \frac{\partial \Gamma(x, y)}{\partial y_j}\, \Phi_1(y)\, dy.$$

Now, we show that the last integral can be taken by parts. Eq. (12) defines the function $\Phi_1(x)$ everywhere in $\mathbb{R}^m$ except $x = 0$ and $x = \infty$, in particular, that function is defined in a ball layer $\Sigma_2 \supset \Sigma$. We form the mollification $\Phi_{1h}(x)$ of it in that ball layer. Then $\Phi_{1h}(x) \to \Phi_1(x)$ as $h \to 0$ in the metric of the space $\mathbf{W}_2^{(1)}(\Sigma)$, and this implies $\Phi_{1h}(x) \to \Phi_1(x)$ as $h \to 0$ in the metric of the space $\mathbf{L}_2(\partial\Sigma)$. Put

$$G_j(x, h) = -\int_{\Sigma} \frac{\partial \Gamma(x, y)}{\partial y_j}\, \Phi_{1h}(y)\, dy.$$

Obviously, $G_j(x, h) \to \partial F_1/\partial x_j$ in $\mathbf{L}_2(\partial\Sigma)$ as $h \to 0$. Integrating by parts and letting $h \to 0$ we obtain

$$\frac{\partial F}{\partial x_j} = - \int_{\partial\Sigma} \Gamma(x, y)\, \Phi_1(y)\, \mathrm{d}(\partial\Sigma) + \int_{\Sigma} \Gamma(x, y) \frac{\partial \Phi_1(y)}{\partial y_j} \mathrm{d}y .$$

The first integral right has derivatives of arbitrary order in Σ' and the second one belongs due to the results of subsection 5.1 to the space $\mathbf{W}_2^{(2s)}(\Sigma)$. Thus, we finally arrive at $F(\theta) \in \mathbf{W}_2^{(2s+1)}(\Sigma') = \mathbf{W}_2^{(l)}(S)$. ▮

6.2. Lemma 6.1. *Let α be an arbitrary multi-index. Then the inequality*

$$|D^\alpha Y_{n,m}(\theta)| \leqq C n^{(m/2)-1+|\alpha|} \| Y_{n,m} \|_{\mathbf{L}_2(S)}, \tag{14}$$

holds true where $Y_{n,m}$ stands for an arbitrary spherical function of order n in the m-dimensional space $\mathbf{R}^m$.

In case $|\alpha| = 0$ the Lemma was proved in the book [12] by Mikhlin such that we may restrict ourselves here to the case $|\alpha| \neq 0$.

Let $x \in \mathbf{R}^m$, $\varrho = |x|$, $\theta = x/\varrho$. The product $Q_n(x) = \varrho^n Y_{n,m}(\theta)$ is a harmonic function in x, and by the Green formula

$$\int_{\varrho<1} |\operatorname{grad} Q_n(x)|^2 \,\mathrm{d}x = \int_S \bar{Q}_n \frac{\mathrm{d}Q_n}{\mathrm{d}\varrho} \,\mathrm{d}S$$

or

$$\frac{1}{2n+m-2} \int_S |\operatorname{grad} Q_n|^2 \,\mathrm{d}S = n \int_S |Q_n|^2 \mathrm{d}S .$$

Thence,

$$\left\| \frac{\partial Q_n}{\partial x_j} \right\| \leqq C_1 n \, \|Q_n\| = C_1 n \| Y_{n,m}\|, \qquad j = 1, 2, \ldots, m .$$

The expression $\left.\dfrac{\partial Q_n}{\partial x_j}\right|_{\varrho=1}$ provides an m-dimensional spherical function of order $n-1$; if $|\alpha| = 0$ equation (14) yields

$$| Y_{n,m}(\theta) | \leqq C n^{(m/2)-1} \| Y_{n,m} \|, \tag{15}$$

therefore,

$$\left| \frac{\partial Q_n}{\partial x_j} \right| \leqq C(n-1)^{(m/2)-1} \left\| \frac{\partial Q_n}{\partial x_j} \right\| \leqq C_2 n^{m/2} \| Y_{n,m} \| .$$

But, on the other hand,

$$\frac{\partial Y_{n,m}}{\partial x_j} = \left.\frac{\partial Q_n}{\partial x_j}\right|_{\varrho=1} - n \frac{\partial \varrho}{\partial x_j} Y_{n,m} .$$

Applying (15) once again we obtain

$$\left| \frac{\partial Y_{n,m}}{\partial x_j} \right| \leqq C_3 n^{m/2} \| Y_{n,m} \| ,$$

and this is exactly the inequality (14) for $|\alpha| = 1$. Now the general case follows easily by induction. ▮

Inequality (15) was obtained first by Mikhlin [7], whereas the inequality (14) is due to Calderon and Zygmund [2].

6.3. Theorem 6.2. *Let l be a natural number, $l \geqq m - 1$, and $f(\theta) \in \mathbf{W}_2^{(l)}(S)$. The series* (1) *and all those series which are obtained from* (1) *by differentiating it* $(l - m + 1)$ *times with respect to the Cartesian coordinates of the point θ, converge absolutely and uniformly.*

Proof. Due to (2) and (15) the series

$$C_1 \sum_{n=1}^{\infty} \sum_{k=1}^{k_{n,m}} |\beta_n^{(k)}| \, n^{-(l-(m/2)+1)} \leqq \frac{C_1}{2} \sum_{n=1}^{\infty} \sum_{k=1}^{k_{n,m}} |\beta_n^{(k)}|^2 + \frac{C_1}{2} \sum_{n=1}^{\infty} k_{n,m} n^{-(2l-m+2)}$$

provides a majorant for the series (1). The first series in the right-hand side is convergent. Furthermore, $k_{n,m} = O(n^{m-2})$, and the second series right converges if $2l - 2m + 4 > 1$ for which the condition $l \geqq m - 1$ is sufficient.

In order to prove the assertion concerning the derivatives of the series (1) we make use of the inequality (14). The series which is obtained from (1) by differentiating it r times has the majorant

$$C_2 \sum_{n=1}^{\infty} \sum_{k=1}^{k_{n,m}} |\beta_n^{(k)}| \, n^{-(l-(m/2)+1-r)}.$$

Repeating the foregoing arguments we may convince ourselves that this majorant is convergent if $r \leqq l - m + 1$. ∎

Theorem 6.3. *Let $f(\theta) \in \mathbf{W}_2^{(l)}(S)$. The series which is obtained from the series* (1) *by differentiating it r times ($r \leqq l$) with respect to the Cartesian coordinates of the point θ is convergent in the metric of the space $\mathbf{L}_p(S)$ if $p = 2(m - 1)\,[m - 1 - 2(l - r)]^{-1}$.*

Proof. By Theorem 6.1 the function f belongs to $D(\delta^{l/2})$. If f is represented by the series (1) then

$$\varphi(\theta) = \delta^{l/2} f(\theta) = \sum_{k=1}^{\infty} \sum_{n=1}^{k_{n,m}} n^{l/2}(n + m - 2)^{l/2} \, a_n^{(k)} Y_{n,m}^{(k)}(\theta)$$

holds where this series is convergent in $\mathbf{L}_2(S)$. Therefore, we have with the notations

$$\varphi_N(\theta) = \sum_{n=1}^{N} \sum_{k=1}^{k_{n,m}} n^{l/2}(n + m - 2)^{l/2} \, a_n^{(k)} Y_{n,m}^{(k)}(\theta)$$

and

$$f_N(\theta) = \sum_{n=0}^{N} \sum_{k=1}^{k_{n,m}} a_n^{(k)} Y_{n,m}^{(k)}(\theta)$$

the convergencee $f_N(\theta) \to f(\theta)$ and $\varphi_N(\theta) \to \varphi(\theta)$ in the $\mathbf{L}_2(S)$ metric.

For a given function $\varphi(\theta)$ the function $f(\theta)$ is determined uniquely up to a constant summand such that the derivatives of $f(\theta)$ are determined uniquely. Therefore, $D^l f(\theta)$ can be considered as the value of a certain operator G at $\varphi(\theta)$ in the space $\tilde{\mathbf{L}}_2(S)$ what we shall write as $G\varphi = D^l f$. Due to Theorem 6.1 the operator G is defined on the whole space $\tilde{\mathbf{L}}_2(S)$; we show it to be closed. Let $\Phi_\nu(\theta) \in \tilde{\mathbf{L}}_2(S)$ with $\Phi_\nu \to \Phi$. Put $\delta^{-l/2}\Phi_\nu = F_\nu$, and let $D^l F_\nu(\theta) \to \omega(\theta)$. The function F_ν will be completely defined if $F_\nu(\theta) \in \tilde{\mathbf{L}}_2(S)$ is required. Then the operator $\delta^{-l/2}$ is bounded, and $F_\nu(\theta)$ converges in the $\mathbf{L}_2(S)$ metric to the function $F(\theta) = \delta^{-l/2}\Phi(\theta)$. Hence, $F_\nu(\theta) \to F(\theta)$ and $D^l F_\nu(\theta) \to \omega(\theta)$. Since the operator of generalized differentiation is closed we obtain $D^l F(\theta) = \omega(\theta)$ which was to be demonstrated.

The closed operator G which is defined on the whole space is bounded such that $G\varphi_N \to G\varphi$ or $D^l f_N \to D^l f$. The last relation shows that the series (1) may be differentiated

l times and the series obtained that way are convergent in $\mathbf{L}_2(S)$. Thus, the theorem is proved if $r = l$; in case $r < l$ it is an immediate consequence of the Sobolev imbedding theorem (cf. SOBOLEV [3]). ∎

6.4. Lemma 6.2. *The set of spherical functions is dense in the space* $\mathbf{W}_2^{(l)}(S)$ *for arbitrary real* $l \geq 0$.

Proof. Let l be a natural number. In that case, Theorem 6.3 asserts that series (1) is convergent in $\mathbf{W}_2^{(l)}(S)$, and this implies the assertion of the lemma. If l is not integer we put $l_1 = [l] + 1$. In the metric $\mathbf{W}_2^{(l_1)}(S)$ the set of spherical functions is dense, but then it dense also in the weaker metric $\mathbf{W}_2^{(l)}(S)$. ∎

Theorem 6.4. *The assertions of Theorem 6.1 are true for any real* $l > 0$.

Proof. Let $f \in \mathbf{W}_2^{(l)}(S)$. Due to Lemma 6.2 there is a sequence $\{\Phi_N(\theta)\}$ of spherical functions such that $\|f - \Phi_N\|_{\mathbf{W}_2^{(l)}(S)} \to 0$ as $N \to \infty$. Then also $\|\Phi_N - \Phi_\nu\|_{\mathbf{W}_2^{(l)}(S)} \to 0$ as $N, \nu \to \infty$.

For arbitrary l the spherical functions belong to the domain of the operator $\delta^{l/2}$, but then, due to Theorem 4.1, Chap. VIII, $\|\delta^{l/2}\Phi_N - \delta^{l/2}\Phi_\nu\|_{\mathbf{L}_2(S)} \to 0$. Therefore, in $\mathbf{L}_2(S)$ there exist the limits $g_0 = \lim \delta^{l/2} \Phi_N$ and $f = \lim \Phi_N$ as well. Since the operator $\delta^{l/2}$ is closed we obtain $f \in D(\delta^{l/2})$.

Now we suppose, conversely, $f \in D(\delta^{l/2})$. Then the expansions

$$f(\theta) = \sum_{n=0}^{\infty} \sum_{k=1}^{k_{n,m}} a_n^{(k)} Y_{n,m}^{(k)}(\theta),$$

$$\delta^{l/2} f(\theta) = \sum_{n=1}^{\infty} \sum_{k=1}^{k_{n,m}} n^{l/2}(n + m - 2)^{l/2} a_n^{(k)} Y_{n,m}^{(k)}(\theta)$$

hold where both the series are convergent in $\mathbf{L}_2(S)$. Put

$$f_N(\theta) = \sum_{n=0}^{N} \sum_{k=1}^{k_{n,m}} a_n^{(k)} Y_{n,m}^{(k)}(\theta)$$

then $\|\delta^{l/2} f_N - \delta^{l/2} f_\nu\|_{\mathbf{L}_2(S)} \to 0$ as $N, \nu \to \infty$. Again due to Theorem 4.1, Chap. VIII, $\|f_N - f_\nu\|_{\mathbf{W}_2^{(l)}(S)} \to 0$ as $N, \nu \to \infty$. Then there exists a function $f_0 \in \mathbf{W}_2^{(l)}(S)$ such that $\|f_N - f_0\|_{\mathbf{W}_2^{(l)}(S)} \to 0$ as $N \to \infty$. As we already proved, then $f_0 \in D(\delta^{l/2})$. Another application of Theorem 4.1, Chap. VIII, yields $\|\delta^{l/2} f_N - \delta^{l/2} f_0\|_{\mathbf{L}_2(S)} \to 0$ as $N \to \infty$. Hence, $f(\theta) = f_0(\theta) + \text{const}$ such that $f \in \mathbf{W}_2^{(l)}(S)$. ∎

Note that Theorem 6.4 allows to generalize Theorem 6.3 to the case of arbitrary real positive l.

Theorem 6.4 was proved by AGRANOVICH [1] and MIKHAILOVA-GUBENKO [2]; the first paper proves the theorem also for negative l.

6.5. In the present section we shall consider the case important for several applications where the functions of the variable θ depend, in addition, on a certain parameter. Such a parameter may be, in particular, a point of the space $\mathbb{R}^m$ or of any other set. Speaking of differentiation we shall always mean differentiation with respect to the Cartesian coordinates of the point $\theta \in S$.

We shall say that a function $f(\lambda, \theta)$ belongs to a certain Banach space B *uniformly with respect to the parameter* λ iff $\|f(\lambda, \cdot)\|_B \leq C$ with a constant C not depending on λ; in that case we shall write $f(\lambda, \theta) \hat{\in} B$. In the following we shall be concerned, in particular, with the case $B = \mathbf{W}_p^{(l)}(S)$, $\lambda = x \in \mathbb{R}^m$; here the relation $f(x, \theta) \hat{\in} \mathbf{W}_p^{(l)}(S)$ means that

$$\int_S |D_\theta^\alpha f(x, \theta)|^p \, dS \leqq C, \qquad 0 \leqq |\alpha| \leqq l \tag{16}$$

with C not depending on x.

For functions belonging to the space $\mathbf{W}_2^{(l)}(S)$ uniformly with respect to a parameter the following theorems hold true which are analogues or modifications of previous theorems on series in spherical functions.

Theorem 6.5. *A function $f(x, \theta)$ belongs to $\mathbf{W}_2^{(l)}(S)$ uniformly with respect to x if and only if the coefficients of the series*

$$f(x, \theta) = \sum_{n=0}^{\infty} \sum_{k=1}^{k_{n,m}} a_n^{(k)}(x) \, Y_{n,m}^{(k)}(\theta) \tag{17}$$

satisfy the inequalities

$$|a_0^{(1)}(x)| \leqq C' = \text{const}, \tag{18a}$$

$$\sum_{n=1}^{\infty} \sum_{k=1}^{k_{n,m}} n^{2l} \, |a_n^{(k)}(x)|^2 \leqq C'' = \text{const}. \tag{18b}$$

Proof. *Necessity.* If $f(x, \theta) \,\hat{\in}\, \mathbf{W}_2^{(l)}(S)$ then

$$\int_S |\delta^{l/2} f|^2 \, dS \leqq C_1 = \text{const} \tag{19}$$

or

$$\sum_{n=1}^{\infty} \sum_{k=1}^{k_{n,m}} n^l (n + m - 2)^l \, |a_n^{(k)}|^2 \leqq C_1.$$

This implies because of $(n + m - 2)/n \leqq m - 1$ the inequality (18b). The inequality (18a) coincides with (16) if $p = 2$ and $|\alpha| = 0$.

Sufficiency. The inequality (19) follows from (18b) by Theorem 4.1, Chap. VIII. When proving Theorem 6.3 we showed that the operator G is bounded, $\|D^l f\| \leqq \|G\| \, \|\delta^{l/2} f\|$. Thence,

$$\int_S |D^l f|^2 \, dS \leqq (m - 1)^l \, C'' \, \|G\|^2,$$

and (16) is proved for $|\alpha| \neq 0$. In order to show that inequality for $|\alpha| = 0$ we note that the operator $\delta^{-l/2}$ is bounded in $\tilde{\mathbf{L}}_2(S)$. Put $\delta^{l/2} f = g(x, \theta)$, then

$$f(x, \theta) = a_0^{(1)}(x) + \delta^{-l/2} g(x, \theta),$$

which implies the estimate

$$|f(x, \theta)|^2 \leqq 2 \, |a_0^{(1)}(x)|^2 + 2 \, |\delta^{-l/2} g(x, \theta)|^2.$$

Integrating over S we obtain

$$\int_S |f(x, \theta)|^2 \, dS \leqq \frac{4\pi^{m/2} C'^2}{\Gamma(m/2)} + 2 \, \|\delta^{-l/2}\|^2 \, \|g(x, \theta)\|^2 \leqq \text{const},$$

which was to be demonstrated. ▮

Theorem 6.6. *Let $l \geqq m - 1$ and $f(x, \theta) \,\hat{\in}\, \mathbf{W}_2^{(l)}(S)$. The series* (17) *as well as those series which are obtained from* (17) *by differentiating it r times* ($r \leqq l - m + 1$) *are absolutely and uniformly conergent.*

Proof. We differentiate the series (17) r times. The remainder of the corresponding series of the moduli

$$\sum_{n=N}^{\infty} \sum_{k=1}^{k_{n,m}} |a_n^{(k)}(x)| \, |D^\alpha Y_{n,m}^{(k)}(\theta)|, \qquad |\alpha| = r,$$

is not greater than the product of some constant with the quantity

$$2 \sum_{n=N}^{\infty} \sum_{k=1}^{k_{n,m}} n^{(m/2)-1+r} \, |a_n^{(k)}(x)| \leq \sum_{n=N}^{\infty} \sum_{k=1}^{k_{n,m}} n^{2l-\eta} \, |a_n^{(k)}(x)|^2 + \sum_{n=N}^{\infty} k_{n,m} n^{-2l+\eta+m-2+2r},$$

where η is an arbitrary positive number. The sum of the first series right is not greater than

$$N^{-\eta} \sum_{n=1}^{\infty} \sum_{k=1}^{k_{n,m}} n^{2l} \, |a_n^{(k)}(x)|^2 \leq C'' N^{-\eta}.$$

The general summand of the second series right is of order $O(n^{-2l+\eta+2m-4+2r})$ such that this series is convergent and the remainder becomes arbitrarily small if N is sufficiently large provided that the condition $2l - \eta - 2m + 4 - 2r > 1$ is fulfilled for which $r \leq l - m + 1$ is sufficient. ▌

Theorem 6.7. *Let $l \geq 1$ and $f(x, \theta) \hat{\in} \mathbf{W}_2^{(l)}(S)$. The series which are obtained from the series (17) by differentiating it r times ($r \leq l - 1$) are convergent in $\mathbf{L}_p(S)$ with the power p being*

$$\frac{2(m-1)}{m - 1 - 2(l - 1 - r)}. \tag{20}$$

Proof. Consider the series

$$\delta^{(l-1)/2} f = \sum_{n=1}^{\infty} \sum_{k=1}^{k_{n,m}} n^{(l-1)/2} (n + m - 2)^{(l-1)/2} \, a_n^{(k)}(x) \, Y_{n,m}^{(k)}(\theta) \tag{21}$$

and estimate the remainder as follows,

$$\left\| \sum_{n=N}^{\infty} \sum_{k=1}^{k_{n,m}} n^{(l-1)/2} (n + m - 2)^{(l-1)/2} \, a_n^{(k)}(x) \, Y_{n,m}^{(k)}(\theta) \right\|_{\mathbf{L}_2(S)}^2$$

$$\leq (m-1)^{l-1} \sum_{n=N}^{\infty} \sum_{k=1}^{k_{n,m}} n^{2l-2} \, |a_n^{(k)}(x)|^2$$

$$\leq \frac{(m-1)^{l-1}}{N^2} \sum_{n=1}^{\infty} \sum_{k=1}^{k_{n,m}} n^{2l} \, |a_n^{(k)}(x)|^2 \leq \frac{(m-1)^l C''}{N^2}.$$

Now you can see that the series (21) converges in the $\mathbf{L}_2(S)$ metric uniformly with respect to x. Therefore, Theorem 6.7 follows for $r = l - 1$ because the operator G is bounded. For $r < l - 1$ it is a consequence of the Sobolev imbedding theorem. ▌

Remark. In order to be sure that the series which is obtained from the series (17) by differentiating it r times is convergent in the metric of the space $\mathbf{L}_p(S)$, $p > 2$, it is sufficient to require $f(x, \theta) \hat{\in} \mathbf{W}_2^{(l)}(S)$ and according to (20)

$$l \geq r + (m-1)\left(\frac{1}{2} - \frac{1}{p}\right) + 1. \tag{22}$$

In what follows we need the case $r \leqq (m+1)/2$ for which the condition

$$l > \frac{m-1}{p'} + 2 \tag{23}$$

is sufficient. In case if $1 < p < 2$ it is sufficient to require $l > (m+3)/2$.

§ 7. Differentiability properties of the symbol and the characteristic

Let us be given a singular integral with a characteristic $f(x, \theta) \hat{\in} \mathbf{W}_2^{(l)}(S)$, $l \geqq 0$. We expand the characteristic in a series with respect to the spherical functions (6.17). Then the corresponding symbol has the series expansion

$$\Phi(x, \theta) = \sum_{n=1}^{\infty} \sum_{k=1}^{k_{n,m}} \gamma_{n,m} a_n^{(k)}(x)\, Y_{n,m}^{(k)}(\theta), \qquad \gamma_{n,m} = \frac{i^n \pi^{m/2} \Gamma(n/2)}{\Gamma((m+n)/2)}.$$

Equation (6.18) gives

$$\sum_{n=1}^{\infty} \sum_{k=1}^{k_{n,m}} n^{2l} \, |a_n^{(k)}(x)|^2 \leqq \text{const} < \infty.$$

By Stirling's formula we have $\gamma_{n,m} = O(n^{-m/2})$, $\gamma_{n,m}^{-1} = O(n^{m/2})$. Therefore, we may write

$$\gamma_{n,m} a_n^{(k)}(x) = n^{-l-m/2} b_n^{(k)}(x); \qquad \sum_{n=1}^{\infty} \sum_{k=1}^{k_{n,m}} |b_n^{(k)}(x)|^2 \leqq \text{const} < \infty.$$

Theorem 6.4 now implies

Theorem 7.1. *The symbol of a singular integral satisfies the relation* $\Phi(x, \theta) \hat{\in} \mathbf{W}_2^{(\lambda)}(S)$ *if and only if the characteristic of this integral fulfills the condition* $f(x, \theta) \hat{\in} \mathbf{W}_2^{(l)}(S)$ *where* $l = \lambda - m/2$.

That theorem was proved by MIKHLIN [11] if l and λ are integers; here the corresponding assertions are

$$f \hat{\in} \mathbf{W}_2^{(l)}(S) \Rightarrow \Phi \hat{\in} \mathbf{W}_2^{(\lambda)}(S), \qquad \lambda = \left[l + \frac{m}{2}\right],$$

$$\Phi \hat{\in} \mathbf{W}_2^{(\lambda)}(S) \Rightarrow f \hat{\in} \mathbf{W}_2^{(l)}(S), \qquad l = \left[\lambda - \frac{m}{2}\right].$$

Theorem 7.1 as given above was proved independently by AGRANOVICH [1] and MIKHAILOVA-GUBENKO [2].

Theorem 7.2. *Let* $p > 1$ *and assume that the characteristic* $f(x, \theta)$ *belongs to* $\mathbf{L}_p(S)$ *for a certain point* x. *Then, for the same* x, *the symbol* $\Phi(x, \theta)$ *is continuous in* θ.

Proof. Since the family of spherical functions is complete in $\mathbf{L}_p(S)$ there exists for a given $\varepsilon > 0$ a polynomial of the form

$$S_N(x, \theta) = \sum_{n=1}^{N} \sum_{k=1}^{k_{n,m}} \alpha_n^{(k)}(x)\, Y_{n,m}^{(k)}(\theta)$$

with $\|f - S_N\|_{\mathbf{L}_p(S)} < \varepsilon$. Let $\sigma_N(x, \theta) = \operatorname{Smb} S_N(x, \theta)$. By (1.14),

$$\Phi(x, \theta) - \sigma_N(x, \theta) = \int_S [f(x, \theta') - S_N(x, \theta')] \left[\ln \frac{1}{\cos \gamma} + \frac{i\pi}{2} \operatorname{sign} \cos \gamma\right] dS'.$$

Using Hölder's inequality and taking into account (1.18) we obtain $|\Phi(x, \theta) - \sigma_N(x, \theta)| \leqq C_{p'}\, \varepsilon$. Thus, the function $\Phi(x,\theta)$ is approximated uniformly by continuous functions $\sigma_N(x, \theta)$ and is, therefore, continuous itself. ▮

Theorem 7.2 is, essentially, contained in a paper by Calderon and Zygmund [1]. For $p = 2$ the theorem was proved earlier and in a different way by Mikhlin [7].

Remark. Recently Gadzhiev [1, 2] gave the following generalization of Theorem 7.1.

Let $1 < p < \infty$ and $l = (m - 2)\left|\frac{1}{p} - \frac{1}{2}\right|$. Then

$$f \in \mathbf{L}_p(S) \Rightarrow \Phi \in \mathbf{W}_p^{(m/2-l)}(S)$$

and

$$\Phi \in \mathbf{W}_p^{(m/2+l)}(S) \Rightarrow f \in \mathbf{L}_p(S).$$

Similar results are true for weighted spaces. The implications given are sharp.

Here $\mathbf{W}_p^{(q)}(S)$, $0 < q < \infty$, means the completion of the set $\mathbf{C}^\infty(S)$ with respect to the norm $\|(I + \delta)^{q/2} f\|_{\mathbf{L}_p(S)}$ $(f \in \mathbf{C}^\infty(S))$, I denotes the identity operator. If $f \in \mathbf{C}^\infty(S)$ is orthogonal to 1 then this norm is equivalent to the norm $\|\delta^{q/2} f\|_{\mathbf{L}_p(S)}$(cf. Theorem 4.1, Chap. VIII, in case if $p = 2$).

§ 8. The symbol ring

8.1. Lemma 8.1 (Peetre [1]). *The space $\mathbf{W}_2^{(s)}(\mathbb{R}^m)$, $s > m/2$, is a ring with respect to usual addition and multiplication of functions.*

Proof. It is sufficient to show that for any two functions $u, v \in \mathbf{W}_2^{(s)}(\mathbb{R}^m)$ the inequality

$$\|uv\|_s \leqq K \|u\|_s \|v\|_s \tag{1}$$

holds where K is a constant and $\|\cdot\|_s$ stands for the norm in $\mathbf{W}_2^{(s)}(\mathbb{R}^m)$ which is defined as follows,

$$\|f\|_s^2 := \int_{\mathbb{R}^m} (1 + |\xi|)^{2s} |\hat{f}(\xi)|^2 \, d\xi; \qquad \hat{f}(\xi) = F_{x\to\xi} f. \tag{2}$$

For $w = uv$ we have

$$\hat{w}(\xi) = \int_{\mathbb{R}^m} \hat{u}(\xi - \eta)\, \hat{v}(\eta)\, d\eta$$

and then

$$\|w\|_s^2 = \int_{\mathbb{R}^m}\int_{\mathbb{R}^m} (1 + |\xi|)^{2s}\, \hat{u}(\xi - \eta)\, \hat{v}(\eta)\, \overline{\hat{w}(\xi)}\, d\xi\, d\eta.$$

Applying Schwarz-Bunyakovski inequality we obtain

$$\|w\|_s^2 \leqq \left[\int_{\mathbb{R}^m}\int_{\mathbb{R}^m} (1 + |\eta|)^{2s} (1 + |\xi - \eta|)^{2s} |\hat{u}(\xi - \eta)|^2 |\hat{v}(\eta)|^2 d\xi\, d\eta\right]^{1/2}$$

$$\times \left[\int_{\mathbb{R}^m}\int_{\mathbb{R}^m} \left(\frac{1 + |\xi|}{(1 + |\xi - \eta|)(1 + |\eta|)}\right)^{2s} (1 + |\xi|)^{2s} |\hat{w}(\xi)|^2 d\xi\, d\eta\right]^{1/2}. \tag{3}$$

We replace in the first integral of the right-hand side $\xi - \eta$ by ξ such that it becomes

$$\int_{\mathbb{R}^m} (1 + |\xi|)^{2s} |\hat{u}(\xi)|^2 \, d\xi \int_{\mathbb{R}^m} (1 + |\eta|)^{2s} |\hat{v}(\eta)|^2 \, d\eta = \|u\|_s^2 \, \|v\|_s^2. \tag{4}$$

In the second integral we have

$$\left(\frac{1 + |\xi|}{(1 + |\xi - \eta|)(1 + |\eta|)}\right)^{2s} < \left(\frac{(1 + |\xi - \eta|) + (1 + |\eta|)}{(1 + |\xi - \eta|)(1 + |\eta|)}\right)^{2s}$$

$$= \left(\frac{1}{1 + |\xi + \eta|} + \frac{1}{1 + |\eta|}\right)^{2s} \leqq 2^{2s-1}\left(\frac{1}{(1 + |\xi - \eta|)^{2s}} + \frac{1}{(1 + |\eta|)^{2}}\right)$$

such that this integral becomes less than

$$2^{2s-1} \int_{\mathbb{R}^m} \left(\int_{\mathbb{R}^m} \frac{d\eta}{(1 + |\xi - \eta|)^{2s}} + \int_{\mathbb{R}^m} \frac{d\eta}{(1 + |\eta|)^{2s}} \right) (1 + |\xi|)^{2s} |\hat{w}(\xi)|^2 \, d\xi$$

$$= K \|w\|_s^2, \qquad K = 2^s \int_{\mathbb{R}^m} \frac{d\eta}{(1 + |\eta|)^{2s}}, \tag{5}$$

where the constant K is finite because of $2s > m$. Inserting (4) and (5) into the inequality (3) we obtain the relation (1). ∎

Lemma 8.1 implies that the space $W_2^{(l)}(\mathbb{R}^{m-1})$, $l > (m - 1)/2$, forms a ring with respect to usual addition and multiplication of functions. Now, we consider an arbitrary ball σ in $\mathbb{R}^{m-1}$ and the space $W_2^{(l)}(\sigma)$. Every function $u(x)$ belonging to that space is extended to the whole space $\mathbb{R}^{m-1}$ according to a theorem proved by Calderon [1] on the extension of functions preserving the class. For the extension $u^*(x)$ we have $u^* \in W_2^{(l)}(\mathbb{R}^{m-1})$ and $\|u^*\|_{W_2^{(l)}(\mathbb{R}^{m-1})} \leqq C \|u\|_{W_2^{(l)}(\sigma)}$, $C = \text{const}$. Using the concept of such an extension we may easily carry over Lemma 8.1 to the space $W_2^{(l)}(\sigma)$. From the definition of the space $W_2^{(l)}(S)$ (see § 2, Chap. VIII) it is now immediately clear that this space for $l > (m - 1)/2$ forms a ring with respect to the operations mentioned above. This implies, together with Theorem 7.1,

Theorem 8.1. *The symbols of singular integral operators with characteristics $f(x, \theta) \,\hat{\in}\, W_2^{(-1/2+\varepsilon)}(S)$, $\varepsilon > 0$, form a ring with respect to addition and multiplication of functions. The functions of that ring are continuous in θ.*

Now we narrow the ring mentioned in Theorem 8.1 taking those functions only which are continuous with respect to the aggregate variables $x \in \overline{\mathbb{R}}^{m-1}$ and $\theta \in S$ where $\overline{\mathbb{R}}^{m-1}$ denotes that compact which is obtained from $\mathbb{R}^{m-1}$ by adding infinity. The resulting smaller ring will be denoted by $\mathfrak{R}_m$.

8.2. Theorem 8.2. *Every maximal ideal of the ring $\mathfrak{R}_m$ is provided by the set of those functions which vanish at a certain point (x_0, θ_0), $x_0 \in \overline{\mathbb{R}}^m$, $\theta_0 \in S$.*

Proof. As it is well-known, the functions with the property as mentioned in the theorem form a maximal ideal in any ring of continuous functions with the usual algebraic operations. Thus, it is sufficient to prove the converse assertion.

A point $\theta \in S$ is determined in spherical coordinates by the angles $\vartheta_1, \vartheta_2, \ldots, \vartheta_{m-1}$ and they correspond to the Cartesian coordinates $\xi_1, \xi_2, \ldots, \xi_{m-1}$ of the point θ via the usual formulae

$$\left.\begin{aligned}
\xi_1 &= \cos\vartheta_1,\\
\xi_2 &= \sin\vartheta_1\cos\vartheta_2,\\
&\cdots\cdots\cdots\cdots\cdots\cdots\\
\xi_k &= \sin\vartheta_1\sin\vartheta_2\ldots\sin\vartheta_{k-1}\cos\vartheta_k,\\
&\cdots\cdots\cdots\cdots\cdots\cdots\\
\xi_{m-1} &= \sin\vartheta_1\sin\vartheta_2\ldots\sin\vartheta_{m-2}\cos\vartheta_{m-1},\\
\xi_m &= \sin\vartheta_1\sin\vartheta_2\ldots\sin\vartheta_{m-2}\sin\vartheta_{m-1}.
\end{aligned}\right\} \tag{6}$$

Due to the definition of the ring $\mathfrak{R}_m$, every function $f(x, \theta) \in \mathfrak{R}_m$ is continuous with respect to the aggregate coordinates $x_1, x_2, \ldots, x_m, \xi_1, \xi_2, \ldots, \xi_m$ where the x_j may assume any finite or infinite value whereas $\xi_1, \ldots, \xi_m$ satisfy the sphere equation $\xi_2^1 + \xi_2^2 + \ldots + \ldots + \xi_m^2 = 1$. Equation (6) shows that the ξ_j are continuous functions of the angular coordinates. Hence, every function of $\mathfrak{R}_m$ is continuous with respect to the aggregate variables $x, \vartheta_1, \vartheta_2, \ldots, \vartheta_{m-1}$; $x \in \overline{\mathrm{R}}^m$, $-\pi \leqq \vartheta_{m-1} \leqq \pi$, $0 \leqq \vartheta_j \leqq \pi$, $1 \leqq j \leqq m-2$. These functions are periodic with respect to the angle ϑ_{m-1} and, in general, non-periodic with respect to the other angular coordinates.

We extend the functions of the ring $\mathfrak{R}_m$ to periodic functions of the coordinates $\vartheta_1, \vartheta_2, \ldots, \vartheta_{m-1}$.

Let $f(x, \theta) \in \mathfrak{R}_m$ .We expand this functions in a series with respect to spherical functions,

$$f(x, \theta) = \sum_{n=0}^{\infty} \sum_{k=1}^{k_{n,m}} a_n^{(k)}(x)\, Y_{n,m}^{(k)}(\theta).$$

These spherical functions are trigonometric polynomials in angular coordinates and, therefore, we may assume them to be defined for any value of those coordinates. Hence, the general summand of the series (7) may be considered as extended to the interval $\vartheta_j \in [-\pi, \pi]$, $1 \leqq j \leqq m-2$. We show that under this extension the series (7) remains to be convergent and, therefore, defines an extension of the function $f(x, \theta)$.

Suppose that an angle ϑ_k, $1 \leqq k \leqq m-2$, varies in the segment $-\pi \leqq \vartheta_k \leqq 0$. That means that the angle ϑ_k alters the sign. As the formulae (6) show, the coordinates $\xi_1, \ldots, \xi_k$ remain untouched whereas the following coordinates alter merely their sign. Thus, the substitution $\vartheta_k \to -\vartheta_k$ is equivalent to the coordinate transformation

$$\xi_1' = \xi_1, \ldots, \xi_k' = \xi_k, \qquad \xi_{k+1}' = -\xi_{k+1}, \ldots, \xi_m' = -\xi_m. \tag{8}$$

This is a unitary transformation of the space R^m, it maps every polynomial into a polynomial of the same degree and does not affect on orthogonality and norm of functions in $\mathbf{L}_2$. Therefore, the substitution $\vartheta_k \to -\vartheta_k$ transforms a complete orthonormal system of spherical functions $\{Y_{n,m}^{(k)}(\theta)\}$ into a similar one. But then the series (7) converges after such a substitution in the metric of $\mathbf{W}_2^{(l)}(S)$ uniformly with respect to x.

The extension of a function $f(x, \theta) \in \mathfrak{R}_m$ as constructed here will be denoted by $f^*(x, \vartheta_1, \ldots, \vartheta_{m-1})$. The funtion f^* is periodic (with a period 2π) with respect to every angular coordinate and, in any case, continuous with respect to the aggregate variables x and ϑ_j. If the function f^* is expanded into a $(m-1)$-dimensional Fourier series then this series is uniformly summable by the $(C, 1)$-method (cf. Zygmund [3], vol. 2, Chap: XVII, Theorem 1.20). The set of all functions f^* with $f \in \mathfrak{R}_m$ is also a ring which will be denoted by $\mathfrak{R}_m^*$.

8.3. Now, let $\mathfrak{M}^*$ be an arbitrary maximal ideal of the ring $\mathfrak{R}_m^*$. We show that all functions belonging to that ideal vanish at a certain point (x_0, θ_0), $x_0 \in \overline{\mathbb{R}}^m$, $\theta_0 \in S$.

The function $\exp(\mathrm{i}\vartheta_k)$ belongs to the ring $\mathfrak{R}_m^*$. The canonical homomorphism of the factor ring of $\mathfrak{R}_m^*$ by the maximal ideal $\mathfrak{M}^*$ attaches a certain number a_k to that function and the number a_k^n to the function $\exp(\mathrm{i}n\vartheta_k)$, n being an arbitrary integer. Due to the property of the canonical homomorphism,

$$|a_k|^n \leqq \|\mathrm{e}^{\mathrm{i}n\vartheta_k}\|_{W_2^{(l)}(S_k)} \leqq C(1 + |n|)^l,$$

where S_k denotes the sphere which is the image of the sphere S under the transformation (8). Extracting the n-th root and letting $n \to \infty$ we find $|a_k| = 1$. Therefore, we may put $a_k = \exp(\mathrm{i}\vartheta_k^0)$ where the constant ϑ_k^0 belongs to the segment $-\pi < \vartheta_k^0 \leqq \pi$.

Now, we consider all those functions of the ring $\mathfrak{R}_m^*$ which depend only on x. They form a subring in $\mathfrak{R}_m^*$ which is isomorphic to the ring $C(\overline{\mathbb{R}}^m)$. The restriction of the maximal ideal $\mathfrak{M}^*$ to that subring is a maximal ideal in $C(\overline{\mathbb{R}}^m)$ and, consequently, consists of functions which are continuous in $\overline{\mathbb{R}}^m$ and vanish at a certain fixed point $x_0 \in \overline{\mathbb{R}}^m$. The canonical homomorphism of the factor ring of $\mathfrak{R}_m^*$ by the maximal ideal $\mathfrak{M}^*$ attaches to a function a the value $a(x_0)$ where to any finite sum

$$\sum_n a_n(x) \exp\left[\mathrm{i} \sum_{k=1}^{m-1} n_k \vartheta_k\right], \qquad n = (n_1, \ldots, n_{m-1}) \tag{9}$$

a sum

$$\sum_n a_n(x_0) \exp\left[\mathrm{i} \sum_{k=1}^{m-1} n_k \vartheta_k^0\right]. \tag{10}$$

is attached.

Let f^* be an arbitrary function of the ring $\mathfrak{R}_m^*$. We expand it into a $(m-1)$-dimensional Fourier series and take the arithmetic mean of the partial sums of this series as a sum (9). Since the Fourier series is summable by the $(C, 1)$-method the sums (9) converge uniformly to f^*. Thence, the canonical homomorphism mentioned above attaches to the function f^* its value at the point $(x_0, \vartheta_1^0, \vartheta_2^0, \ldots, \vartheta_{m-1}^0)$. Equations (6) and (8) show that to every point $(\vartheta_1^0, \vartheta_2^0, \ldots, \vartheta_{m-1}^0)$, $-\pi \leqq \vartheta_j^0 \leqq \pi$, corresponds exactly one point $\theta_0 \in S$. Thus, the canonical homomorphism of the factorring of $\mathfrak{R}_m^*$ by the maximal ideal $\mathfrak{M}^*$ attaches to every function $f^* \in \mathfrak{R}_m^*$ the number $f(x_0, \theta_0)$ where f is that function which is extended as described in 8.2 to f^*. The canonical homomorphism maps the functions belonging to the maximal ideal onto zero. But this means that $\mathfrak{M}^*$ consists of functions f^* such that the corresponding functions $f(x, \theta)$ vanish at $x = x_0$ and $\theta = \theta_0$.

When the functions of the ring $\mathfrak{R}_m$ are extended to elements of the ring $\mathfrak{R}_m^*$ every ideal of the ring $\mathfrak{R}_m$ corresponds to some ideal of the ring $\mathfrak{R}_m^*$. Hence, every maximal ideal of the ring $\mathfrak{R}_m$ consists of functions which vanish at a certain fixed point (x_0, θ_0). Thus, Theorem 8.2 is proved. ▮

Corollary 8.1. *Let* $f(x, \theta) \in \mathfrak{R}_m$. *If* $\inf |f(x, \theta)| > 0$ *then* $f(x, \theta)$ *has in* $\mathfrak{R}_m$ *an inverse element, and this is equal to* $[f(x, \theta)]^{-1}$.

Remark The theorems of the present section are proved in a somewhat different way by Mikhailova-Gubenko [3] and Khvoles [3]. The arguments in section 8.3 are taken, essentially, from the book by Gelfand, Raikov and Shilov [1]; the corresponding part of that book was also used in the papers mentioned by Mikhailova-Gubenko [3] and Khvoles [3].

8.4. In case of singular matrix operators the symbol ring is not commutative, and the foregoing arguments do no longer apply. Nevertheless, it is not difficult to prove the following theorem which will be important for the following.

Theorem 8.3. *Let A be an $N \times N$ matrix of singular operators and $\Phi(x, \theta)$ the corresponding symbol matrix. If the matrix of the characteristics belongs to $\mathbf{W}_2^{(-1/2+\varepsilon)}(S)$ uniformly with respect to $x \in \overline{\mathbb{R}}^m$ for some $\varepsilon > 0$ then $\Phi(x, \theta) \hat{\in} \mathbf{W}_2^{(l)}(S)$ for $l = (m-1)/2 + \varepsilon$, and the symbol matrices corresponding to all such singular matrix operators form a non-commutative ring with respect to the usual operations of addition and multiplication of matrices.*

Let $\mathfrak{R}_m^{(N)}$ denote the restriction of the ring mentioned which consists of all symbol matrices that are continuous on $\overline{\mathbb{R}}^m \times S$. A matrix $\Phi \in \mathfrak{R}_m^{(N)}$ has an inverse in that ring if and only if

$$\inf |\det \Phi(x, \theta)| > 0. \tag{11}$$

Only the last assertion of the theorem needs a proof. That the condition (11) is necessary is obvious, so we show the sufficiency. Theorem 8.1 implies $\det \Phi(x, \theta) \in \mathbf{W}_2^{(l)}(S)$, and since this determinant is continuous on $\overline{\mathbb{R}}^m \times S$ it belongs to the ring $\mathfrak{R}_m$. Then from condition (11) and Corollary 8.1 it follows that $[\det \Phi(x, \theta)]^{-1} \in \mathfrak{R}_m$. If condition (11) is fulfilled then the inverse matrix $[\Phi(x, \theta)]^{-1}$ exists, and the same Theorem 8.1 implies now $[\Phi(x, \theta)]^{-1} \in \mathfrak{R}_m^{(N)}$. ∎

Chapter XI
Singular integral operators in spaces with integral metric

§ 1. An extension of the notion of a singular integral

1.1. Let us consider the singular operator

$$(Au)(x) = au(x) + \int_{R^m} K(x-y)\, u(y)\, dy, \qquad K(x-y) = r^{-m} f(\theta), \tag{1}$$

the symbol $\Phi(\xi) = a + \hat{K}(\xi)$ of which does not depend on the pole. Remember that $\Phi(\xi)$ is a positive-homogeneous function of degree zero in ξ. As we saw in Chapter X (see (1.11)), the operator (1) can be represented in the form

$$A = F^{-1}_{\xi\to x} \Phi(\xi)\, F_{x\to\xi}. \tag{2}$$

If there is no danger of misunderstanding we shall briefly write

$$A = F^{-1}\, \Phi(\xi)\, F \tag{2'}$$

A formula which is analogous to (2) can be constructed also for singular operators the symbol of which depends on the pole. Let such an operator have the form

$$(Au)(x) = a(x)\, u(x) + \int_{R^m} K(x, x-y)\, u(y)\, dy, \qquad K(x, x-y) = r^{-m} f(x, \theta); \tag{3}$$

its symbol will be denoted by $\Phi(x, \xi)$. Consider the singular operator

$$(A_t u)(x) = a(t)\, u(x) + \int_{R^m} K(t, x-y)\, u(y)\, dy \tag{4}$$

that depends on a parameter $t \in \mathrm{R}^m$ and the symbol of which, $\mathrm{Smb}\, A_t = \Phi_t(\xi)$, obviously, is independent of the pole. Furthermore,

$$\Phi_t(\xi) = a(t) + \hat{K}(t, \xi), \qquad A = A_x, \qquad \Phi(x, \xi) = \Phi_x(\xi).$$

For the operator A_t (2) holds true,

$$A_t = F^{-1}_{\xi\to x}\, \Phi_t(\xi)\, F_{x\to\xi}.$$

Put $t = x$, and you will obtain the relation wanted,

$$A = F^{-1}_{\xi\to x}\, \Phi(x, \xi)\, F_{x\to\xi}. \tag{5}$$

Equations (2) and (5) give reason for the extension of the notion of a singular integral operator: This will be an operator of the form (2) or (5) where the functions $\Phi(\xi)$ or $\Phi(x, \xi)$ are measurable with respect to x and ξ and, with respect to ξ, are positive-homogeneous of degree zero. These functions will be called *symbol* of the corresponding singular integral operator.

Remark. One can consider operators of the form (2) or (5) without the assumption that the symbol is a positive-homogeneous function of degree zero in ξ. Under certain rather general assumptions on the symbol this will lead us to the notion of *pseudo-differential operators* (cf. KOHN and NIRENBERG [1], HÖRMANDER [2]). Pseudodifferential operators are the subject of § 8 in the next chapter.

1.2. A symbol that does not depend on the pole is sometimes called a *constant* one. Accordingly, we shall call operators of the form (2) *singular operators with constant symbol*. Such an operator is bounded in $\mathbf{L}_2(\mathbb{R}^m)$ if and only if its symbol is essentially bounded, and its norm is then equal to the essential supremum of the symbol modulus. This assertion is the subject of Theorem 5.1, Chap. X. Equation (2) shows that a singular operator with a constant symbol in the space $\mathbf{L}_2(\mathbb{R}^m)$ is unitarily equivalent to the operator of multiplication with the symbol. This statement implies a number of simple and important consequences which we shall explicate now. One of those consequences is Theorem 5.1, Chap. X. We stress once again that the following theorems concern operators with a constant symbol in the space $\mathbf{L}_2(\mathbb{R}^m)$.

Theorem 1.1. *The singular operator is selfadjoint if its symbol is real-valued, and it is unitary if its symbol has modulus* 1.

Theorem 1.2. *The sepctrum of the singular operator consists of all numbers with the following property. On a certain point set of the unit sphere S having a positive measure the symbol assumes values that are arbitrarily close to a given number. That number is an eigenvalue of multiplicity infinity if the symbol assumes that value on a set of positive measure, and it is a point of the continuous spectrum if the symbol assumes that values on a set of measure zero.*

Theorem 1.3. *Singular operators with constant symbol commute.*

1.3. Let a point θ have the angular coordinates $\vartheta_1, \vartheta_2, \ldots, \vartheta_{m-2}, \vartheta_{m-1}$ where $0 \leqq \vartheta_j \leqq \pi$, $1 \leqq j \leqq m-2$, and $-\pi \leqq \vartheta_{m-1} \leqq \pi$. Let us introduce the singular operators H_j, $j = 1, 2, \ldots, m-1$, which have the symbols $\exp(\mathrm{i}\vartheta_j)$. The Theorems 1.1–1.3 show that the operators H_j are unitary in $\mathbf{L}_2(\mathbb{R}^m)$ and have a purely continuous spectrum which coincides, in case $j = m-1$, with the circle $|\lambda| = 1$, and, in case $1 \leqq j \leqq m-2$, with the half-circle $|\lambda| = 1$, $\operatorname{Im} \lambda \geqq 0$, in the λ-plane. Finally, the operators commute with each other. By $E_j(\vartheta)$ we denote the spectral function of the operator H_j such that

$$H_j = \int_0^\pi e^{\mathrm{i}\vartheta}\, dE_j(\vartheta), \qquad 1 \leqq j \leqq m-2,$$

$$H_{m-1} = \int_{-\pi}^{\pi} e^{\mathrm{i}\vartheta}\, dE_{m-1}(\vartheta).$$

Next, we introduce operators h_j of multiplication by the function $\exp(\mathrm{i}\vartheta_j)$ ($j = 1, 2, \ldots, m-1$), their spectral function will be denoted by $E_j'(\vartheta)$. The symbol $\Phi(\theta)$ is a unique function of $\exp(\mathrm{i}\vartheta_1), \ldots, \exp(\mathrm{i}\vartheta_{m-1})$,

$$\Phi(\theta) = \Phi(e^{\mathrm{i}\vartheta_1}, e^{\mathrm{i}\vartheta_2}, \ldots, e^{\mathrm{i}\vartheta_{m-1}}),$$

and that shows that the operator of multiplication by the symbol $\Phi(\theta)$ is a function of the operators h_j which is equal to $\Phi(h_1, h_2, \ldots, h_{m-1})$. But then the spectral representation holds true,

$$\Phi(\theta)\, u = \int_\Pi \Phi(e^{\mathrm{i}\vartheta_1^*}, e^{\mathrm{i}\vartheta_2^*}, \ldots, e^{\mathrm{i}\vartheta_{m-1}^*})\, dE'(\theta^*)\, u = \int_\Pi \Phi(\theta^*)\, dE'(\theta^*)\, u,$$

where Π stands for the parallelepiped $0 \leqq \vartheta_j^* \leqq \pi$ $(1 \leqq j \leqq m-2)$, $-\pi \leqq \vartheta_{m-1}^* \leqq \pi$, and

$$dE'(\theta^*) = \prod_{j=1}^{m-1} dE_j'(\vartheta_j^*).$$

Due to (2), $\mathrm{d}E_j(\vartheta_j) = F^{-1}\,\mathrm{d}E_j'(\vartheta_j)\,F$, and therefore

$$Au = \int_{\Pi} \Phi(\theta)\,\mathrm{d}E(\theta)\,u \tag{6}$$

where

$$\mathrm{d}E(\theta) = \prod_{j=1}^{m-1} \mathrm{d}E_j(\vartheta_j). \tag{7}$$

Equation (6) provides another representation of a singular operator by its symbol. Using this representation we may easily obtain the theorems of the present section once again.

Let us now consider a singular operator the symbol of which depends on the pole and is a polynomial in the spherical functions,

$$\Phi(x, \theta) = \sum_{n=0}^{N} \sum_{k=1}^{k_{n,m}} a_n^{(k)}(x)\, Y_{n,m}^{(k)}(\theta).$$

Due to (6), this operator (let us call it A) can be represented in the form

$$Au = \sum_{n=0}^{N} \sum_{k=1}^{k_{n,m}} a_n^{(k)}(x) \int_{\Pi} Y_{n,m}^{(k)}(\theta)\,\mathrm{d}E(\theta)\,u.$$

Taking the coefficients under the integral and changing the order of integration and summation, we obtain

$$Au = \int_{\Pi} \Phi(x, \theta)\,\mathrm{d}E(\theta)\,u. \tag{8}$$

Equation (8) can be substantiated for an essentially wider class of symbols. We shall not do so but take (8) as the definition of a singular operator with a variable symbol. Such an operator is defined in $\mathbf{L}_2(\mathbb{R}^m)$ on the set of all those functions that are taken by the integral operator (8) to functions of the same space $\mathbf{L}_2(\mathbb{R}^m)$.

§ 2. Criteria for boundedness in $\mathbf{L}_2(\mathbb{R}^m)$

One such criterion is provided by Theorem 5.1, Chap. X. In the present section we shall give some criteria for the boundedness of singular operators with a variable symbol. Using the results and notations of the previous section, we shall consider here singular operators of type (1.8).

Theorem 2.1. *Assume that the symbol $\Phi(x, \theta)$ and its derivatives*

$$\frac{\partial \Phi}{\partial \vartheta_1}, \frac{\partial \Phi}{\partial \vartheta_2}, \ldots, \frac{\partial^2 \Phi}{\partial \vartheta_1\, \partial \vartheta_2}, \ldots, \frac{\partial^{m-2} \Phi}{\partial \vartheta_1\, \partial \vartheta_2 \ldots \partial \vartheta_{m-2}} \tag{1}$$

are measurable with respect to (x, θ), continuous for fixed x, and independently of x bounded. Furthermore, let Φ have the distributional (or generalized) derivative

$$\frac{\partial^{m-1} \Phi}{\partial \vartheta_1\, \partial \vartheta_2 \ldots \partial \vartheta_{m-1}} \tag{2}$$

satisfying the inequality

$$\int_0^{\pi} \int_0^{\pi} \cdots \int_{-\pi}^{\pi} \left| \frac{\partial^{m-1} \Phi}{\partial \vartheta_1\, \partial \vartheta_2 \ldots \partial \vartheta_{m-1}} \right|^2 \mathrm{d}\vartheta_1\, \mathrm{d}\vartheta_2 \ldots \mathrm{d}\vartheta_{m-1} \leqq C_0^2 = \text{const}. \tag{3}$$

Then the operator (1.8) *is bounded in* $L_2(\mathbb{R}^m)$, *and its norm is not greater than*

$$C \max \left[\sup |\Phi|, \sup \left| \frac{\partial \Phi}{\partial \vartheta_1} \right|, \ldots, \sup \left| \frac{\partial^{m-2} \Phi}{\partial \vartheta_1 \, \partial \vartheta_2 \ldots \partial \vartheta_{m-2}} \right|, C_0 . \right] \tag{4}$$

Proof. We integrate (1.8) by parts. Taking into account the relations $E_1(0) = 0$ and $E_1(\pi) = 1$ we find

$$Au = -\int_0^\pi \int_0^\pi \ldots \int_0^\pi \int_{-\pi}^\pi \frac{\partial \Phi}{\partial \vartheta_1} E_1(\vartheta_1) \, d\vartheta_1 \prod_{j=2}^{m-1} dE_j(\vartheta_j) \, u$$

$$+ \int_0^\pi \ldots \int_0^\pi \int_{-\pi}^\pi \Phi \prod_{j=2}^{m-1} dE_j(\vartheta_j) \, u .$$

Further integration by parts yields finally

$$Au = \sum_{k=0}^{m-1} \pm \int \ldots \int \frac{\partial^k \Phi}{\partial \vartheta_{l_1} \ldots \partial \vartheta_{l_k}} \prod_{r=1}^{k} E_{l_r}(\vartheta_{l_r}) \, u \, d\vartheta_{l_1} \ldots d\vartheta_{l_k} . \tag{5}$$

Under the integral only derivatives of the form (1) or (2) occur where derivatives of order k are just under a k-fold integral. Denote by Bu an arbitrary summand of the sum (5), then, due to the assumptions of the theorem,

$$|Bu|^2 \leqq C^2 \int \ldots \int \left| \prod_{r=1}^{k} E_{l_r}(\vartheta_{l_r}) \, u \right|^2 d\vartheta_{l_1} \ldots d\vartheta_{l_r} . \tag{6}$$

Taking into account the equation $\|E_j\| = 1$ we obtain integrating over $\mathbb{R}^m$ the estimate $\|Bu\| \leqq C' \|u\|$, $C' = \text{const}$. Hence, $\|A\| \leqq NC'$ where N denotes the number of summands in (5). Thus, Theorem 2.1 is proved. ∎

Remark 1. The conditions in Theorem 2.1 are not invariant with respect to rotations of the coordinate system. Such conditions may be called *anisotropic* ones.

Remark 2. The conditions in Theorem 2.1 are, by far, not necessary. The operator (1.8) is, for example, bounded in $L_2(\mathbb{R}^m)$ if its symbol can be expanded in a series

$$\Phi(x, \theta) = \sum_{n=1}^{\infty} a_n(x) \, \Phi_n(\theta) \tag{7}$$

such that

$$\sum_{n=1}^{\infty} \sup |a_n(x)| \sup |\Phi_n(\theta)| < \infty . \tag{8}$$

The norm of that operator is then not greater than the sum of the last series. Let us give another similar criterion. Assume that the symbol of the operator (1.8) has the expansion (7). Put $A_n = F^{-1} \Phi_n(\theta) F$, then $A = \sum_{n=1}^{\infty} a_n(x) A_n$, and therefore

$$\|Au\|^2 = \int_{\mathbb{R}^m} \left| \sum_{n=1}^{\infty} a_n(x) \, (A_n u) \, (x) \right|^2 dx \leqq \int_{\mathbb{R}^m} \sum_{n=1}^{\infty} |a_n(x)|^2 \sum_{n=1}^{\infty} |(A_n u) \, (x)|^2 \, dx$$

$$\leqq \sup_x \sum_{n=1}^{\infty} |a_n(x)|^2 \sum_{n-1}^{\infty} \|A_n u\|^2 .$$

($\|\cdot\|$ means the norm in $L_2(\mathbb{R}^m)$). Furthermore, due to the Plancherel theorem,

$$\|A_n u\|^2 = \|\Phi_n(\theta) \, \hat{u}\|^2 = \int_{R^m} |\Phi_n(\theta)|^2 \, |\hat{u}(\xi)|^2 \, d\xi, \qquad \theta = \xi/|\xi| ,$$

such that finally

$$\sum_{n=1}^{\infty} \|A_n u\|^2 = \int |\hat{u}(\xi)|^2 \sum_{n=1}^{\infty} |\Phi_n(\theta)|^2 \, d\xi \leqq \sup_{\theta} \sum_{n=1}^{\infty} |\Phi_n(\theta)|^2 \, \|u\|^2.$$

Thus, the operator (1.8) is bounded in $\mathbf{L}_2(\mathbb{R}^m)$ if the conditions

$$P^2 := \sup_{x} \sum_{n=1}^{\infty} |a_n(x)|^2 < \infty, \qquad Q^2 := \sup_{\theta} \sum_{n=1}^{\infty} |\Phi_n(\theta)|^2 < \infty \tag{9}$$

are fulfilled. If this is the case, we have

$$\|A\| \leqq PQ. \tag{10}$$

Theorem 2.2. *Let* $\Phi(x, \theta) \hat{\in} \mathbf{W}_2^{(l)}(S)$, $l > (m-1)/2$. *Then the operator* (1.8) *is bounded in* $\mathbf{L}_2(\mathbb{R}^m)$.

Proof. We expand the symbol in a series with respect to spherical functions and give the series the following form,

$$\Phi(x, \theta) = \sum_{n=0}^{\infty} a_n(x) \sum_{k=1}^{k_{n,m}} \alpha_n^{(k)}(x) \, Y_{n,m}^{(k)}(\theta)$$

where

$$\sum_{k=1}^{k_{n,m}} |\alpha_n^{(k)}(x)|^2 = 1.$$

From the proof of Theorem 6.1, Chap. X, the relation

$$\sum_{n=1}^{\infty} \sum_{k=1}^{k_{n,m}} |a_n(x) \, \alpha_n^{(k)}(x)|^2 n^{2l} = \sum_{n=1}^{\infty} |a_n(x)|^2 \, n^{2l} \leqq C_0^2 M^2 \tag{11}$$

follows where $M = \sup_{x} \|\Phi(x, \cdot)\|_{\mathbf{W}_2^{(l)}(S)}$ and $C_0 = \text{const}$. Denote by A_n the singular operator with the symbol

$$\Phi_n(x, \theta) = \sum_{k=1}^{k_{n,m}} \alpha_n^{(k)}(x) \, Y_{n,m}^{(k)}(\theta).$$

According to the inequality (10) we have

$$\|A_n\|^2 \leqq \max_{\theta} \sum_{k=1}^{k_{n,m}} |Y_{n,m}^{(k)}(\theta)|^2. \tag{12}$$

We shall show that

$$\max_{\theta} \sum_{k=1}^{k_{n,m}} |Y_{n,m}^{(k)}(\theta)|^2 = \max_{\beta,\theta} \left| \sum_{k=1}^{k_{n,m}} \beta_k Y_{n,m}^{(k)}(\theta) \right|^2 \tag{13}$$

where β_k are constants and $\sum_{k=1}^{k_{n,m}} |\beta_k|^2 = 1$. By the Cauchy inequality,

$$\left| \sum_{k=1}^{k_{n,m}} \beta_k Y_{n,m}^{(k)}(\theta) \right|^2 \leqq \sum_{k=1}^{k_{n,m}} |Y_{n,m}^{(k)}(\theta)|^2,$$

thence,

$$\max_{\beta,\theta} \left| \sum_{k=1}^{k_{n,m}} \beta_k Y_{n,m}^{(k)}(\theta) \right|^2 \leqq \max_{\theta} \sum_{k=1}^{k_{n,m}} |Y_{n,m}^{(k)}(\theta)|^2. \tag{14}$$

Denote by θ_0 that point at which the sum in the right-hand side of inequality (14) assumes its maximum, and put

$$\beta_k = \overline{Y_{n,m}^{(k)}(\theta_0)} \left[\sum_{k=1}^{k_{n,m}} |Y_{n,m}^{(k)}(\theta_0)|^2\right]^{-1/2}.$$

Then

$$\left|\sum_{k=1}^{k_{n,m}} \beta_k Y_{n,m}^{(k)}(\theta_0)\right|^2 = \sum_{k=1}^{k_{n,m}} |Y_{n,m}^{(k)}(\theta_0)|^2 = \max_\theta \sum_{k=1}^{k_{n,m}} |Y_{n,m}^{(k)}(\theta)|^2,$$

hence,

$$\max_{\beta,\theta}\left|\sum_{k=1}^{k_{n,m}} \beta_k Y_{n,m}^{(k)}(\theta)\right|^2 \geqq \max_\theta \sum_{k=1}^{k_{n,m}} |Y_{n,m}^{(k)}(\theta)|^2$$

and this gives, together with (14), the equality (13).

As usually, we assume the functions $Y_{n,m}^{(k)}(\theta)$ to be orthonormal on S. Then

$$Y_n(\theta) = \sum_{k=1}^{k_{n,m}} \beta_k Y_{n,m}^{(k)}(\theta)$$

provides a normed spherical function of order n. Due to (6.14), Chap. X, $|Y_n(\theta)| \leqq Cn^{m/2-1}$, but then by means of (12), $\|A_n\| \leqq Cn^{m/2-1}$. Furthermore,

$$\left\|\sum_{n=1}^{\infty} A_n u\right\|^2 = \int_{\mathbb{R}^m} \left|\sum_{n=1}^{\infty} a_n(x) A_n u\right|^2 \mathrm{d}x \leqq \sup_x \sum_{n=1}^{\infty} |a_n(x)|^2 n^{2l} \sum_{n=1}^{\infty} n^{-2l} \|A_n u\|^2$$

$$\leqq CM^2 \|u\|^2 \sum_{n=1}^{\infty} n^{m-2-2l}.$$

Since $m - 2 - 2l < -1$ the last series is convergent, and the operator A is bounded. ∎

Remark 3. As one can see from the proof, the norm of the operator (1.8) is bounded by the quantity $CM = C \sup_x \|\Phi(x, \cdot)\|_{\mathbf{W}_2^{(l)}(S)}$. Furthermore, the assumption in Theorem 2.2 is, due to Theorem 7.1, Chap. X, equivalent to the following condition: $\exists \varepsilon > 0 : f \in \mathbf{W}_2^{(\varepsilon-1/2)}(S)$.

Remark 4. Theorem 2.2 is contained in the paper by AGRANOVICH [1] but proved there in a different way. Another proof to Theorem 2.2 is given in the paper by MAZYA and HAIKIN [1]. Similar but somewhat weaker results are contained in MIKHLIN [2] (if $m = 2$) and in CALDERON and ZYGMUND [4, 5] (for m arbitrary). The proof presented here uses essentially the arguments of the last two papers.

The theorems proved in the present section remain to hold true if the singular integral is taken over a measurable set $G \subseteq \mathbb{R}^m$ and the space $\mathbf{L}_2(\mathbb{R}^m)$ is replaced by $\mathbf{L}_2(G)$. Indeed, let

$$(Au)(x) = a(x)\, u(x) + \int_G r^{-m} f(x, \theta)\, u(y)\, \mathrm{d}y,$$

and the symbol of the operator A satisfy the assumptions of any theorem derived here. Define the functions $a(x)$, $u(x)$, and $f(x, \theta)$ on the whole $\mathbb{R}^m$ by putting them zero for $x \notin G$. The conditions of the theorems on boundedness are not violated, and if $u \in \mathbf{L}_2(G)$ then the extended function belongs to $\mathbf{L}_2(\mathbb{R}^m)$. Now, $\|Au\|_{\mathbf{L}_2(\mathbb{R}^m)} \leqq C \|u\|_{\mathbf{L}_2(\mathbb{R}^m)}$, or

$$\int_{\mathbb{R}^m} |Au|^2 \,\mathrm{d}x \leqq C^2 \int_{\mathbb{R}^m} |u|^2 \,\mathrm{d}x = C^2 \int_G |u|^2 \,\mathrm{d}x.$$

The more,

$$\int_G |Au|^2 \,\mathrm{d}x \leqq C^2 \int_G |u|^2 \,\mathrm{d}x,$$

which was to be demonstrated. A similar remark applies also to theorems on boundedness in $\mathbf{L}_p$ (see the following § 3).

§ 3. The theorem of Calderon and Zygmund

3.1. Theorem of Calderon and Zygmund (CALDERON and ZYGMUND [3]). *If the characteristic $f(x, \theta)$ of the singular integral*

$$(A_0 u)(x) = \int_{\mathbb{R}^m} K(x, x-y)\, u(y)\, \mathrm{d}y, \qquad K(x, y) = r^{-m} f(x, \theta), \tag{1}$$

satisfies the inequality

$$\int_S |f(x, \theta)|^{p'} \,\mathrm{d}S \leqq C_0 = \mathrm{const} \tag{2}$$

then the operator (1) *is bounded in* $\mathbf{L}_p(\mathbb{R}^m)$ *and*

$$\|A_0\|_{\mathbf{L}_p(\mathbb{R}^m)} \leqq C \sup_x \|f(x, \cdot)\|_{\mathbf{L}_{p'}(S)}. \tag{3}$$

Proof. The kernel $K(x, z)$ can be split into $K(x, z) = K_1(x, z) + K_2(x, z)$ where K_1 is an odd and K_2 an even function of z. Accordingly, we shall prove the theorem for odd and even kernels separately.

Let the charateristic $f(x, \theta)$ have, in addition to (2), the property $f(x, -\theta) = -f(x, \theta)$. Put

$$\tilde{u}_\varepsilon(x) = \int_{r>\varepsilon} r^{-m} f(x, \theta)\, u(y)\, \mathrm{d}y \tag{4}$$

and $\bar{u}(x) = \sup_\varepsilon |\tilde{u}_\varepsilon(x)|$, $u \in \mathbf{L}_p(\mathbb{R}^m)$. We show

a) The inequality

$$\|\bar{u}\|_{\mathbf{L}_p(\mathbb{R}^m)} \leqq C \|u\|_{\mathbf{L}_p(\mathbb{R}^m)} \tag{5}$$

holds true.

b) The limit $\tilde{u}(x) = \lim_{\varepsilon\to 0} \tilde{u}_\varepsilon(x)$ exists a.e. in $\mathbb{R}^m$, or, in other words, the integral (1) exists a.e. in $\mathbb{R}^m$.

c) The operator A_0 is bounded in $\mathbf{L}_p(\mathbb{R}^m)$.

Let t be a real variable and $g(t) \in \mathbf{L}_p(\mathbb{R})$. Consider the one-dimensional integral operator

$$\tilde{g}_\varepsilon(s) = \int_{|s-t|>\varepsilon(s)} \frac{g(t)}{s-t}\,\mathrm{d}t \tag{*}$$

with a bounded and measurable function $\varepsilon(s)$. As shown by CALDERON and ZYGMUND [1] (Chap. II, Theorem 1), the norm of this operator in $\mathbf{L}_2(\mathbb{R})$ is bounded by a constant that does not depend on the function $\varepsilon(s)$. Due to the Fatou lemma (F. RIESZ and SZ.-NAGY [1], No. 20) you can show

$$\|\bar{g}\|_{\mathbf{L}_p} \leqq C \|g\|_{\mathbf{L}_p}, \qquad \bar{g}(s) = \sup_\varepsilon |\tilde{g}_\varepsilon(s)|. \tag{6}$$

With y' being an arbitrary unit vector we put

$$\tilde{u}_\varepsilon(x, y') = \int_{|t|>\varepsilon} u(x - ty') \frac{\mathrm{d}t}{t}, \tag{7}$$

$$\bar{u}(x, y') = \sup_\varepsilon |\tilde{u}_\varepsilon(x, y')|; \tag{8}$$

both the functions (7) and (8) are measurable. Suppose that the end point of the vector x varies on a straight line parallel to y', i.e. $x = sy' + x_0$, $x_0 = \text{const}$. Then the integral (7) is a special case of the integral (*) for $g(t) = u(ty' + x_0)$. By inequality (6) we obtain

$$\int_{\mathrm{R}} [\bar{u}(x - ty')]^p \mathrm{d}t \leqq C \int_{\mathrm{R}} |u(x - ty')|^p \, \mathrm{d}t.$$

Integrating over the space of the straight lines parallel to the vector y' we get

$$\int_{\mathrm{R}^m} [\bar{u}(x, y')]^p \mathrm{d}x \leqq C \int_{\mathrm{R}^m} |u(x)|^p \, \mathrm{d}x. \tag{9}$$

Now, we put

$$u_0(x) = \frac{1}{2} \int_S \bar{u}(x, y') \, |K(x, y')| \, \mathrm{d}y', \tag{10}$$

$$\check{u}_\varepsilon(x) = \frac{1}{2} \int_S \tilde{u}_\varepsilon(x, y') \, K(x, y') \, \mathrm{d}y'. \tag{11}$$

It is easy to see that $\check{u}_\varepsilon(x) = \tilde{u}_\varepsilon(x)$ where $\tilde{u}_\varepsilon(x)$ is defined by (4). To that aim it is sufficient to replace $\tilde{u}_\varepsilon(x, y')$ in (11) by the right-hand side of (7). The integral obtained coincides with the integral (4) if in the latter spherical coordinates with the point x as center are introduced. Now, due to (8), $|\tilde{u}_\varepsilon(x)| \leqq u_0(x)$.

Further, we find

$$\int_{\mathrm{R}^m} [u_0(x)]^p \, \mathrm{d}x = 2^{-p} \int_{\mathrm{R}_m} \Big[\int_S \bar{u}(x, y') \, |K(x, y')| \, \mathrm{d}y' \Big]^p \mathrm{d}x$$
$$\leqq 2^{-p} \int_{\mathrm{R}^m} \mathrm{d}x \int_S [\bar{u}(x, y')]^p \, \mathrm{d}y' \Big[\int_S |K(x, y')|^p \mathrm{d}y' \Big]^{p/p'}.$$

The inner integral coincides with the integral (2), therefore,

$$\int_{\mathrm{R}^m} [u_0(x)]^p \, \mathrm{d}x \leqq 2^{-p} C \int_S \mathrm{d}y' \int_{\mathrm{R}^m} [\bar{u}(x, y')]^p \, \mathrm{d}x$$

and, by the inequality (9),

$$\|u_0\|_{\mathbf{L}_p(\mathrm{R}^m)} \leqq C \, \|u\|_{\mathbf{L}_p(\mathrm{R}^m)}. \tag{12}$$

Thus, we have $u_0 \in \mathbf{L}_p(\mathbb{R}^m)$. Furthermore, also $\tilde{u}_\varepsilon \in \mathbf{L}_p(\mathbb{R}^m)$, and its $\mathbf{L}_p(\mathbb{R}^m)$ norm satisfies the same inequality (12). Now it is easy to show that $\bar{u} \in \mathbf{L}_p(\mathbb{R}^m)$, where the $\mathbf{L}_p(\mathbb{R}^m)$ norm satisfies also the inequality (12), what is equivalent to (5).

Next, we show that the function $\tilde{u}_\varepsilon(x)$ has a limit as $\varepsilon \to 0$ a.e. in $\mathbb{R}^m$ as well as in the $\mathbf{L}_p(\mathbb{R}^m)$ norm.

Let $\varrho(t)$ be an even and continuously differentiable function such that $\varrho(0) = 1$ and $\varrho(t) = 0$, $|t| \geqq 1$. Since the one-dimensional singular integral $\int_{-\infty}^{\infty} \varrho(t)\, t^{-1} \mathrm{d}t$ is convergent the integral $\int_{|t|>\varepsilon} \varrho(t)\, t^{-1} \, \mathrm{d}t$ has a finite limit as $\varepsilon \to 0$. Introducing spherical coordinates we can show that the integral $\int_{r>\varepsilon} K(x, x - y)\, \varrho(r) \, \mathrm{d}y$ has a finite limit as $\varepsilon \to 0$.

Suppose that the function u is continuously differentiable and vanishes outside of a certain ball. Then

$$\tilde{u}_\varepsilon(x) = \int_{r>\varepsilon} K(x, x-y)\,[u(y) - u(x)\,\varrho(r)]\,dy + u(x) \int_{r>\varepsilon} K(x, x-y)\,\varrho(r)\,dy.$$

Since $\varrho(0) = 1$ the integrand in the first integral is absolutely integrable over the whole space $\mathbb{R}^m$, and it is clear that the limit $\tilde{u}_\varepsilon(x) = \lim \tilde{u}_\varepsilon(x)$ exists everywhere in $\mathbb{R}^m$.

In the general case, if $u \in \mathbf{L}_p(\mathbb{R}^m)$, one can put $u = v + w$ with v being continuously differentiable and w of arbitrarily small norm. Then $\tilde{u}_\varepsilon(x) = \tilde{v}_\varepsilon(x) + \tilde{w}_\varepsilon(x)$ and $|\tilde{w}_\varepsilon|(x)| \leqq \overline{w}(x)$. The limit $\lim\limits_{\varepsilon\to 0} \tilde{v}_\varepsilon(x)$ exists, hence

$$\overline{\lim}\, \tilde{u}_\varepsilon(x) - \underline{\lim}\, \tilde{u}_\varepsilon(x) \leqq 2\overline{w}(x).$$

Since the norm $\|\overline{w}\| \leqq C\,\|w\|$ is arbitrarily small the last inequality shows that the limit $\tilde{u}(x) = \lim\limits_{\varepsilon\to 0} \tilde{u}_\varepsilon(x)$ exists a.e. in $\mathbb{R}^m$. As mentioned above, we have $|\tilde{u}_\varepsilon(x)| \leqq u_0(x)$ and $u_0 \in \mathbf{L}_p(\mathbb{R}^m)$. Thence, $\|\tilde{u}\| \leqq \|u_0\| \leqq C\,\|u\|$, and this completes the proof of Theorem 3.1 for the case of an odd kernel.

3.2. Now, we consider the case of an even kernel. To begin with, we introduce the so-called *Riesz kernel* or vector singular kernel due to M. Riesz,

$$R(x) = \pi^{-(m+1/2)}\,\Gamma\left(\frac{m+1}{2}\right) \frac{x}{|x|^{m+1}}\,. \tag{13}$$

It is an odd kernel such that for $u \in \mathbf{L}_p(\mathbb{R}^m)$ and

$$v(x) = \int_{\mathbb{R}^m} R(x-y)\,u(y)\,dy \tag{14}$$

the inequality $\|v\| \leqq C\,\|u\|$ holds. We shall show now that

$$u(x) = -\int_{\mathbb{R}^m} R(x-y)\,v(y)\,dy. \tag{15}$$

In order to do so, we apply Fourier transformation to (14), then by (1.3), Chap. X, $Fv = FR \cdot Fu$. The function FR is the symbol of the Riesz kernel. On the other hand, the characteristic $\pi^{-(m+1)/2}\,\Gamma((m+1)/2)\,(x-y)/r$ of that kernel turns out to be a spherical function of order 1 and its symbol is obtained by multiplying it by $\gamma_{m,1}$ (see eq. (2.6), Chap. X). Hence,

$$FR = \frac{i\pi^{m/2}\,\Gamma\left(\frac{1}{2}\right)}{\Gamma\left(\frac{m+1}{2}\right)}\,\pi^{-(m+1)/2}\,\Gamma\left(\frac{m+1}{2}\right)\frac{x}{|x|} = i\frac{x}{|x|}$$

and $Fv = ixFu/|x|$. Multiplying by $x/|x|$ we find $Fu = -ixFv/|x| = -FR \cdot Fv$. Now, inverse Fourier transformation yields (15).

For $\tilde{u}(x) = \int_{\mathbb{R}^m} K(x, x-y)\,u(y)\,dy$ from the relations (14) and (15) the equation $\tilde{u}(x) = \int_{\mathbb{R}^m} L(x, x-y)\,v(y)\,dy$ follows where $L(x, x-y) = \int_{\mathbb{R}^m} K(x, z)\,R(y-z)\,dz$. It is easy to see that L is an odd singular kernel. It is possible to show that the characteristic of that kernel satisfies inequality (2) (cf. Calderon and Zygmund [3]; we shall omit this proof since it is rather cumbersome). But then $\|\tilde{u}\| \leqq C\,\|v\| \leqq C\,\|u\|$, and Theorem 3.1 is proved completely. ∎

3.3. In Calderon and Zygmund [3, 6] a theorem is proved which is somewhat more general than Theorem 3.1: If the characteristic $f(x, \theta)$ satisfies the inequality

$$\int_S |f(x, \theta)|^q \, dS \leq C_0 = \text{const} \tag{16}$$

for some $q > 1$ then the operator (1) is bounded in $\mathbf{L}_p(\mathbb{R}^m)$ whenever $p \geq q/(q-1)$ (more precisely, if $p/q < (p-1)\, m/(m-1)$, $1 < p \leq 2$, or $p/q < m/(m-1) + p - 2$, $2 \leq p < \infty$, and the result is sharp for $1 < p \leq 2$).

Without proof we quote another interesting theorem from Calderon and Zygmund [3]. If the characteristic of the singular integral (1) does not depend on the pole, i.e. $f(x, \theta) = f(\theta)$ then the operator (1) is bounded in $\mathbf{L}_p(\mathbb{R}^m)$ for any $p \in (1, \infty)$ provided that

$$\int_S |f(\theta)| \, dS < \infty, \qquad \int_S |f(\theta) + f(-\theta)| \ln |f(\theta) + f(-\theta)| \, dS < \infty. \tag{17}$$

3.4. Theorem 3.2. *The singular operator*

$$(Au)(x) = a(x)\, u(x) + \int_{\mathbb{R}^m} r^{-m} f(x, \theta)\, u(y)\, dy \tag{18}$$

is bounded in $\mathbf{L}_p(\mathbb{R}^m)$ *if its symbol* Φ *satisfies the condition* $\Phi(x, \theta) \hat{\in} \mathbf{W}_2^{(\lambda)}(S)$ *where* $\lambda \geq (m-1)/p + 1/2$ *if* $1 < p \leq 2$ *and* $\lambda \geq m/2$ *if* $p > 2$. *The norm of the operator* (18) *is bounded by*

$$C\,[\sup_x |a(x)| + \sup_x \|\Phi(x, \cdot)\|_{\mathbf{W}_2^{(\lambda)}(S)}]. \tag{19}$$

Proof. By the assumptions of the theorem, the coefficient $a(x)$ is bounded, and then so is the operator of multiplication by this coefficient in $\mathbf{L}_p(\mathbb{R}^m)$. Further, the relation $\Phi(x, \theta) \hat{\in} \mathbf{W}_2^{(\lambda)}(S)$ implies (cf. Theorem 7.1, Chap. X) that $f(x, \theta) \hat{\in} \mathbf{W}_2^{(\lambda - m/2)}(S)$. Let $p \geq 2$. Because of $\lambda \geq m/2$ it follows that $f(x, \theta) \hat{\in} \mathbf{L}_2(S)$ and, the more, $f(x, \theta) \hat{\in} \mathbf{L}_{p'}(S)$. Due to Theorem 3.1, the operator (18) is bounded. Now, let $1 < p < 2$. By the Sobolev imbedding theorem, it is sufficient for $f(x, \theta) \hat{\in} \mathbf{L}_{p'}(S)$ that

$$\frac{2(m-1)}{m - 1 - 2(\lambda - m/2)} \geq p'$$

or $\lambda \geq (m-1)/p + 1/2$. If λ is so then, again by Theorem 3.1, the operator (18) is bounded in $\mathbf{L}_p(\mathbb{R}^m)$. The estimate (19) for the norm of the operator (18) follows from Theorem 3.1 and the imbedding theorem for Sobolev spaces. ▮

Theorem 3.3. *Let the symbol of the operator* (18) *satisfy the condition mentioned in Theorem 3.2. The expansion of the symbol in a series with respect to spherical functions*

$$\Phi(x, \theta) = \sum_{n=0}^{\infty} \sum_{k=1}^{k_{n,m}} a_n^{(k)}(x)\, Y_{n,m}^{(k)}(\theta) \tag{20}$$

corresponds to the expansion of the operator (18)

$$(Au)(x) = a_1^{(0)}(x)\, u(x) + \sum_{n=1}^{\infty} \sum_{k=1}^{k_{n,m}} \frac{a_n^{(k)}(x)}{\gamma_{n,m}} \int_{\mathbb{R}^m} \frac{Y_{n,m}^{(k)}(\theta)}{r^m} u(y)\, dy \tag{21}$$

which is convergent in the norm of the space $\mathbf{L}_p(\mathbb{R}^m)$.

Proof. By Theorem 6.7, Chap. X, the remainder of the series (20) converges to zero in the metric of the space $\mathbf{W}_{(\lambda)}^2(S)$ uniformly with respect to x. By Theorem 3.2, the $\mathbf{L}_p(\mathbb{R}^m)$ norm of the corresponding remainder of the series (21) tends to zero. ▮

§ 4. Some further results

In the present section we shall formulate without proof some new results concerning the boundedness of singular integral operators in spaces with integral metrics.

4.1. Mazya and Haikin [2] consider the singular integral operator

$$A = F^{-1}_{\xi\to x}\,\Phi(x, \theta)\,F_{x\to\xi}, \qquad \theta = \xi/|\xi| \tag{1}$$

in rather general Banach spaces. Let $B = B(\mathbb{R}^m)$ and $M = M(\mathbb{R}^{m-1})$ be Banach spaces of measurable functions defined on $\mathbb{R}^m$ and $\mathbb{R}^{m-1}$, resp. The norms in these spaces are supposed to be monotoneous (i.e. $u, v \in B$ and $|u(x)| \leqq |v(x)|$, $\forall\, x \in \mathbb{R}^m$, implies $\|u\|_B \leqq \|v\|_B$). Furthermore, the spaces B and M are supposed to have the following property: If the function $u(x, t)$ is defined on $\mathbb{R}^m \times \mathbb{R}^{m-1}$ and the two norms $\mu_u(t) = \|u(\cdot, t)\|_B$, $\nu_u(x) = \|u(x, \cdot)\|_M$ are defined and measurable functions then

$$\|\nu_u\|_B \leqq \|\mu_u\|_M. \tag{2}$$

Let us introduce the space $Q = Q(S)$ (S denotes as usually the unit sphere in $\mathbb{R}^m$). Let φ_k, $k = 1, \ldots, q$, be real-valued functions belonging to the class $C^\infty(S)$ which form on S a partition of unity subordinate to a covering of this sphere by sets U_k, $k = 1, 2, \ldots, q$, and $\varkappa_k$ be diffeomorphisms of U_k into $\mathbb{R}^{m-1}$. A function $\Phi(\theta)$ defined on S belongs to the space Q iff for every k, $1 \leqq k \leqq q$, $\varphi_k F\Phi \circ \varkappa_k^{-1} \in M^*$. Here, F is the Fourier transformation, M^* the dual to M, and $\circ$ denotes the following coordinate transformation: For $\eta \in \mathbb{R}^{m-1}$ and $\eta = \varkappa_k(\theta)$ it is $\Phi \circ \varkappa_k^{-1} = \Phi(\varkappa_k^{-1}(\eta))$. The norm in Q is defined by the formula

$$\|\Phi\|_Q = \max_k \|\varphi_k F\Phi \circ \varkappa_k^{-1}\|_{M^*}. \tag{3}$$

An example is provided by $M = \mathbf{W}_2^{(l)}(\mathbb{R}^{m-1})$, l real, and $Q = \mathbf{W}_2^{(-l)}(S)$.

In the paper cited the estimate

$$\|A\|_B \leqq \sup_x \|\Phi(x, \cdot)\|_Q \sum_{k=1}^{q} \|G_k\|_M \tag{4}$$

is proved where $G_k(\tau) = \|D_k(\tau)\|_B$, and $D_k(\tau)$ is the singular operator with the symbol $\varphi_k(\theta)\exp(2\pi i\tau\varkappa_k(\theta))$.

As a corollary to the estimate (4) the following criterion for the boundedness of the operator (1) in $\mathbf{L}_p(\mathbb{R}^m)$, $p \in (1, \infty)$, is obtained. Let $\mu(t)$, $t \in \mathbb{R}^{m-1}$, be a weight function of tempered growth (for the definition see Hörmander [3], No. 17). Let $H = \mathbf{H}^\mu_{p'}(\mathbb{R}^{m-1})$ denote the space with the norm

$$\|u\|_H = \left[\int_{\mathbb{R}^m} |\mu(t)\,Fu(t)|^{p'}\,dt\right]^{1/p'}$$

and $H^1 = \mathbf{H}^\mu_{p'}(S)$ the space of functions defined on S with the norm

$$\|\Phi\|_{H^1} = \max_k \|\varphi_k \Phi_0 \varkappa_k^{-1}\|_H.$$

Then the operator (1) is bounded in $\mathbf{L}_p(\mathbb{R}^m)$ provided that the condition

$$c_0^p := \int_{\mathbb{R}^{m-1}} \frac{[(1 + |t|)^{m/2}\ln(1 + |t|)]^{|p-2|}}{[\mu(t)]^p}\,dt < \infty \tag{5}$$

is fulfilled, and

$$\|A\|_{\mathbf{L}_p(\mathbb{R}^m)} \leqq C c_0 \sup_x \|\Phi(x, \cdot)\|_H. \tag{6}$$

4.2. Birman and Solomyak [1, 2] gave a representation of the singular integral operator as a double operator Stieltjes integral

$$A = \int_{\mathbb{R}^m} \int_S \Phi(x, \theta) \, \mathrm{d}E(x) \, \mathrm{d}E'(\theta) \tag{7}$$

with certain spectral measures $E(x)$ and $E'(\theta)$. The representation (7) gives reason for the following criterion on the boundedness of a singular operator in $\mathbf{L}_2(\mathbb{R}^m)$.

Let, as before, Σ denote the Riemann sphere and y be a variable point on Σ. By $x = x(y)$ we denote the inverse stereographic projection of Σ onto $\mathbb{R}^m$. Put $\Phi(x(y), \theta) = \Phi(y, \theta)$. If $\Phi(y, \theta) \mathrel{\hat{\in}} \mathbf{W}_2^{(\alpha)}(\mathbb{R}^m)$ (the hat means here uniformly with respect to θ) and $\alpha > m/2$ then the operator (1) is bounded in $\mathbf{L}_2(\mathbb{R}^m)$.

This new criterion is interesting, in particular, for the symbol is assumed to be smooth with respect to the variable x but not with respect to the variable θ.

4.3. Besov, Ilyin, and Lizorkin [1] consider a certain class of singular integrals which they call *anisotropic* ones. They prove several theorems on the boundedness of these operators in $\mathbf{L}_p(\mathbb{R}^m)$.

Let $a = (a_1, a_2, \ldots, a_m)$, $a_j \geqq 0$, $\sum_{j=1}^m a_j = n$. A function $K(x) = K(x_1, x_2, \ldots, x_m)$ is called *a-homogeneous* of degree n iff for any $t > 0$

$$K(t^a x) = K(t^{a_1} x_1, t^{a_2} x_2, \ldots, t^{a_m} x_m) = t^n K(x).$$

Let $K(x)$ be a function which is a-homogeneous of degree $-m$ satisfying the conditions

$$\int_S K(x) \sum_{j=1}^m a_j x_j^2 \, \mathrm{d}S = 0,$$

$$|K(x) - K(y)| \leqq \omega(x - y), \qquad \forall x, y \in S; \qquad \int_0^1 \frac{\omega(t)}{t} \mathrm{d}t < \infty.$$

Put

$$(K_\varepsilon u)(x) = \int_{\mathbb{R}^m \setminus \sigma_\varepsilon} K(x - y) \, u(y) \, \mathrm{d}y, \tag{8}$$

where the domain σ_ε is defined by the inequality

$$\sum_{j=1}^m \frac{a_j (x_j - y_j)^2}{\varepsilon^{2a_j}} < 1.$$

In the paper cited it is proved that the operator K_ε is bounded in $\mathbf{L}_p(\mathbb{R}^m)$ independently of ε, and it has a limit Ku in $\mathbf{L}_p(\mathbb{R}^m)$ as $\varepsilon \to 0$. This limit serves as a definition of an anisotropic singular integral,

$$\lim_{\varepsilon \to 0} K_\varepsilon u = Ku := \int_{\mathbb{R}^m} K(x - y) \, u(y) \, \mathrm{d}y. \tag{9}$$

Obviously, the operator K is bounded in $\mathbf{L}_p(\mathbb{R}^m)$.

Finally, we mention a paper by Rozin[1]) [1] where the following statement is proved. Let K be the singular operator defined by eq. (9) the characteristic of which belongs to the class $\mathbf{Lip}_1(S)$. If $v(t)$, $0 \leqq t < \infty$, is a monotoneously decreasing function then the

[1]) A. L. Rozin was a young Leningrad mathematician who died in a tragic accident when mountaineering.

estimate

$$\int_{\mathbb{R}^m} |(Ku)(x)|^p v(x)\, dx \leq C \Big(\int_{\mathbb{R}^m} |u(x)|\, dx\Big)^p, \qquad 0 < p < 1,$$

holds if and only if

$$\int_0^1 v(t)\, t^{n-1-np}\, dt < \infty .$$

§ 5. Singular integrals in weighted spaces. Stein's theorem

5.1. By $L_p(\Omega, \varrho)$ we shall denote the space of functions defined a.e. on Ω for which the norm

$$\|u\|_{L_p(\Omega,\varrho)} = \Big[\int_\Omega \varrho^p(x)\, |u(x)|^p\, dx\Big]^{1/p} \tag{1}$$

is finite.

The function $\varrho(x)$ is called usually *weight function*, and it is always assumed to be non-negative.

STEIN [2] proved the following

Theorem 5.1. *The singular operator*

$$(Ku)(x) = \int_{\mathbb{R}^m} r^{-m} f(x, \theta)\, u(y)\, dy \tag{2}$$

is bounded in $L_p(\mathbb{R}^m, |x|^\alpha)$, $1 < p < \infty$, *if the characteristic* $f(x, \theta)$ *is measurable and bounded, and the exponent* α *satisfies the condition*

$$-\frac{m}{p} < \alpha < \frac{m}{p'} . \tag{3}$$

We shall give here Stein's proof in a simplified fashion by SAMKO [1].

Obviously, the boundedness of the operator K in $L_p(\mathbb{R}^m, \varrho)$ is equivalent to the boundedness of the operator $\varrho K \varrho^{-1}$ in $L_p(\mathbb{R}^m)$. Under the assumption of Theorem 5.1 the operator K is bounded in $L_p(\mathbb{R}^m)$ (Theorem 3.1). Therefore, it is sufficient to show that the operator $L = |x|^\alpha K |x|^{-\alpha} - K$ is bounded in $L_p(\mathbb{R}^m)$. We have

$$(Lu)(x) = \int_{\mathbb{R}^m} r^{-m} f(x, \theta) \left[\frac{|x|^\alpha}{|y|^\alpha} - 1\right] u(y)\, dy . \tag{4}$$

By M we denote the operator

$$(Mu)(x) = \int_{\mathbb{R}^m} r^{-m} \left|\frac{|x|^\alpha}{|y|^\alpha} - 1\right| u(y)\, dy . \tag{5}$$

Since the characteristic $f(x, \theta)$ is bounded we have $\|L\|_{L_p(\mathbb{R}^m)} \leq C \|M\|_{L_p(\mathbb{R}^m)}$. Thus, it is sufficient to show that the operator M is bounded in $L_p(\mathbb{R}^m)$.

5.2. The proof of the last assertion is based upon the following theorem which was proved by SCHUR, HARDY and LITTLEWOOD (cf. HARDY, LITTLEWOOD, and POLYA [1]) $m = 1$ and for arbitrary m by MIKHAILOV [1].

Theorem 5.2. *Let $\nu(x, y)$ be a non-negative homogeneous function of degree $-m$ defined on $\mathbb{R}^m \times \mathbb{R}^m$ which is invariant under rotations in $\mathbb{R}^m$. If*

$$\nu_0 = \int_{\mathbb{R}^m} |y|^{-m/p} \nu(e_1, y)\, dy < \infty, \qquad \nu_1 = \int_{\mathbb{R}^m} |x|^{-m/p'} \nu(x, e_1)\, dx < \infty,$$

$$e_1 = (1, 0, \ldots, 0),$$

then the operator

$$(Nu)(x) = \int_{\mathbb{R}^m} \nu(x, y)\, u(y)\, dy$$

is bounded in $\mathbf{L}_p(\mathbb{R}^m)$ and $\|N\|_{\mathbf{L}_p(\mathbb{R}^m)} \leqq \nu_0^{1/p'} \nu_1^{1/p}$.

Proof. By the Hölder inequality,

$$|(Nu)(x)| = \left| \int_{\mathbb{R}^m} \nu^{1/p'}(x, y) \, |y|^{-m/pp'} \cdot \nu^{1/p}(x, y)\, |y|^{m/pp'} u(y)\, dy \right|$$

$$\leqq \left[\int_{\mathbb{R}^m} |y|^{-m/p} \nu(x, y)\, dy \right]^{1/p'} \left[\int_{\mathbb{R}^m} |y|^{m/p'} \nu(x, y)\, |u(y)|^p\, dy \right]^{1/p'}$$

$$= |x|^{-m/pp'} \nu_0^{1/p'} \left[\int_{\mathbb{R}^m} \nu(x, y)\, |y|^{m/p'}\, |u(y)|^p\, dy \right]^{1/p}.$$

Raising this into the p-th power and integrating with respect to x we find

$$\|Nu\|^p_{\mathbf{L}_p(\mathbb{R}^m)} \leqq \nu_0^{p/p'} \int_{\mathbb{R}^m} |u(y)|^p \, |y|^{m/p'}\, dy \int_{\mathbb{R}^m} \nu(x, y)\, |x|^{-m/p'}\, dx$$

$$= \nu_0^{p/p'} \int_{\mathbb{R}^m} |u(y)|^p\, dy \int_{\mathbb{R}^m} \nu\left(x, \frac{y}{|y|}\right) |x|^{-m/p'}\, dx = \nu_0^{p/p'} \nu_1 \|u\|^p_{\mathbf{L}_p(\mathbb{R}^m)}. \quad \blacksquare$$

Remark. For finite ν_0 and ν_1 one can show that $\nu_0 = \nu_1$, and, therefore, $\|N\| \leqq \nu_0$ (cf. Samko [1]).

5.3. Now let us come back to the operator M. Its kernel $\nu(x, y) = r^{-m} \left|1 - |x|^\alpha |y|^{-\alpha}\right|$ is homogeneous of degree $-m$ and invariant under rotations. We shall show that the quantity ν_0 for this kernel is finite; for ν_1 the proof goes analogously. We have

$$\nu_0 = \int_{\mathbb{R}^m} |y - e_1|^{-m} \left|1 - |y|^{-\alpha}\right| |y|^{-m/p}\, dy = \int_{\mathbb{R}^m} |y - e_1|^{-m} \left| |y|^{-m/p} - |y| \right|^{-m/p-\alpha} dy.$$

The integrand is singular at the points $y = 0, \infty, e_1$. Close to the point $y = 0$ the singularity is of the form $|y|^{-m/p}$ if $\alpha < 0$ and $|y|^{-m/p-\alpha}$ if $\alpha > 0$. In the first case the singularity is summable, and in the second one we have due to (3) $-m/p - \alpha > -m$ such that the singularity is summable, too. The singularity at the point $y = \infty$ is treated analogously. Finally, close to the point $y = e_1$ it suffices to consider the function

$$|y - e_1|^{-m} \left| |y|^{-\alpha} - 1 \right| = |y - e_1|^{-m} \left| |y|^{-\alpha} - |e_1|^{-\alpha} \right| \leqq C\, |y - e_1|^{1-m}.$$

Again the singularity turns out to be summable. Now, Theorem 5.2 yields $\|M\|_{\mathbf{L}_p(\mathbb{R}^m)} < \infty$. As argued in 5.1, then also $\|K\|_{\mathbf{L}_p(\mathbb{R}^m, |x|^\alpha)} < \infty$ and Theorem 5.1 is proved. $\blacksquare$

Some generalizations of Stein's theorem can be found in Yu. S. Nikolski [1] and Pereyra [1]. In the first paper anisotropic singular integrals (§ 4) in weighted spaces are considered. Pereyra's paper is concerned with the case $p = 1$. In general, the singular operator

$$(Au)(x) = \int_{\mathbb{R}^m} r^{-m} f(x, \theta)\, u(y)\, dy \tag{6}$$

is not bounded in $\mathbf{L}_1(\mathbb{R}^m)$. However, under certain conditions on the characteristic $f(x, \theta)$ the operator (6) is of the weak type (1, 1). This is to be understood as follows. For some $s \in (0, 1)$ and any set $X \subset \mathbb{R}^m$ of a finite measure $|X|$ the inequality

$$\Big(\int_X |(Au)(x)|^s \, dx\Big)^{1/s} \leq C\, |X|^{1/s-1} \, \|u\|_{\mathbf{L}_1(\mathbb{R}^m)} \tag{7}$$

holds true.

Pereyra proved the following theorem. *If the operator* (6) *is of the weak type* (1, 1) *in* $\mathbf{L}_1(\mathbb{R}^m)$ *and the characteristic* $f(x, \theta)$ *is bounded then*

$$\Big(\int_X |(Au)(x)|^s \, |x|^\beta \, dx\Big)^{1/s} \leq C\, |X|^{1/s-1} \, \|u\|_{\mathbf{L}_1(\mathbb{R}^m, |x|^\beta)} \tag{8}$$

where $X \subset \mathbb{R}^m$ *is a set of finite measure, and* s *and* β *are arbitrary numbers satisfying the conditions* $-m < \beta < 0$ *and* $(m-1)/m < s < 1$.

§ 6. Singular integrals in weighted spaces. The theorems of Plamenevski and Haikin

In connection with the Theorem of Stein the question arises how far it can be extended for α values outside of the interval (5.3). Important results in that direction were obtained independently by Plamenevski [1, 2] and, somewhat later, by Haikin [1]. In the following we make, essentially, use of Plamenevski [2].

6.1. We shall consider a singular operator with the constant symbol $\Phi(\theta) = \Phi(\xi)$, $\theta = \xi/|\xi|$,

$$A = F^{-1}_{\xi \to x} \, \Phi(\theta) \, F_{x \to \xi} \, . \tag{1}$$

In the present subsection we want to prove the following formula ($\varrho = |x|$, $\varphi = x/\varrho$),

$$Au = (2\pi)^{-m} \, e^{im\pi/2} \int_{-\infty + i(m-2)/2}^{+\infty + i(m-2)/2} \varrho^{i\lambda - 1} \, \Gamma(i\lambda + m - 1) \, \Gamma(1 - i\lambda) \, d\lambda$$
$$\times \int_S (-\varphi\omega + i0)^{i\lambda - 1} \, \Phi(\omega) \, d_\omega S \cdot \int_S (\omega\psi + i0)^{-i\lambda - m - 1} \, \tilde{u}(\lambda, \psi) \, d_\psi S \, . \tag{2}$$

Here, $u = u(x) = u(\varrho, \omega)$ is an arbitrary sufficiently smooth function with a compact support, and $\tilde{u}(\lambda, \varphi)$ denotes its Mellin transform with respect to ϱ. The one-dimensional case is not excluded, if $m = 1$ we suppose only that the variables φ, ω, and ψ assume the values ± 1, and the integral over the sphere S is replaced by a corresponding sum of two summands.

In (2) and likewise in the sequel an expression $(t \pm i0)^\nu$ with a real variable t and a complex exponent ν is understood in the distributional sense (cf. Gelfand and Shilov [1], Chap. I, § 3, No. 6), namely

$$(t + i0)^\nu = t_+^\nu + e^{i\pi\nu} \, t_-^\nu \, , \qquad (t - i0)^\nu = t_+^\nu + e^{-i\pi\nu} \, t_-^\nu \, , \qquad \nu \neq -1, -2, \ldots,$$

$$(t + i0)^{-k} = t^{-k} - \frac{i\pi(-1)^{k-1}}{(k-1)!} \, \delta^{(k-1)}(t) \, ,$$

$$(t - i0)^{-k} = t^{-k} + \frac{i\pi(-1)^{k-1}}{(k-1)!} \, \delta^{(k-1)}(t) \, , \qquad k = 1, 2, \ldots,$$

with δ being the Dirac distribution.

In the following we shall frequently use the Mellin transformation. Let us recall some of its properties. Let $v(r)$ be a function defined on the positive half-axis, its Mellin transform is defined as

$$\tilde{v}(\lambda) = \frac{1}{2\pi} \int_0^\infty r^{-i\lambda} v(r)\, dr, \qquad \lambda \in (-\infty + i\tau, +\infty + i\tau).$$

The following inversion formula holds,

$$v(r) = \int_{-\infty+i\tau}^{+\infty+i\tau} r^{i\lambda-1} \tilde{v}(\lambda)\, d\lambda,$$

and so does the Parseval equality

$$\int_{-\infty+i\tau}^{+\infty+i\tau} |\tilde{v}(\lambda)|^2\, d\lambda = \int_0^\infty |v(r)|^2\, r^{2\tau-1}\, dr.$$

Now let us go in for the proof of (2). To begin with, we show that the Fourier transformation admits of the representation

$$(Fu)(\varrho, \omega) = (2\pi)^{-m/2} \int_{-\infty+i(m-2)/2}^{+\infty+i(m-2)/2} e^{i(i\lambda+m-1)\pi/2}\, \Gamma(i\lambda + m - 1)\, \varrho^{-i\lambda-m+1}\, d\lambda$$
$$\times \int_S \tilde{u}(\lambda, \psi)\, (\psi\omega + i0)^{-i\lambda-m+1}\, d_\psi S,$$

$$\tilde{u}(\lambda, \psi) = \frac{1}{2\pi} \int_0^\infty r^{-i\lambda} u(r, \psi)\, dr, \qquad \operatorname{Im} \lambda = m - 2, \tag{3}$$

where the Fourier transformation in the present section is defined as

$$F_{x\to\xi}\, u = (2\pi)^{-m/2} \int_{\mathbb{R}^m} e^{i(x,\xi)}\, u(x)\, dx.$$

Using the inversion formula of the Mellin transformation we may rewrite the Fourier transformation as

$$(Fu)(\varrho, \omega) = (2\pi)^{-m/2} \int_{\mathbb{R}^m} e^{i\varrho r(\omega,\psi)}\, r^{m-1}\, dr\, d_\psi S \int_{-\infty+i(m-2)/2}^{+\infty+i(m-2)/2} r^{i\lambda-1}\, \tilde{u}(\lambda, \psi)\, d\lambda.$$

We show that we are allowed to change the order of integration in the last formula. To that aim we replace that formula by the following one which contains a parameter $\tau > 0$,

$$(Fu)(\varrho, \omega) = (2\pi)^{-m/2} \lim_{\tau\to+0} \int_{\mathbb{R}^m} e^{i\varrho r(\omega,\psi)-r\tau}\, r^{m-1}\, dr\, d_\psi S \int_{-\infty+i(m-2)/2}^{+\infty+i(m-2)/2} r^{i\lambda-1}\, \tilde{u}(\lambda, \psi)\, d\lambda$$
$$= (2\pi)^{-m/2} \lim_{\tau\to+0} \int_S \int_{-\infty+i(m-2)/2}^{+\infty+i(m-2)/2} \tilde{u}(\lambda, \psi)\, d\lambda\, d_\psi S \int_0^\infty e^{i\varrho r(\omega,\psi)-r\tau}\, r^{i\lambda+m-2}\, dr. \tag{4}$$

Further, the following identity is well-known (cf. e.g. Bateman et al. [1]),

$$\int_0^\infty e^{i\varrho r(\omega,\psi)-r\tau}\, r^{i\lambda+m-2}\, dr = e^{i(\pi/2)(i\lambda+m-1)}\, (\varrho(\omega, \psi) + i\tau)^{\,i\lambda-m+1}\, \Gamma(i\lambda + m - 1),$$

using it we may give (4) the form

$$(Fu)(\varrho, \omega) = (2\pi)^{-m/2} \lim_{\tau \to +0} \int_{-\infty+i(m-2)/2}^{+\infty+i(m-2)/2} \Gamma(i\lambda + m - 1)\, e^{i(\pi/2)(i\lambda+m-1)}$$

$$\times \varrho^{-i\lambda-m+1}\, d\lambda \int_S \tilde{u}(\lambda, \psi)\, ((\omega, \psi) + i\tau)^{-i\lambda-m+1}\, d_\psi S$$

$$= (2\pi)^{-m/2} \int_{-\infty+i(m-2)/2}^{+\infty+i(m-2)/2} \Gamma(i\lambda + m - 1)\, e^{i(\pi/2)(i\lambda+m-1)}\, \varrho^{-i\lambda-m+1}\, d\lambda$$

$$\times \int_S \tilde{u}(\lambda, \psi)\, ((\omega, \psi) + i0)^{-i\lambda-m+1}\, d_\psi S.$$

Taking into account the relation $((\omega, \psi) + i\tau)^\mu \to ((\omega, \psi) + i0)^\mu$ as $\tau \to +0$ (in the distributional sense) we can convince ourselves easily that in the last formula the limit under the integral is admissible. Thus, (3) is proved.

Now, let us consider the operator (1). Introducing again a parameter $\tau > 0$ and applying (3) we may rewrite it as

$$Au = \lim_{\tau \to +0} (2\pi)^{-m} \int_{R^m} e^{-i\varrho r(\varphi,\omega) - \varrho r}\, \varrho^{m-1}\, \Phi(\omega)\, d\varrho\, d_\omega S$$

$$\times \int_{-\infty+i(m-2)/2}^{+\infty+i(m-2)/2} \Gamma(i\lambda + m - 1)\, e^{i(\pi/2)(i\lambda+m-1)}\, \varrho^{-i\lambda-m+1}\, d\lambda$$

$$\times \int_S \tilde{u}(\lambda, \psi)\, ((\omega, \psi) + i0)^{-i\lambda-m+1}\, d_\psi S.$$

Changing the order of integration we get

$$Au = \lim_{\tau \to +0} (2\pi)^{-m} \int_{-\infty+i(m-2)/2}^{+\infty+i(m-2)/2} \int_S \Phi(\omega)\, \Gamma(i\lambda + m - 1)\, e^{i(\pi/2)(i\lambda+m-1)}\, d\lambda\, d_\omega S$$

$$\times \int_S \tilde{u}(\lambda, \psi)\, ((\omega, \psi) + i0)^{-i\lambda-m+1}\, d_\psi S \int_0^\infty e^{-ir\varrho(\omega,\varphi) - \varrho\tau}\, \varrho^{-i\lambda}\, d\varrho.$$

We substitute in that formula the value of the integral

$$\int_0^\infty e^{-ir\varrho(\omega,\varphi) - \varrho\tau}\, \varrho^{-i\lambda}\, d\varrho = e^{i(\pi/2)(1-i\lambda)}\, \Gamma(1 - i\lambda)\, ((-\varphi, \omega) + i0)^{i\lambda-1}$$

take the limit as $\tau \to +0$ and obtain the relation (2).

6.2. Here, we consider the operator

$$E_\lambda v = (2\pi)^{-m/2}\, e^{i(\pi/2)(i\lambda+m-1)}\, \Gamma(i\lambda + m - 1) \int_S ((\omega, \psi) + i0)^{-i\lambda-m+1}\, v(\psi)\, d_\psi S. \tag{5}$$

In the following we shall use several results from the theory of spherical function, and the reader is referred to e.g. the book of Bateman et al. [1], vol. 2. In particular, we make use of the following fact. Let n be a given natural number. We choose integers k_j, $1 \leq j \leq m - 2$, such that $n \geq k_1 \geq k_2 \geq \ldots \geq k_{m-2} \geq 0$, and put $\bar{k} = (k_1, k_2, \ldots, \ldots, k_{m-3}, \pm k_{m-2})$. Furthermore, let $r_\mu^2 = x_{\mu+1}^2 + x_{\mu+2}^2 + \ldots + x_m^2$. The family of functions

$$Y_{n,m}^{(\bar{k})}(\theta) = B_{n,m}^{(\bar{k})}\, e^{\pm i k_{m-2} \vartheta_{m-1}} \prod_{j=0}^{m-3} (\sin \vartheta_{j+1})^{k_{j+1}}\, C_{k_j - k_{j+1}}^{(k_{j+1} + (m-j-2)/2)} (\cos \vartheta_{j+1}), \tag{6}$$

corresponding to all possible vectors $\bar{k}$ forms an orthonormal basis for the spherical functions of order n. In (6) $C_\nu^{(p)}$ denote the Gegenbauer polynomials, and the constant $B_{n,m}^{(\bar{k})}$ is determined by the condition $\| Y_{n,m}^{(\bar{k})} \|_{L_2(S)} = 1$.

Let $v(\psi)$ be a smooth function on the sphere S. We expand this function in a series with respect to spherical functions,

$$v(\psi) = \sum_{n=0}^{\infty} \sum_{\bar{k}} v_n^{(\bar{k})} Y_{n,m}^{(\bar{k})}(\psi)$$

such that

$$E_\lambda v = \sum_{n=0}^{\infty} \sum_{\bar{k}} (2\pi)^{-m/2} e^{i(\pi/2)(i\lambda+m-1)} v_n^{(\bar{k})} \Gamma(i\lambda + m - 1)$$

$$\times \int_S ((\omega, \psi) + i0)^{-i\lambda-m+1} Y_{n,m}^{(\bar{k})}(\psi) \, d_\psi S.$$

Consider the integral

$$\int_S ((\omega, \psi) + i0))^{-i\lambda-m+1} Y_{n,m}^{(\bar{k})}(\psi) \, d_\psi S.$$

Let N denote the north pole of the sphere S and g_ω be a rotation of the sphere which takes the point N to the point ω. Then the last integral can be written as

$$\int_S Y_{n,m}^{(\bar{k})}(\psi) ((g_\omega N, \psi) + i0)^{-i\lambda-m+1} \, d_\psi S$$

$$= \int_S Y_{n,m}^{(\bar{k})}(g_\omega \psi) ((N, \psi) + i0)^{-i\lambda-m+1} \, d_\psi S$$

$$= \int_S Y_{n,m}^{(\bar{k})}(g_\omega \psi) (\cos \psi_1 + i0)^{-i\lambda-m+1} \, d_\psi S.$$

The spherical functions are transformed under a rotation of S according to the following rule (cf. VILENKIN [1], p. 464),

$$Y_{n,m}^{(\bar{k})}(g_\omega \psi) = \sum_{\bar{p}} t_{\bar{p}\bar{k}}^{n}(g_\omega^{-1}) \, Y_{n,m}^{(\bar{p})}(\psi). \tag{7}$$

Here $t_{\bar{p}\bar{k}}^{n}(g)$ denote the matrix elements of the irreducible representation of the rotation group $SO(m)$, $g \in SO(m)$. We shall use the following property of the functions $t_{\bar{p}\bar{k}}^{n}(g)$ only,

$$t_{\bar{0}\bar{k}}^{n}(g_\omega^{-1}) = \sqrt{\frac{n! \, \Gamma(m-1)}{\Gamma(m+n-2)\,(m+2n-2)}} \, Y_{n,m}^{(\bar{k})}(\omega); \qquad \bar{0} = (0, 0, \ldots, 0). \tag{8}$$

Using (7) we get

$$\int_S Y_{n,m}^{(\bar{k})}(g_\omega \psi) (\cos \psi_1 + i0)^{-i\lambda-m+1} \, d_\psi S$$

$$= \sum_{\bar{p}} t_{\bar{p}\bar{k}}^{n}(g_\omega^{-1}) \int_S Y_{n,m}^{(\bar{p})}(\psi) (\cos \psi_1 + i0)^{-i\lambda-m+1} \, d_\psi S. \tag{9}$$

By means of (6) the integral on the right-hand side of the last equation is transformed to

$$\int_S Y_{n,m}^{(\bar{p})}(\psi) (\cos \psi_1 + i0)^{-i\lambda-m+1} \, d_\psi S$$

$$= A_{n\bar{p}} \int_S e^{\pm i k_{m-2} \psi_{m-1}} \prod_{j=0}^{m-3} (\sin \psi_{j+1})^{k_{j+1}} C_{k_j - k_{j+1}}^{(k_j+(m-2-j))/2} (\cos \psi_{j+1})$$

$$\times (\cos \psi_1 + i0)^{-i\lambda-m+1} \sin^{m-2} \psi_1 \ldots \sin \psi_{m-2} \, d\psi_1 \ldots d\psi_{m-1}.$$

This integral is the product of the following integrals

$$\int_0^{2\pi} e^{\pm i k_{m-2}\psi_{m-1}}\, d\psi_{m-1}; \tag{10_1}$$

$$\int_0^{\pi} (\sin \psi_{j+1})^{k_{j+1}+m-j-2}\, C_{k_j-k_{j-1}}^{(k_{j+1}+(m-j-2)/2)} (\cos \psi_{j+1})\, d\psi_{j+1}, \qquad 1 \leqq j \leqq m-3; \tag{10_2}$$

$$\int_0^{\pi} (\sin \psi_1)^{k_1+m-2}\, C_{k_1-k_2}^{(k_1+(m-2)/2)} (\cos \psi_1)\, (\cos \psi_1 + i0)^{-i\lambda-m+1}\, d\psi_1. \tag{10_3}$$

We show that the integrals (10) can be simultaneously non-zero only if all integers $k_1, k_2, \ldots, k_{m-2}$ are zero.

The integral (10_1) is non-zero if and only if $k_{m-2} = 0$. Now, let $k_{j+1} = 0$. We show that the integral (10_2) does not vanish if and only if $k_j = 0$. The integral (10_2) is as one can easily see equal to

$$\int_0^{\pi} \sin^{m-2-j} \psi\, C_{k_j}^{((m-2-j)/2)} (\cos \psi)\, d\psi .$$

The Gegenbauer polynomials are orthogonal in the sense

$$\int_0^{\pi} C_l^{(p)} (\cos \theta)\, C_n^{(p)} (\cos \theta) \sin^2 \theta\, d\theta = 0, \qquad l \neq n. \tag{11}$$

Because of $C_0^{(p)}(t) = 1$ the integral (10_2) for $k_j \neq 0$ equals zero.

Now, we put $\bar{k} = \bar{0}$ and compute the integral

$$\int_S Y_{n,m}^{(\bar{0})}(\psi)\, (\cos \psi_1 + i0)^{-i\lambda-m+1}\, d_\psi S .$$

It turns out to be equal to the product of the integrals (10) with $k_j = 0$ ($j = 1, 2, \ldots, \ldots, m-2$). In that case all the integrals (10) except the last one are different from zero and depend only on the dimension of the space such that their product yields a certain constant. Next, we consider the integral

$$\int_0^{\pi} \sin^{m-2} \psi\, C_n^{((m-2)/2)} (\cos \psi)\, (\cos \psi + i0)^{\mu}\, d\psi$$

$$= \int_{-1}^{1} (1-t^2)^{(m-3)/2}\, C_n^{((m-2)/2)} (t)\, (t+i0)^{\mu}\, dt .$$

Integrating it n times by part we find it to be equal to

$$\frac{\mu(\mu-1)\cdot \ldots \cdot (\mu-n+1)}{2^n\, n!\, \Gamma(m-2)\, \Gamma(n+(m-1)/2)}\, \Gamma(m+n-2) \sqrt{\Gamma((m-1)/2)}$$

$$\times \int_{-1}^{1} (1-t^2)^{n+(m-3)/2}\, (t+i0)^{\mu-n}\, dt .$$

Because of

$$(t+i0)^{\mu-n} = \begin{cases} t_+^{\mu-n}, & t > 0, \\ t_-^{\mu-n}, & t < 0, \end{cases}$$

we have

$$\int_{-1}^{1} (1-t^2)^{n+(m-3)/2} (t+\mathrm{i}0)^{\mu-n} \,\mathrm{d}t = (1+\mathrm{e}^{\mathrm{i}\pi(\mu-n)}) \int_0^1 t^{\mu-n} (1-t^2)^{n+(m-3)/2} \,\mathrm{d}t$$

$$= \frac{1+\mathrm{e}^{\mathrm{i}\pi(\mu-n)}}{2} \cdot \frac{\Gamma(n+(m-1)/2)\,\Gamma((\mu-n+1)/2)}{\Gamma((m+n+\mu)/2)}. \tag{12}$$

Using the relation (cf. VILENKIN [1])

$$A_{n\bar{0}} = \sqrt{\frac{n!\,\Gamma(m-2)\,(2n+m-2)}{\Gamma(n+m-2)\,(m-2)}}$$

and the formulae (8), (9), and (12) we obtain the following representation of the operator E_λ

$$(E_\lambda v)(\omega) = \sum_{n=0}^{\infty} \sum_{\bar{k}} v_{n\bar{k}} \mu_n(\lambda)\, Y_{n,m}^{(\bar{k})}(\omega).$$

Here, $\mu_n(\lambda)$ are meromorphic functions of the form

$$\mu_n(\lambda) = c_m \frac{\Gamma(\mathrm{i}\lambda+m+n-1)\,\Gamma((-\mathrm{i}\lambda-m-n+2)/2)}{2^m \Gamma((n-\mathrm{i}\lambda+1)/2)}$$
$$\times \mathrm{e}^{-\pi\lambda/2} (1+\exp(\mathrm{i}\pi(-\mathrm{i}\lambda-m-n+1))).$$

By c_m we denote always in the present section a certain factor the modulus of which depends only on the space dimension.

We apply the formula $\Gamma(z)\,\Gamma(1-z) = \pi/\sin \pi z$ for $z = (\mathrm{i}\lambda+m+n)/2$ and get

$$\Gamma((-\mathrm{i}\lambda-m-n+2)/2)\,\mathrm{e}^{-\pi\lambda/2} (1-\exp(-\mathrm{i}\pi(\mathrm{i}\lambda+m+n)))$$
$$= \gamma_{mn} [\Gamma((\mathrm{i}\lambda+m+n)/2)]^{-1}, \qquad |\gamma_{mn}| = 2\pi.$$

Thence, the $\mu_n(\lambda)$ become

$$\mu_n(\lambda) = c_m \frac{\Gamma(\mathrm{i}\lambda+m+n-1)}{2^n \Gamma((n-\mathrm{i}\lambda+1)/2)\,\Gamma((\mathrm{i}\lambda+m+n)/2)}.$$

By the doubling formula $\Gamma(2z) = 2^{2z-1}\sqrt{\pi}\cdot\Gamma(z)\cdot\Gamma(z+\frac{1}{2})$ we obtain for $z = (\mathrm{i}\lambda+m+n-1)/2$ the relation

$$\mu_n(\lambda) = 2^{\mathrm{i}\lambda} c_m \frac{\Gamma((\mathrm{i}\lambda+m+n-1)/2)}{\Gamma((n-\mathrm{i}\lambda+1)/2)}.$$

Thus, we proved the following

Lemma 6.1. *Let $v(\psi)$ be an arbitrary sufficiently smooth function on the sphere S. Then the operator $E_\lambda v$ admits to the representation*

$$(E_\lambda v)(\psi) = \sum_{n=0}^{\infty} \sum_{\bar{k}} \mu_n(\lambda)\, v_{n\bar{k}}\, Y_{n,m}^{(\bar{k})}(\psi). \tag{13}$$

Here, $\mu_n(\lambda)$ are meromorphic functions of λ of the form

$$\mu_n(\lambda) = 2^{-\tau} c_m \frac{\Gamma((\mathrm{i}\lambda+m+n-1)/2))}{\Gamma((n-\mathrm{i}\lambda+1)/2)}, \qquad \lambda = \sigma + \mathrm{i}\tau, \tag{14}$$

c_m are factors introduced above, and $v_{n\bar{k}}$ are the coefficients in the expansion of $v(\psi)$ in a series with respect to spherical functions,

$$v(\psi) = \sum_{n=0}^{\infty} \sum_{\bar{k}} v_{n\bar{k}} Y_{n,m}^{(\bar{k})}(\psi).$$

6.3. Now let us go back to (2). We replace the integration along the straight line $(-\infty + \mathrm{i}(m-2)/2, +\infty + \mathrm{i}(m-2)/2)$ by the integration along the line $(-\infty + \mathrm{i}\tau, +\infty + \mathrm{i}\tau)$, $\tau = (\alpha + m - 2)/2$,

$$Au = \int_{-\infty+\mathrm{i}\tau}^{+\infty+\mathrm{i}\tau} r^{\mathrm{i}\lambda-1} A_\lambda \tilde{u} \, \mathrm{d}\lambda + 2\pi \mathrm{i} R(r, \varphi). \tag{15}$$

Here, $A_\lambda \tilde{u}$ stands for the operator

$$A_\lambda v = (2\pi)^{-m} \mathrm{e}^{\mathrm{i}m\pi/2} \Gamma(1 - \mathrm{i}\lambda) \Gamma(m - 1 + \mathrm{i}\lambda) \times \int_S (-(\varphi, \omega) + \mathrm{i}0)^{\mathrm{i}\lambda-1} \Phi(\omega) \, \mathrm{d}_\omega S \int_S ((\omega, \psi) + \mathrm{i}0)^{-\mathrm{i}\lambda-m+1} v(\psi) \, \mathrm{d}_\psi S$$

applied to the function $\tilde{u}(\lambda, \varphi)$, and $R(r, \psi)$ is the sum of the residuals of the function $r^{\mathrm{i}\lambda-1} A_\lambda \tilde{u}$ in the stripe

$$\frac{m-2}{2} < \operatorname{Im} \lambda < \tau, \qquad \alpha > 0,$$

$$\tau < \operatorname{Im} \lambda < \frac{m-2}{2}, \qquad \alpha < 0.$$

Our next aim is to investigate the function $R(r, \varphi)$. If $\Phi(\omega) \equiv 1$ the operator A is the identity. In that case also the operator A_λ on the sphere S for $\operatorname{Im} \lambda = (m-2)/2$ equals identity,

$$(2\pi)^{-m/2} \mathrm{e}^{\mathrm{i}m\pi/2} \Gamma(\mathrm{i}\lambda + m - 1) \Gamma(1 - \mathrm{i}\lambda) \int_S (-(\varphi, \omega) + \mathrm{i}0)^{\mathrm{i}\lambda-1} \mathrm{d}_\omega S \times \int_S ((\omega, \psi) + \mathrm{i}0)^{-\mathrm{i}\lambda-m+1} v(\psi) \, \mathrm{d}_\psi S = v(\varphi). \tag{16}$$

Applying the principle of analytical extension we obtain (16) for all complex λ except, possibly, the poles of the left-hand side function. Note that these points coincide with the poles of the functions $\Gamma(m - 1 + \mathrm{i}\lambda)$ and $\Gamma(1 - \mathrm{i}\lambda)$. Therefore, the integral operators E_λ and E_λ^{-1},

$$E_\lambda^{-1} w = (2\pi)^{-m/2} \mathrm{e}^{\mathrm{i}\pi(1-\mathrm{i}\lambda)/2} \Gamma(1 - \mathrm{i}\lambda) \int_S (-(\varphi, \omega) + \mathrm{i}0)^{\mathrm{i}\lambda-1} w(\omega) \, \mathrm{d}_\omega S,$$

are mutually inverse on the complex λ-plane except, possibly, the poles mentioned above. Hence, applying Lemma 6.1 we obtain for the operator E_λ^{-1} the representation

$$E_\lambda^{-1} v = \sum_{n=0}^{\infty} \sum_{\bar{k}} \nu_n(\lambda) v_{n\bar{k}} Y_{n,m}^{(\bar{k})}(\omega),$$

where $v_{n\bar{k}}$ are the coefficients of the expansion of v with respect to spherical functions and $\nu_n(\lambda)$ meromorphic functions,

$$\nu(\lambda) = 2^{-\mathrm{i}\lambda} c_m \frac{\Gamma((m - \mathrm{i}\lambda + 1)/2)}{\Gamma((m + \mathrm{i}\lambda + n - 1)/2)}.$$

As you can see the zeros and the poles of the functions $\nu_n(\lambda)$ coincide with the poles and zeros, resp., of the functions $\mu_n(\lambda)$, and these are $\lambda = \mathrm{i}(m + n + 2s - 1)$, $s = 0, 1, 2, \ldots$, and $\lambda = -\mathrm{i}(m + 2s + 1)$, $s = 0, 1, 2, \ldots$, resp.

Let, first, $\alpha > 0$ and $(\alpha - m)/2 \neq k$, $k = 0, 1, 2, \ldots$. Only the functions $\mu_n(\lambda)$, $0 \leqq n \leqq k$, $n \equiv k \pmod 2$, $k = 0, 1, \ldots, [(\alpha - m)/2]$, have poles of first order in the stripe $(m-2)/2 < \operatorname{Im} \lambda < \tau$, $\tau = (\alpha + m - 2)/2$. Since $r^{i\lambda - 1} A_\lambda \tilde{u} = r^{i\lambda} E_\lambda^{-1} \Phi E_\lambda \tilde{u}$, for an arbitrary symbol none of the poles of $\mu_n(\lambda)$ is, in general, "erased" by a zero of $\nu_l(\lambda)$. Therefore, the function $R(r, \varphi)$ is of the form

$$R(r, \varphi) = \sum_{k=0}^{[(\alpha-m)/2]} \sum_{\substack{n=0 \\ n \equiv k (\mathrm{mod} 2)}}^{k} \tilde{u}_{n\bar{k}}(\mathrm{i}(k + m - 1))\, r^{-k-m} f_{n\bar{k}}(\varphi).$$

Here the $f_{n\bar{k}}(\varphi)$ are functions that depend on φ only, they can be determined easily. This, together with (15), implies that the singular operator (1) applied to functions satisfying the conditions

$$\tilde{u}_{n\bar{k}}(\mathrm{i}(k + m - 1)) = 0, \quad 0 \leqq n \leqq k, \quad n \equiv k \pmod 2, \quad k = 0, 1, \ldots, [(\alpha - m)/2] \tag{17}$$

can be represented in the form

$$Au = \int_{-\infty + \mathrm{i}\tau}^{+\infty + \mathrm{i}\tau} r^{\mathrm{i}\lambda - 1} A_\lambda \tilde{u}\, \mathrm{d}\lambda. \tag{18}$$

Analogous arguments lead us in case of $\alpha < -m$, $(\alpha + m)/2 \neq k$, $k = 0, -1, -2, \ldots$, to the following result. If the function $u(r, \varphi)$ satisfies the condition

$$\tilde{u}_{l\bar{k}}(\mathrm{i}(q - 1)) = 0 \tag{19}$$

with $l > -q$ or $0 \leqq l \leqq -q$, $l \equiv q \pmod 2$, $q = 0, -1, \ldots, [(\alpha + m)/2]$, then (18) holds true. Here as elsewhere in the present section, $[a]$ means that integer which is closest to a and $|[a]| \leqq |a|$.

We may give the conditions (17) and (19) a more handy form. Condition (17) means that

$$\int_{\mathrm{R}^m} u(r, \psi)\, r^k Y_{n,m}^{(\bar{k})}(\psi)\, r^{m-1}\, \mathrm{d}r\, \mathrm{d}_\psi S = 0$$

for $0 \leqq n \leqq k$, $n \equiv k \pmod 2$, $k = 0, 1, \ldots, [(\alpha - m)/2]$. In other words,

$$\int_{\mathrm{R}^m} u(x)\, h_k(x)\, \mathrm{d}x = \int_{\mathrm{R}^m} u(x)\, h_{k-2}(x)\, r^2\, \mathrm{d}x = \ldots = \int_{\mathrm{R}^m} u(x)\, h_{k-2s}(x)\, r^{2s}\, \mathrm{d}x = 0,$$

$$s = [k/2], \qquad k = 0, 1, \ldots, [(\alpha - m)/2], \tag{20}$$

where $h_l(x)$ are homogeneous harmonic polynomials of degree l. Now, we make use of the following fact that is easy to prove. If $f(x)$ is a homogeneous polynomial of degree k then there exist harmonic polynomials $h_l(x)$ such that

$$f(x) = \sum_{\nu=0}^{[k/2]} r^{2\nu} h_{k-2\nu}(x).$$

This gives together with (20) the equivalence of (17) and

$$\int_{\mathrm{R}^m} u(x)\, x^l\, \mathrm{d}x = 0, \qquad |l| \leqq [(\alpha - m)/2]. \tag{21}$$

In the same way it is clear that the conditions (19) can be written as

$$\int_0^\infty u(r, \psi)\, r^{q-1}\, \mathrm{d}r = \sum_{n=0}^{-q} \sum_{\bar{k}} c_{n\bar{k}} Y_{n,m}^{(\bar{k})}(\psi).$$

$$n \equiv q \pmod 2, \qquad q = 0, -1, \ldots, -[(\alpha + m)/2]. \tag{22}$$

6.4. Let us introduce two sets M_+ and M_-. By M_+ we denote the set of all sufficiently smooth functions satisfying the conditions

$$\int_{\mathbb{R}^m} u(x)\, x^l\, dx = 0, \qquad |l| \leqq [(\alpha - m)/2], \qquad \alpha \geqq m;$$

their support is supposed to be compact and does not contain the origin.

In case if $\alpha \leqq -m$ we denote by M_- the set of all sufficiently smooth functions satisfying the condition (22), their support is also supposed to be compact and does not contain the origin.

We shall show now that the sets M_+ and M_- are dense in $\mathbf{L}_2(\mathbb{R}^m, |x|^\alpha)$.

Let p be an integer. By M_p^+, $p \geqq 0$ (and M_p^-, $p > 0$, resp.) we denote the set of all sufficiently smooth functions satisfying the condition

$$\int_0^\infty v(r, \psi)\, r^q\, dr = 0, \qquad q = 0, 1, \ldots, p \quad (\text{and } q = -1, -2, \ldots, -p, \text{ resp.}),$$

and we shall show that these sets are dense in $\mathbf{L}_2(\mathbb{R}^m, |x|^\alpha)$. This will imply the density of the sets $M_\pm$.

First, we show the density of the set M_0^+. To that aim it is sufficient to show that any bounded measurable function $v(r, \psi)$ with a compact support that does not contain the origin is the $\mathbf{L}_2(\mathbb{R}^m, |x|^\alpha)$ limit of a sequence of functions belonging to M_0^+. Let us denote by $v_n(r, \psi)$, $n = 1, 2, \ldots$, the functions

$$v_n(r, \psi) = \begin{cases} n v(nr, \psi), & \alpha > 2 - m, \\ n^{-1} v(n^{-1} r, \psi), & \alpha < 2 - m. \end{cases}$$

Obviously, $v(r, \psi) - v_n(r, \psi) \in M_0^+$ and $\|v - v_n\|_{\mathbf{L}_2(\mathbb{R}^m, |x|^\alpha)} \to 0$ as $n \to \infty$. Thus, it is clear that M_0^+ is dense in $\mathbf{L}_2(\mathbb{R}^m, |x|^\alpha)$ provided that $\alpha \neq 2 - m$. If $\alpha = 2 - m$ we define the functions $v_n(r, \psi)$ by the formula

$$v_n(r, \psi) = \begin{cases} \dfrac{c_n w(\psi)}{r \ln r}, & n^{-1} < r < 1/2, n = 3, 4, \ldots, \\ 0, & r \notin [n^{-1}, 1/2] \end{cases}$$

with

$$w(\psi) = -\int_0^\infty v(r, \psi)\, dr, \qquad c_n = [\ln \ln 2 - \ln \ln n]^{-1}.$$

Now, we proceed by induction. Suppose the set M_{p-1}^+ to be dense, then we show that the set M_p^+ is also dense in the space considered. To that aim it is sufficient to show that any function of M_{p-1}^+ is limit of a sequence of functions belonging to M_p^+. Put

$$v_n(r, \psi) = \begin{cases} n^{p+1} v(nr, \psi), & \alpha > 2(p+1) - m, \\ n^{-p-1} v(n^{-1} r, \psi), & \alpha < 2(p+1) - m. \end{cases}$$

Obviously, $v - v_n \in M_p^+$, and $v_n \to 0$ as $n \to \infty$ in the $\mathbf{L}_2(\mathbb{R}^m, |x|^\alpha)$ metric. Hence, if $\alpha \neq 2(p+1) - m$ any function $v \in M_{p-1}^+$ is limit of the sequence $\{v - v_n\} \subset M_p^+$, what we wanted to show.

Next, let $\alpha = 2(p+1) - m$. By N_p we denote the set of functions satisfying the only condition

$$\int_0^\infty v(r, \psi)\, r^p\, dr = 0, \tag{23}$$

and we show that the set N_p is dense in the space $\mathbf{L}_2(\mathbb{R}^m, |x|^\alpha)$. Let $v(r, \psi)$ be an arbitrary sufficiently smooth function with a compact support supp $v \not\ni 0$. In order to show that v is limit of a sequence of functions belonging to N_p we put

$$v_n(r, \psi) = \begin{cases} \dfrac{c_n w(\psi)}{r^{p+1} \ln r}, & n^{-1} < r < 1/2, \qquad n = 3, 4, \ldots, \\ 0 & , r \notin [n^{-1}, 1/2], \end{cases}$$

where

$$w(\psi) = -\int_0^\infty v(r, \psi)\, dr, \qquad c_n = [\ln \ln 2 - \ln \ln n]^{-1}.$$

The functions $v + v_n$ satisfy the condition (23). Since $v_n \to 0$ as $n \to \infty$ in the $\mathbf{L}_2(\mathbb{R}^m, |x|^\alpha)$ metric it is immediately clear that the set N_p is dense.

Now, let $v(r, \psi)$ be an arbitrary function of N_p. In order to prove the density of the set $N_p \cap N_{p-1}$ we construct a sequence of functions belonging to that set which has the limit v. For example, the sequence $v(r, \psi) - v_n(r, \psi)$ with $v_n(r, \psi) = n^p v(n\, r, \psi)$ has the properties required.

Proceeding in this way we can show the density of the set $N_p \cap N_{p-1} \cap \ldots \cap N_0 = M_p^+$ if $\alpha = 2(p + 1) - m$. Analogously, the density of the set M_p^- is proved.

Finally, we note the following. The conditions (21) and (22) exactly define the sets M_+ and M_- whereas eq. (18) is equivalent to the relation (2). Thus, (2) is valid for functions belonging to the sets M_+ and M_-.

6.5. Here we shall formulate and prove the basic theorems of Plamenevski for singular operators in weighted spaces.

Theorem 6.1. *Assume that the operator A is defined on the set M_+ by* (1), *and* $\Phi(\xi) \not\equiv$ const. *The operator A is bounded in* $\mathbf{L}_2(\mathbb{R}^m, |x|^\alpha)$ *if and only if*

a) $(\alpha - m)/2 \neq k$, $k = 0, 1, \ldots$;

b) *the symbol is a multiplier in* $\mathbf{W}_2^{(\alpha/2)}(S)$.

If the conditions a) *and* b) *are fulfilled then for the closure $\bar{A}$ of the operator A in* $\mathbf{L}_2(\mathbb{R}^m, |x|^\alpha)$ *the formula* (2) *is valid.*

Theorem 6.2. *Let the singular operator with the symbol* $\Phi(\theta) = \Phi(\xi) \not\equiv$ const, $\theta = \xi/|\xi|$, *be defined by* (1) *on the set M_-. That operator is bounded in* $\mathbf{L}_2(\mathbb{R}^m, |x|^\alpha)$ *if and only if*

$$(\alpha + m)/2 \neq k, \qquad k = 0, -1, -2, \ldots \tag{25}$$

and $\Phi \in \mathbf{W}_2^{(\alpha/2)}(S)$. *If these conditions are fulfilled then for the operator $\bar{A}$ the formula* (2) *is valid.*

Proof of both the Theorems 6.1 and 6.2. The results of subsection 6.3 imply that the operator (1) on the sets M_+ and M_- can be represented as

$$Au = \int_{-\infty + i\tau}^{+\infty + i\tau} r^{i\lambda - 1} E_\lambda^{-1}\, \Phi E_\lambda \tilde{u}\, d\lambda; \qquad \tau = \frac{\alpha + m - 2}{2}. \tag{26}$$

Since the sets M_+ and M_- are dense in $\mathbf{L}_2(\mathbb{R}^m, |x|^\alpha)$ it is sufficient to prove that under the assumptions of the Theorems 6.1 and 6.2 the operator defined by the right-hand side of (26) is bounded in that space. Due to the Parseval equality (see 6.1) this is the case if the operator $A_\lambda = E_\lambda^{-1}\, \Phi E_\lambda$, $\lambda \in (-\infty + i\tau, +\infty + i\tau)$, is bounded in the space $\mathbf{L}_2(S)$ uniformly with respect to λ.

In the space $\mathbf{W}_2^{(l)}(S)$ of functions defined on the sphere S we introduce the norm

$$\|u\|^2_{\mathbf{W}_2^{(l)}(S)} = \sum_{n=0}^{\infty} \sum_{\bar{k}} (1 + |\lambda|^2 + n^2)^l \, |u_{n\bar{k}}|^2,$$

$$u(\psi) = \sum_{n=0}^{\infty} \sum_{\bar{k}} u_{n\bar{k}} Y_{n,m}^{(\bar{k})}(\psi).$$

For an arbitrary fixed λ this norm is equivalent to the usual norm in the Sobolev-Slobodetski space $\mathbf{W}_2^{(l)}(S)$.

We show that the operator E_λ from $\mathbf{W}_2^{(l)}(S)$ into $\mathbf{W}_2^{(l-s)}(S)$ is bounded if $\lambda \neq \mathrm{i}(m + k - 1)$, $k = 0, 1, \ldots$, and $s = m/2 - \tau - 1$, $\lambda = \sigma + \mathrm{i}\tau$, where on every line $(-\infty + \mathrm{i}\tau, +\infty + \mathrm{i}\tau)$, $\tau \neq m + k - 1$, the estimate

$$\|E_\lambda\| \leqq c_\tau \tag{27}$$

holds, the constant c_τ depends only on τ.

To that aim, we use the representations (13) and (14) of E_λ. There is an asymptotic formula for the quotient of two Γ-functions (cf. BATEMAN et al. [1]), and this gives

$$\left|\frac{\Gamma((\mathrm{i}\lambda + m + n - 1)/2)}{\Gamma((n - \mathrm{i}\lambda + 1)/2)}\right| = \left|\frac{\mathrm{i}\sigma + n}{2}\right|^{m/2 - \tau - 1} |1 + O((\mathrm{i}\sigma + n)^{-1})|. \tag{28}$$

On the line $(-\infty + \mathrm{i}(\alpha + m - 2)/2, +\infty + \mathrm{i}(\alpha + m - 2)/2)$, $\alpha - m \neq 2k$, $k = 0, 1, 2, \ldots$, none of the functions $\mu_n(\lambda)$ has poles. Hence, the eqs. (14) and (28) imply the boundedness of the operator E_λ from $\mathbf{W}_2^{(l)}(S)$ into $\mathbf{W}_2^{(l-s)}(S)$ as well as the estimate (27). Similarly, one can show that the operator E_λ^{-1} from $W_2^{(l)}(S)$ into $W_2^{(l+s)}(S)$ is bounded provided that $\alpha + m \neq 2k$, $k = 0, -1, \ldots$, and $\lambda = \sigma + \mathrm{i}\tau$, $2\tau = \alpha + m - 2$, $s = m/2 - \tau - 1$. Then on every line $(-\infty + \mathrm{i}\tau, +\infty + \mathrm{i}\tau)$ the estimate

$$\|E_\lambda^{-1}\| \leqq c_\tau \tag{29}$$

holds, and the constant c_τ again depends only on τ.

Since the Fourier transformation is unitary (3) shows that the operator E_λ, $\operatorname{Im} \lambda = (m - 2)/2$, is also unitary in $\mathbf{L}_2(S)$.

Under the assumption that the symbol is a multiplier in $\mathbf{W}_2^{(\alpha/2)}(S)$ the estimates (27) and (29) imply the inequality $\|A_\lambda v\|_{\mathbf{L}_2(S)} \leqq c_\tau \|v\|_{\mathbf{L}_2(S)}$. Thus, the boundedness of the operator

$$\int_{-\infty + \mathrm{i}\tau}^{+\infty + \mathrm{i}\tau} r^{\mathrm{i}\lambda - 1} A_\lambda \tilde{u} \, \mathrm{d}\lambda, \qquad \tau = \frac{\alpha + m - 2}{2},$$

in $\mathbf{L}_2(\mathbb{R}^m, |x|^\alpha)$ is proved provided that

$$\frac{\alpha - m}{2} \neq k, \qquad \frac{\alpha + m}{2} \neq -k, \qquad k = 0, 1, 2, \ldots,$$

and that means that the sufficiency part of the Theorems 6.1 and 6.2 is proved, too.

Now, let us come to the necessity part. Suppose that e.g. condition a) in Theorem 6.1 is violated, i.e. $\alpha = 2k + m$ for some non-negative integer k. Put $v(r, \psi) = u(r) Y_{k,m}^{(\bar{n})}(\psi)$ with $Y_{k,m}^{(\bar{n})}$ being a spherical function of order k. Then expand the product $\Phi(\omega) Y_{k,m}^{(\bar{n})}(\omega)$ in a series with respect to spherical functions,

$$\Phi(\omega) \, Y_{k,m}^{(\bar{n})}(\omega) = \sum_{l=0}^{\infty} \sum_{\bar{k}} b_{l\bar{k}} Y_{l,m}^{(\bar{k})}(\omega).$$

Because of $\Phi(\omega) \not\equiv$ const, for a suitable choice of the function $Y_{k,m}^{(\bar{n})}$, there is at least one coefficients $b_{l_0\bar{k}} \neq 0$ for some $l_0 > k$. The function

$$w(r,\varphi) = \int_{-\infty+i\tau}^{+\infty+i\tau} r^{i\lambda-1} E_\lambda^{-1} \Phi E_\lambda \tilde{v} \, d\lambda, \qquad \tau = m + k - 1,$$

admits of a representation

$$w(r,\varphi) = \sum_{l=0}^{\infty} \sum_{\bar{k}} b_{l\bar{k}} \int_{-\infty+i\tau}^{+\infty+i\tau} r^{i\lambda-1} \mu_k(\lambda) \tilde{u}(\lambda) \nu_l(\lambda) \, d\lambda \cdot Y_{l,m}^{(\bar{k})}(\varphi). \tag{30}$$

Thence,

$$\|w\|^2_{L_2(\mathbb{R}^m,|x|^\alpha)} = \sum_{l=0}^{\infty} \sum_{\bar{k}} \int_{-\infty+i\tau}^{+\infty+i\tau} |\mu_k(\lambda) \tilde{u}(\lambda)|^2 \, |\nu_l(\lambda)|^2 \, |b_{l\bar{k}}|^2 \, d\lambda$$

$$\geqq |b_{l_0\bar{k}}|^2 \int_{-\infty+i\tau}^{+\infty+i\tau} |\mu_k(\lambda)|^2 |\nu_{l_0}(\lambda)|^2 \, |\tilde{u}(\lambda)|^2 \, d\lambda. \tag{31}$$

Remember that (see 6.3) the function $\mu_k(\lambda)$ has a pole of order 1 at $\lambda = i(k + m - 1)$ whereas the function $\nu_{l_0}(\lambda)$ does not vanish at that point. Therefore, the product $\mu_k(\lambda) \cdot \nu_{l_0}(\lambda)$ has a pole at the point mentioned. Taking this into account and using the inequality (31) we can easily construct a sequence of functions $v_n(r, \psi) \in M_+$ such that $\|v_n\|_{L_2(\mathbb{R}^m,|x|^\alpha)} = 1$ and $w_n \to 0$ as $n \to \infty$ in the $L_2(\mathbb{R}^m, |x|^\alpha)$ norm where w_n is constructed by v_n according to eq. (30). Now, it remains to note merely that the singular operator (1) on the set M_+ coincides with the operator (30).

If the condition a) is fulfilled then the necessity of the condition b) follows from comparing these two facts:

1. The operator (1) admits to the representation

$$Au = \int_{-\infty+i\tau}^{+\infty+i\tau} r^{i\lambda-1} A_\lambda \tilde{u} \, d\lambda, \qquad \tau = \frac{\alpha + m - 2}{2}.$$

2. The operator $A_\lambda = E_\lambda^{-1} \Phi E_\lambda$, $\lambda = \sigma + i\tau$, is similar to the operator of multiplication by the symbol $\Phi(\omega)$ in the space $\mathbf{W}_2^{(-\alpha/2)}(S)$. Consequently, for the boundedness of the operator A_λ in $\mathbf{L}_2(S)$ it is necessary that the operator of multiplication by $\Phi(\omega)$ is bounded in $\mathbf{W}_2^{(-\alpha/2)}(S)$, i.e. the symbol $\Phi(\omega)$ is a multiplier in $\mathbf{W}_2^{(\alpha/2)}(S)$.

Thus, the necessity part of Theorem 6.1 is proved. Similarly, the necessity part of Theorem 6.2 can be shown. ∎

6.6. Haikin [2, 3] considered singular operators with a symbol that depends on the pole. Without proof we quote here the most important result by Haikin for such operators in weighted $\mathbf{L}_2$ spaces.

Suppose that the symbol of the singular operator satisfies the condition $\Phi(x, \theta) \hat{\in} \mathbf{W}_2^{(l)}(S)$. We expand it in a series with respect to spherical functions,

$$\Phi(x,\theta) = \sum_{n=0}^{\infty} \sum_{k=1}^{k_{n,m}} a_n^{(k)}(x) \, Y_{n,m}^{(k)}(\theta).$$

As we did before, by $A_n^{(k)}$ we denote the singular operator with the symbol $Y_{n,m}^{(k)}(\theta)$,

$$A_n^{(k)} = F_{\xi\to x}^{-1} \, Y_{n,m}^{(k)}(\theta) \, F_{x\to\xi}, \qquad \theta = \frac{\xi}{|\xi|},$$

and we put

$$A_\nu = \sum_{n=0}^{\nu} \sum_{k=1}^{k_{n,m}} a_n^{(k)}(x) \, A_n^{(k)}.$$

Theorem 6.3. *Let* $\Phi(x,\theta)\,\hat{\in}\,\mathbf{W}_2^{(l)}(S)$, $l>(m-1)/2+|\alpha|$ *and* $|\alpha| \neq m/2+j$, $j=0,1,\ldots$ *Then the limit* $A=\lim\limits_{\nu\to\infty} A_\nu$ *exists in the operator convergence of the space* $\mathbf{L}_2(\mathbb{R}^m\,|x|^\alpha)$. *The norm of the operator* A *in that space has the estimate*

$$\|A\| \leqq C \sup_{x\in\mathbb{R}^m} \|\Phi(x,.)\|_{\mathbf{W}_2^{(l)}(S)}. \tag{32}$$

6.7. We shall briefly consider here singular operators in weighted spaces $\mathbf{L}_p(\mathbb{R}^m, |x|^\alpha)$, $p \neq 2$. They are considered in Haikin [2]. Let the singular operator be given on the set M_+ or (corresponding to the sign of the parameter α) on M_-. Here we exclude the α-values

$$\alpha = \frac{m}{p'}+j, \qquad -\frac{m}{p}+j; \qquad j=0,1,\ldots \tag{33}$$

If α is not exceptional then, under certain assumptions on the kernel of the singular integral, the corresponding operator turns out to be bounded in $\mathbf{L}_p(\mathbb{R}^m, |x|^\alpha)$. For example, if $\alpha > m/p'$ the following conditions are sufficient for the boundedness,

$$|f(x,\theta)| \leqq C; \qquad \int\limits_{R/2<|x|<4R} |D_z^\beta K(x,z)|_{z=x}|^p\,\mathrm{d}x \leqq CR^{-m(p-1)-|\beta|p}$$

$$\forall\beta, |\beta| \leqq s := \left[\alpha - \frac{m}{p'}\right], \qquad \forall R\in(0,\infty);$$

$$\int\limits_{R/2<|x|<4R} |K^+(x,y)|^p\,\mathrm{d}x \leqq CR^{-m(p-1)-sp}\,|y|^{sp}, \tag{34}$$

where

$$K^+(x,y) = K(x,x-y) - \sum_{|\beta|=0}^{s} D_z^\beta K(x,z)|_{z=x}\frac{y^\beta}{\beta!}.$$

Here, $K(x,x-y)=r^{-m}f(x,\theta)$ is the kernel of the singular operator under consideration.

In Haikin [2] several other weighted spaces are also considered.

6.8. Another criterion on the boundedness of the singular operator in $\mathbf{L}_2(\mathbb{R}^m, |x|^\alpha)$ is contained in Mazya and Haikin [2]. As before, let the singular operator be given on M_+ or M_- corresponding to the sign of the parameter α. Furthermore, let μ be a weight function of tempered growth, and

$$c^2 = \int\limits_{\mathbb{R}^m} \frac{(1+|x|)^{2\alpha}}{\mu^2(x)}\,\mathrm{d}x < \infty.$$

The singular operator A with the symbol $\Phi(x,\theta)$ admits to the estimate

$$\|A\|_{\mathbf{L}_2(\mathbb{R}^m,|x|^\alpha)} \leqq C\sup_x \|\Phi(x,.)\|_{\mathbf{H}_2^{(\mu)}(S)}. \tag{35}$$

provided that α is not exceptional.

§ 7. A multiplication rule for symbols

7.1. In what follows we shall consider operators of the form

$$(Au)(x) = a(x)\,u(x) + \int\limits_{\mathbb{R}^m} r^{-m}f(x,\theta)\,u(y)\,\mathrm{d}y + (Tu)(x) \tag{1}$$

where the operator

$$(A_0u)(x) = a(x)\,u(x) + \int_{\mathbb{R}^m} r^{-m} f(x,\theta)\,u(y)\,dy \tag{2}$$

is assumed to be bounded in some Banach space and the operator T is completely continuous in the same space. It is reasonable to call operators of type (1) *general singular operators* whereas operators of the form (2) will be called *simple singular operators.*

In Chapter X we introduced the notion of the symbol of a simple singular operator. Here, we extend this concept to general singular operators setting the symbol of an arbitrary completely continuous operator identically equal to zero and taking the symbol of the corresponding simple operator (2) as the symbol of the operator (1). Such a definition makes sense if any simple singular operator is completely continuous in the Banach space considered if and only if its symbol is identically equal to zero. A proof to this statement for $\mathbf{L}_p(\mathbb{R}^m)$ spaces will be given in § 10 of the present chapter.

For the definition given here, the sum of general singular operators corresponds to the sum of their symbols, moreover, the general singular operator is determined by its symbol up to an arbitrary completely continuous summand.

In the present section we shall show that under certain additional assumptions the product of general singular operators corresponds to the product of their symbols. Thus, the definition of the symbol of a singular operator is in accordance with the general concept of a symbol as given in § 5, Chap. I.

7.2. Lemma 7.1. *If a function $b(x)$ is continuous on $\overline{\mathbb{R}}^m$ then the operator*

$$(Mu)(x) = \int_{\mathbb{R}^m} [b(y) - b(x)]\, r^{-m} Y_{n,m}^{(k)}(\theta)\, u(y)\,dy \tag{3}$$

is completely continuous in $\mathbf{L}_p(\mathbb{R}^m)$ for any $p \in (1,\infty)$.

Proof. Under the stereographic projection of the space $\mathbb{R}^m$ onto the Riemann sphere Σ the function $b(x)$ is taken to a certain function $b'(\xi)$, $\xi \in \Sigma$, that is continuous on Σ. If, in addition, $b'(\xi) \in \mathbf{Lip}_\lambda(\Sigma)$, $\lambda > 0$, then the lemma is a consequence of Theorem 3.4, Chap. VIII. In the general case we approximate the function $b'(\xi)$ uniformly by a sequence of functions $b_j' \in \mathbf{Lip}_1(\Sigma)$ and put $b_j'(\xi) = b_j(x)$. Then $\max_{x\in\overline{\mathbb{R}}^m} |b_j(x) - b(x)| \to 0$ as $j \to \infty$ and the operators

$$(M_ju)(x) = \int_{\mathbb{R}^m} [b_j(y) - b_j(x)]\, r^{-m} Y_{n,m}^{(k)}(\theta)\, u(y)\,dy$$

are completely continuous in $\mathbf{L}_p(\mathbb{R}^m)$. For estimating the norm $\|M - M_j\|$ we denote by a the norm of the singular operator with the characteristic $Y_{n,m}^{(k)}(\theta)$. We get

$$\|Mu - M_ju\| \leqq \Big\| \int_{\mathbb{R}^m} [b(y) - b_j(y)]\, r^{-m} Y_{n,m}^{(k)}(\theta)\, u(y)\,dy \Big\|$$

$$+ \Big\| [b(x) - b_j(x)] \int_{\mathbb{R}^m} r^{-m} Y_{n,m}^{(k)}(\theta)\, u(y)\,dy \Big\| \leqq 2a \max_{x\in\overline{\mathbb{R}}^m} |b(x) - b_j(x)| \cdot \|u\|$$

and, hence,

$$\|M - M_j\| \leqq 2a \max_{x\in\overline{\mathbb{R}}^m} |b(x) - b_j(x)| \xrightarrow[j\to\infty]{} 0.$$

Thus, the operator M is completely continuous. ∎

7.3. Theorem 7.1. *Assume that the symbols $\Phi_A(x,\theta)$ and $\Phi_B(x,\theta)$ of two singular operators A and B are continuous functions of $x \in \overline{\mathbb{R}}^m$ uniformly with respect to θ. If $\Phi_A(x,\theta)$,*

$\Phi_B(x,\theta) \hat{\in} \mathbf{W}_2^{(l)}(S)$ where, for a certain $p \in (1, \infty)$,

$$l \geqq \begin{cases} (m-1)/2 & , p = 2, \\ (m-1)/p + 1/2, & p < 2, \\ m/2 & , p > 2, \end{cases}$$

then the operator $AB - BA$ is completely continuous in $\mathbf{L}_p(\mathbb{R}^m)$, and the symbol of the product AB is equal to the product $\Phi_A(x,\theta)\,\Phi_B(x,\theta)$.

Proof. By D we denote the singular operator with the symbol $\Phi_A(x,\theta)\,\Phi_B(x,\theta)$. By Theorem 3.2 (if $p \neq 2$) or 2.2 (if $p = 2$) the operators A and B as well as the operator D are bounded in $\mathbf{L}_p(\mathbb{R}^m)$. Due to Theorem 3.3 we can expand these operators into norm convergent series that correspond to the expansions of their symbols with respect to spherical functions. Let

$$\Phi_A(x,\theta) = \sum_{n=0}^{\infty} \sum_{k=1}^{k_{n,m}} a_n^{(k)}(x)\, Y_{n,m}^{(k)}(\theta),$$

$$\Phi_B(x,\theta) = \sum_{n=0}^{\infty} \sum_{k=1}^{k_{n,m}} b_n^{(k)}(x)\, Y_{n,m}^{(k)}(\theta), \tag{4}$$

then

$$(Au)(x) = \sum_{n=0}^{\infty} \sum_{k=1}^{k_{n,m}} a_n^{(k)}(x)\, \gamma_{m,n}^{-1} \int_{\mathbb{R}^m} r^{-m}\, Y_{n,m}^{(k)}(\theta)\, u(y)\, dy + (T_1 u)(x), \tag{5}$$

$$(Bu)(x) = \sum_{n=0}^{\infty} \sum_{k=1}^{k_{n,m}} b_n^{(k)}(x)\, \gamma_{m,n}^{-1} \int_{\mathbb{R}^m} r^{-m}\, Y_{n,m}^{(k)}(\theta)\, u(y)\, dy + (T_2 u)(x).$$

The first terms in the series (5) are, in fact, $a_0^{(1)}(x) \cdot u(x)$ and $b_0^{(1)}(x) \cdot u(x)$, resp., but that remark does not affect to further arguments.

Now, we divide the series (4) into finite sums Φ_A', Φ_B', and their remainders Φ_A'', Φ_B''. Accordingly, we divide the series (5) into $A = A' + A''$ and $B = B' + B''$, resp. This splitting is performed in such a manner that the norms of A'' and B'' are sufficiently small. Then, $AB = A'B' + (A'B'' + A''B' + A''B'') + (AT_2 + T_1 B)$, and the norm of the first parentheses is arbitrarily small whereas the operator in the second parentheses is completely continuous. Finally, $A'B'u$ is a finite sum of expressions as

$$\frac{a_n^{(k)}(x)}{\gamma_{n,m}} \int_{\mathbb{R}^m} \frac{Y_{n,m}^{(k)}(\theta_{xy})}{r_{xy}^m} \cdot \frac{b_q^{(s)}(y)}{\gamma_{q,m}}\, dy \int_{\mathbb{R}^m} \frac{Y_{q,m}^{(s)}(\theta_{yz})}{r_{yz}^m}\, u(z)\, dz$$
$$= \frac{a_n^{(k)}(x)\, b_q^{(s)}(x)}{\gamma_{n,m}\gamma_{q,m}} \int_{\mathbb{R}^m} \frac{Y_{n,m}^{(k)}(\theta_{xy})}{r_{xy}^m}\, dy \int_{\mathbb{R}^m} \frac{Y_{q,m}^{(s)}(\theta_{yz})}{r_{yz}^m}\, u(z)\, dz$$
$$+ \frac{a_n^{(k)}(x)}{\gamma_{n,m}\gamma_{q,m}} \int_{\mathbb{R}^m} \frac{b_q^{(s)}(y) - b_q^{(s)}(x)}{r_{xy}^m}\, Y_{n,m}^{(k)}(\theta_{xy})\, dy \int_{\mathbb{R}^m} \frac{Y_{q,m}^{(s)}(\theta_{yz})}{r_{yz}^m}\, u(z)\, dz. \tag{6}$$

The inner integral in the second summand on the right-hand side of (6) turns out to be a bounded operator whereas the outer integral is, by Lemma 7.1, completely continuous. Hence the summand as a whole is completely continuous and, therefore, so is the sum of the corresponding terms which belong to the product $A'B'$. Further, the iterated integral in the first summand on the right-hand side of (6) is a product of singular operators the characteristics of which, $Y_{n,m}^{(k)}(\theta)$ and $Y_{q,m}^{(s)}(\theta)$, do not depend on the pole. Such a product turns out to be a singular operator with the symbol $\gamma_{n,m}\gamma_{q,m}\, Y_{n,m}^{(k)}(\theta) \times$

$Y_{q,m}^{(s)}(\theta)$. Hence, the first summand right in (6) has the symbol $a_n^{(k)}(x)\, b_q^{(s)}(x)\, Y_{n,m}^{(k)}(\theta)\, Y_{q,m}^{(s)}(\theta)$. The sum of the corresponding terms forms a finite section D' of the series for the operator D, obviously, the norm of the remainder $D'' = D - D'$ is arbitrarily small. On the other hand, $A'B' = D' + T_3$ where the completely continuous operator T_3 is the sum of second summands in expressions (6). Therefore,

$$AB = D + (A'B'' + A''B' + A''B'' - D'') + (AT_2 + T_1B + T_3).$$

The difference $AB - D$ is the sum of two operators, the first one is of arbitrarily small norm, and the second one is completely continuous. That means, the difference $T = AB - D$ is completely continuous, and the symbols of the operators AB and D coincide. But this means, in turn, that the symbol of the product AB is equal to the product of the symbols of the operators A and B.

The same arguments apply when proving the difference $T' = BA - D$ to be completely continuous. But then the difference $AB - BA = T - T'$ is completely continuous, too. ▌

Now, let A be a singular operator satisfying the conditions of Theorem 7.1. By B we denote the operator of multiplication by a function $b(x)$ that is continuous on $\overline{\mathbb{R}}^m$. Then, by Theorem 7.1, the commutator $[A, B] = AB - BA$ of the singular operator and the multiplication operator is completely continuous in $\mathbf{L}_p(\mathbb{R}^m)$. There is an interesting theorem by CALDERON [3] according to which the following estimate holds,

$$\|\nabla([A, B]\, u)\|_{\mathbf{L}_p(\mathbb{R}^m)} \leqq c\, \|u\|_{\mathbf{L}_p(\mathbb{R}^m)}, \qquad c = \text{const}.$$

Here, u is a finite sufficiently smooth function, the characteristic of the operator A is supposed to be of a certain smoothness, and $b \in \mathbf{Lip}_1(\mathbb{R}^m)$; ∇ denotes the gradient with respect to x.

ROZIN [2] obtained the following supplement to Calderon's theorem. Let n be a natural number, $n < m$. Assume that the characteristic f of the singular operator A satisfies the condition $f(x, \theta) \hat{\in} \mathbf{W}_2^{(l)}(S)$, $l > (n-1)/p$ if $1 < p \leqq 2$, $l > (n-1)/2$ if $2 < p < \infty$. If $|\nabla f(x, \theta)| \leqq c/\varrho_n$ where $\varrho_n^2 = x_1^2 + x_2^2 + \dots + x_n^2$, and $-n/p < \alpha < n/p'$ then there is the estimate

$$\|\varrho_n^{1+\alpha}\, \nabla([A, \varrho_n^{-\alpha}]\, u)\|_{\mathbf{L}_p(\mathbb{R}^m)} + \|\varrho_n^{\alpha}[A, \varrho_n^{-\alpha}]\, u\|_{\mathbf{L}_p(\mathbb{R}^m)} \leqq c\, \|u\|_{\mathbf{L}_p(\mathbb{R}^m)}.$$

§ 8. The adjoint singular operator

The adjoint of a singular operator will be constructed in the spaces $\mathbf{L}_p(\mathbb{R}^m)$, $1 < p < \infty$.

Theorem 8.1. *Let A be a simple singular operator the symbol $\Phi(\theta)$ of which does not depend on the pole. Then its adjoint A^* is a simple singular operator, too, and its symbol is the conjugate complex $\overline{\Phi(\theta)}$ of the symbol $\Phi(\theta)$.*

Proof. We have $A = F^{-1}\, \Phi(\theta)\, F$, and the adjoint to the Fourier transformation is $F^* = F^{-1}$. Hence,

$$(Au, v) = (F^{-1}\, \Phi(\theta)\, Fu, v) = (u, F^{-1}\, \overline{\Phi(\theta)}\, Fv)$$

and, consequently, $A^* = F^{-1}\, \overline{\Phi(\theta)}\, F$ which was to be proved. ▌

Theorem 8.2. *Let A be a general singular operator in $\mathbf{L}_p(\mathbb{R}^m)$. Assume that its symbol $\Phi(x, \theta)$ satisfies the following conditions:*

1. $\Phi(x, \theta) \hat{\in} \mathbf{W}_2^{(l)}(S)$

where

$$l > \begin{cases} (m-1)/2, & p = 2, \\ (m-1)/q + 1/2, & p \neq 2, \quad q = \min(p, p'); \end{cases} \tag{1}$$

2. *the symbol is continuous on the Riemann sphere (or, what is the same, on* $\overline{\mathbb{R}}^m$*) uniformly with respect to* θ.

Then the adjoint to A *is a general singular operator with the conjugate complex symbol* $\overline{\Phi(x, \theta)}$.

Proof. To begin with, we show that the given operator A and the operator with the symbol $\overline{\Phi(x, \theta)}$ are bounded in the spaces $\mathbf{L}_p(\mathbb{R}^m)$ and $\mathbf{L}_{p'}(\mathbb{R}^m)$, resp. For $p = 2$ this is a direct consequence of the relation (1). Let, e.g., $p > 2$, then $p' < 2$ and $l > (m-1)/p' + 1/2$. By Theorem 3.2, the singular operator with the symbol $\overline{\Phi(x, \theta)}$ is bounded in $\mathbf{L}_{p'}(\mathbb{R}^m)$. On the other hand,

$$l - \frac{m}{2} > \frac{1}{2} - \frac{m-1}{p'} - \frac{m}{2} = (m-1)\left(\frac{1}{p'} - \frac{1}{2}\right) > 0$$

and, by the same Theorem 3.2, the operator A is bounded in $\mathbf{L}_p(\mathbb{R}^m)$. In case $p < 2$ the reasoning is the same.

Next, we expand $\Phi(x, \theta)$ in a series with respect to spherical functions,

$$\Phi(x,\theta) = \sum_{n=0}^{\infty} \sum_{k=1}^{k_{n,m}} a_n^{(k)}(x)\, Y_{n,m}^{(k)}(\theta). \tag{2}$$

Denoting by $A_n^{(k)}$ the simple singular operator with the symbol $Y_{n,m}^{(k)}(\theta)$ we get

$$A = \sum_{n=0}^{\infty} \sum_{k=1}^{k_{n,m}} a_n^{(k)} A_n^{(k)} + T. \tag{3}$$

By Theorem 6.3, Chap. X, the series (2) converges in $\mathbf{W}_2^{(l)}(S)$ uniformly with respect to x, and by Theorem 3.3 the series (3) is convergent in the $\mathbf{L}_p(\mathbb{R}^m)$ metric. According to Theorem 8.1 it is clear that $(A_n^{(k)})^*$ is a simple singular operator with the symbol $\overline{Y_{n,m}^{(k)}(\theta)}$, thence

$$A^* u = \sum_{n=0}^{\infty} \sum_{k=1}^{k_{n,m}} A_n^{(k)*} (\overline{a_n^{(k)}} u) + T^* u = \sum_{n=0}^{\infty} \sum_{k=1}^{k_{n,m}} \overline{a_n^{(k)}(x)}\, A_n^{(k)*} u$$
$$+ \sum_{n=0}^{\infty} \sum_{k=1}^{k_{n,m}} \int_{\mathbb{R}^m} \frac{\overline{a_n^{(k)}(y)} - \overline{a_n^{(k)}(x)}}{r^m} \cdot \frac{\overline{Y_{n,m}^{(k)}(\theta)}}{\gamma_{n,m}}\, u(y)\, dy + T^* u. \tag{4}$$

The first series in the right-hand side of (4) turns out to be a singular operator with the symbol $\overline{\Phi(x, \theta)}$, this operator is bounded in $\mathbf{L}_{p'}(\mathbb{R}^m)$. Every term of the second series in (4) is, by Lemma 7.1, completely continuous in the same space. We show that this series is norm convergent. Its sum is equal to

$$\sum_{n=1}^{\infty} \sum_{k=1}^{k_{n,m}} A_n^{(k)*} (\overline{a_n^{(k)}} u) - \sum_{n=1}^{\infty} \sum_{k=1}^{k_{n,m}} \overline{a_n^{(k)}}\, A_n^{(k)*} u. \tag{5}$$

The minuend in (5) is a series of operators which are adjoint to the operators under the sum in (3). The convergence of the series (3) in $\mathbf{L}_p(\mathbb{R}^m)$ implies the convergence of the first series in (5) in the space $\mathbf{L}_{p'}(\mathbb{R}^m)$. The subtrahend in (5) is the expansion of a singular operator with the symbol $\overline{\Phi(x,\theta)} - \overline{a_1^{(0)}(x)}$, and the assumptions of the theorem to be proved imply that this series is convergent in $\mathbf{L}_{p'}(\mathbb{R}^m)$, too. ∎

§ 9. Singular operators in Sobolev spaces

The results of the present section are contained in MIKHLIN [19].

9.1. Theorem 9.1. *If the symbol $\Phi_A(\theta)$ of the singular operator*

$$(Au)(x) = au(x) + \int_{\mathrm{R}^m} r^{-m} f(\theta)\, u(y)\, dy \tag{1}$$

is essentially bounded then the operator A is bounded in $\mathbf{W}_2^{(l)}(\mathbb{R}^m)$ for any natural number l, and

$$\| A \|_{\mathbf{W}_2^{(l)}(\mathrm{R}^m)} \leqq C \sup \mathrm{ess}\, | \Phi_A(\theta) |, \qquad C = \mathrm{const}. \tag{2}$$

Proof. The norm in $\mathbf{W}_2^{(l)}(\mathbb{R}^m)$ will be defined by

$$\| u \|^2_{\mathbf{W}_2^{(l)}(\mathrm{R}^m)} = \sum_{|\alpha|=0}^{l} \| D^\alpha u \|^2_{\mathrm{L}_2(\mathrm{R}^m)}. \tag{3}$$

By (1.5) we have

$$(Au)(x) = \int_{\mathrm{R}^m} \mathrm{e}^{2\pi\mathrm{i}(x,\xi)}\, \Phi_A(\theta)\, \hat{u}(\xi)\, \mathrm{d}\xi, \qquad \theta = \frac{\xi}{|\xi|}.$$

Consider a function $u \in \mathbf{W}_2^{(l)}(\mathbb{R}^m)$ and a multi-index α, $|\alpha| < l$. Differentiating we find

$$(D^\alpha Au)(x) = (2\pi\mathrm{i})^{|\alpha|} \int_{\mathrm{R}^m} \mathrm{e}^{2\pi\mathrm{i}(x,\xi)}\, \Phi_A(\theta)\, \xi^\alpha \hat{u}(\xi)\, \mathrm{d}\xi.$$

Taking into account $(2\pi\mathrm{i})^\alpha \xi^\alpha \hat{u}(\xi) = \widehat{D^\alpha u(\xi)}$ we get

$$(D^\alpha Au)(x) = \int_{\mathrm{R}^m} \mathrm{e}^{2\pi\mathrm{i}(x,\xi)} \Phi_A(\theta)\, \widehat{D^\alpha u}(\xi)\, \mathrm{d}\xi. \tag{4}$$

By the same formula (1.5), $D^\alpha Au$ is a singular operator with the symbol $\Phi_A(\theta)$ and the density $D^\alpha u$. Due to Theorem 5.1, Chap. X, $D^\alpha Au \in \mathbf{L}_2(\mathbb{R}^m)$, and

$$\| D^\alpha Au \|^2_{\mathrm{L}_2(\mathrm{R}^m)} \leqq \sup \mathrm{ess}\, | \Phi(\theta) |^2 \cdot \| D^\alpha u \|^2_{\mathrm{L}_2(\mathrm{R}^m)}. \tag{5}$$

Summation over α, $|\alpha| \leqq l$, yields (2) with $C = 1$. If the norm (3) is replaced by another one which is equivalent to it then this may, at most, cause a constant factor in (2). ∎

9.2. In the same space $\mathbf{W}_2^{(l)}(\mathbb{R}^m)$ we consider now the singular operator with a variable symbol,

$$(Au)(x) := a(x)\, u(x) + \int_{\mathrm{R}^m} r^{-m} f(x, \theta)\, u(y)\, \mathrm{d}y = \int_{\mathrm{R}^m} \mathrm{e}^{2\pi\mathrm{i}(x,\xi)}\, \Phi_A(x, \theta)\, \hat{u}(\xi)\, \mathrm{d}\xi. \tag{6}$$

In order to simplify notations we introduce a class $\mathfrak{R}_{l,\lambda}$ of those symbols that satisfy the condition

$$D_x^\alpha \Phi(x, \theta) \mathrel{\hat{\in}} \mathbf{W}_2^{(\lambda)}(S), \qquad \forall \alpha: |\alpha| \leqq l. \tag{7}$$

Theorem 9.2. *If $\Phi_A(x, \theta) \in \mathfrak{R}_{l,\lambda}$, $\lambda > (m-1)/2$, the operator (6) is bounded in $\mathbf{W}_2^{(l)}(\mathbb{R}^m)$.*

If $\Phi_A(x, \theta) \in \mathfrak{R}_{l,\lambda}$, $\lambda > (m-1)/p + 1/2$ if $p < 2$, and $\lambda > m/2$ if $p > 2$, then the operator (6) is bounded in $\mathbf{W}_p^{(l)}(\mathbb{R}^m)$.

Proof. We differentiate (6) and obtain, $|\alpha| \leqq l$,

$$(D^\alpha Au)(x) = \sum_{\beta \leqq \alpha} \frac{\alpha!}{\beta!\,(\alpha-\beta)!} (2\pi\mathrm{i})^{|\beta|} \int_{\mathrm{R}^m} \mathrm{e}^{2\pi\mathrm{i}(x,\xi)} D_x^{\alpha-\beta} \Phi_A(x, \theta)\, \xi^\beta \hat{u}(\xi)\, \mathrm{d}\xi$$

$$= \sum_{\beta \leqq \alpha} \frac{\alpha!}{\beta!\,(\alpha-\beta)!} \int_{\mathrm{R}^m} \mathrm{e}^{2\pi\mathrm{i}(x,\xi)}\, D_x^{\alpha-\beta} \Phi_A(x, \theta)\, \widehat{D^\beta u}(\xi)\, \mathrm{d}\xi.$$

Each term in the last sum turns out to be a singular operator with the symbol $D_x^{\alpha-\beta}\Phi_A(x, \theta)$ that satisfies the conditions of Theorem 2.2 or 3.2 whereas its density belongs to the space $\mathbf{L}_2(\mathbb{R}^m)$ or $\mathbf{L}_p(\mathbb{R}^m)$, resp. If the norm in $\mathbf{W}_p^{(l)}(\mathbb{R}^m)$ is e.g. defined by the formula

$$\|u\|^p_{\mathbf{W}_p^{(l)}(\mathbb{R}^m)} = \sum_{|\alpha|=0}^{l} \|D^\alpha u\|^p_{\mathbf{L}_p(\mathbb{R}^m)} \tag{8}$$

then the operator (6) is bounded in $\mathbf{L}_p(\mathbb{R}^m)$ according to the theorems mentioned. ∎

9.3. We shall call an operator in $\mathbf{L}_p(\mathbb{R}^m)$ *smoothing* of order l iff it takes any function of $\mathbf{W}_p^{(j)}(\mathbb{R}^m)$, $0 \leqq j \leqq l-1$, to a function belonging to the class $\mathbf{W}_p^{(j+1)}(\mathbb{R}^m)$. For the sake of simplicity, we restrict ourselves to the case $p = 2$.

Theorem 9.3. *Assume that the symbol $\Phi_0(x, \theta)$ of the singular integral operator*

$$(A_0 u)(x) = \int_{\mathbb{R}^m} r^{-m} f(x, \theta)\, u(y)\, dy \tag{9}$$

belongs to the class $\mathfrak{R}_{l+1,\lambda}$ for some $\lambda > (m-1)/2$, and its characteristic be continuously differentiable with respect to the Cartesian coordinates of x and θ. Then the operator

$$(Mu)(x) = \int_{\mathbb{R}^m} r^{-m} [f(x, \theta) - f(y, \theta)]\, u(y)\, dy \tag{10}$$

is smoothing of order l.

Proof. We expand the symbol into a series with respect to spherical functions,

$$\Phi_0(x, \theta) = \sum_{n=1}^{\infty} \sum_{k=1}^{k_{n,m}} a_n^{(k)}(x)\, Y_{n,m}^{(k)}(\theta). \tag{11}$$

Then

$$f(x, \theta) = \sum_{n=1}^{\infty} \sum_{k=1}^{k_{n,m}} b_n^{(k)}(x)\, Y_{n,m}^{(k)}(\theta), \qquad b_n^{(k)}(x) = \frac{a_n^{(k)}(x)}{\gamma_{n,m}}, \tag{12}$$

and

$$\begin{aligned}(Mu)(x) &= \sum_{n=1}^{\infty} \sum_{k=1}^{k_{n,m}} \int_{\mathbb{R}^m} [b_n^{(k)}(x) - b_n^{(k)}(y)]\, r^{-m}\, Y_{n,m}^{(k)}(\theta)\, u(y)\, dy \\ &=: \sum_{n=1}^{\infty} \sum_{k=1}^{k_{n,m}} (M_n^{(k)} u)(x).\end{aligned} \tag{13}$$

Let us introduce the notations $(Mu)(x) = v(x)$, $\partial/\partial x_q = D_q$, $r^{-m} Y_{n,m}^{(k)}(\theta) = K_n^{(k)}(x-y)$, then

$$\begin{aligned}D_q v(x) = \lim_{h\to 0} \sum_{n=1}^{\infty} \sum_{k=1}^{k_{n,m}} \frac{1}{h} \int_{\mathbb{R}^m} \{[b_n^{(k)}(x+\bar{h}) - b_n^{(k)}(y)]\, K_n^{(k)}(x+\bar{h}-y) \\ - [b_n^{(k)}(x) - b_n^{(k)}(y)]\, K_n^{(k)}(x-y)\}\, u(y)\, dy\end{aligned}$$

where $\bar{h} = hl_q$, and l_q denotes the unit vector along the x_q axis. This equation is brought to the form

$$\begin{aligned}D_q v(x) = \sum_{n=1}^{\infty} \sum_{k=1}^{k_{n,m}} \lim_{h\to 0} \frac{1}{h} \int_{\mathbb{R}^m} K_n^{(k)}(x-y)\, [b_n^{(k)}(x+\bar{h})\, u(y+\bar{h}) - b_n^{(k)}(x)\, u(y)]\, dy \\ - \sum_{n=1}^{\infty} \sum_{k=1}^{k_{n,m}} \lim_{h\to 0} \frac{1}{h} \int_{\mathbb{R}^m} K_n^{(k)}(x-y)\, [b_n^{(k)}(y+\bar{h})\, u(y+\bar{h}) - b_n^{(k)}(y) u(y)]\, dy.\end{aligned} \tag{14}$$

The first limit in (14) is transformed as follows,

$$\lim_{h\to 0} \frac{b_n^{(k)}(x+\bar{h}) - b_n^{(k)}(x)}{h} \int_{\mathbb{R}^m} K_n^{(k)}(x-y)\, u(y)\, dy$$

$$+ \lim_{h\to 0} b_n^{(k)}(x+\bar{h}) \int_{\mathbb{R}^m} K_n^{(k)}(x-y) \frac{u(y+\bar{h}) - u(y)}{h} dy$$

$$= D_q b_n^{(k)}(x) \int_{\mathbb{R}^m} K_n^{(k)}(x-y)\, u(y)\, dy + b_n^{(k)}(x) \int_{\mathbb{R}^m} K_n^{(k)}(x-y)\, D_q u(y)\, dy.$$

Similarly, the second limit in (14) becomes

$$\int_{\mathbb{R}^m} K_n^{(k)}(x-y)\, D_q b_n^{(k)}(y)\, u(y)\, dy + \int_{\mathbb{R}^m} K_n^{(k)}(x-y)\, b_n^{(k)}(y)\, D_q u(y)\, dy.$$

Inserting these expressions into (14) we obtain

$$D_q v(x) = \int_{\mathbb{R}^m} r^{-m} D_q f(x, \theta)\, u(y)\, dy - \int_{\mathbb{R}^m} r^{-m} D_q f(y, \theta)\, u(y)\, dy + (M\, D_q u)\,(x). \tag{15}$$

Here $D_q f(x, \theta)$ and $D_q f(y, \theta)$ mean derivatives with respect to x and y resp., computed under the assumption that θ does not depend on x or y; an analogous notation will be used in similar cases, too.

The first term in the right-hand side of (15) is a singular integral operator with the symbol $D_q \Phi(x, \theta) \in \mathfrak{R}_{l,\lambda}$ which is bounded in $\mathbf{W}_2^{(j)}(\mathbb{R}^m)$, $j \leq l$, by Theorem 9.2. The second term in (15) right is also bounded in $\mathbf{W}_2^{(j)}(\mathbb{R}^m)$. We do not want to interrupt the main line of reasoning so that assertion will be proved in subsection 9.4 in a slightly more general framework. Thus, it remains to consider the last term in (15). If we succeed in proving that the operator M is smoothing of order $l - 1$ then this last term belongs to $\mathbf{W}_2^{(j)}(\mathbb{R}^m)$ provided that $u \in \mathbf{W}_2^{(j)}(\mathbb{R}^m)$. But then $D_q v \in \mathbf{W}_2^{(j)}(\mathbb{R}^m)$, and hence $v \in \mathbf{W}_2^{(j+1)}(\mathbb{R}^m)$ which was to be demonstrated.

Thus, it is sufficient to show that M is a smoothing operator of order $l - 1$. For that it is, in turn, sufficient to show that it is smoothing of order $l - 2$. Finally, we have to show that M is smoothing of order 1.

The integral in (10) has a weak singularity at $x = y$. Differentiating it we obtain using (7.2), Chap. IX,

$$D_q v(x) = \int_{\mathbb{R}^m} \frac{\partial}{\partial x_q} \{r^{-m}[f(x, \theta) - f(y, \theta)]\}\, u(y)\, dy + u(x) \int_S (\nabla_x f(x, \theta), \theta)\, \theta_q\, dS; \tag{16}$$

the gradient ∇_x is, again, taken under the assumption that θ does not depend on x.

The second summand in (16) belongs to $\mathbf{L}_2(\mathbb{R}^m)$ if u belongs to it. The first summand is a general singular operator that is bounded in $\mathbf{L}_2(\mathbb{R}^m)$. Hence, $D_q v \in \mathbf{L}_2(\mathbb{R}^m)$. On the other hand, it is easy to see that the operator M is bounded in $\mathbf{L}_2(\mathbb{R}^m)$ such that $v \in \mathbf{L}_2(\mathbb{R}^m)$ and, finally, $v \in \mathbf{W}_2^{(1)}(\mathbb{R}^m)$. ∎

Remark. In the special case $l = 1$ and $f(x, \theta) = a(x)\, \varphi(\theta)$ Theorem 9.3 is proved under weaker assumptions on the smoothness of the characteristic by Calderon [2].

9.4. Now, we show that the singular integral

$$(Bu)\,(x) = \int_{\mathbb{R}^m} r^{-m} f(y, \theta)\, u(y)\, dy \tag{17}$$

is an operator that is bounded in $\mathbf{W}_2^{(l)}(\mathbb{R}^m)$ if $\Phi(x, \theta) \in \mathfrak{R}_{l,\lambda}$, $\lambda > m/2 - 1$. In case $l = 0$ the assertion is obvious, since B is then adjoint to the bounded singular operator with the characteristic $\overline{f(x, -\theta)}$.

Let $l > 0$. We expand the characteristic in a series (12). Denote by $B_n^{(k)}$ the singular operator with the characteristic $Y_{n,m}^{(k)}(\theta)$ and put $Bu = w$, then

$$D^\alpha w = \sum_{n=1}^{\infty} \sum_{k=1}^{k_{n,m}} D^\alpha B_n^{(k)}(b_n^{(k)} u) \qquad \forall \alpha \colon |\alpha| \leqq l,$$

or, by Theorem 9.1,

$$D^\alpha w = \sum_{n=1}^{\infty} \sum_{k=1}^{k_{n,m}} B_n^{(k)}(D^\alpha(b_n^{(k)} u)) = \sum_{\beta \leqq \alpha} \sum_{n=1}^{\infty} \sum_{k=1}^{k_{n,m}} \frac{\alpha!}{\beta!\,(\alpha-\beta)!} B_n^{(k)}(D^{\alpha-\beta} b_n^{(k)}\, D^\beta u). \tag{18}$$

We separate in the right-hand side the sum for some fixed multiindex β,

$$w_\beta := \frac{\alpha!}{\beta!\,(\alpha-\beta)!} \sum_{n=1}^{\infty} \sum_{k=1}^{k_{n,m}} B_n^{(k)}(D^{\alpha-\beta} b_n^{(k)}\, D^\beta u). \tag{19}$$

The sum in the right-hand side of (19) turns out to be an operator of type (17) with the characteristic $D_y^{\alpha-\beta} f(y, \theta)$, its adjoint operator in $\mathbf{L}_2(\mathbb{R}^m)$ has the characteristic $D_x^{\alpha-\beta} \overline{f(x, -\theta)}$ and, in view of the assumptions of Theorem 9.3, is bounded in $\mathbf{L}_2(\mathbb{R}^m)$. Thence, $\|w\|_{\mathbf{L}_2(\mathbb{R}^m)} \leqq C \|D^\beta u\|_{\mathbf{L}_2(\mathbb{R}^m)}$ and, therefore, also $\|D^\alpha w\|_{\mathbf{L}_2(\mathbb{R}^m)} \leqq C \|u\|_{\mathbf{W}_2^{(\alpha)}(\mathbb{R}^m)}$. Summing over all multi-indices α, $|\alpha| \leqq l$, we find that the operator B is bounded in $\mathbf{W}_2^{(l)}(\mathbb{R}^m)$.

9.5. In conclusion we note the following. The reasoning in 9.3 and 9.4 implies that the operator M from $\mathbf{W}_2^{(j)}(\mathbb{R}^m)$ into $\mathbf{W}_2^{(j+1)}(\mathbb{R}^m)$, $j = 0, 1, \ldots, l-1$, is bounded. Its operator norm can be estimated by the quantity

$$C \sup_{x \in \mathbb{R}^m, |\alpha| \leqq l+1} \|D_x^\alpha \Phi(x, .)\|_{\mathbf{W}_2^{(\lambda)}(S)}. \tag{20}$$

9.6. Theorem 9.3 remains to hold true also in the space $\mathbf{L}_p(\mathbb{R}^m)$, $p \neq 2$, provided that

$$\Phi(x, \theta) \in \mathfrak{R}_{l+1,\lambda}, \qquad \lambda > \begin{cases} (m-1)/p + 1/2, & p < 2, \\ m/2, & p > 2, \end{cases}$$

and the characteristic is continuously differentiable with respect to the Cartesian coordinates of the points x and θ.

§ 10. Factor ring of singular operators

Le us consider the set of all general singular operators the symbols of which satisfy the assumptions of §§ 7 and 8 such that, in particular, these operators are bounded in a corresponding space $\mathbf{L}_p(\mathbb{R}^m)$ and commute up to a summand that is completely continuous in that space. The adjoint operators have the same properties.

10.1. Theorem 7.1 implies that the operators of the set mentioned form a ring. In this ring, the set of all completely continuous operators in $\mathbf{L}_p(\mathbb{R}^m)$ is a two-sided ideal. The factor ring of the general singular operators with respect to the ideal of completely continuous operators is a commutative ring; it was studied in papers by Gohberg [2], Seeley [2], and Krupnik [1]. In the following we shall develop some of their results on the factor ring mentioned.

10.2. Let A be a simple singular operator the symbol of which, $\Phi_A(x, \theta)$, satisfies for some $p \in (1, \infty)$ the conditions of § 7. We expand the symbol into a series with respect to spherical functions,

$$\Phi_A(x, \theta) = \sum_{n=0}^{\infty} \sum_{k=1}^{k_{m,m}} a_n^{(k)}(x)\, Y_{n,m}^{(k)}(\theta). \tag{1}$$

By Theorem 3.3, the expansion (1) corresponds to the expansion

$$A = \sum_{n=0}^{\infty} \sum_{k=1}^{k_{n,m}} a_n^{(k)} A_n^{(k)} \tag{2}$$

which is convergent in the norm of $\mathbf{L}_p(\mathbb{R}^m)$; $A_n^{(k)} = F^{-1}\, Y_{n,m}^{(k)}(\theta)\, F$ is the simple singular operator with the symbol $Y_{n,m}^{(k)}(\theta)$. In order to simplify notations we shall write $\|\cdot\|_q$ instead of $\|\cdot\|_{\mathbf{L}_q(\mathbb{R}^m)}$.

Lemma 10.1. *Let $p \geqq 2$ and (x_0, θ_0) be an arbitrary point of $\mathbb{R}^m \times S$. To any natural number N and arbitrary positive numbers $\varepsilon_1, \varepsilon_2, \varrho$ there exists a function $\psi \in \mathbf{L}_p(\mathbb{R}^m)$, $\|\psi\|_p = 1$, such that*

$$\int_{|x-x_0| \geqq \varrho} |\psi(x)|^p \,\mathrm{d}x < \varepsilon_1, \qquad \|A_n^{(k)}\psi - Y_{n,m}^{(k)}(\theta_0)\,\psi\|_p < \varepsilon_2. \tag{3}$$

Proof. Let K be a circular cone with vertex at the origin, the apex angle 2δ, and the axis passing the point θ_0. The positive number δ is chosen small enough such that any function $\chi \in \mathbf{L}_{p'}(\mathbb{R}^m)$ vanishing outside of K satisfies the inepuality

$$\left\| \left[Y_{n,m}^{(k)}\left(\frac{x}{|x|}\right) - Y_{n,m}^{(k)}(\theta_0) \right] \chi(x) \right\|_{p'} < \frac{\varepsilon_2}{2} \|\chi\|_{p'}. \tag{4}$$

Let us denote by M the set of all functions belonging to $\mathbf{L}_p(\mathbb{R}^m)$ which vanish outside of K and outside of a certain ball where this ball may depend on the function under consideration.

For $p > 2$ the Fourier transformation F turns out to be a bounded operator from $\mathbf{L}_{p'}(\mathbb{R}^m)$ into $\mathbf{L}_p(\mathbb{R}^m)$ (see Titchmarsh [1], § 4.1). It is easy to see that F^{-1} is also bounded in that pair of spaces. Thence,

$$c_0 := \sup_{\chi \in M} \frac{\|F^{-1}\,\chi\|_p}{\|\chi\|_{p'}} < \infty,$$

and there is a function $\chi_0 \in M$ such that

$$\|F^{-1}\chi_0\|_p \geqq \frac{c_0}{2} \|\chi_0\|_{p'}. \tag{5}$$

Put $\varphi(x) = \chi_0(x)/\|F^{-1}\chi_0\|_p$, then $\|F^{-1}\varphi\|_p = 1$ and $\|\varphi\|_{p'} < 2/c_0$. Consider the sequence $\varphi_s(x) = s^{-m/p'}\varphi(x/s)$, $s = 1, 2, \ldots$ Obviously, $\varphi_s \in M$ and $\|\varphi_s\|_{p'} = \|\varphi\|_{p'} < 2/c_0$. It is not difficult to find the Fourier transform F^{-1} of φ_s: With the notations $F^{-1}\varphi_s = \Phi_s$, $F^{-1}\varphi = \Phi$, we have

$$\Phi_s(x) = s^{-m/p'} \int_{\mathbb{R}^m} \mathrm{e}^{2\pi\mathrm{i}(x,\xi)}\, \varphi\left(\frac{\xi}{s}\right) \mathrm{d}\xi = s^{m/p} \int_{\mathbb{R}^m} \mathrm{e}^{2\pi\mathrm{i}(sx,\xi)} \varphi(\xi)\,\mathrm{d}\xi = s^{m/p}\Phi(sx).$$

Now,

$$\lim_{s\to\infty} \int_{|x|>\varrho} |\Phi_s(x)|^p \,\mathrm{d}x = \lim_{s\to\infty} \int_{|x|>\varrho s} |\Phi(t)|^p \,\mathrm{d}t = 0,$$

and there exists a number s_0 such that $\int_{|x|>\varrho} |\Phi_{s_0}(x)|^p\,\mathrm{d}x < \varepsilon_1$. Put $\psi(x) = \Phi_{s_0}(x - x_0)$, then

$$\int_{|x-x_0|>\varrho} |\psi(x)|^p \mathrm{d}x < \varepsilon_1, \qquad \|\psi\|_p = \|\Phi_{s_0}\|_p = s_0^{m/p}\|\Phi(s_0x)\|_p = \|F^{-1}\varphi\|_p - 1.$$

Let us consider the norm

$$\|A_n^{(k)}\psi - Y_{n,m}^{(k)}(\theta_0)\,\psi\|_p = \|F^{-1}[Y_{n,m}^{(k)}(\theta) - Y_{n,m}^{(k)}(\theta_0)]\,F\psi\|_p.$$

Because of $F\psi \in M$ the product $[Y_{n,m}^{(k)}(\theta) - Y_{n,m}^{(k)}(\theta_0)]\,F\psi$ belongs to M, too, and by the inequalities (4) and (5) we find

$$\|F^{-1}[Y_{n,m}^{(k)}(\theta) - Y_{n,m}^{(k)}(\theta_0)]\,F\psi\|_p \leqq c_0\,\|[Y_{n,m}^{(k)}(\theta) - Y_{n,m}^{(k)}(\theta_0)]\,F\psi\|_{p'}$$

$$\leqq \frac{c_0}{2}\,\varepsilon_2\,\|F\psi\|_{p'} = \frac{c_0}{2}\,\varepsilon_2\,\|\varphi\|_{p'} < \varepsilon_2. \quad \blacksquare$$

10.3. Theorem 10.1. *If the symbol $\Phi_A(x,\theta)$ of a general singular operator A satisfies the conditions of § 7 then for any completely continuous operator T in $\mathrm{L}_p(\mathrm{R}^m)$ the inequality*

$$\|A + T\|_p \geqq \max_{(x,\theta)\in\overline{\mathrm{R}}^m\times S} |\Phi_A(x,\theta)| \tag{6}$$

holds true.

Proof. Let, first, $p \geqq 2$. We expand the symbol in a series (1), and by $\Phi_s(x,\theta)$ we denote its partial sum

$$\Phi_s(x,\theta) = \sum_{n=0}^{s}\sum_{k=1}^{k_{n,m}} a_n^{(k)}(x)\,Y_{n,m}^{(k)}(\theta);$$

similarly, we denote by A_s the singular operator with the symbol $\Phi_s(x,\theta)$. We shall prove the inequality

$$\|A_s + T\|_p \geqq \max_{(x,\theta)\in\overline{\mathrm{R}}^m\times S} |\Phi_s(x,\theta)|. \tag{7}$$

Let $(x_0,\theta_0) \in \mathrm{R}^m\times S$. Due to Lemma 10.1, there is a sequence of functions $\psi_l \in M$ such that

$$\|A_n^{(k)}\psi_l - Y_{n,m}^{(k)}(\theta_0)\,\psi_l\|_p \leqq \frac{1}{l}, \qquad \|\psi_l\|_p = 1, \quad \int_{|x-x_0|>1/l} |\psi_l(x)|^p\,dx < \frac{1}{l^p}, \tag{8}$$

$n = 0, 1, 2, \ldots, s$; $l = 1, 2, \ldots$; $k = 1, 2, \ldots, k_{n,m}$. Thence,

$$\|(A_s + T)\,\psi_l\|_p \geqq \left|\sum_{n=0}^{s}\sum_{k=1}^{k_{n,m}} a_n(x_0)\,Y_{n,m}^{(k)}(\theta_0)\right|$$

$$- \left\|\sum_{n=0}^{s}\sum_{k=1}^{k_{n,m}} [a_n^{(k)}(x)\,A_n^{(k)}\psi_l - a_n^{(k)}(x_0)\,Y_{n,m}^{(k)}(\theta_0)]\,\psi_l\right\|_p - \|T\psi_l\|_p. \tag{9}$$

By b_l we denote the second term in the right-hand side of (9) where

$$b_l \leqq \left\|\sum_{n=0}^{s}\sum_{k=1}^{k_{n,m}} a_n^{(k)}(x)\,[A_n^{(k)}\psi_l - Y_{n,m}^{(k)}(\theta_0)\,\psi_l]\right\|_p + \left\|\sum_{n=0}^{s}\sum_{k=1}^{k_{n,m}} [a_n^{(k)}(x) - a_n^{(k)}(x_0)]\,Y_{n,m}^{(k)}(\theta_0)\,\psi_l\right\|_p,$$

and, due to the first relation in (8),

$$b_l \leqq \frac{1}{l}\sum_{n=0}^{s}\sum_{k=1}^{k_{n,m}} \sup_{x\in\mathrm{R}^m} |a_n^{(k)}(x)| + \left\|\sum_{n=0}^{s}\sum_{k=1}^{k_{n,m}} [a_n^{(k)}(x) - a_n^{(k)}(x_0)]\,Y_{n,m}^{(k)}(\theta_0)\,\psi_l\right\|_p. \tag{10}$$

Now, we introduce the functions

$$\varphi_l(x) = \begin{cases} \psi_l(x), & |x - x_0| \geqq 1/l, \\ 0, & |x - x_0| < 1/l, \end{cases}$$

and $\omega_l(x) = \psi_l(x) - \varphi_l(x)$. The norm in (10) right can be estimated by

$$\left\| \sum_{n=0}^{s} \sum_{k=1}^{k_{n,m}} [a_n^{(k)}(x) - a_n^{(k)}(x_0)]\, Y_{n,m}^{(k)}(\theta_0)\, \omega_l \right\|_p + \left\| \sum_{n=0}^{s} \sum_{k=1}^{k_{n,m}} [a_n^{(k)}(x) - a_n^{(k)}(x_0)]\, Y_{n,m}^{(k)}(\theta_0)\, \varphi_l \right\|_p$$

$$\leqq \sum_{n=0}^{s} \sum_{k=1}^{k_{n,m}} \max_{|x-x_0| \leqq l^{-1}} |a_n^{(k)}(x) - a_n^{(k)}(x_0)| \max_{\theta \in S} |Y_{n,m}^{(k)}(\theta)|$$

$$+ \sum_{n=0}^{s} \sum_{k=1}^{k_{n,m}} \frac{2}{l} \sup_{x \in \mathbb{R}^m} |a_n^{(k)}(x)| \max_{\theta \in S} |Y_{n,m}^{(k)}(\theta)|.$$

This implies together with the inequality (10) $b_l \to 0$ as $l \to \infty$.

10.4. Here we show that the sequence ψ_l converges weakly to zero in $\mathbf{L}_p(\mathbb{R}^m)$ as $l \to \infty$. Let g be an arbitrary function of $\mathbf{L}_{p'}(\mathbb{R}^m)$. Using (8) we obtain

$$|(g, \psi_l)| \leqq \int_{|x-x_0|>l^{-1}} |g(x)\, \psi_l(x)|\, dx + \int_{|x-x_0|<l^{-1}} |g(x)\, \psi_l(x)|\, dx$$

$$\leqq \frac{1}{l} \|g\|_{p'} + \left\{ \int_{|x-x_0|<l^{-1}} |g(x)|^{p'} dx \right\}^{1/p'} \xrightarrow[l \to \infty]{} 0.$$

As a consequence, we obtain that $\|T\psi_l\|_p \to 0$ as $l \to \infty$. Replacing in (9) the left-hand side by its supremum and the right-hand side by its limit as $l \to \infty$ we arrive at

$$\sup_l \|(A_s + T)\, \psi_l\|_p \geqq |\Phi_s(x_0, \theta_0)|.$$

Because of $\|\psi_l\|_p = 1$ we have $\|(A_s + T)\, \psi_l\|_p \leqq \|A_s + T\|_p$, $\|A_s + T\|_p \geqq |\Phi_s(x_0, \theta_0)|$. If we take as (x_0, θ_0) that point at which the function $\Phi_s(x, \theta)$ assumes its maximum we obtain the inequality (7).

10.5. Now, we fix the operator T and take in (7) the limit as $s \to \infty$. Then, by Theorem 3.3, $\|A_s + T\|_p \to \|A + T\|_p$. Furthermore, the function $\Phi_A(x, \theta)$ is continuous on the compact set $\overline{\mathbb{R}}^m \times S$ such that $\Phi_s(x, \theta) \to \Phi_A(x, \theta)$ as $s \to \infty$ uniformly on that compactum. Hence,

$$\max_{(x,\theta) \in \overline{\mathbb{R}}^m \times S} |\Phi_A(x, \theta)| = \|\Phi_A\|_{C(\overline{\mathbb{R}}^m \times S)} = \lim_{s \to \infty} \|\Phi_s\|_{C(\overline{\mathbb{R}}^m \times S)} = \lim_{s \to \infty} \max_{(x,\theta) \in \overline{\mathbb{R}}^m \times S} |\Phi_s(x, \theta)|.$$

As a result, we obtain the inequality (6), and Theorem 10.1 is proved in case $p \geqq 2$.

10.6. Now, let $1 < p < 2$. The adjoint operator A^* is bounded in $\mathbf{L}_{p'}(\mathbb{R}^m)$ where $p' > 2$. As shown in § 8, we have $A^* = A_1 + T_1$ where A_1 is a singular operator with the symbol $\overline{\Phi_A(x, \theta)}$, and T_1 is completely continuous in $\mathbf{L}_{p'}(\mathbb{R}^m)$. By the inequality (6),

$$\|A^* + T^*\|_{p'} = \|A_1 + (T_1 + T^*)\|_{p'} \geqq \max_{(x,\theta) \in \overline{\mathbb{R}}^m \times S} |\Phi_A(x, \theta)|.$$

Since $\|A^* + T^*\|_{p'} = \|A + T\|_p$ the last inequality is equivalent to (6), and Theorem 10.1 is proved completely. ∎

10.7. We denote by $\mathfrak{R}$ the ring of the general singular operators in $\mathbf{L}_p(\mathbb{R}^m)$ the symbols of which satisfy the conditions of § 8, and by $\mathfrak{J}$ the two-sided ideal of completely continuous operators in $\mathbf{L}_p(\mathbb{R}^m)$, finally, $\check{\mathfrak{R}} = \mathfrak{R}/\mathfrak{J}$ denotes the factor ring of $\mathfrak{R}$ by the ideal $\mathfrak{J}$. The ring $\check{\mathfrak{R}}$ is commutative, it is, moreover, a normed ring if

$$\|\check{A}\|_p = \inf_{T \in \mathfrak{J}} \|A + T\|_p \tag{11}$$

is taken as a norm where $\check{A}$ denotes the equivalence class containing the operator A. Then, inequality (6) means

$$\|\check{A}\|_p \geqq \max_{(x,\theta)\in\bar{\mathrm{R}}^m\times S} |\Phi_A(x,\theta)|. \tag{12}$$

In case $p = 2$ the inequality (12) becomes under certain assumptions on the symbol an equality

$$\|\check{A}\|_2 = \max_{(x,\theta)\in\bar{\mathrm{R}}^m\times S} |\Phi_A(x,\theta)|. \tag{13}$$

A proof of (13) will be given in § 4 of the next chapter.

§ 11. Higher derivatives of the volume potential

11.1. Before we come to the subject of the present section, we shall prove a theorem which turns out to be a stronger variant of the fundamental result in § 7, Chap. IX.

Theorem 11.1. *Let Ω be a finite domain in the space R^m and*

$$v(x) = \int_\Omega \frac{\varphi(x,\theta)}{r^{m-1}} u(y)\,\mathrm{d}y, \tag{1}$$

$u \in L_p(\Omega)$ for some $p \in (1, \infty)$. If the function $\varphi(x,\theta)$ has continuous first derivatives with respect to the Cartesian coordinates of the points $x \in \Omega$ and $\theta \in S$ then the integral (1) *has generalized* (*distributional*) *derivatives $\partial v/\partial x_k \in \mathbf{L}_p(\Omega)$ that can be computed by* (7.2), *Chap. IX,*

$$\frac{\partial v}{\partial x_k} = \int_\Omega u(y) \frac{\partial}{\partial x_k}\left[\frac{\varphi(x,\theta)}{r^{m-1}}\right] \mathrm{d}y - u(x) \int_S \varphi(x,\theta) \cos(r, x_k)\,\mathrm{d}S. \tag{2}$$

Proof. The characteristic of the singular integral (2) is continuous and, due to the results of § 3, the operator on the right-hand side in (2) is bounded in $\mathbf{L}_p(\Omega)$.

The function u can be represented as an $\mathbf{L}_p(\Omega)$ limit of a sequence of functions u_n satisfying a Lipschitz condition. Then put

$$v_n(x) = \int_\Omega \frac{\varphi(x,\theta)}{r^{m-1}} u_n(y)\,\mathrm{d}y.$$

According to the results of § 7, Chap. IX, the function $v_n(x)$ has first derivatives which satisfy a Lipschitz condition on every compactum with respect to Ω,

$$\frac{\partial v_n}{\partial x_k} = \int_\Omega u_n(y) \frac{\partial}{\partial x_k}\left[\frac{\varphi(x,\theta)}{r^{m-1}}\right] \mathrm{d}y - u_n(x) \int_S \varphi(x,\theta) \cos(r, x_k)\,\mathrm{d}S. \tag{3}$$

The functions v_n as well as the derivatives $\partial v_n/\partial x_k$ belong, obviously, to the space $\mathbf{L}_p(\Omega)$. Since the operator (2) is bounded these derivatives converge in the $\mathbf{L}_p(\Omega)$ metric to a limit as $n \to \infty$, and that limit is the right-hand side of (2). By the well-known theorem on the closedness of the generalized differentiation operator, the first derivatives $\partial v/\partial x_k$, do exist and are determined by (2). ∎

Theorem 11.1 can be extended to unbounded domains. Let Ω be such a domain and assume that

1. for $u \in \mathbf{L}_p(\Omega)$ given also $v \in \mathbf{L}_p(\Omega)$;
2. if the functions u_n are sufficiently fast decreasing at infinity then the corresponding function v_n belongs to $\mathbf{L}_p(\Omega)$.

Under these assumptions the reasoning can be repeated provided that finite and continuously differentiable functions $u_n(x)$ are taken as approximating ones.

11.2. Now, let us consider a differential operator of second order,

$$(Lu)(x) = -\sum_{i,j=1}^{m} A_{ij}(x)\frac{\partial^2 u(x)}{\partial x_i\,\partial x_j} + \sum_{j=1}^{m} B_j(x)\frac{\partial u(x)}{\partial x_j} + C(x)\,u(x), \tag{4}$$

where $A_{ij} = A_{ji}$, B_j, and C are functions defined on a domain D. We shall assume the operator L to be elliptic in D such that for every $x \in D$ and any reals $t_1, t_2, \ldots, t_m$

$$\sum_{i,j=1}^{m} A_{ij}(x)\,t_i t_j \geqq \mu \sum_{j=1}^{m} t_j^2 \tag{5}$$

with a certain positive constant μ. Furthermore, the coefficients of the operator (4) are supposed to satisfy a Lipschitz condition in the domain D with the exponent $\lambda > 0$. As it is well-known (see e.g. MIRANDA [1]), then under rather general assumptions on the domain there exists a singular solution (the so-called fundamental solution) to the equation $Lu = 0$ and it can be represented as

$$H(x, y) = \psi(x, y) + \int_D \psi(x, z)\,f(z, y)\,\mathrm{d}y =: \psi(x, y) + \psi_1(x, y). \tag{6}$$

For $m > 2$, the function $\psi(x, y)$ is defined by the formula

$$\psi(x, y) = \frac{1}{(m-2)\,\omega_m\sqrt{A(x)}}\left\{\sum_{i,j=1}^{m} C_{ij}(x)\,(x_i - y_i)\,(x_j - y_j)\right\}^{-(m-2)/2} \tag{7}$$

where $\omega_m = 2\pi^{m/2}\,\Gamma(m/2)$ denotes the measure of the surface of the unit ball in $\mathbb{R}^m$, $A = \det(A_{ij})$, and C_{ij} denote the elements of the inverse to the matrix (A_{ij}). Finally, $f(z, y)$ is a certain function which is continuous if $y \neq z$ and has a pole if $z \to y$ of the order $\leqq m - \lambda$. By a well-known theorem on the composition of weakly singular integrals, the function $\psi_1(x, y)$ has a pole as $x \to y$ of order $\leqq m - 2 - \lambda$ such that the main part of the singular solution is given by (7). It is worth noting that in case $m = 2$ equation (7) is to be replaced by the following one,

$$\psi(x,y) = -\frac{1}{2\pi}\ln\left\{\sum_{i,j=1}^{2} C_{ij}(x_i - y_i)\,(x_j - y_j)\right\}^{1/2}. \tag{7a}$$

Let $\Omega \subset D$ be an arbitrary domain, and let $f \in \mathbf{L}_p(\Omega)$ for some $p \in (1, \infty)$. We shall call the integral

$$g(x) = \int_\Omega H(x, y)\,f(y)\,\mathrm{d}y \tag{8}$$

volume potential.

11.3. Theorem 11.2. *The volume potential* (8), $f \in \mathbf{L}_p(\Omega)$, *has generalized derivatives of second order which are summable in* Ω *to the p-th power, and*

$$Lg = f \tag{9}$$

a.e. in Ω.

Proof. We have

$$g(x) = \int_\Omega \psi(x, y)\,f(y)\,\mathrm{d}y + \int_\Omega \psi_1(x, y)\,f(y)\,\mathrm{d}y. \tag{10}$$

The second summand in (10) has, obviously, generalized derivatives of first and second order. The first summand in (10) can be differentiated once under the integral and another time according to (2). Hence, the second derivatives $\partial^2 g/\partial x_i\,\partial x_j \in \mathbf{L}_p(\Omega)$ do exist. Thus, it remains to show that g satisfies (9).

Performing the computations necessary we obtain

$$\frac{\partial^2 g}{\partial x_i\,\partial x_j} = \int_\Omega \frac{\partial^2 H}{\partial x_i\,\partial x_j} f(y)\,\mathrm{d}y - \frac{f(x)}{\omega_m \sqrt{A(x)}} \int_S \left\{ \sum_{k,l=1}^m C_{kl}(x) \cos(\nu, x_k) \cos(\nu, x_l) \right\}^{-m/2}$$

$$\times \sum_{l=1}^m C_{il}(x) \cos(\nu, x_l) \cos(\nu, x_j)\,\mathrm{d}S$$

such that

$$Lg = \int_\Omega f(y)\,LH\,\mathrm{d}y + \frac{f(x)\,E(x)}{\omega_m \sqrt{A(x)}} = \frac{f(x)\,E(x)}{\omega_m \sqrt{A(x)}} \tag{11}$$

where

$$E(x) = \int_S \left\{ \sum_{i,j=1}^m C_{ij}(x) \cos(\nu, x_i) \cos(\nu, x_j) \right\}^{-m/2} \sum_{i,j,k=1}^m A_{ij}(x)\,C_{ik}(x) \cos(\nu, x_j) \cos(\nu, x_k)\,\mathrm{d}S. \tag{12}$$

Equation (12) shows that $E(x)$ is invariant under rotations of the coordinate system. We perform a rotation such that the coordinate axes coincide with the principal axes of the form (5). Then $A_{ij} = C_{ij} = 0$, $i \neq j$, $C_{ii} = A_{ii}^{-1}$ and, therefore,

$$E(x) = \int_S \left\{ \sum_{i=1}^m \frac{\cos^2(\nu, x_i)}{A_{ii}} \right\}^{-m/2} \mathrm{d}S. \tag{13}$$

We want to compute this integral. Denote $A_{ii}^{-1} = a_i^2$. Translating the origin to the point x we may assume $\cos(\nu, x_i) = y_i$, then

$$E(x) = \int_S \left\{ \sum_{i=1}^m a_i^2 y_i^2 \right\}^{-m/2} \mathrm{d}S.$$

Now, we introduce spherical coordinates such that $y_1 = \cos\vartheta_1$ and the other coordinates contain the factor $\sin\vartheta_1$. Then, we may put

$$\sum_{i=2}^m a_i^2 y_i^2 = \alpha^2 \sin^2\vartheta_1$$

where α does not depend on a_1 and ϑ_1. Finally, taking into account $\mathrm{d}S = \prod_{k=1}^{m-1} \sin^{m-k-1}\vartheta_k\,\mathrm{d}\vartheta_k$, we get

$$E(x) = \int_0^\pi \dots \int_0^{2\pi} \sin^{m-3}\vartheta_2 \dots \sin\vartheta_{m-2}\,\mathrm{d}\vartheta_2 \dots \mathrm{d}\vartheta_{m-1}$$

$$\times \int_0^\pi (a_1^2 \cos^2\vartheta_1 + \alpha^2 \sin^2\vartheta_1)^{-m/2} \sin^{m-2}\vartheta_1\,\mathrm{d}\vartheta_1.$$

The inner integral is equal to

$$2 \int_0^{\pi/2} (a_1^2 \cos^2\vartheta_1 + \alpha^2 \sin^2\vartheta_1)^{-m/2} \sin^{m-2}\vartheta_1\,\mathrm{d}\vartheta_1. \tag{14}$$

Now, we make use of the well-known formula

$$\int_0^{\pi/2} \frac{\sin^{2r-1}x \cos^{2s-1}x}{(p^2 \sin^2 x + q^2 \cos^2 x)^{r+s}}\,\mathrm{d}x = \frac{B(r, s)}{2p^{2r}q^{2s}}.$$

Putting here $r = (m - 1)/2$, $s = 1/2$, $p = \alpha$ and $q = a_1$, we find that the integral (14) is indirectly proportional to a_1, and so is $E(x)$. Hence,

$$E(x) = C(a_1 a_2 \dots a_m)^{-1} = C\sqrt{A_{11}A_{22} \dots A_{mm}}\,.$$

In order to determine the constant C put in (13) $A_{ii} = 1$, then $E(x) = C = \omega_m$. Since the coordinate axes coincide with the principal axes we have $A_{11}A_{22} \dots A_{mm} = A(x)$. Thus, finally $E(x) = \omega_m \sqrt{A(x)}$, and (11) coincides with (9). ▮

§ 12. Weighted norm inequalities for maximal functions and singular integral operators

12.1. Muckenhoupt, Hunt, and Wheeden indicated first a characterization of the boundedness of singular integral operators in the space $\mathbf{L}_p(\mathbb{R}^m, \varrho)$ $(1 < p < \infty$, ϱ being an arbitrary non-negative measurable and locally integrable function) by means of the concept of *maximal functions* introduced by Hardy and Littlewood. A maximal function is defined as

$$f^*(x) := \sup_Q m_Q |f|, \qquad m_Q f := \frac{1}{|Q|} \int_Q f(y)\, \mathrm{d}y, \tag{1}$$

for an arbitrary complex-valued and locally integrable function f (i.e. $f \in \mathbf{L}^1_{\mathrm{loc}}(\mathbb{R}^m)$) where the supremum in (1) is taken over all cubes Q centered at the point x the edges of which are parallel to the coordinate axes (see e.g. Stein [1], Chap. I, § 1).

Theorem 12.1 (Muckenhoupt [1]). *The inequality*

$$\|f^*\|_{\mathbf{L}_p(\mathbb{R}^m,\varrho)} \leqq C\, \|f\|_{\mathbf{L}_p(\mathbb{R}^m,\varrho)}$$

holds true for any $f \in \mathbf{L}_p(\mathbb{R}^m, \varrho)$, $1 < p < \infty$, if and only if the weight function ϱ satisfies the condition

$$\|\varrho\|_{\mathbf{L}_p(Q)}\, \|\varrho^{-1}\|_{\mathbf{L}_q(Q)} \leqq C'\, |Q|; \tag{A_p}$$

here $q = p/(p - 1)$, and C, C' are positive constants that do not depend on f and Q, resp.

Theorem 12.2 (Hunt, Muckenhoupt, and Wheeden [1]). *Let $m = 1$ and T denote the Hilbert transform,*

$$(Tf)(x) := \frac{1}{\pi \mathrm{i}} \int_{-\infty}^{\infty} \frac{f(y)\, \mathrm{d}y}{y - x}\,.$$

Then the operator T is bounded in the space $\mathbf{L}_p(\mathbb{R}, \varrho)$, $1 < p < \infty$, if and only if the weight function ϱ satisfies the condition (A_p).

A simplified proof of the Theorems 12.1 and 12.2 is given by Coifman and Fefferman [1]. We reproduce here the particularly simple necessity part of the proof of Theorem 12.2 given in the paper cited.

To that aim, we denote by Q_1 and Q_2 the two halfs of an arbitrary fixed interval Q_0. Let f be a non-negative function such that $\operatorname{supp} f \subseteq Q_1$, then

$$|(Tf)(x)| \geqq C\, |Q_1|^{-1} \int_{Q_1} f(y)\, \mathrm{d}y, \qquad \forall\, x \in Q_2,$$

hence

$$C\, |Q_1|^{-1}\, \|f\|_{\mathbf{L}_1(Q_1)}\, \|\varrho\|_{\mathbf{L}_p(Q_2)} \leqq \|Tf\|_{\mathbf{L}_p(\mathbb{R},\varrho)} \leqq C_1\, \|f\|_{\mathbf{L}_p(Q_1,\varrho)}\,. \tag{2}$$

For $f \equiv 1$ relation (2) yields

$$C \|\varrho\|_{L_p(Q_2)} \leqq C_1 \|\varrho\|_{L_p(Q_1)}. \tag{3}$$

Putting $f = \varrho^{-q}$ we obtain from (2)

$$C \|\varrho^{-1}\|_{L_q(Q_1)} \|\varrho\|_{L_p(Q_2)} \leqq C_1 |Q_1|,$$

and this gives, together with (3), the condition (A_p).

Note that the proof given here for the condition (A_p) to be necessary for the boundedness of the operator T in the space $L_p(\mathbb{R}^m, \varrho)$ applies also if $m > 1$ and T is the singular integral operator with the Riesz kernel. Similarly, the necessity of condition (A_p) can be proved for Theorem 12.1.

12.2. We indicate some properties of the Muckenhoupt weights A_p.

1. For $\omega := \varrho^p$ condition (A_p) assumes the equivalent form

$$\left(\int_Q \omega(x)\, dx\right) \left(\int_Q [\omega(x)]^{-1/(p-1)}\, dx\right)^{p-1} \leqq C\,|Q|^p$$

(cf. Muckenhoupt [1, 2]).

2. If $1 < p_1 < p_2 < \infty$ then $A_{p_1} \subseteq A_{p_2}$.
3. $|x|^\alpha \in A_p$ if and only if condition (5.3) is satisfied.

The properties 1.–3. are immediate consequences of the definition of (A_p). Much deeper are the following properties (see Muckenhoupt [1, 2] and Coifman and Fefferman [1]).

4. If $\varrho \in A_p$, $1 < p < \infty$, then the reverse Hölder inequality

$$\int_Q [\varrho(x)]^{p\nu}\, dx \leqq C\,|Q|^{1-\nu} \left[\int_Q \varrho^p(x)\, dx\right]^\nu$$

holds where C and ν are constants not depending on Q, and $\nu > 1$.

5. If $\varrho \in A_p$, $1 < p < \infty$, then $\varrho \in A_{p-\varepsilon}$ for some $\varepsilon = \varepsilon(\varrho) > 0$.
6. For some p, $1 < p < \infty$, $\varrho \in A_p$ if and only if ϱ satisfies the following condition (A_∞): To any $\varepsilon > 0$ there is a $\delta > 0$ such that for every cube Q and every measurable set $E \subseteq Q$, $|E| < \delta\,|Q|$, the inequality

$$\int_E \varrho^p(x)\, dx \leqq \varepsilon \int_Q \varrho^p(x)\, dx$$

holds.

Property 5. is a simple consequence of (A_p) and 4. with $\varepsilon = (p-1)(\nu-1)/\nu$. The implication ($A_p$) $\Rightarrow$ (A_∞) follows from 4. using Hölder inequality. It was the reverse Hölder inequality which was used by Coifman and Fefferman [1] in order to prove sufficiency of the condition (A_p) in the Theorems 12.1 and 12.2.

12.3. Coifman and Meyer [1] introduced a class of operators which they call Calderón-Zygmund operators. This class contains the singular operators considered above as well as pseudo-differential operators (cf. Chap. XII, § 8) and operators with Cauchy kernel on Lipschitz curves (cf. Chap. II, § 2). The Calderón-Zygmund operators are, similarly to the Hilbert transform, bounded in the space $L_p(\mathbb{R}^m, \varrho)$ provided that $\varrho \in A_p$ and $1 < p < \infty$. The Lecture Notes by Journé [1] are devoted to those operators. The following reasoning is taken from Chap. 4 of Journé [1].

In what follows we denote by Δ^c the complement of the diagonal $\Delta = \{(x, y) \in \mathbb{R}^m \times \mathbb{R}^m : x = y\}$, and by $C_c^\infty(\mathbb{R}^m)$ ($L_c^\infty(\mathbb{R}^m)$, resp.) the subset of all functions $f \in C^\infty(\mathbb{R}^m)$ ($f \in L_\infty(\mathbb{R}^m)$, resp.) having compact support.

Definition 12.1. A *standard kernel* is a continuous function $K: \Delta^c \mapsto \mathbb{C}$ for which there exists a positive constant C such that for all $(x, y) \in \Delta^c$,

$$|K(x, y)| \leqq C\, |x - y|^{-m} \tag{4}$$

and

$$|\text{grad}_x K(x, y)| + |\text{grad}_y K(x, y)| \leqq C\, |x - y|^{-(m+1)}. \tag{5}$$

The gradients are taken in the distributional sense, and they are assumed to be functions.

Note that Definition 12.1 is a straight-forward generalization of the conditions to the kernel K formulated in Chap. IX, Section 3.2.

Definition 12.2. An operator T taking $\mathbf{C}_c^\infty(\mathbb{R}^m)$ into $\mathbf{L}_{\text{loc}}^1(\mathbb{R}^m)$ is called a *Calderón-Zygmund operator* (CZO) iff

(i) T extends to an operator $T \in \mathscr{L}(\mathbf{L}_2(\mathbb{R}^m))$;

(ii) there exists a standard kernel K such that for every $f \in \mathbf{L}_c^\infty(\mathbb{R}^m)$,

$$(Tf)(x) = \int_{\mathbb{R}^m} K(x, y) f(y)\, dy \quad \text{a.e. on } (\operatorname{supp} f)^c.$$

Now, the following important generalization of Theorem 12.2 is true.

Theorem 12.3. *Let $\varrho \in A_p$, $1 < p < \infty$. If T is a CZO then $T \in \mathscr{L}(\mathbf{L}_p(\mathbb{R}^m, \varrho))$.*

As noted above, for the Riesz transform the reverse of Theorem 12.3 is also true.

For the case of a convolution kernel K Theorem 12.3 is proved by Coifman and Fefferman [1]. When proving Theorem 12.3, Journé [1] makes use of the *Fefferman-Stein inequality:* Let T be a CZO, and $p > 1$. Then

$$(Tf)^\# \leqq C(p, T)\, ((|f|^p)^*)^{1/p}, \quad \forall f \in \mathbf{L}_c^\infty(\mathbb{R}^m). \tag{6}$$

The $\#$-function introduced by Fefferman and Stein [1] which is called *sharp function*, is defined by

$$g^\#(x) := \sup_Q m_Q\, |g - m_Q g|$$

for any $g \in \mathbf{L}_{\text{loc}}^1(\mathbb{R}^m)$. Obviously, $g^\#$ takes $\mathbb{R}^m$ into the segment $[0, \infty]$, and $g^\# \leqq 2g^*$.

The inequality (6) can be shown as follows. Let Q be a cube centered at the point x_0, and $f \in \mathbf{L}_c^\infty(\mathbb{R}^m)$. Denoting, as usually, by $\chi_{\bar{Q}}$ the characteristic function of $\bar{Q}$ we put $f_1 = f\chi_{\bar{Q}}$ and $f_2 = f - f_1$. Estimating similarly as in § 3, Chap. IX, and using (4) and (5) we obtain

$$\frac{1}{|Q|} \int_Q |(Tf_2)(x) - (Tf_2)(x_0)|\, dx \leqq C_1 f^*(x_0). \tag{7}$$

Since the operator T is bounded in $\mathbf{L}_p(\mathbb{R}^m)$ (see the Calderón-Zygmund Theorem),

$$\|Tf_1\|_{\mathbf{L}_p(Q)} \leqq C_2\, \|f_1\|_{\mathbf{L}_p(Q)}. \tag{8}$$

The estimates (7) and (8) together with the Hölder inequality immediately yield

$$\begin{aligned}\frac{1}{|Q|} \int_Q |(Tf)(x) - (Tf_2)(x_0)|\, dx &\leqq C_1 f^*(x_0) + \frac{1}{|Q|} \int_Q |(Tf_1)(x)|\, dx \\ &\leqq C_1 f^*(x_0) + \frac{C_2}{|Q|^{1/p}} \|f_1\|_{\mathbf{L}_p(Q)} \\ &\leqq C((|f|^p)^*)^{1/p}(x_0),\end{aligned}$$

and thence (6).

In order to apply inequality (6) we need, moreover, the estimate

$$\|Tf\|_{L_p(\mathbb{R}^m,\varrho)} \leqq C \,\|(Tf)^{\#}\|_{L_p(\mathbb{R}^m,\varrho)} \tag{9}$$

(for the proof, see JOURNE [1], pp. 41 and 53).

Now, let $\varrho \in A_p$, and $f \in L_c^\infty(\mathbb{R}^m)$. Then $\varrho \in A_{p/r}$ where $r = r(\varrho) > 1$. (cf. property 5.). From (6) and (9) using Theorem 12.1 we get

$$\begin{aligned}\|Tf\|_{L_p(\mathbb{R}^m,\varrho)} &\leqq \|(Tf)^{\#}\|_{L_p(\mathbb{R}^m,\varrho)}\\ &\leqq C\,\|((|f|^r)^*)^{1/r}\|_{L_p(\mathbb{R}^m,\varrho)}\\ &= C\,\|(|f|^r)^*\|^{1/r}_{L_{p/r}(\mathbb{R}^m,\varrho)}\\ &\leqq C'\,\||f|^r\|^{1/r}_{L_{p/r}(\mathbb{R}^m,\varrho)} = C'\,\|f\|_{L_p(\mathbb{R}^m,\varrho)}\end{aligned}$$

which was to be demonstrated. ▮

If ϱ satisfies the condition (A_1): There exists a positive constant C such that $\varrho^*(x) \leqq C\varrho(x)$ a.e., then every CZO in $L_1(\mathbb{R}^m, \varrho)$ is of weak type (1, 1) (see COIFMAN and MEYER [1]).

Furthermore, the following result can be proved (see JOURNE [1]). Let K be the standard kernel of a CZO such that the limit

$$\lim_{\varepsilon\to 0} \int_{\varepsilon<|x-y|<1} K(x, y)\, dy \quad \text{a.e.} \tag{10}$$

exists. If $\varrho \in A_p$, $1 \leqq p < \infty$, and $f \in L_p(\mathbb{R}^m, \varrho)$, then the limit

$$\lim_{\varepsilon\to 0} \int_{\varepsilon<|x-y|} K(x, y)\, f(y)\, dy \quad \text{a.e.} \tag{11}$$

exists.

Obviously, condition (10) is also necessary for the limit (11) to exist.

Definition 12.3. Define the space $\mathbf{BMO}(\mathbb{R}^m)$ (of *functions of bounded mean oscillation*) to consist of those functions $f \in L^1_{\text{loc}}(\mathbb{R}^m)$ such that $f^{\#} \in L_\infty(\mathbb{R}^m)$.

BMO spaces were introduced originally by JOHN and NIRENBERG [1]. The space $\mathbf{BMO}(\mathbb{R}^m)$ is a semi-normed vector space, with the semi-norm $\|f\|^* := \|f^{\#}\|_\infty$ vanishing on the constant functions. FEFFERMAN and STEIN [1] characterized $\mathbf{BMO}(\mathbb{R}^m)$ (more precisely, the quotient space of $\mathbf{BMO}(\mathbb{R}^m)$ by the subspace of constant functions) as the dual of the Hardy space $\mathbf{H}^1(\mathbb{R}^m)$.

An example of an unbounded function belonging to $\mathbf{BMO}(\mathbb{R}^m)$ is provided by $\log |x|$.

The following theorem is proved by COFMAN, ROCHBERG, and WEISS [1].

Theorem 12.4. *Let T be a CZO, and $b \in \mathbf{BMO}(\mathbb{R}^m)$. Then the commutator of T and the operator of multiplication by b is bounded on all $L_p(\mathbb{R}^m, \varrho)$, $1 < p < \infty$, if $\varrho \in A_p$.*

Remark. The reader interested in a more detailed survey on the subject of the present section is referred to the recent work by DYNKIN and OSILENKER [1].

Chapter XII
Multidimensional singular integral equations

In the present chapter we shall consider scalar singular integral equations in certain important function spaces. It is worth noting that the results of §§ 1, 2 carry over to systems of such equations which will be studied in detail in Chap. XIII.

§ 1. The constant symbol case

Let us consider the simple singular equation

$$A_0 u := F^{-1}\,\Phi(\theta)\,Fu = g \tag{1}$$

with a given function $g \in \mathbf{L}_2(\mathbb{R}^m)$. Assume that the symbol $\Phi(\theta)$ is finite on the sphere S a.e., and $\inf |\Phi(\theta)| > 0$. Then the function $[\Phi(\theta)]^{-1}$ is bounded such that, by Theorem 5.1, Chap. X, the operator $B_0 = F^{-1}[\Phi(\theta)]^{-1}\,F$ is bounded in $\mathbf{L}_2(\mathbb{R}^m)$. By the symbol multiplication rule, $B_0 = A_0^{-1}$. Thus, equation (1) has one and only one solution in $\mathbf{L}_2(\mathbb{R}^m)$ which is given by the formula

$$u = F^{-1}[\Phi(\theta)]^{-1}\,Fg. \tag{2}$$

The more general equation,

$$F^{-1}\,\Phi(\theta)\,Fu + Tu = g \tag{3}$$

with a completely continuous operator T in $\mathbf{L}_2(\mathbb{R}^m)$ is, obviously, equivalent to the Riesz-Schauder equation

$$u + B_0 T u = B_0 g. \tag{4}$$

Equation (1) can be considered in the space $\mathbf{L}_p(\mathbb{R}^m)$, $1 < p < \infty$, too. In that case we assume that the symbol $[\Phi(\theta)]^{-1}$ satisfies the conditions of Theorem 3.2, Chap. XI. Then, the operator B_0 is bounded in $\mathbf{L}_p(\mathbb{R}^m)$, and for a given $g \in \mathbf{L}_p(\mathbb{R}^m)$ equation (1) has one and only one solution in that space which is given, again, by equation (2). If the operator T is completely continuous in $\mathbf{L}_p(\mathbb{R}^m)$ then (3) is equivalent to the Riesz-Schauder equation (4).

§ 2. The general case and Noether theorems

Let us consider the general singular equation

$$(Au)(x) := a(x)\,u(x) + \int\limits_{\mathbb{R}^m} r^{-m} f(x, \theta)\,u(y)\,\mathrm{d}y = g(x), \qquad g \in \mathbf{L}_p(\mathbb{R}^m). \tag{1}$$

Assume that the symbol $\Phi_A(x, \theta)$ of the operator A satisfies the conditions of the Theorems 3.2 and 7.1, Chap. XI, and

$$\inf |\Phi_A(x, \theta)| > 0, \tag{2}$$

In that case the operator A admits of a two-sided regularization. Indeed, let B be some singular operator with the symbol $[\Phi_A(x, \theta)]^{-1}$ which also satisfies the conditions of the theorems mentioned. By Theorem 7.1, Chap. XI, the symbols of the operators AB and BA are equal to 1 such that $BA = I + T$, $AB = I + T_1$ where I stands for the identity, and T, T_1 are completely continuous in $L_p(\mathbb{R}_m)$. Hence, the operator B provides a two-sided regularizer to the operator A.

Additionally, we assume that the symbol $\Phi_A(x, \theta)$ satisfies the conditions of Theorem 8.2, Chap. XI. Then the adjoint A^* to the operator A admits also of a two-sided regularization, and a regularizer to A^* is provided by the singular operator with the symbol $[\overline{\Phi_A(x, \theta)}]^{-1}$. Due to the results of § 2, Chap. I, we obtain now the following theorems concerning (1).

Theorem 2.1. *The null spaces of the operators A and A^* are finite-dimensional.*

Theorem 2.2. *The operator A is normally solvable.*

Theorem 2.3. *The index of the operator $A + T$ does not depend on the completely continuous summand T.*

Now, let us consider an operator that depends on a complex parameter λ,

$$(A_\lambda u)(x) = u(x) - \lambda \int_{\mathbb{R}^m} r^{-m} f(x, \theta)\, u(y)\, dy + (Tu)(x). \tag{3}$$

Assume that the symbol $\Phi(x, \theta)$ of the integral in (3) satisfies the conditions as mentioned above for the symbol $\Phi_A(x, \theta)$ except, possibly, condition (2). By σ we denote the set in the complex λ-plane on which the symbol $1 - \lambda\Phi(x, \theta)$ of the operator (3) becomes $\inf |1 - \lambda\Phi(x, \theta)| = 0$. The set σ is a closed one, and its complement Δ is as an open set the finite or countable union of domains, $\Delta = \bigcup_j \Delta_j$.

Theorem 2.4. *In each of the domains Δ_j the index of the operator* (3) *is constant.*[1])

Proof. Let δ_j be a bounded and closed subset of the domain Δ_j. Then, on δ_j, $\inf |1 - \lambda\Phi(x, \theta)| > 0$, and the operator (3) has a two-sided regularizer B_λ the symbol of which is $[1 - \lambda\Phi(x, \theta)]^{-1}$.

Replace λ by $\lambda + \lambda'$ where λ' is chosen such that $\lambda + \lambda' \in \delta_j$ and $|\lambda'|\, \|B_\lambda\|\, \|C\| < 1$ with C being

$$(Cu)(x) = \int_{\mathbb{R}^m} r^{-m} f(x, \theta)\, u(y)\, dy.$$

By Theorem 3.5, Chap. I, the indices of the operators A_λ and $A_{\lambda+\lambda'}$ coincide. Hence, if λ belongs together with some sufficiently small vicinity of it to the domain Δ_j then the index of the operator (3) is constant in that vicinity. But then it is constant in the entire domain Δ_j. ∎

[1]) See Mikhlin [9]; for a one-dimensional singular operator analogous domains were introduced also by Mikhlin [15, 3].

§ 3. Equivalent regularization. The index theorem

3.1. Lemma 3.1. *Let $\Phi(x, \theta)$ satisfy the conditions of Theorem 8.2, Chap. XI, and* $\sup |\Phi(x, \theta)| < 1$. *Then the index of the singular operator*

$$u(x) - \int_{\mathbb{R}^m} r^{-m} f(x, \theta)\, u(y)\, dy + (Tu)(x) \tag{1}$$

is equal to zero.

Proof. Consider the operator

$$u(x) - \lambda \int_{\mathbb{R}^m} r^{-m} f(x, \theta)\, u(y)\, dy + (Tu)(x) \tag{2}$$

with a complex parameter λ, put $\sup |\Phi(x, \theta)| = q$, $q < 1$. In the disk $|\lambda| < q^{-1}$ the symbol of the operator (2) is such that

$$\inf |1 - \lambda\Phi(x, \theta)| > 1 - |\lambda|\, q > 0.$$

By Theorem 2.4, the index of the operator (2) is constant in that disk. But for $\lambda = 0$ the index is equal to zero. ∎

Lemma 3.2. *Let the singular operator A_t with the symbol $\Phi(x, \theta, t)$ depend on a parameter $t \in [0, 1]$. Assume that*

1. $\Phi(x, \theta, t) \,\hat{\in}\, \mathrm{W}_2^{(l)}(S)$, *$l$ as in Theorem 8.2, Chap. XI*;
2. $\|\Phi(x, \theta, t + t') - \Phi(x, \theta, t)\|_{\mathrm{W}_2^{(l)}(S)} \to 0$, $t' \to 0$,

uniformly with respect to x and t;

3. *the symbol $\Phi(x, \theta, t)$ is continuous on $\overline{\mathbb{R}}^m \times [0, 1]$ uniformly with respect to θ;*
4. $\inf |\Phi(x, \theta, t)| > 0$.

Then Ind A_t *does not depend on t.*

Proof. Due to Theorem 3.4, Chap. I, it is sufficient to consider the case of a simple singular operator A_t. Then A_t is bounded independently of t, and $\|A_{t+t'} - A_t\| \to 0$ as $t' \to 0$ uniformly with respect to t. Furthermore, for the simple singular operator B_t with the symbol $[\Phi(x, \theta, t)]^{-1}$, obviously, $\|B_t\| \leqq M$ with some constant M. By Theorem 3.5, Chap. I, Ind $A_{t+t'} =$ Ind A_t for sufficiently small t'. Now, the lemma follows by means of the Heine-Borel Lemma. ∎

3.2. Theorem 3.1. *Let the symbol $\Phi_A(x, \theta)$ of the singular operator A satisfy the conditions of Theorem 8.2, Chap. XI, and* $\inf |\Phi_A(x, \theta)| > 0$. *Then the operator A admits of an equivalent regularization in the corresponding space $\mathrm{L}_p(\mathbb{R}^m)$ and its index is equal to zero.*

Theorem 3.1 was proved 1956 by Mikhlin [9, 11] and 1961 by Seeley [2] in a different way. In the proof it turns out to be necessary to separate the cases of dimension $m > 2$ and $m = 2$. In what follows we shall give Seeley's arguments for $m > 2$ and Mikhlin's demonstration for $m = 2$.

Let $m > 2$ and the symbol satisfy the conditions of the theorem to be proved. We shall show that the operator A is homotopic to the operator of multiplication by a certain function $\psi(x)$ where this function is continuous on $\overline{\mathbb{R}}^m$ and $\inf |\psi(x)| \geqq \inf |\Phi_A(x, \theta)|$. We construct a kernel $L_t(\theta, \theta')$, $t \in (0, 1]$, $\theta, \theta' \in S$, that has the following properties:

a) $L_t(\theta, \theta') \geqq 0$;

b) $L_1(\theta, \theta') = \mathrm{const}$;

c) $L_t(\theta, \theta')$ is infinitely differentiable with respect to all three variables;

d) for any fixed $\varepsilon > 0$ and $|\theta' - \theta| > \varepsilon$, $\sup L_t(\theta, \theta') \to 0$ as $t \to 0$;
e) $\int\limits_S L_t(\theta, \theta')\, \mathrm{d}S = 1$

Such kernels do exist, e.g.

$$L_t(\theta, \theta') = c(t) \exp\left[-\frac{|\theta' - \theta|^2}{t}(1-t)\right],$$

$$c(t) = \int\limits_S \exp\left[-\frac{|\theta' - e_1|^2}{t}(1-t)\right] \mathrm{d}S, \qquad e_1 = (1, 0, \ldots, 0).$$

By the formulae

$$|\Phi_t(x, \theta)| = \int\limits_S |\Phi_A(x, \theta')|\, L_t(\theta, \theta')\, \mathrm{d}_{\theta'}S,$$

$$\arg \Phi_t(x, \theta) = \int\limits_S \arg \Phi_A(x, \theta')\, L_t(\theta, \theta')\, \mathrm{d}_{\theta'}S \tag{3}$$

we define a symbol $\Phi_t(x, \theta)$. Because of $m > 2$ the sphere S is simply connected, and that means that every closed path on S can be contracted on S to a point. Thence, $\arg \Phi_A(x, \theta)$ and $\arg \Phi_t(x, \theta)$ are unique and sufficiently smooth functions in their arguments. Moreover, $\Phi_t(x, \theta) \to \Phi_A(x, \theta)$ as $t \to 0$ uniformly with respect to (x, θ). Let A_1 be the operator with the symbol $\Phi_1(x, \theta)$. By Lemma 3.2, $\operatorname{Ind} A = \operatorname{Ind} A_1$. Now, (3) and property b) of the kernel $L_t(\theta, \theta')$ show that $\Phi_1(x, \theta)$ does not depend on θ. Hence, A_1 is the operator of multiplication by the function $\psi(x) = \Phi_1(x)$ with a strictly positive modulus. Obviously, $\operatorname{Ind} A_1 = 0$ such that also $\operatorname{Ind} A = 0$. By Theorems 2.3 and 2.4, Chap. I, the operator A admits of an equivalent regularization.

The arguments given here do not apply in case if $m = 2$ since the circle is no longer simply connected. This case will be considered separately.

Let $m = 2$. Besides on x, the symbol depends only on one angle ϑ varying in the segment $0 \leqq \vartheta \leqq 2\pi$, and we shall write $\Phi(x, \vartheta)$ instead of $\Phi(x, \theta)$. Continuity with respect to the point θ in that case means continuity with respect to the angular coordinate ϑ, in particular, the function $\exp(\mathrm{i}\vartheta)$ as a function of a variable point on the circle is infinitely differentiable. Remember that the simple singular operator h with the symbol $\exp(\mathrm{i}\vartheta)$ has due to the results of § 1 a bounded inverse with the symbol $\exp(-\mathrm{i}\vartheta)$.

The assumptions of the theorem to be proved are $\Phi(x, \theta) \hat{\in} \mathrm{W}_2^{(\lambda)}(S)$ and, as one can easily see, $\lambda > 1$ if $p \neq 2$. For $p = 2$, $\lambda > (m-1)/2 = 1/2$. Thus, $\lambda > 1/2$ for any $p \in (1, \infty)$.

We expand the symbol into a Fourier series,

$$\Phi(x, \vartheta) = \sum_{n=-\infty}^{\infty} a_n(x)\, \mathrm{e}^{\mathrm{i}n\vartheta}. \tag{4}$$

Next, we want to show that this series is absolutely and uniformly convergent with respect to (x, ϑ). First we show that Theorem 6.1, Chap. X, remains to hold true in the case considered for arbitrary positive λ so that

$$\sum_{n=-\infty}^{\infty} n^{2\lambda}\, |a_n(x)|^2 \leqq C = \text{const}. \tag{5}$$

Let $[\lambda] = l$, $\lambda - l = \mu$. We have for the Fourier coefficients $b_n(x)$ of the derivative $\Phi_\vartheta^{(l)}(x, \vartheta) = \partial^l \Phi(x, \vartheta)/\partial\vartheta^l$ the relation $a_n(x) = (in)^{-l}\, b_n(x)$. Furthermore,

$$C \geqq \|\Phi(x, \cdot)\|^2_{\mathrm{W}_2^{(\lambda)}(S)} = \|\Phi(x, \cdot)\|^2_{\mathrm{W}_2^{(l)}(S)} + \int_{-\pi}^{\pi} \frac{\mathrm{d}h}{|h|^{2\mu+1}} \int_{-\pi}^{\pi} |\Phi_\vartheta^{(l)}(x, \vartheta + h) - \Phi_\vartheta^{(l)}(x, \vartheta)|^2 \mathrm{d}\vartheta. \tag{6}$$

Let us estimate the last integral. It is easy to see that

$$\int_{-\pi}^{\pi} |\Phi_\vartheta^{(l)}(x, \vartheta + h) - \Phi_\vartheta^{(l)}(x, \vartheta)|^2 \, d\vartheta = 8\pi \sum_{n=-\infty}^{\infty} |b_n(x)|^2 \sin^2 \frac{nh}{2}.$$

Hence,

$$\int_{-\pi}^{\pi} \frac{dh}{|h|^{2\mu+1}} \int_{-\pi}^{\pi} |\Phi_\vartheta^{(l)}(x, \vartheta + h) - \Phi_\vartheta^{(l)}(x, \vartheta)|^2 \, d\vartheta = 8\pi \sum_{n=-\infty}^{\infty} |b_n(x)|^2 \int_{-\pi}^{\pi} \frac{\sin^2 \frac{nh}{2}}{|h|^{2\mu+1}} \, dh$$

$$\geqq \frac{\pi}{2^{2\mu-2}} \int_0^{\pi/2} \frac{\sin^2 t}{t^{2\mu+1}} \, dt \sum_{n=-\infty}^{\infty} n^{2\mu} |b_n(x)|^2.$$

Now, relation (6) implies the inequality (5) to hold true.

We form the modulus series to (4) and estimate the remainder

$$R_N(x) = \sum_{|n|>N} |a_n(x)|.$$

We get

$$R_N^2(x) \leqq \sum_{n=-\infty}^{\infty} n^{2\lambda} |a_n(x)|^2 \sum_{|n|>N} n^{-2\lambda} \leqq C \sum_{|n|>N} n^{-2\lambda} \to 0, \quad N \to \infty,$$

such that the series (4) converges absolutely and uniformly with respect to (x, ϑ). Put

$$\tilde{\Phi}(x, \vartheta) = \sum_{k=-n}^{n} a_k(x) \, e^{ik\vartheta}$$

and choose n sufficiently large such that for a given positive ε

$$|a_n(x)| < \frac{\varepsilon}{3}, \qquad |\Phi(x, \vartheta) - \tilde{\Phi}(x, \vartheta)| < \frac{\varepsilon}{3}.$$

Then we introduce the function

$$\Phi_0(x, \vartheta) = \frac{2}{3} \varepsilon \, e^{in\vartheta} + \tilde{\Phi}(x, \vartheta) \tag{7}$$

which turns out to be a trigonometric polynomial with the following properties: The coefficient $\frac{2}{3}\varepsilon + a_n(x)$ at $\exp(in\vartheta)$ is in modulus greater than $\varepsilon/3$, and $|\Phi(x, \vartheta) - \Phi_0(x, \vartheta)| < \varepsilon$.

By A_0 we denote the singular operator with the symbol $\Phi_0(x, \vartheta)$. This symbol can be written as a product

$$\Phi_0(x, \vartheta) = \left[\frac{2}{3} \varepsilon + a_n(x)\right] e^{-in\vartheta} \prod_{k=1}^{2n} [\alpha_k(x) - e^{i\vartheta}].$$

On the Riemann sphere (or, what is the same, on $\overline{\mathbb{R}}^2$) the zeros $\alpha_k(x)$ are continuous but, in general, not unique. All they are in modulus different from 1 such that either $\sup_x |\alpha_k(x)| < 1$ or $\inf_x |\alpha_k(x)| > 1$. The zeros of the first group will be denoted by $\alpha_k'(x)$, $k = 1, 2, \ldots, s$, and those of the second group by $\alpha_k''(x)$, $k = 1, 2, \ldots, \sigma$; $s + \sigma = 2n$. Accordingly,

$$\Phi'(x, \vartheta) = \prod_{k=1}^{s} [\alpha_k'(x) - e^{i\vartheta}], \qquad \Phi''(x, \vartheta) = \prod_{k=1}^{\sigma} [\alpha_k''(x) - e^{i\vartheta}].$$

Suppose now that the point x runs on a closed curve in the plane $\mathbb{R}^2$. The symbol $\Phi_0(x, \vartheta)$ is unique and continuous such that the zeros $\alpha_k(x)$ only permute when x is running. And they do not pass the unit circle such that the two groups mentioned permute among one another. But this means that the polynomials Φ' and Φ'' are unique functions of x.

For $t \in [0, 1]$ we put

$$\Phi(x, \vartheta, t) = \left[\frac{2}{3}\varepsilon + a_n(x)\right] e^{-in\vartheta} \prod_{k=1}^{s} [(1-t)\,\alpha_k'(x) - e^{i\vartheta}] \prod_{k=1}^{\sigma} [\alpha_k''(x) - (1-t)\,e^{i\vartheta}]$$

$$= \left[\frac{2}{3}\varepsilon + a_n(x)\right] (1-t)^s\, e^{in\vartheta}\Phi'\left(x, \frac{e^{i\vartheta}}{1-t}\right) \Phi''(x, (1-t)\,e^{i\vartheta})$$

and denote by A_t the simple singular operator with the symbol $\Phi(x, \vartheta, t)$ where $t = 0$ corresponds to the operator A_0 considered above. The operator A_t satisfies the assumptions of Lemma 3.2 such that $\operatorname{Ind} A_0 = \operatorname{Ind} A_1$. But

$$A_1 = (-1)^s \left[\frac{2}{3}\varepsilon + a_n(x)\right] \prod_{k=1}^{\sigma} \alpha_k''(x)\, h^{s-n},$$

hence, $\operatorname{Ind} A_1 = (s - n) \operatorname{Ind} h = 0$ because the operators h and $h^* = h^{-1}$ have the trivial null space only, i.e. $\operatorname{Ind} h = 0$. Thus, we obtained $\operatorname{Ind} A_0 = 0$.

Choose a positive ε small enough such that $\sup |(\Phi - \Phi_0)/\Phi_0| < 1$. The decomposition $\Phi = \Phi_0(1 - (\Phi_0 - \Phi)/\Phi_0)$ of the symbol corresponds to the operator decomposition $A = A_0(I - A_1) + T_2$ where A_1 denotes the singular operator with the symbol $(\Phi_0 - \Phi)/\Phi_0$ and T_2 is completely continuous. Due to Lemma 3.1, $\operatorname{Ind} A_1 = 0$, and we obtain $\operatorname{Ind} A = \operatorname{Ind} A_0 + \operatorname{Ind} A_1 = 0$. By Theorems 2.3 and 2.4, Chap. I, the operator A admits of an equivalent regularization. Theorem 3.1 is proved completely. ▮

§ 4. A necessary condition for the existence of a regularizer

The main result of § 2 can be formulated in the following manner. Under the assumption on a certain smoothness of the symbol $\Phi_A(x, \theta)$ of the operator A the condition $\inf |\Phi_A(x, \theta)| > 0$ is sufficient for the operator A to be a Fredholm operator in the space $\mathbf{L}_p(\mathbb{R}^m)$, $1 < p < \infty$. In the present section, we shall prove that this condition is also necessary (cf. Gohberg [2], Krupnik [1], Seeley [3]).

Let us denote by $\mathfrak{R}$ the ring of all bounded operators in $\mathbf{L}_p(\mathbb{R}^m)$ and by $\hat{\mathfrak{R}}$ the corresponding factor ring of $\mathfrak{R}$ by the ideal $\mathfrak{J}$ of the completely continuous operators in $\mathbf{L}_p(\mathbb{R}^m)$.

4.1. Lemma 4.1. *A singular operator A is a Fredholm operator if and only if the corresponding coset $\hat{A}$ of the ring $\hat{\mathfrak{R}}$ is invertible in $\hat{\mathfrak{R}}$.*

Proof. If there exists an inverse $\hat{B}$ to $\hat{A}$ in $\hat{\mathfrak{R}}$ then any operator B of the coset $\hat{B}$ provides a two-sided regularizer to A. By Theorem 3.1, Chap. I, A is a Fredholm operator. Conversely, if $A \in \mathfrak{R}$ is a Fredholm operator then it has, by the same Theorem 3.1, Chap. I, a two-sided regularizer $B \in \mathfrak{R}$, i.e. $BA = I + T_1$, $AB = I + T_2$, $T_1, T_2 \in \mathfrak{J}$. Hence $\hat{B}\hat{A} = \hat{A}\hat{B} = \hat{I}$, $\hat{B} = \hat{A}^{-1}$, and the element $\hat{A}$ is invertible in $\hat{\mathfrak{R}} = \mathfrak{R}/\mathfrak{J}$. ▮

Lemma 4.2. *Let $\mathfrak{X}$ be a commutative normed ring over the field of complex numbers, and a, b be elements of the ring $\mathfrak{X}$ having real spectra. If $a + bi$ is invertible in $\mathfrak{X}$ then so is the element $a - bi$.*

Proof. Let $\mathfrak{M}$ denote the set of all maximal ideals of the ring $\mathfrak{X}$. Since the element $a + bi$ is assumed to be invertible the corresponding function $a(M) + b(M)i$ on the maximal ideals $M \in \mathfrak{M}$ (see GELFAND, RAIKOV, and SHILOV [1]) does never vanish. The spectra of the elements a and b are real then the functions $a(M)$, $b(M)$ are real-valued, and the function $a(M) - b(M)\,\mathrm{i}$ does never vanish. Consequently, the element $a - b\mathrm{i}$ is invertible. ∎

Lemma 4.3. *Let the symbol $\Phi_A(x, \theta)$ of a general singular operator A satisfy the conditions of Theorem 8.2, Chap. XI. If A is a Fredholm operator in $\mathrm{L}_p(\mathrm{R}^m)$ then so is the operator $\bar{A}$ with the symbol $\Phi_{\bar{A}}(x, \theta) = \overline{\Phi_A(x, \theta)}$.*

Proof. If λ does not belong to the range of the symbol $\Phi_A(x, \theta)$ then the symbol $\Phi_B(x, \theta) = [\Phi_A(x, \theta) - \lambda]^{-1}$ satisfies the conditions of Theorem 8.2, Chap. XI. Obviously, every singular operator B with the symbol $\Phi_B(x, \theta)$ is a two-sided regularizer to the operator $A - \lambda I$ which is, thus, a Fredholm operator. Therefore, the element $\hat{A} - \lambda \hat{I}$ of the factor ring $\mathfrak{R}/\mathfrak{J}$ is invertible[1]). Hence, we proved that the spectrum of the element $\hat{A} \in \mathfrak{R}/\mathfrak{J}$ is contained in the range of the symbol $\Phi_A(x, \theta)$.

Now, let A be a Fredholm operator. Put $B_1 = (A + \bar{A})/2$, $B_2 = (A - \bar{A})/2\mathrm{i}$; the symbols of the operators B_1, B_2 are real-valued such that the spectra of the elements $\hat{B}_1, \hat{B}_2 \in \mathfrak{R}/\mathfrak{J}$ are real, too. The element $\hat{A} = \hat{B}_1 + \mathrm{i}\hat{B}_2$ is invertible in $\mathfrak{R}/\mathfrak{J}$, then so is (by Lemma 4.2) $\hat{B}_1 - \mathrm{i}\hat{B}_2 = \hat{\bar{A}}$, i.e. $\bar{A}$ is a Fredholm operator. ∎

4.2. Theorem 4.1. *Let the symbol $\Phi_A(x, \theta)$ of a general singular operator A satisfy the conditions of Theorem 8.2, Chap. XI. If A is a Fredholm operator in $\mathrm{L}_p(\mathrm{R}^m)$ then* $\inf\limits_{x \in \overline{\mathrm{R}}^m, \theta \in S} |\Phi_A(x, \theta)| > 0$.

Proof. By Lemma 4.2, the operator $\bar{A}$ with the symbol $\overline{\Phi_A(x, \theta)}$ is a Fredholm operator. The operator $D = A\bar{A}$ belongs to the coset $\hat{D} = \widehat{A\bar{A}} = \hat{A}\hat{\bar{A}}$ of the ring $\hat{\mathfrak{R}}$, and the element $\hat{D}$ is invertible in that ring: $\hat{D}^{-1} = (\hat{A}\hat{\bar{A}})^{-1} = \hat{\bar{A}}^{-1}\hat{A}^{-1}$. By Theorem 7.1, Chap. XI, D is a general singular operator with the symbol $\Phi_D(x, \theta) = |\Phi_A(x, \theta)|^2$. Consider the sequence of operators $D_n = D + (1/n)\, I$, $n = 1, 2, \ldots$, with the symbols $\Phi_{D_n}(x, \theta) = |\Phi_A(x, \theta)|^2 + 1/n$. These are positive and the elements $\hat{D}_n$ are invertible in $\hat{\mathfrak{R}}$. Since $\|\hat{D}_n^{-1} - \hat{D}_n\| \to 0$ the norms $\|\hat{D}_n^{-1}\|$ are uniformly bounded.

Suppose now that the symbol $\Phi_A(x, \theta)$ vanishes at a certain point $(x_0, \theta_0) \in \overline{\mathrm{R}}^m \times S$. The symbols satisfying the conditions of Theorem 8.2, Chap. XI, form a ring. Hence, the symbol $[\Phi_{D_n}(x, \theta)]^{-1} = [|\Phi_A(x, \theta)|^2 + 1/n]^{-1}$ satisfies these conditions, too, and the inequality (10.12), Chap. XI, holds,

$$\|\hat{D}_n^{-1}\| \geqq \max |\Phi_{D_n}(x, \theta)|^{-1} \geqq |\Phi_{D_n}(x_0, \theta_0)|^{-1} = n,$$

but this is in contradiction to the uniform boundedness of the norms $\|\hat{D}_n^{-1}\|$. ∎

4.3. Theorem 4.2. *Let the symbol of a general singular operator A satisfy the conditions of Theorem 8.2, Chap. XI, for $p = 2$. Then*

$$\|\hat{A}\|_{\mathrm{L}_2(\mathrm{R}^m)} = \max_{x \in \overline{\mathrm{R}}^m, \theta \in S} |\Phi_A(x, \theta)|. \tag{1}$$

The proof is based on the following lemma by SEELEY [3], and we reproduce it here with some unimportant changes.

[1]) For notations see § 10, Chap. XI.

Lemma 4.4. *Let A be a selfadjoint singular operator the symbol of which satisfies the conditions of Theorem 8.2, Chap. XI, for $p = 2$. If ε is an arbitrary positive number then there exists a finite-dimensional orthogonal projection P_ε in $\mathbf{L}_2(\mathbb{R}^m)$ such that*

$$\|A(I - P_\varepsilon)\| \leq \max |\Phi_A(x, \theta)| + \varepsilon. \tag{2}$$

Proof. From Theorem 8.2, Chap. XI, it follows that the symbol of a selfadjoint singular operator is real-valued. On the other hand, to any real-valued symbol $\Phi(x, \theta)$ a selfadjoint singular operator can be constructed. Indeed, if H is any singular operator with the symbol $\Phi(x, \theta)$ then its adjoint H^* has the symbol $\overline{\Phi(x, \theta)} = \Phi(x, \theta)$ such that it is sufficient to put $A = (H + H^*)/2$.

Let λ_0 be a real number such that $|\lambda_0| > \max |\Phi_A(x, \theta)|$. The difference $\Phi_A(x, \theta) - \lambda_0$ does never vanish, and by the results of §§ 2, 3, the operator $A - \lambda_0 I$ is normally solvable and $\operatorname{Ind}(A - \lambda_0 I) = 0$. By $E(\lambda_0)$ we denote the finite-dimensional null space of the operator $A - \lambda_0 I$, and by $P(\lambda_0)$ the orthogonal projection of $\mathbf{L}_2(\mathbb{R}^m)$ onto $E(\lambda_0)$. The operator $A - \lambda_0 I - P(\lambda_0)$ turns out to be invertible. Indeed, let $\varphi_1, \ldots, \varphi_\nu$ be an orthonormal basis of the subspace $E(\lambda_0)$ and $[A - \lambda_0 I - P(\lambda_0)]\, u = 0$. Then

$$(A - \lambda_0 I)\, u = \sum_{j=1}^{\nu} (u, \varphi_j)\, \varphi_j$$

so that

$$(u, \varphi_j) = ((A - \lambda_0 I)\, u, \varphi_j) = (u, (A - \lambda_0 I)\, \varphi_j) = 0$$

and $u = 0$. Hence, for any real λ such that $|\lambda_0 - \lambda| < \eta$ with a sufficiently small η the operator $A - \lambda I - P(\lambda_0)$ is invertible, too. Namely, from $(A - \lambda I)\, v = 0$ one has

$$0 = ((A - \lambda I)\, v, \varphi_j) = (v, (A - \lambda I)\, \varphi_j) = (\lambda_0 - \lambda)\, (v, \varphi_j)$$

and then $(v, \varphi_j) = 0$, $P(\lambda_0)\, v = 0$, $(A - \lambda I - P(\lambda_0))\, v = 0$, i.e. $v = 0$. All that reasoning implies that those spectral points of the operator A which belong to the domain $|\lambda| > \max |\Phi_A(x, \theta)|$ are isolated eigenvalues of finite multiplicity.

By M we denote the set of the spectral points of the operator A which belong to the domain $\|A\| \geq |\lambda| \geq \max |\Phi_A(x, \theta)| + \varepsilon$. The set M is compact and consists of isolated points, i.e. it is a finite set, say $M = \{\lambda_1, \ldots, \lambda_N\}$. Put $T = -\sum_{j=1}^{N} \lambda_j P(\lambda_j)$. General theorems on the spectrum of a selfadjoint operator state that the spectrum of the operator $A + T$ belongs to the domain $|\lambda| < \max |\Phi_A(x, \theta)| + \varepsilon$. (In SEELEY [3] you can find an elementary proof of this statament.) Therefore,

$$\|A + T\| < \max |\Phi_A(x, \theta)| + \varepsilon. \tag{3}$$

It remains to show that the operator T can be represented as $T = -AP_\varepsilon$ with some orthogonal projection P_ε. We have

$$Tu = -\sum_{k=1}^{N} \lambda_j P(\lambda_j)\, u, \quad \forall u.$$

Denote by $\varphi_{j1}, \ldots, \varphi_{j\nu_j}$ an orthonormal basis of the space $E(\lambda_j)$ then

$$P(\lambda_j)\, u = \sum_{k=1}^{\nu_j} (u, \varphi_{jk})\, \varphi_{jk},$$

$$AP(\lambda_j)\, u = \sum_{k=1}^{\nu_j} (u, \varphi_{jk})\, A\varphi_{jk} = \lambda_j \sum_{k=1}^{\nu_j} (u, \varphi_{jk})\, \varphi_{jk} = \lambda_j P(\lambda_j)\, u.$$

Hence, $\lambda_j P(\lambda_j) = AP(\lambda_j)$, and $T = -AP_\varepsilon$ where $P_\varepsilon = \sum_{j=1}^{N} P(\lambda_j)$ is an orthogonal projection as a sum of pairwise orthogonal projections. ∎

Proof to Theorem 4.2. Let A be a singular operator with the symbol $\Phi_A(x, \theta)$, and A^* its adjoint. Then A^*A is a selfadjoint operator with the symbol $|\Phi_A(x, \theta)|^2$. By Lemma 4.4, there is an orthogonal projection P_ε such that $\|A^*A - A^*AP_\varepsilon\| < \max|\Phi_A(x, \theta)|^2 + \varepsilon$. Put $T = -AP_\varepsilon$ then

$$\|A + T\|^2 = \|(A^* + T^*)(A + T)\| = \|(I - P_\varepsilon) A^*A(I - P_\varepsilon)\|$$
$$\leqq \|A^*A(I - P_\varepsilon)\| < \max|\Phi_A(x, \theta)|^2 + \varepsilon.$$

The more, $\inf_{T\in\mathfrak{J}}\|A + T\|^2 < \max|\Phi_A(x, \theta)|^2 + \varepsilon$, and, since ε is arbitrary, $\|\hat{A}\| \leqq \max|\Phi_A(x, \theta)|$. Taking into account the inequality (10.12), Chap. XI, we obtain (1). ∎

Remark 1. In Gohberg [2] and Krupnik [1] a more general class of singular operators is considered. It consists of the closure (in the norm) of singular operators the symbols of which are spherical polynomials, i.e. linear combinations of spherical functions. The inequality (10.12), Chap. XI, for polynomial symbols shows that the symbols converge for such a closure uniformly to a certain limit which is taken as the symbol of the corresponding limit operator. For these closures the inequality (10.12), Chap. XI, as well as (1) hold.

Krupnik [1] deals with singular operators in an arbitrary Banach space B consisting of functions defined in $\mathbb{R}^m$ and which satisfy the following conditions: 1. The simple singular operators $A_n^{(k)}$ with the symbols $Y_{n,m}^{(k)}(\theta)$ are bounded in B; 2. there exists a set Q of multipliers in B continuous on $\overline{\mathbb{R}^m}$ with the following property: if $a_\nu^{(k)} \in Q$, and the symbol $\Phi_{A_n}(x, \theta)$ of the operator

$$A_n = \sum_{\nu=0}^{n} \sum_{k=1}^{k_{\nu,m}} a_\nu^{(k)}(x)\, Y_{\nu,m}^{(k)}(\theta) \tag{4}$$

does not vanish on $\overline{\mathbb{R}^m} \times S$ then the simple singular operator with the symbol $[\Phi_{A_n}(x, \theta)]^{-1}$ belongs to the closure of the operators (4) (with respect to the operator norm in B); 3. if $a(x) \in Q$ then so does the conjugate complex function $\overline{a(x)}$; 4. the operators of type (4) commute up to a completely continuous operator. For singular operators obtained as closures in such a space the basic results of § 10, Chap. XI, and of the present section are valid.

Remark 2. Theorem 4.1 implies inequality (10.6), Chap. XI. This is easily seen repeating the arguments employed in the proof to Theorem 3.3, Chap. III.

§ 5. Singular equations in Sobolev spaces

5.1. Let us consider the singular equation

$$(A + T)\, u(x) := a(x)\, u(x) + \int_{\mathbb{R}^m} \frac{f(x, \theta)}{r^m}\, u(y)\, dy + (Tu)(x) = g(x) \tag{1}$$

under the following assumptions:

1. $g \in \mathbf{W}_2^{(l)}(\mathbb{R}^m)$ for some natural l;
2. the operator T is completely continuous in $\mathbf{L}_2(\mathbb{R}^m)$ and smoothing of order l;
3. for the symbol of equation (1), $\Phi(x, \theta) \in \mathfrak{R}_{l+1,\lambda}$, $\lambda > (m - 1)/2$;
4. $\inf|\Phi(x, \theta)| > 0$.

Theorem 5.1. *Under the assumptions 1.–4. every solution* $u \in \mathbf{L}_2(\mathbb{R}^m)$ *to equation* (1) *belongs to the space* $\mathbf{W}_2^{(l)}(\mathbb{R}^m)$.

5.2. Suppose, for the time being, that λ is sufficiently large, say $\lambda = \infty$. By B we denote the simple singular operator with the symbol $[\Phi(x, \theta)]^{-1}$. Obviously, $[\Phi(x, \theta)]^{-1} \in \mathfrak{R}_{l+1,\infty}$. By Theorem 9.2, Chap. XI, the operators A and B are bounded in $\mathbf{W}_2^{(l)}(\mathbb{R}^m)$. Moreover, B is a regularizer to the operator A. Applying B to (1) we get

$$u(x) + (T_1 u)(x) + (BTu)(x) = (Bg)(x), \tag{2}$$

where T_1 is a completely continuous operator in $\mathbf{L}_2(\mathbb{R}^m)$. What we shall do next is to show that T_1 is smoothing of order l in that space. The operator B is of the form

$$(Bu)(x) = b(x)\, u(x) + \int_{\mathbb{R}^m} \frac{h(x, \theta)}{r^m}\, u(y)\, dy.$$

With the notations $r_{yz} = |z - y|$, $\theta_{yz} = (z - y)/r_{yz}$ we obtain

$$u(x) + (T_1 u)(x) = b(x)\, a(x)\, u(x) + \int_{\mathbb{R}^m} \frac{h(x, \theta_{xy})}{r_{xy}^m}\, a(y)\, u(y)\, dy$$

$$+ b(x) \int_{\mathbb{R}^m} \frac{f(x, \theta_{xy})}{r_{xy}^m}\, u(y)\, dy + \int_{\mathbb{R}^m} \frac{h(x, \theta_{xy})}{r_{xy}^m}\, dy \int_{\mathbb{R}^m} \frac{f(y, \theta_{yz})}{r_{yz}^m}\, u(z)\, dz$$

$$= b(x)\, a(x)\, u(x) + b(x) \int_{\mathbb{R}^m} \frac{f(x, \theta_{xy})}{r_{xy}^m}\, u(y)\, dy + a(x) \int_{\mathbb{R}^m} \frac{h(x, \theta_{xy})}{r_{xy}^m}\, u(y)\, dy$$

$$+ \int_{\mathbb{R}^m} \frac{h(x, \theta_{xy})}{r_{xy}^m}\, dy \int_{\mathbb{R}^m} \frac{f(x, \theta_{yz})}{r_{yz}^m}\, u(z)\, dz + \int_{\mathbb{R}^m} \frac{h(x, \theta_{xy})}{r_{xy}^m}\, [a(y) - a(x)]\, u(y)\, dy$$

$$+ \int_{\mathbb{R}^m} \frac{h(x, \theta_{xy})}{r_{xy}^m}\, dy \int_{\mathbb{R}^m} \frac{f(y, \theta_{yz}) - f(x, \theta_{yz})}{r_{yz}^m}\, u(y)\, dy. \tag{3}$$

The first four terms in the right-hand side of (3) as a sum give $u(x)$. In order to see this consider the operators A_t and B_t with the symbols $\Phi(t, \theta_{xy})$ and $[\Phi(t, \theta_{xy})]^{-1}$, resp. $t \in \mathbb{R}^m$. Obviously, $(B_t A_t u)(x) = u(x)$, and

$$(A_t u)(x) = a(t)\, u(x) + \int_{\mathbb{R}^m} \frac{f(t, \theta_{xy})}{r_{xy}^m}\, u(y)\, dy,$$

$$(B_t u)(x) = b(t)\, u(x) + \int_{\mathbb{R}^m} \frac{h(t, \theta_{xy})}{r_{xy}^m}\, u(y)\, dy.$$

Hence,

$$u(x) = (B_t A_t u)(x) = b(t)\, a(t)\, u(x) + b(t) \int_{\mathbb{R}^m} \frac{f(t, \theta_{xy})}{r_{xy}^m}\, u(y)\, dy$$

$$+ a(t) \int_{\mathbb{R}^m} \frac{h(t, \theta_{xy})}{r_{xy}^m}\, u(y)\, dy + \int_{\mathbb{R}^m} \frac{h(t, \theta_{xy})}{r_{xy}^m}\, dy \int_{\mathbb{R}^m} \frac{f(t, \theta_{yz})}{r_{yz}^m}\, u(z)\, dz.$$

Now, the last equality proves the assertion made if $t = x$ is put herein. Therefore,

$$(T_1u)(x) = \int_{\mathbb{R}^m} \frac{h(x, \theta_{xy})}{r_{xy}^m} [a(y) - a(x)]\, u(y)\, dy$$

$$+ \int_{\mathbb{R}^m} \frac{h(x, \theta_{xy})}{r_{xy}^m} dy \int_{\mathbb{R}^m} \frac{f(y, \theta_{yz}) - f(x, \theta_{yz})}{r_{yz}^m} u(z)\, dz =: (T_{11}u)(x) + (T_{12}u)(x). \quad (4)$$

5.3. By means of the same arguments as used in the proof to Lemma 7.1, Chap. XI, we can show that the operator T_{11} (and likewise the operator T_{12}) is completely continuous in $L_2(\mathbb{R}^m)$. Thus, we shall prove next that these operators are smoothing of order l. Here we shall do so for the operator T_{11}, the other operator T_{12} will be studied in Section 5.4.

Obviously, $h(x, \theta) \in \mathfrak{R}_{l+1,\infty}$. We expand $h(x, \theta)$ into a series with respect to spherical functions,

$$h(x, \theta) = \sum_{n=1}^{\infty} \sum_{k=1}^{k_{n,m}} b_n^{(k)}(x)\, Y_{n,m}^{(k)}(\theta). \quad (5)$$

The corresponding operator expansion is

$$(T_{11}u)[(x) = \sum_{n=1}^{\infty} \sum_{k=1}^{k_{n,m}} b_n^{(k)}(x) \int_{\mathbb{R}^m} \frac{a(y) - a(x)}{r^m} Y_{n,m}^{(k)}(\theta)\, u(y)\, dy =: \sum_{n=1}^{\infty} \sum_{k=1}^{k_{n,m}} b_n^{(k)}(x)\, (M_n^{(k)}u)(x). \quad (6)$$

Obviously, $a \in C^{l+1}(\overline{\mathbb{R}}^m)$. By Theorem 9.3, Chap. XI, the operator $M_n^{(k)}$ is smoothing of order l, moreover,

$$\| M_n^{(k)} \|_{W_2^{(j)}(\mathbb{R}^m) \to W_2^{(j+1)}(\mathbb{R}^m)} \leqq \frac{C}{n}, \qquad 0 \leqq j \leqq l; \quad (7)$$

this last inequality means

$$\| D^\alpha M_n^{(k)} u \|_{L_2(\mathbb{R}^m)} \leqq \frac{C}{n} \sum_{|\beta|=1}^{j} \| D^\beta u \|_{L_2(\mathbb{R}^m)} \leqq \frac{C}{n} \| u \|_{W_2^{(j)}(\mathbb{R}^m)}, \quad (8)$$

$|\alpha| \leqq j + 1$. Differentiate (6),

$$(D^\alpha T_{11}u)(x) = \sum_{\beta \leqq \alpha} \sum_{n=1}^{\infty} \sum_{k=1}^{k_{n,m}} \frac{\alpha!}{\beta!\,(\alpha - \beta)!} D^{\alpha-\beta} b_n^{(k)}(x)\, (D^\beta M_n^{(k)} u)(x). \quad (9)$$

Separate from (9) the sum with constant β,

$$v_\beta(x) := \sum_{n=1}^{\infty} \sum_{k=1}^{k_{n,m}} \frac{\alpha!}{\beta!\,(\alpha - \beta)!} D^{\alpha-\beta} b_n^{(k)}(x)\, (D^\beta M_n^{(k)} u)(x). \quad (10)$$

Then we estimate the norm $\| v_\beta \|_{L_2(\mathbb{R}^m)}$. With $D^\beta M_n^{(k)} u = w_n^{(k)}$ we get

$$\| v_\beta \|_{L_2(\mathbb{R}^m)}^2 = \left(\frac{\alpha!}{\beta!\,(\alpha - \beta)!} \right)^2 \int_{\mathbb{R}^m} \left| \sum_{n=1}^{\infty} \sum_{k=1}^{k_{n,m}} D^{\alpha-\beta} b_n^{(k)}(x)\, w_n^{(k)}(x) \right|^2 dx$$

$$\leqq C \int_{\mathbb{R}^m} \sum_{n=1}^{\infty} \sum_{k=1}^{k_{n,m}} |D^{\alpha-\beta} b_n^{(k)}(x)|^2\, n^{2\mu} \sum_{n=1}^{\infty} \sum_{k=1}^{k_{n,m}} n^{-2\mu}\, |w_n^{(k)}(x)|^2\, dx$$

$$\leqq C \sup_{\substack{|\alpha| \leqq l+1 \\ x \in \mathbb{R}^m}} \| D^\alpha h(x, \cdot) \|_{W_2^{(\mu)}(S)}^2 \sum_{n=1}^{\infty} \sum_{k=1}^{k_{n,m}} n^{-2\mu}\, \| w_n^{(k)} \|_{L_2(\mathbb{R}^m)}^2$$

$$\leqq C \| u \|_{W_2^{(j)}(\mathbb{R}^m)}^2 \sum_{n=1}^{\infty} \sum_{k=1}^{k_{n,m}} n^{-2\mu-2} \leqq C \| u \|_{W_2^{(j)}(\mathbb{R}^m)}^2 \sum_{n=1}^{\infty} n^{-2\mu-4+m},$$

μ being a sufficiently large positive number. Therefore, the last series is convergent so that $\|v_\beta\|_{L_2(R^m)} \leqq C\,\|u\|_{W_2^{(j)}(R^m)}$. Summing up over all α, β, $|\alpha| \leqq j+1$, we obtain $\|T_{11}u\|_{W_2^{(j+1)}(R^m)} \leqq C\,\|u\|_{W_2^{(j)}(R^m)}$ which was to be demonstrated.

5.4. Now let us come to the operator T_{12}. We expand the characteristic $f(x, \theta)$ into a series with respect to spherical functions,

$$f(x, \theta) = \sum_{\nu=1}^{\infty} \sum_{\varkappa=1}^{k_{\nu,m}} a_\nu^{(\varkappa)}(x)\, Y_{\nu,m}^{(\varkappa)}(\theta).$$

Put

$$v_\nu^{(\varkappa)}(x) = \int_{R^m} \frac{Y_{\nu,m}^{(\varkappa)}(\theta)}{r^m}\, u(y)\, dy$$

and use the expansion (5). Then

$$(T_{12}u)(x) = \int_{R^m} \sum_{n=1}^{\infty} \sum_{k=1}^{k_{n,m}} \sum_{\nu=1}^{\infty} \sum_{\varkappa=1}^{k_{\nu,m}} \frac{b_n^{(k)}(x)}{r^m} [a_\nu^{(\varkappa)}(y) - a_\nu^{(\varkappa)}(x)]\, Y_{n,m}^{(k)}(\theta)\, v_\nu^{(\varkappa)}(y)\, dy.$$

With the notation

$$\varphi_{n,\nu}^{(k,\varkappa)}(x) = \int_{R^m} \frac{a_\nu^{(\varkappa)}(y) - a_\nu^{(\varkappa)}(x)}{r^m}\, Y_{n,m}^{(k)}(\theta)\, v_\nu^{(\varkappa)}(y)\, dy \tag{12}$$

we arrive at the representation

$$(T_{12}u)(x) = \sum_{n=1}^{\infty} \sum_{k=1}^{k_{n,m}} \sum_{\nu=1}^{\infty} \sum_{\varkappa=1}^{k_{\nu,m}} b_n^{(k)}(x)\, \varphi_{n,\nu}^{(k,\varkappa)}(x). \tag{13}$$

Thence,

$$(D^\alpha T_{12}u)(x) = \sum_{\beta \leqq \alpha} \frac{\alpha!}{\beta!\,(\alpha-\beta)!} \sum_{n=1}^{\infty} \sum_{k=1}^{k_{n,m}} \sum_{\nu=1}^{\infty} \sum_{\varkappa=1}^{k_{\nu,m}} D^{\alpha-\beta} b_n^{(k)}(x)\, D^\beta \varphi_{n,\nu}^{(k,\varkappa)}$$

$$=: \sum_{\beta \leqq \alpha} \frac{\alpha!}{\beta!\,(\alpha-\beta)!}\, \psi_{\alpha,\beta}(x).$$

Here,

$$\|\psi_{\alpha,\beta}\|_{L_2(R^m)}^2 \leqq C \int_{R^m} \sum_{n=1}^{\infty} \sum_{k=1}^{k_{n,m}} |D^{\alpha-\beta} b_n^{(k)}(x)|^2\, n^{2\mu} \sum_{n=1}^{\infty} \sum_{k=1}^{k_{n,m}} \left| \sum_{\nu=1}^{\infty} \sum_{\varkappa=1}^{k_{\nu,m}} D^\beta \varphi_{n,\nu}^{(k,\varkappa)} \right|^2 n^{-2\mu}\, dx$$

$$\leqq C \sup_{x \in R^m, |\alpha| \leqq l} \|h(x, \cdot)\|_{W_2^{(\mu)}(R^m)}^2 \sum_{n=1}^{\infty} \sum_{k=1}^{k_{n,m}} n^{-2\mu} \left\| \sum_{\nu=1}^{\infty} \sum_{\varkappa=1}^{k_{\nu,m}} D^\beta \varphi_{n,\nu}^{(k,\varkappa)} \right\|_{L_2(R^m)}^2$$

$$= C \sum_{n=1}^{\infty} \sum_{k=1}^{k_{n,m}} n^{-2\mu} \left\| \sum_{\nu=1}^{\infty} \sum_{\varkappa=1}^{k_{n,m}} D^\beta \varphi_{n,\nu}^{(k,\varkappa)} \right\|_{L_2(R^m)}.$$

By Theorem 9.3 and Theorem 9.1, Chap. XI,

$$\|\varphi_{n,\nu}^{(k,\varkappa)}\|_{W_2^{(j+1)}(R^m)} \leqq \frac{C}{n} \max_{x \in \overline{R}^m} |a_\nu^{(\varkappa)}(x)| \cdot \|v_\nu^{(\varkappa)}\|_{W_2^{(j)}(R^m)} \leqq \frac{C}{n\nu} \|u\|_{W_2^{(j)}(R^m)}.$$

Hence,

$$\|D^\beta \varphi_{n,\nu}^{(k,\varkappa)}\|_{L_2(R^m)} \leqq \frac{C}{n\nu} \|u\|_{W_2^{(j)}(R^m)} \max_{x \in \overline{R}^m} |a_\nu^{(\varkappa)}(x)|$$

and therefore

$$\left\| \sum_{\nu=1}^{\infty} \sum_{\varkappa=1}^{k_{\nu,m}} D^\beta \varphi_{n,\nu}^{(k,\varkappa)} \right\|_{L_2(R^m)} \leqq \frac{C}{n} \sum_{\nu=1}^{\infty} \sum_{\varkappa=1}^{k_{\nu,m}} \max_{x \in \overline{R}^m} |a_\nu^{(\varkappa)}(x)|\, \nu^{-1} \|u\|_{W_2^{(j)}(R^m)}. \tag{14}$$

Since $f \hat{\in} \mathbf{W}_2^{(\mu)}(S)$ we have

$$\sum_{\nu=1}^{\infty} \sum_{\varkappa=1}^{k_{\nu,m}} |a_\nu^{(\varkappa)}(x)|^2 \nu^{2\mu} \leqq C.$$

This gives $\max |a_\nu^{(\varkappa)}(x)| \leqq C\nu^{-\mu}$, and the sum in the right-hand side of (14) can be estimated by

$$C \sum_{\nu=1}^{\infty} \sum_{\varkappa=1}^{k_{\nu,m}} \nu^{-\mu-1} \leqq C \sum_{\nu=1}^{\infty} \nu^{-\mu+m-3},$$

and this series is convergent since μ is supposed to be sufficiently large. Hence,

$$\|\psi_{\alpha,\beta}\|_{\mathbf{L}_2(\mathbb{R}^m)} \leqq C\|u\|_{\mathbf{W}_2^{(j)}(\mathbb{R}^m)} \sum_{n=1}^{\infty} \sum_{k=1}^{k_{n,m}} n^{-2\mu-1} \leqq C\|u\|_{\mathbf{W}_2^{(j)}(\mathbb{R}^m)} \sum_{n=1}^{\infty} n^{-2\mu+m-3} \leqq C\|u\|_{\mathbf{W}_2^{(j)}(\mathbb{R}^m)}. \tag{15}$$

Summing up over all β, $\beta \leqq \alpha$, we find

$$\|D^\alpha T_{12} u\|_{\mathbf{L}_2(\mathbb{R}^m)} \leqq C \|u\|_{\mathbf{W}_2^{(j)}(\mathbb{R}^m)}.$$

Summing up this last inequality over all α, $|\alpha| \leqq j + 1$, we obtain

$$\|T_{12} u\|_{\mathbf{W}_2^{(j+1)}(\mathbb{R}^m)} \leqq C \|u\|_{\mathbf{W}_2^{(j)}(\mathbb{R}^m)}.$$

Thus, the operator T_{12} is smoothing of order l.

Now the reasoning in 5.3 and 5.4 implies that the operator $T = T_1 + BT$ is smoothing of order l, too.

5.5. Let now $\Phi(x, \theta) \in \mathfrak{R}_{l+1,\lambda}$, $\lambda > (m - 1)/2$. We expand $\Phi(x, \theta)$ in a series with respect to spherical functions. Its partial sum of order N will be denoted by $\Phi_1(x, \theta)$ and the corresponding remainder by $\Phi_2(x, \theta)$. We choose N large enough such that $\inf |\Phi_1(x, \theta)| > 0$ and $\|\Phi_2\|_{\mathfrak{R}_{l+1,\lambda}} < \varepsilon$ where ε is a sufficiently small positive number which will be specified later. By A_k, $k = 1, 2$, we denote the simple singular operator with the symbol Φ_k. Then $\Phi_1 \in \mathfrak{R}_{l+1,\infty}$ and $\|A_2\|_{\mathbf{W}_2^{(j)}(\mathbb{R}^m)} \leqq C_0\varepsilon$, $0 \leqq j \leqq l$, with a certain constant C_0.

We rewrite (1) as

$$A_1 u + A_2 u + Tu = g. \tag{16}$$

Denoting by B_1 the simple singular operator with the symbol $[\Phi_1(x, \theta)]^{-1}$ and applying it to (16) we get

$$u + B_1 A_2 u = g - BTu - T_1 u. \tag{17}$$

As shown in 5.2 and 5.3, the operator T_1 is smoothing of order l. Now we choose the number ε such that

$$\|B_1 A_2\|_{\mathbf{W}_2^{(j)}(\mathbb{R}^m)} < 1, \qquad 0 \leqq j \leqq l. \tag{18}$$

Suppose that (1) has a solution $u_0 \in \mathbf{L}_2(\mathbb{R}^m)$. Then $B_1 T u_0$, $T_1 u_0 \in \mathbf{W}_2^{(1)}(\mathbb{R}^m)$. Put $B_1 A_2 = D$ and $g(x) - (BTu_0)(x) - (T_1 u_0)(x) = h(x)$ such that u_0 satisfies the equation

$$u_0 + Du_0 = h \tag{19}$$

Inequality (18) implies that the operator $(I - D)^{-1}$ is bounded in $\mathbf{W}_2^{(j)}(\mathbb{R}^m)$, $0 \leqq j \leqq l$, and $u_0 = (I - D)^{-1} h \in \mathbf{W}_2^{(1)}(\mathbb{R}^m)$. But then $h \in \mathbf{W}_2^{(2)}(\mathbb{R}^m)$ and $u_0 \subset \mathbf{W}_2^{(2)}(\mathbb{R}^m)$. Proceeding in this way we arrive, finally, at $u_0 \in \mathbf{W}_2^{(l)}(\mathbb{R}^m)$. ∎

5.6. Theorem 5.1 remains to hold true for spaces $\mathbf{W}_p^{(l)}(\mathbb{R}^m)$, $p \neq 2$, provided that $\Phi(x, \theta) \in \mathfrak{R}_{l+1,\lambda}$, $\lambda > \max\{(m-1)/p + 1/2, m/2\}$ is assumed.

Remark. The results presented here are contained in Mikhlin [19]. A result similar to Theorem 5.1 can be obtained from a paper by Krupnik [2] (see Lemma 2 therein) assuming, however, stronger conditions to the symbol.

§ 6. Singular equations in test and in distribution spaces

This topic is the subject of Krupnik [2]. The material given here is similar to that contained in the paper cited but it is not the same: We restrict the space of test functions and assume stronger conditions on the equation.

6.1. Let us consider singular operators of the form

$$(Au)(x) = a(x)\,u(x) + \int_{\mathbb{R}^m} r^{-m} f(x, \theta)\,u(y)\,dy, \tag{1}$$

$$(Bu)(x) = \overline{a(x)}\,u(x) + \int_{\mathbb{R}^m} r^{-m}\overline{f(y, -\theta)}\,u(y)\,dy. \tag{2}$$

For an arbitrary multi-index q we shall denote by $f^{(q)}(x, \theta)$ the derivative of order $|q|$ of $f(x, \theta)$ computed under the assumption that θ does not depend on x. A similar meaning is given to $f^{(q)}(x, -\theta)$. By V we shall denote the countably normed space of infinitely differentiable functions in $\mathbb{R}^m$ the norms of which

$$\|u\|_s = \sup_{|q| \leq s, x \in \mathbb{R}^m} (1 + |x|)^m\,|D^q u(x)| \tag{3}$$

are finite. Obviously, the space V is complete. Remember that in countably normed spaces the notions of continuity and boundedness for linear operators are equivalent (see § 4, Chap. I).

Theorem 6.1. *Let the functions $a(x)$ and $f(x, \theta)$ be infinitely differentiable with respect to $x \in \mathbb{R}^m$. If*

$$\forall q\colon |a^{(q)}(x)| \leq N_q, \qquad |f^{(q)}(x, \theta)| \leq N_q, \qquad N_q = \text{const}, \tag{4}$$

and $r^{-m} f^{(q)}(x, \theta) \in \mathcal{A}$ (see § 5.2, Chap. IX) then the operators A and B are bounded in V.

Proof. Put

$$(A_0 u)(x) = a(x)\,u(x), \qquad (A_1 u)(x) = \int_{\mathbb{R}^m} r^{-m} f(x, \theta)\,u(y)\,dy. \tag{5}$$

The operator A_0 is, obviously, bounded. With $y - x = t$, $r = |t|$, and $\theta = t/r$ we get

$$(A_1 u)(x) = \int_{\mathbb{R}^m} r^{-m} f(x, \theta)\,u(x + t)\,dt$$

and then

$$(D^q A_1 u)(x) = \sum_{k+l=q} \frac{q!}{k!\,l!} \int_{\mathbb{R}^m} r^{-m} f^{(k)}(x, \theta)\,u^{(l)}(x + t)\,dt$$

$$= \sum_{k+l=q} \frac{q!}{k!\,l!} \int_{\mathbb{R}^m} r^{-m} f^{(k)}(x, \theta)\,u^{(l)}(y)\,dy. \tag{6}$$

The assumptions of the theorem imply $u^{(l)} \in \mathbf{A}_\alpha$ for all $\alpha \in (0, 1)$ and $r^{-m} f^{(k)}(x, \theta) \in \mathcal{A}$. By Theorem 5.2, Chap. IX, $D^q A_1 u \in \mathbf{A}_\alpha$. But then $|D^q A_1 u| \leq C_q(1 + |x|^2)^{-m/2}$, $C_q =$ const, and $A_1 u \in V$.

Hence, the space V is invariant under the operator A_1. It remains to show A_1 to be bounded. Let M be a bounded set in V, $\forall\, u \in M$: $\|u\|_s \leqq C_s$, $s = 1, 2, \ldots$ Consider an integral from the right-hand side of (6),

$$v_{k,l}(x) = \int_{\mathbb{R}^m} r^{-m} f^{(k)}(x, \theta)\, u^{(l)}(y)\, \mathrm{d}y. \tag{7}$$

We map the space $\mathbb{R}^m$ onto the Riemann sphere Σ stereographically and put $(1 + |x|^2)^{m/2}\, u^{(l)}(x) =: \tilde{u}_l(\xi)$, $(1 + |x|^2)^{m/2}\, v_{k,l}(x) =: \tilde{v}_{k,l}(\xi)$. The kernel $r^{-m} f^{(k)}(x, \theta)$ is then multiplied by $[(1 + |x|^2)\,(1 + |y|^2)]^{m/2}$, and that new kernel can be written as $\varrho^{-m}\, \varphi_k(\xi, \Lambda)$ such that φ_k is the new characteristic (see § 2, Chap. IX). Thus, we have

$$\tilde{v}_{k,l}(\xi) = \int_{\Sigma} \varrho^{-m} \varphi_k(\xi, \Lambda)\, \tilde{u}_l(\eta)\, \mathrm{d}_\eta \sigma.$$

Due to Remark 1, from § 1 Chap. IX, the function $\tilde{v}_{k,l}(\xi)$ is continuous and, consequently, bounded on Σ so that $|v_{k,l}(x)| \leqq C_q(1 + |x|^2)^{-m/2}$, $C_q = \text{const}$. Hence, such an estimate is true for the complete right-hand side of (6). Therefore, the operator A_1 (and likewise the operator A) is bounded in V. The proof goes analogously for the operator B. ∎

6.2. By V' we denote the distribution space (i.e. the space of continuous linear functionals) over V. As usually, the convergence in V' is the weak convergence of functionals: $f_n \to f$ in V', $f_n, f \in V'$, means the convergence $(f_n, \varphi) \to (f, \varphi)$ for any $\varphi \in V$.

Suppose that the operators A and B satisfy the conditions of Theorem 6.1. By Theorem 2.2, Chap. XI, the operator A is bounded in $\mathbf{L}_2(\mathbb{R}^m)$. The operator B is the adjoint to A in $\mathbf{L}_2(\mathbb{R}^m)$ and bounded, too. For any two functions $u, v \in \mathbf{L}_2(\mathbb{R}^m)$, $(Au, v) = (u, Bv)$. Due to Theorem 6.1, the right-hand member of this equality makes sense for any $u \in V'$ and $v \in V$. Thus, the operator A can be defined on the space V'. Similarly, the operator B is defined on V'. In other words, the operator A (and B) is defined on V' as the dual of the operator B (and A, resp.) on V. This immediately implies

Theorem 6.2. *If $a(x)$ and $f(x, \theta)$ satisfy the conditions of Theorem 6.1 then the operators A and B are continuous in V'.*

6.3. Theorem 6.3. *Let the functions $a(x)$ and $f(x, \theta)$ satisfy the conditions of Theorem 6.1. If, in addition,* $\inf |\Phi(x, \theta)| > 0$ *for the symbol Φ of the operator A and $a \in \mathbf{C}^\infty(\overline{\mathbb{R}}^m)$, $f(x, \theta) \hat{\in} \mathbf{C}^\infty(\mathbb{R}^m)$, then the index of the operators A and B in the spaces V and V' is equal to zero.*

Proof. To begin with, let us consider the operator A in the space V. In the space $\mathbf{L}_2(\mathbb{R}^m)$ this operator has the index zero (Theorem 3.1). We introduce a basis $\varphi_1, \ldots, \varphi_n$ and $\psi_1, \ldots, \psi_n$ in the null spaces of A and B, resp. Condition (4) means that $\Phi(x, \theta) \in \mathfrak{K}_{\infty,\lambda}$, $\lambda > m/2$. Theorem 5.1 (which can be also proved for operators (2)) implies that the functions φ_j, ψ_j are infinitely differentiable. We show that these functions belong to the space V. We have $A\varphi_j = 0$. The simple singular operator R with the symbol $[\Phi(x, \theta)]^{-1}$ provides a two-sided regularizer to A, i.e. $RA\varphi_j = \varphi_j + T\varphi_j = 0$ where T is completely continuous in $\mathbf{L}_2(\mathbb{R}^m)$. It is easy to see that

$$(Tu)(x) = \int_{\mathbb{R}^m} \{[a(y) - a(x)]\, L(x, x - y) + \int_{\mathbb{R}^m} L(x, x - z)\, [K(z, z - y) - K(x, z - y)]\, \mathrm{d}z\}\, u(y)\, \mathrm{d}y \tag{8}$$

where $K(x, x - y)$ and $L(x, x - y)$ are the kernels of the singular operators A and R, resp.

We introduce the notations

$$\left(\frac{1+|x|^2}{2}\right)^{m/2} \varphi_j(x) = \tilde{\varphi}_j(\xi), \qquad \left(\frac{1+|x|^2}{2}\right)^{m/2} T = \tilde{T}$$

and map the space $\mathbb{R}^m$ onto the sphere Σ stereographically. The equation for φ_j is transformed to $\tilde{\varphi}_j(\xi) + \tilde{T}\tilde{\varphi}_j = 0$. The operator $\tilde{T}$ is completely continuous in $\mathbf{L}_2(\Sigma)$,

$$(\tilde{T}\tilde{u})(\xi) = \int_\Sigma \Bigg\{[a(y) - a(x)]\frac{\varphi(x, \theta_{xy})}{\varrho^m} + \int_\Sigma \frac{[f(z, \theta_{yz}) - f(x, \theta_{yz})]}{|\xi - \eta|^m\, |\eta - \zeta|^m} \varphi(x, \theta_{xz})\, \mathrm{d}_\zeta\sigma\Bigg\} \tilde{u}(\eta)\, \mathrm{d}_\eta\sigma. \tag{9}$$

Here $\varphi(x, \theta)$ is the characteristic of the kernel L, and ξ, η, ζ are points on Σ, namely the images of the points $x, y, z \in \mathbb{R}^m$ under stereographic projection.

The first summand in the kernel of the operator $\tilde{T}$ (eq. (9)) admits to an estimate $O(\varrho^{-m+1})$ and is, therefore, weakly singular. The second summand is also weakly singular; in order not to interrupt the demonstration this will be shown in § 6.5.

Thus, the kernel of the operator T can be written as $\varrho^{-\lambda} N(\xi, \eta)$ where $0 < \lambda < m$, and $N(\xi, \eta)$ is a bounded function. The functions $\tilde{\varphi}_j$ turn out to be eigenfunctions to a weakly singular operator. Such a function is bounded (see Mikhlin [1], § 16, Theorem 3) such that $\varphi_j = O(1 + |x|^2)^{-m/2}$. Because of $\varphi_j \in \mathbf{C}^\infty$ we have, particularly, $\varphi_j \in \mathbf{A}_\alpha$.

The equality

$$(A\varphi_j)(x) = a(x)\, \varphi_j(x) + \int_{\mathbb{R}^m} \frac{f(x, \theta)}{r^m} u(y)\, \mathrm{d}y = 0$$

is differentiated with respect to x_ν. Here we can make use of (6) with $q = (0, \ldots, 1, \ldots, 0)$ with 1 at place ν,

$$a(x)\, \varphi_j^{(q)}(x) + \int_{\mathbb{R}^m} r^{-m} f(x, \theta)\, \varphi_j^{(q)}(y)\, \mathrm{d}y = -a^{(q)}(x)\, \varphi_j(x) - \int_{\mathbb{R}^m} r^{-m} f^{(q)}(x, \theta)\, \varphi_j(y)\, \mathrm{d}y =: g(x). \tag{10}$$

Since $\varphi_j \in \mathbf{A}_\alpha$ we have by Theorem 5.2, Chap. IX, $g \in \mathbf{A}_\alpha$. We apply to (10) the regularizer R. The same theorem gives $Rg \in \mathbf{A}_\alpha$. Equation (10) becomes $\varphi_j^{(q)}(x) + (T\varphi_j^{(q)})(x) = (Rg)(x)$ or (on the Riemann sphere Σ) $\widetilde{\varphi_j^{(q)}}(\xi) + (\widetilde{T\varphi_j^{(q)}})(\xi) = (\widetilde{Rg})(\xi)$. The function $(\widetilde{Rg})(\xi) = ((1 + |x|^2)/2)^{m/2}\, (Rg)(x)$ is bounded, but then so is the function $\widetilde{\varphi_j^{(q)}}(\xi)$ (cf. Mikhlin [1], § 16, Theorem 3). Hence, $\varphi_j^{(q)}(x) = O((1 + |x|^m))$, $|q| = 1$. Similarly we can find such an estimate for any multi-index q such that, finally, $\varphi_j \in V$. Analogously, $\psi_j \in V$.

6.4. Put

$$A_* u = Au - \sum_{j=1}^{n} (u, \varphi_j)\, \psi_j.$$

It is easy to see that the null space of A_* in $\mathbf{L}_2(\mathbb{R}^m)$ is trivial. Indeed, $A_* v = 0$ gives

$$Av = \sum_{j=1}^{n} (v, \varphi_j)\, \psi_j.$$

This equation is solvable such that necessarily

$$\sum_{j=1}^{n} (v, \varphi_j)(\psi_j, \psi_k) = (v, \varphi_k) = 0, \quad k = 1, 2, \ldots, n,$$

and hence $Av = 0$. But then v is a linear combination of the functions φ_j which is orthogonal to all these functions, i.e. $v = 0$. Furthermore, $\operatorname{Ind} A_* = \operatorname{Ind} A = 0$ in $\mathbf{L}_2(\mathbb{R}^m)$. Hence, the adjoint to the operator A_* in $\mathbf{L}_2(\mathbb{R}^m)$ has only the trivial null space. But then the equation

$$(A_* u)(x) = g(x), \qquad \forall g \in \mathbf{L}_2(\mathbb{R}^m) \tag{11}$$

has in $\mathbf{L}_2(\mathbb{R}^m)$ one and only one solution u. We show that this solution belongs to V if g does so.

Note that, by Theorem 5.1, $u \in \mathbf{C}^\infty(\mathbb{R}^m)$.

Equation (11) can be written as

$$(Au)(x) = g(x) + \sum_{j=1}^{n} (u, \varphi_j)\, \psi_j =: g_1(x). \tag{12}$$

Obviously, $g_1 \in V$. We apply the regularizer R to (12) and then transform it stereographically. With the same notation as used above we find $\tilde{u}(\xi) + (\tilde{T}\tilde{u})(\xi) = (\widetilde{Rg})(\xi)$. Repeating the arguments of section 6.3 we obtain $u^{(q)}(x) = O(1 + |x|^m)$ for any multiindex q and, therefore, $u \in V$.

Thus, (11) is uniquely solvable in V for any $g \in V$. Hence, $\operatorname{Ind} A_* = 0$ in V. The operator $\sum_{j=1}^{n} (u, \varphi_j)\, \psi_j$ is bounded and finite-dimensional, i.e. completely continuous in V. Now, Theorem 3.4, Chap. I (which is true in linear topological spaces, too, cf. § 4.4, Chap. I), implies $\operatorname{Ind} A = 0$ in V. The index of the adjoint operator is also zero, i.e. $\operatorname{Ind} B = 0$ in V'. In the same way $\operatorname{Ind} B = 0$ in V and $\operatorname{Ind} A = 0$ in V' can be shown.

6.5. It remains to show that the second summand in the kernel of the operator (9) is weakly singular. Let the points ξ and η be sufficiently close together. A certain point $\xi_0 \in \Sigma$ is taken as the center of two m-dimensional spheres Σ' and Σ'', their intersections with Σ form two $(m-1)$-dimensional spheres σ_1 and σ_2. Then Σ can be represented as the union of two spherical domains Σ_1 and Σ_2 which are bounded by the spheres σ_1 and σ_2 (see Fig. 11). Put $t = |\zeta - \xi_0|$; obviously, $0 \leqq t \leqq 2$.

By t', t'' we denote the radii of the spheres Σ', Σ''. We form a function $\omega \in \mathbf{C}^\infty[0, 2]$,

$$\omega(t) = \begin{cases} 1, & 0 \leqq t \leqq t'', \\ 0, & t' \leqq t \leqq 2, \end{cases} \qquad 0 < \omega(t) < 1, \qquad t'' < t < t'.$$

The kernel under consideration can be represented as

$$\int_\Sigma \frac{[\tilde{f}(\zeta, \theta_{yz}) - \tilde{f}(\xi, \theta_{yz})]\, \tilde{\varphi}(\xi, \theta_{xz})\, \omega(t)\, d_\zeta\sigma}{|\xi - \zeta|^m \cdot |\eta - \zeta|^m} + \int_\Sigma \frac{[\tilde{f}(\zeta, \theta_{yz}) - \tilde{f}(\xi, \theta_{yz})]\, \tilde{\varphi}(\xi, \theta_{xz})\, (1 - \omega(t))\, d_\zeta\sigma}{|\xi - \zeta|^m\, |\eta - \zeta|^m}. \tag{13}$$

Note that Σ in the first integral can be replaced by Σ_1 and in the second one by Σ_2. Furthermore, we assume $|\xi - \xi_0| \leqq t'' - \delta$, $|\eta - \xi_0| \leqq t'' - \delta$ with some constant $\delta > 0$. Then the second integral in (13) is bounded.

We have

$$\tilde{f}(\zeta, \theta_{yz}) - \tilde{f}(\xi, \theta_{yz}) = (\xi - \zeta, \tilde{F}(\xi, \zeta, \theta_{yz})),$$

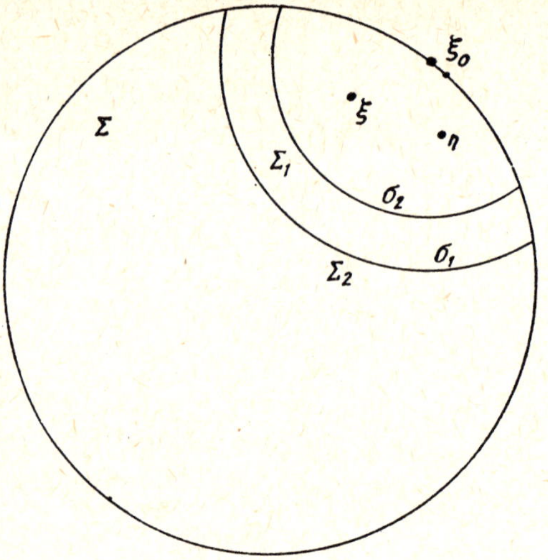

Fig. 11

where the function $\tilde{F}$ is infinitely differentiable with respect to ξ and ζ and is bounded as a function of all three arguments. The first integral in (13) becomes

$$\int_{\Sigma_1} \frac{\tilde{F}(\xi, \zeta, \theta_{yz})}{|\eta - \zeta|^m} \cdot \frac{\tilde{\varphi}(\xi, \theta_{xz})\, \omega(t)\, (\xi - \zeta)}{|\xi - \zeta|^m}\, d_\zeta \sigma. \tag{14}$$

By an inverse stereographic projection with center at ξ_0 we map Σ into $\mathbb{R}^m$. Then te integral (14) is transformed to

$$\left(\frac{1 + |x|^2}{2}\right)^{(m-1)/2} \left(\frac{1 + |y|^2}{2}\right)^{m/2} \int_D \frac{F(x, z, \theta_{yz})}{|y - z|^m} \cdot \frac{\psi(x, z, \theta_{xz})}{|x - z|^{m-1}} \frac{\omega_0(z)}{\sqrt{(1 + |z|^2)/2}}\, dz \tag{15}$$

(D denotes a bouuded domain in $\mathbb{R}^m$).

A composite kernel is the kernel of a product of two corresponding operators. With such a terminology the integral (15) turns out to be the composition of two kernels, namely the weakly singular kernel $|x - z|^{-m+1} \psi(x, z, \theta_{xz})\, \omega_0(z)/((1 + |z|^2)/2)^{1/2} =:$ $|x - z|^{-m+1} \omega(x, z)$ and the singular kernel $F(x, z, \theta_{yz})\, |y - z|^{-m}$. Let us study this composition. Denote the integral in (15) by $G(x, y)$. Then

$$\begin{aligned} G(x, y) &= \lim_{\varepsilon \to 0} \int_{D \cap (|y - z| > \varepsilon)} \frac{F(x, z, \theta_{yz})}{|y - z|^m} \frac{\omega(x, z)}{|x - z|^{m-1}}\, dz \\ &= \int_D \frac{F(x, z, \theta_{yz}) - F(x, y, \theta_{yz})}{|y - z|^m} \cdot \frac{\omega(x, z)}{|x - z|^{m-1}}\, dz \\ &\quad + \int_D \frac{F(x, y, \theta_{yz})\, [\omega(x, z) - \omega(x, y)]}{|y - z|^m\, |x - z|^{m-1}}\, dz \\ &\quad + \omega(x, y) \lim_{\varepsilon \to 0} \int_S F(x, y, \theta_{yz})\, d_z S \int_\varepsilon^R \frac{dr_{yz}}{r_{yz} r_{yz}^{m-1}}. \end{aligned} \tag{16}$$

In the last integral we put $R = \max |y - z|$ for a fixed θ_{yz}.

In the right-hand side of (16) the first two integrals are compositions of weakly singular kernels such that they are weakly singular kernels themselves (cf. SOBOLEV [1] or MIKHLIN [1]). Consider the third integral. From the assumptions $|\xi - \xi_0| \leqq t'' - \delta$ and $|\eta - \xi_0| \leqq t'' - \delta$ it follows that the distances $\varrho(x, \partial D)$ and $\varrho(y, \partial D)$ are bounded from below by some number $2\delta' > 0$ what, in particular, implies $R > \delta'$. Furthermore,

$$\int_{\varepsilon}^{R} \frac{\mathrm{d}r_{yz}}{r_{yz} r_{xz}^{m-1}} = \int_{\varepsilon}^{\infty} \frac{\mathrm{d}r_{yz}}{r_{yz} r_{xz}^{m-1}} - \int_{R}^{\infty} \frac{\mathrm{d}r_{yz}}{r_{xz} r_{xz}^{m-1}}. \tag{17}$$

Because of $R > \delta'$ the second integral in (17) is bounded and, therefore, leads to a bounded summand in the expression for $G(x, y)$. In the first integral of (17) we substitute $r_{yz} = r\tau$, $r = |x - y|$ with γ denoting the angle between the vectors $x - y$ and $z - y$. Then

$$\int_{\varepsilon}^{\infty} \frac{\mathrm{d}r_{yz}}{r_{yz} r_{xz}^{m-1}} = \frac{1}{r^{m-1}} \int_{\varepsilon/r}^{\infty} \frac{\mathrm{d}\tau}{\tau(1 - 2\tau\cos\gamma + \tau^2)^{(m-1)/2}}$$

$$= \frac{1}{r^{m-1}} \left[\int_{\varepsilon/r}^{1/2} \frac{\mathrm{d}\tau}{\tau(1 - 2\tau\cos\gamma + \tau^2)^{(m-1)/2}} + \int_{1/2}^{\infty} \frac{\mathrm{d}\tau}{\tau(1 - 2\tau\cos\gamma + \tau^2)^{(m-1)/2}} \right]. \tag{18}$$

The first integral in the right-hand side of (18) is, obviously, equal to $\ln(r/\varepsilon) + B(r, \gamma) + O(\varepsilon)$ with a bounded function $B(r, \gamma)$, $O(\varepsilon)$ stands for a function that tends to zero uniformly with respect to γ as $\varepsilon \to 0$ for $r \neq 0$. Since $F(x, y, \theta_{xy})$ is the characteristic of a convergent singular integral, we have necessarily

$$\int_S F(x, y, \theta)\,\mathrm{d}S = 0$$

and the integral under consideration leads in $G(x, y)$ to a summand of order $O(r^{1-m})$.

Now, let us consider the second integral in the right-hand side of (18). In case if $m = 2$ this integral is estimated by $O(\ln|\sin\gamma/2|)$, and the third integral in (16) admits to an estimate $O(r^{-1})$, i.e. it is weakly singular. Therefore, let $m > 2$. In the denominator of the integrand we replace the factor τ by its smallest value $1/2$ and then substitute $\tau = \cos\gamma + \sigma\sin\gamma$. Thence,

$$\int_{1/2}^{\infty} \frac{\mathrm{d}\tau}{\tau(1 - 2\tau\cos\gamma + \tau^2)^{(m-1)/2}} < \frac{K}{\sin^{m-2}\gamma}; \qquad K = 2 \int_{-\infty}^{\infty} \frac{\mathrm{d}\sigma}{(\sigma^2 + 1)^{(m-1)/2}}.$$

Thus, the second integral in (17) leads in $G(x, y)$ to a summand which is estimated by

$$\frac{K|\omega(x, y)|}{r^{m-1}} \int_S |F(x, y, \theta_{yz})| \sin^{2-m}\gamma\,\mathrm{d}S \leqq \frac{C}{r^{m-1}} \int_S \sin^{2-m}\gamma\,\mathrm{d}S.$$

We choose a coordinate system with origin at the point y and its first axis directed along the vector $x - y$. The angular coordinates of the point θ_{yz} are denoted by $\vartheta_1, \vartheta_2, \ldots, \ldots, \vartheta_{m-1}$, where $\vartheta_1 = \gamma$ such that

$$\int_S \sin^{2-m}\gamma\,\mathrm{d}S = \int_0^{\pi} \mathrm{d}\gamma \int_0^{\pi} \sin^{m-3}\vartheta_2\,\mathrm{d}\vartheta_2 \ldots \int_0^{\pi} \sin\vartheta_{m-2}\,\mathrm{d}\vartheta_{m-2} \int_{-\pi}^{\pi} \mathrm{d}\vartheta_{m-1},$$

which gives a finite constant. Thus, finally the second integral in (18) leads to a weakly singular summand for $G(x, y)$. The assertion of the present subsection is proved completely. ∎

§ 7. Polysingular integral operators

7.1. Let $\alpha = (\alpha_1, \alpha_2, \ldots, \alpha_m)$ be a multi-index of dimension m all the components of which are either zero or one, and $|\alpha| = k$. By $j_1 < j_2 < \cdots < j_k$ we denote those indices for which $\alpha_{j_i} = 1$. For a vector $x = (x_1, x_2, \ldots, x_m) \in \mathbb{R}^m$ we denote by x' the k-dimensional vector $x' = (x_{j_1}, x_{j_2}, \ldots, x_{j_k})$ and by $\bar{x}$ the corresponding $(m - k)$-dimensional vector of the remaining components of the vector x in natural order. Frequently, we shall write $x = (\bar{x}, x')$.

Let us introduce the *polysingular operator*

$$(S_\alpha u)(x) = \frac{1}{(\pi \mathrm{i})^k} \int_{\mathbb{R}^k} u(\bar{x}, y) \prod_{\nu=1}^{k} (y_\nu - x_{j_\nu})^{-1} \, dy \tag{1}$$

if $k > 0$. In case if $k = 0$, i.e. $\alpha = (0, 0, \ldots, 0)$ we put by definition $S_\alpha = I$, the identity. The integral in (1) is to be understood as a singular one in the following sense,

$$\int_{\mathbb{R}^k} u(\bar{x}, y) \prod_{\nu=1}^{k} (y_\nu - x_{j_\nu})^{-1} \, dy$$

$$= \lim_{\varepsilon \to 0} \int_{|y_1 - x_{j_1}| > \varepsilon} \cdots \int_{|y_k - x_{j_k}| > \varepsilon} u(\bar{x}, y) \prod_{\nu=1}^{k} (y_\nu - x_{j_\nu})^{-1} \, dy .$$

The polysingular operator S_α, $|\alpha| > 0$, is a convolution operator with the kernel

$$K(x) = (-1)^k (\pi \mathrm{i})^{-k} \prod_{\nu=1}^{k} x_{j_\nu}^{-1} \qquad (k > 0). \tag{2}$$

As you can see from the results of § 6.2, Chap. I, the Fourier transform of the kernel (2) is

$$(FK)(\xi) = \prod_{\nu=1}^{k} \operatorname{sign} \xi_{j_\nu}, \qquad \forall\, \xi \in \mathbb{R}^m \setminus \{0\}, \tag{3}$$

such that the symbol of the operator S_α can be defined by the right-hand side of (3),

$$\operatorname{Smb} S_\alpha = \prod_{\nu=1}^{k} \operatorname{sign} \xi_{j_\nu} . \tag{4_1}$$

The symbol of this operator S_α can be also defined as

$$\operatorname{Smb} S_\alpha = \theta_{j_1} \theta_{j_2} \cdots \theta_{j_k} \tag{4_2}$$

where the θ_l are independent variables assuming the values ± 1 only.

To a more general polysingular operator

$$A = \sum_{|\alpha|=0}^{m} a_\alpha S_\alpha, \qquad a_\alpha = \text{const}, \tag{5}$$

we may assign the symbol

$$\operatorname{Smb} A = a_0 + \sum_{|\alpha|=1}^{m} a_\alpha \theta_{j_1} \theta_{j_2} \cdots \theta_{j_k}, \qquad k = |\alpha| . \tag{6}$$

Note that the operator (5) is determined by its symbol uniquely.

Polysingular integrals arise in solving boundary value problems from the theory of functions of several complex variables, and these problems, in turn, are connected with certain problems in quantum mechanics (see e.g. VLADIMIROV [1], § 29).

7.2. Theorem 7.1. *The polysingular operator* (5) *is bounded in* $\mathbf{L}_p(\mathbb{R}^m)$, $1 < p < \infty$.

We shall give here two proofs of Theorem 7.1. One of them is based on a theorem that the one-dimensional singular operator is bounded in $\mathbf{L}_p(\mathbb{R})$ (cf. § 2, Chap. II). Obviously, such a one-dimensional singular operator can be considered as a special case of a polysingular operator if $m = 1$. The other proof uses a theorem on multipliers of Fourier integrals.

7.3. Proof 1. We have to consider operators S_α, $|\alpha| = k \geq 1$, only. Let $k = 1$. The Cauchy operator is bounded in $\mathbf{L}_p(\mathbb{R})$, $1 < p < \infty$, hence

$$\int_{\mathbb{R}} \left| \int_{\mathbb{R}} \frac{u(\bar{x}, y)}{y_1 - x_{j_1}} \, dy_1 \right|^p dx_{j_1} \leq A_p^p \int_{\mathbb{R}} |u(\bar{x}, x_{j_1})|^p \, dx_{j_1},$$

where A_p stands for the norm of the Cauchy operator in $\mathbf{L}_p(\mathbb{R})$. Integrating this inequality with respect to $\bar{x}$ we obtain

$$\int_{\mathbb{R}^m} |S_\alpha u|^p \, dx \leq A_p^p \int_{\mathbb{R}^m} |u(x)|^p \, dx, \qquad |\alpha| = 1,$$

such that the theorem is proved if $|\alpha| = 1$, where $\|S_\alpha\|_{\mathbf{L}_p(\mathbb{R}^m)} \leq A_p$.

Suppose that Theorem 7.1 is proved for $|\alpha| \leq k - 1$ where $\|S_\alpha\|_{\mathbf{L}_p(\mathbb{R}^m)} \leq A_p^{(\alpha)}$. Then we show it to be true also for $|\alpha| = k$, and $\|S_\alpha\|_{\mathbf{L}_p(\mathbb{R}^m)} \leq A_p^k$, $k = |\alpha|$. Let the multi-index α have 1's at places $j_1, j_2, \ldots, j_k$. By β we denote a multi-index of dimension m with 1's at places $j_2, \ldots, j_k$ whereas the other components are zero so that $|\beta| = k - 1$. Put $\gamma = \alpha - \beta$ then, obviously, $S_\alpha = S_\gamma S_\beta$, and since the Cauchy operator is bounded we have

$$\int_{\mathbb{R}} |(S_\alpha u)(x)|^p \, dx_{j_1} \leq A_p^p \int_{\mathbb{R}} |(S_\beta u)(x)|^p \, dx_{j_1}.$$

Integrating over the remaining coordinates we obtain

$$\int_{\mathbb{R}^m} |S_\alpha u|^p \, dx \leq A_p^p \int_{\mathbb{R}^m} |S_\beta u|^p \, dx \leq A_p^{kp} \int_{\mathbb{R}^m} |u(x)|^p \, dx. \quad \blacksquare$$

7.4. Proof 2. The symbol $\Phi(\xi)$, $\xi \in \mathbb{R}^m \setminus \{0\}$, of the operator (5) can be written as

$$\Phi(\xi) = a_0 + \sum_{|\alpha|=1}^{m} a_\alpha \prod_{\nu=1}^{|\alpha|} \operatorname{sign} \xi_{j_\nu}, \tag{7}$$

and $A = F^{-1} \Phi(\xi) F$. Let us consider the simpler case $p = 2$ separately. The function $\Phi(\xi)$ is, obviously, bounded and the Fourier transformation is a unitary operator in $\mathbf{L}_2(\mathbb{R}^m)$. Hence, starting from $FAu = \Phi(\xi) \cdot Fu$ we get

$$\|FAu\| \leq \max |\Phi(\xi)| \cdot \|Fu\|,$$

and taking into account $\|FAu\| = \|Au\|$, $\|Fu\| = \|u\|$ we arrive at

$$\|A\|_{\mathbf{L}_2(\mathbb{R}^m)} \leq \max |\Phi(\xi)|. \tag{8}$$

The proof for the case $p \neq 2$ is more difficult. Mikhlin [32, 33] proved the following theorem on multipliers of Fourier integrals.

Theorem 7.2. *Let us be given an operator A of the form*

$$A = F_{\xi \to x}^{-1} \Phi(\xi) F_{x \to \xi} \tag{9}$$

where F denotes the Fourier transform in $\mathbb{R}^m$, and $\Phi(\xi)$, the so-called multiplier is a function subject to the following conditions.

1. *it is continuous for any* $\xi \neq 0$;
2. *the derivative*

$$D_\xi^\alpha \Phi, \quad \alpha = (1, 1, \ldots, 1), \tag{10}$$

exists at any point $\xi \neq 0$, *and all lower derivatives are continuous for* $\xi \neq 0$;

3. *there is an estimate*

$$|\xi|^k \left| \frac{\partial^k \Phi}{\partial \xi_{i_1} \partial \xi_{i_2} \ldots \partial \xi_{i_k}} \right| \leqq M = \text{const}, \tag{11}$$

$k = 0, 1, \ldots, m;\ 1 \leqq i_1 < i_2 < \ldots < i_k \leqq m$.

Then the operator (9) *is bounded in* $\mathbf{L}_p(\mathbb{R}^m)$ *for any* $p \in (1, \infty)$, *and* $\|A\| \leqq C_{p,m} M$ *where* $C_{p,m}$ *depends only on* p *and* m.

After the papers Mikhlin [32, 33] were published several colleagues called the author's attention to the fact that his proof demonstrates, in fact, a more general theorem which will be given below. This theorem and its proof is published together with further results in Lizorkin [1].

Theorem 7.3. *Let* $\xi = (\xi_1, \xi_2, \ldots, \xi_m)$ *be an arbitrary vector in* $\mathbb{R}^m$, *and suppose that in* (9) *the multiplier* $\Phi(\xi)$ *is continuous everywhere except at the coordinate planes* $\xi_1 = 0$, $\xi_2 = 0, \ldots, \xi_m = 0$. *Furthermore, assume that the derivative* (10) *exists at every point outside of these coordinate planes and that the lower derivatives are continuous there, finally let*

$$|\xi_{j_1} \xi_{j_2} \ldots \xi_{j_k}| \left| \frac{\partial^k \Phi}{\partial \xi_{j_1} \partial \xi_{j_2} \ldots \partial \xi_{j_k}} \right| \leqq M = \text{const}, \tag{12}$$

$k = 0, 1, 2, \ldots, m;\ 1 \leqq j_1 < j_2 < \ldots < j_k \leqq m$. *Then the operator* (9) *is bounded in* $\mathbf{L}_p(\mathbb{R}^m)$, $1 < p < \infty$, *and* $\|A\| \leqq \gamma_{m,p} M$ *where* $\gamma_{m,p}$ *depends only on* p *and* m.

The operator (9) is, obviously, defined on $\mathbf{C}_0^\infty(\mathbb{R}^m)$ which is a dense set in $\mathbf{L}_p(\mathbb{R}^m)$. If this operator satisfies the conditions of Theorem 7.2 or Theorem 7.3 then it is bounded in $\mathbf{L}_p(\mathbb{R}^m)$ and extends continuously to the whole space $\mathbf{L}_p(\mathbb{R}^m)$, $1 < p < \infty$.

Now, we prove the polysingular operator (5) to be bounded in $\mathbf{L}_p(\mathbb{R}^m)$, $1 < p < \infty$. It suffices to show that the symbol (7) satisfies the conditions (12). This symbol is piecewise constant: it is equal to some constant in every domain bounded by the coordinate planes. In each such domain the symbol (7) has all derivatives vanishing identically. But then the symbol (7) does satisfy the conditions (12) where M can be taken as the maximum modulus of the symbol. ∎

7.5. Now let us consider a polysingular integral equation

$$Au = f, \ f \in \mathbf{L}_p(\mathbb{R}^m), \ 1 < p < \infty, \tag{13}$$

with A being the operator (5). Applying Fourier transform to this equation we obtain the new equation $\operatorname{Smb} A \cdot Fu = Ff$, and (13) has the unique solution

$$u = A^{-1} f. \tag{14}$$

If $\operatorname{Smb} A \neq 0$ then as you will easily see, $(\operatorname{Smb} A)^{-1}$ can be given in the form (6) so that it can be considered as a symbol of a certain polysingular operator, and this operator is, obviously, the operator A^{-1}. Thus, if the symbol of a polysingular operator A does not vanish then its inverse A^{-1} exists and is also a polysingular operator and bounded in $\mathbf{L}_p(\mathbb{R}^m)$.

Let us consider the equation

$$Au + Tu = f, \ f \in \mathbf{L}_p(\mathbb{R}^m), \ 1 < p < \infty, \tag{15}$$

where A is a polysingular and T a completely continuous operator in $\mathbf{L}_p(\mathbb{R}^m)$. If the symbol of the operator A does not vanish then this equation becomes equivalent to the Fredholm equation

$$u + A^{-1}\,Tu = A^{-1}f. \tag{16}$$

7.6. Let $a_\alpha(x)$ be measurable and bounded functions in $\mathbb{R}^m$. Then the operator

$$(Bu)\,(x) = \sum_{|\alpha|=0}^{m} a_\alpha(x)\,(S_\alpha u)\,(x) \tag{17}$$

is bounded in $\mathbf{L}_p(\mathbb{R}^m)$, $1 < p < \infty$. An operator (17) and likewise the adjoint to it,

$$(B^*u)\,(x) = \sum_{|\alpha|=0}^{m} S_\alpha(\bar{a}_\alpha u)\,(x) \tag{18}$$

are also called polysingular operators.

7.7. For one-dimensional singular operators

$$a(x_1)\,u(x_1) + \frac{b(x_1)}{\pi i}\int_{\mathbb{R}}\frac{u(y)}{y - x_1}\,dy + (Tu)\,(x_1) = f(x_1), \tag{19}$$

with $x_1 \in \mathbb{R}$ and T being completely continuous in $\mathbf{L}_p(\mathbb{R}^m)$, there is a theorem (see § 3, Chap. III) according to which the operator (19) is a Fredholm operator provided that $a^2(x_1) - b^2(x_1)$ never vanishes. That theorem does not carry over to polysingular equations (17). In order to see this let us consider the operator

$$a(x_1)\,u(x_1, x_2) + \frac{b(x_1)}{\pi i}\int_{\mathbb{R}}\frac{u(y, x_2)}{y - x_1}\,dy \tag{20}$$

in $\mathbf{L}_p(\mathbb{R}^2)$ with coefficients a and b as in (19). Suppose that $a^2 - b^2$ never vanishes so that (19) is a Fredholm operator in $\mathbf{L}_p(\mathbb{R})$, and that the index of that operator is positive. Then there is a function $u_0 \in \mathbf{L}_p(\mathbb{R})$ not identically zero which satisfies the equation

$$a(x_1)\,u_0(x_1) + \frac{b(x_1)}{\pi i}\int_{\mathbb{R}}\frac{u_0(y)}{y - x_1}\,dy = 0, \forall\, x_1 \in \mathbb{R}. \tag{21}$$

Take an arbitrary sequence $v_j \in \mathbf{L}_p(\mathbb{R})$, $j = 1, 2, \ldots$, such that the functions $v_j(x_2)$, $1 \leqq j \leqq n$, are linearly independent for any n. Put $u_j(x_1, x_2) := u_0(x_1)\,v_j(x_2)$, then each of the functions $u_j(x_1, x_2)$ is annihilated by the operator (20). Hence, this operator has a countable number of eigenfunctions corresponding to the zero value of the spectral parameter. Such an operator is not a Fredholm one: In the case considered the operator (20) is a Φ_--operator.

Remark. Criterions for the Fredholm property of polysingular operators were given by Simonenko [8] (for $\mathbf{L}_2$) and by Pilidi [1] (for $\mathbf{L}_p$, $1 < p < \infty$).

§ 8. Pseudo-differential operators

The concept of *pseudo-differential operators* (*PDO*) arises as a natural generalization of the notion of singular integral operators. The fundamentals of a PDO theory are given in Kohn and Nirenberg [1], Hörmander [2], and Seeley [3]. A rather exhaustive treatment of PDO theory is provided by the monographs Shubin [1] and Treves [1].

Elements of this theory are given briefly and at the same time distinctly in the book by Agranovich and Vishik [1]. It is this book which we used widely when compiling the present section, and there you can find various simple facts concerning PDO.

8.1. Let $a(x, \xi)$, $x \in \mathbb{R}^m$, $\xi \in \mathbb{R}^m \setminus \{0\}$, be a positively homogeneous function with respect to ξ of degree $\sigma \geqq 0$,

$$a(x, t\xi) = t^\sigma a(x, \xi), \quad t \geqq 0; \tag{1}$$

assume that the limit $\lim_{|x| \to \infty} a(x, \xi) = a(\infty, \xi)$ exists where the difference

$$a'(x, \xi) = a(x, \xi) - a(\infty, \xi) \tag{2}$$

is supposed to belong as a function of x to the space $S(\mathbb{R}^m)$ uniformly with respect to ξ. Here, we denote by $S(\mathbb{R}^m)$ the space of functions which decrease together with all its derivatives with respect to x as $|x| \to \infty$ faster than any power of x.

As usually, we denote by $F_{x\to\xi}\, u$ or $\hat{u}(\xi)$ the Fourier transform of a function $u(x)$.

Definition 8.1. An operator

$$\begin{aligned} Au &= F^{-1}_{\xi\to x}\, a(x, \xi)\, F_{x\to\xi}\, u \\ &= (2\pi)^{-m/2} \int_{\mathbb{R}^m} e^{i(x,\xi)}\, a(x, \xi)\, \hat{u}(\xi)\, d\xi \end{aligned} \tag{3}$$

defined for all functions $u \in S(\mathbb{R}^m)$ is called *homogeneous pseudo-differential operator* with the *symbol* $a(x, \xi)$. The number σ mentioned in the characterization of the symbol is called the *degree* of the PDO (3).

For the sake of simplicity in notations we shall omit henceforth in this section the sign $\mathbb{R}^m$ in integrals taken over that space.

It is not difficult to obtain the following formula for the PDO (3),

$$(\widehat{Au})(\xi) = a(\infty, \xi)\, \hat{u}(\xi) + (2\pi)^{-m/2} \int \hat{a}'(\xi - \eta, \eta)\, \hat{u}(\eta)\, d\eta \tag{4}$$

where $\hat{a}'(\xi, \eta) = F_{x\to\xi}\, a'(x, \eta)$.

We note two simple special cases of a PDO. 1. for $\sigma = 0$ the PDO (3) is a singular integral operator with the symbol $a(x, \xi)$; 2. if $a(x, \xi)$ is a polynomial of degree σ with respect to ξ,

$$a(x, \xi) = \sum_{|\alpha| = \sigma} a_\alpha(x)\, \xi^\alpha,$$

then the PDO becomes a differential operator,

$$(Au)(x) = \sum_{|\alpha| = \sigma} a_\alpha(x)\, D^\alpha u(x).$$

8.2. In the PDO theory the spaces $\mathbf{W}_2^{(l)}(\mathbb{R}^m)$ play an important role, they will be denoted in the sequel by $\mathbf{H}_l$, $l > 0$. A norm in $\mathbf{H}_l$ is provided by the formula

$$\|u\|_l^2 := \int (1 + |\xi|^2)^l\, |\hat{u}(\xi)|^2\, d\xi; \tag{5}$$

if l is a positive integer then we may put

$$\|u\|_l^2 := \int \sum_{|\alpha|=0}^{l} |D^\alpha u|^2\, dx, \tag{6}$$

the two norms (5) and (6) are then equivalent.

If some linear operator A acts from $\mathbf{H}_l$ into $\mathbf{H}_{l-\sigma}$ and is bounded in that pair of spaces ($l > 0$ arbitrary) then it is said to have the *order* σ. The infimum of all these orders of the operator A is called its *essential order*.

The definition of the order and of the essential order of an operator does not require the order to be non-negative, one can consider operators of negative order, too. Of particular interest in the PDO theory are operators the essential order of which is $-\infty$. In some sense they play a similar role as the completely continuous operators in the theory of singular integral operators.

Theorem 8.1 (Boundedness Theorem). *A homogeneous PDO of degree $\sigma \geqq 0$ has the order σ.*

Proof. We make use of (4). Put

$$(\widehat{A_1 u})(\xi) = a(\infty, \xi)\, \hat{u}(\xi),$$
$$(\widehat{A_2 u})(\xi) = (2\pi)^{-m/2} \int \hat{a}'(\xi - \eta, \eta)\, \hat{u}(\eta)\, d\eta .$$

From (1) one can easily derive the estimate $|a(\infty, \xi)| \leqq C\, |\xi|^\sigma$, $C = \text{const}$. Hence, by (5)

$$\|A_1 u\|_{l-\sigma}^2 = \int (1 + |\xi|^2)^{l-\sigma}\, |a(\infty, \xi)|^2\, |\hat{u}(\xi)|^2\, d\xi$$
$$\leqq C^2 \int (1 + |\xi|^2)^l\, |\hat{u}(\xi)|^2\, d\xi = C^2\, \|u\|_l^2 . \tag{7}$$

Due to the assumptions made in § 8.1 we have

$$|a'(\xi - \eta, \eta)| \leqq C_p \frac{|\eta|^\sigma}{(1 + |\xi - \eta|^2)^p},$$

where p is an arbitrarily large number, and the constant C_p depends on p only. Using this estimate we get

$$(1 + |\xi|^2)^{l-\sigma}\, |(\widehat{A_2 u})(\xi)|^2$$
$$\leqq C \left(\int \frac{(1 + |\xi|^2)^{(l-\sigma)/2}}{(1 + |\eta|^2)^{(l-\sigma)/2}} \frac{(1 + |\eta|^2)^{l/2}}{(1 + |\xi - \eta|^2)^p} |\hat{u}(\eta)|\, d\eta \right)^2 . \tag{8}$$

From the inequality $|\xi|^2 \leqq 2\, |\xi - \eta|^2 + 2\, |\eta|^2$ we get $1 + |\xi|^2 \leqq 2(1 + |\xi - \eta|^2) \times (1 + |\eta|^2)$ and, therefore, for any $k > 0$,

$$\left(\frac{1 + |\xi|^2}{1 + |\eta|^2} \right)^k \leqq 2^k (1 + |\xi - \eta|^2)^k .$$

Put in (8) $p = (m + 1)/2 + (l - \sigma)/2$ then

$$(1 + |\xi|^2)^{l-\sigma}\, |(\widehat{A_2 u})(\xi)|^2$$
$$\leqq C \left(\int (1 + |\xi - \eta|^2)^{-(m+1)/2} (1 + |\eta|^2)^{l/2}\, |\hat{u}(\eta)|\, d\eta \right)^2 . \tag{9}$$

Let us introduce the abbreviations

$$(1 + |\xi - \eta|^2)^{-(m+1)/2} =: \varphi(\xi - \eta),$$
$$(1 + |\eta|^2)^{l/2}\, |\hat{u}(\eta)| =: v(\eta) .$$

We estimate the integral in the right-hand side of (9) by means of Schwarz's inequality,

$$\left| \int \varphi(\xi - \eta)\, v(\eta)\, d\eta \right|^2 \leqq \int \varphi(\xi - \eta)\, d\eta \cdot \int \varphi(\xi - \eta)\, v^2(\eta)\, d\eta .$$

Integrating this with respect to ξ we find

$$\left\| \int \varphi(\xi - \eta)\, v(\eta)\, d\eta \right\|_{L_2(R^m)}^2 \leqq C\, \|v\|_{L_2(R^m)}^2,$$

and this is equivalent to

$$\|Au\|_{l-\sigma}^2 \leqq C\, \|u\|_l^2 : \quad \blacksquare$$

8.3. Now let us define the PDO of negative degree. Assume that $a(x, \xi)$ satisfies the conditions as given in § 8.1 with one exception: the exponent σ is assumed to be negative. Then this function becomes infinity at $\xi = 0$, and (3) makes no longer sense. Therefore, we define a PDO of negative degree as follows. We introduce a non-negative function $\zeta(\xi) \in \mathbf{C}^\infty(\mathbb{R}^m)$ which is zero in a certain finite vicinity of zero and 1 outside of a larger but also finite vicinity. Put

$$(A_\zeta u)(x) = (2\pi)^{-m/2} \int e^{i(x,\xi)} \zeta(\xi)\, a(x, \xi)\, \hat{u}(\xi)\, d\xi. \tag{10}$$

The operator A_ζ is taken as the definition of a *homogeneous PDO of negative degree* σ with the *symbol* $a(x, \xi)$.

Unlike in case $\sigma > 0$, the operator A_ζ is determined by its symbol not uniquely. However, it is easy to see that the difference of two operators A_{ζ_1} and A_{ζ_2} with one and the same symbol is a PDO of the essential order $-\infty$. This assertion is true also if $\sigma > 0$, moreover, in that case the operator $A - A_\zeta$ is also of the essential order $-\infty$. For that reason we may take the operator A_ζ as a homogeneous PDO with the symbol $a(x, \xi)$ also if $\sigma > 0$. Theorem 8.1 on the boundedness of PDO remains to hold also for PDO of negative degree.

8.4. Theorem 8.2. (PDO multiplication theorem). *Let A and B be homogeneous PDO with the symbols $a(x, \xi)$ and $b(x, \xi)$, resp. and of degree σ_a and σ_b, resp. Then for any non-negative integer ϱ there is the representation*

$$AB = C_0 + C_1 + \dots + C_{\varrho-1} + T_\varrho \tag{11}$$

where C_i is a PDO of degree $\sigma_a + \sigma_b - i$ with the symbol

$$c_i(x, \xi) = \sum_{|\alpha|=i} \frac{1}{\alpha!} \partial^\alpha a(x, \xi)\, D^\alpha b(x, \xi), \tag{12}$$

and the operator T_ϱ has the order $\sigma_a + \sigma_b - \varrho$.

Here, ∂^α means differentiation with respect to ξ, and D^α with respect to x.

We shall omit the proof to that theorem which is rather cumbersome. We mention only that the multiplication rule for differential operators as given in § 6, Chap. I, is a special case of the theorem.

8.5. Let A be a homogeneous PDO of degree σ with the symbol $a(x, \xi)$,

$$(Au)(x) = (2\pi)^{-m/2} \int e^{i(x,\xi)} \zeta(\xi)\, a(x, \xi)\, \hat{u}(\xi)\, d\xi,$$

where $\zeta(\xi)$ is the function introduced in § 8.3. The adjoint A^* to it with respect to the duality pairing in $\mathbf{L}_p(\mathbb{R}^m)$ is given by

$$(\widehat{A^*v})(\xi) = (2\pi)^{-m/2} \int e^{-i(x,\xi)} \zeta(\xi)\, a^*(x, \xi)\, v(x)\, dx \tag{13}$$

where the function $a^*(x, \xi)$ is defined as follows. If $a(x, \xi)$ is a scalar function then $a^*(x, \xi) := \overline{a(x, \xi)}$, if $a(x, \xi)$ is a matrix then $a^*(x, \xi)$ is the corresponding adjoint matrix.

Suppose that the symbol $a = a(\xi)$ does not depend on x, and let

$$(Bu)(x) = (2\pi)^{-m/2} \int e^{i(x,\xi)} \zeta(\xi)\, \overline{a(\xi)}\, \hat{u}(\xi)\, d\xi.$$

Then, as you can easily see, $B = A^*$. In the general case, i.e. if the symbol does depend on x, there is the following

Theorem 8.3. *Let A be a homogeneous PDO of degree σ with the symbol $a(x, \xi)$. Then for any positive integer ϱ the adjoint A^* admits of the representation*

$$A^* = \sum_{i=1}^{\varrho-1} B_i + T_\varrho \tag{14}$$

where B_i is a PDO with the symbol

$$\sum_{|\alpha|=i} \frac{1}{\alpha!} D^\alpha \partial^\alpha a^*(x, \xi) \tag{15}$$

and T_ϱ is an operator of order $\sigma - \varrho$.

8.6. Now let us give the definition of a *general* PDO in $\mathbb{R}^m$. Let $\{\sigma_k\}$ be a strictly decreasing finite or infinite sequence of real numbers. If it is infinite then we shall assume $\sigma_k \to -\infty$. Let A_k be a homogeneous PDO of degree σ_k and $a_k(x, \xi)$ its symbol. Assume to be given an operator A on the function space $S(\mathbb{R}^m)$ such that the difference

$$A - \sum_{k=0}^{N} A_k$$

is of order less than σ_N for any positive integer N. We shall say that A is a PDO with the *asymptotic expansion*

$$A \sim \sum_{k=0}^{\infty} A_k. \tag{16}$$

The *symbol* $a(x, \xi)$ of the operator A is taken to be the sequence $\{a_k(x, \xi)\}$ or as it is frequently done, the formal series

$$a(x, \xi) = \sum_{k=0}^{\infty} a_k(x, \xi); \tag{17}$$

the summand $a_0(x, \xi)$ of the formal series (17) is called the *major part* of the symbol.

From the definition given here it follows that a general PDO is determined by its symbol up to an additive term the essential order of which is $-\infty$. Hence, an operator with the essential order $-\infty$ can be considered as a special case of a general PDO the symbol of which is zero.

Theorem 8.4. *Let $\sigma_0, \sigma_1, \ldots$ be a sequence of strictly decreasing real numbers which tends to $-\infty$. Assume that the general term in the formal series* (17) *is a homogeneous function in ξ of degree σ_k satisfying the conditions of § 8.1. Then there exists a PDO with the symbol* (17).

Proof (Hörmander [2]). Let $\varphi(\xi) \in C^\infty$ be a function such that $\varphi(\xi) = 0$, $|\xi| < 1/2$, and $\varphi(\xi) = 1$, $|\xi| \geqq 1$, and t_j be a sequence of numbers rapidly converging to ∞ (this requirement will be specified later). Define a function

$$\tilde{a}(x, \xi) = \sum_{j=0}^{\infty} \varphi\left(\frac{\xi}{t_j}\right) a_j(x, \xi); \tag{18}$$

such a definition makes sense because the sum in (18) contains for any ξ only a finite number of non-zero terms. The operator A can be defined as

$$(Au)(x) = (2\pi)^{-m/2} \int e^{i(x,\xi)} \tilde{a}(x, \xi)\, \hat{u}(\xi)\, d\xi. \tag{19}$$

If we put

$$(A_j u)(x) = (2\pi)^{-m/2} \int e^{(ix,\xi)} \varphi\left(\frac{\xi}{t_j}\right) a_j(x, \xi)\, \hat{u}(\xi)\, d\xi \tag{20}$$

then the asymptotic expansion (16) becomes an equality,

$$A = \sum_{j=1}^{\infty} A_j.$$

The numbers t_j are choosen in the following way. Put $t_0 = 1$ then the following t_j are selected such that for any x, ξ and any multi-indices α, β with $|\alpha| > 0$, $|\alpha| + |\beta| + k \leqq j$, the inequality

$$\left|\partial^\beta \varphi\left(\frac{\xi}{t_j}\right) a_j(x, \xi)\right| + (1 + |x|^k) \left| D^\alpha \partial^\beta \varphi\left(\frac{\xi}{t_j}\right) a_j(x, \xi)\right|$$
$$\leqq 2^{-j} (1 + |\xi|)^{\sigma_j - 1 - |\beta|} \tag{21}$$

is fulfilled. It is easy to show that the t_j's can be chosen that way.

Now we choose a sufficiently large positive integer N and put

$$\tilde{a}(x, \xi) = \sum_{j=0}^{N+1} \varphi\left(\frac{\xi}{t_j}\right) a_j(x, \xi) + \tau_{N+1}.$$

Obviously, the operator A_j (eq. (20)) is of order σ_j. From (21) the inequality

$$|\partial^\beta \tau_{N+1}(x, \xi) + (1 + |x|)^k \, |D^\alpha \partial^\beta \tau_{N+1}(x, \xi)| \leqq (1 + |\xi|)^{\sigma_{N+1} - |\beta|}$$

follows provided that $|\alpha| > 0$ and $|\alpha| + |\beta| + k \leqq N + 1$. Therefrom it can be deduced that the PDO T_{N+1} with the symbol $\tau_{N+1}(x, \xi)$ is subject to the estimate

$$\|T_{N+1} u\|_{l - \sigma_{N+1}} \leqq C \, \|u\|_l$$

where the subscript l varies in some interval the endpoints of which tend to $-\infty$ and ∞ as $N \to \infty$. Hence, the operator

$$A - \sum_{j=0}^{k} A_j$$

has the order σ_{k+1}, and it is also clear that the operator A has the order σ_0. ∎

If A and B are PDO the symbols of which are the formal series

$$a(x, \xi) = \sum_{j=0}^{\infty} a_j(x, \xi), \quad b(x, \xi) = \sum_{j=0}^{\infty} b_j(x, \xi)$$

then it is easy to see that AB is a PDO with the symbol $c(x, \xi)$ which can be represented as a formal series

$$c(x, \xi) = \sum_{|\alpha|=0}^{\infty} \frac{1}{\alpha!} \partial^\alpha a(x, \xi) \, D^\alpha b(x, \xi). \tag{22}$$

The equality in (22) is to be understood in that way that terms of one and the same power in the right and in the left-hand side are equal. In the same sense the following formula is true which defines the symbol $d(x, \xi)$ of the adjoint PDO A^* to A,

$$d(x, \xi) = \sum_{|\alpha|=0}^{\infty} \frac{1}{\alpha!} D^\alpha \partial^\alpha a^*(x, \xi). \tag{23}$$

Equation (22) can be considered as a multiplication rule for formal series (16) since such a definition preserves the main properties of a multiplication, namely to be associative and distributive. Furthermore, the product of PDO's corresponds to the product of their symbols as we demonstrated above; the analogous correspondence of sums is obvious. Consequently, the definition of a symbol to a PDO given here is correct.

8.7. A PDO A with the symbol (16) is called *elliptic* if the major part of the symbol does not vanish for any x and ξ, $x \in \mathbb{R}^m$, $\xi \in \mathbb{R}^m \setminus \{0\}$. In the case if A is a matrix operator this condition is to be replaced by Det $a_0(x, \xi) \neq 0$.

Theorem 8.5 (Hörmander [2]). *For an elliptic PDO A there is a PDO B such that the differences $AB - I$ and $BA - I$ (I being the identity) are operators of the essential order $-\infty$.*

Proof. Suppose that the operator B exists, and its symbol is

$$b(x, \xi) = \sum_{k=0}^{\infty} b_k(x, \xi).$$

The symbol of the product AB is 1. Putting in (22) $c(x, \xi) = 1$ and equating terms with one and the same power left and right we obtain a recurrent system of equations of the following structure. The first equation is

$$a_0(x, \xi)\, b_0(x, \xi) = 1,$$

from which $b_0(x, \xi)$ can be determined. Further, the k-th equation ($k > 1$) contains in the left-hand side a term $a_0(x, \xi) \cdot b_k(x, \xi)$ and, in general, some terms depending on the functions $b_j(x, \xi)$, $j < k$, which have been already determined by means of preceding equations; the right-hand side of the k-th equation is zero. Therefore, the functions $b_k(x, \xi)$ can be determined for any $k \geq 0$ such that the symbol $b(x, \xi)$ can be constructed. Due to Theorem 8.4 there exists a PDO B having the symbol $b(x, \xi)$.

The symbol of the difference $T_1 = AB - I$ is identically zero, but then T_1 is a PDO of essential order $-\infty$.

Analogously a PDO B' is constructed such that $T_2 = B'A - I$ has the essential order $-\infty$. Now,

$$B'AB = B'(I + T_1) = (I + T_2)\, B,$$

hence, $B' - B = T_2 B - B'T_1$. It is easy to see that both the operators B and B' have the order $-\sigma$, and then it is clear that $T = B - B'$ is an operator of essential order $-\infty$, and

$$BA = B'A + TA = I + (T_2 + TA).$$

The operator $T_2 + TA$ has the essential order $-\infty$, and the theorem is completely proved. ∎

8.8. It is possible to develop a theory of PDO acting in spaces of functions which are defined on an inifinitely smooth manifold of finite dimension, cf. Hörmander [2], Seeley [3].

A further generalization of singular integral operators is provided by the so-called *Fourier operators*. The reader is referred to Hörmander [5], Shubin [1], Treves [1].

8.9. Finally, we mention the papers by Plamenevski [3, 4] and Plamenevski and Senichkin [1, 2] where they studied algebras of PDO on manifolds without boundary and having singular "conic" points.

Plamenevski [1, 2] considered, in particular, algebras of PDO with order zero on smooth manifolds the symbols of which may have isolated singularities. For such type PDO he introduced an operator symbol, and he proved the factor algebra of those PDO by the ideal of compact operators to be isomorphic to the algebra of operator symbols. The spectrum of this PDO factor algebra was studied in Plamenevski and Senichkin [1, 2]. Thus, they produced at the same time a certain multi-dimensional analogy to the theory of one-dimensional singular integral operators with piecewise continuous coefficients (cf. Chaps. IV and V).

Chapter XIII
Singular equations on smooth manifolds without boundary

The subject of the present chapter is considered in Mikhlin [6, 7, 8], Gegelia [1–7], Seeley [1, 2], Mikhailova-Gubenko [1–3], and Khvoles [1–3]. The results of Seeley on singular operators in $\mathbf{L}_p$ spaces and those of Khvoles on singular operators in $\mathbf{Lip}_\alpha$ seem to be the most complete ones. The main results of these authors will be given here with some unimportant modifications. Seeley's results will be given in §§ 2, 3, whereas §§ 6, 7 are devoted to the results of Khvoles. The chapter will be opened by a section where we introduce the notion of a manifold and some of its properties.

§ 1. Manifolds

The notion of a manifold and its basic properties are discussed at some length e.g. in the books by Steenrod [1], Sternberg [1], Bakelman, Werner, and Kantor [1]. For the convenience of the reader we recall here some of the fundamentals connected with manifolds.

1.1. Let Γ be a point set in the Euclidean space $\mathbb{R}^{m'}$. Any point $\xi \in \Gamma$ is taken to be a center of a ball of sufficiently small radius (which may depend on ξ). By U_ξ we denote the set of those points of Γ which belong to the corresponding ball. The set Γ is called *manifold of dimension m* iff there is a homeomorphic (i.e. a 1-to-1 and continuous) mapping of the set U_ξ onto a certain bounded open set D_ξ of the m-dimensional Euclidean space $\mathbb{R}^m$ for any $\xi \in \Gamma$. Obviously, $m \leqq m'$. In the present chapter we shall consider only closed manifolds (or compact manifolds without boundary). These are characterized by the following property: There is a finite set $\{\xi^1, \ldots, \xi^N\} \subset \Gamma$ such that the corresponding sets $U_j = U_{\xi^j}$, $j = 1, 2, \ldots, N$, form a covering of the set Γ.

The homeomorphic mapping $D_{\xi^j} = D_j \to U_j$ can be given by a system of equations

$$\xi_i = \omega_{ij}(x) = \omega_{ij}(x_1, \ldots, x_m), \qquad i = 1, 2, \ldots, m'. \tag{1}$$

Let $\lambda \in (0, 1]$, and $s \geqq 0$ be an integer. We shall say $\Gamma \in \mathbf{C}^s$ (or $\Gamma \in \mathbf{C}^{s,\lambda}$) if the corresponding homeomorphisms ω_{ij} belong to the class $\mathbf{C}^s(D_j)$ (or to the class $\mathbf{C}^{s,\lambda}(D_j)$), $j = 1, 2, \ldots, m$; $i = 1, 2, \ldots, m'$. Manifolds of the class $\mathbf{C}^1$ will be called *smooth* ones, and those of the class $\mathbf{C}^{1,\lambda}$ *Lyapunov manifolds*. Instead of (1) we shall write briefly

$$\xi = \omega(x) \quad \text{or} \quad \xi = \omega_j(x), \qquad \xi \in U_j. \tag{1'}$$

Let us give another definition *of a manifold which is essentially equivalent to the* previously given one: A Hausdorff topological space with a countable basis is called *manifold of dimension m* iff any point of that space possesses a neighborhood which is homeomorphic to a certain domain in the Euclidean space $\mathbb{R}^m$.

1.2. Suppose that in (1') the variable x is a smooth function of a real parameter t, $x = x(t)$. Then the equation $\xi = \omega_j(x(t))$ determines a smooth curve on $U_j \subset \Gamma$. The tangent to that curve at the point ξ^0 is given by the equations

$$\frac{\xi_1 - \xi_1^0}{d\xi_1^0} = \frac{\xi_2 - \xi_2^0}{d\xi_2^0} = \dots = \frac{\xi_{m'} - \xi_{m'}^0}{d\xi_{m'}^0}. \tag{2}$$

Let P be a plane in $\mathbb{R}^{m'}$ containing the point ξ^0; its equation can be given in the form

$$\sum_{i=1}^{m'} p_i(\xi_i - \xi_i^0) = 0.$$

The line (2) is contained in that plane if and only if

$$\sum_{i=1}^{m'} p_i \, d\xi_i^0 = 0 \quad \text{or} \quad \sum_{k=1}^{m} dx_k \sum_{i=1}^{m'} p_i \frac{\partial \omega_{ij}(x^0)}{\partial x_k} = 0$$

where x^0 corresponds to the point ξ^0 under the mapping (1'). The tangents (2) passing through one and the same point ξ^0 are determined by the corresponding direction vector $(dx_1, dx_2, \dots, dx_m)$; all these tangents are contained in the plane P if

$$\sum_{i=1}^{m'} p_i \frac{\partial \omega_{ij}(x^0)}{\partial x_k} = 0, \qquad k = 1, 2, \dots, m. \tag{3}$$

In the relations (3) there are $m' - m$ parameters p_i free. Therefore, all tangents (2) corresponding to various direction vectors are contained in a certain m-dimensional plane. And vice versa: Every straight line in the m-dimensional plane mentioned passing through the point ξ^0 is tangent to a certain curve on Γ at that point.

The m-dimensional plane introduced here is called *tangential plane to the manifold* Γ at the point ξ^0.

In the tangential plane we consider all unit vectors with origin at the point ξ^0, these are the tangential unit vectors to Γ at the point ξ^0. The end points of these vectors form a $(m - 1)$-dimensional unit sphere centered at the point ξ^0. A pair consisting of the point $\xi \in \Gamma$ and one of the tangential unit vectors is called *linear element of the manifold* Γ. It is easy to see that the set of all linear elements of a manifold $\Gamma \in \mathbf{C}^s$ forms a new $(2m - 1)$-dimensional manifold of the class $\mathbf{C}^{s-1}$.

1.3. On a manifold Γ Sobolev spaces $\mathbf{W}_p^{(s)}(\Gamma)$ (in particular, $\mathbf{L}_p(\Gamma)$) can be defined. To that aim let $\{U_j\}$ be a finite covering of Γ and $\{\varphi_j\}$ the corresponding partition of unity, i.e.

1. $\varphi_j \in \mathbf{C}^\infty(\mathbb{R}^m)$, $\operatorname{supp} \varphi_j \subset D_j$;
2. $\varphi_j(x) \geqq 0$;
3. $\Sigma\, \varphi_j(x) = 1$.

For a function $u(x)$ defined on Γ we put $u_j(x) = u(\varphi_j(x))$. Then the space $\mathbf{W}_p^{(s)}(\Gamma)$ consists of all functions u the norm

$$\|u\|_{\mathbf{W}_p^{(s)}(\Gamma)} := \left[\sum_{|\alpha|=0}^{s} \sum_j \|u_j^{(\alpha)}\|_{\mathbf{L}_p(D_j)}^p\right]^{1/p} \tag{4}$$

of which is finite. In particular,

$$\|u\|_{\mathbf{L}_p(\Gamma)} = \left[\sum_j \|u_j\|_{\mathbf{L}_p(D_j)}^p\right]^{1/p}. \tag{5}$$

Analogously the other usual function spaces can be introduced. For instance, in the space $\mathbf{C}^s(\Gamma)$ consisting of functions that are s times continuously differentiable, a norm is provided by

$$\|u\|_{\mathbf{C}^s(\Gamma)} = \max_{|\alpha| \leq s} \sup_{x \in D_j} |u_j^{(\alpha)}(x)|. \tag{6}$$

§ 2. Singular operators on manifolds. The symbol

2.1. A definition for a singular operator on a Lyapunov manifold is given in § 2, Chap. IX. We shall give here another definition due to SEELEY [1]. This new definition which is essentially equivalent to the one given previously is particularly suitable for studying singular operators in the spaces $\mathbf{L}_p(\Gamma)$ and in some other function spaces as well.

An operator A is called *singular operator* in $\mathbf{L}_p(\Gamma)$ iff

a) For arbitrary sufficiently smooth functions $\varphi(\xi)$ and $\psi(\xi)$ with disjoint supports the operator $\varphi A \psi$ is completely continuous. Here, φ and ψ stand for the operator of multiplication by the functions $\varphi(\xi)$ and $\psi(\xi)$, resp., i.e. $(\varphi A \psi u)(\xi) = \varphi(\xi)(A(\psi u))(\xi)$.

b) If the supports of the functions $\varphi(\xi)$ and $\psi(\xi)$ are contained in one and the same domain U_j of the covering for the manifold Γ introduced above, then $\varphi A \psi = \varphi_j A_j \psi_j + T$ where $\varphi_j(x) = \varphi(\omega_j(x))$, $\psi_j(x) = \psi(\omega_j(x))$, and T is completely continuous in $\mathbf{L}_p(\Gamma)$. Finally, A_j denotes the simple singular operator in $\mathbb{R}^m$,

$$(A_j u)(x) = a_j(x)\, u(x) + \int_{\mathbb{R}^m} K_j(x, x-y)\, u(y)\, dy;$$
$$K_j(x, x-y) = r^{-m} f_j(x, \theta). \tag{1}$$

The *symbol* of the operator A is defined as that function of the linear elements of the manifold Γ which coincides for $\xi \in U_j$ with the symbol of the operator (1) as defined in § 1, Chap. X.

A linear element (which will be denoted by τ) is a pair of vectors $\xi \in \Gamma$ and $\theta \in S$. In general, these vectors are not independent from each other, therefore the symbol of a singular operator on a manifold cannot be, in general, considered as a function of two independent variables ξ and θ. An exception is provided by the case if a regular coordinate system can be introduced on Γ. So we can do, in particular, on each set U_j: For ξ belonging to one of the sets U_j the variable θ is defined independently of ξ. Accordingly, we shall use the following notation. In the general case the symbol of a singular operator on a manifold will be denoted by $\Phi(\tau)$ (or by $\Phi_A(\tau)$ if it is necessary to stress the corresponding operator). The symbol for $\xi \in U_j$ with j fixed will be denoted by $\Phi(\xi, \theta)$ (or by $\Phi_A(\xi, \theta)$).

It can be shown (see SEELEY [1]) that the symbol of a singular operator does not depend on the choice of the covering $\{U_j\}$ and the mappings ω_j. We will resign to give the proof here, it is rather cumbersome, and we shall not make use of this assertion in the following.

The investigation of the operator A is reduced to studying the operators A_j such that the following theorems can be easily derived.

Theorem 2.1. *The singular operator A is bounded in $\mathbf{L}_p(\Gamma)$ if its symbol for $\xi \in U_j$ (j arbitrary but fixed) satisfies the condition $\Phi_A(\xi, \theta) \mathrel{\hat{\in}} \mathbf{W}_2^{(\lambda)}(S)$ where*

$$\lambda > \begin{cases} (m-1)/p + 1/2, & 1 < p < 2, \\ (m-1)/2, & p = 2, \\ m/2, & 2 < p < \infty. \end{cases} \tag{2}$$

Theorem 2.2. *Let A_1, A_2 be singular operators the symbols of which satisfy the conditions of Theorem 2.1. If these symbols depend continuously on the linear element τ on Γ then the products A_1A_2 and A_2A_1 are singular operators with the common symbol $\Phi_{A_1}(\tau)\,\Phi_{A_2}(\tau)$. The difference $A_1A_2 - A_2A_1$ is completely continuous in the corresponding space $\mathbf{L}_p(\Gamma)$.*

Theorem 2.3. *Let the symbol of a singular operator A be continuous on Γ as a function of the linear element τ. If for $\xi \in U_j$ (j arbitrary but fixed) $\Phi_A(\xi, \theta) \hat{\in} \mathbf{W}_2^{(\lambda)}(S)$ where*

$$\lambda > \begin{cases} (m-1)/q + 1/2, & q = \min(p, p'), \quad p \neq 2, \\ (m-1)/2, & p = 2, \end{cases} \tag{3}$$

then the adjoint A^* to the operator A is a singular operator with the symbol $\Phi_{A^*}(\tau) = \overline{\Phi_A(\tau)}$.

§ 3. Singular equations in $\mathbf{L}_p(\Gamma)$

3.1. We shall write $\Phi(\tau) \hat{\in} \mathbf{W}_2^{(\lambda)}(S)$ if an analogous relations is true on each set U_j.

Theorem 3.1. *Let $\Phi(\tau)$ be an arbitrary function of τ such that $\Phi(\tau) \hat{\in} \mathbf{W}_2^{(\lambda)}(S)$. If λ satisfies the inequality* (2.2) *then there is a singular operator with the symbol $\Phi(\tau)$ which is bounded in $\mathbf{L}_p(\Gamma)$.*

Proof. Let $\{\varphi_j(\xi)\}$ be a partition of unity on Γ subordinate to the covering $\{U_j\}$. On the set U_j the function Φ of the linear element τ can be considered as a function of two independent variables $\xi \in U_j$ and $\theta \in S$, $\Phi(\tau) = \Phi(\xi, \theta)$. The set U_j is mapped by (1.1′) onto the domain $D_j \subset \mathbb{R}^m$. Then the function $\Phi(\xi, \theta)$ is taken to some function $\Phi_j(x, \theta)$. The product $\tilde{\Phi}_j(x, \theta) = \varphi_j(\omega_j(x))\,\Phi_j(x, \theta)$ defined for $x \in D_j$ is extended to the whole space $\mathbb{R}^m$ by $\tilde{\Phi}_j(x, \theta) = 0$, $x \in \mathbb{R}^m \setminus D_j$. Obviously, $\tilde{\Phi}_j(x, \theta) \hat{\in} \mathbf{W}_2^{(\lambda)}(S)$, and there is a simple singular operator $\tilde{A}_j$ with the symbol $\tilde{\Phi}_j(x, \theta)$ which is bounded in $\mathbf{L}_p(\mathbb{R}^m)$. By definition, this operator corresponds to a singular operator A_j defined on U_j. This last operator extends by zero to $\Gamma \setminus U_j$ since $\varphi_j(\xi) = 0$ in this set. Due to the definition given in § 2 the symbol of the operator A_j is $\varphi_j(\xi)\,\Phi_j(x, \theta) = \varphi_j(\tau)\,\Phi(\xi)$. But then $\Phi(\tau) = \sum_{j=1}^{N} \varphi_j(\xi)\,\Phi(\tau)$ is the symbol of the singular operator $A = \sum_{j=1}^{N} A_j$. ∎

3.2. Theorem 3.2. *Let $\Phi(\tau) \hat{\in} \mathbf{W}_2^{(\lambda)}(S)$ where λ satisfies the inequality* (2.3), *and $\Phi(\tau)$ as a function of ξ, $\xi \in U_j$ (j arbitrary but fixed), be continuous uniformly with respect to θ. If* $\min |\Phi(\tau)| > 0$ *then the singular operator A with the symbol $\Phi(\tau)$ admits of an equivalent regularization, and its index is equal to zero.*

That theorem is an analogue to Theorem 3.1, Chap. XII, for manifolds.

Proof. Here, too, we have to separate the cases $m > 2$ and $m = 2$. In case $m > 2$ the arguments of the proof to the Theorem 3.1. Chap, XII, mentioned before can be repeated without any changes. For the case $m = 2$ we reproduce the demonstration by Mikhlin [11]. Here we have to consider separately manifolds which are homeomorphic to a torus and those which are not so.

a) Let Γ be a manifold that is not homeomorphic to a torus. In that case there does not exist a regular coordinate system on Γ. In what follows we need a formula for the transformation of the symbol under rotations of a local coordinate system. Such a formula can be derived from the general results in § 3, Chap. X; its direct derivation is much more simpler. Denote by ξ_1', ξ_2' axes of the new coordinate system which enclose the angle α with the old axes. In that new coordinate system the characteristic $f'(\xi, \theta')$ is related to the original one $f(x, \theta)$ by the equation $f'(\xi, \theta') = f(\xi, \theta' + \alpha)$. Equations

(2.7) and (2.9) of Chap. X then imply the relation wanted between the symbols,

$$\Phi'(\xi, \theta) = \Phi(\xi, \theta' + \alpha). \tag{1}$$

As before, let $\{U_j\}$ be a finite covering of the manifold Γ. For a fixed j the symbol can be considered as a function of the two independent variables ξ and θ, $\Phi(\tau) = \Phi_j(\xi, \theta)$. Expand Φ_j in a Fourier series,

$$\Phi_j(\xi, \theta) = \sum_{n=-\infty}^{\infty} a_{jn}(\xi)\, \mathrm{e}^{in\theta}.$$

This series is convergent in the $\mathrm{W}_2^{(\lambda)}(S)$ metric uniformly with respect to ξ. Because the imbedding operator of $\mathrm{W}_2^{(\lambda)}(S)$ into $\mathrm{C}(S)$ is bounded the remainder of that series tends to zero uniforlmly. We choose a number n_0 such that

$$\Big|\sum_{|n|>n_0} a_{jn}(\xi)\, \mathrm{e}^{in\theta}\Big| < \varepsilon \tag{2}$$

for a positive $\varepsilon < \inf |\Phi(\tau)|$. Put

$$\Phi_0(\xi, \theta) = \sum_{k=-n_0}^{n_0} a_k(\xi)\, \mathrm{e}^{ik\theta}$$

then, obviously,

$$\inf |\Phi_0(\xi, \theta)| > 0. \tag{3}$$

With

$$\mathrm{e}^{i\theta} = \frac{z + \mathrm{i}}{z - \mathrm{i}} \tag{4}$$

we have

$$\Phi_0(\xi, \theta) = P(\xi, z)\, (z^2 + 1)^{-n_0}$$

where

$$P(\xi, z) = \sum_{k=-n_0}^{n_0} a_k(\xi)\, (z + \mathrm{i})^{n_0+k}\, (z - \mathrm{i})^{n_0-k}. \tag{5}$$

The inequality (3) shows that the polynomial $P(\xi, x)$ has no real roots. Note that the highest coefficient of the polynomial (5) is

$$p_0(\xi) = \sum_{k=-n_0}^{n_0} a_k(\xi) = \Phi_0(\xi, 0)$$

and, therefore, satisfies the inequality (3) itself.

Denote the roots of the polynomial (5) by $z_k(\xi)$, $k = 1, 2, \ldots, 2n_0$, so that

$$P(\xi, z) = p_0(\xi) \prod_{k=1}^{2n_0} [z - z_k(\xi)]. \tag{6}$$

It is clear that the coefficient $p_0(\xi)$ is continuous on Γ.

Now let us find out the effect of a coordinate system rotation to (6). Suppose the axes to be rotated by an angle α. By θ' we denote the value of θ in the new coordinate system, i.e. $\theta' = \theta + \alpha$. Generally, we shall mark by a prime all quantities computed in the new coordinate system. Equation (4) implies

$$z' = \frac{z \cos\dfrac{\alpha}{2} + \sin\dfrac{\alpha}{2}}{\cos\dfrac{\alpha}{2} - z \sin\dfrac{\alpha}{2}}. \tag{7}$$

Similarly, the roots $z_k(\xi)$ change,

$$z_k'(\xi) = \frac{z_k(\xi)\cos\frac{\alpha}{2} + \sin\frac{\alpha}{2}}{\cos\frac{\alpha}{2} - z_k(\xi)\sin\frac{\alpha}{2}}. \tag{8}$$

Furthermore, by (1),

$$p_0'(\xi) = \sum_{k=-n_0}^{n_0} a_k(\xi)\, e^{ik\alpha}, \tag{9}$$

what shall be used later.

Equation (1) can be written as

$$P'(\xi, z')\,(z'^2 + 1)^{-n_0} = P(\xi, z)\,(z^2 + 1)^{-n_0}$$

or, by (6),

$$p_0'(\xi)\,(z'^2 + 1)^{-n_0} \prod_{k=1}^{2n_0} [z' - z_k'(\xi)] = p_0(\xi)\,(z^2 + 1)^{-n_0} \sum_{k=1}^{2n_0} [z - z_k(\xi)]. \tag{10}$$

Expressing z and $z_k(\xi)$ by z' and $z_k'(\xi)$, resp., we see from (10) that

$$p_0'(\xi) \prod_{k=1}^{2n_0} \left[\cos\frac{\alpha}{2} + z_k'(\xi)\sin\frac{\alpha}{2}\right] \tag{11}$$

does not depend on the choice of the coordinate axes. In particular, we get

$$p_0'(\xi) = p_0(\xi) \left\{\prod_{k=1}^{2n_0} \left[\cos\frac{\alpha}{2} + z_k'(\xi)\sin\frac{\alpha}{2}\right]\right\}^{-1}.$$

For $t \in [0, 1]$ we put

$$z_k(\xi, t) = \frac{(2 - t)\, z_k(\xi) \pm it}{\pm\frac{t}{i}\, z_k(\xi) + (2 - t)}, \tag{12}$$

where the sign herein is taken according to the sign of $\operatorname{Im} z_k(\xi)$. Obviously, $z_k(\xi, 0) = z_k(\xi)$, $z_k(\xi, 1) = \pm i$. Moreover, it is easy to see that the roots $z_k(\xi, t)$ are not real for $t \in [0, 1]$. Note that (12) is invariant under a coordinate system rotation.

Accordingly, we introduce the polynomial $P(\xi, z, t)$ which is defined in any local coordinate system by

$$P'(\xi, z', t) = p_0'(\xi, t) \prod_{k=1}^{2n_0} [z' - z_k'(\xi, t)] \tag{13}$$

where

$$p_0'(\xi, t) = p_0(\xi) \left\{\prod_{k=1}^{2n_0} \left[\cos\frac{\alpha}{2} + z_k'(\xi, t)\sin\frac{\alpha}{2}\right]\right\}^{-1}. \tag{14}$$

We show the polynomial (13) to be continuous with respect to ξ and t. That it is continuous with respect to t is obvious. The continuity with respect to ξ is reduced to uniqueness which will be shown now.

Suppose the point ξ runs along a closed curve on Γ. After one revolution the polynomial (5) assumes its starting value again, therefore the roots of that polynomial may, at most, undergo a certain permutation. Such a permutation does not cause the roots to move from the upper to the lower half-plane and vice versa, because the polynomial

(5) has no real roots. The same is true for the roots (12), too. But then (13) and (14) show that the polynomial (13) also returns to its starting value.

Thus, we constructed a polynomial (13) that has no real roots, that depends continuously on a parameter $t \in [0, 1]$ such that it coincides for $t = 0$ with the polynomial $P(\xi, z)$ and for $t = 1$ with a polynomial of the form

$$Q(\xi, z) = b_0(\xi) \, e^{i\gamma(\xi)} \, (z - i)^s \, (z + i)^{2n_0 - s} \tag{15}$$

(s means the number of the roots of polynomial (5) contained in the upper half-plane).

Equation (9) shows that the function $\gamma(\xi)$ is subject to an increment $(n_0 - s)\,\alpha$ when the coordinate system is rotated by the angle α. Thence, all directions that form an angle

$$\omega_j(\xi) = \frac{2\pi j - \gamma(\xi)}{n_0 - s}, \qquad j = 0, 1, \ldots, |n_0 - s| - 1$$

with the x_1 axis are invariant under rotations of the coordinate axes. Thus, we defined on the manifold Γ a $|n_0 - s|$-valued field of tangential directions which depends continuously on the point $\xi \in \Gamma$. We make that field unique by means of covering Γ by a multileafed surface Γ_1. Since Γ is not homeomorphic to a torus so does not Γ_1 as well (cf. Seifert and Threlfall [1]).

It is well-known, however, that on closed surfaces that are not homeomorphic to a torus there does not existe a continuous field of tangential directions (cf. Alexandroff and Hopf [1], Kap. XIV, § 4, p. 552). Thus, the assumption $n_0 \neq s$ leads us to a contradiction. Therefore, $n_0 = s$, and the polynomial (15) is of the form

$$Q(\xi, z) = b_0(\xi) \, e^{i\gamma(\xi)} \, (z^2 + 1)^{n_0}.$$

It corresponds to the symbol $\tilde{\Phi}_0(\xi, \theta) = b_0(\xi) \, e^{i\gamma(\xi)}$ and to the singular operator

$$\tilde{A}_0 = b_0(\xi) \, e^{i\gamma(\xi)} \, I + T \tag{16}$$

with a completely continuous operator T in $\mathbf{L}_p(\Gamma)$.

Thus, the operator A_0 with the symbol $\Phi_0(\xi, \theta)$ is homotopic to the operator (16) the index of which is zero. Therefore, $\operatorname{Ind} A_0 = 0$.

b) Now, let Γ be homeomorphic to a torus. Then the arguments derived under a) remain to hold true up to the construction of the polynomial (15) incl. However, the assertion $n_0 = s$ is no longer true, and so we are forced to continue the demonstration differently.

Obviously, $Q(\xi, z) = b_0(\xi) \, e^{i\gamma(\xi)} \, (z^2 + 1)^{n_0} \, e^{i(n_0 - s)\theta}$, and the corresponding symbol is $b_0(\xi) \, e^{i\gamma(\xi)} \, e^{i(n_0 - s)\theta}$. It is sufficient to show that the operator with that symbol has the index zero. For this, in turn, it is sufficient to prove that the index of the operator with the symbol $e^{i\theta}$ is zero.

The position of a point ξ on Γ is determined by two parameters, say φ and ψ. The coordinate axes can be assumed to be orthogonal. The corresponding local coordinates will be denoted by x_1 and x_2.

Denote the singular operator with the symbol $e^{i\theta}$ by q. In the singular equation $qu = 0$ we transform the coordinates according to $\varphi = \tilde{\varphi}$, $\psi = -\tilde{\psi}$, and put $u(\varphi, \psi) = v(\tilde{\varphi}, \tilde{\psi})$. Such a transformation does not touch the index of the equation, whereas θ is transformed to $-\theta$. Hence, the operators with the symbols $e^{i\theta}$ and $e^{-i\theta}$ have a common index. On the other hand, the operator with the symbol $e^{-i\theta}$ is the adjoint q^* to the operator q. But then $\operatorname{Ind} q = \operatorname{Ind} q^* = -\operatorname{Ind} q$ implies $\operatorname{Ind} q = 0$. ∎

§ 4. On the gradient of a harmonic function

4.1. Let Ω be a (bounded or unbounded) domain in the m-dimensional Euclidean space $\mathbb{R}^m$ the boundary of which is a closed Lyapunov surface Γ. Let us consider a harmonic function $u(x)$ in Ω; for the sake of simplicity we shall assume this function to be continuously differentiable in the closed region $\bar{\Omega} = \Omega \cup \Gamma$. By ν we denote the normal vector to Γ, and by $\operatorname{grad}_\Gamma u$ that component of the gradient of the function u which is parallel to the tangential surface to Γ at the point x. We shall prove the inequality

$$\int_\Gamma |\operatorname{grad}_\Gamma u|^2 \,\mathrm{d}\Gamma \leqq C_p \int_\Gamma \left|\frac{\partial u}{\partial \nu}\right|^p \mathrm{d}\Gamma \tag{1}$$

for any $p \in (1, \infty)$ where the constant C_p depends only on p and Γ.

A harmonic function can be represented as a simple-layer potential

$$u(x) = \int_\Gamma r^{2-m} \mu(y) \,\mathrm{d}_y\Gamma, \qquad r = |y - x|, \tag{2}$$

the density $\mu(y)$ of which is determined by the integral equation

$$\pm \frac{\omega_m}{2} \mu(x) + (m-2) \int_\Gamma r^{1-m} \cos(r, \nu)\, \mu(y) \,\mathrm{d}_y\Gamma = \frac{\partial u}{\partial \nu}, \tag{3}$$

here $\omega_m = 2\pi^{m/2}/\Gamma(m/2)$ is the surface measure of the unit sphere in $\mathbb{R}^m$.

Let us consider the tangential plane to Γ at the point x. Denote by $\lambda = \lambda(x)$ a direction in that plane such that $\partial u/\partial\lambda = |\operatorname{grad}_\Gamma u|$. We introduce a local coordinate system $x_1', x_2', \ldots, x_m'$ with origin at the point x such that the x_m' axis is directed along the normal to Γ whereas the other axes are contained in the tangential plane. Due to the differentiation rule for weakly singular integrals (§ 6, Chap. IX),

$$|\operatorname{grad}_\Gamma u| = \sum_{j=1}^{m-1} \frac{\partial u}{\partial x_j'} \cos(\lambda, x_j') = \sum_{j=1}^{m-1} \cos(\lambda, x_j') \int_\Gamma \frac{\partial r^{2-m}}{\partial x_j'} \mu(y) \,\mathrm{d}_y\Gamma. \tag{4}$$

The right-hand side of this equation provides a bounded operator on $\mu(y)$ in $\mathbf{L}_p(\Gamma)$, hence

$$\int_\Gamma |\operatorname{grad}_\Gamma u|^p \,\mathrm{d}\Gamma \leqq C_p' \int_\Gamma |\mu(x)|^p \,\mathrm{d}\Gamma, \qquad C_p' = \text{const}. \tag{5}$$

If Ω is an outer domain with respect to Γ then (3) has a unique solution (we assume $m \geqq 3$), and this is a bounded operator on $\partial u/\partial \nu$,

$$\int_\Gamma |\mu(x)|^p \,\mathrm{d}\Gamma \leqq C_p'' \int_\Gamma \left|\frac{\partial u}{\partial \nu}\right|^p \mathrm{d}\Gamma, \qquad C_p'' = \text{const},$$

such that inequality (1) is proved with $C_p = C_p' C_p''$.

4.2. Now let Ω be an inner domain. Because of

$$\int_\Gamma \frac{\partial u}{\partial \nu} \,\mathrm{d}\Gamma = 0$$

(3) is solvable, but its solution is not unique: it is of the form

$$\mu(x) = \mu_0(x) + c\mu_1(x), \qquad c = \text{const}$$

(for the sake of simplicity we assume the boundary Γ to be connected); here $\mu_0(x)$ is a certain particular solution of the inhomogeneous equation (3) and $\mu_1(x)$ a non-trivial solution of the corresponding homogeneous equation. As it is well-known, this last solution is determined up to a constant factor. As $\mu_0(x)$ we take the solution of minimal norm to (3); such a solution does exist due to a well-known theorem of Riesz. The integral operator in (3) is completely continuous in $\mathbf{L}_p(\Gamma)$, hence

$$\|\mu_0\| \leqq \tilde{C} \left\| \frac{\partial u}{\partial \nu} \right\|.$$

We introduce

$$u_0(x) = \int_\Gamma \mu_0(y)\, r^{2-m}\, \mathrm{d}_y \Gamma.$$

It is well-known that $u(x) - u_0(x)$ is a constant such that the gradient of the function $u_0(x)$ can be estimated. Analogously to (4) we find

$$|\mathrm{grad}_\Gamma u_0| = \sum_{j=1}^{m-1} \cos(\lambda, x_j') \int_\Gamma \frac{\partial r^{2-m}}{\partial x_j'} \mu(y)\, \mathrm{d}_y \Gamma$$

and, as above, this gives

$$\int_\Gamma |\mathrm{grad}_\Gamma u|^p\, \mathrm{d}\Gamma \leqq \tilde{C}' \int_\Gamma |\mu_0(x)|^p\, \mathrm{d}\Gamma \leqq C_p \int_\Gamma \left| \frac{\partial u}{\partial \nu} \right|^p \mathrm{d}\Gamma, \qquad C_p = \tilde{C}' \tilde{C}^p.$$

The inequality (1) extends also to more general elliptic equations of second order, but then, of course, the normal derivative is to be replaced by the co-normal one.

4.3. Inequality (1) was obtained by Mikhlin [23] in case $p = 2$ and for arbitrary p in his book [11]. In the paper [23] he observed that inequality (1) becomes an equality with $C_2 = 1$ in case of a half-space and if $p = 2$, $m = 3$. This result was generalized by Horváth [1] to the case of an arbitrary dimension m. Horváth proves his assertion by taking the function $u(x)$ which is harmonic in the half-space as a simple-layer potential the density of which differs from the normal derivative only by a certain scalar factor. Then $\mathrm{grad}_\Gamma u$ can be represented as a Riesz singular integral (cf. § 3, Chap. XI). Applying Fourier transformation Horváth obtained the identity under consideration as a consequence of the Plancherel equality.

4.4. Vishik [2] proved for $p = 2$ an inequality that is inverse to (1),

$$\int_\Gamma \left(\frac{\partial u}{\partial \nu} \right)^2 \mathrm{d}\Gamma \leqq \int_\Gamma (\mathrm{grad}_\Gamma u)^2\, d\Gamma,$$

where the function $u(x)$ is harmonic in a ball bounded by the sphere Γ. He mentioned in his paper [2] that this inequality extends to more general domains and more general elliptic equations.

§ 5. The oblique derivative problem

5.1. Let us consider the elliptic differential equation

$$-\sum_{i,j=1}^{m} A_{ij} \frac{\partial^2 u}{\partial x_i\, \partial x_j} + \sum_{i=1}^{m} B_i \frac{\partial u}{\partial x_i} + Cu = f \tag{1}$$

with sufficiently smooth coefficients. The oblique derivative problem for (1) is posed as follows. Let Ω be a domain in the space $\mathbb{R}^m$ bounded by a surface Γ. Assume that to every point of Γ a direction λ is attached. We seek a solution of (1) satisfying the boundary condition

$$\frac{\partial u}{\partial \lambda} + \sigma(x)\, u = \varphi(x), \qquad x \in \Gamma, \tag{2}$$

with a function $\varphi(x)$ given on Γ.

The boundary value problem (1), (2) is a higher dimensional analogue to the Poincaré problem considered in § 2.4, Chap. VII.

In the following we restrict ourselves to the case of a sufficiently smooth and closed surface Γ; the direction λ is assumed to form an acute angle with the outer normal at any point of this surface, and this angle be a sufficiently smooth function of the point $x \in \Gamma$.

Under these assumptions the oblique derivative problem was studied by GIRAUD [2, 3]; his results are also reported in the book by MIRANDA [1]. The following investigation of the oblique derivative problem is essentially simpler than Giraud's approach, and that thanks to the concept of the symbol.

For the oblique derivative problem there are the following uniqueness theorems (MIRANDA [1]).

1. Let Ω be a bounded domain. If $C \geqq 0$, $\sigma \geqq 0$, and at least one of these functions is not identically zero then the oblique derivative problem has not more than one solution.

2. If the domain Ω is unbounded, $C \geqq 0$, $\sigma \geqq 0$, and the solution $u(x)$ is subject to the condition $u(x) \to 0$ as $x \to \infty$, then the solution is unique.

3. For identically vanishing C and σ and Ω being a bounded domain two solutions to the oblique derivative problem may differ only by an additive constant.

5.2. We shall solve the oblique derivative problem in the following way. To begin with, we may assume $f(x) \equiv 0$ – this can be achieved easily by subtracting from the unknown function $u(x)$ the volume potential with density $f(x)$ (cf. § 11, Chap. XI). Furthermore, we assume that the surface Γ is such that any solution of the equation (which is of a certain smoothness)

$$-\sum_{i,j=1}^{m} A_{ij} \frac{\partial^2 u}{\partial x_i\, \partial x_j} + \sum_{i=1}^{m} B_i \frac{\partial u}{\partial x_i} + Cu = 0 \tag{3}$$

can be represented as a simple-layer potential

$$u(x) = \int_\Gamma H(x, y)\, \mu(y)\, \mathrm{d}_y\Gamma. \tag{4}$$

Inserting (4) into the boundary conditions (2) we obtain a singular integral equation for the density $\mu(y)$. Let us derive this equation.

Using common methods from potential theory we obtain easily the following formula given first by Giraud,

$$\frac{\partial u}{\partial \lambda} = \pm \frac{\mu(x)}{2a^{(\lambda)}(x)} + \int_\Gamma \frac{\partial H}{\partial \lambda}\, \mu(y)\, \mathrm{d}_y\Gamma, \qquad x \in \Gamma. \tag{5}$$

(a full derivation of (5) is given in the book by MIRANDA [1]). The minus and plus signs refer to the limits from inside and from outside, resp. Further, we have

$$a^{(\lambda)}(x) = \frac{1}{\cos(\nu, \lambda)} \sum_{i,j=1}^{m} A_{ij}(x) \cos(\nu, x_i) \cos(\nu, x_j)$$

where ν denotes the outward normal to Γ. The equation for μ is of the form

$$\pm\frac{\mu(x)}{2a^{(\lambda)}(x)}+\int_\Gamma\frac{\partial H}{\partial\lambda}\mu(y)\,\mathrm{d}_y\Gamma+\sigma(x)\int_\Gamma H(x,y)\,\mu(y)\,\mathrm{d}_y\Gamma=\varphi(x). \tag{6}$$

5.3. We show that the symbol of (6) does never vanish provided that the direction λ is never tangential to Γ. Remember that due to the results of § 3, Chap. X, the range of the symbol is invariant under a non-singular coordinate transformation. Then we may perform such a transformation with the result that for x under consideration we get $A_{ij}=0$, $i\neq j$, $A_{ii}=1$, and the tangential plane to Γ coincides with the plane $x_m=0$. Thence, $[a^{(\lambda)}(x)]^{-1}=\cos(\nu,\lambda)$, and the major part of H (see eq. (11.3), (11.4), Chap. XI) becomes

$$\psi(x,y)=\frac{1}{(m-2)\,\omega_m r^{m-2}}.$$

At the point $x\in\Gamma$ under consideration equation (6) has now the form

$$\pm\frac{\cos(\nu,\lambda)}{2}\mu(x)+\frac{1}{\omega_m}\int_\Gamma\frac{\cos(r,\lambda)}{r^{m-1}}\mu(y)\,\mathrm{d}_y\Gamma+\int_\Gamma M(x,y)\,\mu(y)\,\mathrm{d}_y\Gamma=\varphi(x)$$

$$y=x,\qquad \mathrm{d}_y\Gamma=\mathrm{d}y, \tag{7}$$

with a weakly singular kernel $M(x,y)$. Further,

$$\cos(r,\lambda)=\sum_{k=1}^{m-1}\frac{y_k-x_k}{r}\cos(\lambda,x_k)=\sum_{k=1}^{m-1}Y_{1,m-1}^{(k)}(\theta)\cos(\lambda,x_k)$$

where $Y_{1,m-1}^{(k)}$ are spherical functions in the $(m-1)$-dimensional space. The symbol of the singular operator (7) is equal to

$$\Phi(x,\theta)=\pm\frac{\cos(\nu,\lambda)}{2}+\frac{1}{\omega_m}\gamma_{1,m-1}\cos(r,\lambda)$$

with r being an arbitrary direction in the tangential plane. Because of

$$\gamma_{1,m-1}=\frac{\mathrm{i}\pi^{(m-1)/2}\,\Gamma\left(\frac{1}{2}\right)}{\Gamma\left(\frac{m}{2}\right)}=\frac{\mathrm{i}\pi^{m/2}}{\Gamma\left(\frac{m}{2}\right)}=\mathrm{i}\frac{\omega_m}{2}$$

we have

$$\Phi(x,\theta)=\frac{\pm\cos(\nu,\lambda)+\mathrm{i}\cos(r,\lambda)}{2}$$

such that

$$\inf|\Phi(x,\theta)|\geqq\inf|\cos(\nu,\lambda)|/2. \tag{8}$$

For (6) the Fredholm theorems are valid such that uniqueness implies solvability of the oblique derivative problem. If the uniqueness is violated then the number k of solutions to the homogeneous problem is finite, and the corresponding inhomogeneous problem is solvable if and only if the function $\varphi(x)$ is orthogonal to the (also k) solutions of the adjoint homogeneous equation.

Remark. If the direction λ becomes tangential to the surface Γ at a certain point x then(8) shows that the symbol of the equation (7) vanishes for $\cos(r,\lambda)=0$. In the case $m=2$ this equation cannot hold because the direction r differs only slightly from the tangential direction for points on the curve

Γ close to x. For λ being tangential to the curve in at least one point the oblique derivative problem leads in that case ($m = 2$) to a thoroughly studied one-dimensional singular integral equation the symbol of which does not degenerate but, in general, with an index different from zero (cf. § 2.4, Chap. VII).

If in case $m > 2$ the direction λ becomes tangential to the boundary of the domain at a certain point then the symbol may vanish at that point. Then (6) does not have a bounded regularizer. In that case this equation is either not normally solvable in any $\mathbf{L}_p$-space or its index is not finite or both of them occur. This case is studied in a great many of contributions, and we mention here particularly the paper of Mazya [1] which seems to be the most complete one (see also the references cited there).

§ 6. On the boundedness of singular operators in Lipschitz spaces

A sufficient condition for the singular operator to be bounded in spaces of functions satisfying a Lipschitz condition with positive exponent is provided by Giraud's theorem (§ 4, Chap. IX). This theorem was essentially improved by Mikhailova-Gubenko [1, 2], Shevchenko [1], and Khvoles [1, 3], see also the paper by Taibleson [1]. In the present section we give the results of Khvoles which seem to be most general ones. Without going into further details we mention only that Khvoles considered more general spaces.

6.1. Let Γ be a closed compact Lyapunov manifold (i.e. a manifold of the class $\mathbf{C}^{1,\beta}$, $0 < \beta \leqq 1$) of dimension m, $\{U_j\}$ be a finite covering of Γ and $\{\varphi_j\}$ the corresponding partition of unity; finally, let ω_j be the diffeomorphism $U_j \to D_j$ (see § 1) and ω_j^{-1} its inverse. A function $u(\xi)$ defined on Γ belongs to the space $\mathbf{Lip}_\alpha(\Gamma)$ iff $\varphi_j(\omega_j(x))\, u(\omega_j(x)) \in \mathbf{Lip}_\alpha(D_j)$ for every j. A norm in $\mathbf{Lip}_\alpha(\Gamma)$ can be defined by

$$\|u\|_{\mathbf{Lip}_\alpha(\Gamma)} = \sum_j \|(\varphi_j u) \circ \omega_j\|_{\mathbf{Lip}_\alpha(D_j)} \tag{1}$$

where $\circ$ indicates the substitution $v \circ \omega_j = v(\omega_j(x))$.

Let us give a further definition. Let D' be a domain in the Euclidean space $\mathbb{R}^m$, $x' \in D'$, and $w(x')$ be a function defined on D'. Due to S. M. Nikolski [1] we shall say $w \in \mathbf{H}_p^{(\nu)}(D')$, $1 < p < \infty$, iff the norm

$$\|w\|_{\mathbf{H}_p^{(\nu)}(D')} = \|w\|_{\mathbf{L}_p(D')} + \sup_{|h'| \leqq 1} |h'|^{-\nu} \|w(x + h') - w(x)\|_{\mathbf{L}_p(D')} \tag{2}$$

is finite.

Now, let $\{\psi_j(\theta)\}$ be a partition of unity subordinate to a finite covering $\{S_j\}$ of the $(m - 1)$-dimensional unit sphere S, and τ_j be a diffeomorphism $S_j \to D_j' \subset \mathbb{R}^{m-1}$. We shall say that a function $f(\theta)$ defined on the sphere S belongs to the space $\mathbf{H}_p^{(\nu)}(S)$ iff $(\psi_j f) \circ \tau_j \in \mathbf{H}_p^{(\nu)}(D_j')$ for every j. A norm in $\mathbf{H}_p^{(\nu)}(S)$ can be defined by

$$\|f\|_{\mathbf{H}_p^{(\nu)}(S)} = \sum_j \|(\psi_j f) \circ \tau_j\|_{\mathbf{H}_p^{(\nu)}(D_j')} \,. \tag{3}$$

It can be shown (see Hörmander [3]) that different coverings of S lead to equivalent norms in $\mathbf{H}_p^{(\nu)}(S)$.

6.2. Lemma 6.1. *Let G be a bounded domain in $\mathbb{R}^k$,* supp $F \subset G$, *and* $\|F\|_{\mathbf{H}_p^{(\nu)}(G)} \leqq B$. *If $\varphi_h(x)$ is a smooth regular mapping $\mathbb{R}^k \to \mathbb{R}^k$ the Jacobian determinant of which is continuous with respect to h and not vanishing, such that $|\varphi_h(x) - x| \leqq c\delta$ then*

$$\int_G |F(\varphi_h(x)) - F(x)|^p \, dx \leqq cB^p \, \delta^{\nu p} . \tag{4}$$

Proof. Denote the integral in the left-hand side of (4) by J; it can be written as

$$J = \int_G |F(\xi_1, \xi_2, \ldots, \xi_k) - F(x_1, x_2, \ldots, x_k)|^p \, dx_1 \, dx_2 \ldots dx_k$$

with ξ_j, $j = 1, 2, \ldots, k$, being the coordinates of the point $\varphi_h(x)$.

Additionally we introduce the notations $\Xi = (\xi_1, \ldots, \xi_j)$, $X_j = (x_{j+1}, \ldots, x_k)$, and put

$$(\Xi_j, X_j) := (\xi_1, \ldots, \xi_j, x_{j+1}, \ldots, x_k), \qquad j = 1, 2, \ldots, k-1;$$

$$(\Xi_0, X_0) := X_0 = (x_1, x_2, \ldots, x_m); \qquad (\Xi_k, X_k) := \Xi_k = (\xi_1, \xi_2, \ldots, \xi_k).$$

Then

$$J = \int_G |F(\Xi_k) - F(X_0)|^p \, \mathrm{d}x$$

$$\leqq k^{p-1} \sum_{j=1}^{k} \int_G |F(\Xi_k, X_k) - F(\Xi_{k-1}, X_{k-1})|^p \, \mathrm{d}x =: k^{p-1} \sum_{j=1}^{k} J_j. \tag{5}$$

Now let us consider the matrix $(D_j \xi_i)$, $i, j = 1, 2, \ldots, k$, $D_j = \partial/\partial x_j$; by the assumption of our lemma we have $\det (D_j \xi_i) \neq 0$ on $\bar{G}$. Since the mapping φ_h is a diffeomorphism we have on $\bar{G}$ the inequality

$$|\det (D_j \xi_i)_{i,j=1}^{l}| \leqq C = \text{const} \tag{6}$$

where l is an arbitrary integer such that $1 \leqq l \leqq k$.

In order to estimate the integral J_1 we get with the notations $G^1 =: \{X^1 \colon (x_1, X_1) \in G\}$, $F(\xi_1, X_1) = f(\xi_1)$, $F(x_1, X_1) = f(x_1)$

$$J_1 = \int_{G'} \mathrm{d}X_1 \int_{x_0(X_1)}^{x_1(X_1)} |f(\xi_1) - f(x_1)|^p \, \mathrm{d}x_1$$

where the meaning of the notations used is obvious. It is easy to see that

$$\int_0^\delta J_1 \, \mathrm{d}y = \int_0^\delta \mathrm{d}y \int_{G'} \mathrm{d}X_1 \int_{x_0(X_1)}^{x_1(X_1)} |f(\xi_1) - f(x_1)|^p \mathrm{d}x_1$$

$$\leqq \int_0^\delta \mathrm{d}y \int_{G'} \mathrm{d}X_1 \int_{x_0(X_1)}^{x_1(X_1)} |f(\xi_1) - f(x_1 + y)|^p \, \mathrm{d}x_1$$

$$+ \int_0^\delta \mathrm{d}y \int_{G'} \mathrm{d}X_1 \int_{x_0(X_1)}^{x_1(X_1)} |f(x_1 + y) - f(x_1)|^p \, \mathrm{d}x_1$$

$$\leqq \int_{G'} \mathrm{d}X_1 \int_{x_0(X_1)}^{x_1(X_1)} \mathrm{d}x_1 \int_0^\delta |f(\xi_1) - f(x_1 + y)|^p \, \mathrm{d}y + B \int_0^\delta y^{\nu p} \, \mathrm{d}y \leqq i_1 + B\, \delta^{\nu p + 1}; \tag{7}$$

here i_1 denotes the last but one of the integrals. Put $z = \xi_1 - x_1 - y$ and write

$$i_1 = \int_{G'} \mathrm{d}X_1 \int_{x_0(X_1)}^{x_1(X_1)} \mathrm{d}x_1 \int_{\xi_1 - x_1 - \delta}^{\xi_1 - x_1} |f(\xi_1) - f(\xi_1 - z)|^p \, \mathrm{d}z.$$

Because of $|\xi_1 - x_1| \leqq |\varphi_h(x) - x| \leqq c\,\delta$ we may continue

$$i_1 \leqq \int_{G'} \mathrm{d}X_1 \int_{x_0(X_1)}^{x_1(X_1)} \mathrm{d}x_1 \int_{-c\delta}^{c\delta} |f(\xi_1) - f(\xi_1 - z)|^p \, \mathrm{d}z$$

$$= \int_{-c\delta}^{c\delta} \mathrm{d}z \int_{G'} \mathrm{d}X_1 \int_{x_0(X_1)}^{x_1(X_1)} |f(\xi_1) - f(\xi_1 - z)|^p \, \mathrm{d}x_1.$$

Taking into account condition (6) we obtain

$$i_1 \leqq C'B \int_{-c\delta}^{c\delta} z^{\nu p} dz = C''B\, \delta^{\nu p+1}.$$

Inserting this into (7) we find

$$\int_0^\delta J_1\, dy \leqq C'''B\, \delta^{\nu p+1},$$

and that is equivalent to the estimate $J_1 \leqq C'''B\delta^{\nu p}$. By means of (6) you can obtain similarly estimates for the integrals J_j, $j = 2, 3, \ldots, m$; $J_j \leqq C'''B\, \delta^{\nu p}$. Due to (5), this will imply the inequality (4). ∎

6.3. Theorem 6.1. *Let A be a singular operator,*

$$(Au)(\xi) = \int_\Gamma K(\xi, \eta)\, u(\eta)\, d_\eta \Gamma, \tag{8}$$

the kernel $K(\xi, \eta)$ of which is in every domain D_j (see § 6.1) of the form

$$K(\xi, \eta) = \frac{f_j(x, \theta)}{r^m} + K_j'(x, y) \tag{9}$$

where $r = |y - x|$, $\theta = (y - x)/r$, and $K_j'(x, y)$ is a weakly singular kernel. If for every j the conditions

$$\|f_j(x + h, \cdot) - f_j(x, \cdot)\|_{\mathbf{L}_1(S)} \leqq B\, |h|^\alpha, \qquad 0 < \alpha < 1,$$
$$\|f_j(x, \cdot)\|_{\mathbf{H}_1^{(\nu)}(S)} \leqq B, \qquad B = \text{const}, \tag{10}$$

are satisfied then the operator (8) *is bounded in* $\mathbf{Lip}_\alpha(\Gamma)$.

Proof. Obviously, it is sufficient to prove the following assertion: For the operator

$$A_j u = \int_{D_j} r^{-m} f_j(x, \theta)\, u(y)\, dy$$

let $u \in \mathbf{Lip}_\alpha(D_j)$, supp $u \subset D_j$, and f_j satisfy the conditions (10). Then

$$\|A_j u\|_{\mathbf{Lip}_\alpha(D_j)} \leqq cB\, \|u\|_{\mathbf{Lip}_\alpha(D_j)}, \qquad c = \text{const}.$$

In order to prove this assertion it is, in turn, sufficient to prove a still simpler assertion: let

$$v(x) = \int_{\mathbb{R}^m} r^{-m} f(x, \theta)\, u(y)\, dy \tag{11}$$

where the charactristic f satisfies the conditions (10), and the function $u \in \mathbf{Lip}_\alpha(\mathbb{R}^m)$ has a compact support in $\mathbb{R}^m$. Then

$$\|v\|_{\mathbf{Lip}_\alpha(\mathbb{R}^m)} \leqq cB\, \|u\|_{\mathbf{Lip}_\alpha(\mathbb{R}^m)}, \qquad c = \text{const}. \tag{12}$$

Put $r^{-m} f(x, \theta) = K(x, y - x)$ and substitute $y = x + z + h$ in the integral (11), then

$$v(x) = \int_{\mathbb{R}^m} K(x, z + h)\, u(x + z + h)\, dz.$$

Let G be a bounded domain in $\mathbb{R}^m$ containing supp u, and $x \in G$. Then there is a sufficiently large number δ such that

$$v(x) = \int_{|z| < \delta} K(x, z + h)\, [u(x + z + h) - u(x + h)]\, dz.$$

For a fixed vector h, $|h| < \delta/6$, we put $r_h = |y - x - h|$, $\theta_h = (y - x - h)/r_h$, and consider the difference

$$\begin{aligned} &v(x + 2h) - 2v(x) + v(x - 2h) \\ &= \int_{|z|<\delta} K(x + 2h, z - h)\,[u(x + z + h) - u(x + h)]\,\mathrm{d}z \\ &\quad - 2 \int_{|z|<\delta} K(x, z - h)\,[u(x + z - h) - u(x - h)]\,\mathrm{d}z \\ &\quad + \int_{|z|<\delta} K(x - 2h, z + h)\,[u(x + z +- h) - u(x - h)]\,\mathrm{d}z. \end{aligned}$$

Each of these integrals is divided into three summands corresponding to the integration domains $|z| < 3\,|h|$, $3\,|h| < |z| < \delta - 2\,|h|$, and $\delta - 2\,|h| < |z| < \delta$. We start with estimating the first integral over $|z| < 3\,|h|$ and $\delta - 2\,|h| < |z| < \delta$,

$$\begin{aligned} &\Bigg| \int_{|z|<3|h|} K(x + 2h, z - h)\,[u(x + z + h) - u(x + h)]\,\mathrm{d}z \\ &\quad + \int_{\delta-2|h|<|z|<\delta} K(x + 2h, z - h)\,[u(x + z + h) - u(x + h)]\,\mathrm{d}z \Bigg| \\ &\leqq B \int_{|z|<3|h|} r_{-h}^{-m}\,|f(x + 2h, \theta_{-h})|\,|z|^\alpha\,\mathrm{d}z \\ &\quad + B \int_{\delta-2|h|<|z|<\delta} r_{-h}^{-m}\,|f(x + 2h, \theta_{-h})|\,|z|^\alpha\,\mathrm{d}z \\ &\leqq CB \int_0^{3|h|} t^{\alpha-1}\,\mathrm{d}t + CB \int_{\delta-2|h|}^{\delta} t^{\alpha-1}\,\mathrm{d}t \leqq CB\,|h|^\alpha, \qquad C = \text{const}. \end{aligned} \tag{13}$$

Similarly, the second and the third integral can be estimated over these domains.

Now, we come to the integral over the domain $3\,|h| < |z| < \delta - 2\,|h|$ and consider the expression

$$\begin{aligned} J &= \int_{3|h|<|z|<\delta-2|h|} K(x + 2h, z - h)\,[u(x + z + h) - u(x + h)]\,\mathrm{d}z \\ &\quad - 2 \int_{3|h|<|z|<\delta-2|h|} K(x, z - h)\,[u(x + z - h) - u(x - h)\,]\mathrm{d}z \\ &\quad + \int_{3|h|<|z|<\delta-2|h|} K(x - 2h, z + h)\,[u(x + z - h) - u(x - h)]\,\mathrm{d}z \\ &= \int_{3|h|<|z|<\delta-2|h|} [K(x + 2h, z - h) - K(x, z - h)][u(x + z + h) - u(x + h)]\,\mathrm{d}z \\ &\quad + \int_{3|h|<|z|<\delta-2|h|} K(x, z - h)\,[u(x + z + h) - u(x + z - h)]\,\mathrm{d}z \\ &\quad + \int_{3|h|<|z|<\delta-2|h|} K(x - 2h, z + h)\,[u(x + z - h) - u(x - h)]\,\mathrm{d}z \\ &\quad + \int_{3|h|<|z|<\delta-2|h|} K(x, z + h)\,[u(x - h) - u(x + z - h)]\,\mathrm{d}z + g_1(x, h) \\ &=: J_1 + J_2 + J_3 + J_4 + g_1(x, h), \end{aligned}$$

where we put

$$g_1(x, h) = \int_{3|h|<|z|<\delta-2|h|} [K(x, z - h) - K(x, z + h)][u(x + z - h) - u(x - h)]\,\mathrm{d}z.$$

The integral J_4 differs from the integral

$$\int_{3|h|<|z|<\delta-2|h|} K(x, z - h)\,[u(x + z - h) - u(x - h)]\,\mathrm{d}z$$

by a quantity which is less than $CB\,|h|^\alpha$. Hence, $|g_1(x, h)| \leqq CB\,|h|^\alpha$. For the integral J_1 the estimate

$$|J_1| \leqq \int\limits_{3|h|<|z|<\delta-2|h|} r_h^{-m}\,|f(x+2h, \theta_h) - f(x, \theta_h)|\,|z|^\alpha\,dz$$

$$\leqq CB\,|h|^\alpha \int\limits_0^\delta t^{\alpha-1}\,dt \leqq CB\,|h|^\alpha \tag{14}$$

is obtained.

Now let us consider

$$J_2 + J_3 + J_4 = \int\limits_{3|h|<|z|<\delta-2|h|} K(x, z-h)\,[u(x+z+h) - u(x+z-h)]\,dz$$
$$+ \int\limits_{3|h|<|z|<\delta-2|h|} [K(x-2h, z+h) - K(x, z+h)][u(x+z-h) - u(x-h)]\,dz$$
$$+ \int\limits_{3|h|<|z|<\delta-2|h|} K(x, z+h)\,[u(x+z-h) - u(x+z+h)]\,dz$$
$$= \int\limits_{3|h|<|z|<\delta-2|h|} [K(x, z-h) - K(x, z+h)]\,[u(x+z+h) - u(x+z-h)]\,dz$$
$$+ \int\limits_{3|h|<|z|<\delta-2|h|} [K(x-2h, z+h) - K(x, z+h)]$$
$$\times [u(x+z-h) - u(x-h)]\,dz =: J_5 + J_6.$$

Inserting the estimates obtained in the expression for J we get $|J| \leqq CB\,|h|^\alpha$, and this leads together with (13) to the inequality $|v(x+2h) - 2v(x) + v(x-2h)| \leqq CB\,|h|^\alpha$. Furthermore,

$$\sup_{x\in G} |v(x)| \leqq C \sup_{x\in G} \int\limits_{r<\delta} |f(x, \theta)|\,r^{\alpha-m}\,dy \leqq C \sup_{x\in G} \int\limits_0^\delta r^{\alpha-1}\,dr \int\limits_S |f(x, \theta)|\,dS \leqq CB. \tag{15}$$

The following assertion is a special case of a theorem by Marchoud (cf. A. F. Timan [1]): If $|v(x+2h) - 2v(x) + v(x-2h)| \leqq C\omega(|h|)$ then

$$|v(x+h) - v(x)| \leqq C\,|h| \left[\int\limits_h^2 \tau^{-2}\,\omega(\tau)\,d\tau + \max_{x\in G} |v(x)|\right].$$

This yields together with (15) the estimate $|v(x+h) - v(x)| \leqq CB\,|h|^\alpha$ which gives together with the same inequality (15) the final result. ∎

Theorem 6.2. *Let A be a singular operator in $\mathbb{R}^m$,*

$$Au = F^{-1}_{\xi\to x}\,\Phi(x, \theta)\,F_{x\to\xi}\,u, \qquad \theta = \xi/|\xi|,$$

the symbol of which is subject to the the conditions

$$\Phi \in \mathbf{W}_2^{(l)}(S), \quad l > \frac{m}{2},$$

$$\|\Phi(x+h, \cdot) - \Phi(x, \cdot)\|_{\mathbf{W}_2^{(l)}(S)} \leqq B\,|h|^\alpha.$$

Then the expansion of the symbol into a series with respect to spherical functions

$$\Phi(x, \theta) = \sum_{n=0}^\infty \sum_{k=1}^{k_{n,m}} b_n^{(k)}(x)\,Y_{n,m}^{(k)}(\theta)$$

corresponds to the series

$$A = \sum_{n=0}^\infty \sum_{k=1}^{k_{n,m}} B_n^{(k)}$$

of the operator A which is convergent in the space $\mathbf{Lip}_\alpha(\mathbb{R}^m)$; *here*

$$(B_n^{(k)}u)(x) = \frac{b_n^{(k)}(x)}{\gamma_{n,m}} \int_{\mathbb{R}_m} \frac{Y_{n,m}^{(k)}(\theta)}{r^m} u(y)\, \mathrm{d}y, \qquad n > 0,$$

$$\gamma_{n,m} = \frac{\mathrm{i}^n \pi^{m/2} \Gamma\left(\frac{n}{2}\right)}{\Gamma\left(\frac{m+n}{2}\right)}, \qquad (B_0^{(1)}u)(x) = b_0^{(1)}(x)\, u(x).$$

Proof. It is sufficient to show that for functions $u \in \mathbf{Lip}_\alpha(\mathbb{R}^m)$ with a support contained in a fixed ball the estimate

$$\left\| \sum_{n=N}^{L} \sum_{k=1}^{k_{n,m}} B_n^{(k)} u \right\|_{\mathbf{Lip}_\alpha(\mathbb{R}^m)} \leqq \varepsilon_{NL} \|u\|_{\mathbf{Lip}_\alpha(\mathbb{R}^m)}, \qquad \varepsilon_{NL} \xrightarrow[N,L\to\infty]{} 0,$$

holds.

The sum $\sum_{n=N}^{L} \sum_{k=1}^{k_{n,m}} B_n^{(k)}$ is a singular operator with the characteristic

$$f_{NL}(x, \theta) = \sum_{n=N}^{L} \sum_{k=1}^{k_{n,m}} \frac{b_n^{(k)}(x)}{\gamma_{n,m}} Y_{n,m}^{(k)}(\theta).$$

By Theorem 6.1,

$$\left\| \sum_{n=N}^{L} \sum_{k=1}^{k_{n,m}} B_n^{(k)} \right\|_{\mathbf{Lip}_\alpha(\mathbb{R})^m} \leqq C \Big\{ \sup_x \|f_{NL}(x, \cdot)\|_{\mathbf{H}_1^{(\alpha)}(S)}$$

$$+ \sup_{x,h} |h|^{-\alpha} \|f_{NL}(x+h, \cdot) - f_{NL}(x, \cdot)\|_{\mathbf{L}_2(S)} \Big\}, \qquad \alpha = l - \frac{m}{2}. \tag{16}$$

Denote by Φ_{NL} the symbol that corresponds to the characteristic f_{NL}, then

$$\sup_{x,h} |h|^{-2\alpha} \|f_{NL}(x+h, \cdot) - f_{NL}(x, \cdot)\|^2_{\mathbf{L}_2(S)}$$

$$\leqq C \sup_{x,h} |h|^{-2\alpha} \sum_{n=N}^{L} \sum_{k=1}^{k_{n,m}} |b_n^{(k)}(x+h) - b_n^{(k)}(x)|^2 n^m$$

$$\leqq CN^{m-2l} \sup_{x,h} |h|^{-2\alpha} \sum_{n=N}^{L} \sum_{k=1}^{k_{n,m}} |b_n^{(k)}(x+h) - b_n^{(k)}(x)|\, n^{2l}$$

$$\leqq CN^{m-2l} \sup_{x,h} |h|^{-2\alpha} \|\Phi_{NL}(x+h, \theta) - \Phi_{NL}(x, \theta)\|^2_{\mathbf{W}_2^{(l)}(S)}.$$

Similarly the quantity $\sup \|f_{NL}(x, \cdot)\|_{\mathbf{H}_1^{(\alpha)}(S)}$ can be estimated. Inserting the estimates obtained into (16) we arrive at the inequality to be proved. ∎

§ 7. Singular integral equations in Lipschitz spaces

7.1. To begin with we shall prove three lemmas concerning the integral

$$(Hu)(x) = v(x) := \int_{\mathbb{R}^m} r^{-m}[a(y) - a(x)]\, Y(\theta)\, u(y)\, \mathrm{d}y. \tag{1}$$

These lemmas are contained in their final form in SHEVCHENKO [1]. Here, $a \in \mathbf{Lip}_\alpha(\mathbb{R}^m)$, $0 < \alpha < 1$, and $Y(\theta)$ is an arbitrary spherical polynomial (i.e. a linear combination of spherical functions) that is orthogonal to 1.

Lemma 7.1. *For $u \in \mathbf{Lip}_\alpha(\mathbb{R}^m)$ and $p > m/\alpha$ we have*

$$|v(x)| \leqq C \|u\|_p \|a\|_{\mathbf{Lip}_\alpha} \|Y\|_{\mathbf{C}(S)}, \tag{2}$$

$$|v(x) - v(\xi)| \leqq C \|u\|_p \|a\|_{\mathbf{Lip}_\alpha} \|Y\|_{\mathbf{C}^1(S)} |x - \xi|^\beta,$$

$$|x - \xi| < 1, \qquad \beta = \alpha - \frac{m}{p}. \tag{3}$$

Here and in the following lemmas the notation $\mathbb{R}^m$ in the norm subscript will be suppressed.

Proof. The estimate (2) is an immediate consequence of Hölder's inequality if we put $y = x + z$ and the integral is divided in two ones over the domains $|z| < 1$ and $|z| > 1$. Let us come to the estimate (3). We have

$$v(x) - v(\xi) = \int\limits_{\mathbb{R}^m} \left[\frac{a(x) - a(y)}{|x - y|^m} Y(\theta_{xy}) - \frac{a(\xi) - a(y)}{|\xi - y|^m} Y(\theta_{\xi y})\right] u(y)\, dy. \tag{4}$$

The integration domain in the right-hand side herein is divided into the two domains $|y - x| < 2$ and $|y - x| > 2$. The integral over the second domain can be written in the form

$$[a(x) - a(\xi)] \int\limits_{|x-y|>2} \frac{Y(\theta_{xy})}{|y - x|^m} u(y)\, dy + \int\limits_{|y-x|>2} [a(\xi) - a(y)] \times$$

$$\left[\frac{Y(\theta_{xy})}{|y - x|^m} - \frac{Y(\theta_{\xi y})}{|\xi - y|^m}\right] u(y)\, dy. \tag{5}$$

By the Calderón-Zygmund theorem the first integral in (5) is estimated by $C\,|x - \xi|^\alpha \|Y\|_{\mathbf{C}(S)} \|a\|_{\mathbf{Lip}_\alpha} \|u\|_p$. In the second integral the term $a(\xi) - a(y)$ is bounded, and the expression $|y - x|^{-m}\, Y(\theta_{xy})$ is continuously differentiable with respect to x. Applying the Lagrange formula and using Hölder's inequality we obtain the estimate $C\,|x - \xi|\, \|a\|_{\mathbf{Lip}_\alpha} \|Y\|_{\mathbf{C}^1(S)} \|u\|_p$ for the integral in question.

Hence, for the integral over the domain $|y - x| > 2$ the estimate (3) is true. The corresponding estimate for the integral over the ball $|y - x| < 2$ is obtained in the same way as it was done when deriving Zygmund's inequality (§ 3, Chap. IX). ∎

Lemma 7.2. *If $u \in \mathbf{L}_p$, $1 < p \leqq m/\alpha$, then $v \in \mathbf{L}_\gamma$ for any*

$$\gamma \in (p, mp/(m - p\alpha)).$$

Moreover,

$$\|v\|_\gamma \leqq C \|a\|_{\mathbf{Lip}_\alpha} \|Y\|_{\mathbf{C}(S)} \|u\|_p. \tag{6}$$

Proof. Again we put $y = z + x$ and divide the integration domain in (1) into the two domains $|z| < 1$ and $|z| > 1$; the corresponding integrals are denoted by $v_1(x)$ and $v_2(x)$, resp. We have

$$|v_1(x)| \leqq \|a\|_{\mathbf{Lip}_\alpha} \|Y\|_{\mathbf{C}(S)} \int\limits_{|z|<1} |z|^{\alpha - m} |u(x + z)|\, dz. \tag{7}$$

The function $|z|^{\alpha-m}$ is summable in the q-th power for any $q \in (1, m/(m - \alpha))$. Let $p < m/\alpha$ and determine γ by the equation $\gamma^{-1} = p^{-1} + q^{-1} - 1$. Obviously, $p^{-1} >$

$\gamma^{-1} > p^{-1} - \alpha m^{-1}$. With the abbreviation $g = |z|^{\alpha-m}$ the integrand in (7) can be written as a product of $|u|^{p/\gamma} g^{q/\gamma}$, $|u|^{p(p^{-1}-\gamma^{-1})}$, and $g^{q(q^{-1}-\gamma^{-1})}$. Now we apply Hölder's inequality with the exponents γ, $p\gamma/(\gamma - p)$, and $q\gamma/(\gamma - q)$ and obtain

$$|v_1(x)| \leqq C \|u\|_p^{(\gamma-p)/p} \left\{ \int\limits_{|z|<1} |z|^{(\alpha-m)q} |u(x+z)|^p \, dz \right\}^{1/\gamma} \|a\|_{\mathbf{Lip}_\alpha} \|Y\|_{C(S)}.$$

Raising this to the power γ and integrating over $\mathbb{R}^m$ we can easily see that $v_1(x)$ satisfies the inequality (6). Analogously, the integral $v_2(x)$ and the case $p = m/\alpha$ can be treated, and this last one is even still simpler. ∎

Lemma 7.3. *Let* $u \in \mathbf{L}_p(\mathbb{R}^m) \cap \mathbf{Lip}_\beta(\mathbb{R}^m)$, $p > 1$, $\beta < \alpha$, *and* $a \in \mathbf{Lip}_\alpha(\mathbb{R}^m)$. *Then* $v \in \mathbf{Lip}_\alpha(\mathbb{R}^m)$ *and*

$$\|v\|_{\mathbf{Lip}_\alpha} \leqq C \|a\|_{\mathbf{Lip}_\alpha} \|Y\|_{C^1(S)} (\|u\|_p + \|u\|_{\mathbf{Lip}_\beta}). \tag{8}$$

Proof. Consider the difference

$$\begin{aligned} &|v(x) - v(\xi)| \\ &= \left| \int\limits_{\mathbb{R}^m} \frac{a(x) - a(y)}{|x-y|^m} Y(\theta_{xy}) u(y) \, dy - \int\limits_{\mathbb{R}^m} \frac{a(\xi) - a(y)}{|\xi - y|^m} Y(\theta_{\xi y}) u(y) \, dy \right| \\ &\leqq \left| \int\limits_{|x-y|<\delta} \left[\frac{a(x) - a(y)}{|x-y|^m} Y(\theta_{xy}) - \frac{a(\xi) - a(y)}{|\xi - y|^m} Y(\theta_{\xi y}) \right] u(y) \, dy \right| \\ &\quad + \left| \int\limits_{|y-x|>\delta} \left[\frac{a(x) - a(y)}{|x-y|^m} Y(\theta_{xy}) - \frac{a(\xi) - a(y)}{|\xi - y|^m} Y(\theta_{\xi y}) \right] u(y) \, dy \right|. \end{aligned} \tag{9}$$

The integral over the domain $|x - y| > \delta$ can be written as

$$\begin{aligned} &[a(x) - a(\xi)] \int\limits_{|x-y|>\delta} \frac{Y(\theta_{xy})}{|x-y|^m} u(y) \, dy + \int\limits_{|x-y|>\delta} [a(\xi) - a(y)] \\ &\times \left[\frac{Y(\theta_{xy})}{|x-y|^m} - \frac{Y(\theta_{\xi y})}{|\xi - y|^m} \right] u(y) \, dy. \end{aligned} \tag{10}$$

The first integral in (10) can be estimated by $C \|a\|_{\mathbf{Lip}_\alpha} \|Y\|_{C(S)} \|u\|_p |x - \xi|^\alpha$. In the second integral the term $a(\xi) - a(y)$ is bounded whereas the function $Y(\theta_{xy}) |x - y|^{-m}$ is continuously differentiable with respect to x in the domain $|x - y| > \delta$. By the Lagrange formula and Hölder's inequality this integral is estimated by $C \|a\|_{\mathbf{Lip}_\alpha} \times \|Y\|_{C^1(S)} \|u\|_p |x - \xi|$.

Now let us consider the first integral in the right-hand side of (9). We divide the integration domain into $|x - y| < 2|h|$ and $2|h| < |x - y| < \delta$ where $h = x - \xi$. As in § 3, Chap. XI, we obtain that the integral over the ball $|x - y| < 2|h|$ is of order $C \|a\|_{\mathbf{Lip}_\alpha} \|Y\|_{C(S)} \|u\|_p |x - \xi|$. The inequality

$$\left| [a(\xi) - a(y)] \left[\frac{Y(\theta_{xy})}{|x-y|^m} - \frac{Y(\theta_{\xi y})}{|\xi - y|^m} \right] \right| \leqq C \|a\|_{\mathbf{Lip}_\alpha} \|Y\|_{C^1(S)} \frac{|h| r^\alpha + |h|^{\alpha+1}}{r^{m+1}}$$

and the assumptions of the lemma imply that the second integral in (10) over the domain $2|h| < |x - y| < \delta$ admits of the estimate wanted. It remains to consider the first integral in (10) over the domain $2|h| < |x - y| < \delta$,

$$[a(x) - a(\xi)] \int\limits_{2|h|<|x-y|<\delta} \frac{Y(\theta_{xy})}{|x-y|^m} u(y) \, dy.$$

We get

$$|[a(x) - a(\xi)]| \left| \int\limits_{2|h|<|x-y|<\delta} \frac{Y(\theta_{xy})}{|x-y|^m} u(y)\, \mathrm{d}y \right|$$

$$= \left| \int\limits_{2|h|<|x-y|<\delta} \frac{Y(\theta_{xy})}{|x-y|^m} [u(y) - u(x)]\, \mathrm{d}y \right| |[a\,(x) - a(\xi)]|$$

$$\leqq C\, \|a\|_{\mathbf{Lip}_\alpha} \|Y\|_{\mathbf{C}} \|u\|_{\mathbf{Lip}_\beta} \int\limits_{r<\delta} \frac{\mathrm{d}y}{|x-y|^{m-\beta}}$$

such that the assertion is proved. ∎

7.2. Now let us consider the singular equation

$$a(\xi)\, u(\xi) + \int\limits_{\Gamma} K(\xi, \eta)\, u(\eta)\, \mathrm{d}_\eta \Gamma = g(\xi) \tag{11}$$

the symbol of which will be denoted by $\Phi(\tau)$. Assume that the manifold Γ satisfies the conditions of the present chapter. The notations of the current section are preserved.

Theorem 7.1. *Let* $\inf |\Phi(\tau)| > 0$ *and* $g \in \mathbf{Lip}_\alpha(\Gamma)$. *If the characteristic of the integral in* (11) *satisfies on the set* U_j *(from the finite covering of the manifold* Γ*) the conditions*

$$f_j(x, \cdot) \mathrel{\hat{\in}} \mathbf{W}_2^{(\varepsilon)}(S) \cap \mathbf{L}_{p'}(S), \qquad p > 1,$$

$$\|f_j(x+h, \cdot) - f_j(x, \cdot)\|_{\mathbf{W}_2^{(\varepsilon)}(S)} \leqq B\, |h|^\alpha, \qquad 0 < \alpha < 1, \qquad B = \text{const}, \tag{12}$$

with ε *being some positive number then every solution* $u \in \mathbf{L}_p(\Gamma)$ *to* (11) *belongs also to the space* $\mathbf{Lip}_\alpha(\Gamma)$.

Proof. As in the proof to Theorem 6.1 it suffices to consider the singular equation

$$(Au)\,(x) = a(x)\, u(x) + \int\limits_{\mathbb{R}^m} r^{-m} f(x, \theta)\, u(y)\, \mathrm{d}y + (Tu)\,(x)$$

$$=: (A_0 u)\,(x) + (Tu)\,(x) = g(x) \tag{13}$$

in $\mathbb{R}^m$, and that under the assumptions that $f(x, \theta)$ satisfies the conditions (12), the symbol $\Phi(x, \theta)$ of the operator A does not degenerate, and $a, g \in \mathbf{Lip}_\alpha(\mathbb{R}^m)$. Moreover, T is assumed to be a completely continuous operator in $\mathbf{L}_p(\mathbb{R}^m)$ such that the assertions of Lemmas 7.1 to 7.3 are valid. Let G be a bounded domain in $\mathbb{R}^m$. It is sufficient to show that any solution $u \in \mathbf{L}_p(\mathbb{R}^m)$ to (13) such that $\operatorname{supp} u \subset G$ belongs also to $\mathbf{Lip}_\alpha(\mathbb{R}^m)$.

In order to do so let

$$\Phi(x, \theta) = \sum_{n=0}^{\infty} \sum_{k=1}^{k_{n,m}} b_n^{(k)}(x)\ Y_{n,m}^{(k)}(\theta)$$

and consider the sequence of functions

$$\Phi_N(x, \theta) = \sum_{n=0}^{N} \sum_{k=1}^{k_{n,m}} b_n^{(k)}(x)\ Y_{n,m}^{(k)}(\theta).$$

Due to the results of § 7, Chap. X,

$$\|\Phi(x, \cdot) - \Phi_N(x, \cdot)\|_{\mathbf{W}_2^{(l)}(S)} \leqq \varepsilon_N, \tag{14}$$

$$\|\Phi_N(x+h, \cdot) - \Phi(x, \cdot)\|_{\mathbf{W}_2^{(l)}(S)} \leqq C\, |h|^\alpha, \tag{15}$$

and

$$\|\Phi_N(x,\cdot)-\Phi_N(x+h,\cdot)-[\Phi(x+h,\cdot)-\Phi(x,\cdot)]\|_{\mathrm{W}_2^{(l)}(S)}\leqq C\varepsilon_N\,|h|^\alpha \tag{16}$$

where $l=m/2+\varepsilon$ and $\varepsilon_N\to 0$ as $N\to\infty$.

By Theorem 6.3, Chap. X, the sequence $\Phi_N(x,\theta)$ converges uniformly to $\Phi(x,\theta)$. Therefore, we may assume $|\Phi_N(x,\theta)|>\delta$, $\delta=\text{const}>0$, for sufficiently large N.

Now let $u\in \mathbf{L}_p(\mathbb{R}^m)$ be a solution to (13). Rewrite this equation as

$$(A_0-A_N)\,u+A_Nu+Tu=g \tag{17}$$

where A_N is the operator with the symbol $\Phi_N(x,\theta)$. Applying to that equation (17) the operator R_N with the symbol $[\Phi_N(x,\theta)]^{-1}$ we get

$$R_N(A_0-A_N)\,u+R_NA_Nu+R_NTu=R_Ng. \tag{18}$$

We have $R_NA_N=I+T_1$ with some completely continuous operator T_1 such that (18) becomes

$$u+T_1u+R_NTu+R_N(A_0-A_N)\,u=R_Ng. \tag{19}$$

By the Calderón-Zygmund theorem and by Theorem 6.1 we have for sufficiently large N the inequalities

$$\|R_N(A_0-A_N)\|_{\mathbf{L}_p(\mathbb{R}^m)}\leqq c_p<1$$

and

$$\|R_N(A_0-A_N)\|_{\mathbf{Lip}_\alpha(\mathbb{R}^m)}\leqq c_0<1.$$

Thence, the operator $R=[I+R_N(A_0-A_N)]^{-1}$ does exist, and that operator is bounded in $\mathbf{L}_p(\mathbb{R}^m)$ as well as in $\mathbf{Lip}_\alpha(\mathbb{R}^m)$. We apply this operator to (19) and obtain

$$u+RT_1u+RR_NTu=g_1 \tag{20}$$

with $g_1=RR_Ng\in\mathbf{L}_p(\mathbb{R}^m)\cap\mathbf{Lip}_\alpha(\mathbb{R}^m)$. Equation (20) will be written as

$$u=-RR_NTu-RT_1u+g_1. \tag{21}$$

Put as usually

$$A_n^{(k)}u=\int_{\mathbb{R}^m}\frac{Y_{n,m}^{(k)}(\theta)}{r^m}\,u(y)\,\mathrm{d}y$$

then the operator A_N can be given as the sum

$$A_N=\sum_{n=0}^{N}\sum_{k=1}^{k_{n,m}}b_n^{(k)}(x)\,A_n^{(k)}. \tag{22}$$

Inequality (12) and the theorems of § 6, Chap. X, imply the estimates

$$|b_n^{(k)}(x)-b_n^{(k)}(\xi)|\leqq cn^{-l}\,|x-\xi|^\alpha, \tag{23}$$

$$|b_n^{(k)}(x)|\leqq cn^{-l}. \tag{24}$$

The operator $A_n^{(k)}$ is bounded in the spaces $\mathbf{L}_p(\mathbb{R}^m)$ and $\mathbf{Lip}_\alpha(\mathbb{R}^m)$, and its norms in these spaces can be estimated as

$$\|A_n^{(k)}\|_{\mathbf{L}_p}<C,\qquad \|A_n^{(k)}\|_{\mathbf{Lip}_\alpha}\leqq Cn^{m/2}.$$

The operator R_N can be represented in the following form,

$$R_N=\sum_{n=0}^{\infty}\sum_{k=1}^{k_{n,m}}c_n^{(k)}(x)\,A_n^{(k)}. \tag{25}$$

Now let us consider the operator T in the space $\mathbf{L}_p(\mathbb{R}^m)$. By the multiplication rule for symbols we have $R_N A_N = I + T'_N$ with T'_N being some completely continuous operator. The estimates (23) and (24) allow to multiply the operators A_N and R_N term by term,

$$R_N A_N = \sum_{i=0}^{\infty} \sum_{j=0}^{k_{i,m}} c_i^{(j)} A_i^{(j)} \sum_{n=0}^{N} \sum_{k=1}^{k_{n,m}} b_n^{(k)} A_n^{(k)} = \sum_{i=0}^{\infty} \sum_{j=1}^{k_{i,m}} c_i^{(j)} \sum_{n=0}^{N} \sum_{k=1}^{k_{n,m}} A_i^{(j)} b_n^{(k)} A_n^{(k)}$$

$$= \sum_{i=0}^{\infty} \sum_{j=1}^{k_{i,m}} \sum_{n=0}^{N} \sum_{k=1}^{k_{n,m}} c_i^{(j)} b_n^{(k)} A_i^{(j)} A_n^{(k)} + \sum_{i=0}^{\infty} \sum_{j=1}^{k_{i,m}} \sum_{n=0}^{N} \sum_{k=1}^{k_{n,m}} c_i^{(j)} [A_i^{(j)} b_n^{(k)} - b_n^{(k)} A_i^{(j)}] A_n^{(k)}.$$

The first multiple sum yields the identity, and all other terms are completely continuous. Hence,

$$T'_N = \sum_{i=0}^{\infty} \sum_{j=1}^{k_{i,m}} \sum_{n=0}^{N} \sum_{k=1}^{k_{n,m}} c_i^{(j)}(x) \, [A_i^{(j)} b_n^{(k)} - b_n^{(k)}(x) A_i^{(j)}] \, A_n^{(k)}.$$

The function $g_1(x)$ belongs simultaneously to the spaces $\mathbf{L}_p(\mathbb{R}^m)$ and $\mathbf{Lip}_\alpha(\mathbb{R}^m)$. Let $u \in \mathbf{L}_q(\mathbb{R}^m)$, $q \leqq m/\alpha$. Then $v_n^{(k)} = A_n^{(k)} u \in \mathbf{L}_q(\mathbb{R}^m)$, too. Fix a sufficiently small number $\delta > 0$ such that $\delta_1 = \alpha/m - \delta > 0$. Then there is an integer ν such that $\delta_1 \nu \leqq p^{-1} < (\nu + 1)\, \delta_1$. Consider the integral

$$w_{ij,nk} = \int_{\mathbb{R}^m} \frac{b_n^{(k)}(x) - b_n^{(k)}(y)}{|x - y|^m} \, Y_{i,m}^{(j)}(\theta) \, v_n^{(k)}(y) \, \mathrm{d}y .$$

With the notation

$$g_{nk}(x) = \int_{\mathbb{R}^m} |x - y|^{-m} \, |b_n^{(k)}(x) - b_n^{(k)}(y)| \, |v_n^{(k)}(y)| \, \mathrm{d}y$$

we obtain $|w_{ij,nk}| \leqq c i^{m/2-1} \, g_{nk}(x)$. The estimate (23) and Lemma 7.2 imply $g_{nk} \in \mathbf{L}_{q_1}(\mathbb{R}^m)$, $q_1^{-1} = p^{-1} + \delta$. But then also $g \in \mathbf{L}_{q_1}(\mathbb{R}^m)$.

Similar arguments lead us to $T'_N u \in \mathbf{L}_{q_1}(\mathbb{R}^m)$. By the Calderón-Zygmund theorem the operator R is bounded in $\mathbf{L}_{q_1}(\mathbb{R}^m)$, hence $RT'_N u \in \mathbf{L}_{q_1}(\mathbb{R}^m)$. Now, (21) implies $u \in \mathbf{L}_{q_1}(\mathbb{R}^m)$, too. Repeating these arguments ν times we arrive finally at $u \in \mathbf{L}_{q_\nu}(\mathbb{R}^m)$, $q_\nu^{-1} = 1/p - \nu\,\delta < \delta$, where $q_\nu > \delta^{-1} > m/\beta$. Now we can apply Lemma 7.1 and obtain $Tu \in \mathbf{Lip}_\beta(\mathbb{R}^m)$. Since the operator R is bounded in $\mathbf{Lip}_\beta(\mathbb{R}^m)$ we arrive at $u \in \mathbf{Lip}_\beta(\mathbb{R}^m)$.

Analogously we obtain due to Lemma 7.1 that the sum $\sum_{n=0}^{N} \sum_{k=1}^{k_{n,m}} w_{ij,nk} =: d_{ij}$ belongs to the space $\mathbf{Lip}_\beta(\mathbb{R}^m)$ where

$$|d_{ij}(x)| \leqq C i^{m\,2-1}, \qquad |d_{ij}(x) - d_{ij}(\xi)| \leqq C n^{-l} \, |x - \xi|^\alpha, \qquad |x - \xi| \leqq 1 .$$

With the notation $t_N(x) = (T'_N u)\,(x)$ we have

$$t_N(x) - t_N(\xi) = \sum_{i=0}^{\infty} \sum_{j=1}^{k_{i,m}} [b_i^{(j)}(x) - b_i^{(j)}(\xi)] \, d_{ij}(x) + \sum_{i=0}^{\infty} \sum_{j=1}^{k_{i,m}} b_i^{(j)}(\xi) \, [d_{ij}(x) - d_{ij}(\xi)],$$

and we conclude $t_N \in \mathbf{Lip}_\alpha(\mathbb{R}^m)$. Now the operator R is bounded in $\mathbf{Lip}_\alpha(\mathbb{R}^m)$ such that (21) finally yields $u \in \mathbf{Lip}_\alpha(\mathbb{R}^m)$. ∎

Chapter XIV
Systems of multidimensional singular equations

§ 1. General remarks

In the present chapter, Γ means either the Euclidean space $\mathbb{R}^m$ or a compact, sufficiently smooth m-dimensional manifold without boundary which is imbedded in a higher dimensional Euclidean space. We shall be concerned with systems of singular equations of the form

$$\sum_{k=1}^{n} (A_{jk}u_k)(x) = g_j(x), \qquad j = 1, 2, \ldots, n. \tag{1}$$

Here, A_{jk} means a general singular operator as defined in § 2, Chap. XIII. The symbol of the operator A_{jk} will be denoted by $\Phi_{jk}(\tau)$ where τ stands for a linear element of the manifold Γ. Remember that we can write $\Phi_{jk}(\tau) = \Phi_{jk}(x, \theta)$ $(x \in \Gamma, \theta \in S)$ provided that Γ can be endowed with a regular coordinate system.

Let us introduce the matrix

$$A = \begin{pmatrix} A_{11} & A_{12} & \ldots & A_{1n} \\ A_{21} & A_{22} & \ldots & A_{2n} \\ \ldots & \ldots & \ldots & \ldots \\ A_{n1} & A_{n2} & \ldots & A_{nn} \end{pmatrix} \tag{2}$$

and the column vectors u and g with components $u_1, u_2, \ldots, u_n$ and $g_1, g_2, \ldots, g_n$, resp. Then the system (1) can be written briefly as one equation

$$Au = g. \tag{1a}$$

The matrix A will be called a *singular matrix operator*. Furthermore, let us introduce the *symbol matrix* of the operator A (or of the system (1)),

$$\Phi(\tau) = \begin{pmatrix} \Phi_{11}(\tau) & \Phi_{12}(\tau) & \ldots & \Phi_{1n}(\tau) \\ \Phi_{21}(\tau) & \Phi_{22}(\tau) & \ldots & \Phi_{2n}(\tau) \\ \ldots & \ldots & \ldots & \ldots \\ \Phi_{n1}(\tau) & \Phi_{2n}(\tau) & \ldots & \Phi_{nn}(\tau) \end{pmatrix}; \tag{3}$$

its determinant is called the *symbol determinant* of the operator A (or of the corresponding system (1)). We shall assume that the symbols $\Phi_{jk}(\tau)$ satisfy the conditions of Theorem 8.2, Chap. XI.

The multiplication of singular matrix operators is performed according to the usual multiplication rule for matrices; of course, the product is not commutative. The theorems of § 7, Chap. XI, imply that the product of singular matrix operators corresponds to the product of their symbol matrices. Furthermore, it is obvious that the same is true for sums.

Theorem 1.1. *Under the assumption*

$$\inf |\det \Phi(\tau)| > 0 \tag{4}$$

the system (1) *is normally solvable in the corresponding space* $\mathbf{L}_p(\Gamma)$, *and its index is finite.*

The proof follows immediately from the fact that the operator A and its adjoint A^* admit of a regularization: the singular operators with the symbol matrices $[\Phi(x, \theta)]^{-1}$ and $[\Phi^*(x, \theta)]^{-1}$, resp., provide regularizers, where Φ^* means the adjoint to the matrix Φ.

Unlike the case of the singular equation, the index of a system can be non-zero also if the inequality (4) is obeyed and the symbols $\Phi_{jk}(\tau)$ satisfy the conditions of the theorems in §§ 7, 8, Chap. XI. This problem will be discussed detailed in the subsequent sections of this chapter. We mention here only that the index of singular matrix operators is studied in the papers Wolpert [1, 2], Seeley [3, 4], Dynin [1, 2], Atiyah and Singer [1, 2], Mikhlin [20, 21], Agranovich [1], Fedosov [1, 2] et al. In Wolpert's papers first it was shown that the index of a singular matrix operator can be different from zero. Atiyah and Singer obtained an index formula for elliptic operators of very general form; the singular matrix operators are a rather special subclass of operator considered by them. The papers of Atiyah and Singer initiated a reasonable number of investigations which are summarized in their vast majority in the books by Palais [1] and in the Séminaire Henri Cartan [1] (cf. also Booss [1], Rempel and Schulze [1]). The papers by Fedosov mentioned above aim also at that direction.

Finally, we mention a very simple class of systems (1) which is obtained for $\Gamma = \mathbb{R}^m$ with a symbol matrix that depends only on $\theta \in S$. In that case the simple singular operators with the symbols $\Phi^{-1}(\theta)$ and $[\Phi^*(\theta)]^{-1}$ provide equivalent regularizers for the operators A and A^*, resp. The index of such a system is then, obviously, zero.

§ 2. The index problem. Reduction to a more special case

2.1. Let Γ be a sufficiently smooth, compact m-dimensional manifold without boundary. Suppose that Γ consists of a finite number N connected sub-manifolds $\Gamma_1, \Gamma_2, \ldots, \Gamma_N$ which are pairwise disjoint. One of these sub-manifolds may coincide with the extended space $\overline{\mathbb{R}^m}$.

Let us consider the system of singular equations

$$\sum_{k=1}^{n} \left\{a_{jk}(x)\, u_k(x) + \int_\Gamma K_{jk}(x, y)\, u(y)\, \mathrm{d}_y\Gamma + (T_{jk}u)\,(x)\right\} = g_j(x); \qquad j = 1, 2, \ldots, n. \tag{1}$$

Here, x and y are points on the manifold Γ, and $K_{jk}(x, y)$ are singular kernels satisfying the usual conditions; the operators T_{jk} are completely continuous in the corresponding function space. Furthermore, we assume that the symbol determinant of the system (1) never vanishes.

Let us introduce some notations. The manifold of linear elements of Γ will be denoted by $\tilde{\Gamma}$, and τ is a generic element of $\tilde{\Gamma}$. As in § 1 we denote by $\Phi(\tau) = (\Phi_{jk}(\tau))_{j,k=1}^{n}$ the symbol matrix of the system (1). The index of a singular system will be called also *index of the symbol matrix*. If a symbol matrix is given on the manifold of linear elements of a manifold Γ_i we shall speak of the index of that matrix on Γ_i. For a non-degenerate symbol matrix Ψ defined on the manifold $\tilde{\Delta}$ of linear elements of a manifold Δ we shall denote the index of that matrix on Δ by Ind (Ψ, Δ) (or simply by Ind Ψ if there is no danger of ambiguity).

2.2. Theorem 2.1. *Let* Γ, Γ_j, and Φ *have the meaning as described in Section 2.1. Then*

$$\operatorname{Ind}(\Phi, \Gamma) = \sum_{j=1}^{N} \operatorname{Ind}(\Phi, \Gamma_j). \tag{2}$$

Proof. To begin with, we derive an auxiliary relation.

Let Φ be an arbitrary element of a topological group G and denote by $l_1(\Phi), l_2(\Phi), \ldots, \ldots, l_\nu(\Phi)$ homomorphisms of G into the group of integers such that there are elements $\Phi_1, \Phi_2, \ldots, \Phi_\nu \in G$ with $l_j(\Phi_k) = \delta_{jk}$. Furthermore, assume that there exists a homomorphism $l_0(\Phi)$ of G into the group of integer numbers with the following property: If for some $\Phi^* \in G$ the equations $l_j(\Phi^*) = 0$, $j = 1, 2, \ldots, \nu$, hold then $l_0(\Phi^*) = 0$. Under these assumptions we have

$$l_0(\Phi) = \sum_{j=1}^{\nu} \mu_j l_j(\Phi) \tag{3}$$

for any $\Phi \in G$, where μ_j are integers.

In order to prove this assertion put $\mu_j = l_0(\Phi_j)$ and $l_j(\Phi) = q_j$ for an arbitrary but fixed element $\Phi \in G$. Obviously,

$$l_j(\Phi\Phi_1^{-q_1}\Phi_2^{-q_2} \ldots \Phi_\nu^{-q_\nu}) = q_j - \sum_{k=1}^{\nu} \delta_{jk} q_k = 0, \qquad j = 1, 2, \ldots, \nu,$$

and hence

$$l_0(\Phi\Phi_1^{-q_1}\Phi_2^{-q_2} \ldots \Phi_\nu^{-q_\nu}) = 0,$$

what is equivalent to the relation (3).

Now we shall show that $\operatorname{Ind}(\Phi, \Gamma_j) = 0$, $j = 1, 2, \ldots, N$, implies $\operatorname{Ind}(\Phi, \Gamma) = 0$. Let us write the system (1) as a matrix equation,

$$a(x)\, u(x) + \int_{\Gamma} K(x, y)\, u(y)\, \mathrm{d}_y\Gamma + (Tu)(x) = g(x). \tag{4}$$

For the vector functions $u(x)$ and $g(x)$ on Γ_j we shall use the notation $u^{(j)}(x)$ and $g^{(j)}(x)$, resp. Then, for $x \in \Gamma$ (4) assumes the form

$$a(x)\, u^{(j)}(x) + \int_{\Gamma_j} K(x, y)\, u^{(j)}(y)\, \mathrm{d}_y\Gamma + \sum_{k \neq j} \int_{\Gamma_k} K(x, y)\, u^{(k)}(y)\, \mathrm{d}_y\Gamma + (Tu)(x) = g^{(j)}(x), \quad x \in \Gamma_j. \tag{5}$$

In (5) only the integral over Γ_j is singular, the other integrals are Fredholm operator for the corresponding vectors $u^{(k)}$. Hence, (5) is singular with respect to $u^{(j)}(x)$, and its symbol matrix is $\Phi(\tau)$ restricted to $\tilde{\Gamma}_j$.

Now, the proof goes as follows. Let A and B be two operators in the Banach spaces E and F, resp., $E \cap F = \{0\}$, which admit of a two-sided regularization. For $H = E \dotplus F$, $C = A \dotplus B$ the following formula is true,

$$\operatorname{Ind} C = \operatorname{Ind} A + \operatorname{Ind} B. \tag{6}$$

In order to prove it we count the solution numbers of the corresponding equations. Let $u_1, u_2, \ldots, u_k$ and $v_1, v_2, \ldots, v_l$ be bases in the null spaces of the operators A and B resp., $\alpha(A) = k$, $\alpha(B) = l$. The equation $Cw = 0$, $w = (u, v)$, is equivalent to the pair of equations $Au = 0$, $Bv = 0$; the null space of he operator C has, obviously, the basis $(u_i, 0)$, $(0, v_j)$, $i = 1, 2, \ldots, k$; $j = 1, 2, \ldots, l$. Hence, $\alpha(C) = k + l = \alpha(A) + \alpha(B)$. Similarly, $\alpha(C^*) = \alpha(A^*) + \alpha(B^*)$ such that (6) is proved.

We neglect in (5) the completely continuous terms Tu and $\sum_{k \neq j} \int_{\Gamma_k} K(x, y)\, u^{(k)}(y)\, \mathrm{d}_y\Gamma$ which do not affect to the index of the system. Then (5) separates into independent systems according to the values of j, and Theorem 2.1 follows immediately from (6). The index of (5) equals Ind (Φ, Γ_j) and is, by assumption, zero. In that case this equation admits of an equivalent regularization: There is a singular operator $R^{(j)}$ that makes (5) equivalent to an equation of the form $u^{(j)}(x) + (T^{(j)}u)(x) = g^{(j)}(x)$ where $T^{(j)}$ is a completely continuous operator. That means that (4) admits also of an equivalent regularization: Any regularizer the restriction of which to functions of $x \in \Gamma_j$ coincides with $R^{(j)}$ is, at the same time, an equivalent regularizer. But if (4) admits of an equivalent regularization then Ind $(\Phi, \Gamma) \geqq 0$ (cf. Theorem 2.3, Chap. I, or also Mikhlin [22]).

The symbol matrix of the adjoint system to (4) equals $\Phi^*(\tau)$, the adjoint matrix to $\Phi(\tau)$. Due to the assumption, we have Ind $(\Phi^*, \Gamma_j) = -$Ind $(\Phi. \Gamma_j) = 0$. Repeating the foregoing arguments we obtain Ind $(\Phi^*, \Gamma) \geqq 0$. These two inequalities imply Ind $(\Phi, \Gamma) = 0$.

Now let $\Phi_j(\tau)$ be a matrix defined on the manifold $\tilde{\Gamma}_j$ of linear elements of Γ_j the index of which assumes the smallest possible for Γ_j positive value, say $\varkappa_j$. It is easy to see that the index of an arbitrary matrix Ψ defined on $\tilde{\Gamma}_j$ is an integer multiple of $\varkappa_j$. Indeed, suppose Ind $(\Psi, \Gamma_j) = a\varkappa_j + b$, a and b integers, $0 < b < \varkappa_j$. Then Ind $(\Phi_j^{-a}\Psi, \Gamma_j) = b$, and this is in contradiction to the definition of the number $\varkappa_j$.

It is possible that for every j and any matrix $\Phi(\tau)$ the relation Ind $(\Phi, \Gamma_j) = 0$ holds. In that case we showed above that Ind $(\Phi, \Gamma) = 0$, and our theorem is proved.

Now, let Ind $(\Phi, \Gamma_j) \neq 0$, $j = 1, 2, \ldots, s$, and Ind $(\Phi, \Gamma_j) = 0$, $j = s + 1, \ldots, N$. We extend the matrix $\Phi_j(\tau)$ to the whole manifold $\tilde{\Gamma}$ by setting $\Phi_j(\tau)$ the unit matrix for $\tau \in \tilde{\Gamma}_k$, $k \neq j$. In order to show Ind $(\Phi_j, \Gamma_j) =$ Ind (Φ_j, Γ) we consider the corresponding singular system. By $\Phi(\tau) * v$ we denote the result when the singular operator with the symbol matrix $\Phi(\tau)$ is applied to the function $v(x)$. Then the singular system with the extended symbol matrix $\Phi_j(\tau)$ can be written as

$$\Phi_j(\tau) * v = g(x), \qquad x \in \Gamma_j, \tag{7$_1$}$$

$$v = g(x), \qquad x \in \Gamma_k, \qquad k \neq j. \tag{7$_2$}$$

Equation (7_2) determines uniquely the vector $v(x)$ we are seeking, if $k \neq j$. Therefore the solution number of the homogeneous equation (7) and the number of solvability conditions of the inhomogeneous equation (7) coincide with the corresponding numbers of (7_1), i.e. Ind $(\Phi_j, \Gamma) =$ Ind (Φ_j, Γ_j).

The non-degenerate symbol matrices defined on $\tilde{\Gamma}$ form a topological group, and in that group there are defined N homomorphisms into the group of integer numbers,

$$l_j(\Phi) = \frac{1}{\varkappa_j} \text{Ind}\,(\Phi, \Gamma_j), \qquad j = 1, 2, \ldots, N,$$

obviously, $l_j(\Phi_k) = \delta_{jk}$. There is another homomorphism $l_0(\Phi) =$ Ind (Φ, Γ) defined on that group. As we showed above, the relations $l_j(\Phi) = 0$, $j = 1, 2, \ldots, N$, imply $l_0(\Phi) = 0$. Then, by (3),

$$\text{Ind}\,(\Phi, \Gamma) = \sum_{j=1}^{N} \frac{\mu_j}{\varkappa_j} \text{Ind}\,(\Phi, \Gamma_j).$$

Put $\Phi = \Phi_k$ herein and find $\mu_k = \varkappa_k$ such that Theorem 2.1 is proved. ▮

2.3. Theorem 2.2. *If in the system* (1) $n > m$ *then one can construct a singular system of order m the index of which is equal to the index of the system* (1).

In case $n < m$ it suffices to add the equations

$$u_{n+1} = 0, \qquad u_{n+2} = 0, \ldots, u_m = 0$$

to system (1); *it is clear that the index is the same.*

Proof. Let $n > m$. Separating in $\Phi_{jk}(\tau)$ the real and the imaginary part, $\Phi_{jk}(\tau) = \Phi'_{jk}(\tau) + i\Phi''_{jk}(\tau)$, we put

$$\Phi_1(\tau_1) = \sqrt{\sum_{k=1}^{n} [\Phi'^2_{k2}(\tau) + \Phi''^2_{k1}(\tau)]} = \sqrt{\sum_{k=1}^{n} |\Phi_{k1}(\tau)|^2}.$$

Since the matrix $\Phi(\tau)$ is non-singular, we have inf $|\Phi_1(\tau)| > 0$. Furthermore, we introduce a normed vector $\varphi(\tau) = (\varphi_1(\tau), \varphi_2(\tau), \ldots, \varphi_{2n}(\tau))$ of dimension $2n$ by

$$\varphi_{2k-1}(\tau) = \frac{\Phi'_{k1}(\tau)}{\Phi_1(\tau)}, \qquad \varphi_{2k}(\tau) = \frac{\Phi''_k(\tau)}{\Phi_1(\tau)}.$$

The vector $\varphi(\tau)$ provides a mapping of the $(2m-1)$-dimensional space $\tilde{\Gamma}$ into the $(2n-1)$-dimensional sphere S_{2n-1}. Because of $n > m$ all the values $\varphi(\tau)$ are unstable for this mapping (cf. Hurewicz and Wallman [1]), i.e. an arbitrarily small perturbation of that mapping may cause any point of the sphere S_{2n-1}, e.g. the south pole, to fall out of the range of the mapping. Now the space $\tilde{\Gamma}$ is complete and compact, the mapping $\varphi(\tau)$ is continuous, therefore, the range of the mapping $\varphi(\tau)$ is closed. Hence, if the south pole is no image point then also a certain spherical neighborhood does not belong to the range of the mapping $\varphi(\tau)$. It is worth noting that the small perturbation of the mapping $\varphi(\tau)$ can be caused by an appropriately small perturbation of the matrix $\Phi(\tau)$ which leaves the index invariant.

Now, we want to show that to any linear element $\tau \in \tilde{\Gamma}$ there is a unit vector $\lambda(\tau) = (\lambda_1(\tau), \ldots, \lambda_{2n}(\tau))$ in the $2n$-dimensional Euclidean space with the following properties:

1. $|\lambda_1(\tau)| \geqq \eta_1$ and $|(\varphi, \lambda)| \geqq \eta_2$ with η_1, η_2 being two positive constants;
2. $\lambda(\tau)$ is a sufficiently smooth function of τ.

Let us denote the coordinate axes in $\mathbb{R}^{2n}$ by $0z_1, 0z_2, \ldots, 0z_{2n}$. Assume that the $0z_1$ axis is directed from the center of the sphere S_{2n-1} to the north pole of it. We introduce angular coordinates $\vartheta_1, \vartheta_2, \ldots, \vartheta_{2n-1}$ such that ϑ_1 is the angle between the radius vector and the $0z_1$ axis. The neighborhood σ mentioned above can be given by the inequalities $\pi - \alpha < \vartheta_1 \leqq \pi$, $0 < \alpha < \pi/2$. Take an angle $\beta \in (0, \alpha)$ and define $\lambda(\tau)$ as follows: Let the end point of the unit vector $\varphi(\tau)$ have the angular coordinates $\vartheta_1, \vartheta_2, \ldots, \vartheta_{2n-1}$. If that point belongs to the upper hemi-sphere ($\vartheta_1 \leqq \pi/2$) then the end point of the vector $\lambda(\tau)$ is defined by the angular coordinates $(\pi - 2\beta)\, \vartheta_1/\pi, \vartheta_2, \ldots, \vartheta_{2n-1}$. If the end point of the vector $\varphi(\tau)$ lies on the southern hemi-sphere then it does not belong to the neighborhood σ, and therefore $\pi/2 < \vartheta_1 \leqq \pi - \alpha$. In that case we define the end point of the vector $\lambda(\tau)$ by the angular coordinates $\pi/2 - \beta, \vartheta_2, \ldots, \vartheta_{2n-1}$. For the vector $\lambda(\tau)$ defined that way the angle δ between $\lambda(\tau)$ and the $0z_1$ axis satisfies the conditions $0 \leqq \delta \leqq \pi/2 - \beta$ and $|\cos \delta| = |\lambda_1(\tau)| \geqq \sin \beta$. Moreover, the angle between the vectors $\varphi(\tau)$ and $\lambda(\tau)$ is not greater than $\pi/2 - \alpha + \beta$, hence $|(\varphi, \lambda)| \geqq \sin(\alpha - \beta)$. So condition (1) is fulfilled. The vector λ as a function of τ is merely continuous. In order to satisfy condition 2. the function $\lambda(\tau)$ is subject to Sobolev mollification with sufficiently small radius and then normed again. This mollification does not affect to condition 1.

Now,

$$\left|\sum_{k=1}^{2n} \varphi_k \lambda_k\right| \geqq \eta_2$$

or

$$\left|\operatorname{Re} \sum_{k=1}^{n} \Phi_{k1} (\lambda_{2k-1} - i\lambda_{2k})\right| \geqq \eta_2 \inf |\Phi_1| =: \eta_3 > 0,$$

hence

$$\left|\sum_{k=1}^{n} \Phi_{k1}(\lambda_{2k-1} - i\lambda_{2k})\right| \geqq \eta_3 .$$

Put

$$\psi_k(\tau) = \frac{\lambda_{2k-1}(\tau) - i\,\lambda_{2k}(\tau)}{\lambda_1(\tau) - i\lambda_2(\tau)} .$$

The functions $\psi_k(\tau)$ are sufficiently smooth, and

$$\left|\Phi_{11}(\tau) + \sum_{k=2}^{n} \psi_k(\tau)\, \Phi_{k1}(\tau)\right| \geqq \frac{\eta_3}{|\lambda_1 - i\lambda_2|} \geqq \frac{\eta_3}{\eta_1} . \tag{8}$$

We rewrite system (1),

$$L_j u := \sum_{k=1}^{n} \{\Phi_{jk} * u_k + T_{jk} u_k\} = g_j, \qquad j = 1, 2, \dots, n,$$

then it is equivalent to the following one,

$$L_1 u + \sum_{k=2}^{n} \psi_k * L_k u = g_1 + \sum_{k=2}^{n} \psi_k * g_k , \tag{9}$$

$$L_j u = g_j, \qquad j = 2, 3, \dots, n. \tag{10}$$

In (9) the symbol for the operator at u_1 is

$$\Omega(\tau) = \Phi_1(\tau) + \sum_{k=2}^{n} \psi_k(\tau)\, \Phi_{k1}(\tau) .$$

Due to inequality (8) we have $|\Omega(\tau)| \geqq \eta_3/\eta_1$. Taking u_1 for the only unknown we may give (9) the equivalent form

$$u_1 = - \sum_{k=2}^{n} [\Omega(\tau)]^{-1} \psi_k(\tau) * (L_k u - \Phi_{k1} * u_1) + T'u + [\Omega(\tau)]^{-1} * \left(g_1 + \sum_{k=2}^{s} \psi_k * g_k\right) \tag{11}$$

where T' is a certain completely continuous operator. Now, the system (9)–(10) can be replaced by the equivalent one (10)–(11). The index of such a system is not touched by a completely continuous term; hence, it remains the same if in the right-hand side of (11) all terms are omitted which contain u_1 under a completely continuous operator. Then (11) yields an explicit expression for u_1 by the remaining unknowns. Inserting this expression instead of u_1 into (10) we obtain a new system of the same index but with only $n - 1$ unknowns $u_2, u_3, \dots, u_n$. In this manner we proceed until the dimension of the system is m. ▮

We add a remark that will be useful for the following. Let $\Phi(\tau)$ be a non-singular symbol matrix. As it is well-known, such a matrix can be factorized, $\Phi(\tau) = S(\tau)\, U(\tau)$ where S is a symmetric and U a unitary matrix. Due to Atkinson's theorem, $\operatorname{Ind} \Phi = \operatorname{Ind} S + \operatorname{Ind} U$. In § 5 we shall show that the index of a singular operator with a Hermitean symbol matrix is zero, hence $\operatorname{Ind} \Phi = \operatorname{Ind} U$. Thus, the index problem for the general case is reduced to the computation of the index for a singular operator with a unitary symbol matrix.

§ 3. Computation of the index

In this section we shall give without proof the Atiyah-Singer formula for the index of an elliptic operator on a manifold and, moreover, several special formulas for the case of singular matrix operators.

Let A be an elliptic operator (for the definition we refer to Atiyah and Singer [1]) with the symbol Φ which is defined on a sufficiently smooth compact m-dimensional manifold Γ without boundary. By ch Φ we denote the Chern character of the symbol Φ, and by $\mathscr{T}(\Gamma)$ the Todd class of the manifold Γ. Then the Atiyah-Singer formula can be given as

$$\operatorname{Ind} A = (\operatorname{ch} \Phi \cdot \mathscr{T}(\Gamma))\,[\Gamma]. \tag{1}$$

Here the right-hand side means the value of the m-dimensional component of the element ch $\Phi \cdot \mathscr{T}(\Gamma)$ at the fundamental cycle $[\Gamma]$ of the manifold Γ (cf. Atiyah and Singer [1], Palais [1], § 1, Chap. I, or Booss [1], Rempel and Schulze [1]).

Fedosov [1] indicated a simpler formula for the index of pseudo-differential operators and, in particular, of singular integral operators.

Let Γ be again a compact sufficiently smooth m-dimensional manifold without boundary. By $T(\Gamma)$ we denote its co-tangential bundle, and by $S(\Gamma)$ its unit spheres bundle on $T(\Gamma)$. It can be shown that

$$\mathscr{T}(\Gamma) = 1 + \sum_{k=1}^{[m/4]} \mathscr{T}_k \tag{2}$$

where $\mathscr{T}_k$ are certain differential forms of degree $4k$. Furthermore, we introduce the differential forms

$$\psi_k = \frac{1}{(2\pi \mathrm{i})^k} \frac{(k-1)!}{(2k-1)!} \operatorname{sp} (\Phi^{-1}\,\mathrm{d}\Phi)^{2k-1}. \tag{3}$$

It turns out that (1) for pseudo-differential operators (and, in particular, for singular integral operators) is equivalent to the following formula,

$$\operatorname{Ind} A = \frac{(-1)^{m+1}\,(m-1)!}{(2\pi \mathrm{i})^m\,(2m-1)!} \int_{S(\Gamma)} \left\{\psi_m + \sum_{k=1}^{[m/4]} \psi_{m-2k}\mathscr{T}_k\right\}. \tag{4}$$

In case if $\mathscr{T}(\Gamma) = 1$ this formula becomes considerably simpler, namely

$$\operatorname{Ind} A = \frac{(-1)^{m+1}\,(m-1)!}{(2\pi \mathrm{i})^m\,(2m-1)!} \int_{S(\Gamma)} \operatorname{sp} (\Phi^{-1}\,\mathrm{d}\Phi)^{2m-1}, \tag{5}$$

where the multiplication under the integral is to be understood as the outer multiplication of differential forms.

Equation (5) was further simplified by Fedosov [2] for a special case. Let $\Gamma = \mathbb{R}^m$, and for the sake of simplicity we restrict ourselves to the case of singular integral operators. Suppose that the symbol matrix coincides with the unit matrix outside of a certain ball $|x| < a$. Then the integral in (5) can be taken over the $(2n-1)$-dimensional sphere $|x|^2 + |\xi|^2 = a^2$ instead of $S(\Gamma)$.

Finally, we shall formulate some results by Seeley [3]. We begin with some simple remarks. Let Γ be again a compact sufficiently smooth m-dimensional manifold without boundary, and on Γ be given a system of m singular equations with a unitary symbol matrix $\Phi(\tau)$. Consider e.g. the first row of that matrix, $(\Phi_{11}(\tau), \Phi_{12}(\tau), \ldots, \Phi_{1m}(\tau))$, separate real and imaginary part, and put $\Phi_{1j}(\tau) = \varphi_j(\tau) + \mathrm{i}\varphi_{m+j}(\tau)$. Since the matrix

$\Phi(\tau)$ is unitary we have $\sum_{k=1}^{2m} \varphi_k^2(\tau) = 1$. This last relation defines a certain mapping $\Pi(\Phi)$ of the set of all unitary matrices of order m as functions on the set $\tilde{\Gamma}$ of linear elements of Γ, into the $(2m-1)$-dimensional sphere S_{2m-1}. This mapping takes every linear element $\tau \in \tilde{\Gamma}$ to a point of the sphere S_{2m-1} with the Cartesian coordinates $\varphi_k(\tau)$, $k = 1, 2, \ldots, 2m$. The degree of that mapping $\Pi(\Phi)$ will be denoted by $p = p(\Phi)$.

Now we are able to formulate Seeley's fundamental theorem on the index of a system on the manifold Γ[1]):

Theorem 3.1. *Let there be given a system of m singular equations on a oriented compact, sufficiently smooth m-dimensional manifold Γ without boundary with a unitary symbol matrix $\Phi(\tau)$ the elements of which satisfy the conditions of Theorem 8.2, Chap. XI. If on Γ any system of $m-1$ singular integral equations with $m-1$ unknowns and non-degenerate symbol matrix has the index zero then the co-tangential manifold $S^*(\Gamma)$ and the sphere S_{2m-1} can be oriented such that the formula*

$$\operatorname{Ind} \Phi(\tau) = \frac{1}{(m-1)!} p(\Phi) \tag{6}$$

holds. Conversely, if (6) *is true then the index of any system of k singular equations with k unknowns, $k < m$, and non-degenerate symbol matrix is zero.*

Equation (6) is true for any manifold the Todd class of which is equal to 1. In particular, it is true for the Euclidean space, for the sphere, and for every two-dimensional manifold (cf. SEELEY [3]). This last case which is more elementary will be considered in the next section applying different arguments.

§ 4. The case of a two-dimensional manifold

In this section we shall prove in a rather elementary way that for two-dimensional manifolds (3.6) is valid up to a certain integer factor where this factor does neither depend on the system in question nor on the manifold on which the system is considered. The proof is contained in MIKHLIN [20, 21]. That the integer factor does not depend on the manifold, was observed by Wolpert. Using the index formula for a two-dimensional sphere proved by WOLPERT [1, 2] you can easily show that this factor is 1. Thus, formula (3.6) is completely proved for two-dimensional manifolds.

4.1. Let us be given a system of two singular equations in two unknowns on a two-dimensional oriented, sufficiently smooth compact manifold Γ without boundary the symbol matrix of which is unitary.

A unitary matrix of order 2 can be given in the form

$$\Phi = \begin{pmatrix} \alpha & \beta \\ -\bar{\beta} e^{i\gamma} & \bar{\alpha} e^{i\gamma} \end{pmatrix} \tag{1}$$

with a real γ and, in general, complex α and β. In our case these quantities are functions on the set $\tilde{\Gamma}$ of the linear elements of the manifold given.

[1]) This theorem was formulated and proved by SEELEY for a more general class of pseudo-differential operators.

The elements of the matrix (1) are subject to certain smoothness conditions which come from the theorems of §§ 7, 8, Chap. XI. In that case the function $e^{i\gamma}$, the determinant of the matrix (1), satisfies the same smoothness conditions. The matrix (1) can be represented as a product

$$\begin{pmatrix} \alpha & \beta \\ -\bar{\beta}e^{i\gamma} & \bar{\alpha}e^{i\gamma} \end{pmatrix} = \begin{pmatrix} 1 & 0 \\ 0 & e^{i\gamma} \end{pmatrix} \begin{pmatrix} \alpha & \beta \\ -\bar{\beta} & \bar{\alpha} \end{pmatrix}.$$

The index of the first matrix on the right-hand side is zero because the corresponding system separates into two independent equations with the smooth and non-degenerate symbols 1 and $e^{i\gamma}$. Thus, the problem of computing the index is reduced to find the index for the unimodular matrix

$$U = \begin{pmatrix} \alpha & \beta \\ -\bar{\beta} & \bar{\alpha} \end{pmatrix}, \quad |\alpha|^2 + |\beta|^2 = 1. \tag{2}$$

Put $\alpha = \alpha' + i\alpha''$, $\beta = \beta' + i\beta''$ and consider the 4-dimensional vector $A = (\alpha', \alpha'', \beta', \beta'')$ determined by the first row of the matrix (2). Because of $\alpha'^2 + \alpha''^2 + \beta'^2 + \beta''^2 = 1$ this vector defines a certain mapping of the manifold $\tilde{\Gamma}$ into the three-dimensional unit sphere. As in § 3 we denote by $p(U)$ the degree of that mapping which will be called, due to Krasnoselski, *rotation* of the vector A.

We shall show the relation

$$p(U_1U_2) = p(U_1) + p(U_2). \tag{3}$$

To that aim we associate to the vector A the normed quaternion $K(U) = \alpha' + i\alpha'' + j\beta' + k\beta''$. As it is well-known, $K(U_1U_2) = K(U_1)\,K(U_2)$ (this relation can be easily verified immediately). We shall also speak of the rotation of the corresponding quaternion meaning the rotation of the vector A. It suffices to show that the rotations are added if the quaternions are multiplied.

Let a quaternion $K(\tau)$ be a function of the linear element $\tau \in \tilde{\Gamma}$, the rotation of it will be denoted by ν. Let us, first, find the rotation of the quaternion $K^n(\tau)$, n integer. Choose an arbitrary unit quaternion L and consider the equation

$$K^n(\tau) = L. \tag{4}$$

We may represent L in the form $L = \cos\theta + \lambda \sin\theta$ where θ is the so-called argument, and λ the axis of the quaternion L. Then (4) is equivalent to n single equations

$$K(\tau) = \cos\frac{\theta + 2h\pi}{n} + \lambda \sin\frac{\theta + 2h\pi}{n}, \quad h = 0, 1, \ldots, n-1. \tag{5}$$

Let the h-th equation (5) have a_h and b_h solutions, resp., for which the Jacobian determinant of the mapping from $\tilde{\Gamma}$ into the three-dimensional sphere is positive or negative, resp., at a certain point. Obviously, $a_h - b_h = \nu$. Thence, (4) has $\sum_{h=0}^{n-1} a_h$, $\sum_{h=0}^{n-1} b_h$ solutions with positive or negative Jacobian determinant, resp. Therefore, the rotation of the quaternion $K^n(\tau)$ is $\sum_{h=0}^{n-1} (a_h - b_h)$.

It is easy to see that this result is true also if $n \leq 0$. Indeed, for $n = 0$ this is obious. For $n = -m$, $m > 0$, (5) is of the form

$$K(\tau) = \cos\frac{-\theta + 2h\pi}{n} + \lambda \sin\frac{-\theta + 2h\pi}{n}, \quad h = 0, 1, \ldots, n-1.$$

A sign changing in θ causes a sign changing in the Jacobian determinant mentioned above. Hence, the rotation of the quaternion $K^{-1}(\tau)$ is $-\nu$. But then the rotation of the quaternion $K^n(\tau) = [K^{-1}(\tau)]^m$ is, according to the assertion proved above, equal to $-\nu m = n\nu$.

Now, let $K_1(\tau)$ and $K_2(\tau)$ be two quaternions with the rotations ν_1 and ν_2, resp., and $L(\tau)$ be an arbitrary quaternion with rotation 1. Then the quaternions K_1, L^{ν_1} and K_2, L^{ν_2} have, in each case, the same rotation such that they are, by the well-known Hopf theorem (cf. Alexandroff and Hopf [1]), homotopic to each other, $K_1 \sim L^{\nu_1}$, $K_2 \sim L^{\nu_2}$. Hence, $K_1K_2 \sim L^{\nu_1+\nu_2}$, and since homotopic quaternions have the same rotation we obtain that the rotation of the quaternion K_1K_2 is equal to $\nu_1 + \nu_2$ which was to be shown.

As a consequence, we obtain the result that the rotation $p(U)$ of the vector A provides a homomorphism of the topological group of unimodular symbol matrices into the group of integer numbers.

Assume that for the matrix U we have $p(U) = 0$. By the Hopf theorem cited above the vector A is homotopic to the vector $(1, 0, 0, 0)$. That means the existence of two continuous functions $\varphi_1(\tau, t)$, $\varphi_2(\tau, t)$, $\tau \in \tilde{\Gamma}$, $t \in [0, 1]$, (which can be assumed to be sufficiently smooth) with the following properties: $\inf \{|\varphi_1|^2 + |\varphi_2|^2\} > 0$, $\varphi_1(\tau, 0) = 1$, $\varphi_1(\tau, 1) = \alpha$, $\varphi_2(\tau, 0) = 0$, $\varphi_2(\tau, 1) = \beta$. But then the matrix

$$\begin{pmatrix} \varphi_1(\tau, t) & \varphi_2(\tau, t) \\ -\overline{\varphi_2(\tau, t)} & \overline{\varphi_1(\tau, t)} \end{pmatrix}$$

joins homotopically the matrix (2) with the unit matrix in the class of sufficiently smooth non-singular matrices. Hence, if the rotation of the first row in the matrix (2) is zero then this matrix is homotopic to the unit matrix and has, therefore, the index zero. Equation (2.3) implies $\operatorname{Ind} U = \mu p(U)$ where the constant μ may depend only on Γ. Because of $\operatorname{Ind} U = \operatorname{Ind} \Phi$ we obtain

$$\operatorname{Ind} \Phi = \mu p(U). \tag{6}$$

4.2. Now we shall show that the coefficient μ in (6) does not depend on Γ.

To that aim we choose on Γ a part K_1 that is homeomorphic to a disk: the interior of K_1 will be denoted by K (and K is also homeomorphic to the disk). Furthermore, we form a manifold Γ_1 that is homeomorphic to a two-dimensional sphere and contains the set K_1.

On the set of linear elements $\tau_1 \in \tilde{\Gamma}_1$ we define a matrix $\Phi_1(\tau_1)$ subject to the following conditions:

1. $p(\Phi_1) = 1$;
2. $\Phi_1(\tau_1) = I$ (I is the unit matrix of order 2) if the linear element τ_1 is associated with the point $x \in \Gamma_1 \setminus K$.

Now we construct on the manifold Γ_1 a homogeneous singular system with the symbol matrix $\Phi_1(\tau_1)$,

$$a_1(x)\, u(x) + \int_{\Gamma_1} b_1(x, y)\, u(y)\, \mathrm{d}_y\Gamma = 0. \tag{7}$$

Obviously,

$$a_1(x) = I, \qquad b_1(x, y) = 0; \qquad x \in \Gamma_1 \setminus K. \tag{8}$$

Next we define on the manifold Γ_1 a sufficiently smooth function $\omega(x)$ such that

$$\omega(x) = \begin{cases} 1, & x \in K, \\ 0, & x \in \Gamma_1 \setminus K_1, \end{cases} \qquad 0 \leqq \omega(x) \leqq 1,$$

and put

$$c_1(x, y) = \omega(y)\, b_1(x, y);$$

$$a(x) = \begin{cases} a_1(x), & x \in K_1, \\ 0, & x \in \Gamma \setminus K_1; \end{cases}$$

$$c(x, y) = \begin{cases} c_1(x, y), & x \in K, \quad y \in K_1, \\ 0, & x \in \Gamma \setminus K_1 \quad \text{or} \quad y \in \setminus K_1. \end{cases}$$

We consider the two singular operators

$$(Au)(x) = a(x)\, u(x) + \int_\Gamma c(x, y)\, u(y)\, \mathrm{d}_y\Gamma, \qquad x \in \Gamma;$$

$$(A_1 u_1)(x) = a_1(x)\, u_1(x) + \int_{\Gamma_1} c_1(x, y)\, u_1(y)\, \mathrm{d}_y\Gamma_1, \qquad x \in \Gamma_1,$$

and show that their indices coincide. Let $Au = 0$. If $x \in \Gamma \setminus K_1$ the operator A is the identity, i.e. $u(x) = 0$ for $x \in \Gamma \setminus K_1$, and the function $u(x)$ satisfies the equation

$$a(x)\, u(x) + \int_{K_1} c(x, y)\, u(y)\, \mathrm{d}_y\Gamma = 0, \qquad x \in K_1.$$

Put

$$u_1(x) = \begin{cases} u(x), & x \in K_1, \\ 0, & x \in \Gamma_1 \setminus K_1, \end{cases}$$

then $(A_1 u_1)(x) = 0$, hence $\alpha(A_1) \geqq \alpha(A)$. Interchanging A and A_1 we obtain the opposite inequality such that $\alpha(A) = \alpha(A_1)$. In the same way we find $\alpha(A^*) = \alpha(A_1^*)$, and our assertion is proved.

The kernel $c_1(x, y)$ differs from $b_1(x, y)$ only by a weakly singular kernel,

$$c_1(x, y) = b_1(x, y)\, [\omega(y) - \omega(x)] + b_1(x, y)\, \omega(y);$$

the first term is weakly singular, and the second one is

$$b_1(x, y)\, \omega(y) = \begin{cases} b_1(x, y), & x \in K, \\ 0, & x \in \Gamma_1 \setminus K, \end{cases}$$

i.e. equals $b_1(x, y)$. But then the indices of the operators A_1 and (7) and, consequently, those of the operators A and (7) coincide.

Let us denote by μ_0 the value of the factor μ for the sphere. Since Γ_1 is homeomorphic to the sphere, we have $\operatorname{Ind} A_1 = \mu_0 p(\Phi_1)$. It is easy to see that for the symbol matrix $\Phi(\tau)$ of the operator A

$$\Phi(\tau) = \begin{cases} \Phi_1(\tau), & x \in K_1, \\ I, & x \in \Gamma \setminus K_1. \end{cases}$$

Hence, $p(\Phi) = 1 = p(\Phi_1)$, and

$$\mu = \mu p(\Phi) = \operatorname{Ind} A = \operatorname{Ind} A_1 = \mu_0 p(\Phi_1) = \mu_0,$$

which was to be proved.

It remains to note that by Wolpert's theorem mentioned above, $\mu_0 = 1$ such that for any two-dimensional manifold $\mu = 1$.

§ 5. Elementary cases with index zero

The cases to be studied here were considered by MIKHLIN [11, 21]. The order n of the symbol matrix is arbitrary, and $m \neq n$ is allowed.

Let us consider the singular system

$$Au = g \tag{1}$$

where

$$A = \begin{pmatrix} A_{11} & A_{12} \dots A_{1n} \\ A_{21} & A_{22} \dots A_{2n} \\ \dots & \dots \\ A_{n1} & A_{n2} \dots A_{nn} \end{pmatrix}$$

and

$$(A_{jk}u)(x) = a_{jk}u(x) + \int_\Gamma K_{jk}(x, y)\, u(y)\, \mathrm{d}_y\Gamma + (T_{jk}u)(x). \tag{2}$$

Here Γ is a compact, sufficiently smooth m-dimensional manifold without boundary, $K_{jk}(x, y)$ denote singular kernels and T_{jk} completely continuous operators in the corresponding space where the system (1) is to be considered. The symbol of the operator A_{jk} will be denoted by $\Phi_{jk}(\tau)$, and

$$\Phi(\tau) = \begin{pmatrix} \Phi_{11}(\tau) & \Phi_{12}(\tau) \dots \Phi_{1n}(\tau) \\ \Phi_{21}(\tau) & \Phi_{22}(\tau) \dots \Phi_{2n}(\tau) \\ \dots & \dots \\ \Phi_{n1}(\tau) & \Phi_{n2}(\tau) \dots \Phi_{nn}(\tau) \end{pmatrix}. \tag{3}$$

5.1. The simplest example of a singular system with index zero is provided if $\Gamma = \mathbb{R}^m$ and if the symbol matrix is constant (i.e. independent of x). In that case the simple singular operators with the symbol matrices Φ^{-1} and Φ^{*-1} provide equivalent regularizers for the operators A and A^*, resp. By Theorem 2.3, Chap. I, Ind $A = 0$. We mentioned that case previously.

5.2. Theorem 5.1. *Assume that the symbol matrix of the system* (1) *has the form* $\Phi(\tau) = I - \Psi(\tau)$ *(I being the unit matrix). If the characteristic values of the matrix Ψ are in modulus less than* 1 *for any* $\tau \in \tilde{\Gamma}$ *then system* (1) *has the index zero.*

Proof. Let $\Psi(\tau)$ be an arbitrary symbol matrix the elements of which satisfy the conditions of Theorem 8.2, Chap. XI. By A_λ we denote the singular matrix operator with the symbol matrix $I - \lambda\Psi(\tau)$. In the comlex λ-plane we find an open set Δ where the characteristic values of that matrix are different from zero. As in the proof to Theorem 2.4, Chap. XII, we obtain that the index of the operator A_λ is constant on every connected part of Δ. By the assumptions of our theorem, the disk $|\lambda| \leqq 1$ is completely contained in one of these connected parts. Hence, Ind $A_1 =$ Ind A_0. But $\lambda = 1$ corresponds to the system (1) in question, and $\lambda = 0$ to a system with a completely continuous operator the index of which is zero. ∎

5.3. Theorem 5.2. *If the moduli of the minors*

$$\delta_1 = \Phi_{11}, \qquad \delta_2 = \begin{vmatrix} \Phi_{11} & \Phi_{12} \\ \Phi_{21} & \Phi_{22} \end{vmatrix}, \dots, \qquad \delta_n = \begin{vmatrix} \Phi_{11} & \Phi_{12} \dots \Phi_{1n} \\ \Phi_{21} & \Phi_{22} \dots \Phi_{2n} \\ \dots & \dots \\ \Phi_{n1} & \Phi_{n2} \dots \Phi_{nn} \end{vmatrix} \tag{4}$$

are bounded from below by a positive constant then the index of the system (1) *is zero.*

Proof. We take the first equation of the system and consider it as an equation for the unknown u_1. It admits of an equivalent regularization such that it can be transformed into the equivalent equation

$$u_1 = \sum_{k=2}^{n} A_k^{(1)} u_k + \sum_{k=1}^{n} T_k^{(1)} u_k + g^{(1)}. \tag{5}$$

Here $g^{(1)}$ is a certain known function, and $T_k^{(1)}$ are completely continuous, $A_k^{(1)}$ singular operators with the symbols Φ_{1k}/Φ_{11}. Inserting (5) into the remaining equations (1) we obtain a system of $n - 1$ equations containing the unknowns $u_2, u_3, \ldots, u_n$ under singular integrals. In the first of these equations the unknown u_2 is contained under a singular integral with the non-degenerate symbol δ_2/δ_1. Hence, this equation can be given by the equivalent form

$$u_2 = \sum_{k=1}^{n} A_k^{(2)} u_k + \sum_{k=1}^{n} T_k^{(2)} u_k + g^{(2)} \tag{6}$$

(the meaning of the notations is obvious). Again we insert this expression into the remaining $n - 2$ equations. Proceeding in this manner we arrive finally at the system

$$u_j = \sum_{k=j+1}^{n} A_k^{(j)} u_k + \sum_{k=1}^{n} T_k^{(j)} u_k + g^{(j)}, \tag{7}$$

which is equivalent to (1); for $j = n$ the first sum in (7) is to be omitted.

System (7) can be, obviously, transformed into an equivalent one with completely continuous operators. But that means that system (1) admits of an equivalent regularization. By Theorem 2.3, Chap. I, the index of system (1) is non-negative. On the other hand, the conditions of our theorem are valid also for the adjoint system such that its index is also non-negative. Thus, the index of system (1) is zero. ∎

5.4. Theorem 5.3. *Let A be a singular operator with the symbol matrix $\Phi(\tau)$ defined on the set $\tilde{\Gamma}$ of linear elements of the manifold Γ. If in the complex ζ-plane there exists a smooth curve L connecting the points $\zeta = 0$ and $\zeta = \infty$ such that it has no common points with the set of characteristic values of the matrix $\Phi(\tau)$ then* Ind $A = 0$.

Proof. Put $\zeta = (\lambda - 1)/\lambda$ then the curve L is transformed to some curve L_0 in the λ-plane, and it connects the points $\lambda = 0$ and $\lambda = 1$. We may assume that the curve L does not pass the point $\zeta = 1$ so that the curve L_0 has finite length.

We consider the system

$$u - \lambda(I - A)\, u = g \tag{8}$$

with the symbol matrix $I - \lambda(I - \Phi) = \lambda(\Phi - \zeta I)$ and show that its determinant is non-zero everywhere. Indeed, for small $|\lambda|$ the determinant assumes values close to 1. Hence, there are numbers $\eta > 0$ and $q > 0$ such that $|\lambda| < \eta$ implies the inequality $|\det [(1 - \lambda)\, I + \lambda\Phi]| > q$.

Now, let $|\lambda| \geqq \eta$. Then

$$|\det [(1 - \lambda)\, I + \lambda\Phi]| \geqq \eta^n\, |\det (\zeta I - \Phi)| = \eta^n \prod_{k=1}^{n} |\zeta - \zeta(\tau)|,$$

where ζ_k are the characteristic values of the matrix Φ. They depend continuously on the parameter τ which varies in the compact manifold $\tilde{\Gamma}$ such that the set of the ζ_k-values, $k = 1, 2, \ldots, n$, is compact. The assumption on the existence of a curve L with the properties as described above implies the existence of a constant $\gamma > 0$ such that

$|\zeta - \zeta_k(\tau)| \geqq \gamma$ if $\zeta \in L$. Thus, we obtain

$$|\det [(1 - \lambda) I + \lambda \Phi]| \geqq (\gamma\eta)^n; \qquad \lambda \in L_0, \qquad |\lambda| \geqq \eta,$$

which proves our assertion.

Now it is obvious that the singular operator H_λ with the symbol matrix $[(1 - \lambda) I + \lambda\Phi(\tau)]^{-1} [I - \Phi(\tau)]$ is bounded in the space where the singular system is considered, and that independently of λ, $\|H_\lambda\| \leqq C$.

If $\lambda = 0$ we obtain $\|I - A\| \leqq C$. Hence, for $\lambda_0 \in L_0$, $|\lambda_0| \leqq C/2$, there exists the bounded and everywhere defined operator $[I - \lambda_0(I - A)]^{-1}$ and its index is, obviously, zero. Applying this operator to eq. (8) we obtain an equivalent to (8) equation with the same index. Its symbol matrix is

$$[(1 - \lambda_0) I + \lambda_0\Phi]^{-1} [(1 - \lambda) I + \lambda\Phi] = I - (\lambda - \lambda_0) [(1 - \lambda_0) I + \lambda_0\Phi]^{-1} (I - \Phi).$$

Now apply to the equation obtained the operator

$$[I - (\lambda_1 - \lambda_0) H_{\lambda_0}]^{-1}, \qquad \lambda_1 \in L_0, \qquad |\lambda_1 - \lambda_0| \leqq C/2,$$

then we obtain an equation that is again equivalent to (8) and has the same index. Its symbol matrix is

$$I - (\lambda - \lambda_1) [(1 - \lambda_1) I + \lambda_1\Phi]^{-1} (I - \Phi).$$

Proceeding that way we obtain after k steps a system that is equivalent to (8) and has the symbol matrix

$$I - (\lambda - \lambda_k) [(1 - \lambda_k) I + \lambda_k\Phi]^{-1} (I - \Phi). \tag{9}$$

The differences $\lambda_{j+1} - \lambda_j$ can be chosen sufficiently large in modulus, if only the inequality $|\lambda_{j+1} - \lambda_j| \leqq C/2$ is kept. Therefore, we may assume $\lambda_k = 1$ for sufficiently large k. For $\lambda = 1$ system (8) becomes system (1). Putting in (9) $\lambda = \lambda_k = 1$ you can see that system (1) is transformed into an equivalent one the symbol matrix of which is the unit matrix, i.e. into a system with a completely continuous operator. Since the index remains in every step the same, and the index of a system with a completely continuous operator is zero, we obtain finally Ind $A = 0$. ▮

Corollary 5.1. *If the symbol determinant is non-zero everywhere and the symbol matrix is symmetric or skew-symmetric then the index of the corresponding singular system is zero.*

We made use of this corollary in §§ 3, 4 of the present chapter.

5.5. *Let* $\Phi(\tau) = \Phi_1(\tau) + i\Phi_2(\tau)$ *with Hermitean matrices* Φ_1 *and* Φ_2. *If at least one of these matrices is definite then* Ind $\Phi = 0$.

Proof. Let e.g. the matrix Φ_1 be definite. Then $\Phi = i\Phi_1(\Phi_1^{-1}\Phi_2 - iI)$ and Ind $\Phi =$ Ind $(\Phi_1^{-1}\Phi_2 - iI)$ with I denoting again the unit matrix. The eigenvalues of the matrix $\Phi_1^{-1}\Phi_2$ are the roots of the equation $\det (\Phi_2 - \lambda\Phi_1) = 0$ and, hence, real. But then the eigenvalues of the matrix $\Phi_1^{-1}\Phi_2 - iI$ lie on the straight line Im $\lambda = -1$. By Theorem 5.3, Ind $(\Phi_1^{-1}\Phi_2 - I) = 0$. ▮

5.6. Let Γ be either a two-dimensional plane or the two-dimensional torus so that Γ may be endowed with a regular coordinate system. In that case $\tilde{\Gamma}$ is the direct product of the surface Γ and the unit circle: Every linear element is determined by a point $x \in \Gamma$ and a point $\zeta = e^{i\theta}$ on the unit circle. Thence, the symbol matrix can be considered as a function $\Phi(x, \zeta)$ of two independent variables. For fixed x the matrix $\Phi(x, \zeta)$ admits to the left factorization (cf. § 3, Chap. V),

$$\Phi(x, \zeta) = \Phi_+(x, \zeta)\, \Phi_0(x, \zeta)\, \Phi_-(x, \zeta), \tag{10}$$

where the matrices $\Phi_+(x, \zeta)$ and $\Phi_-(x, \zeta)$ can be extended analytically into the interior and into the exterior, resp., of the unit circle in the ζ-plane; the determinants of these matrices are non-zero everywhere in the corresponding domains. The matrix $\Phi_0(x, \zeta)$ is a diagonal one,

$$\Phi_0(x, \zeta) = (\zeta^{\varkappa_j} \delta_{jk})_{j,k=1}^n,$$

where the integers $\varkappa_j$ (the left partial indices of the matrix Φ) do not depend on ζ. Of course, the matrix Φ admits to a right factorization, too.

If the left (or the right) partial indices of the matrix $\Phi(x, \zeta)$ do not depend on x then Ind $\Phi = 0$.

Indeed, by the Atkinson theorem (cf. § 3, Chap. I),

$$\text{Ind } \Phi = \text{Ind } \Phi_+ + \text{Ind } \Phi_0 + \text{Ind } \Phi_-$$

where Ind $\Phi_0 = 0$ since this matrix does not depend on x. Let us consider the matrix $\Phi_+(x, \zeta)$. It can be approximated in the disk $|\zeta| \leqq 1$ by a non-singular matrix $P_+(x, \zeta)$ which is a polynomial in ζ and has the same index. Denote the roots of the polynomial $\det P_+(x, \zeta)$ by $\alpha_k(x)$, $k = 1, 2, \ldots, N$, so that $|\alpha_k(x)| > 1$ for $x \in \Gamma$. Replace ζ in the matrix $P_+(x, \zeta)$ by $t\zeta$, $t \in [0, 1]$. Then the roots $\alpha_k(x)$ become $t^{-1} \alpha_k(x)$, $|t^{-1} \alpha_k(x)| > 1$. Hence, we obtain a homotopy between the matrices $P_+(x, \zeta)$ and $P_+(x, 0)$ independently of ζ such that its index is zero. Therefore, Ind $\Phi_+ = 0$.

The matrix $\Phi_-(x, \zeta)$ can be approximated by a matrix $P_-(x, \zeta)$ which is a polynomial in ζ^{-1} and has the same index. Denote by $\beta_k(x)$ those values of ζ for which $\det P_-(x, \zeta) = 0$. Obviously, $|\beta_k(x)| < 1$. Replace in the matrix $P_-(x, \zeta)$ the argument ζ^{-1} by $t\zeta^{-1}$, $t \in [0, 1]$. The roots of the determinant of the new matrix are $t\beta_k(x)$, $|t\beta_k(x)| < 1$. The matrix $P_-(x, \zeta)$ is homotopic to the matrix $P_-(x, \infty)$ that does not depend on ζ. As in the previous paragraph we find Ind $\Phi_- = 0$. ▮

In particular, on the torus the index of a matrix that does not depend on x is equal to zero.

5.7. The factorization technique applies sometimes to multidimensional symbol matrices, too. Assume that the manifold Γ can be endowed with a regular coordinate system. Then the symbol matrix can be given in the form $\Phi(x, \theta)$, $x \in \Gamma$, and $\theta = (\vartheta_1, \vartheta_2, \ldots, \vartheta_{m-1})$ being a point on the $(m - 1)$-dimensional sphere S, $\vartheta_{m-1} \in [0, 2\pi]$. Put $\zeta = \exp(i\vartheta_{m-1})$ and fix the other arguments then the matrix $\Phi(x, \theta)$ can be factorized as done before. If all left (or all right) partial indices vanish then Ind $\Phi = 0$.

Indeed, in that case $\Phi(x, \theta) = \Phi_+(x, \theta)\, \Phi_-(x, \theta)$ where the expansion of $\Phi_+(x, \theta)$ with respect to spherical function contains non-negative powers of ζ only, whereas the corresponding expansion of $\Phi_-(x, \theta)$ contains only non-positive powers of ζ. Approximating Φ_+ and Φ_- by spherical polynomials (i.e. linear combinations of spherical functions) and replacing ζ in Φ_+ by $t\zeta$, in Φ_- by $t^{-1} \zeta$, $t \in [0, 1]$, we obtain to each of the matrices $\Phi_\pm$ a homotopic matrix which does not depend on ζ. Then it can be shown (cf. Mikhlin [11]) that the index of such a matrix is zero. Hence, Ind $\Phi_+ =$ Ind $\Phi_- = 0$ and, by the Atkinson theorem, Ind $\Phi = 0$.

§ 6. Problems in static elasticity theory

6.1. The present section is devoted to applications of the theory of singular systems to basic problems in elasticity theory. A large number of applications of the theory of singular equations to a wide range of problems in elasticity theory was given by Kupradze and his collaborators. The corresponding results are collected in the book by Kupradze, Gegelia, Bashelishvili, and Burchuladze [1], furthermore we refer to the earlier book by Kupradze [2]. Kupradze [1] applied in 1953 the concept of singular equations to the simplest problems in elasticity theory. Somewhat simpler with respect to the technique used is the paper by Kinoshita and Mura [1] (1956) which will be used in the current section. While Kupradze used the so-called antenna potentials inroduced by H. Weyl the Japanese authors mentioned apply the simple-layer and double-layer potentials which are completely analogous to the classic potentials. They study simple properties of these potentials, obtain limit formulas for them, and derive integral equations for the first two basic boundary value problems (see below). However, the authors consider their equations as Fredholm equations with discontinuous kernels. In order to obtain continuous kernels they suggest an iterative method. But, as a matter of fact, there is no need to do so: It is the question here of singular equations which cannot be transformed into Fredholm equations by means of a finite number of iterations. Nevertheless, the consequences drawn by Kinoshita and Mura are valid, although their arguments are not quite exact: As we shall show in the following, these integral equations form a system with index zero, and for such systems the fundamental Fredholm theorems are valid.

The errors committed by Kinoshita and Mura are corrected in the book by Mikhlin [11].

6.2. Assume that an isotropic elastic body occupies a (bounded or unbounded) domain Ω in the space $\mathbb{R}^3$ with coordinates x_1, x_2, x_3. Let the boundary $\partial\Omega$ be a connected[1]) closed Lyapunov surface Γ. By $u = (u_1, u_2, u_3)$ we denote the elastic displacement vector, and by $\tau_{ik} = \tau_{ik}(u)$ the components of the corresponding stress tensor. We remember the well-known *Lamé equations*

$$\tau_{ik}(u) = \lambda \operatorname{div} u \, \delta_{ik} + \mu \left(\frac{\partial u_i}{\partial x_k} + \frac{\partial u_k}{\partial x_i}\right) \tag{1}$$

where δ_{jk} is Kronecker's δ, and λ and μ are Lamé's elastic constants. For the sake of simplicity we assume that there are no volume forces, then we have for the displacement vector in an equilibrium state the equation

$$\mu \Delta u + (\lambda + \mu) \operatorname{grad} \operatorname{div} u = 0. \tag{2}$$

This equation (2) will be considered together with boundary conditions of the following type:

Problem I. On the surface Γ the displacement vector

$$u|_\Gamma = g(x), \qquad x \in \Gamma, \tag{3}$$

is given.

Problem II. On Γ the surface stress

$$p(u) = \tau_{jk}(u)\, \alpha_k e_j = h(x), \qquad x \in \Gamma, \tag{4}$$

is given.

For the present section we agree upon the following notations and symbols. Equal indices in an expression imply summation over that index between 1 and 3; α_k denote the direction cosines of the outward normal to Γ; e_j is the unit vector on the x_j axis; the vectors $g(x)$ and $h(x)$ are given on Γ.

---

[1]) This condition is not essential.

As usually we shall distinguish the interior from the exterior problem I, II according to whether the body occupies the interior or the exterior domain with respet to Γ.

Now the potential energy of an elastic deformation is positive so that we have the well-known inequalities $\lambda + 2\mu/3 > 0$, $\mu > 0$. For such values of Lamé's constants the following uniqueness theorems are known:

1. The interior problem I has not more than one solution.
2. Under the additional condition $u = O(|x|^{-1})$, $\tau_{jk}(u) = O(|x|^{-2})$ as $|x| \to \infty$ the exterior problems I, II have not more than one solution.
3. If the interior problem II is solvable then its solution is defined up to a "rigid" displacement $a + b \times x$ with arbitrary constant vectors a and b; the symbol $\times$ means the vector product.

For the interior problem II to be solvable it is necessary that the main vector and the main momentum of the vector $h(x)$ vanish,

$$\int_\Gamma h(x)\,\mathrm{d}\Gamma = 0, \qquad \int_\Gamma x \times h(x)\,\mathrm{d}\Gamma = 0. \tag{5}$$

Let us recall the Green formula for the equations of elasticity theory which is better known under the name of Betti's formula. Let Ω be a bounded domain for which the Gauss-Ostrogradski formula is valid, $\partial\Omega = \Gamma$. For two vectors u and v belonging to $\mathbf{W}_p^{(2)}(\Omega)$, $p > 1$, we put

$$Au = -\mu\,\Delta u - (\lambda + \mu)\,\mathrm{grad}\,\mathrm{div}\,u \tag{6}$$

then *Betti's formula* is

$$\int_\Omega (v \cdot Au - u \cdot Av)\,\mathrm{d}x = \int_\Gamma [u \cdot p(v) - v \cdot p(u)]\,\mathrm{d}\Gamma. \tag{7}$$

6.3. In what follows, the singular solution for the equations of elasticity theory plays an important role. This is a symmetric tensor $V = V(x, y) = (v_{ik})$,

$$v_{ik}(x, y) = \frac{1}{16\pi\mu(1 - \sigma)}\left\{\frac{3 - 4\sigma}{r}\,\delta_{ik} + \frac{(y_i - x_i)\,(y_k - x_k)}{r^3}\right\}; \tag{8}$$

here, $r = |x - y|$, and $\sigma = \lambda/2(\lambda + \mu)$ is the Poisson constant of the elastic material under consideration. For x fixed and $y \neq x$ every column $v_i = (v_{1i}, v_{2i}, v_{3i})$ satisfies (2), more precisely, for any x and y, $Av_i = \delta(x - y)$. The tensor V is called *Somiglian's tensor*.

Let u be a vector which is as smooth as necessary in $\bar{\Omega}$, and x be a point interior with respect to Γ. We take x as the center of a ball with sufficiently small radius and remove that ball from the integration domain Ω. Applying Betti's formula over the domain obtained that way and letting the radius of the ball tend to zero, we obtain

$$u(x) = \int_\Omega V(x, y) \cdot Au(y)\,\mathrm{d}y - \int_\Gamma [P(x, y) \cdot u(y) - V(x, y) \cdot p(u(y))]\,\mathrm{d}_y\Gamma. \tag{9}$$

By $P(x, y)$ we denote the tensor the i-th column of which is $p(v_i)$.

If the point x is exterior with respect to Γ the right-hand side of (9) vanishes identically.

As in the usual potential theory, (9) gives reason to introduce the three potentials

$$\int_\Omega V(x, y) \cdot \psi(y)\,\mathrm{d}y, \qquad \int_\Gamma P(x, y) \cdot \varkappa(y)\,\mathrm{d}_y\Gamma, \qquad \int_\Gamma V(x, y) \cdot \varrho(y)\,\mathrm{d}_y\Gamma,$$

which will be called *volume potential, double-layer potential,* and *simple-layer potential,* resp. Using the results on differentiation of weakly singular integrals one can easily show that the volume potential satisfies the equation $Au = \psi$ in the interior domain with respect to Γ, and $Au = 0$ in the exterior one. The volume potential can be used (in case of volume forces being present) for reducing the inhomogeneous equation of elasticity theory to the homogeneous one in the usual manner.

The simple-layer and double-layer potentials satisfy the homogeneous equation (2) in the interior and likewise in the exterior domain of Γ. The simple-layer potential decreases as $O(|x|^{-1})$ $(|x| \to \infty)$, and the corresponding stress as $O(|x|^{-2})$. For the double-layer potential the corresponding orders are $O(|x|^{-2})$ and $O(|x|^{-3})$, resp. The simple-layer potential is continuous in the entire space provided that its density is continuous on Γ.

The potentials introduced here have limit theorems which are analogous to the well-known theorems of the classical potential theory. We shall derive here such a theorem concerning the limits of the double-layer potential where we assume that its density satisfies a Lipschitz condition with a positive exponent.

Simple calculations lead to the following formula for the components P_{ij} of the tensor $P = P(x, y)$,

$$P_{ij} = \frac{1}{8\pi(1-\sigma)} \left[\frac{1-2\sigma}{r^3} (\xi_i \, \delta_{jk} - \xi_j \delta_{ik} - \xi_k \, \delta_{ij}) - \frac{3}{r^5} \xi_i \xi_j \xi_k \right] \alpha_k, \qquad \xi_i = y_i - x_i. \quad (10)$$

Let $\varkappa$ be a constant vector. We want to compute the double-layer potential

$$u_0(x) = \int_\Gamma P(x, y) \cdot \varkappa \, \mathrm{d}_y \Gamma .$$

If x belongs to the exterior domain of Γ we obtain, after applying Betti's formula to the vectors $\varkappa$ and v_i, the relation $u_0(x) = 0$ because of $p(\varkappa) = 0$ and $A\varkappa = 0$.

Now let x belong to the interior domain of Γ. We surround the point x by a sphere S_ε with radius ε. By Betti's formula,

$$u_0(x) = \int_{S_\varepsilon} P(x, y) \cdot \varkappa \, \mathrm{d}S_\varepsilon$$

(the normal involved in $P(x, y)$ has radial direction). Replacing $y - x$ by εy we give the last equation the form

$$u_0(x) = \frac{1}{8\pi(1-\sigma)} \int_S \varkappa_i [(1-2\sigma)(y_i \, \delta_{jk} - y_j \, \delta_{ik} - y_k \, \delta_{ij}) - 3 y_i y_j y_k] \, \alpha_k e_j \, \mathrm{d}S$$

S denoting the unit sphere. Let ω denote the unit ball with boundary S. Taking into account the obvious relation $\partial y_i / \partial y_j = \delta_{ij}$ we obtain by means of the Gauss-Ostrogradski formula

$$u_0(x) = \frac{-1}{8\pi(1-\sigma)} \int_\omega \varkappa_i [(1-2\sigma) \, \delta_{ij} \delta_{kk} + 3(y_j y_k \, \delta_{ik} + y_i y_j \, \delta_{jk} + y_i y_j \, \delta_{kk})] \, e_j \, \mathrm{d}y =$$

$$= - \frac{1}{8\pi(1-\sigma)} \left\{ (3-2\sigma) \int_\omega \varkappa_j e_j \, \mathrm{d}y + 15 \int_\omega \varkappa_i y_i y_j e_j \, \mathrm{d}y \right\}.$$

The first integral in the right-hand side equals $4\pi\varkappa/3$. The second one is a sum of integrals corresponding to various combinations of the indices i and j. For $i \neq j$ these integrals are zero, and for $i = j$ their sum amounts to

$$\varkappa_j e_j \int_\omega y_i^2 \, \mathrm{d}y = \varkappa \int_\omega y_i^2 \, \mathrm{d}y = \frac{4\pi}{15} \varkappa .$$

Finally, we obtain $u_0(x) = -\varkappa$ for x being interior with respect to Γ.

Now, let $x \in \Gamma$. In that case $u_0(x)$ is a singular integral. Consider the sphere S_ε centered at the point x, and denote by Γ_ε the set $\Gamma \setminus S_\varepsilon$, by S'_ε the set $S_\varepsilon \setminus \Gamma$. As shown above,

$$\int_{\Gamma_\varepsilon} P(x, y) \cdot \varkappa \, \mathrm{d}_y\Gamma + \int_{S'_\varepsilon} P(x, y) \cdot \varkappa \, \mathrm{d}S_\varepsilon = -\varkappa .$$

Hence,

$$u_0(x) = -\varkappa - \lim_{\varepsilon \to 0} \int_{S'_\varepsilon} P(x, y) \cdot \varkappa \, \mathrm{d}S_\varepsilon =$$

$$= -\varkappa - \frac{1}{8\pi(1-\sigma)} \int_{S'} \varkappa_i[(1-2\sigma)(y_i \delta_{jk} - y_j \delta_{ik} - y_k \delta_{ij}) - 3y_i y_j y_k] \alpha_k e_j \, \mathrm{d}S,$$

where S' denotes a hemi-sphere of the unit sphere S. By symmetry reasons, the integral over S' equals half the integral over S so that $u_0(x) = -\frac{1}{2}\varkappa$, $x \in \Gamma$.

Thus, we obtain the analogue to the Gauss formula in classical potential theory,

$$\int_{\Gamma} P(x, y) \cdot \varkappa \, \mathrm{d}_y\Gamma = \begin{cases} -\varkappa, & \text{if } x \text{ is inside of } \Gamma, \\ -\frac{1}{2}\varkappa, & \text{if } x \in \Gamma, \\ 0, & \text{if } x \text{ is outside of } \Gamma. \end{cases} \tag{11}$$

6.4. Now it is easy to obtain the limit formula for the double-layer potential. Let

$$u(x) = \int_{\Gamma} P(x, y) \cdot \varkappa(y) \, \mathrm{d}_y\Gamma \tag{12}$$

and $x_0 \in \Gamma$. We have

$$u(x) = \int_{\Gamma} P(x, y) \cdot [\varkappa(y) - \varkappa(x_0)] \, \mathrm{d}_y\Gamma + \int_{\Gamma} P(x, y) \cdot \varkappa(x_0) \, \mathrm{d}_y\Gamma .$$

Under the assumption made before that $\varkappa(y)$ satisfies a Lipschitz condition with positive exponent, the first integral on the right-hand side of this equation is continuous at $x = x_0$. The value of the second integral is provided by (11). This leads easily to the limit formulae

$$u_i(x_0) = -\tfrac{1}{2}\varkappa(x_0) + \int_{\Gamma} P(x_0, y) \cdot \varkappa(y) \, \mathrm{d}_y\Gamma,$$

$$u_e(x_0) = \tfrac{1}{2}\varkappa(x_0) + \int_{\Gamma} P(x_0, y) \cdot \varkappa(y) \, \mathrm{d}_y\Gamma, \tag{13}$$

where the subscripts i and e indicate the interior and the exterior limit, resp.

Let us give the limit formula for the simple-layer potential. To that aim let

$$u(x) = \int_{\Gamma} V(x, y) \cdot \varrho(y) \, \mathrm{d}_y\Gamma \tag{14}$$

and $\varrho \in \mathbf{Lip}_\alpha(\Gamma)$, $\alpha > 0$. For $x_0 \in \Gamma$,

$$p(u)_i = \tfrac{1}{2}\varrho(x_0) + \int_{\Gamma} P^*(y, x_0) \cdot \varrho(y) \, \mathrm{d}_y\Gamma,$$

$$p(u)_e = -\tfrac{1}{2}\varrho(x_0) + \int_{\Gamma} P^*(y, x_0) \cdot \varrho(y) \mathrm{d}_y\Gamma. \tag{15}$$

6.5. By means of the formulae (13) and (15) problems I and II can be transformed to singular integral equations. We seek the solution to problem I in the shape of a double-layer potential (12) and the solution to problem II as a simple-layer potential (14). The boundary conditions (3) and (4) together with the relations (13) and (15) yield the integral equations

$$\varkappa(x) - 2 \int_\Gamma P(x, y) \cdot \varkappa(y) \, \mathrm{d}_y\Gamma = -2g(x), \tag{16}$$

$$\varkappa(x) + 2 \int_\Gamma P(x, y) \cdot \varkappa(y) \, \mathrm{d}_y\Gamma = 2g(x), \tag{17}$$

$$\varrho(x) + 2 \int_\Gamma P^*(y, x) \cdot \varrho(x) \, \mathrm{d}_y\Gamma = 2h(x), \tag{18}$$

$$\varrho(x) - 2 \int_\Gamma P^*(y, x) \cdot \varrho(y) \, \mathrm{d}_y\Gamma = -2h(x), \tag{19}$$

which correspond, in the order given, to the interior and to the exterior problem I and to the interior and exterior problem II, resp. The equations (16) and (19) and likewise (17) and (18) are adjoint to each other.

Each of the equations (16)–(19) is, in fact, a system of three singular equations. Let us consider the symbol matrices of these systems. Taking into account the relation $\xi_k \alpha_k = r \cos(r, \nu) = O(r^{1+\gamma})$ (γ is the Lyapunov exponent of the surface Γ) we get from (10)

$$P_{ij}(x, y) = \frac{1 - 2\sigma}{8\pi(1 - \sigma)} \cdot \frac{\xi_i \, \delta_{jk} - \xi_j \, \delta_{ik}}{r^3} \alpha_k + O(r^{\gamma - 2}) =$$

$$= \frac{1 - 2\sigma}{8\pi(1 - \sigma)} \cdot \frac{\xi_i \alpha_j - \xi_j \alpha_i}{r^3} + O(r^{\gamma - 2}). \tag{20}$$

The singular part of this kernel is given by the first term in (20).

Under a coordinate transformation the argument of every symbol matrix element is subject to a linear mapping (cf. § 3, Chap. X). Note that the same mapping applies also to the argument of the symbol determinant, and its range is invariant under the transformation of coordinates. Taking all that into account, we introduce local coordinates at every point $x \in \Gamma$ where the x_3 axis is directed along the outward normal to Γ. As unknowns we introduce the components of the vectors $\varkappa$ and ϱ in local coordinates what does, obviously, not affect to the index of the singular system.

In local coordinates we have $\alpha_1 = \alpha_2 = 0$, $\alpha_3 = 1$. With these relations in mind let us consider e.g. system (16). Applying (20) we can give it in local coordinates the form

$$\varkappa_1(x) + \frac{1 - 2\sigma}{4\pi(1 - \sigma)} \int_\Gamma \frac{\xi_1}{r^3} \varkappa_3(y) \, \mathrm{d}_y\Gamma + T_1(x) = 2g_1(x),$$

$$\varkappa_2(x) + \frac{1 - 2\sigma}{4\pi(1 - \sigma)} \int_\Gamma \frac{\xi_2}{r^3} \varkappa_3(y) \, \mathrm{d}_y\Gamma + T_2(x) = 2g_2(x),$$

$$\varkappa_3(x) - \frac{1 - 2\sigma}{4\pi(1 - \sigma)} \int_\Gamma \frac{1}{r^3} [\xi_1 \varkappa_1(y) + \xi_2 \varkappa_2(y)] \, \mathrm{d}_y\Gamma + T_3(x) = 2g_3(x)$$

where T_k are weakly singular operators. The characteristics of the singular integrals in that system are $\xi_1/r = \cos\theta$ and $\xi_2/r = \sin\theta$. The symbols of these integrals are obtained from the characteristics by multiplying them by the factor $2\pi\mathrm{i}$. With the abbreviation $\delta = (1 - 2\sigma)/2(1 - \sigma)$ we obtain for this system the symbol matrix

$$\begin{pmatrix} 1 & 0 & \mathrm{i}\,\delta\cos\theta \\ 0 & 1 & \mathrm{i}\,\delta\sin\theta \\ -\mathrm{i}\,\delta\cos\theta & -\mathrm{i}\,\delta\sin\theta & 1 \end{pmatrix}. \tag{21}$$

The determinant of it – the symbol determinant – equals $(3 - 4\sigma)/4(1 - \sigma)^2$, and it is non-zero for $\sigma \neq 3/4, \infty$. Analogous relations are true for the other systems (17)–(19).

Equation (21) shows that the symbol matrix is Hermitean. By the Corollary 5.1 the index of system (16) is zero, and, therefore, the Fredholm theorems apply to that system. The same is true for the other systems (17)–(19).

Now let the Poisson constant σ belong to the interval $(-1, 1/2)$. Arguments as in the classical potential theory lead to the following results: Equations (16) and (19) are always uniquely solvable. Equation (18) of the interior Problem II is solvable if and only if condition (5) is fulfilled. (17) is, in general, not solvable, and that due to the fact that the solution to the exterior Problem I cannot be, in general, represented as a (rapidly decaying) double-layer potential.

Now let us come back again to (1). As mentioned above, by physical reasons we have to assume $\lambda + 2\mu/3 > 0$, $\mu > 0$, or, what is the same, $\sigma \in (-1, 1/2)$. It can be shown that under these assumptions the system of the static elasticity theory equations

$$(1 - 2\sigma)\,\Delta u_k + \frac{\partial \operatorname{div} u}{\partial x_k} = 0, \qquad k = 1, 2, 3,$$

is equivalent to the vector equation (1) which is, in the sense of VISHIK [1], strongly elliptic. We want to show now that the system (1) is elliptic in the sense of Petrovski for all σ except $\sigma = 1/2$ and $\sigma = 1$. To that aim it suffices to show that for any real numbers ξ_1, ξ_2, ξ_3 such that $\xi_1^2 + \xi_2^2 + \xi_3^2 = 1$ the determinant

$$\begin{vmatrix} 1 - 2\sigma + \xi_1^2 & \xi_1\xi_2 & \xi_1\xi_3 \\ \xi_1\xi_2 & 1 - 2\sigma + \xi_2^2 & \xi_2\xi_3 \\ \xi_1\xi_3 & \xi_2\xi_3 & 1 - 2\sigma + \xi_3^2 \end{vmatrix}$$

is non-zero. We can easily compute this determinant; it is equal to $2(1 - 2\sigma)^2(1 - \sigma)$, hence it is non-zero if $\sigma \neq 1, 1/2$. Equation (8) shows that for such σ values there exists a singular solution to the system of the static elasticity theory. Furthermore, it is obvious that the technique developed in the present section (i.e. potentials and their properties, singular equations associated to boundary value problem) carries over without any changes to all those σ values which are different from 1 and 1/2. If, in addition, $\sigma \neq 3/4$, ∞ then the symbols of the singular systems (16)–(19) do not degenerate. Hence, for all σ values different from 1/2, 3/4, 1, ∞ the index of Problems I and II is zero, and both the problems are normally solvable. For the exceptional values of σ the corresponding systems are either not normally solvable, or they do not have a finite index, or both of these properties occur.

We mention, in particular, Problem I for $\sigma = \infty$, which turns out to be the well-known Dirichlet problem for the vector Laplace equation. That problem has a unique solution. It may happen that the symbol determinant of the system (16) (or (19)) becomes zero, and that means that the solution to the Dirichlet problem in question either cannot be represented as a double-layer potential (12), or that the subspace of densities of this potential for the homogeneous Dirichlet problem is of infinite dimension.

Chapter XV
The localization principle. Singular operators on manifolds with boundary

This chapter is mainly devoted to the so-called localization principle introduced by SIMONENKO [1, 2], and its applications to singular integral operators. We shall apply this principle, in particular, to singular equations on manifolds with boundary. This last problem is the subject of the profound papers by VISHIK and ESKIN [1–6] which contain a considerable number of results. A brief survey of some of these results is given in § 10.

§ 1. Operators of local type

1.1. Let us introduce some definitions which will be used in the following frequently.

Let X be a compact Hausdorff space endowed with a non-negative measure which is defined on a certain σ-algebra of subsets of X. We shall assume this σ-algebra to contain all open and all closed subsets of X.

As it is well-known, two closed sets with no common points in a Hausdorff space can be separated by open sets. Moreover, for any two closed sets F_0 and F_1 with no common points there exists a continuous function $w(x)$, $0 \leqq w(x) \leqq 1$, such that $w(x) = 0$, $x \in F_0$, and $w(x) = 1$, $x \in F_1$. For these notions and their properties we refer to the book NEUMARK [1]; concerning the notion of dimension we refer to P. S. ALEXANDROV [1], and concerning the fundamentals of measure theory to HALMOS [1].

As usually, we denote by $\mathbf{L}_p(X)$ the space of a.e. on X defined functions that are summable to the p-th power. In the current chapter we shall consider operators only which act in $\mathbf{L}_p(X)$. By $\|\cdot\|$ we shall always denote the norm in that space. The *essential norm* of an operator A is defined as $|||A||| := \inf \|A - T\|$, the infimum taken over the set of completely continuous operators T. Two operators A and B will be called *equivalent*, $A \sim B$, iff $|||A - B||| = 0$.

A sequence of operators $\{A_n\}$ is called a *fundamental sequence* with respect to the essential norm iff $|||A_m - A_n||| \to 0$ as $n, m \to \infty$. As it is well-known, such a sequence is convergent in the essential norm, i.e. there is an operator A such that $|||A_n - A||| \to 0$ as $n \to \infty$.

For a measurable set $M \subset X$ we define an operator P_M by

$$(P_M u)(x) = \begin{cases} u(x), & x \in M, \\ 0, & x \notin M. \end{cases}$$

An operator A is called *operator of local type* iff for any two disjoint closed sets F_1, F_2 the operator $P_{F_1} A P_{F_2}$ is completely continuous.

It is worth noting that a singular integral operator on a compact manifold is an operator of local type. This is an immediate consequence of Seeley's definition (cf. § 2, Chap. XIII).

There is another definition which is equivalent to the given one: An operator A is called operator of local type iff for any two sets U_1, U_2, $U_1 \subset \operatorname{int} U_2$, the relations $P_{U_1} A P_{U_2} \sim P_{U_1} A$ and $P_{U_2} A P_{U_1} \sim A P_{U_1}$ hold.

We mention some obvious properties of operators of local type.

a) If A_n are operators of local type and $|||A_n - A||| \to 0$ then the operator A is of local type, too.

b) The sum of two operators of local type is again an operator of local type, and the product of bounded operators of local type is also of local type.

c) If A is an operator of local type and $p \in (1, \infty)$ then the adjoint A^* is an operator of local type, too.

Example 1. The operator of multiplication by a measurable and bounded function is an operator of local type.

Example 2. The singular operator

$$(Au)(x) = \int_{\mathbb{R}^m} \frac{f(x, \theta)}{r^m} u(y) \, dy,$$

the characteristic of which satisfies the conditions of the Calderón-Zygmund theorem (§ 3, Chap. XI) is an operator of local type in $X = \overline{\mathbb{R}^m}$. This fact was mentioned above.

1.2. Theorem 1.1. *Let A be an operator of local type, U be an open set, and F_1, F_2 two disjoint closed sets such that $F_1 \subset U$. If two operators R_d, R_s satisfy the conditions $P_U A R_d \sim P_U$, $R_s A P_U \sim P_U$ then the operators $P_{F_2} R_d P_{F_1}$ and $P_{F_1} R_d P_{F_2}$ are completely continuous.*

Proof. We shall give the proof for the operator $P_{F_1} R_d P_{F_2}$, for the other operator it goes analogously. Suppose that the operator $P_{F_1} R_d P_{F_2}$ is not completely continuous. Then there is a norm bounded sequence $\{\varphi_n\}$ of functions that vanish outside of F_2 such that the sequence $\{P_{F_1} R_d \varphi_n\}$ is not compact. With $R_d \varphi_n =: \psi_n$ the sequence $\{P_{F_1} \psi_n\}$ is not compact, i.e. there is a number $\varepsilon_0 > 0$ and a sub-class $\{\psi_{n_k}\}$ such that $\|P_{F_1} \psi_{n_k} - P_{F_1} \psi_{n_l}\| \geqq \varepsilon_0$, $k \neq l$. In order to simplify notations, we shall write briefly ψ_k instead of ψ_{n_k} so that $\|P_{F_1} \psi_k - P_{F_1} \psi_l\| \geqq \varepsilon_0$. Since the two closed sets F_1 and F_2 are disjoint, there exists a continuous, real-valued function $\Phi(x)$ defined on X with the following properties:

1. $\Phi(x) = 0$ for $x \in F_2 \cup (X \setminus U)$;
2. $\Phi(x) = 1$ for $x \in F_1$;
3. $0 \leqq \Phi(x) \leqq 1$.

Let us denote by V_{r_1, r_2} the set of all x such that $r_1 < \Phi(x) < r_2$.

The proof to our theorem is based upon the following assertion:

(A) If, for $k \neq l$, $\|P_{F_1} \psi_k - P_{F_1} \psi_l\| \geqq \varepsilon_0$ then for any set V_{r_1, r_2} there is the estimate

$$\overline{\lim_{k \to \infty}} \int_{V_{r_1, r_2}} |\psi_k|^p \, d\mu > \frac{1}{2^p} \left(\frac{\varepsilon_0}{2} \right)^p \frac{1}{\|A\|^p \, \|R_s\|^p} =: \delta^p \tag{1}$$

(μ denotes the measure in X).

We shall prove this assertion indirectly. Suppose that there is a set $V = V_{r_1,r_2}$, $0 < r_1 < r_2 < 1$, such that

$$\overline{\lim_{k\to\infty}} \int_V |\psi_k|^p \,\mathrm{d}\mu \leqq \delta^p. \tag{2}$$

Then choose numbers $\varrho_1, \varrho_2, r_1 < \varrho_1 < \varrho_2 < r_2$, and put $W := V_{\varrho_1,\varrho_2}$. Furthermore, we introduce the notations

$$W_+ = \{x: \Phi(x) \geqq \varrho_2\}, \qquad W_- = \{x: \Phi(x) \leqq \varrho_1\},$$
$$V_+ = \{x: \Phi(x) \geqq r_2\}, \qquad V_- = \{x: \Phi(x) \leqq r_1\}.$$

The sets W, W_+, W_- and likewise the sets V, V_+, V_- have no common points, and, obviously, cover the entire space X.

We have

$$P_U A \psi_k = P_U A R_d \varphi_k = P_U \varphi_k + \zeta_k, \tag{3}$$

where $\{\zeta_k\}$ is a compact sequence. Further we put

$$\omega_k = P_U A P_V \psi_k,$$
$$\xi_k = [P_{W_-} P_U A P_{V_+} + P_{W_+} P_U A P_{V_-} + P_W P_U A (P_{V_+} + P_{V_-})]\, \psi_k. \tag{4}$$

By (2) we have $\overline{\lim}\, \|\omega_k\| \leqq \|A\|\, \delta$. Moreover, the sequence $\{\xi_k\}$ is compact, too, since the operator A is of local type, and the intersections $\overline{V}_+ \cap \overline{W}_-$, $\overline{W}_+ \cap \overline{V}_-$, $\overline{W} \cap (\overline{V}_+ \cup \overline{V}_-)$ are all empty.

With the notations (4) equation (3) can be written as

$$(P_{W_+} P_U A P_{V_+} + P_{W_-} P_U A P_{V_-})\, \psi_k = P_U \varphi_k - \omega_k - \xi_k + \zeta_k.$$

Applying the operator P_{W_+} to this equation and taking into account the relations $W_+ \subset U$ and $\varphi_k(x) = 0$, $x \in W_+$, we find

$$P_{W_+} A P_{V_+} \psi_k = -P_{W_+} \omega_k - \xi_k', \qquad \xi_k' = P_{W_+}(\xi_k - \zeta_k). \tag{5}$$

Because of $V_+ \subset \operatorname{int} W_+$ we have $P_{W_+} A P_{V_+} \sim A P_{V_+}$, hence, the relation (5) can be written in the form

$$A P_{W_+} \psi_k = -P_{W_+} \omega_k - \xi_k'' \tag{6}$$

with a compact sequence $\{\xi_k''\}$.

To (6) we apply the operator R_s. Because of $R_s A P_U \sim P_U$ and $V_+ \subset U$ we find $R_s A P_{V_+} \sim P_{V_+}$. Thence, $P_{V_+} \psi_k = -R_s P_{W_+} \omega_k - \eta_k$ with a compact sequence $\{\eta_k\}$. From that sequence we select a convergent sub-sequence which will also be denoted by $\{\eta_k\}$. Because of $V_+ \supset F_1$ we have, by assumption, $\|P_{V_+} \psi_k - P_{V_+} \psi_{k+1}\| > \varepsilon_0$, and, on the other hand, by the inequality (2),

$$\overline{\lim_{k\to\infty}}\, \|P_{V_+} \psi_k - P_{V_+} \psi_{k+1}\| = \overline{\lim_{k\to\infty}}\, \|R_s P_{W_+} (\omega_{k+1} - \omega_k)\|$$

$$\leqq 2\, \overline{\lim_{k\to\infty}}\, \|R_s P_{W_+} \omega_k\| \leqq 2\, \|R_s\|\, \|A\|\, \delta = \frac{\varepsilon_0}{2}.$$

The contradiction obtained in that way proves assertion (A).

Now, let us come back to the proof of our theorem. We choose a sequence of pairwise disjoint sets V_n each of which is of the form V_{r_1,r_2}, $0 < r_1 < r_2 < 1$. Due to the relation (1) we can select from the sequence $\{\psi_k\} =: N_0$ a new sequence N_1 such that its elements ψ_n satisfy the inequality

$$\int_{V_1} |\psi_n|^p \,\mathrm{d}\mu > \delta^p.$$

Proceeding in this way, we obtain a sequence of sequences N_n with the properties: a) $N_{n+1} \subset N_n$; b) $\psi_k \in N_n$ implies

$$\int_{V_j} |\psi_k|^p \, d\mu > \delta^p, \qquad j \leqq n.$$

Then, for $\psi_k \in N_n$ by b) we have the inequality $\|\psi_k\| \geqq n^{1/p}\delta$, and this is in contradiction to the boundedness of the sequence $\{\psi_k\}$. ∎

Theorem 1.2. *If an operator A of local type is invertible, then its inverse A^{-1} is of local type, too.*

Theorem 1.3. *If an operator of local type is a Fredholm operator, then its left and right regularizers are also operators of local type.*

These theorems are simple consequences of Theorem 1.1.

§ 2. Equivalence at a point and locally Fredholm operators

2.1. Two operators A and B will be called *equivalent at a point* $x_0 \in X$ iff for any $\varepsilon > 0$ there exists a neighborhood $U \ni x_0$ such that $|||AP_U - BP_U||| < \varepsilon$ and $|||P_UA - P_UB||| < \varepsilon$. In that case we shall briefly write $A \overset{x_0}{\sim} B$. The equivalence at a point is, obviously, reflexive, commutative, and transitive. Here are some further properties of this notion.

a) Let A, A', B, B' be operators of local type. If $A \overset{x_0}{\sim} A'$ and $B \overset{x_0}{\sim} B'$ then $AB \overset{x_0}{\sim} A'B'$.

b) For two operators A and B of local type we have $A \overset{x_0}{\sim} B$ if and only if there exists to any $\varepsilon > 0$ a neighborhood $U \ni x_0$ such that either $|||(A - B) P_U||| < \varepsilon$ or $|||P_U(A - B)||| < \varepsilon$.

c) If the operators A and B are of local type and $A \overset{x_0}{\sim} B$ then $P_{U_1}AP_{U_2} \overset{x_0}{\sim} P_{U_3}AP_{U_4}$ for arbitrary neighborhoods U_i $(i = 1, 2, 3, 4)$ of the point x_0.

d) If $|||A_n - A||| \to 0$, $|||B_n - B||| \to 0$, and $A_n \overset{x_0}{\sim} B_n$ then $A \overset{x_0}{\sim} B$.

The three examples given hereafter are important for the following later applications.

Example 1. Let the function $a(x)$ be continuous on X. Then the operator of multiplication by $a(x)$ is equivalent at the point $x_0 \in X$ to the operator of multiplication by the constant $a(x_0)$.

Example 2. Let us consider again the singular integral from Example 2, § 1 under the additional assumption

$$\lim_{x \to x_0} \int_S |f(x, \theta) - f(x_0, \theta)|^{p'} \, dS = 0.$$

It is easy to see that $A \overset{x_0}{\sim} A_{x_0}$ with the operator A_{x_0} defined by

$$(A_{x_0}u)(x) = \int_{\mathbb{R}^m} \frac{f(x_0, \theta)}{r^m} u(y) \, dy. \tag{1}$$

It is worth noting that the characteristic of the operator (1) does not depend on the pole.

Example 3. Again we consider the singular integral in Example 2, § 1. Assume that a smooth $(m - 1)$-dimensional surface Γ passes through the point x_0 such that Γ divides a certain neighborhood U of x_0 into two parts U^+ and U^-. Let there exist two functions $f^\pm(\theta)$ which are positively homogeneous of degree zero (with respect to θ) and satisfy the condition

$$\lim_{\substack{x \to x_0 \\ x \in U^\pm}} \int_S |f^\pm(\theta) - f(x, \theta)|^{p'} \, dS = 0.$$

Put

$$A_{x_0}^{\pm} u = \int_{\mathbf{R}^m} \frac{f^{\pm}(\theta)}{r^m} u(y)\,dy$$

then it is easy to see that

$$A_{x_0}^{+} P_{U^+} + A_{x_0}^{-} P_{U^-} \overset{x_0}{\approx} A. \tag{2}$$

2.2. An operator R_s is called *local left regularizer* of the operator A at the point x_0 iff there exists a neighborhood $U \ni x_0$ such that $R_s A P_U \sim P_U$. A *local right regularizer* is defined similarly.

Let A be an operator of local type, and R_s, R_d its local left and right, resp., regularizers at the point $x_0 \in X$.

Lemma 2.1. *Let V be an arbitrary neighborhood of the point x_0. Then the operators $R_s P_V$ and $P_V R_s$ are local left regularizers whereas $R_d P_V$ and $P_V R_d$ are local right regularizers of the operator A.*

Proof. Choose a neighborhood W of x_0 such that $\overline{W} \subset V \cap U$. Then $P_V A P_W \sim A P_W$, and therefore $R_s A P_W \sim P_W$ and $R_s A P_W \sim R_s P_V A P_W$, simultaneously. Hence, $(R_s P_V)\, A P_W \sim P_W$, and the operator $R_s P_V$ provides a local left regularizer of A at the point x_0. The other operators are considered similarly. ∎

We shall call an operator A *locally Fredholm operator at a point* x_0 iff it has local left and right regularizers at that point.

Theorem 2.1. *Let A, B be operators in $\mathbf{L}_p(X)$, and $A \overset{x_0}{\approx} B$, $x_0 \in X$. If A has a left (or right) local regularizer at the point x_0 then so does B.*

Proof. We shall prove the theorem for the case of a local left regularizer – the other case goes similarly.

Let R_s be a local left regularizer of the operator A at the point x_0. Then there is a neighborhood $U_1 \ni x_0$ such that $R_s A P_{U_1} \sim P_{U_1}$. Because of $A \overset{x_0}{\approx} B$ there is another neighborhood $U_0 \ni x_0$ such that for any neighborhood U_2 of the point x_0 contained in U_0

$$|||(A - B)\, P_{U_2}||| \leqq \frac{1}{4\,\|R_s\|}. \tag{3}$$

Let us consider the intersection $U := U_1 \cap U_0$. By the definition of the essential norm there exists a completely continuous operator T_1 such that

$$\|(A - B)\, P_U - T_1 P_U\| \leqq \frac{1}{2\,\|R_s\|}.$$

We put $C := R_s[(A - B)\, P_U - T_1 P_U]$ and find

$$R_s B P_U = (I - C)\, P_U - P_U T P_U \tag{4}$$

with some completely continuous operator T.

Obviously, $\|C\| \leqq 1/2$ so that the inverse operator $(I - C)^{-1}$ exists. Applying this operator from the left to (4) we obtain

$$[(I - C)^{-1}\, P_U R_s]\, B P_U = P_U + T_2$$

with a completely continuous operator T_2. Therefore, the operator $(I - C)^{-1}\, P_U R_s$ provides a local left regularizer of B at the point x_0. ∎

Remark. If the operators A, B, and R_s are of local type then so is the operator $(I - C)^{-1} P_U R_s$. This statement implies the following one: If under the assumptions of Theorem 2.1 the operators A and B are of local type, then there exist at the point x_0 local left (right) regularizers either for both of them or for none of them, and they are (in case of existence) of local type, too.

Theorem 2.2. *Let the operators A and B satisfy the conditions of Theorem 2.1. If A is a locally Fredholm operator at the point x_0 then so is B.*

The theorem follows, obviously, from Theorem 2.1.

§ 3. The envelope of an operator family

3.1. Let us be given a family of operators A_x acting in $\mathbf{L}_p(X)$ and depending on the point $x \in X$. Such an operator family will be called *locally continuous* iff for every point $x_0 \in X$ and any $\varepsilon > 0$ there exists a neighborhood $U \ni x_0$ such that $|||(A_{x_0} - A_x) P_U||| < \varepsilon$ and $|||P_U(A_{x_0} - A_x)||| < \varepsilon$ whenever $x \in U$.

If the operators A_x are of local type it suffices to require only one of these two inequalities.

The operator A is called *envelope* of the family A_x iff for any $x \in X$ the relation $A \overset{x}{\sim} A_x$ holds.

Theorem 3.1. *Let A_x be a locally continuous family of operators of local type. Then there exists an envelope operator of local type, and it is determined up to a completely continuous additive term.*

Before proving our theorem we formulate two lemmas.

Lemma 3.1. *Let $\{\tilde{U}_i\}$ be a finite family of measurable sets with an intersection multiplicity $k + 1$, and U_i be measurable and pairwise disjoint sets. Then for any operators C_i bounded in $\mathbf{L}_p(X)$ there are the inequalities*

$$\Big\|\sum_i P_{\tilde{U}_i} C_i P_{U_i}\Big\| \leqq (k + 1) \max_i \|C_i\|, \qquad \Big\|\sum_i P_{U_i} C_i P_{\tilde{U}_i}\Big\| \leqq (k + 1) \max_i \|C_i\|, \tag{1}$$

$$\Big|\Big|\Big|\sum_i P_{\tilde{U}_i} C_i P_{U_i}\Big|\Big|\Big| \leqq (k + 1) \max_i |||C_i|||, \qquad \Big|\Big|\Big|\sum_i P_{U_i} C_i P_{\tilde{U}_i}\Big|\Big|\Big| \leqq (k + 1) \max_i |||C_i|||. \tag{2}$$

Lemma 3.2. *If an operator of local type is equivalent at every point to the zero operator then it is completely continuous.*

Proof of Lemma 3.1. For $k = 0$ the sets $\tilde{U}_i$ are pairwise disjoint, and therefore we have the inequalities

$$\Big\|\sum_i P_{\tilde{U}_i} C_i P_{U_i} u\Big\|^p \leqq \sum_i \|C_i P_{U_i} u\|^p \leqq \sum_i \|C_i P_{U_i}\|^p \|P_{U_i} u\|^p$$
$$\leqq \max_i \|C_i P_{U_i}\|^p \sum_i \|P_{U_i} u\|^p \leqq \max_i \|C_i\|^p \|u\|^p,$$

and (1) is proved for $k = 0$.

Suppose now the inequality (1) to be true for $k \leqq l - 1$. We shall show it to be true also for $k = l$. The relations

$$V_1 = \tilde{U}_1, \quad V_2 = \tilde{U}_2 \setminus \tilde{U}_1, \ldots, \quad V_{n+1} = \tilde{U}_{n+1} \setminus \bigcup_{i=1}^{n} \tilde{U}_i; \quad W_i = \tilde{U}_i \setminus V_i$$

define measurable sets V_i and W_i. The sets V_i are, obviously, pairwise disjoint, $V_i \cap W_i = \emptyset$, $V_i \cup W_i = \tilde{U}_i$, and the intersection multiplicity of the family $\{W_i\}$ is not greater than l. Hence,

$$\left\| \sum_i P_{\tilde{U}_i} C_i P_{U_i} \right\| \leqq \left\| \sum_i P_{V_i} C_i P_{U_i} \right\| + \left\| \sum_i P_{W_i} C_i P_{U_i} \right\|$$
$$\leqq \max_i \|C_i\| + l \max_i \|C_i\| = (l+1) \max_i \|C_i\|,$$

and the inequality (1) is proved. From (1) one can easily obtain the inequality (2). ∎

Proof of Lemma 3.2. Let A be the operator in question, and $\varepsilon > 0$ be arbitrary. To any point x_0 there exists a neighborhood $U_\varepsilon \ni x_0$ such that $|||AP_{U_\varepsilon}||| < \varepsilon$. These neighborhoods form a covering of X. Then we can select a finite covering $\{\tilde{U}_i\}_{i=1}^N$ the intersection multiplicity of which is $m+1$ (remember: m is the dimension of the space X). Let $\{F_k\}$ be a finite covering of the space X by closed sets each of which is contained in one of the sets $\tilde{U}_i$. Then we construct a covering $\{U_i\}_{i=1}^N$ consisting of measurable and pairwise disjoint sets: As U_1 we take the union of all F_k contained in $\tilde{U}_1$. Proceeding we put by induction

$$U_{n+1} = \bigcup_{F_k \subset \tilde{U}_{n+1}} F_k \setminus \bigcup_{k=1}^{n} U_k .$$

Obviously, $\overline{U}_k \subset \operatorname{int} \tilde{U}_k$. The operator A is represented as $A = \sum_{i=1}^N P_{U_i} A$. Since it is of local type, we have $P_{U_i} A \sim P_{U_i} A P_{\tilde{U}_i}$, $A \sim \sum_{i=1}^N P_{U_i} A P_{\tilde{U}_i}$. Now, inequality (2) gives

$$|||A||| = \left|\left|\left| \sum_{i=1}^N P_{U_i} A P_{\tilde{U}_i} \right|\right|\right| \leqq (m+1)\, |||AP_{U_i}||| \leqq (m+1)\, \varepsilon$$

and since ε is arbitrary we obtain $|||A||| = 0$. ∎

Proof of Theorem 3.1. We shall construct the envelope A as the limit of an operator sequence A_n defined as follows: Let $\varepsilon_n > 0$, $\varepsilon_n \to 0$ as $n \to \infty$. For every ε_n we choose a finite covering $\{\tilde{U}_i^{(n)}\}_{i=1}^{N_n}$ of an intersection multiplicity $m+1$ such that for any x, $x_0 \in \tilde{U}_i^{(n)}$

$$|||(A_x - A_{x_0}) P_{\tilde{U}_i^{(n)}}||| < \varepsilon_n .$$

To every covering $\{\tilde{U}_i^{(n)}\}$ we construct as in the proof to Lemma 3.2 a covering $\{U_i^{(n)}\}_{i=1}^{N_n}$ consisting of measurable and pairwise disjoint sets such that $\overline{U}_i^{(n)} \subset \operatorname{int} \tilde{U}_i^{(n)}$. Then the operator A_n is defined by

$$A_n = \sum_{i=1}^{N_n} P_{\tilde{U}_i^{(n)}} A_{x_i^{(n)}} P_{U_i^{(n)}},$$

where $x_i^{(n)}$ is a point in the set $U_i^{(n)}$.

Now we shall show that the sequence $\{A_n\}$ is a Cauchy sequence with respect to the essential norm, and, therefore is convergent in that norm. In order to do so we estimate the difference $|||A_n - A_k|||$. First, note that

$$A_n = \sum_{j=1}^{N_k} \sum_{i=1}^{N_n} P_{\tilde{U}_i^{(n)}} A_{x_i^{(n)}} P_{U_i^{(n)}} P_{U_j^{(k)}} \sim \sum_{j=1}^{N_k} \sum_{i=1}^{N_n} P_{\tilde{U}_j^{(k)} \cap \tilde{U}_i^{(n)}} A_{x_i^{(n)}} P_{U_i^{(n)} \cap U_j^{(k)}} . \qquad (3)$$

Here, the equivalence symbol $\sim$ makes sense, since

$$\overline{U_i^{(n)} \cap U_j^{(k)}} = \overline{U_i^{(n)}} \cap \overline{U_j^{(k)}} \subset \operatorname{int} \tilde{U}_i^{(n)} \cap \operatorname{int} \tilde{U}_j^{(k)} = \operatorname{int} (\tilde{U}_i^{(n)} \cap \tilde{U}_j^{(k)})$$

and A_x is an operator of local type. The relation (3) remains to hold true with n and k interchanged. Using Lemma 3.1, we find

$$|||A_n - A_k||| = \left|\left|\left| \sum_{j=1}^{N_k} \sum_{i=1}^{N_n} P_{\tilde{U}_i^{(n)} \cap \tilde{U}_j^{(k)}} (A_{x_i^{(n)}} - A_{x_i^{(k)}}) P_{U_i^{(n)} \cap U_j^{(k)}} \right|\right|\right|$$
$$< 2(m+1)^2 \max(\varepsilon_n, \varepsilon_k) \xrightarrow[n,k\to\infty]{} 0$$

(the factor $(m+1)^2$ occurs because the multiplicity of the covering $\{U_i^{(n)} \cap U_j^{(k)}\}$ is not greater than that factor).

Thus, there is an operator A of local type such that $|||A - A_n||| \to 0$. We show this operator A to be the envelope of the family $\{A_x\}$. To that aim let U be a neighborhood of the point $x_0 \in X$ such that $|||(A_x - A_{x'}) P_U||| < \varepsilon$ for any $x, x' \in U$. Now,

$$|||(A_{x_0} - A) P_U||| \leqq |||(A_{x_0} - A_n) P_U||| + |||(A_n - A) P_U|||.$$

For sufficiently large n the second term becomes arbitrarily small. Concerning the first term, we have

$$\Delta := \left|\left|\left| \left(A_{x_0} - \sum_{i=1}^{N_n} P_{\tilde{U}_i^{(n)}} A_{x_i^{(n)}} P_{U_i^{(n)}}\right) P_U \right|\right|\right| = \left|\left|\left| \left(A_{x_0} \sum_{i=1}^{N_n} P_{U_i^{(n)}} - \sum_{i=1}^{N_n} P_{\tilde{U}_i^{(n)}} A_{x_i^{(n)}} P_{U_i^{(n)}}\right) P_U \right|\right|\right|$$
$$= \left|\left|\left| \sum_{i=1}^{N_n} P_{\tilde{U}_i^{(n)}} (A_{x_0} - A_{x_i^{(n)}}) P_{U_i^{(n)} \cap U} \right|\right|\right|.$$

Choose a point $x_i \in U \cap U_i^{(n)}$ (provided that intersection is not empty), then

$$\Delta \leqq \left|\left|\left| \sum_{i=1}^{N_n} P_{\tilde{U}_i^{(n)}} (A_{x_0} - A_{x_i}) P_{U_i^{(n)} \cap U} \right|\right|\right| + \left|\left|\left| \sum_{i=1}^{N_n} P_{\tilde{U}_i^{(n)}} (A_{x_i} - A_{x_i^{(n)}}) P_{U_i^{(n)} \cap U} \right|\right|\right| \leqq 2(m+1)\varepsilon,$$

i.e. $A \overset{x_0}{\sim} A_{x_0}$. That the operator A is uniquely determined up to completely continuous additive terms, follows from Lemma 3.2. Thus, Theorem 3.1 is proved. Finally, we note that

$$|||A||| \leqq (m+1) \sup_x |||A_x|||. \quad \blacksquare$$

§ 4. A theorem on the connection between Fredholm and locally Fredholm operators

Theorem 4.1. *An operator of local type acting in $\mathbf{L}_p(X)$ is a Fredholm operator if and only if it is a locally Fredholm operator at every point $x \in X$.*

Theorem 4.2. *If an operator of local type acting in $\mathbf{L}_p(X)$ has at every point $x \in X$ a local left (right) regularizer of local type, then it has also a usual left (right) regularizer.*

We shall prove Theorem 4.2 first. It suffices to restrict ourselves to the case of a left regularizer. To any point $x \in X$ there is an operator $R_s^{(x)}$ and a neighborhood $U(x) \ni x$ such that $R_s^{(x)} A P_{U(x)} = P_{U(x)} + T_x$ with a completely continuous operator T_x. Now choose a neighborhood $V(x)$ of the point x such that $\overline{V(x)} \subset U(x)$. For x varying in the set X the neighborhoods $V(x)$ form a covering of X. From that covering we select a finite covering $\{V_i\}$, $V_i = V(x_i)$, $i = 1, 2, \ldots, n$, and then we form a covering consisting of pairwise disjoint sets W_i:

$$W_1 = V_1, \ldots, W_k = \left(\bigcup_{i=1}^{k} V_i\right) \setminus \left(\bigcup_{i=1}^{k-1} V_i\right).$$

We shall show that the operator $R_s = \sum_{i=1}^{n} P_{W_i} R_s^{(x_i)}$ provides the left regularizer in question. Due to Lemma 3.2 it suffices to prove the relation $R_s A \overset{x}{\sim} I$ for any $x \in X$.

We construct a neighborhood $W(x)$ of x as follows: We collect all sets V_i the closures of which contain the point x; let us denote them by V_{i_k}. Then the neighborhood $W(x)$ is chosen such that $\overline{W(x)} \subset \bigcap V_{i_k}$, and $\overline{W(x)}$ has no common points with the sets $V_i \neq V_{i_k}$. Consider the operator

$$R_s A P_{W(x)} = \sum_{i=1}^{n} P_{W_i} R_s^{(x_i)} A P_{W(x)} \sim \sum_{k} P_{W_{i_k}} R_s^{(x_{i_k})} A P_{W(x)}.$$

Because of $W(x) \subset V_{i_k}$ we have $R_s^{(x_{i_k})} A P_{W(x)} \sim P_{W(x)}$, hence $R_s A P_{W(x)} \sim \sum_{k} P_{W_{i_k}} P_{W(x)}$ $= P_{W(x)}$. Thus, Theorem 4.2 is proved. ∎

Now we are able to prove Theorem 4.1. Its necessity part is obvious. The sufficiency part is a consequence of the Theorems 4.2 and 1.1: By Theorem 1.1 the existence of the two local regularizers at the point x implies the existence of local regularizers at that point which are of local type. Now, Theorem 4.2 states that the operator A possesses right and left regularizers and, therefore, is a Fredholm operator. ∎

§ 5. Homogeneous operators and translation invariant operators

5.1. To begin with, we recall some results from the theory of Banach algebras.

For commutative Banach algebras without radical and with symmetric involution it is well-known that every non-invertible element is a generalized zero divisor (cf. Gelfand, Raikov, and Shilov [1]). This property is true also for the algebra of matrices the elements of which are from a commutative algebra.

Let $\mathfrak{R}$ be a (not necessarily commutative) Banach algebra. An element $x \in \mathfrak{R}$ is called a *generalized right zero divisor* of the algebra $\mathfrak{R}$ iff there is a sequence $\{y_n\} \subset \mathfrak{R}$ such that 1. $\inf \|y_n\| > 0$, and 2. $\|y_n x\| \to 0$. The notion of a *generalized left zero divisor* is defined analogously.

Let $\mathfrak{R}$ be a commutative Banach algebra without radical and with symmetric involution, furthermore, assume that $\|xy\| \leq \|x\| \, \|y\|$ and $\|e\| = 1$ (e stands for the unit element of the algebra). By $\mathfrak{R}_n$ we denote the algebra of square matrices of order n with elements from the algebra $\mathfrak{R}$.

Theorem 5.1. *If a matrix $Z \in \mathfrak{R}_n$ has no inverse then it is either left or right generalized zero divisor in $\mathfrak{R}_n$. The matrix Z is invertible if and only if* $\det Z(M) \neq 0$, $M \in \mathfrak{M}$, *where $\mathfrak{M}$ denotes the set of maximal ideals of the algebra $\mathfrak{R}$.*

The proof to Theorem 5.1 is essentially the same as in the case of a commutative algebra (cf. Gelfand, Raikov, and Shilov [1]).

5.2. Let α be a real number, $\alpha \neq 0$, $h \in \mathbb{R}^m$, and $u(x)$ be a function defined in $\mathbb{R}^m$. By $*$ and by $\square$ we denote the operators defined by the relations $(\alpha * u)(x) := u(\alpha x)$ and $(h \square u)(x) := u(x + h)$, resp.

A linear operator (defined e.g. in $\mathbf{L}_p(\mathbb{R}^m)$) is called *homogeneous* iff $A(\alpha * u) = \alpha * (Au)$ for arbitrary real $\alpha \neq 0$. This operator is said to be *translation invariant* iff $A(h \,\square\, u) = h \,\square\, (Au)$ for arbitrary $h \in \mathbb{R}^m$.

Lemma 5.1. *Let C be a homogeneous operator. Then for any neighborhood U_1 of the origin or of infinity,*

$$|||CP_{U_1}||| = |||P_{U_1}C||| = \|C\|. \tag{1}$$

Proof. Let U be an arbitrary neighborhood of the origin, then we show that

$$\|CP_U\| = \|C\|. \tag{2}$$

Obviously, $\|CP_U\| \leqq \|C\|$. On the other hand, for any $\varepsilon > 0$ there is a function $u_0 \in \mathbf{L}_p(\mathbb{R}^m)$ with compact support such that $\|Cu_0\| \geqq (\|C\| - \varepsilon)\,\|u_0\|$. Choose a number $\alpha > 0$ sufficiently large such that $u_0(\alpha x) = 0$ whenever $x \notin U$. Then

$$\begin{aligned} \|CP_U(\alpha * u_0)\| &= \|C(\alpha * u_0)\| = \|\alpha * Cu_0\| = \alpha^{-m/p}\,\|Cu_0\| \\ &\geqq \alpha^{-m/p}\,(\|C\| - \varepsilon)\,\|u_0\| = (\|C\| - \varepsilon)\,\|\alpha * u_0\|. \end{aligned}$$

Hence, $\|CP_U\| \geqq \|C\| - \varepsilon$, and $\|CP_U\| \geqq \|C\|$ since ε can be chosen arbitrarily small. Together with the estimate $\|CP_U\| \leqq \|C\|$ this yields the equality (2).

Next, we shall show the equality $|||CP_{U_1}||| = \|C\|$; the second equality in (1) can be proved similarly. The inequality $|||CP_{U_1}||| \leqq ||C||$ is obvious. We come to the opposite estimate. To any $\varepsilon > 0$ there is a completely continuous operator T such that $\|CP_{U_1} - T\| \leqq |||CP_{U_1}||| + \varepsilon$. We choose a zero neighborhood U contained in U_1 such that $\|TP_U\| < \varepsilon$. In order to do so it is sufficient to consider a sequence $\{U_n\}$ of zero neighborhoods monotoneously contracting to the origin. Then the sequence $\{P_{U_n}\}$ strongly converges to zero, and, therefore, the product TP_{U_n} tends to zero in the norm.

It is easy to see that $P_U = P_{U_1}P_U$. Thence, $CP_U - TP_U = (CP_{U_1} - T)\,P_U$, so that $\|CP_U - TP_U\| \leqq \|CP_{U_1} - T\|\,\|P_U\| = \|CP_{U_1} - T\|$. Thus, we obtain

$$\|CP_{U_1} - T\| \geqq \|CP_U - TP_U\| \geqq \|CP_U\| - \varepsilon = \|C\| - \varepsilon.$$

On the other hand, $\|CP_{U_1} - T\| \leqq |||CP_{U_1}||| + \varepsilon$. So we have $|||CP_{U_1}||| \geqq ||C|| - 2\varepsilon$, and finally $|||CP_{U_1}||| = ||C||$. ∎

Corollary 5.1. *If homogeneous operators are equivalent at the point $x = 0$ then they coincide.*

Lemma 5.2. *If the operator C is homogeneous and translation invariant then, for any neighborhood U,*

$$|||CP_U||| = |||P_UC||| = \|C\|. \tag{3}$$

The proof is obvious.

Theorem 5.2. *Let C be a bounded in $\mathbf{L}_2(\mathbb{R}^m)$ homogeneous operator of local type. If it is locally Fredholm at the point $x = 0$ or at the point $x = \infty$ then it is a Fredholm operator.*

Proof. Suppose that the operator C is not Fredholm. We denote by C^* as usually its adjoint. At least one of the operators C^*C and CC^* does not have a bounded inverse, say e.g. CC^*. Then we shall show that there is a sequence $\{B_n\}$ of homogeneous operators of local type acting in $\mathbf{L}_2(\mathbb{R}^m)$ with the following properties: $\inf \|B_n\| > 0$ and either $\|B_nC\| \to 0$ or $\|CB_n\| \to 0$. To that aim consider the operator sequence

$$D_n = \frac{(D + \lambda_n I)^{-1}}{\|(D + \lambda_n I)^{-1}\|}, \qquad D = CC^*,$$

where $\lambda_n > 0$ and $\lambda_n \to 0$. The operators C^*, CC^*, $CC^* + \lambda_n I$, and $(CC^* + \lambda_n I)^{-1}$ are homogeneous and of local type, moreover $\|DD_n\| \to 0$. Indeed, by assumption the operator D does not have a bounded inverse. Hence, there is a sequence of functions $u_n \in \mathbf{L}_2(\mathbb{R}^m)$, $\|u_n\| = 1$, such that $Du_n \to 0$. Since the operator D is non-negative, we have

$$\inf \frac{\|(D + \lambda_n I)\, u\|}{\|u\|} \leqq \lambda_n .$$

For $v := (D + \lambda_n I)^{-1}\, u$ we obtain

$$\|(D + \lambda_n I)^{-1}\| = \sup_{v \in \mathbf{L}_2(\mathbb{R}^m)} \frac{\|(D + \lambda_n I)^{-1}\, v\|}{\|v\|} \geqq \lambda_n^{-1} .$$

On the other hand, $D + \lambda_n I \geqq \lambda_n I$ such that $(D + \lambda_n I)^{-1} \leqq \lambda_n^{-1}\, I$, and $\|(D + \lambda_n I)^{-1}\| \leqq \lambda_n^{-1}$. Comparing these inequalities, we find $\|(D + \lambda_n I)^{-1}\| = \lambda_n^{-1}$ and $D_n = \lambda_n (D + \lambda_n I)^{-1}$. Therefore, $\|DD_n\| = \lambda_n\, \|D(D + \lambda_n I)^{-1}\|$. In the right-hand side of that equation there is the norm of a function of the self-adjoint operator D. This norm is not greater than the maximum modulus of that function on the spectrum of D. The operator D is non-negative, hence

$$\|D(D + \lambda_n I)^{-1}\| \leqq \max_{\lambda \geqq 0} \frac{\lambda}{\lambda + \lambda_n} = 1 .$$

Thus, we obtained the relation $\|DD_n\| = \|CC^*\, D_n\| \leqq \lambda_n \to 0$.

Now, we come to the construction of the operator B_n. If it happens that $\inf \|C^*\, D_n\| > 0$ then put $B_n := C^* D_n$. Otherwise there exists a sequence $\{k_n\}$ of integers such that $\|C^*\, D_{k_n}\| \to 0$ as $n \to \infty$. In that case put $B_n := D_{k_n}$.

Let us consider the case if $\|CB_n\| \to 0$; the other case can be treated similarly. The operator C is a locally Fredholm operator at the point $x = 0$ (or at $x = \infty$) such that there exist a local left regularizer R_s and a zero neighborhood U with $R_s C P_U \sim P_U$. Take a zero neighborhood U_1, $\overline{U_1} \subset U$. Then $\|R_s C B_n P_{U_1}\| \to 0$, but on the other hand

$$R_s C B_n P_{U_1} \sim R_s C P_U B_n P_{U_1} \sim P_U B_n P_{U_1} \sim B_n P_{U_1} .$$

By Lemma 5.1 this implies

$$|||R_s C B_n P_{U_1}||| = |||B_n P_{U_1}||| = \|B_n\| > \inf \|B_n\| .$$

The contradiction obtained proves our theorem. ▮

All the results of § 5.2 carry over, obviously, to operators in the space $\mathbf{L}_p(\mathbb{R}^m)$.

5.3. Now let us come to some properties of translation invariant operators. The contribution by Hörmander [1] is devoted to these operators, and we shall quote some facts from this paper paraphrasing them according to our purposes.

We shall denote the set of bounded in $\mathbf{L}_p(\mathbb{R}^m)$ and translation invariant operators by $\mathcal{N}_p^p$. In Hörmander's paper mentioned it is shown that $A \in \mathcal{N}_p^p$ implies $A \subset \mathcal{N}_{p'}^{p'}$ and likewise $A \in \mathcal{N}_q^q$ for any q between p and p'; in particular, $A \in \mathcal{N}_2^2$. Under the Fourier transformation the operator $A \in \mathcal{N}_2^2$ becomes the operator of multiplication by a certain bounded measurable function that can be considered as the *symbol* of the operator A. It will be denoted by Φ_A.

The set of symbols for all operators belonging to $\mathcal{N}_p^p$ forms a function space denoted by $\mathcal{M}_p^p$. A norm in that space is provided by the formula

$$\|\Phi_A\|_{\mathcal{M}_p^p} = \|A\| .$$

From the above arguments it follows that $\mathcal{M}_p^p = \mathcal{M}_{p'}^{p'} \subset \mathcal{M}_2^2$. Moreover, we have the relations $\|\Phi\|_{\mathcal{M}_2^2} \leqq \|\Phi\|_{\mathcal{M}_p^p} = \|\Phi\|_{\mathcal{M}_{p'}^{p'}}$.

Let us denote by $\mathcal{H}_p$ the subspace of $\mathcal{M}_p^p$ obtained as the closure (in the metric of that space) of the set of positively homogeneous functions of degree zero which are infinitely differentiable everywhere except the points $x = 0$ and $x = \infty$. The set $\mathcal{H}_2$ coincides, obviously, with the set of positively homogeneous functions which are continuous on the unit sphere.

5.4. Let $\mathfrak{Q}_p$ denote the space of translation invariant operators the symbols of which belong to the class $\mathcal{H}_p$. The spaces $\mathfrak{Q}_p$ and $\mathcal{H}_p$ are, obviously, isomorphic Banach algebras. Between the unit sphere S and the set of maximal ideals $\mathfrak{M}_p$ of the algebra $\mathfrak{Q}_p$ there is a 1-to-1 correspondence such that for $l \in \mathfrak{M}_p$ and $\theta \in S$ in correspondence, we have the equation $l(A) = \Phi_A(\theta)$ for any $A \in \mathfrak{Q}_p$. It can be shown that an operator $A \in \mathfrak{Q}_p$ has an inverse in $\mathbf{L}_p$ if and only if its symbol is non-zero everywhere on S.

We mention an important property of the class $\mathfrak{Q}_p$: An operator $A \in \mathfrak{Q}_p$ is of local type. Indeed, let $A \in \mathfrak{Q}_p$ with a symbol Φ_A that is infinitely differentiable on S. Then it is easy to see by constructing the characteristic f_A corresponding to the symbol Φ_A that A is a simple singular operator with the characteristic f_A and therefore, is an operator of local type. It is worth noting that every operator belonging to $\mathfrak{Q}_p$ is limit (in the sense of norm convergence) of a sequence consisting of operators of the same class $\mathfrak{Q}_p$ with infinitely differentiable symbols.

In $\mathfrak{Q}_p$ an involution can be introduced by assigning to every operator $A \in \mathfrak{Q}_p$ its adjoint A^*. It is easy to see that $\Phi_{A^*} = \overline{\Phi_A}$, and this implies the symmetry of the involution defined.

By $\mathfrak{Q}_p^n$ we denote the Banach algebra of operators A defined in the space $\mathbf{L}_p^n(\mathbb{R}^m)$ (cf. § 1.1, Chap. V) by the formula

$$(Au)_j = \sum_{k=1}^{n} A_{jk} u_k, \qquad j = 1, 2, \ldots, n.$$

Here, $A_{jk} \in \mathfrak{Q}_p$, u_k are the components of the vector u, and $(Au)_j$ the components of the vector Au. The matrix formed by the symbols of the operator A_{jk} will be called as done before the *symbol matrix* of the operator A. The Banach algebra $\mathfrak{Q}_p^n$ is isomorphic to the Banach algebra of matrices of order n with elements belonging to $\mathfrak{Q}_p$. Applying Theorem 5.1 we obtain the following

Theorem 5.3. *If the operator $A \in \mathfrak{Q}_p^n$ has no inverse then it is either a left or a right generalized zero divisor of the algebra $\mathfrak{Q}_p^n$. An operator $A \in \mathfrak{Q}_p^n$ is invertible if and only if its symbol determinant is non-zero everywhere on the unit sphere.*

Theorem 5.4. *Let $C \in \mathfrak{Q}_p^n$, and x_0 be an arbitrary point of $\overline{\mathbb{R}^m}$. Then the following assertions are equivalent:*

(i) *The operator C is locally Fredholm at the point x_0.*

(ii) *The operator C is invertible.*

(iii) *The symbol determinant of the operator C is non-zero everywhere on the unit sphere S.*

Proof. By Theorem 5.3, (iii) implies (i) and (ii). Therefore, it suffices to show that (i) implies (iii). We prove it indirectly. Suppose (i), and for some point $\theta_0 \in S$ be $\det \Phi_C(\theta_0) = 0$. By Theorem 5.1 there is a sequence of operators $B_k \in \mathfrak{Q}_p^n$, $\|B_k\| = 1$, such that $\|CB_k\| \to 0$ or $\|B_kC\| \to 0$. Now C is a locally Fredholm operator at the point x_0, therefore there exists a local left regularizer R_s at that point and a neighborhood $U \ni x_0$ such that $R_s CP_U \sim P_U$. Then for a neighborhood $U_1 \ni x_0$, $\overline{U}_1 \subset U$, we have the relations

$$R_s CB_k P_{U_1} \sim R_s CP_U B_k P_{U_1} \sim P_U B_k P_{U_1} \sim B_k P_{U_1}.$$

Hence, $|||B_k P_{U_1}||| \to 0$. On the other hand, by Lemma 5.1, $|||B_k P_{U_1}||| = ||B_k|| = 1$. The contradiction obtained proves our theorem. ▮

Finally, we mention an important type of operators belonging to the class $\mathfrak{Q}_p$, namely the singular integral operators of the form

$$Au = \int_{\mathbb{R}^m} r^{-m} f(\theta)\, u(y)\, \mathrm{d}y$$

the characteristic of which is subject to the inequality

$$\int_S |f(\theta)|^q \,\mathrm{d}S < \infty, \qquad q = \text{const} > 1.$$

This assertion easily follows from the Calderón-Zygmund theorem and the properties given in § 5.3.

§ 6. Canonical singular integrals with piecewise continuous symbols

In this section we shall consider operators B and D acting in $\mathbf{L}_2(\mathbb{R}^m)$ which can be represented in the form

$$B = B^+ P_{\mathbb{R}^m_+} + B^- P_{\mathbb{R}^m_-}, \tag{1}$$

$$D = P_{\mathbb{R}^m_+} D^+ + P_{\mathbb{R}^m_+} D^-. \tag{1'}$$

Here, $B^\pm$ and $D^\pm$ are operators belonging to $\mathfrak{Q}_2$, and $\mathbb{R}^m_+, \mathbb{R}^m_-$ denote the half-spaces $x_1 > 0$ and $x_1 < 0$, resp.

6.1. Equations with operators (1), (1′) are related to each other as follows: If the operator D^+ is continuously invertible then the equation

$$Du = g \tag{2}$$

is equivalent to the equation

$$\tilde{D}\psi = D^-(D^+)^{-1} P_{\mathbb{R}^m_-}\psi + P_{\mathbb{R}^m_+}\psi = P_{\mathbb{R}^m_-} g - P_{\mathbb{R}^m_-} D^-(D^+)^{-1} P_{\mathbb{R}^m_+} g \tag{2'}$$

in the sense that the solutions to these equations are related to each other by the relations

$$u = (D^+)^{-1}(P_{\mathbb{R}^m_+} g + P_{\mathbb{R}^m_-}\psi),$$

$$\psi = P_{\mathbb{R}^m_-} D^+ u - D^-(D^+)^{-1} P_{\mathbb{R}^m_-} D^+ u + P_{\mathbb{R}^m_-} g - P_{\mathbb{R}^m_-} D^-(D^+)^{-1} P_{\mathbb{R}^m_+} g. \tag{3}$$

Due to this correspondence we may restrict ourselves to considering operators of type (1).

6.2. If the symbols $\Phi_{B^\pm}(\theta)$ of the operators $B^\pm$ are non-zero everywhere then we put $G(\theta) := -\Phi_{B^-}(\theta)\,[\Phi_{B^+}(\theta)]^{-1}$. By L we denote the set of half great circles on the sphere $|\theta| = 1$ connecting the points $(-1, 0, \ldots, 0)$ and $(1, 0, \ldots, 0)$. Note that for $m > 2$ all half circles are homotopic to each other, whereas in case $m = 2$ the set L consists of two non-homotopic half circles.

The increment of the argument of the function $G(\theta)$ along the curve $l \in L$ will be denoted by G_l. For $m > 2$ the increment G_l does not depend on l, and we shall write simply $G_l := d(G)$. In case $m = 2$, G_l assumes two values, $G_{l^\pm} =: d^\pm(G)$ where l^+ and l^- denote the upper and lower half circle, resp.

Theorem 6.1. *The following assertions are equivalent:*

(i) *The operator B is a locally Fredholm operator at the point $x = 0$.*

(ii) *The operator B is continuously invertible.*

(iii) *The symbols $\Phi_{B^\pm}(\theta)$ are non-zero everywhere, and $|d(G)| < \pi$ if $m > 2$ or $|d^\pm(G)| < \pi$ if $m = 2$.*

Theorem 6.2. *Assume that the symbols $\Phi_{B^\pm}(\theta)$ are non-zero everywhere. If the number $d(G)$ (or $d^\pm(G)$ if $m = 2$) is not an odd multiple of π and $|d(G)| > \pi$ (or $|d^\pm(G)| > \pi$ if $m = 2$) then the following assertions are true:*

a) *For $d(G) > 0$ (or $d^\pm(G) > 0$ if $m = 2$) the operator B possesses a right inverse of local type, and $\alpha(B) = \infty$, $\alpha(B^*) = 0$.*

b) *In case of $d(G) < 0$ (or $d^\pm(G) < 0$ if $m = 2$) the operator B possesses a left inverse if local type, and $\alpha(B) = 0$, $\alpha(B^*) = \infty$.*

c) *If $m = 2$ and the numbers $d^\pm(G)$ are of different sign then $\alpha(B) = \alpha(B^*) = \infty$.*

We shall prove these two theorems simultaneously. The idea of the proof is to reduce the problem in question to a certain Riemann boundary value problem.

6.3. We shall consider the operator $\hat{B} := F^{-1} BF$ where F denotes Fourier transformation. Let

$$Fu = v, \qquad FP_{\mathbb{R}^m_+} u = v^+, \qquad FP_{\mathbb{R}^m_-} u = v^-,$$

such that $\hat{B}u = \hat{B}^+ v^+ + \hat{B}^- v^-$. Consider the equation

$$\hat{B}v = \Phi_{\hat{B}^+}(\xi_1, \xi_2, \ldots, \xi_m)\, v^+(\xi_1, \xi_2, \ldots, \xi_m) + \Phi_{\hat{B}^-}(\xi_1, \xi_2, \ldots, \xi_m)\, v^-(\xi_1, \xi_2, \ldots, \xi_m)$$
$$= \psi(\xi_1, \xi_2, \ldots, \xi_m). \tag{4}$$

The symbols $\Phi_{\hat{B}^\pm}(\theta)$ are assumed to be non-zero everywhere so that this equation can be written as

$$v^+(\xi_1, \xi_2, \ldots, \xi_m) = G(\xi_1, \xi_2, \ldots, \xi_m)\, v^-(\xi_1, \xi_2, \ldots, \xi_m) + \psi_1(\xi_1, \xi_2, \ldots, \xi_m) \tag{4'}$$

with $\psi_1 := \psi/\Phi_{\hat{B}^+}$. Problem (4′) is closely related to the Riemann boundary value problem

$$v_1^+(x) = G(x, \xi_2, \ldots, \xi_m)\, v_1^-(x) + g(x) \tag{5}$$

where $g \in \mathbf{L}_2$ is a given function, and the functions v_1^+, v_1^- are unknown. They are Fourier transforms of certain functions belonging to $\mathbf{L}_2(\mathbb{R})$ and vanishing on the left or right half axis, resp. The coordinates $\xi_2, \ldots, \xi_m$ play the role of parameters.

There is an obvious correspondence between problems (4′) and (5): If $v^\pm(\xi_1, \xi_2, \ldots, \xi_m)$ is a solution to problem (4′) then $v_1^\pm(x) = v^\pm(x, \xi_2, \ldots, \xi_m)$ provides a solution to problem (5) with right-hand side $g(x) := \psi_1(x, \xi_2, \ldots, \xi_m)$ for almost all $\xi_2, \ldots, \xi_m$. Conversely, if $v_1^\pm(x, \xi_2, \ldots, \xi_m)$ is a solution to problem (5) with $g(x) = \psi_1(x, \xi_2, \ldots, \xi_m)$ then the function $v^\pm(\xi_1, \ldots, \xi_m) = v_1^\pm(\xi_1, \xi_2, \ldots, \xi_m)$ satisfies (4) and, if it additionally belongs to $\mathbf{L}_2(\mathbb{R}^m)$, equation (4′), too.

6.4. Let us recall some results concerning the Riemann boundary value problem; they are contained in the papers by Simonenko [3, 4].

Consider the boundary value problem

$$v^+(t) = G(t)\, v^-(t) + g(t)$$

with $g \in \mathbf{L}_2$ given. The solutions are to belong to the classes $E_2^\pm$ that are defined as follows. We shall assume that the curve on which the problem is considered is a closed and simply connected Lyapunov curve. By E_2^+ and E_2^- we denote the class of functions that are analytic in the interior and in the exterior domain, resp., of that curve, and which can be represented as a Cauchy integral with a density belonging to $\mathbf{L}_2$. The function $G(t)$ is assumed to be measurable and bounded in modulus from below and from above by positive constants. Furthermore, let there be a number $\delta > 0$ such that to any point t of that curve there exists a neighborhood l on which the values of $G(t)$ lie in the interior of a sector with the apex angle $\pi - \delta$ and with vertex at the origin. Note that the results of Simonenko in the papers cited are true also in the case of a straight line.

For coefficients $G(t)$ satisfying the conditions mentioned above, in SIMONENKO [3] the index ind G is defined, and the following assertions are proved:

a) If ind $G \geqq 0$ the Riemann problem is solvable for every $g \in \mathbf{L}_2$, the dimension of the null space (i.e. the subspace of solutions to the corresponding homogeneous problem) is equal to ind G.

b) If ind $G < 0$ the problem is solvable under certain orthogonality conditions the number of which is $-$ind G. If the problem has a solution then it is unique.

c) In case of ind $G = 0$ the problem is always uniquely solvable.

6.5. Now we show that assertion (iii) implies the operator B to be invertible, and hence (i) and (ii). To begin with, we note that the coefficient $G(x, \xi_2, \ldots, \xi_m)$ in problem (5) is a homogeneous function of degree zero with respect to the variables $x, \xi_2, \ldots, \xi_m$ which is continuous except at the origin. From this remark we may infer the following: If x varies in $\mathbb{R}$ with fixed values $\xi_2, \ldots, \xi_m$ then G assumes the same values as if its arguments vary along the half great circle which lies in the intersection of the unit sphere with the plane passing the origin and the straight line $\xi_2 = \text{const}, \ldots, \xi_m = \text{const}$. For $x = \pm\infty$ we put $G(\pm\infty, \xi_2, \ldots, \xi_m) = G(1, 0, \ldots, 0)$.

Under the assumption (iii) the coefficient G satisfies the conditions of § 6.4, and ind $G = 0$. Then problem (5) is uniquely solvable. Hence, problem (4) is uniquely solvable provided that it has a solution at all. In order to show the existence of a solution we proceed as follows.

We denote by $R_{\xi_2,\ldots,\xi_m}$ the operator that assigns to a function g the function $v_1 = v_1^+ + v_1^-$ where v_1^+, v_1^- solve problem (5). This operator acts in $\mathbf{L}_2(\mathbb{R})$.

Problem (5) can be written as

$$Cv_1 = g, \tag{6}$$

where

$$Cv_1 = FP_0^\infty F^{-1} v_1 - G(x, \xi_2, \ldots, \xi_m)\, FP_{-\infty}^0 F^{-1} v_1, \tag{7}$$

and the operators $P_0^\infty, P_{-\infty}^0$ are defined by tho formulae

$$(P_0^\infty u)(x) = \begin{cases} u(x), & x > 0 \\ 0, & x \leqq 0 \end{cases}, \qquad (P_{-\infty}^0 u)(x) = \begin{cases} 0, & x \geqq 0 \\ u(x), & x < 0 \end{cases}$$

Obviously, $R = C^{-1}$.

For proving the existence of a solution to problem (4) it suffices to show that the norm of the operators $R_{\xi_2,\ldots,\xi_m}$ is bounded independently of $\xi_2, \ldots, \xi_m$; then the function $v(\xi_1, \ldots, \xi_m) = [R_{\xi_2,\ldots,\xi_m} \psi(x, \xi_2, \ldots, \xi_m)](\xi_1)$ belongs to the space $\mathbf{L}_2(\mathbb{R}^m)$. Note that

$G(x, \xi_2, \ldots, \xi_m)$ depends continuously on $\xi_2, \ldots, \xi_m$ on the $(m-2)$-dimensional sphere S^{m-2}, $\xi_2^2 + \ldots + \xi_m^2 = 1$. Therefore, the operators C and R also depend continuously on these parameters (in the sense of norm convergence). This implies the estimate

$$\| R_{\xi_2 \ldots \xi_m} \| \leq M = \text{const} \tag{8}$$

on S^{m-2}. Using the fact that the function G is homogeneous we can show that the estimate (8) is true for arbitrary values of the parameters $\xi_2, \ldots, \xi_m$. Indeed, since G is homogeneous we have

$$R_{\alpha\xi_2 \ldots \alpha\xi_m} = \alpha^{-1} * R_{\xi_2 \ldots \xi_m} * \alpha . \tag{9}$$

From the relation $\|\alpha * g\| = \alpha^{-1/2} \|g\|$ together with (8) and (9) it follows that the norm $\| R_{\xi_2, \ldots, \xi_m} \|$ is uniformly bounded for all values of the parameters $\xi_2, \ldots, \xi_m$.

6.6. Next, we prove the assertion a) of Theorem 6.2. We shall here consider the case $m > 2$ only; the case $m = 2$ is treated similarly. Let $\varkappa$ be an integer, and denote $\xi' := (\xi_2, \ldots, \xi_m)$, $|\xi'|^2 = \xi_2^2 + \ldots + \xi_m^2$. Consider the operator $A_\varkappa$ with the symbol

$$\Phi_{A_\varkappa} = \left(\frac{\xi_1 - \mathrm{i}\,|\xi'|}{\xi_1 + \mathrm{i}\,|\xi'|} \right)^{\varkappa} .$$

That operator considered as an element of the set $\mathfrak{Q}_2$ has the following properties:

1. $\|A_\varkappa\| = 1$;
2. for $\varkappa \geq 0$ ($\varkappa < 0$) it acts in the space of functions vanishing on $\mathbb{R}^m_-$ ($\mathbb{R}^m_+$);
3. $A_{\varkappa_1} A_{\varkappa_2} = A_{\varkappa_1 + \varkappa_2}$;
4. $A_0 = I$.

Let condition a) of Theorem 6.2 be fulfilled. Then there is an integer $\varkappa > 0$ such that $|d(G) - 2\pi\varkappa| < \pi$. Consider the operator

$$B_1 = B^+ P_{\mathbb{R}^m_+} + B^- A_{-\varkappa} P_{\mathbb{R}^m_-} . \tag{10}$$

It is easy to see that this operator satisfies the conditions of Theorem 6.1, therefore, it has an inverse B_1^{-1}. By Theorem 1.1 this inverse is of local type. It is clear that the operator $(P_{\mathbb{R}^m_+} + A_{-\varkappa} P_{\mathbb{R}^m_-}) B_1^{-1}$ is of local type and a right inverse to B. Hence, $\alpha(B^*) = 0$. One can show that $\alpha(B) = \infty$.

6.7. In order to prove the assertion b) of Theorem 6.2 we shall consider the adjoint B^* to the operator B,

$$B^* = P_{\mathbb{R}^m_+} (B^+)^* + P_{\mathbb{R}^m_-} (B^-)^* . \tag{11}$$

As shown in § 6.1, the equation $B^* u = g$ can be transformed into an equation with an operator of the form (1), where instead of B^+, B^- some other operators of $\mathfrak{Q}_2$ are written. Under the assumptions of Theorem 6.2 that new equation satisfies the condition a) of the theorem, and we are able to use the results of § 6.6. Thus, the operator (11) has a right inverse of local type. Then the operator B has a left inverse of local type. But then the range of the operator B is closed and $\alpha(B) = 0$, $\alpha(B^*) = \infty$.

The proof for the assertion c) of Theorem 6.2 is completely similar to those for a) and b).

6.8. In order to complete the proof to Theorem 6.1 we need the following

Lemma 6.1. *If for $m > 2$ the number $d(G)$ (for $m = 2$ at least one of the numbers $d^\pm(G)$) is an odd multiple of π then B is not a Fredholm operator.*

Proof. Suppose that B is a Fredholm operator under the assumptions of the lemma. Then there is a number $\varepsilon > 0$ such that every operator A with $\|A - B\| < \varepsilon$ is also a Fredholm operator. On the other hand, by a small perturbation of the operator B we can make the perturbed operator to satisfy the conditions of Theorem 6.2 such that it is not a Fredholm operator. ∎

Now we show that (i) implies (iii). The operator B is locally Fredholm at the point $x = 0$, hence, by Theorem 5.2, it is a Fredholm operator. By the Theorems 5.4, 4.1, and 3.2, applied for a point x with positive (negative) first coordinate x_1, we obtain that B^+ (B^-) is a Fredholm operator. Thus, we proved the first part of (iii). The other part is proved indirectly: Supposing the contrary we are under the assumption of either Theorem 6.2 or Lemma 6.1. In both cases B is not a Fredholm operator. Thus, Theorem 6.1 is proved completely. ∎

6.9. Theorems similar to those proved in the previous subsections are valid also for matrix operators of type (1) and (1′). We restrict ourselves to formulate them here; they can be proved completely analogously as in the scalar case.

Theorem 6.3. *For matrix operators of the form* (1) *the following assertions are equivalent*:

(i) *The operator B is locally Fredholm at the point $x = 0$.*

(ii) *The operator B is continuously invertible.*

(iii) *The operators B^+ and B^- have the following properties*:

1. *The symbol matrices are non-singular for all $\xi \neq 0$.*

2. *Among the eigenvalues of the matrix $H = G(1, 0, \ldots, 0)\, G^{-1}(-1, 0, \ldots, 0)$ there are no negative real numbers.*

3. *All partial indices $\varkappa_k(G, \xi_2, \ldots, \xi_m)$ of the matrix G are zero independently of the values of the parameters $\xi_2, \ldots, \xi_m$.*

In Theorem 6.3 we put $G(\xi) = -[\Phi_{B^+}(\xi)]^{-1}\, \Phi_{B^-}(\xi)$.

Theorem 6.4. *Assume that the conditions 1. and 2. of part* (iii) *in Theorem 6.3 are fulfilled. If the partial indices mentioned above are all positive* (*negative*) *then $\alpha(B) = \infty$, $\alpha(B^*) = 0$ ($\alpha(B) = 0$, $\alpha(B^*) = \infty$). If in that case the system of partial indices does not depend on $\xi_2, \ldots, \xi_m$ then there is a right inverse* (*left inverse*) *operator of local type.*

§ 7. Generalized singular integrals

A linear operator A (in general, a matrix operator) of local type in $\mathbf{L}_p(\mathbb{R}^m)$, $1 < p < \infty$, will be called a *generalized singular integral operator* iff the operator A is equivalent to a certain operator $A_x \in \mathfrak{Q}_p^n$ ($n \geqq 1$) at each point $x \in \overline{\mathbb{R}^m}$. The operator A_x will be called *simple part* of the operator A at the point x.

The matrix $\Phi_{A_x}(\theta)$ can be taken as the *symbol* of the operator A; as we did before, we shall denote this symbol by $\Phi_A(x, \theta)$.

From Corollary 5.1 it follows that the simple part of a generalized singular operator is determined uniquely. We show A_x to depend continuously on x. To any $\varepsilon > 0$ there is a neighborhood $U_1 \ni x$ such that $|||(A - A_x) P_{U_1}||| < \varepsilon/2$. We choose an arbitrary point $y \in U_1$ and a neighborhood $U_2 \ni y$, $U_2 \subset U_1$, such that $|||(A - A_y) P_{U_2}||| < \varepsilon/2$. Hence, $|||(A_x - A_y) P_{U_2}||| < \varepsilon$, and Lemma 5.1 implies $\|A_x - A_y\| = |||(A_x - A_y) P_{U_2}||| < \varepsilon$ which was to be shown. From what we proved it follows that the symbol $\Phi_A(x, \theta)$ is continuous on $\overline{\mathbb{R}^m} \times S$.

Theorem 3.1 yields the following property of generalized singular operators.

Let $\{A_x\}$ be a family of operators belonging to $\mathfrak{Q}_p^n$ which depend continuously on $x \in \overline{\mathbb{R}}^m$. Then there is a generalized singular operator A the simple part of which is A_x at every point x. The operator A is uniquely determined up to a completely continuous additive term, and

$$|||A||| \leqq (m + 1) \sup_x |||Ax|||.$$

This property, in turn, implies the following one: Every matrix $\Phi(x, \theta)$ which is continuous with respect to $x \in \overline{\mathbb{R}}^m$ and $\theta \in S$, is symbol of a generalized singular operator. This correspondence defines an isomorphism between the algebra of continuous (with respect to x and θ) matrix functions and the factor-algebra of generalized singular operators by the ideal of completely continuous operators.

The main result of this section is

Theorem 7.1. *A generalized singular operator is a Fredholm operator if and only if its symbol determinant* (*i.e. the determinant of the symbol matrix*) *is non-zero everywhere.*

Theorem 7.1 follows, obviously, from the Theorems 4.1, 3.2 and 5.4. It is important in that it strengthens the corresponding theorems of the previous chapters, namely it needs weaker smoothness conditions to the symbol. In particular, if $p = 2$ they are lifted completely.

§ 8. Compound generalized singular operators

8.1. Assume that the compact Euclidean space $\overline{\mathbb{R}}^m$ is divided into a finite number of closed regions D_i that do not have common inner points. The boundary Γ_i of D_i consists of disjoint closed Lyapunov surfaces and do not contain infinity, $\bigcup_i \Gamma_i =: \Gamma$.

Let B be an operator of local type in the space $\mathbf{L}_p^n(\mathbb{R}^m)$. The operator B is called *compound generalized singular operator* iff it can be represented as

$$B = \sum_i A_i P_{D_i} \tag{1}$$

where A_i are generalized singular operators as defined in § 7. In the following we restrict ourselves to the case $p = 2$.

For a compound generalized singular operator we shall introduce a notion of the symbol which, possibly, is not in complete accordance to the symbol concept as introduced in § 5, Chap. I.

The *symbol* of the operator (1) is defined as a matrix $\Phi_B(x, \theta)$, $x \notin \Gamma$, that coincides with $\Phi_{A_i}(x, \theta)$ for $x \in \operatorname{int} D_i$. The symbol defined in that way is, obviously, continuous at the points $x \notin \Gamma$. It is not defined for $x \in \Gamma$, but the limits do exist if $x \in \Gamma$ is approximated from the interior of a region D_i. These limits will be denoted by $\Phi_B^i(x, \theta)$.

We shall assume that every point $x \in \Gamma$ belongs to exactly two closed regions D_i and D_j. By ν^i and ν^j we shall denote the unit normal vector to Γ at a point $x_0 \in \Gamma$ directed into the interior of D_i and D_j, resp. Translating them we may achieve that their origins are at the point $\xi = 0$. Then their end points come to lie at two points e_i and e_j at the unit sphere S. From now on we have to distinguish the scalar case ($n = 1$) from the vector case ($n > 1$).

8.2. The case $n = 1$. We connect the points e_i and e_j by all possible half circles the set of which will be denoted by L.

If $\Phi_B^i(x_0, \theta)$ and $\Phi_B^j(x_0, \theta)$ are non-zero for every θ then we determine the increment of the argument

$$d_{x_0}^l = \left[\arg \frac{\Phi_B^i(x_0, \theta)}{\Phi_B^j(x_0, \theta)}\right]_l,$$

with θ varying along the half circle $l \in L$. For $m > 2$ all curves l are homotopic, and $d_{x_0}^l$ does not depend on l. The common value of $d_{x_0}^l$ is denoted by d_{x_0}. In case $m = 2$ there are two non-homotopic curves, and we obtain two numbers $d_{x_0}^+$ and $d_{x_0}^-$.

We shall give without proof the fundamental theorems on compound operators.

Theorem 8.1. *The operator* (1) *is a Fredholm operator if and only if*

1. *the symbol of this operator and its limits on Γ are non-zero everywhere, and*
2. $|d_x| < \pi$ *(or $|d_x^\pm| < \pi$ in case $m = 2$) for each $x \in \Gamma$.*

Theorem 8.2. *The compound singular operator* (1) *has a right* (*left*) *regularizer and a closed range if the following conditions are fulfilled*:

1. *Condition 1. of Theorem 8.1.*
2. *For no point $x \in \Gamma$ the number d_x is an odd multiple of π.*
3. $d_x > -\pi$ $(d_x < \pi)$ *if* $m > 2$; $d_x^\pm > -\pi$ $(d_x^\pm < \pi)$ *if* $m = 2$.

8.3. Now we come to the case $n > 1$. Let x_0 be a common boundary point of the regions D_i and D_j. If the matrices $\Phi_B^i(x_0, \theta)$ and $\Phi_B^j(x_0, \theta)$ are non-singular for every θ we may introduce the matrices $G_{x_0}^{ij}(\xi) := [\Phi_B^i(x_0, \theta)]^{-1} \Phi_B^j(x_0, \theta)$, $\theta = \xi/|\xi|$, and $H_{x_0}^{ij} := G_{x_0}^{ij}(e_i) \times [G_{x_0}^{ij}(e_j)]^{-1}$. Furthermore, we shall consider the matrix $G_{x_0,\xi}^{ij} := G_{x_0}^{ij}(\nu^i t + \xi)$, $-\infty < t < +\infty$, where ξ belongs to the intersection of the unit sphere with the plane Π_{x_0} that passes the origin and is perpendicular to the straight connection of the points e_i and e_j.

If the matrix $H_{x_0}^{ij}$ has no negative real eigenvalues, then the index

$$\operatorname{ind} G_{x_0,\xi}^{ij}(t) = \frac{1}{2\pi}\left[\{\arg \det G_{x_0,\xi}^{ij}(t)\}_{t=-\infty}^{t=+\infty} + \sum_{k=1}^{n} \arg \lambda_k\right]$$

can be defined, where λ_k denote the eigenvalues of $H_{x_0}^{ij}$, and the branch $\arg \lambda_k$ is to be chosen such that $|\arg \lambda_k| < \pi$. The index defined in that way is, obviously, an integer. The partial indices of the matrix $G_{x_0,\xi}^{ij}$ are ordered in a nondecreasing sequence, and in that order denoted by $\varkappa_1(x_0, \xi), \ldots, \varkappa_n(x_0, \xi)$. Then

$$\operatorname{ind} G_{x_0,\xi}^{ij}(t) = \sum_{k=1}^{n} \varkappa_k(x_0, \xi).$$

The quantity $\operatorname{ind} G_{x_0,\xi}^{ij}$ does not depend on ξ.

Theorem 8.3. *A compound generalized singular operator is Fredholm if and only if*

1. *the symbol matrix $\Phi_B(x, \theta)$ and its limits on Γ are non-singular for every $x \in \mathbb{R}^m$ and $\theta \in S$, and*
2. *the partial indices of the symbol matrix are zero for every $x \in \Gamma$ and $\xi \in \Pi_{x_0} \cap S$ (S is the unit sphere).*

§ 9. Singular integral equations in domains with boundary

Let G be a domain in $\mathbb{R}^m$ bounded by a finite number of pairwise disjoint simple, closed Lyapunov surfaces. Let us consider the (in general, vector) equation

$$(AP_G u)(x) = g(x), \qquad x \in G, \tag{1}$$

with a generalized singular operator A in $\mathbf{L}_2(\mathbb{R}^m)$. Equation (1) can be transformed into an equation in $\mathbb{R}^m$ with a compound generalized singular operator, namely

$$IP_{\mathbb{R}^m\setminus G}v + AP_G v = g_1 . \tag{2}$$

Here I stands for the identity, and

$$g_1(x) = \begin{cases} g(x), & x \in G, \\ 0, & x \notin G. \end{cases}$$

The solutions to eqs. (1) and (2) are related by the formulas

$$u = P_G v, \qquad v = P_G u + P_{\mathbb{R}^m\setminus G} g - AP_G u .$$

Thus, instead of eq. (1) the equivalent eq. (2) with a compound operator can be studied, and the arguments and results of § 8 can be applied.

Remark. Using Simonenko's localization principle, Duduchava [4, 5] and Peterhänsel[1] recently studied singular integral equations on compact manifolds with boundary. Duduchava (l.c.), in particular, investigated the Fredholm properties of such operators in vector Sobolev-Slobodetski spaces.

Finally, we mention the papers by Vasilevski [1, 2] who considered Banach algebras generated by certain two-dimensional singular integral operators on the unit disk with piecewise continuous coefficients; for operators belonging to these algebras a Fredholm theory is derived, and the index is computed.

§ 10. A survey on the papers by Vishik and Eskin

Vishik and Eskin [1–6] studied pseudo-differential operators on manifolds with boundary. A consideration of such pseudo-differential operators would exceed the frame of the present book, and therefore we restrict ourselves to a survey on the papers mentioned. On the other hand, such a survey seems to be necessary, for the singular integral operators provide a special case of pseudo-differential equations such that the papers cited are directly related to the subject of our book.

Their papers [1–3] are devoted to convolution equations on a bounded domain $G \subset \mathbb{R}^m$, i.e. equations of the form

$$P^+ A u_+ = g \tag{1}$$

where u_+ is a function with support in $\overline{G}$, A denotes the convolution operator which can be, in particular, a singular integral operator, and P^+ denotes the restriction of a function to the domain G.

Besides (1) they consider coupled equations of the form

$$P_1 A_1 u = g_1, \qquad P_2 A_2 u = g_2 \tag{2}$$

where the function u is defined on an m-dimensional manifold Γ without boundary. That manifolds is divided into two parts Γ_1 and Γ_2 by a smooth surface such that $\overline{\Gamma}_1 \cup \overline{\Gamma}_2 = \Gamma$. Then P_i denotes the restriction of a function to Γ_i, and A_i are convolution operators on Γ.

If the order α of the operator A is zero then equation (1) is a singular integral equation in G. In case $\alpha < 0$ equation (1) is an integral equation of the first kind. It was shown that the case $\alpha \leqq 0$ fits the framework of an arbitrary "elliptic" operator.

A fundamental role for the problem setting is played by the so-called factorization index of the symbol of the operator A on the boundary Γ of the domain G.

Let the symbol $\Phi_A(x, \xi)$ of the operator A be homogeneous of degree α with respect to ξ, and be given in local coordinates near Γ such that $\xi_m = 0$ is the equation for the corresponding boundary part on Γ. The symbol can be factorized with respect to ξ_m, i.e. it admits a product representation

$$\Phi_A(x, \xi) = \tilde{A}_-(x, \xi', \xi_m) \tilde{A}_+(x, \xi', \xi_m).$$

Here, $\tilde{A}_-$ $(\tilde{A}_+)$ is a homogeneous function with respect to ξ which is analytic in the half plane $\operatorname{Im} \xi_m < 0$ $(\operatorname{Im} \xi_m > 0)$ as a function of ξ_m and has no zeros in the closed half plane for $\xi \neq 0$. The homogeneity degree $\varkappa = \varkappa(x)$ of the function $\tilde{A}_+(x, \xi)$ is called *index* of equation (1).

For the sake of simplicity we suppose $\varkappa$ to be an integer and independent of x. For $\varkappa = 0$ equation (1) behaves like a usual singular integral equation on a manifold without boundary: If $\Phi_A(x, \xi) \neq 0$, $x \in \bar{G}$, $\xi \neq 0$, and, moreover, $\varkappa = 0$ for all points of Γ then (1) is normally solvable in $\mathbf{L}_2(G)$. In case $\varkappa > 0$ (also if $\alpha \leqq 0$) $\varkappa$ boundary conditions of the form

$$\gamma P^+ B_j u_+ = g_j(x'), \qquad x' \in \Gamma, \qquad 1 \leqq j \leqq \varkappa, \tag{3}$$

have to be added to (1) in order to ensure normal solvability of (1) in $\mathbf{L}_2(G)$. Here, B_j are convolution operators, and γ denotes the restriction operator to Γ.

If the symbol $\Phi_A(x, \xi) \neq 0$ for $x \in \bar{G}$ and $\xi \neq 0$ and the symbols Φ_A and Φ_{B_j} satisfy the so-called complementary condition at every point $x' \in \Gamma$ (this is a condition analogous to the well-known Shapiro-Lopatinski condition), then problem (1)–(2) is normally solvable in the corresponding function spaces, and $u_+ \in \mathbf{L}_2(G)$.

In case $\varkappa < 0$ the following problem turns out to be well posed: Find the solution to the equation

$$P^+\left(Au_+ + \sum_{k=1}^{|\varkappa|} C_k \varrho_k\right) = g \tag{4}$$

where C_k are convolution operators, and the new unknown functions ϱ_k depend on points of Γ. Under certain algebraic conditions problem (4) is solvable in the corresponding function spaces, and $u_+ \in \mathbf{L}_2(G)$.

In general, the factorization index can be an arbitrary function on Γ. In such cases the Fredholm properties of the operator can be shown to hold no longer in Sobolev-Slobodetski spaces but in certain other spaces of more complicated structure. It was proved the solvability of equations with an arbitrary number of "potentials" ϱ_k and an arbitrary number of additional boundary conditions provided that the solution is to be found in suitable function spaces that depend on the number of boundary conditions and potentials.

In the papers mentioned the smoothness of the solution in the closed region was studied as well as the behaviour of the solution near the boundary.

In [4] Vishik and Eskin carry over the results of their previous papers to the case of convolution equations systems. Here, the factorization of the symbol matrix plays an important role. In [5, 6] applications to discontinuous boundary value problems for elliptic equations are given.

For further studies in the theory of pseudo-differential operators the reader is referred to Boutet de Monvel [1], Calderón [6], Duistermaat [1], Eskin [1], Friedrichs [1], Hörmander [5], Shubin [1], Schulze [1], Treves [1], and to the collection of papers "Pseudo-differential operators" (in Russian, Mir Publishers, Moscow 1967).

Chapter XVI
Multidimensional singular equations with degenerate symbol

In contrast to the one-dimensional case, the knowledge one has about multidimensional singular integral equations with degenerate symbol is relatively incomplete. In the present chapter we confine our attention to two classes of multidimensional singular equations with degenerate symbol the study of which has achieved a rather advanced stage: convolution equations on the one hand and paired equations involving a higher-dimensional analogue of the Cauchy singular integral operator on the other hand.

§ 1. Convolution equations with degenerate symbol

Multidimensional singular integral equations of convolution type with a degenerate symbol were studied in the works of Mazya and Plamenevski [1, 2] and of Mazya, Plamenevski, and Haikin [1]. In this section we essentially follow the last article, which generalizes the results of the first two papers essentially. In that article it is supposed that the symbol vanishes on the union of a finite number of smooth disjoint submanifolds of the unit sphere S^{n-1} and that on each of these submanifolds the order of the zero is finite and constant. Distribution spaces are constructed so that the equation be solvable for every right-hand side from $\mathbf{L}_2(\mathbb{R}^n)$ and the general form of the solution in these spaces is described. To guarantee the uniqueness of the solution, integrals of the form

$$\int_{S^{n-1}} (B(l)\, u)\, (r, \theta)\, \mathrm{d}S_\theta\,,$$

where $r > 0$ and $B(l)$ is a singular integral operator depending on a parameter $l \in \Gamma$, are required to assume prescribed values $\psi(l, r)$.

1.1. Function spaces

Let $\mathbb{R}^n$ be the n-dimensional Euclidean space and S^{n-1} the $(n-1)$-dimensional unit sphere with center at the origin. Denote by $\mathbf{W}_2^l(S^{n-1})$ the completion of the set $\mathbf{C}^\infty(S^{n-1})$ with respect to the norm

$$|\varphi|_l = \left(\sum_{k=0}^{l} [\delta^k \varphi, \varphi]\right)^{1/2},$$

where $[u, v]$ is the scalar product in $\mathbf{L}_2(S^{n-1})$ and δ the Beltrami operator on S^{n-1} (see §§ 2 and 4, Chapter VIII). Let $\mathbf{W}_2^{-l}(S^{n-1})$ denote the space conjugate to $\mathbf{W}_2^l(S^{n-1})$ with respect to the scalar product $[u, v]$.

Next, denote by $\mathbf{H}(\mathbf{W}_2^l(S^{n-1}))$ (l an integer) the space of $\mathbf{W}_2^l(S^{n-1})$ valued functions of the variable $\varrho \in [0, \infty)$ whose norm is

$$\|u\|_l = \left(\int_0^\infty |u|_l^2 \varrho^{n-1} \, d\varrho\right)^{1/2}.$$

Note that the spaces $\mathbf{H}(\mathbf{W}_2^{-l}(S^{n-1}))$ and $\mathbf{H}(\mathbf{W}_2^l(S^{n-1}))$ are mutually conjugate with respect to the inner product

$$(f, \varphi) = \int_0^\infty [f, \varphi] \varrho^{n-1} \, d\varrho.$$

Let Γ be a smooth closed submanifold of the sphere S^{n-1} and $\mathbf{W}_2^\tau(\Gamma)$ (τ an arbitrary real number) the Sobolev-Slobodetski space over Γ. (Define $\mathbf{W}_2^\tau(\Gamma) = \mathbb{R}^1$ if $\dim \Gamma = 0$). Denote by $\mathbf{H}(\mathbf{W}_2^\tau(\Gamma))$ the space of $\mathbf{W}_2^\tau(\Gamma)$ valued generalized functions on $[0, \infty)$ the norm of which is

$$\|u\|_{\tau,\Gamma} = \left(\int_0^\infty |u(\varrho, \cdot)|_{\tau,\Gamma}^2 \varrho^{n-1} \, d\varrho\right)^{1/2},$$

where $|\cdot|_{\tau,\Gamma}$ is the norm in $\mathbf{W}_2^\tau(\Gamma)$.

In case $l \geqq 0$ the embedding $\mathbf{H}(\mathbf{W}_2^l(S^{n-1})) \subset \mathbf{L}_2(\mathbb{R}^n)$ holds, and so one can define the Fourier transform $\tilde{u} = Fu$ of the functions $u \in \mathbf{H}(\mathbf{W}_2^l(S^{n-1}))$. Since $\mathbf{H}(\mathbf{W}_2^{-l}(S^{n-1}))$ is the conjugate space of $\mathbf{H}(\mathbf{W}_2^l(S^{n-1}))$, the Fourier transform can be defined on the space $\mathbf{H}(\mathbf{W}_2^{-l}(S^{n-1}))$ in the usual fashion. Since the operators δ and F commute (see § 4, Chapter VIII), it is easy to see that the Fourier transform maps each of the spaces $\mathbf{H}(\mathbf{W}_2^l(S^{n-1}))$ and $\mathbf{H}(\mathbf{W}_2^{-l}(S^{n-1}))$ isometrically onto itself.[1]

1.2. Existence and general form of the solution of the singular integral equation

We consider the singular integral operator

$$(Au)(x) = F_{\xi\to x}^{-1} \Phi(\theta) F_{y\to\xi} u(y), \qquad \theta = \xi |\xi|^{-1} \tag{1}$$

with symbol $\Phi \in \mathbf{C}^\infty(S^{n-1})$. This operator is defined and continuous on all spaces $\mathbf{H}(\mathbf{W}_2^l(S^{n-1}))$[2].

Let Γ be the union of a finite number of smooth, closed, and connected submanifolds Γ_j ($1 \leqq j \leqq N$) of the unit sphere S^{n-1}. Suppose $\Gamma_i \cap \Gamma_j = \emptyset$ whenever $i \neq j$ and let $\dim \Gamma_j = d_j$. Further suppose that there exists a continuous field of normal $(n - 1 - d_j)$ legs on each manifold Γ_j. Finally, assume that the symbol Φ takes the value zero only on Γ and that near Γ_j, $1 \leqq j \leqq N$, the estimate

$$c_1 \varrho^{\varkappa_j} \leqq |\Phi(\theta)| \leqq c_2 \varrho^{\varkappa_j}, \qquad \varrho = \operatorname{dist}(\theta, \Gamma_j)$$

holds, where $\varkappa_j$ are certain positive integers.

Theorem 1.1. *For every right-hand side $f \in \mathbf{L}_2(\mathbb{R}^n)$ the equation*

$$Au = f \tag{2}$$

[1]) A detailed proof for the properties of the spaces $\mathbf{H}(\mathbf{W}_2^l(S^{n-1}))$ quoted in this section can be found in the paper of Lorenz [3].

[2]) See also Lorenz [3].

has a solution $u \in \mathbf{H}(\mathbf{W}_2^{-q}(S^{n-1}))$, *where the number* q *is subject to the following condition: let* N_0 *be the set of all subscripts* k, $1 \leq k \leq N$, *such that* $\varkappa_k = \max_{1 \leq j \leq N} \varkappa_j$; *then* $q \geq \max_j \varkappa_j$ *if* $\varkappa_k - \frac{1}{2}(n - 1 - d_k) \neq 0, 1, \ldots$ *for all* $k \in N_0$, *and* $q > \max_j \varkappa_j$ *in all other cases.*

Proof. The equation (2) is equivalent to the equation $\Phi \tilde{u} = \tilde{f}$. Therefore, in order to prove the existence of a solution, it suffices to construct a regularization of the functional $\Phi^{-1} \tilde{f}$.

Let $\{U_i\}$ be a finite covering of the sphere by regions with sufficiently small diameter which are diffeomorphic to the $(n-1)$-dimensional open unit ball D_{n-1}. Denote the corresponding diffeomorphism $U_i \to D_{n-1}$ by P_i and let $\{\chi_i\}$ be a partition of unity subordinate to $\{U_i\}$.

Consider the quadratic functional $[\delta^q \psi, \psi]$ on the set of all functions $\psi \in \mathbf{W}_2^q(S^{n-1})$ satisfying the condition

$$D_\nu^\alpha \psi|_{\Gamma_j} = D_\nu^\alpha \varphi|_{\Gamma_j}$$

on the manifolds Γ_j whose dimension d_j is greater than $n - 1 - 2q$ and denote by $h_q \varphi$ the element in $\mathbf{W}_2^q(S^{n-1})$ at which this functional attains its minimum. Here $D_\nu^\alpha = D_{\nu_1}^{\alpha_1} \ldots D_{\nu_{n-1-d_j}}^{\alpha_{n-1-d_j}}$, and α is an arbitrary multi-index of length $|\alpha| < q - \frac{1}{2}(n - 1 - d_j)$. The existence and uniqueness of the function $h_q \varphi$ was proved by SLOBODETSKI [1].

We claim that the expression

$$\sum_{i_1} \int_0^\infty [\tilde{f}, \chi_{i_1} \bar{\Phi}^{-1} \tilde{\varphi}]\, \varrho^{n-1}\, d\varrho + \sum_{i_2} \int_0^\infty [\tilde{f}, \chi_{i_2} (\tilde{\varphi} - h_q \tilde{\varphi})\, \bar{\Phi}^{-1}]\, \varrho^{n-1}\, d\varrho$$

$(\operatorname{supp} \chi_{i_1} \cap \Gamma_j = \emptyset$, $\operatorname{supp} \chi_{i_2} \cap \Gamma_j = \emptyset)$ is the desired regularization of the functional $\Phi^{-1} \tilde{f}$. Since $\tilde{f} \in \mathbf{L}_2(\mathbb{R}^n)$, we have

$$|[\tilde{f}, \chi_{i_1} \tilde{\varphi} \bar{\Phi}^{-1}]| \leq c\, |\tilde{\varphi}|_0$$

for almost all ϱ. Thus, it remains to show that the estimate

$$|[\tilde{f}, \chi_{i_2} (\tilde{\varphi} - h_q \tilde{\varphi})\, \Phi^{-1}]| \leq c\, |\tilde{\varphi}|_q \tag{3}$$

holds.

In the case where $U_i \cap \Gamma_j = \emptyset$, it can be assumed that the image of Γ_j under the mapping P_i coincides with the intersection of the ball D_{n-1} and the hyper-plane $\mathbb{R}^{d_j} = \{x \in \mathbb{R}^{n-1} : x_{d_j+1} = \ldots = x_{n-1} = 0\}$. The condition imposed upon the symbol Φ implies that near Γ_j the estimate $|\Phi(\theta)| > c\,[\operatorname{dist}(\theta, \Gamma_j)]^{\varkappa_j}$ holds. Hence (3) follows from the well known inequality

$$\|r^{-\varkappa_j} u\|_{\mathbf{L}_2(\mathbb{R}^{n-1})} \leq c\, \|u\|_{\mathbf{W}_2^p(\mathbb{R}^{n-1})}\,.$$

Here $r^2 = x_{d_j+1}^2 + \ldots + x_{n-1}^2$, u is a function which vanishes in some neighborhood of the subspace $\mathbb{R}^{d_j}$, $p \geq \varkappa_j$ if $\varkappa_j - \frac{1}{2}(n - 1 - d_j) \neq 0, 1, \ldots$ and $p > \varkappa_j$ otherwise. ∎

In the following theorem p_j is equal to $q - \frac{1}{2}(n - 1 - d_j)$ and $\langle g, \psi \rangle_{\tau, \Gamma_j}$ denotes the value of the functional $g \in \mathbf{W}_2^{-\tau}(\Gamma_j)$ at the element $\psi \in \mathbf{W}_2^\tau(\Gamma_j)$. We also assume that the manifolds Γ_j are numbered so that $p_j > 0$ for $1 \leq j \leq M$ and $p_j < 0$ for the remaining values of j.

Theorem 1.2. *Let the symbol of the singular integral operator* (1) *satisfy the conditions of Theorem 1.1. Then for a functional* $u \in \mathbf{H}(\mathbf{W}_2^{-q}(S^{n-1}))$ *to be a solution of the equation*

$$Au = 0 \tag{4}$$

it is necessary and sufficient that it be of the form

$$(\tilde{u}, \tilde{\varphi}) = \sum_{j=1}^{M} \sum_{|\alpha|<p_j} \int_0^\infty \langle f_j^\alpha, D_\nu^\alpha \tilde{\varphi} \rangle_{p_j - |\alpha|, \Gamma_j} \varrho^{n-1} \, d\varrho, \tag{5}$$

where f_j^α *are functionals belonging to the space* $\mathbf{H}(\mathbf{W}_2^{|\alpha|-p_j}(\Gamma_j))$ *and satisfying*

$$\sum_{\substack{|\beta|+\varkappa_j \leqq |\alpha| < p_j \\ \alpha \geqq \beta}} \frac{\alpha!}{(\alpha-\beta)!} f_j^\alpha D_\nu^{\alpha-\beta} \Phi|_{\Gamma_j} = 0, \qquad |\beta| < p_j - \varkappa_j, \tag{6}$$

for all subscripts j *such that* $p_j > \varkappa_j$.

Proof. *Necessity.* When proving Theorem 1.1 we observed that

$$\|(\tilde{\varphi} - h_q \tilde{\varphi}) \Phi^{-1}\|_0 \leqq c \|\tilde{\varphi} - h_q \tilde{\varphi}\|_q, \qquad \varphi \in \mathbf{H}(\mathbf{W}_2^q(S^{n-1})).$$

Therefore every solution $u \in \mathbf{H}(\mathbf{W}_2^{-q}(S^{n-1}))$ of the equation (4) satisfies the equation

$$[\tilde{u}, \tilde{\varphi}] = [\Phi \tilde{u}, (\tilde{\varphi} - h_q \tilde{\varphi}) \bar{\Phi}^{-1}] + [\tilde{u}, h_q \tilde{\varphi}].$$

Taking into account that $\Phi \tilde{u} = 0$ we get

$$[\tilde{u}, \tilde{\varphi}] = [\tilde{u}, h_q \tilde{\varphi}].$$

The last equality implies that the Fourier transform of any solution of the equation (4) is supported on the manifolds Γ_j, $1 \leqq j \leqq M$. Further, due to a well-known theorem by SLOBODETSKI [1], we have

$$|[\tilde{u}, \tilde{\varphi}]| \leqq c \, |h_q \tilde{\varphi}|_q \leqq c \sum_{j=1}^{M} \sum_{|\alpha|<p_j} |D_\nu^\alpha \tilde{\varphi}|_{p_j - |\alpha|, \Gamma_j}.$$

This shows that the functional $\tilde{u}$ can be represented in the form (5). Since u is a solution of equation (4), we deduce that

$$0 = [\Phi \tilde{u}, \tilde{\varphi}] = [\tilde{u}, \bar{\Phi} \tilde{\varphi}] = \sum_{j=1}^{M} \sum_{|\alpha|<p_j} \langle f_j^\alpha, D_\nu^\alpha (\bar{\Phi} \tilde{\varphi}) \rangle_{p_j - |\alpha|, \Gamma_j}$$

for almost all ϱ. So we obtain that

$$\sum_{|\alpha|<p_j} \sum_{\alpha \geqq \beta} \frac{\alpha!}{\beta! \, (\alpha-\beta)!} \langle f_j^\alpha D_\nu^{\alpha-\beta} \Phi, D_\nu^\beta \tilde{\varphi} \rangle_{p_j - |\alpha|, \Gamma_j} = 0$$

for $j = 1, 2, \ldots, M$. If $|\alpha - \beta| < \varkappa_j$ then $D_\nu^{\alpha-\beta} \Phi = 0$ and hence condition (6) is also fulfilled.

Sufficiency. Let f_j^α be arbitrary functionals belonging to $\mathbf{H}(\mathbf{W}_2^{|\alpha|-p_j}(\Gamma_j))$ and satisfying the condition (6). Consider the functional (5). For almost all ϱ, the estimate

$$|(\tilde{u}, \tilde{\varphi})| \leqq \sum_{j=1}^{M} \sum_{|\alpha|<p_j} \int_0^\infty |f_j^\alpha|_{|\alpha|-p_j, \Gamma_j} |\tilde{\varphi}|_q \varrho^{n-1} \, d\varrho \leqq \|\varphi\|_q \sum_{j=1}^{M} \sum_{|\alpha|<p_j} \|f_j^\alpha\|_{|\alpha|-p_j, \Gamma_j}$$

holds. Consequently, $u \in \mathbf{H}(\mathbf{W}_2^{-q}(S^{n-1}))$. On replacing in (5) $\tilde{\varphi}$ by $\Phi \tilde{\varphi}$ and carrying out the same transformations as in the proof of the necessity part we arrive at the equality $(\Phi \tilde{u}, \tilde{\varphi}) = 0$. ∎

Remark 1. If all manifolds Γ_j, $1 \leqq j \leqq M$, have the dimension $n - 2$, then the general solution of the equation $Au = 0$ can be described in a simpler way. In that case the conditions (6) take the form

$$\sum_{\alpha=\varkappa_j+\beta}^{[q-1/2]} \frac{\alpha!}{(\alpha-\beta)!} f_j^\alpha \left. \frac{\partial^{\alpha-\beta} \Phi}{\partial \nu^{\alpha-\beta}} \right|_{\Gamma_j} = 0, \qquad \beta = 0, 1, \ldots, [q - 1/2] - \varkappa_j, \qquad 1 \leqq j \leqq M,$$

where ν denotes the normal to Γ_j. For each j, these equations are an algebraic system for the unknowns f_j^α with a triangular matrix the entries on whose diagonal are $D_\nu^{\varkappa_j}\Phi\mid_{\Gamma_j} \neq 0$. Hence $f_j^\alpha = 0$ if $\varkappa_j \leq \alpha \leq [q - 1/2]$ and thus each solution $u \in \mathbf{H}(\mathbf{W}_2^{-q}(S^{n-1}))$ can be written in the form

$$(\tilde{u}, \tilde{\varphi}) = \sum_{j=1}^{M} \sum_{\alpha=0}^{\varkappa_j-1} \int_0^\infty \langle f_j^\alpha, D_\nu^\alpha \tilde{\varphi}\rangle_{q-1/2-|\alpha|,\Gamma_j} \varrho^{n-1}\, d\varrho.$$

Remark 2. In the special case $n = 2$ the general solution of the equation $Au = 0$ is of the form

$$u = \sum_{j=1}^{N} \sum_{k=0}^{\varkappa_j-1} F^{-1}[c_{kj}(\varrho)\, \delta^{(k)}(\theta - \theta_j)].$$

Here ϱ, θ are polar coordinates in $\mathbb{R}^2$, $\theta_j = \Gamma_j$ are the zeros of multiplicity $\varkappa_j$ of the function $\Phi(\theta)$, δ is the Dirac delta function, and $c_{kj}(\varrho) \in \mathbf{L}_2((0, \infty); \varrho^{1/2})$ are arbitrary functions (see Mazya and Plamenevski [1]).

1.3. A correctly posed problem

Denote by $\mathbf{L}_2(\Gamma, \mathbf{W}_2^k(S^{n-1}))$ the space of functions on $\Gamma \times S^{n-1}$ whose norm is

$$\|\sigma\|_{\mathbf{L}_2((\Gamma,\mathbf{W}_2^k(S^{n-1}))} = \left(\int_0^\infty |\sigma(l, \cdot)|_k^2\, dl\right)^{1/2}$$

and put $\mathbf{L}_2(\Gamma, \mathbf{C}^\infty(S^{n-1})) = \bigcap_{k=0}^{\infty} \mathbf{L}_2(\Gamma, \mathbf{W}_2^k(S^{n-1}))$.

We now pose the following problem. Find a solution $u \in \mathbf{H}(\mathbf{W}_2^{-q}(S^{n-1}))$ (q as in Theorem 1.1) of the equation

$$Au = f, \qquad f \in \mathbf{L}_2(\mathbb{R}^n), \tag{7}$$

which satisfies the conditions

$$[B_{mk}u, 1] = \psi_{mk}(l, r), \qquad l \in \Gamma_k; \qquad m = 1, \ldots, s_k - t_k; \qquad 1 \leq k \leq M \tag{8}$$

for almost all $r \in [0, \infty)$. Here B_{mk} are singular integral operators with symbols $\sigma_{mk} \in \mathbf{L}_2(\Gamma_k, \mathbf{C}^\infty(S^{n-1}))$, $\psi_{mk} \in \mathbf{H}(\mathbf{L}_2(\Gamma_k))$ are given functions, s_k is the number of multi-indices whose dimension is $n - 1 - d_k$ and whose length is $|\alpha| < p_k$, and t_k is the number of multi-indices of the same dimension the length of which is $|\alpha| < p_k - \varkappa_k$.

Theorem 1.3. *In order that the problem* (7), (8) *be uniquely solvable for each* $\psi_{mk} \in \mathbf{H}(\mathbf{L}_2(\Gamma_k))$ *and that the solution satisfy the estimate*

$$\|u\|_{-q} \leq c\left(\|f\|_0 + \sum_{j=1}^{M} \sum_{m=1}^{s_k-t_k} \|\psi_{mk}\|_{0,\Gamma_k}\right) \tag{9}$$

it is necessary and sufficient that the system of equations

$$\sum_{j=1}^{M} \sum_{|\alpha|<p_j} \langle f_j^\alpha, D_\nu^\alpha \sigma_{mk}(l, \cdot)\rangle_{p_j-|\alpha|,\Gamma_j} = g_{mk}, \qquad l \in \Gamma_k, \qquad 1 \leq m \leq s_k - t_k,$$
$$1 \leq k \leq M, \tag{10}$$

$$\sum_{\substack{|\beta|+\varkappa_j \leq |\alpha| < p_j \\ \alpha \geq \beta}} f_j^\alpha D_\nu^{\alpha-\beta}\Phi|_{\Gamma_j} = 0, \qquad 0 \leq |\beta| \leq p_j - \varkappa_j \tag{11}$$

be uniquely solvable in the space $\prod_{j=1}^{M} \prod_{|\alpha|<p_j} \mathbf{W}_2^{|\alpha|-p_j}(\Gamma_j)$ *for each* $g_{mk} \in \mathbf{L}_2(\Gamma_k)$ *and that for the solution* $\{f_j^\alpha\}$ *the estimate*

$$\sum_{j=1}^{M} \sum_{|\alpha|<p_j} |f_j^\alpha|_{|\alpha|-p_j,\Gamma_j}^2 \leq c \sum_{k=1}^{M} \sum_{m=1}^{s_k-t_k} |g_{mk}|_{\mathbf{L}_2(\Gamma_k)}^2 \tag{12}$$

hold.

Proof. *Sufficiency.* It is well known that $[g, 1]^{\sim} = [\tilde{g}, 1]$ (see, for instance, Stein and Weiss [1], § 1, Chapter IV). Thus, taking the Fourier transform on both sides of the equation (8) we obtain

$$[\sigma_{mk}(l, \cdot)\, \tilde{u}(\varrho, \cdot), 1] = \tilde{\psi}_{mk}(l, \varrho), \qquad l \in \Gamma_k, \tilde{\psi}_{mk} \in \mathbf{H}(\mathbf{L}_2(\Gamma_k)).$$

Multiply this equality by $\varrho^{n-1}\chi(\varrho)$, where $\chi(\varrho)$ is any function from $\mathbf{C}_0^\infty([0, \infty))$, and then integrate it with respect to ϱ. What results is

$$\int_0^\infty [\sigma_{mk}(l, \cdot)\, \tilde{u}(\varrho, \cdot), 1]\, \chi(\varrho)\, \varrho^{n-1}\, \mathrm{d}\varrho = \int_0^\infty \tilde{\psi}_{mk}(l, \varrho)\, \chi(\varrho)\, \varrho^{n-1}\, \mathrm{d}\varrho \tag{13}$$

for almost all $l \in \Gamma_k$. From the Theorems 1.1 and 1.2 we know that the Fourier transform $\tilde{u} \in \mathbf{H}(\mathbf{W}_2^{-q}(S^{n-1}))$ of the solution of equation (8) can be written in the form

$$(\tilde{u}, \tilde{\varphi}) = (\tilde{v}, \tilde{\varphi}) + \sum_{j=1}^{M} \sum_{|\alpha|<p_j} \int_0^\infty \langle f_j^\alpha, D_\nu^\alpha \tilde{\varphi} \rangle_{p_j - |\alpha|, \Gamma_j}\, \varrho^{n-1}\, \mathrm{d}\varrho, \tag{14}$$

where $\tilde{v} \in \mathbf{H}(\mathbf{W}_2^{-q}(S^{n-1}))$ is the solution of equation (2) constructed in the proof of Theorem 1.1 and f_j^α are functionals belonging to $\mathbf{H}(\mathbf{W}_2^{|\alpha|-p_j}(\Gamma_j))$ and satisfying the conditions (11).

Now let in (14) $\tilde{\varphi} = \sigma_{mk}(l, \cdot)\, \chi$, $l \in \Gamma_k$, $m = 1, \ldots, s_k - t_k$, $k = 1, \ldots, M$. Taking into consideration (13) we then obtain

$$\sum_{j=1}^{M} \sum_{|\alpha|<p_j} \int_0^\infty \langle f_j^\alpha(\varrho, \cdot),\ D_\nu^\alpha \sigma_{mk}(l, \cdot) \rangle_{p_j - |\alpha|, \Gamma_j}\, \chi(\varrho)\, \varrho^{n-1}\, \mathrm{d}\varrho$$
$$= \int_0^\infty \tilde{\psi}_{mk}(l, \varrho)\, \chi(\varrho)\, \varrho^{n-1}\, \mathrm{d}\varrho - (\tilde{v}, \chi\sigma_{mk}(l, \cdot)).$$

Since χ is a function which can be chosen arbitrarily, we conclude that, for almost all $\varrho \geqq 0$ and $l \in \Gamma_k$,

$$\sum_{j=1}^{M} \sum_{|\alpha|<p_j} \langle f_j^\alpha(\varrho, \cdot), D_\nu^\alpha \sigma_{mk}(l, \cdot) \rangle_{p_j - |\alpha|, \Gamma_j} = \zeta_{mk}(l, \varrho)$$

with $\zeta_{mk}(l, \varrho) = \tilde{\psi}(l, \varrho) - [\tilde{v}(\varrho, \cdot), \sigma_{mk}(l, \cdot)]$.

The estimates obtained in Theorem 1.1 for the solution $\tilde{v}$ show that

$$|\zeta_{mk}(l, \varrho)| \leqq |\tilde{\psi}_{mk}(l, \varrho)| + |\sigma_{mk}(l, \cdot)|_q\, |\tilde{v}(\varrho, \cdot)|_{-q}.$$

Consequently,

$$\|\zeta_{mk}\|_{0, \Gamma_k} \leqq c(\|\psi_{mk}\|_{0, \Gamma_k} + \|f\|_0\, \|\sigma_{mk}\|_{\mathbf{L}_2(\Gamma_k, \mathbf{W}_2^q(S^{n-1}))}). \tag{15}$$

Since $\zeta_{mk}(\cdot, \varrho)$ is in $\mathbf{L}_2(\Gamma_k)$ for almost all ϱ, by taking into account the conditions of the theorem we obtain a unique solution $\{f_j^\alpha(\varrho, \cdot)\}$ which belongs to the space $\prod_{j=1}^{M} \prod_{|\alpha|<p_j} \mathbf{W}_2^{|\alpha|-p_j}(\Gamma_j)$ and satisfies the estimate

$$\sum_{j=1}^{M} \sum_{|\alpha|<p_j} |f_j^\alpha(\varrho, \cdot)|^2_{|\alpha|-p_j, \Gamma_j} \leqq c \sum_{k=1}^{M} \sum_{m=1}^{s_k - t_k} |\zeta_{mk}(\cdot, \varrho)|^2_{0, \Gamma_k}.$$

This together with the inequality (15) gives the estimate (9).

Necessity. Let the functions $f_j^\alpha \in \mathbf{W}_2^{|\alpha|-p_j}(\Gamma_j)$ satisfy the homogeneous system (10), (11). Then, for an arbitrary function $\chi \in \mathbf{C}_0^\infty([0, \infty))$, the functional

$$(\tilde{u}, \tilde{\varphi}) = \sum_{j=1}^{M} \sum_{|\alpha|<p_j} \int_0^\infty \langle \chi f_j^\alpha, D_\nu^\alpha \tilde{\varphi} \rangle_{q_j - |\alpha|, \Gamma_j}\, \varrho^{n-1}\, \mathrm{d}\varrho$$

belongs to the space $\mathbf{H}(\mathbf{W}_2^{-q}(S^{n-1}))$ and is a solution of the homogeneous system (7), (8). Hence $f_j^\alpha \equiv 0$.

We now prove that the system (10), (11) is solvable and that the estimate (12) holds. Let $g_{mk} \in \mathbf{L}_2(\Gamma_k)$ and $\chi \in \mathbf{C}_0^\infty([0, \infty))$. Determine the solution of the equation $Au = 0$ which satisfies the additional conditions (8) with $\tilde{\varphi}_{mk}(l, \varrho) = \chi(\varrho)\, g_{mk}(l)$, $l \in \Gamma_k$. Denote by f_j^α the corresponding functionals from the representation (14). Then, for almost all ϱ, the equality

$$\sum_{j=1}^{M} \sum_{|\alpha|<p_j} \langle f_j^\alpha(\varrho, \cdot), D_\nu^\alpha \sigma_{mk}(l, \cdot)\rangle_{p_j-|\alpha|, \Gamma_j} = \chi(\varrho) g_{mk}(l), \qquad l \in \Gamma_k,$$

holds and the conditions (11) are fulfilled. Since the solution of the system (10), (11) is determined uniquely (this was proved above), we deduce that the functionals f_j^α can be represented in the form $\chi(\varrho)\, h_j^\alpha(\theta)$ with $h_j^\alpha \in \mathbf{W}_2^{|\alpha|-p_j}(\Gamma_j)$. The inequality (12) now follows from the estimate (9). ∎

We finally consider the special case where the set of all zeros of the symbol consists of finitely many points $\theta = \theta_j$, $1 \leqq j \leqq N$. Denote by $\varkappa_j$ the order of the zero θ_j and put $\varkappa = \max\limits_j \varkappa_j$. The number q, which fixes the class in which the solution is sought, is now subject to the following condition: $q > \varkappa$ if $2\varkappa \geqq n - 1$ and n is an odd number; $q \leqq \varkappa$ otherwise.

If $2q \leqq n - 1$, then the equation (7) is uniquely solvable in the space $\mathbf{H}(\mathbf{W}_2^{-q}(S^{n-1}))$. In the case where $2q > n - 1$ the problem (7), (8) consists in finding that solution $u \in \mathbf{H}(\mathbf{W}_2^{-q}(S^{n-1}))$ of the equation (7) which satisfies

$$[B_s u, 1] = \psi_s(r), \qquad s = 1, 2, \ldots, \qquad \mu = \sum_{j=1}^{N} \varkappa_j$$

for almost all $r \geqq 0$. Here B_s are singular integral operators with symbol $\sigma_s \in \mathbf{C}^\infty(S^{n-1})$ and $\psi_s \in \mathbf{L}_2([0, \infty); \varrho^{(n-1)/2})$ are given functions.

In view of Theorem 1.3 the problem just formulated is correctly posed if and only if the algebraic system

$$\sum_{j=1}^{M} \sum_{|\alpha|<p} f_j^\alpha D_\theta^\alpha \sigma_s(\theta_j) = g_s,$$

$$\sum_{|\beta|+\varkappa_j \leqq |\alpha| < p} \frac{\alpha!}{(\alpha - \beta)!} f_j^\alpha D_\theta^\alpha \Phi(\theta_j) = 0, \qquad 0 \leqq |\beta| < p - \varkappa_j, \tag{16}$$

where $p = q - \frac{1}{2}(n - 1)$, is uniquely solvable in complex numbers f_j^α.

In particular, if $n = 2$ then the equations (16) are redundant (see Remark 1) and so the criterion for the correctness of the problem assumes the form $\det B \neq 0$, where B is the matrix $(B_1, B_2, \ldots, B_N)$ with B_i given by

$$B_j = \begin{pmatrix} \sigma_1(\theta_i) & \dfrac{d\sigma_1}{d\theta}(\theta_i) & \cdots & \dfrac{d^{\varkappa_i-1}\sigma_1}{d\theta^{\varkappa_i-1}}(\theta_i) \\ \cdots & \cdots & \cdots & \cdots \\ \cdots & \cdots & \cdots & \cdots \\ \sigma_\mu(\theta_i) & \dfrac{d\sigma_\mu}{d\theta}(\theta_i) & \cdots & \dfrac{d^{\varkappa_i-1}\sigma_\mu}{d\theta^{\varkappa_i-1}}(\theta_i) \end{pmatrix}.$$

Remark 3. Note that Mazya and Plamenevski [1] still considered two further classes of additional conditions ensuring the uniqueness of the solution of equation (7).

§ 2. Further results on convolution operators with degenerate symbol

In the present section we formulate (without proofs) some results obtained by Lorenz [1–4] which are closely related to the subject of § 1.

2.1. In [3], Lorenz developed a systematic theory of singular integral operators with infinitely smooth symbols $a(x, \theta)$ on the spaces $\mathbf{H}(\mathbf{W}_2^l(S^{n-1}))$, where l is an integer (recall Section 1.1). He proved the boundedness of the singular integral operator and the compactness of the commutator of two such operators on these spaces and showed that the usual symbol rules hold. Furthermore, he proved that the conditon $|a(x, \theta)| \geqq \text{const} > 0$ is necessary and sufficient for the singular integral operator with symbol $a(x, \theta)$ to be Fredholm on $\mathbf{H}(\mathbf{W}_2^l(S^{n-1}))$. These results have been extended to the case of matrix operators as well as to pseudodifferential operators and pseudomultiplication operators (see Prössdorf [18]).

2.2. In [1, 2], Lorenz studied matrix operators of convolution type of the form (1.1), where $\Phi(\theta) \in [\mathbf{C}^\infty(S^{n-1})]^{n\times n}$ is a square matrix function. He proved that the operator A is invertible on the space $\mathbf{M}^\infty = \bigcap_{l=0}^{\infty} \mathbf{H}(\mathbf{W}_2^l(S^{n-1}))$ if and only if $\det \Phi(\theta) \neq 0$. He also showed that A is correctly solvable[1] on $\mathbf{M}^\infty$ if and only if the operator of multiplication by the function $\det \Phi(\theta)$ has this property when considered on the space $\mathbf{C}^\infty(S^{n-1})$. If $\det \Phi(\theta)$ has at least one zero, then $\dim \operatorname{Coker} A = \infty$.

If the estimate

$$\|\psi\|_l \leqq C \,\|(\det a)\|\psi\|_k \qquad \forall \psi \in \mathbf{C}^\infty(S^{n-1}) \tag{1}$$

holds, then the operator $A : \mathbf{H}(\mathbf{W}_2^l(S^{n-1})) \to \mathbf{H}(\mathbf{W}_2^k(S^{n-1}))$ is continuously left-invertible and the equation $Au = f$ has a solution $u \in \mathbf{H}(\mathbf{W}_2^{-k}(S^{n-1}))$ for every $f \in \mathbf{H}(\mathbf{W}_2^{-l}(S^{n-1}))$.

For the equation $Au = f$ to possess a solution $u \in \mathbf{M}^{-\infty} = \bigcup_{l=0}^{\infty} \mathbf{H}(\mathbf{W}_2^{-l}(S^{n-1}))$ for every $f \in \mathbf{M}^{-\infty}$ it is necessary and sufficient that to every $l \geqq 0$ there correspond a number $k \geqq 0$ such that (1) be true.

Hörmander [4] proved estimates of the form (1) for all functions $b(\theta) = \det a(\theta)$ which are locally polynomials in some neighborhood of their zeros.

2.3. Under the hypothesis (1), Lorenz [2] described the image space $\operatorname{Im} A$ of the operator $A : \mathbf{H}(\mathbf{W}_2^l(S^{n-1})) \to \mathbf{H}(\mathbf{W}_2^k(S^{n-1}))$ and constructed a left-inverse; correspondingly, the kernel of the operator $A : \mathbf{H}(\mathbf{W}_2^{-k}(S^{n-1})) \to \mathbf{H}(\mathbf{W}_2^{-l}(S^{n-1}))$ was decribed and a particular solution of the equation $Au = f$ was given by him.

It is also a result of Lorenz [2] that the inequality (1) holds if $n = 2$ and the function $\det \Phi(\theta)$ has only finitely many zeros of integral orders. Furthermore, in that case he succeeded in constructing the general solution $u \in \mathbf{H}(\mathbf{W}_2^{-(r+s)}(S^{n-1}))$ of the equation $Au = f$ with $f \in \mathbf{H}(\mathbf{W}_2^{-r}(S^{n-1}))$ $(r \geqq 0)$. Every solution of the homogeneous equation $Au = 0$ is a functional of finite order concentrated at the zeros of the function $\det \Phi(\theta)$.

2.4. If $\det \Phi(\theta)$ has at least one zero of infinite order (this occurs, for instance, if $\det \Phi(\theta)$ has infinitely many zeros), then there is no pair of spaces $\mathbf{H}(\mathbf{W}_2^l(S^{n-1}))$, $\mathbf{H}(\mathbf{W}_2^k(S^{n-1}))$ $(l \geqq 0)$ such that the singular integral operator A is left-sided invertible or normally solvable on it. In that case it is also not true that for arbitrary $f \in \mathbf{L}_2(\mathbb{R}^n)$ the equation (7) has a solution in anyone of the spaces $\mathbf{H}(\mathbf{W}_2^{-k}(S^{n-1}))$ (see Lorenz [2, 3]).

---

[1]) An operator is said to be *correctly solvable* if it is normally solvable and has a trivial kernel.

2.5. The results of Lorenz were generalized by SCHÄFER [1], who studied symbols of the form $a(x, \theta) = c(x)\, a_0(x, \theta)\, b(\theta)$ satisfying the following conditions:

1. $c \in \mathbf{C}(\overline{\mathbb{R}}^n) \cap \mathbf{C}^{m-1}(\overline{\Omega})$, where $\Omega \subset \mathbb{R}^n$ is an open and bounded set;
2. $N_0 := \{x \in \mathbb{R}^n : c(x) = 0\} \subset \Omega$ and the Lebesgue measure of N_0 is zero;
3. $|c(x)| \geqq c_0\, d(x, N_0)^m$ with $c_0 > 0$ for all $x \in \Omega$;
4. $(D^\alpha c)(x) = 0$ for all $x \in N_0$ and $|\alpha| \leqq m - 1$;
5. $b \in \mathbf{C}(S^{n-1})$ satisfies the conditions 1–4 locally;
6. $A_0 \in \Phi(\mathbf{H}(\mathbf{W}_2^l(S^{n-1})))$, where A_0 is the singular integral operator (or, more generally, a pseudodifferential operator) with the symbol a_0.

SCHÄFER [1] proved the estimate

$$\|f\|_{\mathbf{L}_2(\Omega)} \leqq C\, \|cf\|_{\mathbf{W}_2^k(\Omega)} \quad \forall f \in \mathbf{C}^\infty(\overline{\Omega})$$

with $k = m + [n/2] + 1$ and then he constructed a Hilbert space $H_1 \subset \mathbf{H}(\mathbf{W}_2^{-l}(S^{n-1}))$ such that the operator $A : H_1 \to H_2 := \mathbf{W}_2^k(\mathbb{R}^n)$ with the symbol a admits a left regularization and hence $A \in \Phi_+(H_1, H_2)$.

§ 3. A multidimensional analogue of the Cauchy singular integral operator and the corresponding paired operators

In this section we present some results of SPRÖSSIG [1–7]. Again the proofs will be omitted. A part of these results was communicated to the authors by Sprössig privately.

3.1. Generalized Cauchy-Riemann systems

Let $G \subset \mathbb{R}^n$ be a bounded region, suppose $\Gamma = \partial G$ is a sufficiently smooth Lyapunov surface, and put $\overline{G} = G \cup \Gamma$. Furthermore, assume $0 \in G$.

Let $(D_i)_{i=0}^{n-1}$ be a family of orthogonal matrices of the order k ($k \geqq n$) whose entries are 0, 1, or -1 only, and which are subject to the conditions

$$D_i' D_j + D_j' D_i = 0, \qquad i \neq j \quad (i, j = 0, 1, \ldots, n-1).$$

Here D_i' denotes the matrix resulting from D_i through transponation. Consider the expression

$$D(a) = \sum_{i=0}^{n-1} a_i\, D_i,$$

where $a = (a_0, \ldots, a_{n-1})$ is a formal vector. Letting, for example, this formal vector be $\nabla = \left(\frac{\partial}{\partial t_0}, \ldots, \frac{\partial}{\partial t_{n-1}}\right)$ we obtain an elliptic differential operator which can be regarded as a higher-dimensional analogue of the Cauchy-Riemann operator (cf. SAAK [1]).

In the case $n = 4$ the operator $D(\nabla)$ can be written in the form $D(\nabla) = \nabla\circ$, where "$\circ$" denotes the quaternionic multiplication, which is defined by

$$\varphi \circ \psi = (\varphi_0\psi_0 - (\overline{\varphi}, \overline{\psi}), \varphi_0\overline{\psi} + \psi_0\overline{\varphi} + \overline{\varphi} \times \overline{\psi})$$

for $\varphi = (\varphi_0, \overline{\varphi})$, $\psi = (\psi_0, \overline{\psi})$, $\overline{\varphi} = (\varphi_1, \varphi_2, \varphi_3)$, $\overline{\psi} = (\psi_1, \psi_2, \psi_3)$. The following examples illustrate what is covered by this formalism:

1. If $\dim G = 3$ and $\varphi_0 = 0$, then

$$\nabla \circ \varphi = (-\operatorname{div} \overline{\varphi}, \operatorname{rot} \overline{\varphi}).$$

This operator was considered by BITSADZE [2].

2. In case $\dim G = 3$ we have

$$\nabla \circ \varphi = (-\operatorname{div} \overline{\varphi}, \operatorname{rot} \overline{\varphi} + \operatorname{grad} \varphi_0)$$

and this operator is called the Moisil-Theodorescu operator (see MOISIL and THEODORESCU [1]).

3. For $\dim G = 4$ and $t = (t_0, t_1, t_2, t_3) \in G$ we get

$$\nabla \circ \varphi = \left(\frac{\partial \varphi_0}{\partial t_0} - \operatorname{div} \overline{\varphi}, \frac{\partial \overline{\varphi}}{\partial t_0} + \operatorname{rot} \overline{\varphi} + \operatorname{grad} \varphi_0\right).$$

This operator was studied by FUETER [1].

Further fundamental results in this direction were obtained by ARONSZAJN [1], DELANGHE [1], MUHAMMED-NASER [1], STEIN and WEISS [2], SUDBERY [1], BRACKX, DELANGHE, and SOMMEN [1], LOUNESTO [1], and others.

A continuously differentiable vector function $\varphi = (\varphi_0, \ldots, \varphi_{n-1})$ satisfying the equation $D(\nabla)\,\varphi = 0$ in G is said to be *analytic* in G (see SPRÖSSIG [3]).

3.2. A generalized Borel-Pompeiu formula

Let T denote the operator defined as follows:

$$(T\varphi)\,(t) = \frac{1}{|S_1|} \int\limits_G \frac{D'(\theta)}{|t-\tau|^{n-1}}\, \varphi(\tau)\, \mathrm{d}G_\tau.$$

Here $\theta = -(\theta_0, \ldots, \theta_{n-1})$, $\theta_i = \dfrac{t_i - \tau_i}{|t-\tau|}$, $D'(\theta) = \sum\limits_{i=0}^{n-1} \theta_i D_i'$, and $|S_1|$ denotes the area of the unit sphere. Furthermore, define

$$(\mathring{S}\varphi)\,(t) = \frac{1}{|S_1|} \int\limits_\Gamma \frac{D'(\theta)\, D(\alpha)}{|t-\tau|^{n-1}}\, \varphi(\tau)\, \mathrm{d}\Gamma_\tau, \qquad t \notin \Gamma,$$

where $\alpha = (\alpha_0, \ldots, \alpha_{n-1})$ denotes the outer normal of length 1 at $\tau \in \Gamma$.

The operators T and $\mathring{S}$ can be viewed as higher-dimensional analogues of the complex *T-operator* (see I. N. VEKUA [3]) and the *Cauchy integral operator*, respectively. They are tied up by the formula

$$T\, D(\nabla)\, \varphi + \mathring{S}\varphi = \begin{cases} \varphi, & t \in G, \\ 0, & t \notin \overline{G}. \end{cases}$$

The last equality is a generalization of the well known Borel-Pompeiu formula, which plays an important role in complex analysis (see SPRÖSSIG [3, 4]).

If the vector function φ is analytic, we immediately obtain the following generalization of the Cauchy integral formula:

$$\mathring{S}\varphi = \begin{cases} \varphi \text{ in } G, \\ 0 \text{ in } \mathbb{R}^n \setminus \overline{G}. \end{cases}$$

If the vector function φ satisfies a Hölder condition on Γ,

$$|\varphi(t) - \varphi(\tau)| \leqq \text{const}\, |t-\tau|^\lambda, \qquad 0 < \lambda \leqq 1,$$

then the integral

$$(S\varphi)\,(t) = \frac{2}{|S_1|} \int\limits_\Gamma \frac{D'(\theta)\, D(\alpha)}{|t-\tau|^{n-1}}\, \varphi(\tau)\, \mathrm{d}\Gamma_\tau$$

is defined for $t \in \Gamma$ in the sense of the Cauchy principal value. In that case for all points $t^0 \in \Gamma$ the *generalized Plemelj-Sokhotski formulae*

$$\lim_{\substack{t\to t^0\\ t\in G}} (\overset{\circ}{S}\varphi)(t) = \left(\frac{I+S}{2}\varphi\right)(t^0), \qquad \lim_{\substack{t\to t^0\\ t\in \mathbf{R}^n\setminus\bar{G}}} (\overset{\circ}{S}\varphi)(t) = \left(\frac{I-S}{2}\varphi\right)(t^0)$$

hold. In analogy to the one-dimensional situation, they imply that

$$(S^2\varphi)(t) = \varphi(t), \qquad t \in \Gamma.$$

Put $\frac{1}{2}(I+S) = P$ and $\frac{1}{2}(I-S) = Q$. It follows at once that $P^2 = P$, $Q^2 = Q$, $P + Q = I$, and $PQ = QP = 0$. The operator S is a *multidimensional analogue of the Cauchy singular integral operator* (see Chapter II, § 1).

3.3. Further properties of the singular integral operator S

1. We denote by $\mathbf{C}^{m,\lambda}(G)$ and $\mathbf{W}_p^{(m)}(G)$ ($m \geqq 0$ an integer, $0 < \lambda < 1$, $1 < p < \infty$), respectively, the Hölder and Sobolev spaces defined above (see Chapter VII, § 3.2) and by $\mathbf{C}_k^{m,\lambda}(G)$ and $\mathbf{W}_{p,k}^{(m)}(G)$ the corresponding spaces of vector functions with k components.

Let $\mathbf{C}_k^{\infty}(G)$ be the space of all vector functions of length k whose components are infinitely differentiable in G. Providing $\mathbf{C}_k^{\infty}(G)$ with the countable system of norms

$$\|\varphi\|_m = \|\varphi\|_{\mathbf{C}_k^m(G)} \qquad (m = 0, 1, 2, \ldots)$$

we make it become a Fréchet space.

Finally, let $\mathbf{C}_k^{m,\lambda}(\Gamma)$, $\mathbf{W}_{p,k}^{(m)}(\Gamma)$, $\mathbf{C}_k^{\infty}(\Gamma)$ denote the corresponding spaces on a Lyapunov manifold Γ (see Chapter XIII, § 1).

2. Fix a point $a \in \Gamma$ and let $U'(a)$ and $U''(a)$ denote two neighborhoods of a such that $U''(a) \subset U'(a) \subset U(a)$, where $U(a)$ denotes the Lyapunov ball. Furthermore, let α_a and β_a be functions in $\mathbf{C}^{\infty}(\Gamma)$ such that $0 \leqq \alpha_a(t), \beta_a(t) \leqq 1$, $\operatorname{supp} \beta_a \subset U'_\Gamma(a) := U'(a) \cap \Gamma$, $\operatorname{supp} \alpha_a \subset U''_\Gamma(a) := U''(a) \cap \Gamma$, $\beta_a(t) = 1$ in $U''_\Gamma(a)$. Then the integral operator

$$(K\varphi)(t) = \alpha_a(t) \int_\Gamma K(t,\tau)\,\varphi(\tau)\,d\Gamma_\tau$$

can be decomposed into the sum $(K\psi_a)(t) + (K\chi_a)(t) = (K\varphi)(t)$, where $\psi_a(t) = \beta_a(t)\,\varphi(t)$ and $\chi_a(t) = (1 - \beta_a(t))\,\varphi(t)$. Let $\tilde{\Gamma}$ be a surface diffeomorphic to Γ and denote by H the corresponding diffeomorphism. Put $H(t) = \tilde{t}$, $t \in \Gamma$; $H(U'_a(\Gamma)) = U'_{\tilde{a}}(\tilde{\Gamma})$, $H(a) = \tilde{a}$. Further, define $\tilde{\varphi}(\tilde{t}) = \varphi(t)$, $\varphi \in X(\Gamma)$, $\tilde{\varphi} \in X(\tilde{\Gamma})$, with $X(\Gamma)$ and $X(\tilde{\Gamma})$ for the time being arbitrary Fréchet spaces of vector functions. Via the formula

$$\tilde{\varphi}(\tilde{t}) = \tilde{\varphi}(H(t)) = \varphi(t) = (H^*\tilde{\varphi})(t)$$

the diffeomorphism H generates a linear homeomorphism $H^* \in \mathscr{L}(X(\tilde{\Gamma}), X(\Gamma))$. With $\tilde{K}$ the operator defined by

$$(\tilde{K}\tilde{\varphi})(\tilde{t}) = \widetilde{(K\varphi)}(\tilde{t})$$

we get

$$(K\varphi)(t) = (\tilde{K}\tilde{\varphi})(\tilde{t})$$

and thus

$$\tilde{K} = H^{*-1} K H^*.$$

It is easy to prove the following auxiliary proposition.

Lemma 3.1. *Let $K \in \mathscr{L}(X(\Gamma), Y(\Gamma))$ and suppose for each point $a \in \Gamma$ there exists a surface $\tilde{\Gamma}$ diffeomorphic to Γ such that*

$$\int_{U_{\tilde{\Gamma}}(\tilde{a})} \tilde{K}(\tilde{t}, \tilde{\tau})\, \tilde{\psi}_a(\tilde{\tau})\, d\tilde{\Gamma}_{\tilde{\tau}} = 0$$

for all $\tilde{t} \in U_{\tilde{\Gamma}}(\tilde{a})$. Then the operator K is compact if and only if $K(1 - \beta_a)$ is so.

Now locate the origin of a local coordinate system $(u_0, \ldots, u_{n-1})$ at the point a and let the u_0-axis have the same direction as the outer normal to Γ. In $U(a)$, the manifold Γ is then given through local coordinates by $u_0 = u_0(u)$, $u = (u_1, \ldots, u_{n-1})$. Put

$$\varphi^*(u) = \varphi(u, u_0(u)), \qquad d\Gamma_t = J(u)\, du, \qquad J(u) = \sqrt{1 + \left(\frac{\partial u_0}{\partial u_1}\right)^2 + \ldots + \left(\frac{\partial u_0}{\partial u_{n-1}}\right)^2}.$$

The following is obvious.

Lemma 3.2. *The operator $K\beta_a$ admits the representation*

$$(K\beta_a\varphi)(t) = (K\beta_\alpha\varphi)^*(u) = \alpha^*(u) \int_{G(a)} K^*(u, v)\, \psi_a^*(v)\, J(v)\, dv,$$

where $G(a)$ is the projection of $U'(a)$ into the u-space.

With the help of Lemma 3.2 the following theorem can be proved.

Theorem 3.1. *One has*

$$S \in \mathscr{L}(\mathbf{C}_k^{m,\lambda}(\Gamma)) \cap \mathscr{L}(\mathbf{C}_k^{\infty}(\Gamma)) \cap \mathscr{L}(\mathbf{W}_{p,k}^{(m)}(\Gamma)).$$

Remark. For the case where S is the Moisil-Theodorescu operator a proof is in Sprössig [2].

3. Now let u and a be in $G(a)$, put $u - a = h$ and $h_0 = h/|h|$. Further let $\varphi \in \mathbf{C}_k^m(G)$. Denote by $(P_m\varphi)(u)$ the vector polynomial

$$(P_m\varphi)(u) = \sum_{l=0}^{m-1} \left(\frac{(h_0, \nabla)^l}{l!}\varphi\right)(a)\, |h|^l.$$

The Taylor formula for directional derivatives with the remainder in integral form gives the following.

Theorem 3.2 (Sprössig [1]) *One has*

$$\frac{I - P_m}{|h|^m} \in \mathscr{L}(\mathbf{C}_k^{m,\lambda}(G), \mathbf{C}_k^{0,\lambda}(G)).$$

Let H denote the diffeomorphism of $U'_\Gamma(a)$ onto $G(a)$. Then through

$$\varphi(t) = \tilde{\varphi}(u) = \tilde{\varphi}(H(t)) = (H^*\tilde{\varphi})(t)$$

a linear homeomorphism is generated. Put $\tilde{P}_m = H^* P_m H^{*\,-1}$.

Here is an immediate consequence of Theorem 3.2.

Corollary 3.1. *Let $\tilde{h} = t - a$. Then*

$$\frac{I - \tilde{P}_m}{|\tilde{h}|^m} \in \mathscr{L}(\mathbf{C}_k^{m,\lambda}(U'_\Gamma(a)), \mathbf{C}_k^{0,\lambda}(U'_\Gamma(a))).$$

Corollary 3.1 on its hand implies the following.

Corollary 3.2. *Let $\varphi \in \mathbf{W}_{p,k}^{(m)}(U'_\Gamma(a))$ and $(D^\alpha\varphi)(a) = 0$ for $|\alpha| \leqq m - 1$. Then*

$$(|\tilde{h}|^{-m}\varphi)(t) \in \mathbf{L}_{p,k}(U'_\Gamma(a)).$$

Theorem 3.3 (Sprössig [1]). *Let $M \in \mathbf{C}^{2m+2}_{k\times k}(\Gamma)$. Then the commutator $K_\Gamma = [S, M]$ is compact on the pairs of spaces $\mathbf{C}^{0,\lambda}_k(\Gamma)$, $\mathbf{C}^{m,\lambda}_k(\Gamma)$ and $\mathbf{L}^k_p(\Gamma)$, $\mathbf{W}^{(m)}_{p,k}(\Gamma)$. If $M \in \mathbf{C}^\infty_{k\times k}(\Gamma)$ then K_Γ is also compact on $\mathbf{C}^\infty_k(\Gamma)$.*

Proof. We merely sketch the proof for the pair of spaces $\mathbf{C}^{0,\lambda}_k(\Gamma)$, $\mathbf{C}^{m,\lambda}_k(\Gamma)$. Let $a \in \Gamma$. Consider first the integral operator $\alpha_a K_{U_\Gamma(a)}$. As already mentioned above, this operator can be written in the form

$$\alpha_a K_{U_\Gamma(a)} = \alpha_a K_{U_\Gamma(a)} \beta_a + \alpha_a K_{U_\Gamma(a)} (1 - \beta_a).$$

It is easily seen that $\alpha_a K_{U_\Gamma(a)}(1 - \beta_a)$ is a compact operator. By virtue of Lemma 3.1, the investigation of the operator $\alpha_a K_{U_\Gamma(a)} \beta_a$ can be reduced to the study of the operator $\tilde{\alpha}_{\tilde{a}} \tilde{K}_{\tilde{U}_{\tilde{\Gamma}}(\tilde{a})} \tilde{\beta}_{\tilde{a}}$, where $\tilde{\Gamma}$ is a surface which is plane in $\tilde{U}_{\tilde{\Gamma}}(\tilde{a})$. Therefore we put $\tilde{U}_{\tilde{\Gamma}}(\tilde{a}) = G(a)$. Then

$$\alpha_a K_{U_\Gamma(a)} \beta_a = \frac{2\alpha^*(u)}{|S_1|} \int_{G(a)} \frac{D'(\theta^*)\, D(\alpha^*)}{|u - v|^{n-1}} [M^*(u) - M^*(v)]\, \varphi^*(v)\, J(v)\, \mathrm{d}v$$

with

$$\theta^* = \left(0, \frac{u_1 - v_1}{|u - v|}, \ldots, \frac{u_{n-1} - v_{n-1}}{|u - v|}\right), \qquad \alpha^* = (1, 0, \ldots, 0).$$

Put

$$\frac{D'(\theta^*)\, D(\alpha^*)}{|u - v|^{n-1}} = S^*(u, v) = S^*(u - v) = S^*(h),$$

and apply Taylor's formula for directional derivatives to get

$$\int_{G(a)} S^*(h)\, \tilde{P}(v)\, \mathrm{d}v = 0$$

for every polynomial $\tilde{P}(v)$. The assertion now follows from the Weierstrass approximation theorem. ▮

3.4. Fredholm properties of certain degenerate paired operators

Consider the paired singular integral operator

$$A = aP + bQ,$$

where $a, b \in \mathbf{C}^{k\times k}(\Gamma)$. This is a multidimensional analogue of the one-dimensional singular integral operator (see Chapter III). Similarly as in the one-dimensional case, the pair of matrix functions $\{a, b\}$ can be taken as a *symbol* of the operator A.

Given a fixed non-negative integer, through

$$\bar{A}\varphi := A\varphi, \qquad \varphi \in D(\bar{A}),$$

$$D(\bar{A}) := \{\varphi \in \mathbf{L}^k_p(\Gamma) : A\varphi \in \mathbf{W}^{(m)}_{p,k}(\Gamma)\}$$

a closed operator with the domain $D(\bar{A})$ acting from $\mathbf{L}^k_p(\Gamma)$ into $\mathbf{W}^{(m)}_{p,k}(\Gamma)$ is defined.[1]) The operator $\bar{A}$ can be analogously defined on the pair of spaces $\mathbf{C}^{0,\lambda}_k(\Gamma)$, $\mathbf{C}^{m,\lambda}_k(\Gamma)$. In this section we assume that the symbol determinants $\det a$ and $\det b$ have at most finitely many point-zeros.

---

[1]) In this connection see also Section 5.2 of Chapter VI or Prössdorf [10].

Put $I_1 = \{1, \ldots, l\}$ and $I_2 = \{l+1, \ldots, s\}$. Suppose that the matrix function a can be written as $a(t) = |t - t^i|^{m_i} \bar{a}_i(t)$ in $U_\Gamma(t^i)$, $i \in I_1$ and that b is of the form $b(t) = |t - t^i|^{m_i} \bar{b}_i(t)$ in $U_\Gamma(t^i)$, $i \in I_2$, where $\det \bar{a}_i(t) \neq 0$, $\det \bar{b}_i(t) \neq 0$, and $t^i \in \Gamma$ $(i \in I_1 \cup I_2)$.

Theorem 3.4 (Sprössig [1]). *Let* $a, b \in \mathbf{C}^{2m+2}_{k\times k}(\Gamma)$. *If*

(i) *the function* $\det(ab)$ *has at most finitely many point-zeros of an order* $\leq m$ *on* Γ,

(ii) *the matrix functions* a *and* b *commute with the matrix* $D'(\theta)\,D(\alpha)$,

then the operator $\bar{A}$ *is a* Φ*-operator on the pairs of spaces* $\mathbf{C}^{0,\lambda}_k(\Gamma)$, $\mathbf{C}^{m,\lambda}_k(\Gamma)$ *and* $\mathbf{L}^k_p(\Gamma)$, $\mathbf{W}^{(m)}_{p,k}(\Gamma)$. *If* a *and* b *are in* $\mathbf{C}^\infty_{k\times k}(\Gamma)$, *then* A *is a* Φ_+*-operator on* $\mathbf{C}^\infty_k(\Gamma)$.

In order to prove this theorem, Sprößig constructs a left regularizer R of the operator A. This regularizer is of the form

$$R = \sum_{i=0}^{s} \alpha_i [R_i^a P + R_i^b Q],$$

where

$$R_i^a = \begin{cases} a^{-1}(t), & t \notin U_\Gamma(t^i), \quad i \in I_1, \\ \dfrac{I - \tilde{P}_{m_i}}{|t - t^i|^{m_i}} \tilde{\bar{a}}_i(t), & t \in U_\Gamma(t^i), \quad i \in I_1, \end{cases}$$

$$R_i^b = \begin{cases} b^{-1}(t), & t \notin U_\Gamma(t^i), \quad i \in I_2, \\ \dfrac{I - \tilde{P}_{m_i}}{|t - t^i|^{m_i}} \tilde{\bar{b}}_i(t), & t \in U_\Gamma(t^i), \quad i \in I_2. \end{cases}$$

Here $\tilde{P}_{m_i}$ denotes the Taylor polynomial corresponding to the zero t^i and $\tilde{\bar{b}}_i$, $\tilde{\bar{a}}_i$ are the matrix functions determined (uniquely) by $\tilde{\bar{b}}_i \cdot \bar{b}_i = \det \bar{b}$ and $\tilde{\bar{a}}_i \cdot \bar{a}_i = \det \bar{a}$.

The pairs of spaces quoted above are "exact" in the following sense.

Theorem 3.5 (Sprössig [2]). *If at least one of the coefficients* a, b *has a point-zero of an order* $\geq m+1$, *then* $\bar{A}$ *is not a* Φ_+*-operator on the pair of spaces* $\mathbf{C}^{0,\lambda}_k(\Gamma)$, $\mathbf{C}^{m,\lambda}_k(\Gamma)$.

An analogous result also holds for the pair of spaces $\mathbf{L}^k_p(\Gamma)$, $\mathbf{W}^{(m)}_{p,k}(\Gamma)$.

3.5. Some generalizations

Definition. A function is said to have a *zero of the order* m on a closed smooth submanifold $\gamma \subset \Gamma$ if, for all $h_0 \in T_a(\Gamma) \setminus T_a(\gamma)$, the directional derivatives $(h_0, \nabla)^l$, $l = 0, \ldots, m-1$, of that function vanish on γ and the m-th directional derivative $(h_0, \nabla)^m$ is distinct from zero on Γ. If the directional derivatives of all orders vanish on γ, then γ is called a *zero of infinite order*. Here $T_a(\Gamma)$ and $T_a(\gamma)$, respectively, are the tangential spaces to Γ and γ at the point a.

Let $\gamma^* \subset G \subset \mathbb{R}^n$ be a closed smooth submanifold which is the boundary of a star-shaped region contained in G. Assume the center of the region coincides with the origin. Put $h_{\gamma^*} = u - \gamma^*(u/|u|)$. It is easy to show that

$$(h_0, \nabla)^l\, |h_{\gamma^*}|^m = m(m-1) \ldots (m-l+1)\, |h_{\gamma^*}|^{m-l}$$

for all $h_0 \notin T_a(\gamma^*)$. Replacing the Taylor polynomial $(P_m\varphi)(u)$ by the vector function

$$(\widehat{P_m\varphi})(u) = \sum_{l=0}^{m} \frac{(h_0, \nabla)^l}{l!} \varphi\left(\gamma^*\left(\frac{u}{|u|}\right)\right) |h_{\gamma^*}|^l,$$

we arrive at the following.

Theorem 3.6 (Sprössig [2]). *Let $a, b \in C^{2m+2}_{k\times k}(\Gamma)$. If the matrix functions a and b commute with the matrix $D'(\theta)\, D(\alpha)$ and if the symbol determinants* $\det a$ *and* $\det b$ *have at most finitely many zeros of the form described above, then the singular integral operator $\bar{A}$ is a Φ_+-operator on the pairs of spaces $C_k^{0,\lambda}(\Gamma)$, $C_k^{m,\lambda}(\Gamma)$ and $L_p^k(\Gamma)$, $W_{p,k}^{(m)}(\Gamma)$. If at least one of the zeros has an order $\geqq m + 1$, then $\bar{A}$ is not a Φ_+-operator on these pairs of spaces.*

Remark 1. The zeros in the sense of the above definition are submanifolds which are pairwise disjoint.

Remark 2. Sprössig also considered more general zero-sets of the symbol determinants, namely, zero-sets consisting of smooth closed Jordan curves which are not necessarily supposed to be pairwise disjoint. In his works [1, 2], for the case $n = 3$ and $k = 4$, he established results on the Φ_+-properties of the corresponding singular integral operator $\bar{A}$.

Remark 3. The operators considered in the Sections 3.1 and 3.2 have applications to certain problems in physics and engineering. In [7], Sprössig proposed an iteration method for the solution of the Maxwell equations, and in [6], Sprössig determined thermal stresses arising in cast steel bodies during the process of heating. For central symmetric bodies the stress τ_α in the direction α is

$$\tau_\alpha = L\beta \frac{m+1}{m-1}\left(\alpha \frac{1}{m-2} t - T \frac{\partial}{\partial\alpha} t - \sum_{i=1}^{3} \alpha_i \mathring{S} n n_i t - (m-1)\, P_1 \mathring{S} t\right),$$

where L denotes the torsion modulus, β the coefficient of heat expansion, t the temperature, m^{-1} the Poisson coefficient, and $P_1\hat{\varphi} := (0, \varphi)$. If the surface temperature is constant this can be simplified to

$$\tau_\alpha = L\beta \frac{m+1}{m-1}\left(\alpha \frac{1}{m-2} t - T \frac{\partial}{\partial\alpha} t\right).$$

Remark 4. For bodies of more complicate shape such simple explicit formulae for the stresses as, for instance, in Sprössig [6] cannot be derived. Gürlebeck [1] showed that three-dimensional boundary value problems of mathematical physics can be globally factorized by means of the operators considered in the Sections 3.1 and 3.2. When applied to the Dirichlet boundary value problem of the linear theory of elasticity,

$$\Delta u + \frac{m}{m-2} \operatorname{grad} \operatorname{div} u = 0 \text{ in } G,$$

$$\gamma_0 u = g \text{ on } \Gamma,$$

that factorization leads to the following two boundary value problems for (generalized) analytic functions:

$$D(\nabla)\, \hat{v} = 0 \text{ in } G,\ \gamma_0 \hat{v} = P\hat{g} \text{ on } \Gamma,$$

and

$$D(\nabla)\, \hat{w} = 0 \text{ in } G,\ \gamma_0 T M^{-1} \hat{w} = Q\hat{g} \text{ on } \Gamma.$$

Here m^{-1} is the Poisson coefficient, $M\hat{f} = \left(\frac{2m-2}{m-2} f_1, f\right)$, $\hat{g} = (0, g)$, and $\hat{u} = (u_1, u) = \hat{v} + TM^{-1}\hat{w}$.

Using some results of Müller [1] and Aleksidze [1], Gürlebeck [1] established a boundary collocation method for the numerical solution of boundary value problems for analytic functions.

He seeks an approximate solution of the problem

$$D(\nabla)\,\hat{u} = 0 \text{ in } G,\ R\hat{u} = g \text{ on } \Gamma$$

in the form

$$\hat{u}_n(x) = \sum_{i=1}^{n} \hat{\varphi}_i(x) \circ \hat{a}_i,$$

where $\{\varphi_i\}_{i=1}^{\infty}$ is a system of functions which is complete in Ker $D(\nabla)$ (G) $\big(\cap H^S(G)\big)$ and where the quaternionic coefficients $\hat{a}_i$ $(i = 1, \ldots, n)$ are to be determined so that the conditions

$$(R\hat{u}_n)\,(x_j) = \hat{g}(x_j) \quad (x_j \in \Gamma, j = 1, \ldots, n)$$

are satisfied.

Chapter XVII
Methods for the approximate solution of one-dimensional singular integral equations

The present chapter is concerned with several methods for the approximate solution of one-dimensional singular integral equations over closed as well as over open Lyapunov curves. Special attention is paid to the method of least squares, the reduction method, the collocation method, and the method of mechanical quadratures. We state conditions for the convergence of these methods in spaces of the type $\mathbf{L}_p$ and $\mathbf{H}^\mu$ and determine the asymptotic convergence order. We begin with some general theorems on the convergence and stability of projection methods (in general with unbounded projections) for linear operator equations in Banach spaces. Thereby stable projection methods for abstract singular equations will be established.

§ 1. General theorems on the convergence of projection methods

In this section we present some results which were obtained by Prössdorf and Silbermann [4], [8–9]. With only minor modifications we follow these authors' works [8, 9].

1.1. Let X and Y be Banach spaces, let $X_n \subset X$ and $Y_n \subset Y$ $(n = 1, 2, \ldots)$ be closed subspaces, and let Q_n be a (not necessarily bounded) projection of Y onto Y_n, that is, suppose

$$Q_n^2 = Q_n, \qquad \operatorname{Im} Q_n = Y_n \qquad (n = 1, 2, \ldots).$$

Consider the equation

$$Ax = y, \tag{1}$$

where $A \in \mathscr{L}(X, Y)$. The *projection method* for the approximate solution of this equation consists of the following. On condition that $y \in D(Q_n)$ $(n = 1, 2, \ldots)$, replace equation (1) by the *approximating equation* ("*projected*" *equation*)

$$Q_n A x_n = Q_n y \tag{2}$$

and find a solution $x_n \in X_n$ of this equation. The problem is to make clear under what conditions and in what a sense x_n can be viewed as an *approximate solution* of equation (1).

In the case of bounded (and, thus, defined on the whole space X) projections Q_n the *projection method* is said to be *applicable to the operator* A, and this will be written as $A \in \Pi\{X_n, Q_n\}$[1]), if there is an n_0 such that for each $y \in Y$ and for all $n \geqq n_0$ the equa-

[1]) Note that the sequences $\{X_n\}$ and $\{Q_n\}$ determine the projection method completely.

tion (2) has a unique solution $x_n \in X$ and if the sequence $\{x_n\}$ converges to a solution $x \in X$ of the equation (1) as $n \to \infty$. If Q_n are bounded projections and if, furthermore, the union $\bigcup_{n=1}^{\infty} X_n$ is dense in X, then it is not difficult to prove the following criterion: $A \in \Pi\{X_n, Q_n\}$ if and only if the operators $A : X \to Y$ and $A_n = Q_nA \mid X_n : X_n \to Y_n$ $(n \geqq n_0)$ are invertible and if $A_n^{-1}Q_n$ converges strongly to A^{-1} as $n \to \infty$.

1.2. In concrete applications one is often confronted with the fact that the projection method does not converge for every $y \in Y$ (see the §§ 3–6 of the present chapter). We have obviously such a situation if the space X is not separable and the subspaces X_n are finite-dimensional or if the projections Q_n are not bounded. It is therefore reasonable to give the following definition.

Definition (Gohberg and Levchenko [1]). Throughout what follows assume that $A(X_n) \subset D(Q_n)$ and that the operators $A_n = Q_nA \mid X_n : X_n \to Y_n$ are continuous and (continuously) invertible for all sufficiently large n, say $n \geqq n_0$. Denote by $\mathfrak{K}(A; \{X_n, Q_n\})$ the set of all elements $y \in Y$ that have the following properties:

1. $y \in D(Q_n)$ for $n \geqq n'(y)$;
2. $x_n = A_n^{-1}Q_ny$ converges in the norm of X to an element $x \in X$ as $n \to \infty$;
3. $Ax = y$.

The set $\mathfrak{K}(A; \{X_n, Q_n\})$ is called the *convergence manifold* of the operator A with respect to the projection method $\{X_n, Q_n\}$.

If we have a sequence of projections $\{P_n\}$ in X such that $\operatorname{Im} P_n = X_n$, then $\mathfrak{K}(A; \{X_n, Q_n\})$ will also be written as $\mathfrak{K}(A; \{P_n, Q_n\})$.

It is clear that the convergence manifold is a linear set contained in the image space $\operatorname{Im} A$. If the operator A is invertible, $\mathfrak{K}(A; \{X_n, Q_n\})$ is the maximal linear set on which the operators $A_n^{-1}Q_n$ converge strongly to A^{-1} as $n \to \infty$. Notice that $\mathfrak{K}(A; \{X_n, Q_n\}) = Y$ if and only if $A \in \Pi\{X_n, Q_n\}$.

In the present chapter we denote by $\mathscr{K}(X, Y)$ $(\subset \mathscr{L}(X, Y))$ the collection of all compact operators between X and Y and by $G\mathscr{L}(X, Y)$ the set of all invertible operators from $\mathscr{L}(X, Y)$.

Theorem 1.1. *Let $A \in \mathscr{L}(X, Y)$ be an invertible operator, let $Z \subset Y$ be a Banach space*[1]) *continuously embedded in Y such that $Z \subset \mathfrak{K}(A; \{X_n, Q_n\})$, and let $T \in \mathscr{K}(X, Z)$. Also suppose that*

1. $\dim \operatorname{Ker} (A + T) = 0$;
2. $Q_n \mid Z \in \mathscr{L}(Z, Y)$.

Then

$$\mathfrak{K}(A; \{X_n, Q_n\}) = \mathfrak{K}(A + T; \{X_n, Q_n\}).$$

Proof. The invertibility of A and condition 1 of the theorem imply that the operator $\tilde{A} = A + T$ and, hence, also the operator $I + A^{-1}T$ are invertible. Note that $\tilde{A}^{-1} = (I + A^{-1}T)^{-1} A^{-1}$.

From the condition 2 of the theorem we deduce that $\operatorname{Im} T \subset D(Q_n)$ and $Q_nT \in \mathscr{L}(X, Y)$ $(n = 1, 2, \ldots)$. Put

$$A_n = Q_nA \mid X_n, \qquad A_n = Q_nA \mid X_n = A_n + Q_nT \qquad (n \geqq n_0).$$

[1]) In what follows the inclusion of two Banach spaces, $Z \subset Y$, will be always understood in the sense of a continuous embedding.

By the hypotheses of the theorem, the operators $A_n^{-1}Q_n \mid Z \in \mathscr{L}(Z, X)$ converge strongly to the operator $A^{-1} \mid Z \in \mathscr{L}(Z, X)$ as $n \to \infty$. Therefore, this convergence is uniform on each relatively compact subset of the space Z. Since $T \in \mathscr{K}(X, Z)$, we conclude that

$$\delta_n = \|A^{-1}T - A_n^{-1}Q_nT\| \to 0 \qquad \text{as } n \to \infty. \tag{3}$$

The operator $I + A_n^{-1}Q_nT \in \mathscr{L}(X)$ is therefore invertible for $n \geqq n_1$. It follows that $\tilde{A}_n \colon X_n \to Y_n$ is also invertible for $n \geqq n_1$ and that

$$\tilde{A}_n^{-1} = (I + A_n^{-1}Q_nT)^{-1} A_n^{-1}. \tag{4}$$

Now take any element $y \in \mathfrak{R}(A; \{X_n, Q_n\})$. We show that $\tilde{A}_n^{-1}Q_ny$ converges in the norm of X to $\tilde{A}^{-1}y$ as $n \to \infty$. Since $A_n^{-1}Q_ny \to A^{-1}y$ as $n \to \infty$, in view of (4) it remains to show that

$$z_n = [(I + A^{-1}T)^{-1} - (I + A_n^{-1}Q_nT)^{-1}]\, A_n^{-1}Q_ny$$

converges to zero as $n \to \infty$. This however, results immediately from the well-known estimate

$$\|z_n\| \leqq \|(I + A^{-1}\, T)^{-1}\|^2 \frac{\delta_n \,\|A_n^{-1}Q_ny\|}{1 - \delta_n \,\|(I + A^{-1}\, T)^{-1}\|} \tag{5}$$

together with (3). Thus,

$$\mathfrak{R}(A; \{X_n, Q_n\}) \subset \mathfrak{R}(\tilde{A}; \{X_n, Q_n\}).$$

Since $A = \tilde{A} - T$ and $Z \subset \mathfrak{R}(\tilde{A}; \{X_n, Q_n\})$, by repeating the preceding arguments with the roles of A and $\tilde{A}$ interchanged we arrive at the inclusion

$$\mathfrak{R}(\tilde{A}; \{X_n, Q_n\}) \subset \mathfrak{R}(A; \{X_n, Q_n\}). \blacksquare$$

Corollary 1.1. (Gohberg and Feldman [1]). *Let $Q_n \in \mathscr{L}(Y)$ $(n = 1, 2, \ldots)$, suppose $A \in \Pi\{X_n, Q_n\}$ is invertible, and let $T \in \mathscr{K}(X, Y)$. Then if the operator $A + T$ is invertible, it is in $\Pi\{X_n, Q_n\}$.*

To see this, apply Theorem 1.1 with $Z = Y$.

Remark. There is an analogue of Theorem 1.1 for perturbations of the operator A by operators with sufficiently small norm, which in particular shows that $\Pi\{X_n, Q_n\}$ is an open subset of the space $\mathscr{L}(X, Y)$ (see Prössdorf and Silbermann [9]). In this connection see also Theorem 1.5 below.

1.3. The following theorem provides an error estimate for a certain class of operators $A \in \mathscr{L}(X, Y)$.

Theorem 1.2. *Suppose $A = CB$, where $B \in \mathscr{L}(X, Y)$ and $C \in \mathscr{L}(Y)$ are invertible operators which satisfy the following two conditions:*

(a) $Q_nB(X_n) = B(X_n), \qquad n = 1, 2, \ldots;$

(b) $Q_nC^{\pm 1}Q_ny = Q_nC^{\pm 1}\, y, \qquad \forall y \in D(Q_n).$

Furthermore, suppose the conditions of Theorem 1.1 are fulfilled.

If $y \in \mathfrak{R}(A; \{X_n, Q_n\})$ and $x \in X$, $x_n \in X_n$ are the solutions of the equations

$$Ax + Tx = y, \qquad Q_n(A + T)\, x_n = Q_ny$$

respectively, then

$$c_1 \,\|Bx - Q_nBx\|_Y \leqq \|x - x_n\|_X \leqq c_2 \,\|Bx - Q_nBx\|_Y \tag{6}$$

and in case the projections Q_n are continuous,

$$\|x - x_n\|_X \leqq c\, \|Q_n\|\, E_n(Y, Bx). \tag{7}$$

Here c_1, c_2, c are positive constants and $E_n(Y, f) := \inf_{y_n \in Y_n} \|f - y_n\|_Y$.

Proof. Put $T_1 = C^{-1}T$. Then the equations

$$Ax + Tx = y \quad \text{and} \quad Bx + T_1x = C^{-1}y$$

are obviously equivalent. By virtue of condition (b), the equations

$$Q_n(A + T)\, x_n = Q_n y \text{ and } Q_n(B + T_1)\, x_n = Q_n C^{-1} y$$

are also so. Taking into account condition (a) we obtain

$$(B + Q_n T_1)\,(x - x_n) = Bx - Q_n Bx. \tag{8}$$

The estimate (6) will follow as soon as we have proved that

$$\|T_1 - Q_n T_1\|_{X \to Y} \to 0 \qquad \text{as } n \to \infty, \tag{9}$$

since (9) implies the uniform convergence

$$B + Q_n T_1 \to B + T_1, \qquad (B + Q_n T_1)^{-1} \to (B + T_1)^{-1},$$

and so (6) is seen to be an immediate consequence of (8).

Thus, let us prove (9). Put $A_n = Q_n A \mid X_n$. From (a) and (b) we get

$$A_n = Q_n C Q_n B \mid X_n, \qquad B^{-1} Q_n C^{-1} A_n = B^{-1} Q_n B \mid X_n = I_n,$$

with I_n the identity operator on X_n. Note that $Q_n C Q_n : Y_n \to Y_n$ is invertible. The invertibility of $A_n : X_n \to Y_n$ therefore implies the invertibility of $B \mid X_n : X_n \to Y_n$ and we have $A_n^{-1} Q_n = B^{-1} Q_n C^{-1}$. Hence,

$$\|T_1 - Q_n T_1\| \leqq \|B\| \, \|B^{-1} C^{-1} T - B^{-1} Q_n C^{-1} T\| = \|B\| \, \|A^{-1} T - A_n^{-1} Q T_n\|$$

and this combined with (3) gives (9).

Finally, let us verify the estimate (7). We have $\|z - Q_n z\|_Y \leqq \|z - z_n\|_Y + \|Q_n(z - z_n)\|_Y \leqq (1 + \|Q_n\|) \, \|z - z_n\|_Y$ for every $z \in D(Q_n)$ and $z_n \in Y_n$. Thus, letting $z = Bx$ it follows that

$$\|Bx - Q_n Bx\|_Y \leqq 2 \, \|Q_n\| \, E_n(Y, Bx),$$

and this together with the second inequality in (6) gives the estimate (7). ∎

Corollary 1.2. *Let the conditions of Theorem 1.2 be fulfilled. Furthermore, suppose $Bx \in Z$, where $Z \subset Y$ are Banach spaces such that*

(c) $\|y - Q_n y\|_Y \to 0, \qquad \forall y \in Z, \qquad n \to \infty.$

Then

$$\|x - x_n\|_X \leqq c E_n(Z, Bx).$$

Proof. Since the embedding $Z \subset Y$ is continuous, we deduce that

$$\|z - Q_n z\|_Y \leqq (c' + \|Q_n\|_{Z \to Y}) \, \|z - z_n\|_Z$$

for every $z \in Z$ and $z_n \in Y_n$. To get the assertion it remains to take into account condition (c) and to make use of the Banach-Steinhaus theorem. ∎

The following theorem states sufficient conditions for a certain set Z to be contained in the converges manifold of an operator.

Theorem 1.3. *Suppose $A = CB$ with invertible operators $B \in \mathscr{L}(X, Y)$ and $C \in \mathscr{L}(Y)$ satisfying the following two conditions:*

(a′) $BX_n = Y_n$ $(n = 1, 2, \ldots)$;

(d) *there is a set $Z \subset Y$ such that $C^{-1}(Z) \subset Z$.*

Also let the above conditions (b) *and* (c) *be fulfilled.*
Then $Z \subset \mathfrak{K}(A; \{X_n, Q_n\})$.

Proof. Condition (b) implies that

$$Q_n C^{-1} Q_n C = Q_n = Q_n C Q_n C^{-1}.$$

This together with (a') shows that the operator $A_n = Q_n A \mid X_n : X_n \to Y_n$ is invertible and that $A_n^{-1} Q_n = B^{-1} Q_n C^{-1}$.

Let $z \in Z$ be an arbitrarily chosen element. Then, by condition (d), $C^{-1} z \in Z$. Taking into consideration condition (c) we conclude that

$$A_n^{-1} Q_n z \to B^{-1} C^{-1} z = A^{-1} z,$$

whence $z \in \mathfrak{K}(A; \{X_n, Q_n\})$. ∎

1.4. The computation of both the operator $A_n = Q_n A \mid X_n$ and the right-hand side $Q_n y$ in the approximating equation (2) is, in general, accompanied with errors (e.g. rounding off errors in the computation of scalar products, integrals etc.). So what we solve actually is not the equation (2), but a certain perturbed equation of the form

$$(A_n + Q_n \Gamma_n)\, \tilde{x}_n = Q_n y + \delta_n. \tag{10}$$

Here one requires of course that $\tilde{x}_n \in X_n$, $\delta_n \in Y_n$, and $\Gamma_n : X_n \to D(Q_n)$. However, the projection method $\{X_n, Q_n\}$ is *stable* in the following sense.[1])

Theorem 1.4. *Let* $A \in \mathscr{L}(X, Y)$ *be an invertible operator and* $Z \subset Y$ *a Banach space such that* $Y_n \subset Z \subset \mathfrak{K}(A; \{X_n, Q_n\})$. *Furthermore, suppose* $Q_n \mid Z \in \mathscr{L}(Z, Y)$ *and let* $y \in Z$.

Then there exist positive constants p, q, γ *which do not depend on* n *and* y *such that for* $\|\Gamma_n\|_{X\to Z} < \gamma$ *the following holds:*

1. *The equation* (10) *has a unique solution for every* $\delta_n \in Y_n$ $(n \geqq n')$.
2. $\|\tilde{x}_n - x_n\|_X \leqq p\, \|y\|_Z\, \|\Gamma_n\|_{X\to Z} + q\, \|\delta_n\|_Z$.

Proof. Since the continuous operators $A_n^{-1} Q_n \mid Z$ converge strongly to $A^{-1} \mid Z$ on the Banach space Z, the Banach-Steinhaus theorem gives the existence of a constant $\gamma_1 > 0$ independent of n such that

$$\|A_n^{-1} Q_n\|_{Z\to X} < \gamma \qquad (n = 1, 2, \ldots).$$

Choose γ so that $\gamma\gamma_1 = \beta < 1$. Then the operator $\tilde{A}_n = A_n + Q_n \Gamma_n$ is invertible if $\|\Gamma_n\|_{X\to Z} < \gamma$, and one has

$$\tilde{A}_n^{-1} = \left[\sum_{k=0}^{\infty} (-1)^k \, (A_n^{-1} Q_n \Gamma_n)^k\right] A_n^{-1}. \tag{11}$$

From

$$x_n = A_n^{-1} Q_n y, \qquad \tilde{x}_n = \tilde{A}_n^{-1} Q_n y + \tilde{A}_n^{-1} Q_n\, \delta_n$$

we get

$$\tilde{x}_n - x_n = (-A_n^{-1} Q_n \Gamma_n) \left[\sum_{k=0}^{\infty} (-1)^k \, (A_n^{-1} Q_n \Gamma_n)^k\right] A_n^{-1} Q_n y$$
$$+ \tilde{A}_n^{-1} Q_n\, \delta_n = -A_n^{-1} Q_n \Gamma_n \tilde{A}_n^{-1}\, Q_n y + \tilde{A}_n^{-1} Q_n \delta_n.$$

The representation (11) shows that

$$\|\tilde{A}_n^{-1} Q_n\|_{Z\to X} < \frac{\gamma_1}{1-\beta}.$$

---

[1]) For the definition of stability see also Mikhlin [25].

Thus,

$$\|\tilde{x}_n - x_n\|_X \leqq \frac{\gamma_1^2}{1-\beta} \|\Gamma_n\|_{X\to Z} \|y\|_Z + \frac{\gamma_1}{1-\beta} \|\delta_n\|_Z . \blacksquare$$

Remark 1. If $y \in \mathfrak{K}(A; \{X_n, Q_n\})$ and $\|A_n^{-1}Q_n\Gamma_n\|_{\mathscr{L}(X)} \leqq \beta < 1$, then it is seen from the preceding proof that

$$\|\tilde{x}_n - x_n\|_X \leqq \frac{1}{1-\beta} [\beta \|A_n^{-1}Q_n y\|_X + \|A_n^{-1}Q_n \delta_n\|_X].$$

Remark 2. Approximation methods of the form (10) and (12) can be interpreted as, in a sense, "*perturbed*" *projection methods.* It turns out that certain approximation methods (for instance, the method of mechanical quadratures, which will be considered below) cannot be written in the form (2) but are of the form (10) or (12).

Remark 3. All results of this section remain true when the subscript n is not restricted to run through the set of positive integers but is allowed to range over an arbitrary unbounded set of positive real numbers.

Remark 4. The results of the present section also yield the well-known theorems of Kantorovich and Vainikko on the Galerkin method with perturbations for Riesz-Schauder equations (see Krasnoselski, Vainikko et al. [1], § 17, Chapter 4).

1.5. The Theorems 1.1 and 1.4 can be easily generalized in a natural way.

We consider the situation where instead of the operators $Q_nA \mid X_n$ an arbitrary sequence of invertible operators $A_n \in \mathscr{L}(X_n, Y_n)$ is given. In that case it is reasonable to denote the collection of all elements $y \in Y$ satisfying the conditions of the definition in Section 1.2 by $\mathfrak{K}(A, A_n; \{X_n, Q_n\})$ and to call it the *convergence manifold* of the operator A with respect to the sequence $\{A_n\}$. If $y \in \mathfrak{K}(A, A_n; \{X_n, Q_n\})$, then the approximation method with the operators $\{A_n\}$ is said to be *convergent* for the element y; if this approximation method converges for all $y \in Y$, then it is simply said to be convergent (or to be *applicable*).

The following theorem involves the Theorems 1.1 and 1.4 as special cases.

Theorem 1.5. *Let A and $A + T$ belong to $G\mathscr{L}(X, Y)$, let $A_n \in G\mathscr{L}(X_n, Y_n)$ $(n = 1, 2, \ldots)$, and let $T_n \in \mathscr{L}(X_n, Y_n)$. Suppose the following conditions are satisfied:*

a) $\operatorname{Im} T \subset D(Q_n)$ *for* $n = 1, 2, \ldots,$

b) $\varrho_n := \|(A^{-1} - A_n^{-1}Q_n)\, T\|_{X\to X} \to 0$ *as* $n \to \infty$,

c) $\varepsilon_n := \|A_n^{-1}(T_n - Q_nT)\|_{X_n\to X} \to 0$ *as* $n \to \infty$.

Then

$$\mathfrak{K}(A, A_n; \{X_n, Q_n\}) = \mathfrak{K}(A + T, A_n + T_n; \{X_n, Q_n\}).$$

If $y \in \mathfrak{K}(A, A_n; \{X_n, Q_n\})$, if $\delta_n \in Y_n$, and if $u \in X$ and $u_n \in X_n$ are the solutions of the equations

$$(A + T)\, u = y \text{ and } (A_n + T_n)\, u_n = Q_n y + \delta_n, \tag{12}$$

respectively, then the estimate

$$\|u - u_n\| \leq c\{\|u - A_n^{-1}Q_nAu\| + \varepsilon_n \|A_n^{-1}Q_n y\| + \|A_n^{-1}\delta\|\}$$

holds. Here c is certain positive constant.

To prove Theorem 1.5 put in the proofs of the Theorems 1.1 and 1.4 $\tilde{A}_n = A_n + Q_nT$ and $\Gamma_n = T_n - Q_nT$ and then repeat the argument of those proofs (see Prössdorf [25], Junghanns and Silbermann [1]). Theorem 1.5 was (without proof) formulated in Prössdorf and Silbermann [12] for the first time.

Remark. Note that, obviously,

$$\|u - A_n^{-1}Q_nAu\| \leqq \varrho_n \|u\| + \|A^{-1}y - A_n^{-1}Q_ny\|.$$

In what follows the sequence $\{A_n\}$ or the approximation method

$$A_nx_n = y_n \quad (x_n \in X_n, y_n \in Y_n) \tag{13}$$

will be called *asymptotically stable* or simply *stable* if $\sup_n \|A_n^{-1}\| < \infty$.

The following theorem, though being very elementary, is of great importance.

Theorem 1.6. *Suppose* $\{A_n\}$ *is stable and let* $P_n \in \mathscr{L}(X, X_n)$ $(n = 1, 2, \ldots)$. *Put* $c = \sup_n \|A_n^{-1}\|$. *Then for the solutions of the equations* (1) *and* (13) *the error estimate*

$$\|x - x_n\| \leqq c(\|Ax - A_nP_nx\| + \|y - y_n\|) + \|x - P_nx\|$$

holds.

This is immediate from the identity

$$P_nx - x_n = A_n^{-1}(A_nP_nx - y_n).$$

§ 2. Projection methods for the solution of abstract singular equations

The results of this section were obtained by Prössdorf and Silbermann [3–6, 10]. Here we essentially follows their works [6] and [10].

Let X be a Banach space, $P \in \mathscr{L}(X)$ a projection, and put $Q = I - P$. Furthermore, assume we are given a subalgebra $\mathfrak{M}$ of the algebra $\mathscr{L}(X)$ which has the property that the inverse a^{-1} of each operator $a \in \mathfrak{M}$ invertible in $\mathscr{L}(X)$ belongs to $\mathfrak{M}$.

Denote by $\mathfrak{M}^+$ (resp. $\mathfrak{M}^-$) the subalgebra of $\mathfrak{M}$ consisting of all operators $a \in \mathfrak{M}$ with $aP = PaP$ (resp. $Pa = PaP$). Let $\mathfrak{M}_l$ (resp. $\mathfrak{M}_r$) denote the collection of all operators $a \in \mathfrak{M}$ representable in the form

$$a = a_+a_- \text{ (resp. } a = a_-a_+),$$

where $a_+^{\pm 1} \in \mathfrak{M}^+$ and $a_-^{\pm 1} \in \mathfrak{M}^-$.

We consider the *abstract singular operator*

$$A = aP + bQ + T, \tag{1}$$

where $a, b \in \mathfrak{M}$ and $T \in \mathscr{K}(X)$. The purpose of this section is to study (under various additional hypotheses) projection methods for the approximate solution of the abstract singular equation $Ax = y$.

Let $\{P_n\}$ and $\{Q_n\}$ be two given sequences of projections on X. Suppose $P_n \in \mathscr{L}(X)$ $(n = 1, 2, \ldots)$. The projections Q_n are required to be defined on certain linear sets $D(Q_n) \subset X$ and to have a closed image space $\operatorname{Im} Q_n \subset X$.

2.1. In the sequel we suppose that there exist two Banach spaces $Z_0 \subset Z \subset X$ with the following properties:

a) $P \mid Z$ is in $\mathscr{L}(Z)$ and $a \mid Z$ is in $\mathscr{L}(Z)$ for all $a \in \mathfrak{M}$.

b) $a \mid Z_0$ is in $\mathscr{L}(Z_0)$ and $Pa - aP$ is in $\mathscr{K}(X, Z_0)$ for all $a \in \mathfrak{M}$.

Furthermore, let the following conditions be fulfilled (for $n \geqq n_0$):

c) $P_n(Pa_- + Qa_+)P_n = (Pa_- + Qa_+)P_n$ for all $a_\pm \in \mathfrak{M}^\pm$.

d) $\operatorname{Im} P_n = \operatorname{Im} Q_n \subset Z$.

e) $Q_n \mid Z_0$ is in $\mathscr{L}(Z_0, Z)$ and $Q_nz \to z$ as $n \to \infty$ in the norm of Z for every $z \in Z_1$, where $Z_1 \subset Z$ is a certain linear set such that $Z_0 \subset Z_1 \subset D(Q_n)$.

Finally, assume at least one of the following two conditions is satisfied:

f) $Q_n a Q_n = Q_n a$ and $a(Z_1) \subset Z_1$ for all $a \in \mathfrak{M}$.

g) $Q_n C Q_n = Q_n C$ and $C(Z_j) \subset Z_j$ $(j = 0, 1)$ for every operator of the form $C = a_+ P + a_- Q$ with $a_\pm \in \mathfrak{M}^\pm$.

Theorem 2.1. *Let the conditions* a)–f) *be fulfilled and let A be an abstract singular operator of the form* (1) *with $T \in \mathcal{K}(X, Z_0)$ and invertible coefficients $a, b \in \mathfrak{M}$. Suppose that $b^{-1}a = c_+ c_- \in \mathfrak{M}_l$ and also that* $\dim \operatorname{Ker} A = 0$.

Then the equation

$$Q_n A x_n = Q_n y \qquad (x_n \in \operatorname{Im} P_n)$$

has a unique solution $x_n \in \operatorname{Im} P_n$ for all $n \geqq n_1$ and for each $y \in Z_1$. As $n \to \infty$, x_n converges in the norm of X to the (unique) solution $x \in X$ of the equation $Ax = y$. Moreover, one has the estimate

$$\|x - x_n\|_X \leqq \gamma \, \|y_0 - Q_n y_0\|_Z, \tag{2}$$

where $\gamma > 0$ is a constant (independent of n and y) and

$$y_0 = (Pc_- + Qc_+^{-1})\, x \in Z_1.$$

If $y \in Z_0$, then $y_0 \in Z_0$ and

$$\|x - x_n\|_X \leqq \gamma' E_n(Z_0, y_0)$$

with

$$E_n(Z_0, y_0) = \inf_{y \in \operatorname{Im} Q_n} \|y_0 - y\|_{Z_0}. \tag{3}$$

Proof. The operator A can be written as

$$A = bc_+(Pc_- + Qc_+^{-1}) + T'$$

with

$$T' = bc_+(Qc_- P + Pc_+^{-1} Q) + T.$$

Condition b) implies that $Qc_- P$ and $Pc_+^{-1}Q$ are in $\mathcal{K}(X, Z_0)$ and therefore T' is in $\mathcal{K}(X, Z_0)$. Thus, since the operator $A_0 = bc_+(Pc_- + Qc_+^{-1})$ is invertible and since $\dim \operatorname{Ker} A = 0$, the operator A is invertible.

Using the conditions c)–f) it is easy to deduce from Theorem 1.3 that

$$Z_1 \subset \mathfrak{R}(A_0; \{P_n, Q_n\}).$$

Due to Theorem 1.1 we have $\mathfrak{R}(A; \{P_n, Q_n\}) = \mathfrak{R}(A_0; \{P_n, Q_n\})$ and this immediately gives the first assertion of the theorem. Finally, from Theorem 1.2 and Corollary 1.2 we obtain the estimates (2) and (3) with

$$y_0 = (Pc_- + Qc_+^{-1})\, x = c_+^{-1} b^{-1} (y - T'x). \ \blacksquare$$

Theorem 2.2. *Suppose the conditions* a)–e) *and the condition* g) *are satisfied. Let A be an abstract singular operator of the form* (1) *with $T \in \mathcal{K}(X, Z_0)$ and invertible coefficients $a, b \in \mathfrak{M}$. Assume $a = a_+ a_- \in \mathfrak{M}_l$ and $b = b_- b_+ \in \mathfrak{M}_r$. Finally, let* $\dim \operatorname{Ker} A = 0$.

Then the assertions of Theorem 2.1 hold with $y_0 = (Pa_- + Qb_+)\, x$.

Proof. We have $A = A_0 + T'$ with $A_0 = (a_+ P + b_- Q)(Pa_- + Qb_+)$ and $T' = (a_+ Q a_- P + b_- P b_+ Q) + T$. Now put

$$y_0 = (Pa_- + Qb_+)\, x = (a_+^{-1} P + b_-^{-1} Q)(y - T'x)$$

and proceed as in the proof of Theorem 2.1. $\blacksquare$

Remark 1. Note that the Theorems 2.1 and 2.2 also hold for abstract singular operators of the form $A = Pa + Qb + T$ with $a, b \in \mathfrak{M}$ and $T \in \mathscr{K}(X, Z_0)$, since these operators can be written as $aP + bQ + T_1$ with $T_1 \in \mathscr{K}(X, Z_0)$.

Remark 2. Under the hypotheses of Theorem 2.1 (or Theorem 2.2) the statement of Theorem 1.4 is true with $Z = Z_0$, that is, the projection method $\{P_n, Q_n\}$ is stable.

2.2. In this subsection we consider abstract singular operators of the form (1) with coefficients of the form

$$a = cr_-, \qquad b = dr_+, \tag{4}$$

where $c, d \in \mathfrak{M}$ are invertible operators and $r_\pm \in \mathfrak{M}^\pm$. Note that r_+ and r_- are not supposed to be invertible. The results of this subsection form the abstract foundation for the approximation methods for the solution of singular integral equations with degenerate coefficients that will be considered in the following sections.

Suppose the operator $B = Pr_- + Qr_+$ satisfies the following two conditions:

h) $\dim \operatorname{Ker} B = 0$.

i) $\operatorname{Im} P_n \subset B(\operatorname{Im} P_n)$.

Remark 3. If the projections P_n are finite-dimensional[1]), condition i) is a consequence of the conditions c) and h).

Indeed, c) implies that $P_nBP_n = BP_n$ and hence $B(\operatorname{Im} P_n) \subset \operatorname{Im} P_n$. Condition i) requires that B maps the subspace $\operatorname{Im} P_n$ onto itself. But if $\operatorname{Im} P_n$ is finite-dimensional this already follows from h).

Theorem 2.3. *Let the conditions* a)–f), h), *and* i) *be fulfilled and let* $Z \subset \operatorname{Im} B$. *Let* A *be an operator of the form* (1) *with* $T \in \mathscr{K}(X, Z_0)$ *and with the coefficients of the form* (4). *Suppose* $c, d \in \mathfrak{M}$ *are invertible and* $d^{-1}c = c_+c_- \in \mathfrak{M}_l$. *Also suppose that* $\dim \operatorname{Ker} A = 0$.

Then all assertions of Theorem 2.1 are true with $y_0 = (Pc_- + Qc_+^{-1})\, Bx \in Z_1$.

Proof. Put $\bar{X} = \{x \in X : Bx \in Z\}$. On defining $|x| = \|Bx\|_Z$ $(x \in \bar{X})$ we make $\bar{X}$ become a Banach space. Note that $\bar{X} \subset X$. Condition a) implies that $Z \subset \bar{X}$. It is obvious that the restriction $\bar{B} = B \mid \bar{X}$ maps the space $\bar{X}$ continuously and one-to-one onto Z.

Let $E : \bar{X} \to X$ be the embedding operator and let $\bar{A}$ denote the restriction $A \mid \bar{X}$. Because $d^{-1}c = c_+c_-$, we have

$$\bar{A} = dc_+(Pc_- + Qc_+^{-1})\, \bar{B} + T' \tag{5}$$

with

$$T' = dc_+(Qc_-r_-P + Pc_+^{-1}r_+Q)\, E + TE. \tag{6}$$

Condition a) gives the invertibility of the operator $A_0 = dc_+(Pc_- + Qc_+^{-1})\, \bar{B} \in \mathscr{L}(\bar{X}, Z)$ and condition b) shows that $T' \in \mathscr{K}(\bar{X}, Z_0)$. Thus, since $\dim \operatorname{Ker} A = 0$, we conclude that $\bar{A} \in \mathscr{L}(\bar{X}, Z_0)$ is invertible. With the help of the closed graph theorem we deduce from d) that the restriction $\bar{P}_n = P_n \mid \bar{X}$ is a continuous projection on the space $\bar{X}$.

It is easily seen from Theorem 1.3 that

$$Z_1 \subset \mathfrak{R}(A_0; \{\bar{P}_n, Q_n\}).$$

Theorem 1.1 gives $\mathfrak{R}(\bar{A}; \{\bar{P}_n, Q_n\}) = \mathfrak{R}(A_0; \{\bar{P}_n, Q_n\})$ and this implies the first assertion of the theorem. Finally, from Theorem 1.2 and Corollary 1.2 we get the estimates (2)

---

[1]) In the concrete situations considered in the following sections of this chapter the projections P_n will always be finite-dimensional.

and (3) with

$$y_0 = (Pc_- + Qc_+^{-1})\, Bx = c_+^{-1}\, d^{-1}(y - T'x)$$

(see equality (5)). ∎

Theorem 2.4. *Let the conditions* a)–e), g), h), *and* i) *be satisfied. Let A be an operator of the form* (1) *with $T \in \mathcal{K}(X, Z_0)$ and the coefficients being of the form* (4). *Suppose $c = c_+c_- \in \mathfrak{M}_l$ and $d = d_-d_+ \in \mathfrak{M}_r$. Assume* dim Ker $A = 0$.

Then all conclusions of Theorem 2.1 are valid with $y_0 = (Pc_- + Q\, d_+)\, Bx$.

Proof. We use the notation introduced in the preceding proof. Under the hypotheses of Theorem 2.3 we have $\bar{A} = A_0 + T'$ with

$$A_0 = (c_+P + d_-Q)\,(Pc_- + Q\, d_+)\, \bar{B}, \tag{7}$$

$$T' = (c_+Qc_-r_-P + d_-P\, d_+r_+\, Q)\, E + TE. \tag{8}$$

Now put

$$y_0 = (Pc_- + Qd_+)\, \bar{B}x = (c_+^{-1}P + d_-^{-1}Q)\,(y - T'x)$$

and repeat the arguments of the proof of Theorem 2.3. ∎

Remark 4. If $d^{-1}\, c = a_-a_+ \in \mathfrak{M}_r$ (resp. $c\, d^{-1} = a_-a_+ \in \mathfrak{M}_r$), then the operator $cP + dQ$ (resp. $Pc + Qd$) is invertible on the space X.

This is immediate from the representations

$$cP + dQ = d(a_-P + Q)\,(a_+P + Q),$$

$$Pc + Q\,d = (Pa_- + Q)\,(Pa_+ + Q)\,d.$$

Remark 5. Under the hypotheses of Theorem 2.3 (or Theorem 2.4) the conclusions of Theorem 1.4 are valid with $Z = Z_0$. In other words, the projection method $\{P_n, Q_n\}$ is then stable.

§ 3. Polynomial approximation methods for the solution of singular integral equations

In this section we consider the reduction method, the collocation method, the method of mechanical quadratures, and the method of least squares for the approximate solution of singular integral equations over the unit circle Γ (Sections 3.1–3.6) or over the interval $[-1, 1]$ (Section 3.7) of the form

$$(A\varphi)\,(t) := c(t)\,\varphi(t) + \frac{d(t)}{\pi i} \int_\Gamma \frac{\varphi(\tau)}{\tau - t}\, d\tau + (T\varphi)\,(t) = f(t) \qquad (t \in \Gamma) \tag{1}$$

in the spaces $\mathbf{L}_p(\Gamma)$ $(1 < p < \infty)$ and $\mathbf{H}^\mu(\Gamma)$ $(0 < \mu < 1)$ (resp. $\mathbf{L}_2([-1, 1], \varrho)$). We suppose that $c, d \in \mathbf{H}^\mu(\Gamma)$ and that T is a linear compact operator on these spaces. Henceforth let

$$a(t) = c(t) + d(t), \qquad b(t) = c(t) - d(t). \tag{2}$$

Furthermore, we shall use the following notations: $C^{l,\lambda}(\Gamma)$ ($l \geqq 0$ an integer, $0 < \lambda \leqq 1$) is the function algebra introduced in Section 6.1 of Chapter II, in case $\lambda = 0$ let $\mathbf{C}^m(\Gamma) := \mathbf{C}^{m,0}(\Gamma)$ and in case $l = 0$ write $\mathbf{C}(\Gamma) := \mathbf{C}^0(\Gamma)$, $\mathbf{H}^\lambda(\Gamma) := \mathbf{C}^{0,\lambda}(\Gamma)$. By $\mathbf{W}_p^{(l)}(\Gamma)$ $(1 < p < \infty)$ we denote the Sobolev space of 2π-periodic functions on the real line which together with their derivates up to the order $l - 1$ are absolutely continuous and whose l-th derivative is absolutely integrable in the p-th power (see also Section 6.1, Chapter II). Let $\mathbf{W}_p^{(l,\lambda)}(\Gamma)$

$(0 < \lambda \leqq 1)$ denote the set of all functions $f \in \mathbf{W}_p^{(l)}(\Gamma)$ whose l-th derivative satisfies a Hölder condition with respect to the $\mathbf{L}_p$-norm, that is,

$$\|f^{(l)}(t+h) - f^{(l)}(t)\|_{\mathbf{L}_p(\Gamma)} \leqq c\,|h|^\lambda \qquad (c = \text{const}).$$

If $\lambda = 0$, put $\mathbf{W}_p^{(l,0)}(\Gamma) = \mathbf{W}_p^{(l)}(\Gamma)$. Then, obviously, $\mathbf{C}^{l,\lambda}(\Gamma) \subset \mathbf{W}_p^{(l,\lambda)}(\Gamma)$. Finally, let $\mathscr{R}(\Gamma)$ be the Banach space of all bounded and Riemann integrable functions on Γ endowed with the norm $\|f\|_\infty := \sup\limits_{t\in\Gamma} |f(t)|$.

3.1. The reduction method

Throughout this subsection we denote by f_j, a_j, b_j, and c_{jk} $(j, k = 0, \pm 1, \ldots)$ the Fourier coefficients of the functions f $(\in \mathbf{L}_p(\Gamma))$, a, b, and Tt^k, respectively.

Theorem 3.1. *Let a and b be functions in $\mathbf{H}^\mu(\Gamma)$ $(0 < \mu < 1)$ which satisfy the following conditions:*

1. $a(t) \neq 0$, $b(t) \neq 0$ $(|t| = 1)$.
2. $\operatorname{ind} a = \operatorname{ind} b = 0$.

If T is a compact operator on the space $\mathbf{L}_p(\Gamma)$ $(1 < p < \infty)$ and if $\dim \operatorname{Ker} A = 0$, *then for each function $f \in \mathbf{L}_p(\Gamma)$ and for all sufficiently large n the system*

$$\sum_{k=0}^{n} a_{j-k}\xi_k + \sum_{k=-n}^{-1} b_{j-k}\xi_k + \sum_{k=-n}^{n} c_{jk}\xi_k = f_j \quad (j = 0, \pm 1, \ldots, \pm n) \tag{3}$$

has exactly one solution $(\xi_k^{(n)})_{k=-n}^n$ and, as $n \to \infty$, the functions

$$\varphi_n(t) = \sum_{k=-n}^{n} \xi_k^{(n)} t^k \tag{4}$$

converge in the norm of the space $\mathbf{L}_p(\Gamma)$ to the solution $\varphi \in \mathbf{L}_p(\Gamma)$ of the equation (1).

If, moreover, $a, b \in \mathbf{C}^{r,\mu}(\Gamma)$, $f \in \mathbf{W}_p^{(r,\nu)}(\Gamma)$ $(r \geqq 0,\ 0 \leqq \nu < 1)$, and T maps $\mathbf{L}_p(\Gamma)$ into $\mathbf{W}_p^{(r,\nu)}(\Gamma)$, then the estimate

$$\|\varphi - \varphi_n\|_{\mathbf{L}_p} = O(n^{-r-\lambda}), \qquad \lambda = \min(\mu, \nu) \tag{5}$$

holds.

Proof. Put $P = \frac{1}{2}(I + S)$ and $Q = \frac{1}{2}(I - S)$, where $S = S_\Gamma$ is the Cauchy singular integral operator. Then the operator A can be written in the form $A = aP + bQ + T$, where a and b denote the operators of multiplication by the functions (2).

Let P_n $(n = 1, 2, \ldots)$ denote the projection on the space $\mathbf{L}_p(\Gamma)$ which sends a function $\varphi(t) = \sum\limits_{j=-\infty}^{\infty} \xi_j t^j \in \mathbf{L}_p(\Gamma)$ to

$$(P_n\varphi)(t) = \sum_{j=-n}^{n} \xi_j t^j \qquad (t \in \Gamma). \tag{6}$$

It is well known (see, for instance, Zygmund [3]) that P_n converges strongly to the identity operator on $\mathbf{L}_p(\Gamma)$ $(1 < p < \infty)$. Finally, to come into the setting of the preceding two sections, put $Q_n = P_n$.

Now apply Theorem 2.2 with $X = Z = Z_0 = Z_1 = \mathbf{L}_p(\Gamma)$ and $\mathfrak{M} = \mathbf{H}^\mu(\Gamma)$ (identify a function $a \in \mathbf{H}^\mu(\Gamma)$ with the corresponding operator of multiplication by a). It is easy to see that the system (3) is nothing else than the projection equation $P_nAP_nf = P_nf$. The conditions a)–e) and condition g) of Section 2.1 can be readily checked in a straightforward way. Furthermore, the hypotheses 1 and 2 of Theorem 3.1 imply that $a \in \mathfrak{M}_l$ and $b \in \mathfrak{M}_r$ and also that A is invertible (see Chapter III). Thus, what results from Theorem 2.2 is the first assertion of Theorem 3.1 together with the estimate

$$\|\varphi - \varphi_n\|_{\mathbf{L}_p} \leqq \gamma \|y_0 - P_n y_0\|_{\mathbf{L}_p} \leqq \gamma' E_n(\mathbf{L}_p, y_0), \tag{7}$$

where $y_0 = (Pa_- + Qb_+)\varphi$ with $a = a_+a_-$, $b = b_-b_+$ (recall that a and b are in $\mathbf{H}^\mu(\Gamma)$).

To get the second part of Theorem 3.1, notice first that the stronger conditions imposed upon a, b, f, T imply that $y_0 \in \mathbf{W}_p^{(r,\lambda)}(\Gamma)$. Indeed, if g is any function in $\mathbf{W}_p^{(0,\nu)}(\Gamma)$, then the Cauchy singular integral Sg is also in $\mathbf{W}_p^{(0,\nu)}(\Gamma)$ (see Butzer and Nessel [1], Proposition 8.2.9, p. 322). So Lemma 6.1 of Chapter II can be applied to see that $Sg \in \mathbf{W}_p^{(r,\nu)}(\Gamma)$ whenever $g \in \mathbf{W}_p^{(r,\nu)}(\Gamma)$. Since both the functions $a_\pm$, $b_\pm$ and their inverses belong to the algebra $\mathbf{C}^{r,\mu}(\Gamma)$ (Corollary 5.3, Chapter III) and since $A\varphi = f - T\varphi \in \mathbf{W}_p^{(r,\nu)}(\Gamma)$, we deduce with the help of Theorem 6.1 of Chapter III that $\varphi \in \mathbf{W}_p^{(r,\lambda)}(\Gamma)$ and thus $y_0 \in \mathbf{W}_p^{(r,\lambda)}(\Gamma)$. It follows that

$$E_n(\mathbf{L}_p, y_0) = O(n^{-r-\lambda}) \tag{7'}$$

(see Butzer and Nessel [1], Corollary 2.2.4, p. 99) and this together with (7) gives the assertion. ∎

Remark 1. In case $T = 0$, the equality dim Ker $A = 0$ is redundant, since it is then a consequence of the conditions 1 and 2.

Remark 2. For the singular integral equation

$$c(t)\,\varphi(t) + \frac{1}{\pi i}\int_\Gamma \frac{d(\tau)\,\varphi(\tau)}{\tau - t}\,d\tau = f(t)$$

the system (3) has to be replaced by the system

$$\sum_{k=-n}^{n} a_{j-k}\xi_k = f_j \qquad (j = 0, 1, \ldots, n), \qquad \sum_{k=-n}^{n} b_{j-k}\xi_k = f_j \qquad (j = -1, \ldots, -n).$$

Remark 3. The first part of Theorem 3.1 is due to Gohberg and Feldman ([1], Theorem 4.1, Chapter VI). They also proved that the conditions 1 and 2 are necessary for the first part of Theorem 3.1 to be valid. The error estimate (5) was probably first given by Prössdorf and Silbermann [7, 10].

The first result about the reduction method goes back to Baxter [1], who proved that under certain conditions the reduction method is applicable to discrete Wiener-Hopf equations. In this context the reduction method is sometimes referred to as the *finite section method*.

Here is the $\mathbf{H}^\mu(\Gamma)$ analogue of Theorem 3.1:

Theorem 3.2. *Let $a, b \in \mathbf{H}^\mu(\Gamma)$ $(0 < \mu < 1)$ satisfy the conditions 1 and 2 of Theorem 3.1. If T is a compact operator from $\mathbf{H}^\nu(\Gamma)$ into $\mathbf{H}^\mu(\Gamma)$ $(0 < \nu < \mu < 1)$ and if* dim Ker $A = 0$, *then for all sufficiently large n and for each $f \in \mathbf{H}^\mu(\Gamma)$ the system* (3) *has precisely one solution $(\xi_k^{(n)})_{k=-n}^n$ and, as $n \to \infty$, the functions* (4) *converge in the norm of $\mathbf{H}^\nu(\Gamma)$ to the solution $\varphi \in \mathbf{H}^\nu(\Gamma)$ of equation* (1). *Furthermore, one has the estimate*

$$\|\varphi - \varphi_n\|_{\mathbf{H}^\nu} = O\left(\frac{\ln n}{n^{\mu-\nu}}\right). \tag{8}$$

If, in addition, $a, b \in \mathbf{C}^{r,\mu}(\Gamma)$, $f \in \mathbf{C}^{r,\mu}(\Gamma)$, and T maps $\mathbf{H}^\nu(\Gamma)$ into $\mathbf{C}^{r,\mu}(\Gamma)$, then

$$\|\varphi - \varphi_n\|_{\mathbf{H}^\nu} = O\left(\frac{\ln n}{n^{r+\mu-\nu}}\right). \tag{9}$$

Proof. Theorem 3.2 follows from Theorem 2.2 applied with $X = Z = \mathbf{H}^\nu(\Gamma)$, $Z_0 = Z_1 = \mathbf{H}^\mu(\Gamma)$, $\mathfrak{M} = \mathbf{H}^\mu(\Gamma)$, $Q_n = P_n$. The conditions a)–d) and g) can be easily checked in a straightforward manner. To see that condition e) is satisfied, note that

$$\|f - P_n f\|_{\mathbf{H}^\nu} \leqq \gamma_1 \frac{\ln n}{n^{\mu-\nu}} \|f\|_{\mathbf{H}^\mu} \tag{10}$$

for every $f \in \mathbf{H}^\mu(\Gamma)$ (see Prössdorf [19] or Prössdorf, Silbermann [11], Chapter 2). So Theorem 2.2 together with the estimate

$$\|\varphi - \varphi_n\|_{\mathbf{H}^\nu} \leqq \gamma \|y_0 - P_n y_0\|_{\mathbf{H}^\nu}, \tag{11}$$

where $y_0 \in \mathbf{H}^\mu(\Gamma)$ has the same meaning as in the proof of Theorem 3.1, gives the first portion of Theorem 3.2.

Now suppose $a, b \in \mathbf{C}^{r,\mu}(\Gamma)$, $f \in \mathbf{C}^{r,\mu}(\Gamma)$, and T maps $\mathbf{H}^\nu(\Gamma)$ into $\mathbf{C}^{r,\mu}(\Gamma)$. The same reasoning as in the proof of the preceding theorem shows that then $y_0 \in \mathbf{C}^{r,\mu}(\Gamma)$. Thus (9) follows from (11) and the also well-known estimate (see Prössdorf, Silbermann [11], Chapter 2)

$$\|y_0 - P_n y_0\|_{\mathbf{H}^\nu} \leqq \gamma_2 \frac{\ln n}{n^{r+\mu-\nu}} \|y_0^{(r)}\|_{\mathbf{H}^\mu}. \blacksquare$$

Remark. The first part of Theorem 3.2 was independently of each other and by different methods proved by Zolotarevski [2–3] and Prössdorf [16]. Zolotarevski gave error estimates which were somewhat rougher than (8), (9), ($\ln^4 n$ in place of $\ln n$). The estimates (8) and (9) are due to Prössdorf and Silbermann [7, 10].

3.2. The collocation method

We now consider the collocation method for the approximate solution of equation (1) with the following collocation points on the unit circle:

$$t_j = t_j^{(n)} = e^{2\pi i j/(2n+1)} \qquad (j = 0, \pm 1, \ldots, \pm n). \tag{12}$$

This method consists in finding an approximate solution of the equation (1) in the form (4), where the coefficients $\xi_k = \xi_k^{(n)}$ are to be determined so that the equation $A\varphi_n = f$ be satisfied at the collocation points (12), that is, so that

$$(A\varphi_n)(t_j) = f(t_j) \qquad (j = 0, \pm 1, \ldots, \pm n). \tag{13}$$

It is easily seen that the equation (13) written down in full takes the form

$$a(t_j) \sum_{k=0}^{n} t_j^k \xi_k + b(t_j) \sum_{k=-n}^{-1} t_j^k \xi_k + \sum_{k=-n}^{n} d_{jk}\xi_k = f(t_j) \quad (j = 0, \pm 1, \ldots, \pm n), \tag{14}$$

where $d_{jk} = (Tt^k)(t_j)$ $(j, k = 0, \pm 1, \ldots, \pm n)$.

Theorem 3.3 *Let a and b be functions in $\mathbf{H}^\mu(\Gamma)$ $(0 < \mu < 1)$ which satisfy the following conditions:*

1. $a(t) \neq 0$, $b(t) \neq 0$ $(|t| = 1)$.
2. $\operatorname{ind} a = \operatorname{ind} b$.

If T is a compact operator from $\mathbf{L}_p(\Gamma)$ $(1/\mu < p < \infty)$ into the space $\mathbf{C}(\Gamma)$ and if $\dim \operatorname{Ker} A = 0$, *then for each function $f \in \mathcal{R}(\Gamma)$ and for all sufficiently large n the system* (14) *has exactly one solution $(\xi_k^{(n)})_{k=-n}^{n}$ and, as $n \to \infty$, the functions* (4) *converge in the norm of $\mathbf{L}_p(\Gamma)$ to the solution $\varphi \in \mathbf{L}_p(\Gamma)$ of the equation* (1).

If, in addition, $a, b \in \mathbf{C}^{r,\mu}(\Gamma), f \in \mathbf{C}^{r,\nu}(\Gamma)$, *and* T *maps* $\mathbf{L}_p(\Gamma)$ *into* $\mathbf{C}^{r,\nu}(\Gamma)$ $(r \geqq 0, 0 \leqq \nu < 1)$, *then the estimate* (5) *holds.*

Proof. Apply Theorem 2.1 with $X = Z = \mathbf{L}_p(\Gamma)$, $Z_0 = \mathbf{C}(\Gamma)$, $Z_1 = \mathscr{R}(\Gamma)$, $\mathfrak{M} = \mathbf{H}^\mu(\Gamma)$, P_n defined as in Section 3.1 (see (6)), and Q_n defined for $f \in \mathscr{R}(\Gamma)$ by

$$(Q_n f)(t) = \sum_{k=-n}^{n} a_k t^k, \qquad a_k = \frac{1}{2n+1} \sum_{j=-n}^{n} f(t_j)\, t_j^{-k}. \tag{15}$$

Thus, $Q_n f$ is the interpolation polynomial for the function $f \in \mathscr{R}(\Gamma)$ with respect to the knots (12). It is easy to see that the system (14) is equivalent to the projection equation $Q_n A P_n \varphi = Q_n f$.

Again the conditions a), c), d), and f) of Section 2.1 can be easily verified in a direct way. The validity of condition b) is a consequence of a well-known theorem on the compactness of integral operators with kernels of potential type (see e.g. Kantorovich and Akilov [1], Theorem 7 (2.X)). Finally, condition e) is a well-known result on the $\mathbf{L}_p$ convergence of interpolation polynomials for functions from $\mathscr{R}(\Gamma)$ (see Zygmund [3], Chapter X.7). Thus, all hypotheses of Theorem 2.1 are satisfied, and we obtain the first assertion of Theorem 3.3 and the estimate

$$\|\varphi - \varphi_n\|_{\mathbf{L}_p} \leqq \gamma E_n(\mathbf{C}, y_0) \tag{16}$$

with $y_0 \in \mathbf{C}(\Gamma)$.

Under the stronger conditions imposed upon a, b, f, and T, a similar reasoning as in the proof of Theorem 3.1 applies: taking into account (11) and (12) we see that $y_0 \in \mathbf{C}^{r,\lambda}(\Gamma)$ with $\lambda = \min(\mu, \nu)$. Hence

$$E_n(\mathbf{C}, y_0) = O(n^{-r-\lambda})$$

(see Butzer and Nessel [1], Corollary 2.2.4, p. 99) and this combined with (16) gives the estimate (5). ∎

The following theorem is the analogue of Theorem 3.2.

Theorem 3.4. *Let* $a, b \in \mathbf{H}^\mu(\Gamma)$ $(0 < \mu < 1)$ *satisfy the conditions 1 and 2 of Theorem 3.3. Then all conclusions of Theorem 3.2 are valid for the system* (14) (*in place of* (3)).

Proof. This follows from Theorem 2.1 applied with $X = Z = \mathbf{H}^\nu(\Gamma)$, $Z_0 = Z_1 = \mathbf{H}^\mu(\Gamma)$, $\mathfrak{M} = \mathbf{H}^\mu(\Gamma)$, and P_n and Q_n defined by (6) and (15), respectively. The reasoning is essentially the same as in the proofs of the Theorems 3.2 and 3.3. The only deviation is that one has now to use the estimate

$$\|\varphi - Q_n\varphi\|_{\mathbf{H}^\nu} = O\left(\frac{\ln n}{n^{r+\mu-\nu}}\right),$$

which holds for any $\varphi \in \mathbf{C}^{r,\mu}(\Gamma)$ (see Prössdorf and Silbermann [11], Chapter 2). ∎

Remark 1. The conditions 1 and 2 of Theorem 3.3 are also necessary for the convergence of the collocation method in the spaces $\mathbf{H}^\nu(\Gamma)$ and $\mathbf{L}_p(\Gamma)$ (see Prössdorf [22] and Prössdorf, Schmidt [1]).

Remark 2. In the form presented here the Theorems 3.3 and 3.4 were stated by Prössdorf and Silbermann [7, 10]. Independently and about at the same time, the first part of Theorem 3.3 was established with the help of other methods by Gabdulhaev and Dushkov [1]. Under the hypotheses of Theorem 3.4, the convergence of the collocation method in a somewhat weaker norm was proved by Ivanov [1] and, subsequently, in the norm of $\mathbf{H}^\nu(\Gamma)$ $(0 < \nu < \mu)$ by Gabdulhaev [1].

3.3. The method of mechanical quadratures

We now consider the singular integral equation (1) with a compact integral operator T:

$$(A\varphi)(t) = c(t)\,\varphi(t) + \frac{d(t)}{\pi i} \int_\Gamma \frac{\varphi(\tau)}{\tau - t}\,d\tau + \int_\Gamma K(t,\tau)\,\varphi(\tau)\,d\tau = f(t), \tag{17}$$

where $K(t,\tau) \in \mathbf{C}(\Gamma\times\Gamma)$. The method of mechanical quadratures again consists in finding an approximate solution of equation (17) in the form (4), but the coefficients $\xi_k = \xi_k^{(n)}$ $(k = 0, \pm 1, , \dots, \pm n)$ are now determined through the system

$$a(t_j) \sum_{k=0}^{n} t_j^k \xi_k + b(t_j) \sum_{k=-n}^{-1} t_j^k \xi_k + \frac{2\pi i}{2n+1} \sum_{k=-n}^{n} \gamma_{jk} \xi_k = f(t_j) \qquad (j = 0, \pm 1, \dots, \pm n), \tag{18}$$

where

$$\gamma_{jk} = \sum_{r=-n}^{n} K(t_j, t_r)\, t_r^{k+1}, \qquad t_j = e^{2\pi i j/(2n+1)}.$$

The equations (18) result from the equations (14) be replacing the integrals

$$d_{jk} = \int_\Gamma K(t_j, \tau)\,\tau^k\,d\tau \qquad (j, k = 0, \pm 1, \dots, \pm n)$$

by their integral sums with respect to the knots $\{t_j\}$.

Theorem 3.5. *Let the functions* (2) *belong to the class* $\mathbf{H}^\mu(\Gamma)$ $(0 < \mu < 1)$ *and let them satisfy the conditions 1 and 2 of Theorem 3.3. Furthermore, suppose* $K(t,\tau) \in \mathbf{C}(\Gamma\times\Gamma)$ *and* dim Ker $A = 0$.

Then for each function $f \in \mathscr{R}(\Gamma)$ *and for all sufficiently large* n *the system* (18) *has precisely one solution* $(\xi_k^{(n)})_{k=-n}^{n}$ *and, as* $n \to \infty$, *the functions* (4) *converge in the norm of* $\mathbf{L}_p(\Gamma)$ $(1/\mu < p < \infty)$ *to the solution* $\varphi \in \mathbf{L}_p(\Gamma)$ *of the equation* (17).

Moreover, if $a, b, f \in \mathbf{C}^{r,\mu}(\Gamma)$ *and* $K(t,\tau) \in \mathbf{C}^{r,\mu}(\Gamma\times\Gamma)$, *then* $\|\varphi - \varphi_n\|_{\mathbf{L}_p} = O(n^{-r-\mu})$.

Proof. It is not difficult to check that the system (18) is equivalent to the equation $A_n\varphi_n + Q_n G_n \varphi_n = Q_n f$, where

$$A_n\varphi_n = Q_n(aP + bQ)\,\varphi_n + Q_{nt} \int_\Gamma K(t,\tau)\,\varphi_n(\tau)\,d\tau,$$

$$G_n\varphi_n = -\int_\Gamma K(t,\tau)\,\varphi_n(\tau)\,d\tau + \int_\Gamma \{\tau^{-1} Q_{n\tau}[\tau K(t,\tau)]\}\,\varphi_n(\tau)\,d\tau.$$

Here Q_{nt} denotes the operator Q_n defined by (15) and applied with respect to the variable t. Hölder's inequality gives

$$\|G_n\varphi_n\|_{\mathbf{C}} \leqq \varepsilon_n \|\varphi_n\|_{\mathbf{L}_p}, \tag{19}$$

$$\varepsilon_n := c \max_{t\in\Gamma} \Big\{ \int_\Gamma |K(t,\tau)\,\tau - Q_{n\tau}[K(t,\tau)\,\tau]|^q\,|d\tau| \Big\}^{1/q}.$$

It is well known that $\varepsilon_n = o(1)$ if $K(t,\tau) \in \mathbf{C}(\Gamma\times\Gamma)$ and that $\varepsilon_n = o(n^{-r-\mu})$ if $K(t,\tau) \in \mathbf{C}^{r,\mu}(\Gamma\times\Gamma)$ (see, for instance, Prössdorf and Silbermann [11], Chapter 2, 3.1–3.2).

We also know that the solution φ of equation (17) belongs to $\mathbf{L}_p(\Gamma)$ for every p such that $1 < p < \infty$ (see Chapter III). Thus, it is enough to consider the case where p is sufficiently large. By virtue of Theorem 3.3, the equation $A_n\tilde\varphi_n = Q_n f$ has exactly one solution $\tilde\varphi_n$ whenever n is large enough, and $\tilde\varphi_n$ converges to φ in the $\mathbf{L}_p$-norm. Now Theorem 3.3, the estimates (5) and (19), and Theorem 1.4 (with $\Gamma_n = G_n$, $\delta_n = 0$) give all assertions of Theorem 3.5. ∎

Theorem 3.6. *Let the hypotheses of Theorem 3.5 be satisfied, suppose* $K(t, \tau) \in \mathbf{H}^{\mu}(\Gamma \times \Gamma)$ *and* $f \in \mathbf{H}^{\mu}(\Gamma)$. *Then the functions* φ_n *given by* (4) *converge in the norm of the space* $\mathbf{H}^{\nu}(\Gamma)$ $(0 < \nu < \mu)$ *to the solution* $\varphi \in \mathbf{H}^{\nu}(\Gamma)$ *of the equation* (17) *and one has* $\|\varphi - \varphi_n\|_{\mathbf{H}^{\nu}} = O(n^{\nu-\mu} \ln n)$.

If, moreover, $a, b, f \in \mathbf{C}^{r,\mu}(\Gamma)$ *and* $K(t,\tau) \in \mathbf{C}^{r,\mu}(\Gamma \times \Gamma)$, *then* $\|\varphi - \varphi_n\|_{\mathbf{H}^{\nu}} = O(n^{\nu-r-\mu} \ln n.)$

Using Theorem 3.4 and the Remark 1 to Theorem 1.4 this can be proved analogously as Theorem 3.5. It only remains to estimate the norm $\|A_n^{-1} Q_n G_n\|_{\mathcal{L}(\mathbf{H}^{\nu})}$. As in the proof of Theorem 2.1 we have $A = A_0 + T_0$ with a compact operator $T_0 : \mathbf{H}^{\nu} \to \mathbf{H}^{\nu}$ and $A_0 = bc_+(Pc_- + Qc_+^{-1})$. The proof of Theorem 1.1 shows that it suffices to consider A_0 instead of A. From (19) we deduce that

$$A_{0n}^{-1} Q_n = (Pc_-^{-1} + Qc_+)\, Q_n c_+^{-1} b^{-1}, \qquad \|A_{0n}^{-1} Q_n G_n \varphi_n\|_{\mathbf{H}^{\nu}}$$
$$\leq c_1 n^{\nu} \, \|Q_n c_+^{-1} b^{-1} G_n \varphi_n\|_{\mathbf{C}} \leq n^{\nu} \ln n \, \varepsilon_n \, \|\varphi_n\|_{\mathbf{L}_p} \leq c_2 n^{\nu} \ln n \, \varepsilon_n \, \|\varphi_n\|_{\mathbf{H}^{\nu}}$$

and this implies that $\|A_n^{-1} Q_n G_n\|_{\mathbf{H}^{\nu}} = O(n^{\nu-r-\mu} \ln n)$. ∎

Remark. Theorem 3.5 was probably first established by PRÖSSDORF and SILBERMANN [11]. GABDULHAEV and DUSHKOV [1] had somewhat earlier already stated the first part of this theorem. Theorem 3.6 was obtained by other methods and with slightly worse error estimates by GABDULHAEV [1], ZOLOTAREVSKI [4], and PRÖSSDORF, SILBERMANN [11]. The error estimate given here was in a somewhat different way proved by JUNGHANNS in his diploma paper (Konvergenzmannigfaltigkeit und reguläre Konvergenz von Folgen linearer Operatoren, Karl-Marx-Stadt, 1979) for the first time. See also PRÖSSDORF and SILBERMANN [12].

3.4. The method of least squares

This method can be described as follows. Consider the equation (1) in the Hilbert space $\mathbf{L}_2(\Gamma)$ and seek the approximate solution in the form (4) with the coefficients $\xi_k = \xi_k^{(n)}$ $(k = 0, \pm 1, \ldots, \pm n)$ determined so that the functional

$$\|f - A\varphi_n\|_{\mathbf{L}_2(\Gamma)} \tag{20}$$

become minimal. It is well known that the functional (20) attains it minimum at that function (4) whose coefficients satisfy the system

$$\sum_{k=-n}^{n} \xi_k (At^k, At^j) = (f, At^j) \qquad (j = 0, \pm 1, \ldots, \pm n), \tag{21}$$

where

$$(f, g) = \int_0^{2\pi} f(e^{is})\, \overline{g(e^{is})}\, ds$$

is the scalar product of the functions $f, g \in \mathbf{L}_2(\Gamma)$ (see MIKHLIN [24], Chapter XII).

Theorem 3.7. *Let a and b be functions on* Γ *which satisfy the conditions 1 and 2 of Theorem 3.3.*

If T *is a compact operator on* $\mathbf{L}_2(\Gamma)$ *and if* $\dim \operatorname{Ker} A = 0$, *then for each function* $f \in \mathbf{L}_2(\Gamma)$ *and for all sufficiently large* n *the system* (21) *has exactly one solution* $(\xi_k^{(n)})_{k=-n}^{n}$ *and, as* $n \to \infty$, *the functions* (4) *converge in the norm of the space* $\mathbf{L}_2(\Gamma)$ *to the solution* $\varphi \in \mathbf{L}_2(\Gamma)$ *of the equation* (1).

If, moreover, $a, b \in \mathbf{C}^{r,\lambda}(\Gamma)$ *and* $f, T\varphi \in \mathbf{W}_2^{(r,\lambda)}(\Gamma)$ $(r \geq 0, 0 \leq \lambda \leq 1)$, *then*

$$\|\varphi - \varphi_n\|_{\mathbf{L}_2} = O(n^{-r-\lambda}). \tag{22}$$

Proof. Let P_n be the projection on $\mathbf{L}_2(\Gamma)$ defined by (6) and put $Q_n = AP_nA^{-1}$. Then the method of least squares is equivalent to the projection method $\{P_n, Q_n\}$ (see MIKHLIN

[24], Chapter XII). Theorem 1.3 applied with $B = A$, $C = I$, and $Z = \mathbf{L}_2(\Gamma)$ shows that $\mathbf{L}_2(\Gamma) = \mathfrak{R}(A; \{P_n, Q_n\})$ (or, equivalently, $A \in \Pi\{P_n, Q_n\}$) and this is the first assertion of Theorem 3.7.

From Theorem 1.2 (see the estimate (1.6)) we get

$$\|\varphi - \varphi_n\|_{\mathbf{L}_2} \leqq c_2 \|A\varphi - Q_n A\varphi\|_{\mathbf{L}_2} \leqq c_2 \|A\| \, \|\varphi - P_n\varphi\|_{\mathbf{L}_2}$$

and thus

$$\|\varphi - \varphi_n\|_{\mathbf{L}_2} \leqq c_3 E_n(\mathbf{L}_2, \varphi). \tag{23}$$

It was shown in the proof of Theorem 3.1 that under the additional hypotheses of Theorem 3.7 φ belongs to $\mathbf{W}_2^{(r,\lambda)}(\Gamma)$, and hence $E_n(\mathbf{L}_2, \varphi) = O(n^{-r-\lambda})$. This together with (23) gives the estimate (22). ∎

Remark. The convergence of the method of least squares for singular integral equations in the space $\mathbf{L}_2$ was proved by Mikhlin [26] for the first time.

3.5. The case of a non-vanishing index

In the hitherto stated theorems of this section the invertibility of the singular integral operator defined by (1) has always been a part of the hypothesis. This subsection is concerned with the case where the functions (2) do not vanish on Γ but are allowed to possess indices

$$\varkappa_1 = \operatorname{ind} a, \qquad \varkappa_2 = \operatorname{ind} b$$

such that $\varkappa = \varkappa_1 - \varkappa_2 \neq 0$. The following argument shows that this situation can be easily reduced to those situations which were considered in the Sections 3.1 to 3.4.

The functions $t^{-\varkappa_1} a(t)$ and $t^{-\varkappa_2} b(t)$ satisfy the conditions of Theorem 3.1, while the functions $t^{-\varkappa} a(t)$ and $b(t)$ fulfil the conditions of Theorem 3.3. Put

$$C = t^{-\varkappa_1} aP + t^{-\varkappa_2} bQ + t^{-\varkappa_2} T(t^{-\varkappa} P + Q),$$

$$D = t^{\varkappa_2} C = t^{-\varkappa} aP + bQ + T(t^{-\varkappa} P + Q).$$

In view of the Theorems 3.1–3.7, the reduction method is applicable to the operator C, while the collocation method, the method of mechanical quadratures, and the method of least squares are applicable to the operator D. Note that D (and hence also C) is invertible if and only if

$$\dim \operatorname{Ker} D = 0 \tag{24}$$

(see Theorem 3.1, Chapter III, and Theorem 3.4, Chapter I). In what follows suppose (24) holds.

It is obvious that

$$A = aP + bQ + T = D(t^{\varkappa} P + Q), \text{ if } \varkappa \geqq 0,$$

and

$$A(t^{-\varkappa} P + Q) = D, \text{ if } \varkappa \leqq 0.$$

Therefore, the operator $(t^{-\varkappa} P + Q) D^{-1}$ is a left inverse of the operator A in case $\varkappa \geqq 0$ and a right inverse of A in case $\varkappa \leqq 0$. Thus, if $\varkappa \leqq 0$, then $\varphi^{(0)} = (t^{-\varkappa} P + Q) D^{-1} f$ is a solution of the equation (1) and if $\varkappa \geqq 0$, then $\varphi^{(0)}$ is the only possible solution of this equation.

Now let $\psi_n = \sum_{k=-n}^{n} \xi_k t^k$ denote both the approximate solution of the equation $C\psi = t^{-\varkappa_2} f(t)$ obtained by applying the reduction method or the approximate solution of the equation $D\psi = f(t)$ resulting from the collocation method (resp. the method of mechanical quadratures, or the method of least squares). From what has been said above it follows that the sequence

$$\varphi_n = (t^{-\varkappa} P + Q)\, \psi_n = \sum_{k=0}^{n} \xi_k t^{k-\varkappa} + \sum_{k=-n}^{-1} \xi_k t^k$$

converges to a solution φ of the equation (1) if only that equation is solvable. For the error $\|\varphi - \varphi_n\|$ the estimates established in the Theorems 3.1–3.7 hold.

Remark. Note that due to Theorem 4.3 of Chapter II a singular integral equation of the form (1) over an arbitrary closed Lyapunov curve Γ can be transformed into a singular integral equation of the same form over the unit circle through an appropriate mapping of the curve Γ onto the unit circle. Concerning the collocation method for the solution of these equations see also the note of Zolotarevski an Seichuk [1].

3.6. Discontinuous coefficients

We now consider singular integral equations whose coefficients are merely supposed to be continuous or even to be piecewise continuous. When studying approximation methods for these equations there arise substantial difficulties, which have been overcome only to a certain degree at the present time. This section is an attempt to survey some important results in this direction.

1. Our first concern is the collocation method. With $\mathbf{L}_2(\Gamma)$ as the underlying space, Theorem 3.3 can be generalized as follows.

Theorem 3.8. *Let a and b be functions belonging to* $\mathbf{PC}(\Gamma)$ *and suppose the singular integral operator* $A_0 := aP + bQ$ *is invertible on* $\mathbf{L}_2(\Gamma)$.

If T is a compact operator from $\mathbf{L}_2(\Gamma)$ *into the space* $\mathcal{R}(\Gamma)$ *and if* $\dim \operatorname{Ker} A = 0$, *then for each function* $f \in \mathcal{R}(\Gamma)$ *and for all sufficiently large n the system* (14) *has exactly one solution* $(\xi_k)_{k=-n}^{n}$ *and, as* $n \to \infty$, *the functions* (4) *converge in the norm of* $\mathbf{L}_2(\Gamma)$ *to the solution* $\varphi \in \mathbf{L}_2(\Gamma)$ *of equation* (1).

Proof. The proof is a natural modification of the proofs above given for the Theorems 3.3 and 2.1.

The hypotheses of the theorem imply that a/b is a 2-regular function and hence

$$a/b = r(1 + g), \tag{25}$$

where r is a rational function without poles or zeros on Γ, where $\operatorname{ind} r = 0$, and where g is a function from $\mathbf{PC}(\Gamma)$ with $\|g\|_\infty < 1$ (see Chapter IV, Theorem 4.2 and the proof of Theorem 4.1).

We first consider the invertible operator $G := I + gP \in \mathcal{L}(\mathbf{L}_2(\Gamma))$ and the corresponding approximating operators $G_n := Q_n G P_n \in \mathcal{L}(\operatorname{Im} P_n)$. Since $\sum_{k=-n}^{n} t_k^m = 0$ for every integer m such that $0 < |m| \leq 2n$, it is easy to deduce from (15) that

$$\|Q_n f\|_{\mathbf{L}_2}^2 = \sum_{k=-n}^{n} |a_k|^2 = (2n+1)^{-1} \sum_{j=-n}^{n} |f(t_j)|^2$$

and, consequently,

$$\|Q_n g f\|_{\mathbf{L}_2} \leq \|g\|_\infty \, \|Q_n f\|_{\mathbf{L}_2}. \tag{26}$$

From (26) and the equality $\|P\| = 1$ it follows that the operators G_n are invertible and that $\|G_n^{-1}\| \leq (1 - \|g\|_\infty)^{-1}$.

The inequality (26) also shows that $\{G_n\}$ is a uniformly bounded sequence of operators on $\mathbf{L}_2(\Gamma)$. For $n \geqq m$ and for arbitrary $u_m \in \operatorname{Im} P_n$ we get

$$G_n u_m = u_m + Q_n g P u_m \to G u_m \quad (n \to \infty),$$

the convergence in the norm of $\mathbf{L}_2(\Gamma)$. Thus, by the Banach-Steinhaus theorem, $G_n P_n u \to Gu$ for every $u \in \mathbf{L}_2(\Gamma)$. This and the obvious inequality

$$\|G_n^{-1} Q_n y - G^{-1} y\|_{\mathbf{L}_2} \leqq \frac{1}{1 - \|g\|_\infty} [\|Q_n y - y\|_{\mathbf{L}_2} + \|Gx - G_n P_n x\|_{\mathbf{L}_2}] + \|P_n x - x\|_{\mathbf{L}_2}, \qquad Gx = y,$$

imply that

$$G_n^{-1} Q_n y \to G^{-1} y \quad (n \to \infty) \tag{27}$$

for every $y \in \mathscr{R}(\Gamma)$.

In view of § 2, Chapter III, the function r admits the factorization $r = r_+ r_-$ and therefore A can be written as

$$A = b r_+ G(P r_- + Q r_+^{-1}) + T'$$

with $T' = b r_+ [(1 + g) Q r_- P + P r_+^{-1} Q] + T$. Since $T' \in \mathscr{K}(\mathbf{L}_2(\Gamma), \mathscr{R}(\Gamma))$, it suffices to prove the assertion of Theorem 3.8 for the operator $B := b r_+ G(P r_- + Q r_+^{-1})$ instead for A (recall Theorem 1.1).

Taking into account the equalities c), d), and f) of Section 2.1, it is easily seen that

$$B_n^{-1} Q_n = (P r_-^{-1} + Q r_+) G_n^{-1} Q_n b^{-1} r_+^{-1}. \tag{28}$$

From (27) and (28) we conclude that

$$B_n^{-1} Q_n y \to (P r_-^{-1} + Q r_+) G^{-1} b^{-1} r_+^{-1} y = B^{-1} y$$

for every $y \in \mathscr{R}(\Gamma)$. ∎

Theorem 3.9. *Let a and b be any continuous functions on Γ subject to the conditions 1 and 2 of Theorem 3.3. Furthermore, let $T \in \mathscr{K}(\mathbf{L}_p(\Gamma), \mathscr{R}(\Gamma))$, $1 < p < \infty$, and suppose* dim Ker $A = 0$. *Then the first assertion of Theorem 3.3 is true.*

Proof. For the spaces $\mathbf{L}_p(\Gamma)$, $1 < p < \infty$, the following weakened version of the inequality (26) holds:

$$\|Q_n g u_n\|_{\mathbf{L}_p} \leqq C_p \|g\|_\infty \|u_n\|_{\mathbf{L}_p} \quad (u_n \in \operatorname{Im} P_n), \tag{29}$$

where $C_p > 0$ is certain constant only depending on p (see Zygmund [3], Theorem X.7.5). The hypotheses of Theorem 3.9 ensure that the continuous function a/b can be written in the form (25) with $\|g\|_\infty < 1/(C_p m_p)$, where m_p denotes the norm of the operator P on the space $\mathbf{L}_p(\Gamma)$. From (29) we deduce that the operators G_n are invertible and that $\|G_n^{-1}\| \leqq (1 - \|g\|_\infty C_p m_p)^{-1}$. The further arguments are the same as those in the proof of Theorem 3.8. ∎

Remark. Theorem 3.9 was established by Boikov [1] for $p = 2$ and by Junghanns [1] for general p. Theorem 3.8 and the proof given here are also due to Junghanns [1]. Recently Junghanns and Silbermann [2] have found a proof based on local methods and have so been able to extend Theorem 3.8 to coefficients $a, b \in \overline{\mathbf{PC}}(\Gamma)$ ($\overline{\mathbf{PC}}(\Gamma)$ is the closure of $\mathbf{PC}(\Gamma)$ with respect to the norm $\|\cdot\|_\infty$). Moreover, they also proved the necessity of the conditions required in the Theorems 3.8 and 3.9. For certain classes of discontinuous coefficients the collocation method in the spaces $\mathbf{L}_p(\Gamma)$ $(1 < p < \infty)$ has been studied by Junghanns [1] and Prössdorf [25]. On the basis of a further development of the ideas of Junghanns and Silbermann [2], Rathsfeld [2] has recently generalized Theorem 3.8 to coefficients $a, b \in \overline{\mathbf{PC}}(\Gamma)$ and the spaces $\mathbf{L}_p(\Gamma)$.

2. An interesting difference between the collocation method and the reduction method is that in the case of discontinuous coefficients an additional condition is needed to ensure the applicability of the reduction method. This was first pointed out by VERBITSKI [1], who established the following result.

Theorem 3.10. *Let $a, b \in \mathbf{PC}(\Gamma)$ and suppose*[1])

$$\left| \arg \frac{a(t+0)}{a(t-0)} - \arg \frac{b(t+0)}{b(t-0)} \right| < \pi \tag{30}$$

for all points t on the unit circle Γ. Then for the applicability of the reduction method (3) *to the operator $aP + bQ$ in the space $\mathbf{L}_2(\Gamma)$ it is necessary and sufficient that both the operators $aP + Q$ and $P + bQ$ be invertible on $\mathbf{L}_2(\Gamma)$.*

Remark 1. It is easy to see that the condition (30) is equivalent to the following condition (compare this with Theorem 4.1, below):

$$\mu \frac{a(t+0)}{a(t-0)} + (1-\mu) \frac{b(t+0)}{b(t-0)} \notin (-\infty, 0], \quad \forall t \in \Gamma, \forall \mu \in [0, 1].$$

Remark 2. VERBITSKI's original proof [1] is hard and rests on a series of subtle function theoretical constructions. It has been considerably simplified by SILBERMANN (unpublished). Moreover, with the help of localization arguments SILBERMANN [12] generalized Theorem 3.10 to the case where the coefficients a, b are taken from $\overline{\mathbf{PC}}(\Gamma)$. The necessity of the condition (30) was proved by VERBITSKI [1] in a special case and by RATHSFELD [1] in the general case. RATHSFELD [1] also extended Theorem 3.10 to the space $\mathbf{Fl}_p$ $(1 < p < \infty)$ of all functions on the unit circle whose sequence of Fourier coefficients is in $\mathbf{l}_p$.

3. Comments. The current state of the theory of approximation methods for singular integral equations with discontinuous coefficients has been essentially inspired by the extensive preparatory work done for Toeplitz operators, that is for singular integral operators over the unit circle which are of the form $aP + Q$. In the Toeplitz case, the situation at the beginning of the eighties was as follows: it was relatively clear what happens (a) for symbols with an arbitrary number of discontinuities when the operator is considered on a Hilbert space and (b) for symbols with only one discontinuity when the operator is considered on certain Banach spaces (say, $\mathbf{l}_p$). In this connection recall that, in contrast to the case of continuous symbols, the properties of Toeplitz operators with discontinuous symbols (resp. singular integral operators with discontinuous coefficients) on the spaces $\mathbf{l}_p$ (resp. $\mathbf{L}_p(\Gamma)$) depend on the value of p essentially (see, for instance, § 4 of Chapter IV).

A first step forward to glueing the results one had for a single discontinuity to get results for symbols with an arbitrary number of discontinuities was made by BÖTTCHER and SILBERMANN [1]. They developed a separation technique to prove that the reduction method is applicable to the Toeplitz operator $T(a_1 \dots a_R)$ on $\mathbf{l}_p$ (equivalently, to the singular integral operator $a_1 \dots a_R P + Q$, considered on the space $\mathbf{Fl}_p$) if it is applicable to each of the Toeplitz operators $T(a_1), \dots, T(a_R)$ (equivalently, the singular integral operators $a_1 P + Q, \dots, a_R P + Q$) and if the singular supports of $a_1, \dots, a_R$ (cf. Section 1.2, Chapter IV) are pairwise disjoint.

The deciding step forwards, however, was made by SILBERMANN [11], who succeeded in establishing a localization technique for the treatment of the reduction method for Toeplitz operators.

The basic idea of localization techniques is to construct an algebra which plays the same role in the theory of an approximation method as the Calkin algebra does in the Fredholm theory. Then the problem of the applicability of a given approximation method can be translated into a problem concerned with the invertibility of certain elements in that algebra. The latter problem, on its hand, admits the application of several local principles, by means of which the questions can be reduced to the study of certain so-called local representatives (compare with Chapter XV). KOZAK [1, 2] (detailed proofs have been recently published in [3]) was the first to realize this program and he obtained a

---

1) Here $\arg z \in (-\pi, \pi]$ for $z \in \mathbb{C} \setminus \{0\}$.

series of remarkable results for operators with continuous coefficients, the depth of which can be rightly appreciated in the higher-dimensional case. However, these techniques failed for discontinuous coefficients, although many attemps had been made to extend them into this direction.

SILBERMANN's contribution to overcome these difficulties was that he understood how to choose the algebra which is intended to replace the Calkin algebra in the right way. He demonstrated this for the case of Toeplitz operators, and so he solved not only a lot of problems which had been open at that time (for example, his approach was just what is needed to attack symbols with countably many discontinuities) but also set up the foundation for the enormous and fruitful development of the last few years. It became clear which peculiarities of the approximation method at hand are the guide to the choice of the corresponding algreba, and the (nevertheless, by no means trivial) problem one is left with in each case is the investigation of the (fortunately or unfortunately chosen) local representatives.

SILBERMANN [12] then also worked out the basic ideas of a local theory of the reduction method for singular integral operators and JUNGHANNS and SILBERMANN [2] established the basis for a local approach to the collocation method. PRÖSSDORF [26–27] and PRÖSSDORF, RATHSFELD [1] finally founded the corresponding theory of finite element approximation methods.

A self-contained and modern theory of the finite section method for Toeplitz operators with symbols having various kinds of discontinuities (or, in our language, of the reduction method for singular integral operators over the unit circle of the special form $aP + Q$) can be found in the monograph of BÖTTCHER and SILBERMANN [2]. However, notice that the passage from the Toeplitz case to singular integral operators of the form $aP + bQ$ is not trivial but is accompanied with a series of essential difficulties. This is also seen from the fact that, for example, the theory of the reduction method for singular integral operators developed up to the present time has not yet achieved that rather advanced stage as the theory for the Toeplitz case.

3.7. Singular integral equations over an interval

In the last few years the interest in numerical analysis for singular integral equations over an interval has been continuously increasing (see IOAKIMIDIS and THEOCARIS [1], DOW and ELLIOTT [1], DZHISHKARIANI [1, 2], ELLIOTT [1], GOLBERG and FROMME [1], HAFTMANN [1], JUNGHANNS and SILBERMANN [1]). Although very good results were obtained in applying certain approximation methods to various concrete problems (see, for example, ERDOGAN, GUPTA, and COOK [1]), many questions about the convergence of the approximation methods that were used have been remained unanswered. However, in recent time a lot of problems concerning the applicability of methods using polynomials as trial functions (Galerkin, collocation, and quadrature methods) were solved completely.

The present section is the slightly modified article of JUNGHANNS and SILBERMANN [3], which gives a survey on essential results and methods along these lines. We also recommend the interested reader to consult the fundamental papers of JUNGHANNS and SILBERMANN [1] and JUNGHANNS [2], the survey by GOLBERG [1], and the Proceedings edited by GERASOULIS and VICHNEVETSKY [1].

3.7.1. The foundation of the aforementioned numerical methods is formed by a sufficiently general and adaptable theory of abstract approximation methods based on the notion of the convergence manifold (see § 1). Such a perturbation theory for projection methods also essentially determines the strategy pursued here: in the first place to study the approximation method for a relatively simple equation, say

$$(Au)(x) := au(x) + \frac{b}{\pi} \int_{-1}^{1} \frac{u(y)\,\mathrm{d}y}{y - x} = f(x), \quad -1 < x < 1, \tag{31}$$

and then to handle the complete problem,

$$(Au)(x) + \int_{-1}^{1} h(x, y)\, u(y)\, dy = f(x), \qquad -1 < x < 1, \tag{32}$$

as a perturbed equation (31).

The relation (33) below is well known from the literature devoted to Galerkin, collocation, and quadrature methods for the equation (32) (see, for example, ERDOGAN, GUPTA, and COOK [1], KARPENKO [1], and GOLBERG [2]). Suppose that a and b are real numbers, that $b \neq 0$, and that $a^2 + b^2 = 1$. Define α so that $-1 < \alpha < 1$ and $a + ib = e^{i\pi\alpha}$. Also choose integers λ and μ so that $-1 < \beta := \lambda + \alpha < 1$ and $-1 < \gamma := \mu - \alpha < 1$. Finally, let $\varkappa := -(\lambda + \mu)$ and $\sigma(x) := \sigma^{(\beta,\gamma)}(x) = (1 + x)^\beta (1 - x)^\gamma$. Then for every $x \in (-1, 1)$ and for all $n = 0, 1, 2, \ldots$

$$a\sigma(x)\, P_n^{(\beta,\gamma)}(x) + \frac{b}{\pi} \int_{-1}^{1} \frac{P_n^{(\beta,\gamma)}(y)}{y - x}\, \sigma(y)\, dy = -\frac{2^{-\varkappa} b}{\sin(\pi\gamma)}\, P_{n-\varkappa}^{(-\beta,-\gamma)}(x), \tag{33}$$

where $P_n^{(\beta,\gamma)}$ is the Jacobi polynomial

$$P_n^{(\beta,\gamma)}(x) = \frac{(-1)^n}{n!} (1 + x)^{-\beta} (1 - x)^{-\gamma} \frac{d^n}{dx^n} (1 + x)^{n+\beta} (1 - x)^{n+\gamma}$$

(put $P_{-1}^{(\beta,\gamma)} \equiv 0$). Let $\tilde{P}_n^{(\beta,\gamma)}$ denote the normalized (with respect to the weight $\sigma^{(\beta,\gamma)}(x)$) Jacobi polynomial:

$$\tilde{P}_n^{(\beta,\gamma)} = P_n^{(\beta,\gamma)} / \| P_n^{(\beta,\gamma)} \|_{2,\sigma^{(\beta,\gamma)}},$$

$$\|u\|_{2,\sigma} = \left(\int_{-1}^{1} |u(x)|^2\, \sigma(x)\, dx \right)^{1/2}.$$

Then one has the equalities

$$A\sigma\tilde{P}_n^{(\beta,\gamma)} = (-1)^\mu\, \tilde{P}_{n-\varkappa}^{(-\beta,-\gamma)} \qquad (n = 0, 1, 2, \ldots). \tag{34}$$

3.7.2. Let us consider the dominant equation (31). We look for an approximate solution of the form

$$u_n(x) = \sigma(x) \sum_{k=0}^{n-1} \xi_k^{(n)} \tilde{P}_k^{(\beta,\gamma)}(x) =: \sigma(x)\, v_n(x).$$

For the sake of simplicity assume that $\varkappa = 0$. Then Theorem 5.1, Chapter IV, implies that A is in $G\mathscr{L}(Y, Y)$ with $Y = \mathbf{L}_2([-1, 1], \sigma^{-1/2})$. From (34) we obtain that

$$Au_n = (-1)^\mu \sum_{k=0}^{n-1} \xi_k^{(n)} P_k^{(-\beta,-\gamma)}.$$

Define the projection P_n by $P_n f := \sum_{k=0}^{n-1} f_k P_k^{(-\beta,-\gamma)}$ with $f_k := \int_{-1}^{1} \sigma^{-1}(x) f(x) P_k^{(-\beta,-\gamma)}(x)\, dx$.

The Galerkin method consists in determining the approximate solution u_n from the equation

$$P_n A u_n = (-1)^\mu \sum_{k=0}^{n-1} \xi_k^{(n)} \tilde{P}_k^{(-\beta,-\gamma)} = \sum_{k=0}^{n-1} f_k \tilde{P}_k^{(-\beta,-\gamma)}.$$

This is equivalent to $\xi_k^{(n)} = (-1)^\mu f_k$ $(k = 0, 1, \ldots, n - 1)$. Consequently, in the case at hand the convergence of the method for all $f \in Y$ is obvious. An approximate solution for the complete equation (32) is determined with the help of the Galerkin equations

$$\xi_k^{(n)} + \sum_{j=0}^{n-1} h_{kj}\xi_j^{(n)} = f_k \quad (k = 0, 1, \ldots, n-1),$$

where $h_{kj} = \int_{-1}^{1} \sigma^{-1}(x)\, \tilde{P}_k^{(-\beta,-\gamma)}(x) \int_{-1}^{1} \sigma(y)\, h(x, y)\, \tilde{P}_j^{(\beta,\gamma)}(y)\, dy\, dx$. If $T \in \mathscr{K}(Y, Y)$ and $A + T \in G\mathscr{L}(Y, Y)$, where $(Tu)(x) := \int_{-1}^{1} h(x, y)\, u(y)\, dy$, then this method converges for all $f \in Y$. Using some results from approximation theory it can be proved that

$$\|u - u_n\|_Y = O(n^{-m-\tau}) \tag{35}$$

if $f \in \mathbf{C}^{m,\tau}$ and $h \in \mathbf{C}_x^{m,\tau}$ (that is, $h(\cdot, y) \in \mathbf{C}^{m,\tau}$ uniformly with respect to $y \in [-1, 1]$).

Another frequently applied approximation method is the quadrature method. This method is based upon a quadrature formula for regular integrals, for example, upon the Gauss quadrature formula

$$\frac{1}{\pi} \int_{-1}^{1} \sigma^{(\beta,\gamma)}(y)\, v(y)\, dy \sim \sum_{k=1}^{n} A_k^{(n)} v(y_{kn}),$$

$$P_n^{(\beta,\gamma)}(y_{kn}) = 0 \quad (k = 1, \ldots, n).$$

Now the approximate solution is sought in the form $u_n(x) = \sigma(x)\, v_n(x)$, where

$$v_n(x) = \sum_{k=1}^{n} v_n(y_{kn}) \frac{P_n^{(\beta,\gamma)}(x)}{(P_n^{(\beta,\gamma)})'(y_{kn})\,(x - y_{kn})}.$$

Let $P_n^{(-\beta,-\gamma)}(x_{jn}) = 0$ $(j = 1, \ldots, n)$. Then $x_{jn} \neq y_{kn}$ $(k, j = 1, \ldots, n)$ and

$$\begin{aligned}
&(A\sigma^{(\beta,\gamma)} v_n)(x_{jn}) \\
&= \sum_{k=1}^{n} \frac{v_n(y_{kn})}{(P_n^{(\beta,\gamma)})'(y_{kn})} \left[a\sigma^{(\beta,\gamma)}(x_{jn}) \frac{P_n^{(\beta,\gamma)}(x_{jn})}{x_{jn} - y_{kn}} + \frac{b}{\pi} \int_{-1}^{1} \frac{\sigma^{(\beta,\gamma)} P_n^{(\beta,\gamma)}(y)\, dy}{(y - y_{kn})(y - x_{jn})} \right] \\
&= \sum_{k=1}^{n} \frac{v_n(y_{kn})}{(P_n^{(\beta,\gamma)})'(y_{kn})} \cdot \frac{1}{y_{kn} - x_{jn}} \left[-(A\sigma^{(\beta,\gamma)} P_n^{(\beta,\gamma)})(x_{jn}) \right. \\
&\qquad \left. + \frac{b}{\pi} \int_{-1}^{1} \frac{\sigma^{(\beta,\gamma)}(y) P_n^{(\beta,\gamma)}(y)\, dy}{y - y_{kn}} \right] \\
&= b \sum_{k=1}^{n} A_k^{(n)} \frac{v_n(y_{kn})}{y_{kn} - x_{jn}} \quad (j = 1, \ldots, n),
\end{aligned}$$

where

$$A_k^{(n)} = \frac{1}{\pi} \int_{-1}^{1} \frac{\sigma^{(\beta,\gamma)}(y)\, P_n^{(\beta,\gamma)}(y)\, dy}{(y - y_{kn})\, (P_n^{(\beta,\gamma)})'(y_{kn})} \quad (k = 1, \ldots, n)$$

are the well-known coefficients in the given quadrature formula. Taking into account this relation the quadrature method for the complete equation (32),

$$\sum_{k=1}^{n} A_k^{(n)} \left[\frac{b}{y_{kn} - x_{jn}} + \pi h(x_{jn}, y_{kn}) \right] \xi_k^{(n)} = f(x_{jn}) \quad (j = 1, \ldots, n),$$

$\xi_k^{(n)} = v_n(y_{kn})$, can be viewed as a perturbed collocation method. On condition that $A + T \in G\mathscr{L}(Y, Y)$, $f \in \mathbf{C}^{m,\tau}$, and $h \in \mathbf{C}_x^{m,\tau} \cap \mathbf{C}_y^{m,\tau}$ Junghanns and Silbermann [1] showed that

$$\|v - v_n\|_X = O(n^{-m-\tau}), \tag{36}$$

where $u = \sigma v$ is the solution of (32) and $X = \mathbf{L}_2([-1, 1], \sigma^{1/2})$. They also proved analogous results for the case where the Radau or Lobatto quadrature formulae are utilized.

The error estimate (36) can be used to show the uniform convergence of the polynomials v_n in case $m + \tau > \nu + 1$, where $\nu := \max(\beta, \gamma)$. To see this, notice first that

$$\|\tilde{P}_n^{(\beta,\gamma)}\|_\infty \leqq c(n + 1)^{\nu+1/2}$$

(see NATANSON [1], Chap. VI, § 3, Theorem 2), which implies that

$$\|v_n\|_\infty \leqq cn^{\nu+1} \|v_n\|_{2,\sigma^{(\beta,\gamma)}}$$

for every polynomial v_n of a degree less than n, and then apply a standard procedure (see KANTOROVICH and AKILOV [1], Chap. XIV, § 4.3) to deduce that $v \in \mathbf{C}[-1, 1]$ and

$$\|v - v_n\|_\infty = O(n^{-m-\tau+\nu+1}). \tag{37}$$

3.7.3. The purpose of what follows is to show how the strategy pointed out above can be applied to generalize the above results to equations with non-constant coefficients of the form

$$(A + T)\, u(x) := a(x)\, u(x) + \frac{1}{\pi} \int_{-1}^{1} b(y) \frac{u(y)\, dy}{y - x}$$
$$+ \int_{-1}^{1} h(x, y)\, u(y)\, dy = f(x), \qquad -1 < x < 1. \tag{38}$$

The basic ideas in this direction are in the paper of JUNGHANNS and SILBERMANN [1]. The functions a and b are assumed to be Hölder continuous and to satisfy $[a(x)]^2 + [b(x)]^2 = 1$ for $x \in [-1, 1]$. For $x \in (-1, 1)$ define the continuous function $g(x)$ as

$$g(x) = \frac{1}{2\pi i} \ln \frac{a(x) - ib(x)}{a(x) + ib(x)}$$

and the function $\sigma(x)$ as

$$\sigma(x) = (1 + x)^\lambda (1 - x)^\mu \exp \int_{-1}^{1} \frac{g(t)\, dt}{t - x},$$

where λ and μ are integers such that

$$-1 < \beta := \lambda - g(-1) < 1 \text{ and } -1 < \gamma := \mu + g(1) < 1.$$

Furthermore, for $z \in \mathbf{C} \setminus [-1, 1]$, define

$$X(z) = (1 - z)^\mu (1 + z)^\lambda \exp \int_{-1}^{1} \frac{g(t)\, dt}{t - z}.$$

Note that $\sigma(x) = (1 + x)^\beta (1 - x)^\gamma w(x)$ with certain Hölder continuous function $w(x) > 0$.

Theorem 3.11. *Let $p_m(x)$ be a polynomial of degree m. Then $q_{m-\varkappa} := A\sigma p_m$ is a polynomial of degree $m - \varkappa$ ($\varkappa := -(\lambda + \mu)$) and*

$$\lim_{z\to\infty} [X(z)\, p_m(z) - q_{m-\varkappa}(z)] = 0.$$

Theorem 3.12. *Suppose the orthogonality relations*

$$\int_{-1}^{1} x^j p_m(x)\, b(x)\, \sigma(x)\, dx = 0 \qquad (j = 0, 1, \ldots, m-1)$$

are satisfied. Then

$$\int_{-1}^{1} x^j q_{m-\varkappa}(x)\, b(x)\, [\sigma(x)]^{-1}\, dx = 0 \qquad (j = 0, 1, \ldots, m - \varkappa - 1).$$

For the sake of simplicity, in the following we shall again restrict ourselves to the case $\varkappa = 0$. Furthermore, let

$$\omega_1(x) = \sigma(x)\, w_1(x), \qquad \omega_2(x) = [\sigma(x)]^{-1}\, w_2(x),$$

$w_j \in \mathbf{C}[-1, 1]$, $w_j(x) > 0$, and let $g_n^{(j)}(x)$ be the orthogonal polynomial with respect to the weight $\omega_j(x)$ of the degree n with the norm $\|g_n^{(j)}\|_{2,\omega_j} = 1$ ($g_n^{(j)}(\infty) = \infty$, for uniqueness).

Galerkin method. The approximate solution is sought in the form

$$u_n(x) = \omega_1(x) v_n(x), \qquad v_n(x) = \sum_{k=0}^{n-1} \xi_k^{(n)} g_k^{(1)}(x), \tag{39}$$

and the coefficients $\xi_k^{(n)}$ are determined from the Galerkin equations

$$\sum_{k=0}^{n-1} \xi_k^{(n)} ((A + T)\, \omega_1 g_k^{(1)} - f, g_j^{(2)})_{2,\omega_2} = 0 \qquad (j = 0, 1, \ldots, n-1). \tag{40}$$

Theorem 3.13. *Let* $Y := \mathbf{L}_2([-1, 1], \sigma^{-1/2})$ *and* $T \in \mathscr{K}(Y, Y)$. *Let* $u(x) = \omega_1(x)\, v(x)$ *be the unique solution of* (38) *in the space* Y. *Then the equations* (40) *are uniquely solvable for all* $f \in Y$ *and for all sufficiently large* n, *and the solutions* (39) *converge to* $u(x)$ *in the norm of the space* Y. *If* $f \in \mathbf{C}^{m,\tau}$, $h \in \mathbf{C}_x^{m,\tau}$, *and* $w_1 \in \mathbf{C}^{m+1,\tau}$ $(0 < \tau < 1)$, *then* (35) *holds, and if, in addition,* $\nu := \max(\beta, \gamma) \geqq -1/2$, *then* (37) *is valid.*

Collocation method. Let $x_{kn}^{(j)}$ be the zeros of $g_n^{(j)}(x)$ $(k = 1, \ldots, n)$. The collocation method consists in determining the approximate solution in the form (39) from the equations

$$(A + T)\, u_n(x_{kn}^{(2)}) = f(x_{kn}^{(2)}) \qquad (k = 1, \ldots, n). \tag{41}$$

Theorem 3.14. *Let the operators* T *and* $w_1 S - S w_1 I$, *with*

$$(w_1 S - S w_1 I)\, u(x) = \frac{1}{\pi} \int_{-1}^{1} \frac{w_1(x) - w_1(y)}{y - x}\, u(y)\, dy,$$

belong to $\mathscr{K}(Y, \mathscr{R}(\sigma^{-1}))$, *where* $\mathscr{R}(\sigma^{-1})$ *denotes the Banach space of all bounded and with respect to* $[\sigma(x)]^{-1}\, dx$ *Riemann-Stieltjes integrable functions* $(\|\cdot\|_{\mathscr{R}} = \|\cdot\|_\infty)$. *Furthermore, let* $u(x) = \omega_1(x)\, v(x)$ *be the unique solution of* (38) *in* Y. *Then for any* $f \in \mathscr{R}(\sigma^{-1})$ *and for all sufficiently large* n *the equations* (41) *are uniquely solvable and all assertions of Theorem 3.13 remain valid.*

Quadrature method. We choose $w_1 = w_2 = 1$. Take, for example, the Gauss quadrature formula

$$\frac{1}{\pi} \int_{-1}^{1} \sigma(x)\, v(x)\, dx \sim \sum_{k=1}^{n} A_k^{(n)} v_n(x_{kn}^{(1)}).$$

Look for an approximate solution in the form

$$u_n(x) = \sigma(x)\, v_n(x),$$

$$v_n(x) = \sum_{k=1}^{n} \xi_k^{(n)} \frac{g_n^{(1)}(x)}{g_n^{(1)\prime}(x_{kn}^{(1)})\,(x - x_{kn}^{(1)})} \tag{42}$$

and determine the coefficients $\xi_k^{(n)}$ from the equations

$$\sum_{k=1}^{n} A_k^{(n)} \left[\frac{b(x_{kn}^{(1)})}{x_{kn}^{(1)} - x_{jn}^{(2)}} + \pi h(x_{jn}^{(2)}, x_{kn}^{(1)})\right] \xi_k^{(n)} = f(x_{jn}^{(2)}) \tag{43}$$

if $x_{jn}^{(2)} \neq x_{kn}^{(1)}$ $(k = 1, \ldots, n)$ and

$$[a(x_{jn}^{(2)})\, \sigma(x_{jn}^{(2)}) + A_k^{(n)}[b'(x_{kn}^{(1)}) + \pi h(x_{jn}^{(2)}, x_{kn}^{(1)})]]\; \xi_k^{(n)}$$

$$+ \sum_{l=1}^{n} A_l^{(n)} \left[\frac{b(x_{ln}^{(1)})}{x_{ln}^{(1)} - x_{jn}^{(2)}} + \pi h(x_{jn}^{(2)}, x_{ln}^{(1)})\right] \xi_l^{(n)} = f(x_{jn}^{(2)}) \tag{44}$$

if $x_{jn}^{(2)} = x_{kn}^{(1)}$. Here the subscript j runs from 1 to n.

Theorem 3.15. *Let $b(x)$ be a polynomial and $T \in \mathcal{K}(Y, \mathcal{R}(\sigma^{-1}))$. Furthermore, let $h(x, \cdot)$ be an element of $\mathcal{R}(\sigma)$ uniformly with respect to $x \in [-1, 1]$. Finally, let $u(x) = \sigma(x)\, v(x)$ be the unique solution of* (38) *in Y and let $f \in \mathcal{R}(\sigma^{-1})$. Then for all sufficiently large n the equations* (43) *or* (44) *are uniquely solvable and the solutions* (42) *converge to $u(x)$ in the norm of the space Y. Moreover, if $f \in \mathbf{C}^{m,\tau}$ and $h \in \mathbf{C}_x^{m,\tau} \cap \mathbf{C}_y^{m,\tau}$, then* (35) *holds, and if, in addition, $\nu \geqq -1/2$, then* (37) *holds.*

§ 4. Spline approximation methods for the solution of singular integral equations

In many practical computations with boundary integral equations for two-dimensional problems or with singular integral equations over an interval it is advantageous to work with spline approximations for the unknown functions. The two most popular discretization schemes are collocation methods and Galerkin procedures. The corresponding mathematical foundation and the error analysis, however, have been developed only rather recently in the work by Arnold and Wendland [1–3], Elschner [5–6], Hsiao and Wendland [1–2], Nedelec [1], Prössdorf [27], Prössdorf and Schmidt [2–3], Prössdorf and Rathsfeld [1–3], Saranen and Wendland [1], Schmidt [1–4], Stephan and Wendland [1], Thomas [1], Wendland [3–5] (see also the remarks on pp. 200–205 in Ivanov [1]). The purpose of the present section is to give a survey of some of these recent developments.

For brevity we shall here only formulate the results for the dominant singular integral equation, since they easily extend to the complete equation (3.1) (see Section 1). Our first concern is one of the simplest collocation methods, the so-called polygonal method. This method will also serve to demonstrate some characteristic features (pertaining to both the results and techniques to prove them) of other spline methods. It will turn out that, roughly speaking, each of the above mentioned spline approximation method is applicable if and only if the singular integral operator is strongly elliptic (in a sense that will be specified below).

4.1. The polygonal method

For simplicity's sake let us suppose Γ is the unit circle and $P = P_\Gamma$, $Q = Q_\Gamma$ are the singular projections on the space $\mathbf{L}_2(\Gamma)$ (see (5.7), Chapter II).

We consider the singular integral equation

$$(Ax)(t) := a(t)(Px)(t) + b(t)(Qx)(t) = f(t) \qquad (t \in \Gamma) \tag{1}$$

with coefficients $a, b \in \overline{\mathbf{PC}}(\Gamma)$ in the space $\mathbf{L}_2(\Gamma)$. The *polygonal method* consists in looking for an approximate solution of the equation (1) in the form of a polygon:

$$x_n = \sum_{k=0}^{n-1} \xi_k \varphi_k^{(n)}. \tag{2}$$

Here the piecewise linear splines $\varphi_k^{(n)}$ are given for $t \in \Gamma$ by

$$\varphi_k^{(n)}(t) = \begin{cases} (t - t_{k-1})/(t_k - t_{k-1}) & \text{if } t \in [t_{k-1}, t_k], \\ (t_{k+1} - t)/(t_{k+1} - t_k) & \text{if } t \in [t_k, t_{k+1}], \\ 0 & \text{otherwise}, \end{cases}$$

where $t_k = t_k^{(n)} = e^{2\pi ik/n}$ $(k = 0, \ldots, n-1)$ and $[t_{k-1}, t_k]$ denotes the arc on Γ whose endpoints are t_{k-1} and t_k. The coefficients $\xi_k = \xi_k^{(n)}$ $(n = 1, 2, \ldots)$ are determined from the collocation equations

$$(Ax_n)(t_j) = f(t_j) \qquad (j = 0, \ldots, n-1). \tag{3}$$

The application of the polygonal method is dated back to a paper of Lavrentyev [1] that was published as early as 1932. Multhopp [1] was probably the first to use the trigonometric collocation for the approximate solution of singular integral equations (his paper appeared in 1938). Only in 1961 Ivanov (see his book [1]) showed that in the case of the Cauchy singular integral operator $A = S_\Gamma$ the collocation system (3) is singular for sufficiently large n. In this way he disproved an earlier statement of Sofronov [1]. Gabdulhaev [2] proved the convergence of the polygonal method for the case of Hölder continuous coefficients, but under the very restrictive extra assumption that $\|a - b\|_\infty$ be sufficiently small.

Necessary and sufficient conditions for the convergence of the polygonal method were first obtained by Prössdorf and Schmidt [2–3] for the case of continuous coefficients and by Prössdorf and Rathsfeld [1] for the case of coefficients from $\overline{\mathbf{PC}}(\Gamma)$. In these authors' works one can find, among other things, the following results.

Theorem 4.1. *The polygonal method* (3) *is stable if and only if the following two conditions are fulfilled:*

(i) $b(t) \neq 0$ *for all* $t \in \Gamma$;

(ii) $\mu \frac{a}{b}(t+0) + (1-\mu)\frac{a}{b}(t-0) \notin (-\infty, 0]$ *for all* $t \in \Gamma$ *and all* $\mu \in [0, 1]$.

If the conditions (i) *and* (ii) *are satisfied, then the system* (3) *has a unique solution* x_n *for all sufficiently large* n *and for each* $f \in \mathcal{R}(\Gamma)$, *and as* $n \to \infty$ *the solutions* x_n *converge in the norm of* $\mathbf{L}_2(\Gamma)$ *to the solution* $x \in \mathbf{L}_2(\Gamma)$ *of the equation* (1).

Remark 1. Note that the operator A is obviously invertible on $\mathbf{L}_2(\Gamma)$ if the conditions (i) and (ii) are satisfied (cf. Theorem 5.1, Chapter IV). In the case of continuous coefficients a and b the conditions (i) and (ii) are equivalent to the condition

$$\nu a(t) + (1-\nu) b(t) \neq 0, \qquad \forall t \in \Gamma, \qquad \forall \nu \in [0, 1]. \tag{4}$$

Remark 2. The conditions (i) and (ii) together are equivalent to each of the following conditions:

(iii) $\mu(\nu_1 a(t+0) + (1-\nu_1)\, b(t+0))\,(\nu_2 a(t-0) + (1-\nu_2)\, b(t-0))$
$+ (1-\mu)\,(\nu_2 a(t+0) + (1-\nu_2)\, b(t+0))\,(\nu_1 a(t-0) + (1-\nu_1)\, b(t-0)) \neq 0,$
$\forall t \in \Gamma,\ \forall \mu, \nu_1, \nu_2 \in [0, 1].$

(iv) There exist both an operator $T \in \mathscr{K}(\mathbf{L}_2(\Gamma))$ and an invertible function $\theta \in \overline{\mathbf{PC}}(\Gamma)$ which is discontinuous at most at the points of discontinuity of a and b such that $A = \theta(D + T)$, where D is an operator with a positive-definite real part, that is,

$$\operatorname{Re}(Df, f) \geq \varepsilon(f, f), \qquad \forall f \in \mathbf{L}_2(\Gamma)$$

with some positive constant ε.

If the condition (iv) is fulfilled, then the operator A is said to be *strongly elliptic*.

The equivalence (i) + (ii) $\Leftrightarrow$(iii) easily follows from the fact that the collection of all straight line segments and circular arcs is preserved by every linear fractional transformation (see also Prössdorf and Rathsfeld [1], Corollary 4.2). It is also a simple matter to show that in the case of continuous coefficients a and b the conditions (4) and (iv) are equivalent (see Prössdorf and Schmidt [2]). For coefficients $a, b \in \mathbf{PC}(\Gamma)$ the equivalence of those conditions was proved in the paper of Prössdorf and Rathsfeld [3] by means of the symbol calculus presented in Chapter V (§§ 7 and 8) and with the help of some methods from convex analysis.

Remark 3. All results of this section can be easily extended to singular integral equations over an arbitrary simple closed Lyapunov curve (see Prössdorf and Rathsfeld [1]).

The proof of Theorem 4.1 will be based upon a localization technique; this technique was (for more general situations) developed in the works by Prössdorf [26–27] and Prössdorf and Rathsfeld [1].

1. We first formulate some basic approximation properties of piecewise linear splines. Detailed proofs can be found in the article of Prössdorf and Schmidt [2].

Let X_n denote the linear span of the splines $\varphi_0^{(n)}, \ldots, \varphi_{n-1}^{(n)}$, that is, the subspace of $\mathbf{L}_2(\Gamma)$ consisting of all polygons of the form (2). Denote by L_n the orthogonal projection from $\mathbf{L}_2(\Gamma)$ onto X_n and by K_n the interpolation projection which assigns the polygon

$$(K_n f)(t) := \sum_{k=0}^{n-1} f(t_k)\, \varphi_k^{(n)}(t) \qquad (\in X_n)$$

to a bounded function f defined everywhere on Γ. Then the system (3) is nothing else than the projection equation $K_n A L_n x_n = K_n f$. Also notice that, obviously, $K_n f = K_n f K_n$.

Further, let $\mathbf{l}_2(n) := \{\xi = (\xi_k)_{k=0}^{n-1} : \xi_k \in \mathbf{C}\}$ be the n-dimensional complex Euclidean space provided with the norm

$$|\xi| = \frac{1}{\sqrt{n}} \left(\sum_{k=0}^{n-1} |\xi_k|^2 \right)^{1/2}.$$

It is well known that the norms $\|\cdot\|_{\mathbf{L}_2(\Gamma)}$ and $|\cdot|$ defined in X_n are equivalent (see, for instance, Aubin [1], Mikhlin [31], or Prössdorf and Schmidt [2]). Thus, we may identify the space X_n with $\mathbf{l}_2(n)$ and, correspondingly, the operator $K_n A L_n$ with the matrix $A_n := ((A\varphi_k^{(n)})\,(t_j^{(n)}))_{j,k=0}^{n-1}$. In particular, $a_n = (aI)_n$ is the diagonal matrix $a_n = (a(t_j)\, \delta_{jk})_{j,k=0}^{n-1}$ and therefore

$$\|K_n a L_n\| \leq \text{const} \cdot \|a\|_\infty \tag{5}$$

for every $a \in \mathscr{R}(\Gamma)$. Here $\|\cdot\|$ denotes the operator norm in $\mathscr{L}(\mathbf{L}_2(\Gamma))$. Furthermore, notice the following properties of the operators K_n and L_n:

(i) $\|K_nf - f\|_{L_2(\Gamma)} \to 0$ for every $f \in \mathscr{R}(\Gamma)$, in particular, $L_n \to I$ as $n \to \infty$.

(ii) If $f \in C(\Gamma)$ then

$$\|K_nf - f\|_{L_2(\Gamma)} \leqq 4\sqrt{\pi}\,\omega\left(f, \frac{2\pi}{n}\right),$$

where $\omega(f, \delta) := \sup_{|h|<\delta} |f(t) - f(e^{ih}t)|$ is the modulus of continuity.

(iii) $K_ngL_n \to gI$ as $n \to \infty$ for every $g \in \mathscr{R}(\Gamma)$.

(iv) If $g \in C(\Gamma)$ then

$$\|(K_n - I)\,gL_n\| = O\left(\omega\left(g, \frac{2\pi}{n}\right)\right).$$

(v) If $g \in \mathbf{Lip}_\mu(\Gamma)$ $(0 < \mu \leqq 1)$ then

$$\|K_nS(I - K_n)\,gL_n)\| = O\left(\frac{\ln n}{n^\mu}\right).$$

(vi) If $a, b, f \in \mathbf{Lip}_\mu(\Gamma)$ $(0 < \mu \leqq 1)$ then one has the following estimate for the singular operator $A = aP + bQ$:

$$\|K_nAL_nf - Af\|_{L_2(\Gamma)} = O\left(\frac{(\ln n)^{[\mu]}}{n^\mu}\right),$$

where $[\mu] = 0$ for $0 < \mu < 1$ and $[1] = 1$.

The properties (i)–(vi) can be easily checked by elementary estimations. Using the formula (1.4) of Chapter II to calculate the singular integrals $S\varphi_k^{(n)}$, it is seen that $(S\varphi_k^{(n)})\,(t_j^{(n)}) = (S\varphi_0^{(n)})\,(t_{j-k}^{(n)})$. Thus, the matrix S_n is a hermitian circulant, since, obviously, $t_{-l}^{(n)} = t_{n-l}^{(n)}$ $(l = 1, \ldots, n)$. Taking into account the well-known formulae for the eigenvalues and eigenvectors of circulants (see e.g. Marcus and Minc [1]) we arrive at the following statements on the eigenvalues $\lambda_k^{(n)}$ $(k = 0, 1, \ldots, n - 1)$ of the matrix S_n:

(vii) $\Lambda_n := (\lambda_j^{(n)}\,\delta_{jk})_{j,k=0}^{n-1} = F_n^{-1}S_nF_n$, where F_n denotes the unitary matrix $\left(\frac{1}{\sqrt{n}}\,e^{i\frac{2\pi}{n}jk}\right)_{j,k=0}^{n-1}$.

(viii) Define f on the interval $[0, 1]$ as

$$f(x) := \frac{2\sin^2 \pi x}{\pi^2} \sum_{k=0}^{\infty} \frac{1}{(x+k)^2} - 1 \quad (0 < x \leqq 1);\ f(0) = 1.$$

Note that f is a monotonically decreasing C^∞-function. Furthermore, put

$$\lambda(e^{2\pi ix}) := f(x) \quad (0 \leqq x \leqq 1).$$

Then $f(1 - x) = -f(x)$, the range of both λ and $f(\in \mathbf{PC}(\Gamma))$ is the interval $[-1, 1]$, and one has the estimate

$$|\lambda_k^{(n)} - \lambda(t_k^{(n)})| \leqq \gamma/n \quad (k = 0, \ldots, n - 1)$$

with some constant $\gamma > 0$.

From (viii) we conclude in particular that $\sup_n \|S_n\| < \infty$, that the set $\bigcup_{k,n} \{\lambda_k^{(n)}\}$ is dense in the interval $[-1, 1]$, and that the estimate

$$\|\Lambda_n - \lambda_n\| \leqq \gamma/n \tag{6}$$

holds. Finally, since $K_nSL_ng \to Sg$ as $n \to \infty$ for every $g \in \mathbf{Lip}_\mu(\Gamma)$ $(0 < \mu \leqq 1)$ (see (vi)) and since $\mathbf{Lip}_\mu(\Gamma)$ is dense in $L_2(\Gamma)$, it follows that $K_nSL_n \to S$. This together with

(iii) implies the strong convergence

$$K_n(aP + bQ)\, L_n \to aP + bQ \qquad (n \to \infty)$$

for every $a, b \in \mathcal{R}(\Gamma)$.

2. The next step of the proof of Theorem 4.1 consists in reducing the problem of the stability of the polygonal method for equation (1) to the case of an equation involving only piecewise constant coefficients.

To do this, we first fix some notation. Given $a, b \in \overline{\mathbf{PC}}(\Gamma)$ and a point $t_0 \in \Gamma$ denote by u the characteristic function of the circular arc $[t_0, -t_0)$, put $v = 1 - u$, and let $a_\pm := a(t_0 \pm 0)$, $b_\pm := b(t_0 \pm 0)$. Finally, let A^{t_0} denote the singular operator with piecewise constant coefficients defined as

$$A^{t_0} := (a_+ u + a_- v)\, P + (b_+ u + b_- v)\, Q\,.$$

Theorem 4.2. *The polygonal method* (3) *is stable for the invertible operator* $A = aP + bQ$ *if and only if for each point* $t_0 \in \Gamma$ *it is stable for the operator* A^{t_0}.

Proof. For the sake of brevity we shall call two sequences of operators B_n and C_n from $\mathcal{L}(X_n)$ equivalent and shall write this as $B_n \simeq C_n$ if $\| B_n - C_n \| \to 0$ as $n \to \infty$.

First suppose A^{t_0} is stable for each $t_0 \in \Gamma$. Let $V_{t_0} \subset U_{t_0} \subset \Gamma$ be sufficiently small neighborhoods of the point t_0 and let $\psi^{t_0} \in \mathbf{C}^\infty(\Gamma)$ be a real-valued function such that $0 \leqq \psi^{t_0} \leqq 1$, $\operatorname{supp} \psi^{t_0} \subset U_{t_0}$, and $\psi^{t_0}(t) \equiv 1$ for $t \in V_{t_0}$. From (5) we obtain that the approximating operators

$$A_n = K_n A L_n, \quad A_n^{t_0} = K_n A^{t_0} L_n, \quad \psi_n^{t_0} = K_n \psi^{t_0} L_n$$

satisfy the relation

$$\psi_n^{t_0}[A_n - A_n^{t_0}] = \psi_n^{t_0} B_n^{t_0}$$

with some $B_n^{t_0} \in \mathcal{L}(X_n)$ such that $\| B_n^{t_0} \|\, \| (A_n^{t_0})^{-1} \| \leqq q < 1$ for all $n \in \mathbb{N}$. Hence

$$\psi_n^{t_0} A_n (A_n^{t_0})^{-1} = \psi_n^{t_0}(I_n + B_n^{t_0}(A_n^{t_0})^{-1})$$

and, thus,

$$\psi_n^{t_0} A_n G_n^{t_0} = \psi_n^{t_0} \tag{7}$$

with $G_n^{t_0} := (A_n^{t_0})^{-1}\, (I_n + B_n^{t_0}(A_n^{t_0})^{-1})^{-1}$. Clearly, $\sup\limits_n \| G_n^{t_0} \| < \infty$.

Since the family of neighborhoods $\{U_\tau\}_{\tau \in \Gamma}$ contains a finite subfamily which covers Γ completely, there exists a finite number of points $t_1, \ldots, t_N \in \Gamma$ such that the function

$$\psi := \sum_{k=1}^{N} \psi^{t_k} \in \mathbf{C}^\infty(\Gamma)$$

is invertible. Define

$$D_n := \sum_{k=1}^{N} \psi_n^{t_k} G_n^{t_k},$$

take into account the property (v), and apply Theorem 4.2 of Chapter II to get

$$A_n D_n \simeq \sum_{k=1}^{N} K_n A \psi^{t_k} G_n^{t_k} = \sum_{k=1}^{N} K_n(\psi^{t_k} + T_k)\, G_n^{t_k}$$

with certain $T_k \in \mathcal{K}(\mathbf{L}_2(\Gamma), \mathbf{C}(\Gamma))$. This together with (7) shows that

$$A_n D_n \cong \psi_n + \sum_{k=1}^{N} K_n T_k G_n^{t_k}. \tag{8}$$

Now consider the operators

$$\tilde{D}_n := D_n - \sum_{k=1}^{N} L_n A^{-1} T_k G_n^{t_k}.$$

From (8) we deduce that

$$A_n \tilde{D}_n \cong \psi_n + W_n$$

with

$$W_n := \sum_{k=1}^{N} (K_n - A_n A^{-1}) T_k G_n^{t_k}.$$

Since $A_n \to A$, since $T_k \in \mathcal{K}(\mathbf{L}_2(\Gamma), \mathcal{R}(\Gamma))$, and because of property (i), it follows that $\|W_n\| \to 0$ as $n \to \infty$ and, consequently, $A_n \tilde{D}_n \cong \psi_n$. Finally, it is obvious that $(\psi_n)^{-1} = (\psi^{-1})_n$, and so we have proved that there exists a sequence of operators $D_n' \in \mathcal{L}(X_n)$ such that

$$A_n D_n' = I_n \text{ and } \sup_n \|D_n'\| < \infty.$$

But A_n is a square matrix, hence $D_n' = A_n^{-1}$ and this proves the stability of the polygonal method for the operator A.

We now prove the converse. Thus, assume the polygonal method for the operator A is stable. A reasoning analogous to the arguments applied above gives

$$\begin{aligned}\psi_n^{t_0} A_n^{t_0} A_n^{-1} &= \psi_n^{t_0} + \psi_n^{t_0}[A_n^{t_0} - A_n]\, A_n^{-1}\\ &\cong \psi_n^{t_0}(I_n + C_n^{t_0}) - \psi_n^{t_0} T^{t_0} A_n^{-1}\end{aligned}$$

with $\|C_n^{t_0}\| \leqq q < 1$ and $T^{t_0} \in \mathcal{K}(\mathbf{L}_2(\Gamma), \mathcal{R}(\Gamma))$. It follows that the operators $I_n + C_n^{t_0} \in \mathcal{L}(X_n)$ are invertible and that, with $D_n^{t_0} := A_n^{-1}(I_n + C_n^{t_0})^{-1}$,

$$\psi_n^{t_0}(A^{t_0} + T^{t_0})_n\, D_n^{t_0} \cong \psi_n^{t_0} \tag{9}$$

and $\sup_n \|D_n^{t_0}\| < \infty$. Now put

$$A_\pm^{t_0} := a_\pm P + b_\pm Q.$$

This are operators with constant coefficients. Both the invertibility of A and the stability of the sequence (A_n) remain untouched when in some sufficiently small right-sided resp. left-sided neighborhoods $U_\pm$ of the point t_0 the coefficients of A are set equal to a_+ and b_+ (for $t \in U_+$) resp. a_- and b_- (for $t \in U_-$). Therefore, we obtain in analogy to (9) that

$$\psi_n^\tau(A_+^{t_0} + T_+^\tau)\, D_{+n}^\tau \cong \psi_n^\tau \tag{10}$$

for all $\tau \in U_+$ and thus for all $\tau \in (t_0, -t_0)$, and also that

$$\psi_n^\tau(A_-^{t_0} + T_-^\tau)\, D_{-n}^\tau \cong \psi_n^\tau \tag{11}$$

for all $\tau \in U_-$ and hence for all $\tau \in (-t_0, t_0)$. Here $T_\pm^\tau$ are operators in $\mathcal{K}(\mathbf{L}_2(\Gamma), \mathcal{R}(\Gamma))$ and we have $\sup_n \|D_{\pm n}^\tau\| < \infty$. Also recall how A^{t_0} was defined to see that obviously

$$\psi_n^{-t_0}(A^{t_0} + T^{-t_0})_n\, D_n^{-t_0} \cong \psi_n^{-t_0}. \tag{12}$$

Now repeat the arguments of the first step of the present proof, but use (9)–(12) in place of (7). This gives the stability of the sequence $(A_n^{t_0})$. ∎

3. Finally, the following theorem answers the question under what conditions the sequence $\{A_n^{t_0}\}$ is stable.

Theorem 4.3. *For the sequence* $\{A_n^{t_0}\}$ *to be stable it is necessary and sufficient that the conditions* (i) *and* (ii) *of Theorem 4.1 be satisfied at the point* $t = t_0$.

Proof. Since $K_n f = K_n f K_n$, it follows that $A_n^{t_0}$ can be written in the form

$$A_n^{t_0} = (a_+ u_n + a_- v_n) P_n + (b_+ u_n + b_- v_n) Q_n.$$

Put

$$U_n := F_n^{-1} u_n F_n, \quad V_n := F_n^{-1} v_n F_n,$$

and let $c := (a + b)/2$, $d := (a - b)/2$. Obviously,

$$U_n^2 = U_n,\ V_n^2 = V_n, \quad U_n + V_n = I_n, \quad U_n V_n = V_n U_n = 0 \tag{13}$$

and (see property (vii))

$$F_n^{-1} A_n^{t_0} F_n = U_n(c_+ I_n + d_+ \Lambda_n) + V_n(c_- I_n + d_- \Lambda_n). \tag{14}$$

First suppose the conditions (i) and (ii) of Theorem 4.1 are fulfilled for $t = t_0$. Then, by virtue of Remark 2,

$$\mu(c_+ + \lambda_1 d_+)(c_- + \lambda_2 d_-) + (1 - \mu)(c_+ + \lambda_2 d_+)(c_- + \lambda_1 d_-) \neq 0 \tag{15}$$

for all $\mu \in [0, 1]$ and $\lambda_1, \lambda_2 \in [-1, 1]$. This implies immediately that the operators $B_n := c_- I_n + d_- \Lambda_n \in \mathscr{L}(X_n)$ are invertible and that $\sup_n \|B_n^{-1}\| < \infty$. Taking into consideration (13) we deduce from (14) (see also formula (1.7′), Chapter III) that

$$F_n^{-1} A_n^{t_0} F_n = (I_n + U_n D_n V_n)(U_n D_n U_n + V_n) B_n,$$

D_n being the diagonal matrix

$$D_n := (c_+ I_n + d_+ \Lambda_n)(c_- I_n + d_- \Lambda_n)^{-1}.$$

Note that, obviously, $(I_n + U_n D_n V_n)^{-1} = I_n - U_n D_n V_n$. Using a well-known property of linear fractional transformations it is easy to check that the diagonal entries of D_n are located on a certain circular arc whose convex hull does by virtue of (15) not contain the origin. Hence, there exist constants $\theta \in \mathbb{C}$, $|\theta| = 1$, and $\delta > 0$ such that $\operatorname{Re}(\theta U_n D_n U_n) \geqq \delta$. The conclusion is that the operators $U_n D_n U_n + V_n$ are invertible and that $\sup_n \|(U_n D_n U_n + V_n)^{-1}\| < \infty$. This proves the stability of the sequence $(A_n^{t_0})$.

Conversely, suppose now that the sequence $(A_n^{t_0})$ is stable. On replacing the coefficients a and b by $\tilde{a}(t) = a(t/t_0)$ and $\tilde{b}(t) = b(t/t_0)$ it is seen that it suffices to consider the case $t_0 = 1$. Since $\lambda_n = K_n \lambda L_n \to \lambda I$ as $n \to \infty$, we deduce from (6) that $\Lambda_n L_n \to \lambda I$. It is easy to verify that

$$U_n K_n t^j = \begin{cases} K_n t^j & \text{if } j \geqq 0, \\ 0 & \text{if } j < 0 \end{cases}$$

for all sufficiently large n. This together with property (ii) shows that $U_n L_n t^j \to P t^j$ for all integers j. Thus, since $\sup_n \|U_n L_n\| < \infty$, we have $U_n L_n \to P$. Combining this with (14) we see that, as $n \to \infty$,

$$F_n^{-1} A_n^{t_0} F_n L_n \to B, \quad F_n^{-1} (A_n^{t_0})^* F_n L_n \to B^*, \tag{16}$$

where $B := P(c_+ + d_+\lambda) + Q(c_- + d_-\lambda)$. From (16) and the stability of the sequence $\{A_{nj}^{t_0}\}$ we conclude that the operator $B \in \mathscr{L}(\mathbf{L}_2(\Gamma))$ is invertible, which implies that the conditions (i) and (ii) of Theorem 4.1 are satisfied for $t = t_0$ (cf. Theorem 4.2, Chapter IV). ∎

The first statement of Theorem 4.1 is now an immediate consequence of the Theorems 4.2 and 4.3, while the second one follows from Theorem 1.6 and property (i). ∎

The Theorem 1.6 and the properties (ii) and (vi) also yield the following error estimate.

Theorem 4.4. *If* $a, b, f \in \mathbf{Lip}_\mu(\Gamma)$ $(0 < \mu \leqq 1)$ *and if the condition* (4) *is satisfied, then*

$$\|x - x_n\|_{\mathbf{L}_2(\Gamma)} = O\left(\frac{(\ln n)^{[\mu]}}{n^\mu}\right).$$

4.2. Collocation methods with splines of arbitrary degree

Let now Γ be an arbitrary simple closed Lyapunov curve in the complex plane and let $z = z(s)$, $0 \leqq s \leqq 1$, be its parametric representation. In what follows we shall identify a function f defined on Γ with the 1-periodic function $f(z(s))$ of the real variable s.

The concern of this section is the so-called ε-point collocation method with spline trial-functions for the approximate solution of the equation (1) with coefficients $a, b \in \mathbf{PC}(\Gamma)$. Denote by $\mathbf{S}_d^n$ $(n \in \mathbb{N})$ the n-dimensional space of all 1-periodic smooth polynomial splines of degree d on the uniform mesh:

$$\mathbf{S}_d^n := \{v \in \mathbf{C}^{d-1}: v\,|_{\left[\frac{j}{n}, \frac{j+1}{n}\right]} \text{ is a polynomial of degree} \leqq d,\ j = 0, \ldots, n-1\}.$$

We let $\mathbf{S}_0^n$ denote the corresponding space of piecewise constant functions.

Fix an arbitrary number $\varepsilon \in [0, 1)$ and put $t_k = (k + \varepsilon)/n$, $k = 0, \ldots, n - 1$. The collocation method which defines $x_n \in \mathbf{S}_d^n$ by (3) with the collocation points $t_k = t_k(\varepsilon, n)$ chosen in this way will be called the *ε-point collocation method*. We henceforth require that

$$d \geqq 1 \text{ for } \varepsilon = 0 \text{ and } d \geqq 0 \text{ for } 0 < \varepsilon < 1;$$

this ensures that the left-hand side of (3) makes sense.

Given any $\tau \in \Gamma$ and $\varepsilon \in (0, 1)$, $\varepsilon \neq 1/2$, define the singular integral operator $B_{\tau,d,\varepsilon}$ over the unit circle $\Gamma_0 = \{t \in \mathbf{C}: |t| = 1\}$ as follows:

$$B_{\tau,d,\varepsilon} := \mathscr{A}_{d,\varepsilon}(\tau + 0, \cdot)\, P_{\Gamma_0} + \mathscr{A}_{d,\varepsilon}(\tau - 0, \cdot)\, Q_\Gamma$$

with

$$\mathscr{A}_{d,\varepsilon}(\tau, e^{2\pi i x}) = a(\tau) \sum_{k=0}^{\infty} (x + k)^{-d-1}\, e^{2\pi i k\varepsilon}$$

$$+ b(\tau) \sum_{k=-1}^{-\infty} (x + k)^{-d-1}\, e^{2\pi i k\varepsilon}, \qquad 0 \leqq x \leqq 1.$$

Then the following generalization of Theorem 4.1 holds.

Theorem 4.5. *Let* $a, b \in \mathbf{PC}(\Gamma)$ *and suppose the operator* $A := aP_\Gamma + bQ_\Gamma \in \mathscr{L}(\mathbf{L}_2(\Gamma))$ *is invertible. Then for the ε-point collocation method to be stable it is necessary and sufficient that*

(i) $\mu \frac{a}{b}(\tau + 0) + (1 - \mu) \frac{a}{b}(\tau - 0) \notin (-\infty, 0]$, $\forall\, \tau \in \Gamma$, $0 \leqq \mu \leqq 1$, *if* $\varepsilon = 0$ *and* d *is odd or if* $\varepsilon = 1/2$ *and* d *is even;*

(ii) $\mu \frac{a}{b}(\tau + 0) + (1 - \mu) \frac{a}{b}(\tau - 0) \notin [0, \infty)$, $\forall \tau \in \Gamma$, $0 \leqq \mu \leqq 1$, *if* $\varepsilon = 0$ *and* d *is even or if* $\varepsilon = 1/2$ *and* d *is odd*;

(iii) *the operator* $B_{\tau,d,\varepsilon}$ *be invertible on* $\mathbf{L}_2(\Gamma_0)$ *for all* $\tau \in \Gamma$ *if* $0 < \varepsilon < 1$ *and* $\varepsilon \neq 1/2$.

If the ε*-point collocation is stable, then it converges in* $\mathbf{L}_2(\Gamma)$ *for every right-hand side* $f \in \mathcal{R}(\Gamma)$.

Note that in the case of continuous coefficients a and b the condition (iii) of Theorem 4.5 is equivalent to the requirement that

$$\mathcal{A}_{d,\varepsilon}(\tau, z) \neq 0, \forall \tau \in \Gamma, \forall z \in \Gamma_0.$$

For a sufficiently smooth curve Γ denote by $\mathbf{H}^s$ ($s \in \mathbb{R}$) the periodic Sobolev space of order $s \in \mathbb{R}$, that is, the closure of the collection of all 1-periodic $\mathbf{C}^\infty$-functions with respect to the norm

$$\|f\|_s := \left\{ |\hat{f}_0|^2 + \sum_{0 \neq k \in \mathbb{Z}} |\hat{f}_k|^2 \, |2\pi k|^{2s} \right\}^{1/2},$$

where

$$\hat{f}_k = \int_0^1 e^{-2\pi i k x} f(x) \, dx.$$

Theorem 4.6. *Let* $a, b, f \in \mathbf{H}^s$, $1/2 < s \leqq d + 1$, *and suppose the conditions* (i)–(iii) *of Theorem 4.5 are fulfilled. Then, for all* t *such that* $0 \leqq t < d + 1/2$ *and* $t \leqq s$, *the error estimate*

$$\|x - x_n\|_t \leqq c n^{t-s} \|f\|_s \tag{17}$$

holds, where c *is a constant independent of* f *and* n.

It should be noted that Theorem 4.6 says that under the mentioned hypotheses the convergence order of the ε-point collocation method is optimal (see, for instance, AUBIN [1], Chapter 2).

The Theorems 4. 5 and 4.6 are due to SCHMIDT [3, 4]. The proofs are based upon the technique utilized in Section 4.1 to prove the Theorems 4.1 and 4.4 and also upon the general localization principle established by PRÖSSDORF [26–27]. That condition (i) with $\varepsilon = 0$ and d odd is sufficient for the convergence of the collocation method in the spaces $\mathbf{H}^t$, $t > 1/2$, as well as the corresponding error estimates in the case of smooth coefficients had been previously proved by ARNOLD and WENDLAND [2]. Their error analysis rests on an equivalence established between the collocation methods and certain nonstandard Galerkin methods (in this connection see also PRÖSSDORF [24]). By applying the same techniques, SARANEN and WENDLAND [1] succeeded for $\varepsilon = 1/2$ and even d in the case of constant coefficients. These results were then generalized to the case of variable coefficients by PRÖSSDORF [26–27] on the basis of his abstract localization principle. By means of a further development of the technique worked out by ARNOLD and WENDLAND [2], SCHMIDT [1, 2] then studied the 0-point collocation method for both even and odd d in the spaces $\mathbf{H}^t$, $t \geqq 0$. Note that in the works cited here pseudo-differential operators over closed smooth curves are considered, too.

4.3. The Galerkin method

Let Γ be a simple closed or open Lyapunov curve, let $z = z(s)$, $0 \leqq s \leqq 1$, be its parametric representation, and let $a, b \in \mathbf{PC}(\Gamma)$. Denote by $s_1, \ldots, s_N \in [0, 1]$ all points of discontinuity of the functions a and b, and choose any (not necessarily equidistant) partition $\Delta := \{0 = x_0 < x_1 < \ldots < x_n = 1\}$ containing all the points s_j ($j = 1, \ldots, N$). Let $\mathbf{PS}_d(\Delta)$ denote the space of all 1-periodic splines which are $d - 1$ times continuously differentiable on $[0, 1] \setminus \{s_1, \ldots, s_N\}$ and whose restriction to $[x_k, x_{k+1}]$ $(k = 0, \ldots, n-1)$ is a polynomial of degree less than or equal to d; $\mathbf{PC}_0(\Delta)$ is defined as the corresponding space of piecewise constant functions. Finally, let $h := \max_k (x_{k+1} - x_k)$.

We consider the following Galerkin method for the equation (1): it is sought an approximate solution $x_\Delta \in \mathbf{PS}_d(\Delta)$ satisfying the Galerkin equations

$$(Ax_\Delta, v) = (f, v), \forall\, v \in \mathbf{PS}_d(\Delta). \tag{18}$$

Here $(\cdot, \cdot)$ is the scalar product in $\mathbf{L}_2(\Gamma)$.

Theorem 4.7. *The Galerkin method* (18) *is stable in the space* $\mathbf{L}_2(\Gamma)$ *(as* $h \to 0$*) if and only if the conditions* (i) *and* (ii) *of Theorem 4.1 are satisfied.*

This theorem was stated by Prössdorf and Rathsfeld [2]. Its sufficiency part is an immediate consequence of the following lemma combined with the Remark 2 to Theorem 4.1 and the results of § 1. Note that the sufficiency portion can also be proved by means of the techniques applied to prove Theorem 4.1, which has the advantage that one avoids the use of the condition (iv) of the mentioned remark (see Prössdorf [27]).

Lemma 4.1. *Let* P_Δ *be the orthogonal projection of* $\mathbf{L}_2(\Gamma)$ *onto the subspace* $\mathbf{PS}_d(\Delta)$ $(d \geqq 0)$. *If the function* $f \in \mathbf{PC}(\Gamma)$ *is continuous on* $[0, 1] \setminus \{s_1, \ldots, s_N\}$, *then as* $h \to 0$

$$\|(I - P_\Delta) f P_\Delta\| \to 0, \quad \|P_\Delta f(I - P_\Delta)\| \to 0.$$

In the case where Γ is the unit circle, $f \in \mathbf{C}(\Gamma)$, and Δ is an equidistant partition, Lemma 4.1 follows immediately from the property (iv) in 4.1.1. For the case where Γ and f are supposed to be sufficiently smooth and Γ is an equidistant partition, Lemma 4.1 (even for the norms $\|\cdot\|_s$) was independently proved by Prössdorf [26] and by Arnold and Wendland [3]; the general case was settled by Prössdorf and Rathsfeld [2].

If Γ is sufficiently smooth, then one has also the following result (see Hsiao and Wendland [2] and Stephan and Wendland [1]).

Theorem 4.8. *Let* a *and* b *be sufficiently smooth and suppose the condition* (4) *is satisfied. Then, if* $-d-1 \leqq t \leqq 0 \leqq s \leqq d+1$ *and* $f \in \mathbf{H}^s$, *for the Galerkin method* (18) *the error estimate*

$$\|x - x_\Delta\|_t \leqq c h^{s-t} \|f\|_s \tag{19}$$

holds. If, in addition, all partitions Δ *are quasiuniform (that is, if* $h \leqq \varrho \min_k (x_{k+1} - x_k)$ *with some* $\varrho > 0$ *independent of* h*) then* (19) *is also true for* $0 \leqq t \leqq s \leqq d+1$ *and* $t < d + 1/2$.

Remark 1. For the case of piecewise smooth datas Prössdorf and Rathsfeld [2] stated error estimates in the $\mathbf{L}_2$ norm for equidistant as well as for certain nonuniform partitions by making use of the complete asymptotic expansion of the solution. In this way they generalized some results of Elschner [5] concerned with the case of an interval (for more about this see also Section 4.4).

Remark 2. The restrictions imposed on the parameters t and s in the Sobolev index inequalities (17) and (19) show that – for the same splines – the collocation method converges at most with the order h^{d+1} $(h = 1/n)$, whereas the Galerkin method converges at most with the order h^{2d+2}. Note that these restrictions are sharp (see Arnold and Wendland [2]). Thus, to obtain the same order of convergence for both methods, we must choose

$$d_C = 2d_G + 1, \tag{20}$$

where d_C denotes the spline degree for the collocation method and d_G that for the Galerkin method. On the other hand, when working with Galerkin's procedure we have to carry out double integrations in order to compute the scalar products in (18), while the collocation method only requires one integration. These effects must be taken into account when choosing a method of numerical integration. After all, the computing times and costs are more or less the same for both methods provided (20) holds (see also Wendland [5] and Arnold and Wendland [1]). In this connection notice that the ε-point collocation method converges for a larger class of equations (compare the Theorems 4.5 and 4.7). The placing of the collocation points is of course decisive for the convergence.

4.4. Singular integral equations over an interval

Considering singular equations over an interval, mention must be made of the papers by Washizu and Ikegawa [1], Dang and Norrie [1], Jen and Srivastav [1], and Gerasoulis [1]. These authors obtained numerical results for Galerkin and collocation methods with splines applied to the Cauchy singular operator S. However, with the exception of Thomas [1], no convergence results have yet been proved for these methods.

In this section we survey some new results on the convergence of Galerkin and collocation methods with splines obtained by Elschner [5] and by Prössdorf and Rathsfeld [1, 2]. The material presented in the following is essentially the paper of Prössdorf and Elschner [1].

1. Consider the singular integral operator

$$A = aI + bS, \qquad (Su)(x) = (\pi \mathrm{i})^{-1} \int_0^1 u(y)\,(x-y)^{-1}\,\mathrm{d}y, \tag{21}$$

where a and b are continuous functions on $[0, 1]$ and I is the identity operator. Let $-1 < \beta_0, \beta_1 < 1$ and $\varrho(x) = x^{\beta_0}(1-x)^{\beta_1}$. Then $\mathbf{L}_2(\varrho^{1/2}) := \varrho^{-1/2}\,\mathbf{L}_2$, where $\mathbf{L}_2 := \mathbf{L}_2(0, 1)$, is a Hilbert space with the scalar product

$$(\varrho u, v) = \int_0^1 \varrho(x)\,u(x)\,\overline{v(x)}\,\mathrm{d}x$$

and A is in $\mathscr{L}(\mathbf{L}_2(\varrho^{1/2}))$ (see § 3, Chapter II). Denote by Γ_0 the oriented boundary of the rectangle $\{0 \leqq x \leqq 1,\ -1 \leqq z \leqq 1\}$ in the x-z-plane. Following Gohberg and Krupnik [4] (Chapter IX, § 5), we associate with the operator (21) the symbol

$$\sigma_{A,\varrho}(x, z) = a(x) + b(x)\,\Omega_\varrho(x, z), \qquad (x, z) \in \Gamma_0, \tag{22}$$

where $\Omega_\varrho(x, z) = 0$ for $x \in (0, 1)$ and

$$\Omega_\varrho(x, z) = [z(1 + \alpha_j^2) - \mathrm{i}(1 - z^2)\,\alpha_j(-1)^j]\,[1 + z^2\alpha_j^2]^{-1}$$

for $x = j$ $(j = 0, 1)$. Here $\alpha_j = \cot \pi(1 + \beta_j)/2$ $(j = 0, 1)$. The range of (22) is a closed oriented curve; for $x = j$ $(j = 0, 1)$, it traces out a circular arc joining $a(j) - b(j)$ and $a(j) + b(j)$. Note that $\Omega_\varrho(x, z) = z$ for all $x \in [0, 1]$ in case $\varrho \equiv 1$. Theorem 5.1, Chapter IV, implies that the operator (21) is Fredholm on $\mathbf{L}_2(\varrho^{1/2})$ if and only if $\sigma_{A,\varrho}(x, z) \neq 0$ for all $(x, z) \in \Gamma_0$ and that then its invertibility is equivalent to the equality $\operatorname{ind}_{\Gamma_0} \sigma_{A,\varrho} = 0$, where ind, as usual, refers to the winding number around the origin.

We say the operator A *satsfies a Gårding inequality* on $\mathbf{L}_2(\varrho^{1/2})$ if A can be represented as $A = A_0 + T$, where $T \in \mathscr{K}(\mathbf{L}_2(\varrho^{1/2}))$ and

$$\operatorname{Re}\,(\varrho A_0 u, u) \geqq c(\varrho u, u), \qquad \forall\, u \in \mathbf{L}_2(\varrho^{1/2}), \tag{23}$$

with a positive constant c independent of u. It is well known that for the operator A the Galerkin method with respect to any orthogonal basis in $\mathbf{L}_2(\varrho^{1/2})$ converges if and only if A satisfies a Gårding inequality (see, e.g. Gohberg and Feldman [1]).

Lemma 4.2. *For the singular integral operator* (21) *to satisfy a Gårding inequality on* $\mathbf{L}_2(\varrho^{1/2})$ *it is necessary and sufficient that*

$$\operatorname{Re}\,\sigma_{A,\varrho}(x, z) > 0, \qquad \forall\,(x, z) \in \Gamma_0. \tag{24}$$

Note that in the case $\varrho \not\equiv 1$ the condition (24) is in general stronger than the condition

$$\operatorname{Re}\{a(x) \pm b(x)\} > 0, \qquad \forall\, x \in [0, 1]. \tag{25}$$

However, if $\varrho \equiv 1$, then the conditions (24) and (25) are equivalent and imply the invertibility of the operator on $\mathbf{L}_2$.

Under the hypotheses (25) we have $-1/2 < \operatorname{Re} \varkappa_j < 1/2$ $(j = 0, 1)$, where

$$\varkappa_0 = \theta(0),\ \varkappa_1 = -\theta(1),\ \theta(x) = \frac{1}{2\pi i} \ln \frac{a(x) + b(x)}{a(x) - b(x)} \tag{26}$$

and ln denotes that branch of the logarithm which is continuous in $\mathbb{C} \setminus (-\infty, 0]$ and takes real values on the positive real axis. The following lemma shows that the conditions (24) and (25) are also equivalent for certain weights $\varrho \not\equiv 1$.

Lemma 4.3. *Let* $\varrho(x) = x^{\beta_0}(1 - x)^{\beta_1}$, *where*

$$\min(0, -2 \operatorname{Re} \varkappa_j) \leqq \beta_j \leqq \max(0, -2 \operatorname{Re} \varkappa_j),\ j = 0, 1. \tag{27}$$

Then (25) *implies* (24).

The operator A is said to be *strongly elliptic* on $\mathbf{L}_2(\varrho^{1/2})$ if there exists a continuous function ϑ on $[0, 1]$ such that the operator ϑA satisfies a Gårding inequality on $\mathbf{L}_2(\varrho^{1/2})$ (compare this with Remark 2 to Theorem 4.1). The Galerkin method with finite elements as trial functions converges for every invertible strongly elliptic operator (see Prössdorf [27]).

Lemma 4.4. *For the singular integral operator* (21) *to be strongly elliptic on* $\mathbf{L}_2(\varrho^{1/2})$ *it is necessary and sufficient that*

$$\begin{aligned} &a(x) + b(x)\,\lambda \neq 0, \\ &\forall\, x \in [0, 1], \qquad \forall\, \lambda \in \operatorname{conv}\{\Omega_\varrho(x, z) : -1 \leqq z \leqq 1\}, \end{aligned} \tag{28}$$

where conv *denotes the convex hull.*

Since the set $\{\Omega_\varrho(x, z) : (x, z) \in \Gamma_0\}$ is just the essential spectrum $\sigma_{\text{ess}}(S)$ of the Cauchy singular operator (see Theorem 5.1, Chapter IV), it follows that the condition

$$a(x) + b(x)\,\lambda \neq 0, \qquad \forall\, x \in [0, 1], \qquad \forall\, \lambda \in \operatorname{conv} \sigma_{\text{ess}}(S) \tag{29}$$

implies the condition (28). Notice however that (28) and (29) are equivalent in the case of constant coefficients a and b. Also notice that condition (28) is equivalent to

$$\begin{aligned} &|2 \operatorname{Re} \varkappa_j + \beta_j| \leqslant 1 \quad (j = 0, 1) \quad \text{and} \\ &a(x) + b(x)\,\lambda \neq 0, \qquad \forall\, x \in [0, 1], \qquad \forall\, \lambda \in [-1, 1]. \end{aligned} \tag{30}$$

2. In this subsection we consider the dominant singular integral equation

$$(Au)(x) := a(x)\,u(x) + b(x)\,(Su)(x) = f(x) \tag{31}$$

on the interval (0, 1) and give a convergence analysis for Galerkin's method with splines for its approximate solution. For a more detailed treatment of these questions see Elschner [5].

We first collect some basic facts concerning the regularity and the asymptotics of the solutions of equation (31). For an arbitrary real number $s \geqq 0$, let $\mathbf{H}^s$ denote the usual Sobolev space of order s on (0, 1) and $\|\cdot\|_s$ the norm in $\mathbf{H}^s$. Let $\varkappa_j$ $(j = 0, 1)$ be the numbers defined in (26).

Theorem 4.9. *Suppose the condition* (25) *is satisfied, let a and b be in $\mathbf{H}^1$, and assume that*

$$0 < s < \operatorname{Re} \varkappa_j + 1/2 \quad (j = 0, 1). \tag{32}$$

Also suppose that $u \in \mathbf{L}_2$ and $Au \in \mathbf{H}^s$. Then $u \in \mathbf{H}^s$.

The following theorem gives the complete asymptotics of the solutions at the endpoints 0 and 1. It also shows that Theorem 4.9 is, in general, no longer true if (32) is violated.

Theorem 4.10. *Suppose* (25) *is fulfilled, let $a, b \in \mathbf{H}^{k+1}$ and $f \in \mathbf{H}^k$, where $k \geqq 1$ is an integer. Then the unique $\mathbf{L}_2$-solution of* (31) *has the representation*

$$u(x) = \varphi_0(x)\, x^{\varkappa_0} \sum_{m=0}^{k-1} \sum_{j=0}^{m} c_{mj} x^m \ln^j x$$

$$+ \varphi_1(x)\,(1-x)^{\varkappa_1} \sum_{m=0}^{k-1} \sum_{j=0}^{m} d_{mj} (1-x)^m \ln^j (1-x) + u_k(x), \tag{33}$$

where $c_{mj}, d_{mj} \in \mathbf{C}$, $u_k \in \mathbf{H}^k$, $u_k^{(j)}(0) = u_k^{(j)}(1) = 0$ $(j = 0, \ldots, k-1)$, $\varphi_0 \in \mathbf{C}_0^\infty(\mathbf{R})$ is some cut-off function such that $\varphi_0 = 1$ for $0 \leqq x \leqq 1/2$ and $\varphi_0 = 0$ for $x \geqq 2/3$, and $\varphi_1(x) := \varphi_0(1-x)$.

In the case of constant coefficients a and b the logarithmic terms in (33) *do not occur.*

Our next concern is the standard Galerkin method with smoothest splines as trial functions. Let $\mathbf{S}_d(\Delta)$ $(d \geqq 0)$ be the space of $d-1$ times continuously differentiable splines of degree d subordinate to the partition $\Delta = \{0 = x_0 < x_1 < \ldots < x_n = 1\}$ of the interval $[0, 1]$. Note that $\mathbf{S}_d(\Delta) \subset \mathbf{H}^s$ if and only if $s < d + 1/2$. Let $\bar{h} = \max (x_k - x_{k-1})$ and $\underline{h} = \min (x_k - x_{k-1})$, $1 \leqq k \leqq n$. A mesh Δ is said to be *γ-quasiuniform* $(\gamma > 0)$ if $\bar{h} \leqq \gamma \underline{h}$. The collection of all γ-quasiuniform meshes will be denoted by $\mathcal{D}_\gamma$. The following properties of spline spaces play an important role in the error analysis of Galerkin methods. (see Babushka and Aziz [1], Arnold and Wendland [2], Elschner and Schmidt [1])

Approximation property (a): Let $0 \leqq s \leqq r \leqq d + 1$ and $s < d + 1/2$. Then for any $u \in \mathbf{H}^r$ and any partition Δ there exists a $u_\Delta \in \mathbf{S}_d(\Delta)$ such that $\|u - u_\Delta\|_s \leqq c(s)\, \bar{h}^{r-s} \|u\|_r$ for all s (here $c(s)$ denotes a constant which does not depend on u and Δ).

Inverse property (i): Let $0 \leqq t \leqq s < d + 1/2$ and $\gamma > 0$. Then there exists a constant c such that $\|v\|_s \leqq c\bar{h}^{t-s} \|v\|_t$ for all $v \in \mathbf{S}_d(\Delta)$ and $\Delta \in D_\gamma$.

The standard Galerkin method with splines for the approximate solution of equation (31) consists in seeking an element $u_\Delta \in \mathbf{S}_d(\Delta)$ satisfying the Galerkin equations

$$(Au_\Delta, v_\Delta) = (f, v_\Delta) \text{ for all } v_\Delta \in \mathbf{S}_d(\Delta). \tag{34}$$

Theorem 4.11. *Let* (25) *be satisfied. Then if $\bar{h}$ is sufficiently small the equations* (34) *are uniquely solvable for any $f \in \mathbf{L}_2$, the approximate solutions u_Δ converge in $\mathbf{L}_2$ to the exact solution u, and the error estimate*

$$\|u - u_\Delta\|_0 \leqq c \min_{v \in \mathbf{S}_d(\Delta)} \|u - v\|_0 \tag{35}$$

holds. If, in addition, $a, b, f \in \mathbf{H}^1$ and $\Delta \in \mathcal{D}_\gamma$ for $\gamma > 0$, then

$$\|u - u_\Delta\|_s \leqq c\bar{h}^{t-s} \|u\|_t \tag{36}$$

for any s and t such that $0 \leqq s \leqq t < \operatorname{Re} \varkappa_j + 1/2$ $(j = 0, 1)$ and $s < d + 1/2$.

The first part of Thorem 4.11 follows from Lemma 4.2 applied with $\varrho \equiv 1$ and from standard theory of Galerkin's method (cf. § 1 and STRANG and FIX [1]). If $a, b, f \in \mathbf{H}^1$, then Theorem 4.9 implies that $u \in \mathbf{H}^t$, and from the estimate (35) and property (a) we get (36) with $s = 0$. Finally, the Aubin-Nitsche duality argument, the estimate (35), and the properties (a) and (i) give (36) for $s > 0$.

Note that in formula (36) $t - s$ is always less than 1 and that, for $d > 1$, the order of convergence cannot be increased, in general. Thus, we have quasioptimal error estimates in a scale of Sobolev spaces which is, however, strictly limited by the lack of regularity of the solution at the endpoints.

We now derive improved error estimates for a variety of Galerkin methods using smoothest splines multiplied by a weight function which reflects the principal term of the asymptotics. These Galerkin methods are nonstandard in the sense that the pairing of test and trial functions is made in the scalar product of weighted $\mathbf{L}_2$ spaces.

Put $\varrho_0(x) = x^{\varkappa_0}(1-x)^{\varkappa_1}$ and $\tilde{\mathbf{S}}_d(\Delta) = \varrho_0 \cdot \mathbf{S}_d(\Delta)$, Further, let $\varrho(x) = x^{\beta_0}(1-x)^{\beta_1}$ be a weight function on $[0, 1]$ which satisfies condition (27). Then (25) implies that the operator (21) is invertible on $\mathbf{L}_2(\varrho^{1/2})$. We approximate the solution of equation (31) with a right-hand side $f \in \mathbf{L}_2(\varrho^{1/2})$ by the solutions $u_\Delta \in \tilde{\mathbf{S}}_d(\Delta)$ of the Galerkin equations

$$(\varrho A u_\Delta, v_\Delta) = (\varrho f, v_\Delta) \text{ for all } v_\Delta \in \tilde{\mathbf{S}}_d(\Delta). \tag{37}$$

Note that the scalar products in (37) are well defined, since $\operatorname{Re} \varkappa_j > -1/2$ and (27) holds, and hence $u_\Delta, v_\Delta \in \mathbf{L}_2(\varrho^{1/2})$. Let

$$\nu = \max_{j=0,1} (0, -\operatorname{Re} \varkappa_j - \beta_j/2), \qquad \mu = \max_{j=0,1} (0, \operatorname{Re} \varkappa_j + \beta_j/2).$$

Theorem 4.12. *Suppose* (25) *and* (27) *are satisfied. Then if* $\bar{h}$ *is sufficiently small, the equations* (37) *are uniquely solvable for each* $f \in \mathbf{L}_2(\varrho^{1/2})$, *the approximate solutions* u_Δ *converge in* $\mathbf{L}_2(\varrho^{1/2})$ *to the exact solution* u, *and the error estimate*

$$\|\varrho^{1/2}(u - u_\Delta)\|_0 \leq c \min_{v \in \tilde{\mathbf{S}}_d(\Delta)} \|\varrho^{1/2}(u - v)\|_0 \tag{38}$$

holds. If, in addition, $a, b \in \mathbf{H}^3$ *and* $f \in \mathbf{H}^2$, *then*

$$\|\varrho^{1/2}(u - u_\Delta)\|_0 \leq c\bar{h}^{t-\nu} \|\varrho_0^{-1} u\|_t \tag{39}$$

and, for $\Delta \in \mathcal{D}_\gamma$,

$$\|\varrho_0^{-1}(u - u_\Delta)\|_s \leq c\bar{h}^{t-s-\nu-\mu} \|\varrho_0^{-1} u\|_t \tag{40}$$

for any s *and* t *satisfying* $\nu \leq t < 3/2$, $0 \leq s \leq t \leq d + 1$, *and* $s < d + 1/2$.

The estimate (38) can be proved with the help of Lemma 4.3 and standard techniques for Galerkin methods. If $a, b \in \mathbf{H}^3$ and $f \in \mathbf{H}^2$, then Theorem 4.10 with $k = 2$ implies that $\varrho_0^{-1} u \in \mathbf{H}^t$ for all $t < 3/2$ and so the estimates (39) and (40) can be verified using (38) and the properties (a) and (i) for $\mathbf{S}_d(\Delta)$.

In (39) and (40) the best asymptotic error estimates are obtained for $\beta_j = -2 \operatorname{Re} \varkappa_j$ $(j = 0, 1)$, since then $\nu = \mu = 0$. If $\varkappa_j \in \mathbb{R}$ and $\beta_j = -\varkappa_j$ $(j = 0, 1)$, the Galerkin method (37) coincides with the method of weighted residuals proposed by WASHIZU and IKEGAWA [1].

Note that, for $d > 1$, we have, in general, no increase of the order of convergence in Theorem 4.12. However, in the case of constant coefficients a and b, one can show that the estimates (39) and (40) hold for any s and t satisfying $\nu \leq t \leq k + 1 - |\operatorname{Re} \varkappa|$, $0 \leq s \leq t \leq d + 1$, $s < d + 1/2$, if only $f \in \mathbf{H}^{k+1}$ $(k \in \mathbb{N})$. Here $\varkappa = \varkappa_0 = -\varkappa_1 = (2\pi i)^{-1} \ln \{(a + b)/(a - b)\}$.

Finally, we shall see that weighted continuous splines on special nonuniform partitions give $O(n^{-d})$ as the asymptotic rate of convergence on $\mathbf{L}_2$, where n is the number of mesh points and $d - 1$ the degree of the splines.

Given a fixed $n \in \mathbb{N}$ define a partition $\Delta_n = \{x_k\}_0^{2n}$ of the interval $[0, 1]$ as follows:

$$x_k = 2^{-1} (k/n)^{(2d+3)/(3+2\operatorname{Re}\varkappa_0)},$$

$$x_{2n-k} = 1 - 2^{-1}(k/n)^{(2d+3)/(3+2\operatorname{Re}\varkappa_1)}, \; k = 0, \ldots, n.$$

Note that the partitions Δ_n $(n \in \mathbb{N})$ are not γ-quasiuniform for any $\gamma > 0$. Let $\mathbf{P}_d(\Delta_n)$ $(d \geqq 1)$ denote the set of piecewise polynomials of degree $\leqq d$ which are continuous on $[0, 1]$ and whose breakpoints are at the x_k, $k = 1, \ldots, 2n - 1$. We introduce the weighted spline spaces

$$\tilde{\mathbf{P}}_d(\Delta_n) = \varrho_0 \cdot \mathbf{P}_d(\Delta_n)$$

and take as an approximate solution of equation (31) the solution $u_n \in \tilde{\mathbf{P}}_d(\Delta_n)$ of the Galerkin equations

$$(Au_n, v_n) = (f, v_n) \text{ for all } v_n \in \tilde{\mathbf{P}}_d(\Delta_n). \tag{41}$$

Theorem 4.13. *Suppose* (25) *holds. Then if n is sufficiently large, the equations* (41) *have a unique solution for each $f \in \mathbf{L}_2$, the approximate solutions u_n converge in $\mathbf{L}_2$ to the exact solution u, and one has the error estimate*

$$\|u - u_n\|_0 \leqq c \min_{v \in \tilde{\mathbf{P}}_d(\Delta_n)} \|u - v\|_0. \tag{42}$$

If, in addition, $a, b \in \mathbf{H}^{d+3}$ and $f \in \mathbf{H}^{d+2}$, then $\|u - u_n\|_0 = O(n^{-d-1})$ as $n \to \infty$.

The estimate (42) is a consequence of Lemma 4.2 for $\varrho \equiv 1$. To prove the second part of Theorem 4.13, apply Theorem 4.10 with $k = d + 1$ and construct elements $v_n \in \tilde{\mathbf{P}}_d(\Delta_n)$ so that $\|u - v_n\|_0 = O(n^{-d-1})$ as $n \to \infty$.

In the paper of Thomas [1] the special case where $a \equiv 1$ and b is purely imaginary was considered. However, the proof of the corresponding result given there is incorrect, since it does not take into account the complete asymptotics of the solutions of equation (31).

With the aid of Lemma 4.4 it can be proved that the Theorems 4.11 and 4.13 remain valid if condition (25) is replaced by (30) and that Theorem 4.12 continues to hold with the conditions (25) and (27) replaced by (28).

3. We finally formulate two convergence theorems for the collocation method utilizing piecewise linear splines. These theorems are immediate consequences of Theorem 4.1 (see Prössdorf and Rathsfeld [1]).

The sequence of collocation points on $[0, 1]$ is given by $x_k = k/n$ $(k = 0, 1, \ldots, n;$ $n = 1, 2, \ldots)$. We look for an approximate solution of equation (31) in the form of a polygon $u_n = \sum_{k=1}^{n-1} \xi_k \varphi_k^{(n)}$, where

$$\varphi_k^{(n)}(x) = \begin{cases} (x - x_{k-1})/(x_k - x_{k-1}) & \text{for } x_{k-1} \leqq x \leqq x_k, \\ (x_{k+1} - x)/(x_{k+1} - x_k) & \text{for } x_k \leqq x \leqq x_{k+1}, \\ 0 & \text{otherwise.} \end{cases}$$

The coefficients $\xi_k = \xi_k^{(n)}$ must be determined so that the equation $Au_n = f$ be satisfied at the point x_k, i.e.

$$a(x_k)\, u_n(x_k) + b(x_k)\, (Su_n)\, (x_k) = f(x_k), \qquad k = 1, \ldots, n - 1. \tag{43}$$

Theorem 4.14. *Let a and b be continuous functions on* $[0, 1]$. *Then the spline collocation method* (43) *converges in* $\mathbf{L}_2$ *for all bounded and Riemann integrable functions f on* $[0, 1]$ *if and only if condition* (30) *is satisfied.*

Theorem 4.15. *Let a and b be piecewise continuous functions on* $[0, 1]$ *and suppose that the following conditions are fulfilled:*

(i) $d(x) := a(x) - b(x) \neq 0, \quad \forall\, x \in [0, 1];$

(ii) $\mu \frac{c}{d}(x + 0) + (1 - \mu) \frac{c}{d}(x - 0) \notin (-\infty, 0]$

for all $0 < x < 1$ *and all* $0 \leq \mu \leq 1$, *where* $c := a + b$;

(iii) $\mu c(0 + 0) + (1 - \mu)\, d(0 + 0) \neq 0,$

$\mu c(1 - 0) + (1 - \mu)\, d(1 - 0) \neq 0$ for all $0 \leq \mu \leq 1$.

Then the spline collocation method (43) *converges in* $\mathbf{L}_2$ *for all bounded Riemann integrable functions f on* $[0, 1]$. *The conditions* (i)–(iii) *are necessary for the stability of the method* (43).

Remark 1. Analogous results can be obtained for the ε-point collocation method (see Section 4.2 as well as the paper of Schmidt [5]).

Remark 2. The convergence and error analysis for more general spline collocation methods, in particular for methods working with trial functions taken from the weighted spline space $\tilde{\mathbf{S}}_d(\Delta)$, is still an open but, as we think, important problem. Some results concerning this problem were recently obtained by Elschner [7].

§ 5. The approximate solution of singular integral equations with degenerate symbol

The present section is concerned with the application of the approximation methods considered in the Sections 3.1–3.4 to singular integral equations of the form (3.1) for the case where the functions (3.2) have a finite number of zeros of integral order on the unit circle Γ. Thus let

$$a(t) = \prod_{i=1}^{k_1} (t^{-1} - \alpha_i^{-1})^{m_i}\, a_1(t), \qquad b(t) = \prod_{j=1}^{k_2} (t - \beta_j)^{n_j}\, b_1(t), \tag{1}$$

where α_i ($i = 1, \ldots, k_1$) and β_j ($j = 1, \ldots, k_2$) are points on the unit circle and the continuous functions a_1 and b_1 do not vanish on Γ. In what follows let

$$l = \max\,(m_1, \ldots, m_{k_1}, n_1, \ldots, n_{k_2}).$$

5.1. The reduction method

Theorem 5.1. *Let a_1 and b_1 (see the representations* (1)) *be functions belonging to the class* $\mathbf{C}^{l,\mu}(\Gamma)$ $(0 < \mu < 1)$ *which satisfy the following conditions:*

1. $a_1(t) \neq 0, \quad b_1(t) \neq 0 \ (|t| = 1).$
2. $\operatorname{ind} a_1 = \operatorname{ind} b_1 = 0.$

Furthermore, let T be a compact operator from $\mathbf{L}_p(\Gamma)$ $(1 < p < \infty)$ *into the space* $\mathbf{W}_p^{(l)}(\Gamma)$ *and suppose* $\dim \operatorname{Ker} A = 0$.

Then for each function $f \in \mathbf{W}_p^{(l)}(\Gamma)$ *and for all sufficiently large n the system* (3.3) *has exactly one solution* $(\xi_k^{(n)})_{k=-n}^{k}$ *and, as* $n \to \infty$, *the functions* (3.4) *converge in the norm of the space* $\mathbf{L}_p(\Gamma)$ *to the solution* $\varphi \in \mathbf{L}_p(\Gamma)$ *of the equation* (3.1).

If $a_1, b_1 \in \mathbf{C}^{l+r,\mu}(\Gamma)$, $(r > 0)$, $f \in \mathbf{W}_p^{(l+r)}(\Gamma)$, and T maps $\mathbf{L}_p(\Gamma)$ into $\mathbf{W}_p^{(l+r)}(\Gamma)$, then

$$\|\varphi - \varphi_n\|_{\mathbf{L}_p} = O(n^{-r}).$$

Proof. Apply Theorem 2.4 with $X = \mathbf{L}_p(\Gamma)$, $Z = Z_1 = Z_0 = \mathbf{W}_p^{(l)}(\Gamma)$, $c = a_1$, $d = b_1$, $\mathfrak{M} = \mathbf{C}^{l,\mu}(\Gamma)$,

$$r_-(t) = \prod_{i=1}^{k_1} (t^{-1} - \alpha_i^{-1})^{m_i}, \qquad r_+(t) = \prod_{j=1}^{k_2} (t - \beta_j)^{n_j}, \tag{2}$$

and $P_n = Q_n$ given by (3.6).

It follows from the results of § 6, Chapter II, and § 4, Chapter VI, that the conditions a), b), h), and i) of § 2 are fulfilled. That the remaining conditions are satisfied can be easily checked in a straightforward fashion. Thus, what results is the first part of Theorem 5.1 and the estimate

$$\|\varphi - \varphi_n\|_{\mathbf{L}_p} \leqq \gamma \, \|y_0 - P_n y_0\|_{\mathbf{W}_p^{(l)}}$$

with $y_0 = (Pc_- + Q\,d_+)\,B\varphi$, where $a_1 = c_+c_-$ and $b_1 = d_-\,d_+$ (note that $a_1, b_1 \in \mathbf{C}^{l,\mu}(\Gamma)$).

Now let $a_1, b_1 \in \mathbf{C}^{l+r,\mu}(\Gamma)$ and $f \in \mathbf{W}_p^{(l+r)}(\Gamma)$. Then both the functions $c_\pm$, $d_\pm$ and their inverses belong to $\mathbf{C}^{l+r,\mu}(\Gamma)$ and from (2.7) and (2.8) we get

$$y_0 = (c_+^{-1}\,P + d_-^{-1}Q)\,(f - T_1\varphi)$$

with $T_1\varphi = (c_+Qc_-r_-P + d_-P\,d_+r_+Q)\,\varphi + T\varphi \in \mathbf{W}_p^{(l+r)}(\Gamma)$. Hence, $y_0 \in \mathbf{W}_p^{(l+r)}(\Gamma)$ and this together with (3.7′) shows that

$$\|y_0 - P_n y_0\|_{\mathbf{W}_p^{(l)}} = \sum_{j=0}^{l} \|f^{(j)} - P_n(f^{(j)})\|_{\mathbf{L}_p} = O(n^{-r}). \;\blacksquare \tag{3}$$

Remark. If the condition 1 of Theorem 5.1 is satisfied and if ind $a_1 =$ ind b_1, then dim Ker $(Pa + Qb) = 0$. If, in addition, the points α_i and β_j $(i = 1, \ldots, k_1; j = 1, \ldots, k_2)$ are pairwise distinct, then also dim Ker $A_0 =$ dim Ker $(aP + bQ) = 0$ (see § 5, Chapter VI).

5.2. The collocation method

Theorem 5.2. *Let a_1 and b_1 (see the representations* (1)) *be functions from $\mathbf{C}^{l+1}(\Gamma)$ which satisfy the following conditions:*

1. $a_1(t) \neq 0, \quad b_1(t) \neq 0 \; (|t| = 1)$.
2. ind $a_1 =$ ind b_1.

Furthermore, suppose T is a compact operator from $\mathbf{L}_p(\Gamma)$ $(1 < p < \infty)$ into the space $\mathbf{C}^{l,\mu}(\Gamma)$ $(0 < \mu < 1)$ and suppose dim Ker $A = 0$.

Then for each function $f \in \mathbf{C}^{l,\mu}(\Gamma)$ and for all sufficiently large n the system (3.14) *has precisely one solution $(\xi_k^{(n)})_{k=-n}^{n}$ and, as $n \to \infty$, the functions* (3.4) *converge in the norm of $\mathbf{L}_p(\Gamma)$ to the solution $\varphi \in \mathbf{L}_p(\Gamma)$ of the equation* (3.1). *One has*

$$\|\varphi - \varphi_n\|_{\mathbf{L}_p} = O(n^{-\mu}). \tag{4}$$

If $a_1, b_1 \in \mathbf{C}^{l+r,\mu}(\Gamma)$ $(r > 0)$, $\quad f \in \mathbf{C}^{l+r,\mu}(\Gamma)$, and T maps $\mathbf{H}^\nu(\Gamma)$ into $\mathbf{C}^{l+r,\mu}(\Gamma)$ $(0 < \nu < 1)$, then

$$\|\varphi - \varphi_n\|_{\mathbf{L}_p} = O(n^{-r-\lambda}) \tag{5}$$

for every λ such that $0 < \lambda < \mu$.

Proof. Apply Theorem 2.3 with $X = \mathbf{L}_p(\Gamma)$, $Z = \mathbf{W}_p^{(l)}(\Gamma)$, $Z_0 = Z_1 = \mathbf{C}^l(\Gamma)$, $\mathfrak{M} = \mathbf{C}^{l+1}(\Gamma)$, $c = a_1$, $d = b_1$, $r_\pm$ given by (2), and the projections given by (3.6) and (3.15).

Recall § 6 of Chapter II and § 4 of Chapter VI to see that the conditions a), h), and i) of § 2 are fulfilled. From Theorem 6.3, Chapter II, and the continuity of the embedding $\mathbf{W}_p^{(l+1)}(\Gamma) \subset \mathbf{C}^{l,1-1/p}$ we deduce that condition b) of § 2 is fulfilled, too. It can be easily checked in a straightforward way that the conditions c), d), and f) are fulfilled. Finally, as for condition e) of § 2, note that the estimate

$$\|g - Q_n g\|_{\mathbf{W}_p^{(l)}} = O(n^{-r-\mu}), \qquad \forall g \in \mathbf{C}^{l+r,\mu}(\Gamma) \tag{6}$$

holds (see, for instance, Prössdorf and Silbermann [11], formula (2.3.5)). Thus, all hypotheses of Theorem 2.3 are fulfilled and we so obtain both the first assertion of Theorem 5.2 and the estimate

$$\|\varphi - \varphi_n\|_{\mathbf{L}_p} \leqq \gamma \, \|y_0 - Q_n y_0\|_{\mathbf{W}_p^{(l)}} \tag{7}$$

with $y_0 = (Pc_- + Qc_+^{-1}) \, B\varphi$ and $b_1^{-1}a_1 = c_+c_-$. From (2.5) and (2.6) we see that y_0 is in $\mathbf{C}^{l,\nu}(\Gamma)$, where $\nu = \min\,(\mu, 1 - 1/p)$. Taking into account the formula for the inverse B^{-1} (see § 4, Chapter VI) we then conclude that $\varphi \in \mathbf{H}^\nu(\Gamma)$. Once again using (2.5) and (2.6) we deduce that $T'\varphi \in C^{l,1-\varepsilon}(\Gamma)$ for any ε, $0 < \varepsilon < 1$, and $y_0 = c_+^{-1}b_1^{-1}(f - T'\varphi) \in \mathbf{C}^{l,\mu}(\Gamma)$. The estimate (4) now follows from (6) and (7) (with $r = 0$).

The preceding arguments show that under the stronger conditions imposed upon a_1, b_1, f, and T the function y_0 is in $\mathbf{C}^{l+r,\lambda}(\Gamma)$ for any λ such that $0 < \lambda < \mu$. So the estimate (5) is also seen to be a consequence of (6) and (7). ∎

5.3. The method of mechanical quadratures

Now consider the equation (3.17) with a kernel $K(t, \tau) \in \mathbf{C}^{l,\mu}(\Gamma \times \Gamma)$. Denote by T the integral operator whose kernel is $K(t, \tau)$.

Theorem 5.3. *Let the conditions of Theorem 5.2 be satisfied and suppose $K(t, \tau) \in \mathbf{C}^{l,\mu}(\Gamma \times \Gamma)$. Then for each function $f \in \mathbf{C}^{l,\mu}(\Gamma)$ and for all sufficiently large n the system (3.18) has exactly one solution $(\xi_k^{(n)})_{k=-n}^{n}$ and, as $n \to \infty$, the functions (3.4) converge in the norm of the space $\mathbf{L}_p(\Gamma)$ to the solution $\varphi \in \mathbf{L}_p(\Gamma)$ of the equation (3.18). Furthermore, the estimate*

$$\|\varphi - \varphi_n\|_{\mathbf{L}_p} = O(n^{-\lambda}) \tag{8}$$

holds for every λ such that $0 < \lambda < \mu$.

If $a_1, b_1 \in \mathbf{C}^{l+r,\mu}(\Gamma)$ $(r > 0)$ and $K(t, \tau) \in \mathbf{C}^{l+r,\mu}(\Gamma \times \Gamma)$, then

$$\|\varphi - \varphi_n\|_{\mathbf{L}_p} = O(n^{-r-\lambda}) \tag{9}$$

for every λ such that $0 < \mu < \lambda$.

Proof. Taking into account the preceding theorem and Theorem 1.4 this can be proved analogously as Theorem 3.5. We retain the notation introduced in the proof of Theorem 3.5.

Due to the results of § 5, Chapter VI, the solution of the equation (3.17) is in $\mathbf{L}_p(\Gamma)$ for all p such that $1 < p < \infty$. Therefore, we can again restrict ourselves to sufficiently large p. Since

$$\|Q_n G_n \varphi_n\|_{\mathbf{C}^l} \leqq \text{const} \cdot \frac{\ln n}{n^{r+\mu-1/p}} \|\varphi_n\|_{\mathbf{L}_p}$$

(see Prössdorf and Silbermann [11], Corollary 2.3.4), by applying Theorem 1.4 with $Z = \mathbf{C}^l(\Gamma)$ and proceeding as in the proof of Theorem 3.5 we then get the assertion. ∎

Remark. The results of the Sections 5.1–5.3 are due to Prössdorf and Silbermann [3–7, 10, 11]. Note that the Theorems 5.1–5.3 have analogues in the Hölder spaces $\mathbf{H}^\nu(\Gamma)$ (see Prössdorf and Silbermann [11], Chapter 6, §§ 6–7).

5.4. The method of least squares

In this section we suppose that the zeros α_i and β_j ($i = 1, \ldots, k_1$; $j = 1, \ldots, k_2$) of the functions a and b on Γ are pairwise distinct. Put (see the equations (1))

$$\boldsymbol{\alpha} = (\alpha_1, \ldots, \alpha_{k_1}), \qquad \boldsymbol{\beta} = (\beta_1, \ldots, \beta_{k_2}),$$
$$\boldsymbol{m} = (m_1, \ldots, m_{k_1}), \qquad \boldsymbol{n} = (n_1, \ldots, n_{k_2}).$$

Theorem 5.4.[1]) *Let $a_1 \in \mathrm{C}(\boldsymbol{\beta}, \boldsymbol{n})$ and $b_1 \in \mathrm{C}(\boldsymbol{\alpha}, \boldsymbol{m})$ be functions which satisfy the conditions 1 and 2 of Theorem 5.2. Furthermore, suppose T is a compact operator from $\mathbf{L}_2(\Gamma)$ into the space $\bar{\mathbf{L}}_2(\boldsymbol{\alpha}, \boldsymbol{m}; \boldsymbol{\beta}, \boldsymbol{n})$ and assume* dim Ker $A = 0$.

Then for any function $f \in \bar{\mathbf{L}}_2(\boldsymbol{\alpha}, \boldsymbol{m}; \boldsymbol{\beta}, \boldsymbol{n})$ and for all sufficiently large n the system (3.21) *has precisely one solution $(\xi_k^{(n)})_{k=-n}^{n}$ and, as $n \to \infty$, the functions* (3.4) *converge in the norm of $\mathbf{L}_2(\Gamma)$ to the solution $\varphi \in \mathbf{L}_2(\Gamma)$ of the equation* (3.1).

If $\varphi \in \mathbf{W}_2^{(r,\lambda)}(\Gamma)$ ($r \geqq 0$, $0 \leqq \lambda \leqq 1$), then

$$\|\varphi - \varphi_n\|_{\mathbf{L}_2} = O(n^{-r-\lambda}).$$

Note that $\mathbf{W}_2^{(l)} \subset \bar{\mathbf{L}}_2(\boldsymbol{\alpha}, \boldsymbol{m}; \boldsymbol{\beta}, \boldsymbol{n})$ (see § 4, Chapter VI).

The proof of Theorem 5.4 is the same as that of Theorem 3.7, since in view of Theorem 5.2, Chapter VI, the operator $A \in \mathscr{L}(\mathbf{L}_2(\Gamma), \bar{\mathbf{L}}_2(\boldsymbol{\alpha}, \boldsymbol{m}; \boldsymbol{\beta}, \boldsymbol{n}))$ is invertible.

Remark 1. Under slightly stronger hypotheses and by means of other methods, Theorem 5.4 was established by Schulz [1]. He also determined the constants comprised in the estimate for the error $\|\varphi - \varphi_n\|_{\mathbf{L}_2}$ and considered a numerical example.

Remark 2. For all the methods considered in this section the arguments of Section 3.5 can be applied to reduce the case where the functions a_1 and b_1 have non-zero or different indices to the case considered above (in this connection see also Prössdorf and Silbermann [4]).

§ 6. The approximate solution of systems of singular integral equations

6.1. Polynomial approximation methods

We now consider systems of singular integral equations of the form (3.1) with $(m \times m)$-matrix coefficients c and d in the spaces $\mathbf{L}_p^m(\Gamma)$ and $[\mathbf{H}^\mu(\Gamma)]^m$. All theorems of the Sections 3.1–3.4 extend to such systems. The only modifications are: in the Theorems 3.1 and 3.3 replace condition 1 by the condition

1'. det $a(t) \neq 0$, det $b(t) \neq 0$ ($|t| = 1$);

in Theorem 3.1 replace the second condition by the condition

2'. *The left indices of a and the right indices of b are equal to zero*;

in Theorem 3.3 replace the second condition by the condition

2''. *The left indices of the matrix function $b^{-1}a$ are equal to zero.*

Using the corresponding factorizations of matrix functions the proofs we have given above for the scalar case can be literally carried over to the system case.

To give an example, we formulate the matrix analogues of the Theorems 3.1 and 3.3.

Theorem 6.1. *Let a and b be matrix functions belonging to $[\mathbf{H}^\mu(\Gamma)]^{m \times m}$ ($0 < \mu < 1$) and satisfying the conditions 1' and 2'. Furthermore, suppose T is a compact operator on $\mathbf{L}_p^m(\Gamma)$ ($1 < p < \infty$) and* dim Ker $A = 0$.

---

[1]) For notation see the Sections 3 and 4 of Chapter VI.

Then for each vector function $f \in \mathbf{L}_p^m(\Gamma)$ *and for all sufficiently large* n *the system* (3.3) *has precisely one solution* $(\xi_k^{(n)})_{k=-n}^n$ *and, as* $n \to \infty$, *the vector functions* (3.4) *converge in* $\mathbf{L}_p^m(\Gamma)$ *to the solution* $\varphi \in \mathbf{L}_p^m(\Gamma)$ *of the system* (3.1).

If, in addition, $a, b \in [\mathbf{C}^{r,\mu}(\Gamma)]^{m\times m}$, $f \in [\mathbf{W}_p^{(r,\nu)}(\Gamma)]^m$ $(r \geqq 0,\ 0 \leqq \nu \leqq 1)$, *and* T *maps* $\mathbf{L}_p^m(\Gamma)$ *into* $[\mathbf{W}_p^{(r,\nu)}(\Gamma)]^m$, *then*

$$\|\varphi - \varphi_n\|_{\mathbf{L}_p^m(\Gamma)} = O(n^{-r-\lambda}),\ \lambda = \min(\mu, \nu). \tag{1}$$

Theorem 6.2. *Let* a *and* b *be matrix functions belonging to* $[\mathbf{H}^\mu(\Gamma)]^{m\times m}$ $(0 < \mu < 1)$ *and satisfying the conditions* 1' *and* 2''. *Suppose* T *is a compact operator from* $\mathbf{L}_p^m(\Gamma)$ $(1/\mu < p < \infty)$ *into* $\mathbf{C}^m(\Gamma)$ *and let* $\dim \operatorname{Ker} A = 0$.

Then for each vector function f *with components from* $\mathcal{R}(\Gamma)$ *and for all sufficiently large* n *the system* (3.14) *has exactly one solution* $(\xi_k^{(n)})_{k=-n}^n$ *and, as* $n \to \infty$, *the vector functions* (3.4) *converge in* $\mathbf{L}_p^m(\Gamma)$ *to the solution* $\varphi \in \mathbf{L}_p^m(\Gamma)$ *of the system* (3.1).

If, in addition, $a, b \in [\mathbf{C}^{r,\mu}(\Gamma)]^{m\times m}$, $f \in [\mathbf{C}^{r,\nu}(\Gamma)]^m$, *and* T *maps* $\mathbf{L}_p^m(\Gamma)$ *into* $[\mathbf{C}^{r,\nu}(\Gamma)]^m$ $(r \geqq 0,\ 0 \leqq \nu < 1)$, *then the estimate* (1) *holds.*

Remark 1. Kozak [1] proved that the conditions 1' and 2' are also necessary for the first part of Theorem 6.1 to be valid. His proof is based upon an essential generalization of Simonenko's local principle (see Chapter XV). For more about this see Kozak [1–3], Prössdorf and Schmidt [1].

Remark 2. The results of this subsection appeared in the form presented here in the article of Prössdorf and Silbermann [7] for the first time. For a more detailed treatment of the problems touched upon in this subsection we refer the reader to the books by Prössdorf [17], (Chapter 11) and Prössdorf, Silbermann [11, 12].

6.2. Discontinuous coefficients

Silbermann [12] and Junghanns and Silbermann [2] have recently extended the results of Section 3.6 to systems with discontinuous coefficients. Here we restrict ourselves to the formulation of the most important results. Γ is again the unit circle.

Theorem 6.3 (Junghanns and Silbermann [2]). *Let* $a, b \in [\overline{\mathbf{PC}}(\Gamma)]^{m\times m}$, $T \in \mathcal{K}(\mathbf{L}_2^m(\Gamma), [\mathcal{R}(\Gamma)]^m)$, *and* $A = aP + bQ + T$. *Then the collocation method* (3.14) *is stable in* $\mathbf{L}_2^m(\Gamma)$ *if and only if both the operators* A *and* $\tilde{a}P + \tilde{b}Q$ *are invertible on* $\mathbf{L}_2^m(\Gamma)$, *where*

$$\tilde{a}(t) := a(t^{-1}) \text{ and } \tilde{b}(t) := b(t^{-1}).$$

If the matrix functions a and b are continuous, then Theorem 6.3 remains true with $\mathbf{L}_2^m(\Gamma)$ replaced by $\mathbf{L}_p^m(\Gamma)$ $(1 < p < \infty)$.

Theorem 6.4 (Silbermann [12]). *Let* $a, b \in [\overline{\mathbf{PC}}(\Gamma)]^{m\times m}$. *Then the reduction method* (3.3) *converges in* $\mathbf{L}_2^m(\Gamma)$ *for the operator* $aP + bQ$ *(resp.* $Pa + Qb$*) if and only if the operators*

$$aP + bQ,\ \tilde{a}P + Q, \text{ and } P + \tilde{b}Q$$

$$(\text{resp. } Pa + Qb,\ P\tilde{a} + Q, \text{ and } P + Q\tilde{b})$$

are invertible on $\mathbf{L}_2^m(\Gamma)$.

Remark 1. Theorem 6.4 was proved by Silbermann [12] under the hypothesis that a and b have no common points of discontinuities. Without that additional hypothesis the theorem was proved by Rathsfeld [1], who also generalized it to the spaces $[\mathbf{Fl}_p]^m$ $(1 < p < \infty)$. On the basis of a further development of some ideas of Junghanns and Silbermann [2], A. Rathsfeld [2] has recently also generalized Theorem 6.3 to the spaces $\mathbf{L}_p^m(\Gamma)$.

Remark 2. From the general point of view, the foundation of approximation methods for systems of singular integral equations on the interval must be included into the collection of the yet unsolved problems. Note however that the approximate solution of some concrete systems arising in certain applications, e.g. in fracture mechanics, led to good numerical results (see, for instance, NIED [1]).

6.3. Degenerate symbols

All theorems of § 3 extend to te corresponding systems of equations. The proofs we have given for the scalar case can be without difficulty carried over to the matrix case almost literally. The matrix analogues of (5.1) are the representations

$$a(t) = a_1(t)\, D_-(t)\, S_-(t), \qquad b(t) = b_1(t)\, D_+(t)\, S_+(t). \tag{2}$$

Here D_- and D_+ are diagonal matrices of the form

$$D_-(t) = \left(\prod_{i=1}^{k_1} (t^{-1} - \alpha_i^{-1})^{\mu_r^{(i)}}\, \delta_{rs}\right)_{r,s=1}^{m}, \qquad D_+(t) = \left(\prod_{j=1}^{k_2} (t - \beta_j)^{\nu_r^{(j)}}\, \delta_{rs}\right)_{r,s=1}^{n},$$

$\mu_1^{(i)} \geqq \mu_2^{(i)} \geqq \dots \geqq \mu_m^{(i)} \geqq 0$ $(i = 1, \dots, k_1)$ and $\nu_1^{(j)} \geqq \nu_2^{(j)} \geqq \dots \geqq \nu_m^{(j)} \geqq 0$ $(j = 1, \dots, k_2)$ are integers, $S_\pm$ polynomial matrices in $t_{\pm 1}$ with constant and non-vanishing determinant, and a_1, b_1 are non-singular matrix functions. By virtue of Theorem 3.1, Chapter VI, the existence of the representation (2) is ensured whenever the matrix functions a and b are sufficiently smooth in some neighborhood of the zeros of their determinants. We now introduce the following conditions:

1′. $\det a_1(t) \neq 0$, $\det b_1(t) \neq 0$ $(|t| = 1)$.

2′. *The left indices of* a_1 *and the right indices of* b_1 *are equal to zero.*

3′. *The left indices of* $b_1^{-1}a_1$ *are equal to zero.*

Then the theorems of § 5 remain valid for systems of singular integral equations of the form (3.1) if they are modified as follows: define the integer l now as

$$l = \max\,(\mu_1^{(1)}, \dots, \mu_1^{(k_1)}, \nu_1^{(1)}, \dots, \nu_1^{(k_2)})$$

and replace the conditions 1 and 2 by the conditions 1′ and 2′ in Theorem 5.1, and by the conditions 1′ and 2″ in Theorem 5.2.

Remark 1. The results of this subsection are due to PRÖSSDORF and SILBERMANN [7].

Remark 2. BÖTTCHER [1] studied the reduction method for singular integral equations of the form (3.17) with $c, d \in [\mathbf{C}^\infty(\Gamma)]^{m\times m}$ and $K(t, \tau) \in [\mathbf{C}^\infty(\Gamma\times\Gamma)]^{m\times m}$ in the space $[\mathbf{C}^\infty(\Gamma)]^m$ (Γ the unit circle). The result he obtained reads as follows: *the system* (3.3) *has exactly one solution* $(\xi_k^{(n)})_{k=-n}^{n}$ *for all* $f \in [\mathbf{C}^\infty(\Gamma)]^m$ *and for all sufficiently large n and the functions* (3.4) *converge in the topology of* $[\mathbf{C}^\infty(\Gamma)]^m$ *to a solution* $\varphi \in [\mathbf{C}^\infty(\Gamma)]^m$ *of the equation* (3.17) *if and only if the operator A defined by* (3.17) *is invertible on* $[\mathbf{C}^\infty(\Gamma)]^m$ *and the operator* $P\tilde{a}P + Q\tilde{b}Q$ *is invertible on* $[\mathbf{C}^{-\infty}(\Gamma)]^m$. Here $\tilde{a}$ and $\tilde{b}$ denote the operators of multiplication by the matrix functions defined as $\tilde{a}(t) = a(1/t)$ and $\tilde{b}(t) = b(1/t)$ $(|t| = 1)$, with a and b given by (3.2). Note that $P\tilde{a}P + Q\tilde{b}Q$ is invertible on $[\mathbf{C}^{-\infty}(\Gamma)]^m$ if and only if both $\tilde{a}P + Q$ and $P + \tilde{b}Q$ are invertible on $[\mathbf{C}^{-\infty}(\Gamma)]^m$. The proof of the above result is based on the explicit construction of two sequences of operators $\{R_n\}$ and $\{C_n\}$ which satisfy the equality $R_n \cdot P_nAP_n = P_n + C_n$ (P_n given by (3.6)) and which have the following properties: for *each* positive integer k there exist a positive constant M_k, a positive integer j_k, and a sequence of real numbers $\delta_n^{(k)} \to 0$ $(n \to \infty)$ such that

$$\|R_n\varphi\|_k \leqq M_k\, \|\varphi\|_{j_k}, \; \|C_n\varphi\|_k \leqq \delta_n^{(k)}\, \|\varphi\|_k$$

for all $\varphi \in [\mathbf{C}^\infty(\Gamma)]^m$ (see Section 10.1 for notation).

6.4. Spline approximation methods

Necessary and sufficient conditions for the stability of the polygonal method (see Section 4.1) for systems of singular integral equations of the form (3.1) in the space $\mathbf{L}_2^m(\Gamma)$ were obtained by PRÖSSDORF and SCHMIDT [3] for the case where the matrix coefficients a and b are continuous and Γ is the unit circle, and by PRÖSSDORF and RATHSFELD [1, 2] for the case of piecewise continuous matrix coefficients and an arbitrary closed Lyapunov curve Γ. If a and b are in $[\mathbf{C}(\Gamma)]^{m\times m}$, the result is: the polygonal method is stable for $A := aP + bQ$ if and only if dim Ker $A = 0$ and

$$\det (\nu a(t) + (1 - \nu)\, b(t)) \neq 0, \qquad \forall\, t \in \Gamma, \forall\, \nu \in [0, 1]. \tag{3}$$

Furthermore, PRÖSSDORF and RATHSFELD [2] showed that for the invertible operator $A = aP + bQ \in \mathscr{L}(\mathbf{L}_2^m(\Gamma))$ with $a, b \in [\mathbf{PC}(\Gamma)]^{m\times m}$ the strong ellipticity is sufficient for the stability of the Galerkin method (4.18) as well as of the polygonal method. Note that the strong ellipticity is also a necessary condition in the case $m = 1$. In the paper of PRÖSSDORF and RATHSFELD [3] criteria for the strong ellipticity of the operator A in terms of the coefficients $a, b \in [\mathbf{PC}(\Gamma)]^{m\times m}$ were established; in the case of continuous coefficients condition (3) is such a criterion.

SCHMIDT [2–3] also studied the convergence of the 0-point collocation with splines of arbitrary degree for systems with smooth coefficients.

Chapter XVIII
Approximate solution of multidimensional singular integral equations

Methods for solving multidimensional singular integral equations approximately are developed not as far as for one-dimensional equations. In the present chapter we shall derive iteration methods, the Bubnov-Galerkin method, and the least squares method. For the latter we shall indicate several modifications which differ in the choice of the coordinate functions.

Every approximate method for singular equations needs the approximate computation of singular integrals with a density given. We open the chapter by a section devoted to this problem.

§ 1. Approximate computation of singular integrals

1.1. Let us be given the singular integral

$$v(\xi) = \int_{\Sigma} \varrho^{-m} f(\xi, \Lambda)\, u(\eta)\, \mathrm{d}_\eta \Sigma, \tag{1}$$

where Σ denotes the m-dimensional sphere $x_0^2 + x_1^2 + \dots + x_m^2 = 1$; in the following we shall make extensive use of the notations introduced in Chap. IX. By (2.10), Chap. IX,

$$v(\xi) = \int_{\Sigma} \varrho^{-m} f(\xi, \Lambda)\, [u(\eta) - u(\xi)]\, \mathrm{d}_\eta \Sigma. \tag{2}$$

We shall assume the function $u(\eta)$ to be sufficiently smooth.

As shown in § 2, Chap. IX, the point $\eta \in \Sigma$ is for given $\xi \in \Sigma$ determined by a number ϱ and a point $\Lambda \in S$ where S denotes as always the $(m - 1)$-dimensional unit sphere in $\mathbb{R}^m$. The coordinates of the point Λ are determined as follows.

Let $OX_0, OX_1, \dots, OX_m$ be the axes of a Cartesian coordinate system in $\mathbb{R}^{m+1}$, and $\xi \in \Sigma$ be given. Then we assign to that coordinate system as usually a system of spherical coordinates such that the point ξ has the angular coordinates $\vartheta_0, \vartheta_1, \dots, \vartheta_{m-1}$. Now we introduce in $\mathbb{R}^{m+1}$ another coordinate system $O\overline{X}_0, O\overline{X}_1, \dots, O\overline{X}_m$ with the same origin. Here, the axis $O\overline{X}_0$ is directed from the origin O to the point ξ, and the axis $O\overline{X}_1$ is put into the two-dimensional plane spanned by the axes OX_0 and $O\overline{X}_0$, perpendicular to $O\overline{X}_0$. The following axes $O\overline{X}_j$, $j = 2, \dots, m$, are chosen so that the axis $O\overline{X}_j$ is perpendicular to the axes $O\overline{X}_i$, $i < j$, and has the following direction cosines (see the subsequent table):

Table of Direction Cosines

	X_0	X_1	X_2	X_3	$\cdots$	X_{m-3}	X_{m-2}	X_{m-1}	X_m
$\overline{X}_0$	α_{00}	α_{01}	α_{02}	α_{03}	$\cdots$	$\alpha_{0,m-3}$	$\alpha_{0,m-2}$	$\alpha_{0,m-1}$	α_{0m}
$\overline{X}_1$	α_{10}	α_{11}	α_{12}	α_{13}	$\cdots$	$\alpha_{1,m-3}$	$\alpha_{1,m-2}$	$\alpha_{1,m-1}$	α_{1m}
$\overline{X}_2$	0	0	0	0	$\cdots$	0	0	$\alpha_{2,m-1}$	α_{2m}
$\overline{X}_3$	0	0	0	0	$\cdots$	0	$\alpha_{3,m-2}$	$\alpha_{3,m-1}$	α_{3m}
$\cdots$	$\cdots$	$\cdots$	$\cdots$	$\cdots$	$\cdots$	$\cdots$	$\cdots$	$\cdots$	$\cdots$
$\overline{X}_{m-2}$	0	0	0	$\alpha_{m-2,3}$	$\cdots$	$\alpha_{m-2,m-3}$	$\alpha_{m-2,m-2}$	$\alpha_{m-2,m-1}$	$\alpha_{m-2,m}$
$\overline{X}_{m-1}$	0	0	$\alpha_{m-1,2}$	$\alpha_{m-1,3}$	$\cdots$	$\alpha_{m-1,m-3}$	$\alpha_{m-1,m-2}$	$\alpha_{m-1,m-1}$	$\alpha_{m-1,m}$
$\overline{X}_m$	0	α_{m1}	α_{m2}	α_{m3}	$\cdots$	$\alpha_{m,m-3}$	$\alpha_{m,m-2}$	$\alpha_{m,m-1}$	α_{mm}

It is easy to see that the non-zero direction cosines are determined by the following formulas:

$$\alpha_{00} = \cos\vartheta_0,$$

$$\alpha_{01} = \sin\vartheta_0 \cos\vartheta_1,$$

....

$$\alpha_{0,m-1} = \sin\vartheta_0 \sin\vartheta_1 \ldots \sin\vartheta_{m-2} \cos\vartheta_{m-1},$$

$$\alpha_{0m} = \sin\vartheta_0 \sin\vartheta_1 \ldots \sin\vartheta_{m-2} \sin\vartheta_{m-1},$$

$$\alpha_{10} = \sin\vartheta_0,$$

$$\alpha_{11} = -\cos\vartheta_0 \cos\vartheta_1,$$

....

$$\alpha_{1,m-1} = -\cos\vartheta_0 \sin\vartheta_1 \ldots \sin\vartheta_{m-2} \cos\vartheta_{m-1},$$

$$\alpha_{1m} = -\cos\vartheta_0 \sin\vartheta_1 \ldots \sin\vartheta_{m-2} \sin\vartheta_{m-1}.$$

Furthermore, for $k + 1 < l \leq m - 1$,

$$\alpha_{m-k,l} = -\cos\vartheta_{k+2} \sin\vartheta_{k+3} \ldots \sin\vartheta_{l-1} \sin\vartheta_l$$

and for $l = k + 1$ we have $\alpha_{m-k,k+1} = \sin\vartheta_{k+1}$. In case $k < m - 2$ we have

$$\alpha_{m-k,m} = -\cos\vartheta_{k+2} \sin\vartheta_{k+3} \ldots \sin\vartheta_{m-2} \sin\vartheta_{m-1}.$$

Finally, $\alpha_{m-2,m} = -\cos\vartheta_{m-1}$.

Next, we introduce spherical coordinates $\varrho, \lambda_0, \lambda_1, \ldots, \lambda_{m-1}$ with origin at the centre of the sphere Σ which are related to the coordinate system $O\overline{X}_0, O\overline{X}_1, \ldots, O\overline{X}_m$. Let the coordinates of the points ξ and η in that Cartesian system be $\xi_0, \xi_1, \ldots, \xi_m$ and $\eta_0, \eta_1, \ldots, \eta_m$, resp. Then

$$\eta_0 = \xi_0 + \varrho \cos\lambda_0,$$

$$\eta_1 = \xi_1 + \varrho \sin\lambda_0 \cos\lambda_1,$$

....

$$\eta_{m-1} = \xi_{m-1} + \varrho \sin\lambda_0 \ldots \sin\lambda_{m-2} \cos\lambda_{m-1},$$

$$\eta_m = \xi_m + \varrho \sin\lambda_0 \ldots \sin\lambda_{m-2} \sin\lambda_{m-1}.$$

For $\eta \in \Sigma$ we have $\varrho = 2 \sin \lambda_0/2$, and as coordinates of that point we take the quantities $\varrho, \lambda_1, \ldots, \lambda_{m-1}$.

By the Taylor formula,

$$u(\eta) - u(\xi) = \frac{\partial u(\xi)}{\partial \varrho}\varrho + R(\xi, \eta); \qquad R(\xi, \eta) = O(\varrho^2).$$

Inserting this into (2) we find

$$v(\xi) = \frac{\partial u(\xi)}{\partial \varrho} \int_{\Sigma} \varrho^{1-m} f(\xi, \Lambda)\, \mathrm{d}_\eta \Sigma + \int_{\Sigma} \varrho^{-m} R(\xi, \eta) f(\xi, \Lambda)\, \mathrm{d}_\eta \Sigma$$

or, applying eq. (2.6), Chap. IX,

$$v(\xi) = \frac{\partial u(\xi)}{\partial \varrho} \int_0^2 \left(1 - \frac{\varrho^2}{4}\right)^{(m-2)/2} \mathrm{d}\varrho \int_0^\pi \ldots \int_0^{2\pi} f(\xi, \Lambda) \sin^{m-2}\lambda_1 \ldots \sin\lambda_{m-2} \mathrm{d}\lambda_1 \ldots \mathrm{d}\lambda_{m-1}$$
$$+ \int_0^2 \int_0^\pi \ldots \int_0^{2\pi} \varrho^{-1} \left(1 - \frac{\varrho^2}{4}\right)^{(m-2)/2} R(\xi, \eta) f(\xi, \Lambda) \sin^{m-2}\lambda_1 \ldots \sin\lambda_{m-2} \mathrm{d}\varrho\, \mathrm{d}\lambda_1 \ldots \mathrm{d}\lambda_{m-1}. \tag{3}$$

The integrals (3) do not have a singularity at $\varrho = 0$, and they can be evaluated by means of usual cubature formulae.

If the singular integral is taken over a manifold Γ that is homeomorphic to the sphere then it can be transformed to an integral taken over that sphere. One can apply, in particular, stereographic projection if the integral is taken over the Euclidean space $\mathbb{R}^m$.

1.2. Perlin (cf. e.g. PERLIN [1]) suggests in a number of papers the following method for the approximate computation of singular integrals. Suppose the integral

$$v(\xi) = \int_\Gamma K(\xi, \eta)\, u(\eta)\, \mathrm{d}_\eta \Gamma \tag{4}$$

is to be computed, where $K(\xi, \eta)$ is a singular kernel, the function u belongs to some Lipschitz space, and Γ is a Lyapunov surface. The integral (4) can be written as

$$v(\xi) = u(\xi) \int_\Gamma K(\xi, \eta)\, \mathrm{d}_\eta \Gamma + \int_\Gamma K(\xi, \eta)\, [u(\eta) - u(\xi)]\, \mathrm{d}_\eta \Gamma. \tag{5}$$

We assume that the first integral in (5) is easily computable–in the elasticity problems studied by Perlin this integral is the unit matrix (cf. § 6, Chap. XIV). In such a case the problem is reduced to computing the second integral in (5), and this is not a singular but, in general, a convergent improper integral.

In order to compute it we divide the surface Γ in polyhedrons Γ_k with curved boundary and of small diameter. The corners η_{jk} of such a polyhedron will be called *nodes*. In the interior of each polyhedron Γ_k we choose a point ξ_k called *supporting node*. Then the second integral in (5) is approximated at the supporting nodes according to the formula

$$\int_\Gamma K(\xi_k, \eta)\, [u(\eta) - u(\xi_k)] \mathrm{d}_\eta \Gamma \approx \sum_l \sum_j K(\xi_k, \eta_{lj})\, [u(\eta_{lj}) - u(\xi_k)]\, |\Gamma_l| \tag{6}$$

where $|\Gamma_i|$ means the measure of the polyhedron Γ_i with nodes η_{ij}.

§ 2. An iteration method

2.1. Let us consider the singular equation

$$u(\xi) - \int_\Gamma K(\xi, \eta)\, u(\eta)\, \mathrm{d}_\eta \Gamma = g(\xi). \tag{1}$$

If the norm of the integral operator from (1) in a certain function space turns out to be less than 1, then (1) can be solved iteratively,

$$u(\xi) = \sum_{n=0}^{\infty} u_n(\xi); \qquad u_0(\xi) = g(\xi), \qquad u_n(\xi) = \int_\Gamma K(\xi, \eta)\, u_{n-1}(\eta)\, \mathrm{d}_\eta \Gamma, \tag{2}$$

and the problem is reduced to compute the integrals (2) the density of which is given. For applying the method of § 1.2 it is necessary to compute the iterate $u_n(\xi)$ not only at the supporting nodes but also at the nodes. This can be done by using some interpolation procedure.

2.2. In PERLIN [1] the equations of elasticity theory are considered which are described in § 6, Chap. XIV. We shall here consider the somewhat more general equations

$$\varkappa(x) - 2\lambda \int_\Gamma P(x, y)\, \varkappa(y)\, \mathrm{d}_y\Gamma = \varphi(x), \tag{3}$$

$$\varrho(x) - 2\lambda \int_\Gamma P^*(y, x)\, \varrho(y)\, \mathrm{d}_y\Gamma = \psi(x). \tag{4}$$

Here, $\varphi(x)$ and $\psi(x)$ stand for the right-hand sides of equations (6.16)–(6.19), Chap. XIV, and λ is a scalar parameter. For the first interior and for the second exterior problem we have $\lambda = 1$, whereas in the case of the second interior and the first exterior problem $\lambda = -1$. PHAM THE LAI [1] showed that for a Poisson coefficient $\sigma \neq -1$, $-\infty < \sigma < 1/2$, the resolvents of (3) and (4) have only simple real poles which lie on the rays $\lambda > 1$ and $\lambda < -1$. The same is true for arbitrary real σ different from -1 and $1/2$; this follows from the results of MIKHLIN [28]. For that reason the simple iteration process for the singular equations in elasticity theory is divergent. But e.g. for $\lambda = 1$ a modified iteration method suggested by Kantorovich (cf. KANTOROVICH and KRYLOV [1]) is convergent. It leads to a representation of the solution as a series

$$\varkappa(x) = \tfrac{1}{2}\varkappa_0(x) + \tfrac{1}{2}[\varkappa_{n-1}(x) + \varkappa_n(x)]. \tag{5}$$

For eq. (3) we have

$$\varkappa_0(x) = \varphi(x), \qquad \varkappa_n(x) = \int_\Gamma P(x, y)\, \varkappa_{n-1}(y)\, \mathrm{d}_y\Gamma. \tag{6}$$

The numerical examples given by Perlin l.c. seem to confirm the method described here to be sufficiently effective.

§ 3. The Bubnov-Galerkin and the least squares methods

A general theory for these methods is given in the monographs by MIKHLIN [24, 25].

Here and in the following sections singular equations will be considered in $\mathbf{L}_2$ spaces only.

3.1. For the sake of simplicity we shall assume the (scalar or vector) singular equation

$$(Au)(x) = a(x)\, u(x) + \int_\Gamma K(x, y)\, u(y)\, \mathrm{d}_y\Gamma + (Tu)(x) = g(x) \tag{1}$$

to be uniquely solvable for any right-hand side. Then the index is zero, and there exists an equivalent regularizer R. Equation (1) is replaced by the equivalent equation

$$(RAu)(x) := u(x) + (T_1 u)(x) = (Rg)(x), \tag{2}$$

and this equation will be solved by means of the Bubnov-Galerkin method. To that aim we take a sequence $\{\varphi_n(x)\}$ of coordinate functions which is complete in $\mathbf{L}_2(\Gamma)$ (for the definition of coordinate functions and their properties see MIKHLIN [24]), we let

$$u_n(x) = \sum_{k=1}^{n} a_k \varphi_k(x), \qquad a_k = \text{const}, \tag{3}$$

and determine the coefficients a_k from the algebraic system $(RAu_n, \varphi_j) = (Rg, \varphi_j)$ or, explicitly,

$$\sum_{k=1}^{n} (RA\varphi_k, \varphi_j)\, a_k = (Rg, \varphi_j); \qquad j = 1, 2, \ldots, n. \tag{4}$$

It is well-known that this system is uniquely solvable for large n, the approximate solution $u_n(x)$ converges to the exact solution $u_0(x)$, and there is the estimate

$$\|u_n - u_0\| \leqq (1 + \varepsilon_n)\, \|(I - P_n)\, u_0\| \tag{5}$$

(cf. VAINIKKO [1]). In (5) the operator P_n denotes orthogonal projection onto the subspace spanned by the functions $\varphi_1, \varphi_2, \ldots, \varphi_n$, and $\varepsilon_n \to 0$ as $n \to \infty$.

The Bubnov-Galerkin method is stable, and the condition number of the matrix in (4) is bounded independently of n provided that the coordinate system $\{\varphi_n\}$ is almost orthonormal in $\mathbf{L}_2(\Gamma)$ (the more, if it is orthonormal in that space).

It is worth noting that system (4) can be written as

$$\sum_{k=1}^{n} (A\varphi_k, R^*\varphi_j)\, a_k = (g, R^*\varphi_j), \qquad j = 1, 2, \ldots, n. \tag{6}$$

In such a form it is no longer necessary to multiply the operators A and R, but it requires to compute the functions $R^*\varphi_j$.

3.2. We preserve the assumptions made in § 3.1 on the properties of the singular equation; the case where these assumptions are violated will be considered later. As before, we want to find an approximate solution in the form (3) where the functions $\varphi_n(x)$ satisfy the conditions mentioned in § 3.1. But here we shall determine the coefficients a_n from the condition $\|Au_n - g\|^2 = \min$ or, what is the same, from the algebraic system

$$\sum_{k=1}^{n} (A\varphi_k, A\varphi_j)\, a_k = (g, A\varphi_j), \qquad j = 1, 2, \ldots, n, \tag{7}$$

and that is the least squares method. System (7) is solvable for arbitrary n, and $u_n \to u_0$ as $n \to \infty$. The numerical process of the least squares method is stable under the same conditions as the Bubnov-Galerkin method, see § 3.1.

To singular integral equations the least squares method was applied first by MIKHLIN [26].

3.3. Let us estimate the rate of convergence in the least squares method (cf. MIKHLIN and RADEVA [1]). Assume that the linear operator A is a 1-to-1 mapping of a separable Hilbert space H onto H such that both the operators A and A^{-1} are bounded. We choose a complete system of coordinate elements $\{\varphi_n\}$ in H and seek the approximate solution $u_n = \sum_{k=1}^{n} a_k \varphi_k$ to the equation $Au = g$ according to the least squares method. Now denote by u_0 the exact solution of that equation, and let $\tilde{u}_n = \sum_{k=1}^{n} b_k \varphi_k$. Assume that for a certain choice of the coefficients b_k there is an estimate

$$\|\tilde{u}_n - u_0\| \leqq \gamma(n) \tag{8}$$

with a known function $\gamma(n)$ ($\|\cdot\|$ means the norm in H). Now, let us estimate $\|u_n - u_0\|$. We have

$$\|u_n - u_0\| = \|A^{-1}(Au_n - g)\| \leqq \|A^{-1}\| \cdot \|Au_n - g\|.$$

Because u_n is the least squares solution, there are the inequalities

$$\|Au_n - g\| \leqq \|A\tilde{u}_n - g\| = \|A(\tilde{u}_n - u_0)\| \leqq \|A\| \cdot \|\tilde{u}_n - u_0\| \leqq \|A\| \gamma(n).$$

Then we obtain the estimate wanted,

$$\|u_n - u_0\| \leqq \text{cond}(A) \cdot \gamma(n),$$

where $\text{cond}(A) := \|A\| \cdot \|A^{-1}\|$ is the condition number of the operator A.

Obviously, $\gamma(n)$ can be taken from the best approximation of u_0 by linear combinations of the elements $\varphi_1, \varphi_2, \ldots, \varphi_n$:

$$\gamma(n) = \inf_{\alpha_k} \left\| u_0 - \sum_{k=1}^{n} \alpha_k \varphi_k \right\|. \tag{10}$$

3.4. In the book [25] by Mikhlin a certain system of coordinate functions is presented such that their singular integrals are computed relatively simple provided that the characteristic is a spherical polynomial, i.e. a linear combination of spherical functions.

3.5. Prössdorf [27] took finite elements which were introduced and called basic functions by Mikhlin [31] as coordinate functions. The results of Prössdorf l.c. (see the sections 4 and 5 there) imply that the Galerkin method with these coordinate functions applied to (1) converges in $\mathbf{L}_2(\Gamma)$ under the assumptions of § 3.1 provided that the singular operator A is strongly elliptic. Here the operator A is said to be strongly elliptic iff there are a nowhere vanishing function $b \in \mathbf{C}(\Gamma)$ and a completely continuous operator T_0 such that the operator $b \cdot (A - T_0)$ has a positive definite real part. Furthermore, error estimates in the Sobolev space scale $\mathbf{W}_2^{(l)}(\Gamma)$ are obtained.

§ 4. Coordinate functions connected with spherical functions

Equations (3.5) and (3.9) show that the problem of estimating the error in the Bubnov-Galerkin and in the least squares methods is reduced to finding an estimate for the best approximation of a given function by linear combinations of the first n coordinate functions. This given function is the solution to the singular equation in question, and the best approximation is understood in the $\mathbf{L}_2$ sense. In the present section we shall show that such estimates for the best approximation can be obtained for a certain class of singular equations provided that a particular coordinate function system is used which is, in some sense, connected with spherical functions.

4.1. Let Σ be the m-dimensional unit sphere in $\mathbb{R}^{m+1}$, and denote the coordinates in $\mathbb{R}^{m+1}$ by $\xi_1, \xi_2, \ldots, \xi_{m+1}$. The sphere Σ is mapped by stereographic projection onto the Euclidean space $\mathbb{R}^m$; the transformation formulae are, as it is well-known,

$$\xi_k = \frac{2x_k}{1 + |x|^2}, \qquad k = 1, 2, \ldots, m; \qquad \xi_{m+1} = \frac{1 - |x|^2}{1 + |x|^2}. \tag{1}$$

Let

$$k_{n,m+1} = (2n + m - 1)\frac{(n + m - 2)!}{(m - 1)!\, n!}, \quad n = 0, 1, 2, \ldots, k = 1, 2, \ldots, k_{n,m+1},$$

$\xi \in \Sigma$, and $\{Y_{n,m+1}^{(k)}\}$ be the system of $(m + 1)$-dimensional spherical functions which is orthonormal and complete in $\mathbf{L}_2(\Sigma)$. In order to simplify notations we shall henceforth omit the subscripts $m + 1$. The transformation

$$u(x) = \left(\frac{1 + |x|^2}{2}\right)^{-m/2} \tilde{u}\,(\xi) = (1 + \xi_m)^{m/2}\, \tilde{u}\,(x) \tag{2}$$

maps the space $\mathbf{L}_2(\Sigma)$ isometrically onto $\mathbf{L}_2(\mathbb{R}^m)$, so that the system of functions

$$\varphi_n^{(k)}(x) = (1 + \xi_{m+1})^{m/2}\; Y_n^{(k)}(\xi) \tag{3}$$

is orthonormal and complete in $\mathbf{L}_2(\mathbb{R}^m)$.

We recall the following fact: For $\tilde{g} \in \mathbf{W}_2^{(l)}(\Sigma)$ and

$$\tilde{g}(\xi) = \sum_{n=0}^{\infty} \sum_{k=1}^{k_n} a_n^{(k)} Y_n^{(k)}(\xi),$$

we have (cf. § 6, Chap. X)

$$\left\| \tilde{g} - \sum_{n=0}^{N} \sum_{k=1}^{k_n} a_n^{(k)} Y_n^{(k)} \right\|_{\mathbf{L}_2(\Sigma)} \leqq \frac{\sigma(N)}{N^l}, \qquad \sigma(N) = o(1). \tag{4}$$

4.2. Let us consider the singular equation

$$(Au)\,(x) := a(x)\, u(x) + \int\limits_{\mathbb{R}^m} r^{-m} f\,(x, \theta)\, u(y)\, \mathrm{d}y + (Tu)\,(x) = g(x) \tag{5}$$

under the following assumptions:

1. $g \in \mathbf{W}_2^{(l)}(\mathbb{R}^m)$;
2. the operator T is completely continuous in $\mathbf{L}_2(\mathbb{R}^m)$ and smoothing of order l;
3. the symbol $\Phi_A(x, \theta)$ of the operator A belongs to the class $\mathfrak{R}_{l+1,\lambda}$, $\lambda > (m - 1)/2$;
4. the characteristic $f(x, \theta)$ is continuously differentiable with respect to the Cartesian coordinates of x and θ.

Due to Theorem 5.1, Chap. XII, every solution $u \in \mathbf{L}_2(\mathbb{R}^m)$ of (5) belongs also to the space $\mathbf{W}_2^{(l)}(\mathbb{R}^m)$.

In addition we shall assume that (in case of a matrix operator A) Ind $A = 0$ and that (5) has not more than one solution. Then this equation is solvable for any right-hand side $g(x)$ of $\mathbf{L}_2(\mathbb{R}^m)$.

Under the stereographic projection of $\mathbb{R}^m$ onto Σ (5) is transformed into a new singular equation with an integral over Σ and the new unknown function $v(\xi) = (1 + \xi_{m+1})^{-m/2}\, u(x)$. Obviously, this new equation has a unique solution $v_0(\xi) = (1 + \xi_{m+1})^{-m/2}\, u_0(x)$, and $v_0 \in \mathbf{L}_2(\Sigma)$ as well as $v_0 \in \mathbf{W}_2^{(l)}(\Sigma \setminus \sigma)$ where σ is an arbitrary neighborhood of the point $\xi = (0, \dots, 0, -1)$ on Σ. Now, put

$$\xi_k = \frac{2x'_k}{1 + |x'|^2}, \qquad k = 1, 2, \dots, m, \qquad \xi_{m+1} = \frac{|x'|^2 - 1}{|x'|^2 + 1}, \qquad |x'|^2 = \sum_{k=1}^{m} x_k'^2. \tag{6}$$

These formulae define a mapping of the sphere Σ onto the space $\overline{\mathbb{R}}^m$ such that the point $x' = \infty$ corresponds to $\xi = (0, \dots, 0, 1)$. With the notation $u'(x') = (1 - \xi_{m+1})^{m/2}\, v(\xi)$ the singular integral equation on the sphere is transformed to a new one in $\mathbb{R}^m$:

$$(A'u')\,(x') := a'(x')\, u'(x') + \int\limits_{\mathbb{R}^m} r'^{-m} f'(x', \theta')\, u'(y')\, \mathrm{d}y' + (T'u')\,(x') = g'(x');$$

$$r' = |y' - x'|, \qquad \theta' = \frac{y' - x'}{r'}. \tag{7}$$

We shall assume that for (7) the same conditions are fulfilled as those required above for (5). Hence, (7) has a unique solution that belongs, due to Theorem 5.1, Chap. XII, to the space $\mathbf{W}_2^{(l)}(\mathbb{R}^m)$. But then for some neighborhood σ', $\sigma \subset \sigma' \subset \Sigma$, we have $v_0 \in \mathbf{W}_2^{(l)}(\sigma')$. Thence, $v_0 \in \mathbf{W}_2^{(l)}(\Sigma)$, and expanding it into a series with respect to spherical functions,

$$v_0(\xi) = \sum_{n=0}^{\infty} \sum_{k=1}^{k_n} b_n^{(k)} Y_n^{(k)}(\xi),$$

we obtain the estimate

$$\left\| v_0 - \sum_{n=0}^{N} \sum_{k=1}^{k_n} b_n^{(k)} Y_n^{(k)} \right\|_{\mathbf{L}_2(\Sigma)} \leqq \frac{\sigma(N)}{N^l}, \qquad \sigma(N) = o(1). \tag{8}$$

By the isometry, according to (2), between the spaces $\mathbf{L}_2(\Sigma)$ and $\mathbf{L}_2(\mathbb{R}^m)$ we find

$$\left\| u_0 - \sum_{n=0}^{N} \sum_{k=1}^{k_n} b_n^{(k)} \varphi_n^{(k)} \right\|_{\mathbf{L}_2(\mathbb{R}^m)} \leqq \frac{\sigma(N)}{N^l}. \tag{9}$$

Let $u_N(x)$ be an approximate solution to (5),

$$u_N(x) = \sum_{n=0}^{N} \sum_{k=1}^{k_n} c_n^{(k)} \varphi_n^{(k)}(x),$$

which is determined e.g. by the least squares method. Then the inequalities (9) and (3.9) yield the estimate wanted,

$$\| u - u_N \|_{\mathbf{L}_2(\mathbb{R}^m)} \leqq \frac{\operatorname{cond}(A)\, \sigma(N)}{N^l}. \tag{10}$$

Let us find the explicit form of the transformation from (5) to (7). Comparing equations (1) and (6) we have $|x'|^2 = 1/|x|^2$ and $x_k' = x_k/|x|^2$, i.e. the transformation from the coordinates x_k to the coordinates x_k' is an inversion with respect to the origin in $\mathbb{R}^m$. Furthermore,

$$u'(x') = \left(\frac{1 - \xi_{m+1}}{1 + \xi_{m+1}}\right)^{m/2} u(x) = |x|^m\, u(x).$$

The relation between $r = |y - x|$ and $r' = |y' - x'|$ is easily found: The four points x, x', y, y' lie in a two-dimensional plane through the origin. Because $|x| \cdot |x'| = |y| \cdot |y'|$, the triangles $0xy$ and $0x'y'$ are similar. Hence, $r = r'/|x'|\,|y'|$. Now, (5) is easily transformed to

$$(A'u')(x') = a\left(\frac{x'}{|x'|^2}\right) u'(x') + \int_{\mathbb{R}^m} r'^{-m} f\left(\frac{x'}{|x'|^2}, \theta' - 2x_0'(\theta', x_0')\right)$$

$$\times u'(y')\, dy' + (T'u')(x') = |x'|^{-m}\, g\left(\frac{x'}{|x'|^2}\right), \qquad x_0' = \frac{x'}{|x'|} \tag{11}$$

and this is essentially the same as (7).

4.3. Finally, let us indicate explicitly those conditions for the symbol of (5) that ensure the estimate (10). The condition $\Phi'(x', \theta') \in \mathfrak{R}_{l+1,\lambda}$ holds for $x' \neq 0, \infty$ if $\lambda \geqq l + 1$. The same condition is true also for $x' = 0$ provided that the symbol $\Phi_0(x, \theta)$ of the integral operator

$$(A_0 u)(x) = \int_{\mathbb{R}^m} r^{-m} f(x, \theta)\, u(y)\, dy$$

is subject to the condition $\Phi_0^{(q)}(\infty, \theta) = 0$ for every q such that $|q| \leqq l$, where the derivative $\Phi_0^{(q)}$ is computed under the assumption that θ does not depend on x. Finally, for $x' = \infty$ the condition $\Phi' \in \mathfrak{R}_{l+1,\lambda}$ is fulfilled if $\Phi_0^{(q)}(0, \theta) = 0$, $|q| \leqq l$.

§ 5. The case of exactly known eigenfunctions

5.1. Let A be a bounded linear operator which is normally solvable in the Hilbert space H. Let us consider the equation

$$Au = g \tag{1}$$

with a given right-hand side $g \in H$.

Assume that all linearly independent solutions $u^{(1)}, u^{(2)}, \ldots, u^{(s)}$ of the homogeneous equation

$$Au = 0 \tag{2}$$

are known; we may assume them to be orthonormal.

By H_1 we denote the solution subspace of (2) (i.e. the null space of the operator A), and by H_1^* that one of the adjoint homogeneous equation

$$A^*u = 0. \tag{3}$$

The orthogonal complements of these spaces will be denoted by H_2 and H_2^*, resp.,

$$H_2 = H \ominus H_1, \qquad H_2^* = H \ominus H_1^*.$$

Equation (1) is solvable if and only if $g \in H_2^*$. The general solution to this equation is of the form $u = u_0 + u'$ where u_0 is a particular solution of (1) belonging to H_2, and u' is arbitrary in H_1. The particular solution u_0 is, obviously, unique.

We shall seek the solution to (1) in H_2.

To that aim we denote by B the restriction of the operator A to H_2. It is a 1-to-1 mapping of the subspace H_2 onto H_2^*, so that both the operators B and B^{-1} are bounded.

Now we choose a complete orthonormal system $\{\varphi_n\}$ in H and consider the projection of the elements φ_n onto H_2,

$$\hat{\varphi}_n = \varphi_n - \sum_{\nu=1}^{s} (\varphi_n, u^{(\nu)})\, u^{(\nu)}. \tag{4}$$

For solving (1) we apply the least squares method. The approximate solution is of the form

$$u_n = \sum_{k=1}^{n} b_k \hat{\varphi}_k, \tag{5}$$

where the coefficients b_k are determined, as usually, from the condition $\|Bu_n - g\|^2 = \min$, and that leads to the linear algebraic system

$$\sum_{k=1}^{n} (B\hat{\varphi}_k, B\hat{\varphi}_j)\, b_k = (g, B\hat{\varphi}_j), \qquad j = 1, 2, \ldots, n \tag{6}$$

Let us estimate the error $\|u_n - u_0\|$. Let u be an arbitrary element of H_2 such that $(u, u^{(\nu)}) = 0$, $\nu = 1, 2, \ldots, s$. Expanding u into a series,

$$u = \sum_{k=1}^{\infty} A_k \varphi_k, \qquad A_k = (u, \varphi_k), \tag{7}$$

we find

$$\sum_{k=1}^{\infty} (\varphi_k, u^{(\nu)}) A_k = 0, \qquad \nu = 1, 2, \ldots, s.$$

The finite section of the series (7),

$$u^{(n)} = \sum_{k=1}^{n} A_k \varphi_k ,$$

is projected onto H_2; its projection will be denoted by $v^{(n)}$. Then

$$u^{(n)} = v^{(n)} + w^{(n)}, \qquad v^{(n)} = \sum_{k=1}^{n} A_k \hat{\varphi}_k, \qquad (v^{(n)}, w^{(n)}) = 0$$

and hence

$$\|u - u^{(n)}\|^2 = \|u - v^{(n)}\|^2 + \|w^{(n)}\|^2.$$

Denote by $\gamma(n)$ the value of the best approximation of the element u_0 by linear combinations of the elements $\hat{\varphi}_k$, $1 \leq k \leq n$. As shown in § 3.3, $\|u_0 - u_n\| \leq \operatorname{cond}(B) \cdot \gamma(n)$.

5.2. Let us consider (4.5) where we assume all conditions formulated in § 4 with respect to that equation to be fulfilled, except the following two ones: The index of that equation may be non-zero, and the equation need not be solvable for any right-hand side. As before we assume all linearly independent solutions $u^{(1)}, u^{(2)}, \ldots, u^{(s)}$ of the corresponding homogeneous equation to be known and orthonormal. Furthermore, we preserve the notations introduced in § 5.1, and put $H = \mathbf{L}_2(\mathbb{R}^m)$.

The complete and orthonormal system of coordinate functions in H,

$$\varphi_n^{(k)} = (1 + \xi_{m+1})^{m/2}\, Y_n^{(k)}(\xi)$$

($\xi \in \Sigma$, Σ is the unit sphere in $\mathbb{R}^{m+1}$) is projected onto H_2.

As described in § 5.1, we find the solution of the singular equation in H_2 by the least squares method. The approximate solution is of the form

$$u_N = \sum_{n=0}^{N} \sum_{k=1}^{k_n} b_n^{(k)} \hat{\varphi}_n^{(k)},$$

where $\hat{\varphi}_n^{(k)}$ is the projection of $\varphi_n^{(k)}$ onto H_2. The solution u_0 of (4.5) belongs to $\mathbf{W}_2^{(l)}(\mathbb{R}^m)$. Mapping $\mathbb{R}^m$ stereographically onto the sphere Σ such that the point $x = \infty$ is taken to the north pole of Σ, we transform the function u_0 into a certain function $v_0 \in \mathbf{W}_2^{(l)}(\Sigma \setminus \sigma)$ where σ is a neighborhood of that north pole. We map, inversely, the sphere Σ onto $\mathbb{R}^m$, but now so that the south pole is taken to $x = \infty$. Here, the function v_0 is transformed into a function u_0' such that $u_0' \in \mathbf{W}_2^{(l)}(\mathbb{R}^m)$. Hence, $v_0 \in \mathbf{W}_2^{(l)}(\Sigma \setminus \sigma')$ with a neighborhood σ' of the south pole. These arguments now imply $v_0 \in \mathbf{W}_2^{(l)}(\Sigma)$.

Thus, we may expand v_0 into a series with respect to spherical functions,

$$v_0 = \sum_{n=0}^{\infty} \sum_{k=1}^{k_n} a_n^{(k)} Y_n^{(k)}(\xi),$$

and obtain

$$\left\| v_0 - \sum_{n=0}^{N} \sum_{k=1}^{k_n} a_n^{(k)} Y_n^{(k)} \right\|_{\mathbf{L}_2(\Sigma)} = o(N^{-l}).$$

The mapping is isometric, hence

$$\left\| u_0 - \sum_{n=0}^{N} \sum_{k=1}^{k_n} a_n^{(k)} \varphi_n^{(k)} \right\|_{\mathbf{L}_2(\mathbb{R}^m)} = o(N^{-l})$$

and then for the projection

$$\left\| u_0 - \sum_{n=0}^{N} \sum_{k=1}^{k_n} a_n^{(k)} \hat{\varphi}_n^{(k)} \right\|_{L_2(R^m)} = o(N^{-l}). \tag{8}$$

Therefore, for the best approximation we obtained $\gamma(n) = o(N^{-l})$, and the least squares approximate solution admits the error estimate

$$\|u_0 - u_N\| \leqq \operatorname{cond}(B) \left\| u_0 - \sum_{n=0}^{N} \sum_{k=1}^{k_n} a_n^{(k)} \hat{\varphi}_n^{(k)} \right\|_{L_2(R^m)} = o(N^{-l}). \tag{9}$$

§ 6. Approximate construction of the eigenfunctions

6.1. Let us consider the homogeneous equation

$$Au = 0 \tag{1}$$

with a bounded linear operator A in a Hilbert space H. The operator A is assumed to be normally solvable and to have a finite-dimensional null-space.

Applying the adjoint operator A^* to (1) we obtain a new equation

$$A^*Au = 0 \tag{2}$$

which is equivalent to (1). The operator A^*A is self-adjoint, and its index is zero, hence it has an equivalent regularizer R. By definition of a regularizer, $RA^*Au = (I + T)\,u$ with a completely continuous operator T (I is the identity). Thus, we obtain the equation

$$u + Tu = 0, \tag{3}$$

which is still equivalent to (1).

We shall find the solutions to (3) as eigenelements of the equation

$$u + \mu Tu = 0 \tag{4}$$

to $\mu = 1$. These eigenelements are determined approximately by the Bubnov-Galerkin method. To that aim let $\{\varphi_n\}$ be a complete orthogonal system of coordinate functions in H. By H_n we denote the subspace spanned by the elements $\varphi_1, \varphi_2, \ldots, \varphi_n$, and by P_n the orthogonal projection from H onto H_n. Put

$$u_n = \sum_{k=1}^{n} a_k \varphi_k \tag{5}$$

and determine the coefficients a_k by means of the Bubnov-Galerkin equation

$$P_n(u_n + \mu T u_n) = 0. \tag{6}$$

The following fact is well-known (cf. Krasnoselski, Vainikko et al. [1]): Let μ_0 be an eigenvalue of (4) with multiplicity s, μ_n be an approximate eigenvalue for that equation, and $\mu_n \to \mu_0$ as $n \to \infty$. Then

$$\varrho(u_n^{(k)}, U_0^{(s)}) \leqq c \sup_{u_0 \in U_0^{(s)}, \|u_0\| = 1} \|(I - P_n)\, u_0\|; \tag{7}$$

$$\varrho(u_n^{(k)}, U_0^{(s)}) = \inf_{u_0 \in U_0^{(s)}} \|u_n^{(k)} - u_0\|; \qquad c = \text{const}.$$

Here, u_0 is the exact solution of (3), and $u_n^{(k)}$ is an arbitrary element of $U_n^{(k)}$, with norm 1,

$$U_n^{(k)} := \{u_n \in H_n : (I + \mu_n P_n T)^k u_n = 0\},$$
$$U_0^{(j)} := \{u_0 \in H : (I + \mu_0 T)^j u_0 = 0\}.$$

6.2. Now, we consider the singular integral equation

$$(Au)(x) := a(x)\, u(x) + \int_{\mathbb{R}^m} r^{-m} f(x, \theta)\, u(y)\, dy = 0. \tag{8}$$

The operator A satisfy all assumptions made in § 5. Applying the general arguments of § 6.1 to that special case we find that (8) is equivalent to some equation (3). Moreover, every solution of (8) and, therefore, of the corresponding equation (3) belongs to the space $\mathbf{W}_2^{(l)}(\mathbb{R}^m)$. We shall show that for any positive integer k the solutions to the equation

$$(I + T)^k u = 0 \tag{9}$$

belong also to that space. This will be done by induction. Suppose that the solutions of the equation $(I + T)^{k-1} u = 0$ belong all to $\mathbf{W}_2^{(l)}(\mathbb{R}^m)$. Put

$$(I + T)\, u = v \tag{10}$$

so that $(I + T)^{k-1} v = 0$ and $v \in \mathbf{W}_2^{(l)}(\mathbb{R}^m)$. Therefore, the function u as a solution of (10) belongs to $\mathbf{W}_2^{(l)}(\mathbb{R}^m)$, and the assertion is proved.

As mentioned before, the functions

$$\varphi_n^{(k)}(x) = (1 + \xi_{m+1})^{m/2}\, Y_n^{(k)}(\xi)$$

form a complete orthonormal system in $\mathbf{L}_2(\mathbb{R}^m)$, and for

$$u \in \mathbf{W}_2^{(l)}(\Sigma), \qquad u(\xi) = \sum_{n=0}^{\infty} \sum_{k=1}^{k_n} a_n^{(k)} Y_n^{(k)}(\xi)$$

one has

$$\left\| u - \sum_{n=0}^{N} \sum_{k=1}^{k_n} a_n^{(k)} Y_n^{(k)} \right\|_{\mathbf{L}_2(\Sigma)} = o(N^{-l}). \tag{11}$$

The estimate (11) is, in particular, valid for the solutions of (8).

Let $u_1, u_2, \ldots, u_p$ be an orthonormal basis in the subspace of solutions to eq. (8). Every normed element of that subspace admits the representation

$$u_0 = \sum_{k=1}^{p} \alpha_k u_k, \qquad \sum_{k=1}^{p} |\alpha_k|^2 = 1.$$

With $P^{(n)} := I - P_n$ one has

$$P^{(n)} u_0 = \sum_{k=1}^{p} \alpha_k P^{(n)} u_k$$

and

$$\|P^{(n)} u_0\|^2 \le \sum_{k=1}^{p} \alpha_k^2 \sum_{k=1}^{p} \|P^{(n)} u_k\|^2 = \sum_{k=1}^{p} \|P^{(n)} u_k\|^2 = o(N^{-2l}).$$

Now, eq. (7) implies the estimate

$$\varrho(u_n^{(k)}, U_0^{(s)}) = o(N^{-l}). \tag{12}$$

§ 7. Constructing the approximations and estimating them

Let us come back again to (5.1). Assume that all linearly independent approximate solutions $\tilde{u}^{(1)}, \tilde{u}^{(2)}, \ldots, \tilde{u}^{(s)}$ of the homogeneous equation (5.2) are known, and assume them to be orthonormal. We preserve the notations of the previous sections. Moreover, by $\tilde{H}_1$ we denote the subspace of approximate solutions to eq. (5.2) i.e. the span of the elements $\tilde{u}^{(1)}, \tilde{u}^{(2)}, \ldots, \tilde{u}^{(s)}$, and $\tilde{H}_2 := H \ominus \tilde{H}_1$. The solution of (5.1) is to be found in $\tilde{H}_2$.

By $\tilde{B}$ we denote the restriction of the operator A to $\tilde{H}_2$. Then we shall show that for sufficiently large N the bounded inverse $\tilde{B}^{-1}$ exists. Suppose $\tilde{B}u = 0$, then $Au = 0$ and hence

$$u = \sum_{j=1}^{s} c_j u^{(j)} . \tag{1}$$

On the other hand, $u \in \tilde{H}_2$, i.e. $(u, \tilde{u}^{(k)}) = 0$, $k = 1, 2, \ldots, s$, and

$$\sum_{j=1}^{s} c_j (u^{(j)}, \tilde{u}^{(k)}) = 0 . \tag{2}$$

By the estimate (6.12) we have for sufficiently large N the inequality $\|\tilde{u}^{(j)} - u^{(j)}\| < \delta$ with an arbitrarily small number δ. If δ were zero then $\tilde{u}^{(j)}$ and $u^{(j)}$ would coincide, and the determinant of system (2) would be 1. Hence, this determinant is non-zero for small δ, and the system (2) has the unique solution $c_j = 0$, i.e. the inverse $\tilde{B}^{-1}$ does exist.

Next, we shall show the operator $\tilde{B}^{-1}$ to be bounded. It is easy to see that $\operatorname{Im} \tilde{B} = H_2^*$. Indeed, let $v \in H_2^*$, then we have to show that there is an element $w \in \tilde{H}_2$ such that $Aw = v$. Obviously, there is an element $u \in H_2$ such that $Au = v$. Put

$$w = u + \sum_{k=1}^{s} a_k u^{(k)} ,$$

$$\sum_{k=1}^{s} a_k (u^{(k)}, \tilde{u}^{(j)}) = -(u, \tilde{u}^{(j)}), \qquad j = 1, 2, \ldots, s .$$

As mentioned above, the determinant of that system is non-zero, and the element w can be constructed. It is obvious that $w \in \tilde{H}_2$ and $Aw = v$.

Thus, the operator $\tilde{B}^{-1}$ is defined on the whole space H_2^* and, as the inverse to the bounded operator $\tilde{B}$, it is closed. Therefore, $\tilde{B}^{-1}$ is bounded.

Now we choose a complete orthonormal system $\{\varphi_n\}$ in H and project it onto $\tilde{H}_2$. So we obtain a new system $\{\tilde{\varphi}_n\}$,

$$\tilde{\varphi}_n = \varphi_n - \sum_{j=1}^{s} (\varphi_n, \tilde{u}^{(j)})\, u^{(j)} . \tag{3}$$

An approximate solution of (5.1) is found by the least squares method in the shape

$$u_n = \sum_{k=1}^{n} b_k \tilde{\varphi}_k , \tag{4}$$

where the coefficients b_k are to be determined from the condition $\|\tilde{B}u_n - g\|^2 = \min$. The error $\|u_n - u_0\|$ is estimated as in § 5. Using the results of § 3 we obtain

$$\|u_n - u_0\| \leqq \operatorname{cond}(\tilde{B}) \cdot \gamma(n) \tag{5}$$

where $\gamma(n)$ denotes the best approximation of the element u_0 by linear combinations of the $\tilde{\varphi}_1, \tilde{\varphi}_2, \ldots, \tilde{\varphi}_n$.

The results of the present section apply also to singular integral equations. So does the estimate

$$\|u_0 - u_N\| = o(N^{-l}) \tag{6}$$

where u_N means the least squares solution constructed by means of the coordinate functions introduced in § 4.

§ 8. Applications to one-dimensional singular equations

The results of §§ 4–7 can be applied to one-dimensional singular equations. Let us consider the equation

$$(Au)(t) := a(t)\,u(t) + \frac{b(t)}{\pi i} \int_{-\infty}^{\infty} \frac{u(\tau)}{\tau - t}\,d\tau + (Tu)(t) = g(t) \tag{1}$$

under the assumptions $a^2(t) - b^2(t) \neq 0$ and $\operatorname{Ind} A = 0$. The operator T is of the form

$$(Tu)(t) = \int_{-\infty}^{\infty} K(t, \tau)\,u(\tau)\,d\tau, \tag{2}$$

it is assumed to be completely continuous and smoothing (this is e.g. the case if $\partial^q K/\partial t^q \in \mathbf{L}_2(\mathbb{R}^2)$, $q = 0, 1, \ldots, l$).

Let $g \in \mathbf{W}_2^{(l)}(\mathbb{R})$, then $g \in \mathbf{L}_2(\mathbb{R})$ and, therefore, also $u \in \mathbf{L}_2(\mathbb{R})$. Furthermore, let $a, b \in \mathbf{C}^{(l)}(\mathbb{R})$. Denoting $h(t) := g(t) - (Tu)(t)$ we consider the function $h(t)$ to be given, and write (1) in the form $a(t)\,u(t) + b(t)\,(Su)(t) = h(t)$. That is a simple one-dimensional singular equation the solution of which is given by

$$u(t) = \frac{a(t)}{a^2(t) - b^2(t)}\,h(t) - \frac{b(t)\,e^{\omega(t)}}{\pi i \sqrt{a^2(t) - b^2(t)}} \int_{-\infty}^{\infty} \frac{e^{-\omega(\tau)}h(\tau)}{\sqrt{a^2(\tau) - b^2(\tau)}}\,\frac{d\tau}{(\tau - t)} \tag{3}$$

with

$$\omega(t) = \frac{1}{2}\,(S\mu)(t), \qquad \mu(t) = \ln \frac{a(t) - b(t)}{a(t) + b(t)}.$$

Taking into account that T is smoothing, we obtain from (3) immediately $u \in \mathbf{W}_2^{(l)}(\mathbb{R})$.

The line $\mathbb{R}$ is mapped by the transformation

$$\xi_1 = \frac{2t}{1 + t^2}, \qquad \xi_2 = \frac{1 - t^2}{1 + t^2} \tag{4}$$

onto the unit circle Σ. For $\xi, \eta \in \Sigma$, $t, \tau \in \mathbb{R}$, and dt, $d\Sigma$ denoting the differential on $\mathbb{R}$ and Σ, resp., we have

$$t^2 + 1 = \frac{2}{1 + \xi_2}, \qquad dt = \frac{1}{1 + \xi_2}\,d\Sigma.$$

The mapping (4) transforms (1) into an integral equation over Σ with the new unknown function $v(\xi) = (1 + \xi_2)^{-1/2}\,u(t)$. That new equation has, obviously, the unique solution $v_0(\xi) = (1 + \xi_2)^{-1/2} u_0(t)$ with $v_0 \in \mathbf{L}_2(\Sigma)$ and $v_0 \in \mathbf{W}_2^{(l)}(\Sigma \setminus \sigma)$ where σ is a neighborhood of the point $(0, -1)$ on Σ.

Now put

$$\xi_1 = \frac{2t'}{1 + t'^2}, \qquad \xi_2 = \frac{t'^2 - 1}{t'^2 + 1}. \tag{5}$$

These formulae define a mapping of the circle Σ onto $\overline{\mathbb{R}}$ such that $t' = \infty$ corresponds to the point $\xi = (0, 1)$. Introducing the new unknown function $u'(t') = (1 - \xi_2)^{1/2}\, v(\xi)$ we transform the singular equation on Σ into a new one,

$$(A'u')\,(t') := a'(t')\, u'(t') + \frac{b'(t')}{\pi i} \int_{-\infty}^{\infty} \frac{u'(\tau')\, d\tau'}{\tau' - t'} + (T'u')\,(t') = g'(t'). \tag{6}$$

We want to give explicitly the transformation from (1) to (6). Comparing (4) and (5) we find $t' = t^{-1}$, i.e. the transformation from t to t' turns out to be an inversion on the real axis with respect to the origin. Now put in (1) $t = 1/t'$, then

$$a(t'^{-1})\, u\,(t'^{-1}) + \frac{b(t'^{-1})}{\pi i} \int_{-\infty}^{\infty} \frac{u(\tau)}{\tau - t'^{-1}}\, d\tau + (Tu)\,(t'^{-1}) = g\,(t'^{-1}). \tag{7}$$

With the notations

$$\tau = \tau'^{-1}, \qquad u\,(t'^{-1}) = u'(t'), \qquad a(t'^{-1}) = a'(t'), \qquad b\,(t'^{-1}) = b'(t'),$$

$$g\,(t'^{-1}) = g'(t'),$$

$$(Tu)\,(t'^{-1}) = -\int_{-\infty}^{\infty} \frac{K(t'^{-1}, \tau'^{-1})}{\tau'}\, u'(\tau')\, d\tau' = (T'u')\,(t'),$$

(7) assumes the form (6). The operator T is smoothing provided that

$$\frac{\partial^q \left[\dfrac{K(t'^{-1}, \tau'^{-1})}{\tau'}\right]}{\partial \tau'^q} \in \mathbf{L}_2(\mathbb{R}^2), \qquad q = 0, 1, \ldots, l.$$

Now we construct the least squares approximate solution to (1) using the coordinate functions defined in § 4. These coordinate functions have in the case considered here the form

$$\varphi_n^{(1)}(t) = \sqrt{1 + \xi_2}\, \cos n\theta, \qquad \varphi_n^{(2)}(t) = \sqrt{1 + \xi_2}\, \sin n\theta$$

or, replacing θ

$$\varphi_n^{(1)}(t) = \sqrt{\frac{2}{1 + t^2}}\, T_n \left(\frac{2t}{1 + t^2}\right), \qquad \varphi_n^{(2)}(t) = \frac{\sqrt{2}\,(1 - t^2)}{(1 + t^2)^{3/2}}\, U_{n-1} \left(\frac{2t}{1 + t^2}\right), \tag{8}$$

where T_k and U_k denote Chebyshev polynomials of first and second kind.

Under the above assumptions that $g' \in \mathbf{W}_2^{(l)}(\mathbb{R})$ and $a'(t'),\ b'(t') \in \mathbf{C}^{(l)}(\mathbb{R})$, the least squares approximate solutions u_N of (1) admit the estimate

$$\|u - u_N\|_{\mathbf{L}_2(\mathbb{R})} = o(N^{-l}). \tag{9}$$

§ 9. Application of Hermite functions

Applying the least squares method for solving singular integral equations we can derive an error estimate under somewhat different conditions as those of the previous sections, when Hermite functions are used as coordinate functions. These are given by the formula

$$\hat{H}_n(x) = \frac{(-1)^n e^{x^2/2}}{\sqrt{2^n n! \sqrt{\pi}}} \frac{d^n e^{-x^2}}{dx^n} := a_n e^{x^2/2} \frac{d^n e^{-x^2}}{dx^n} = e^{-x^2/2} H_n(x) \tag{1}$$

where $H_n(x)$ are Hermite polynomials. Hermite functions form a complete orthonormal system in $\mathbf{L}_2(\mathbb{R})$. In the following we shall denote by $c_n(u)$ the Fourier coefficient of the function $u(x)$ with respect to the system (1),

$$c_n(u) = a_n \int_{-\infty}^{\infty} u(x)\, e^{x^2/2} \frac{d^n e^{-x^2}}{dx^n}\, dx. \tag{2}$$

9.1. Let us estimate the rate of convergence of the coefficients (2) as $n \to \infty$ for $u \in \mathbf{W}_2^{(1)}(\mathbb{R})$. With the identity

$$a_n = -\frac{1}{\sqrt{2n}} a_{n-1}$$

integration by parts gives

$$c_n(u) = \frac{1}{\sqrt{2n}} c_{n-1}(u') + \frac{1}{\sqrt{2n}} \int_{-\infty}^{\infty} x\, u(x)\, \hat{H}_{n-1}(x)\, dx. \tag{3}$$

The recursion formula for Hermite polynomials implies a similar formula for Hermite functions,

$$\hat{H}_n(x) - 2x\hat{H}_{n-1}(x) + 2(n-1)\, \hat{H}_{n-2}(x) = 0.$$

Inserting this into the formula for $c_n(u)$ and replacing n by $n+2$ we obtain

$$c_{n+2}(u) = \frac{1}{\sqrt{2(n+2)} - \frac{1}{2}} c_{n+1}(u') + \frac{n+1}{\sqrt{2(n+2)} - \frac{1}{2}} c_n(u). \tag{4}$$

This relation gives the estimate

$$|c_n(u)| \le \frac{1}{\sqrt{n+1}} g_{1n}(u), \tag{5}$$

where

$$g_{1n}(u) = 2\,|c_{n+2}(u)| + \frac{1}{\sqrt{n+1}} |c_{n+1}(u')|. \tag{6}$$

It is worth noting that

$$\sum_{n=0}^{\infty} g_{1n}^2(u) \le C\, \|u\|^2_{\mathbf{W}_2^{(1)}(\mathbb{R})}, \qquad C = \text{const}. \tag{7}$$

It is easy to obtain similar formulae as (5)–(7) for functions u belonging to $\mathbf{W}_2^{(l)}(\mathbb{R})$, $l > 1$. Indeed, by the formulae mentioned we have

$$|c_n(u')| \le \frac{2}{\sqrt{n+1}} |c_{n+2}(u')| + \frac{1}{n+1} |c_{n+1}(u'')|$$

or, replacing n by $n+1$,

$$|c_{n+1}(u')| \le \frac{2}{\sqrt{n+2}} |c_{n+3}(u')| + \frac{1}{n+2} |c_{n+2}(u'')|.$$

Inserting this into (6) we obtain

$$g_{1n}(u) \leqq 2\,|c_{n+2}(u)| + \frac{2}{\sqrt{n+1}\,(n+2)}\,|c_{n+3}(u')| + \frac{1}{\sqrt{n+1}\,(n+2)}\,|c_{n+2}(u'')|$$

$$\leqq 2\,|c_{n+2}(u)| + \frac{2}{(n+1)^{3/2}}\,|c_{n+3}(u')| + \frac{1}{(n+1)^{3/2}}\,|c_{n+2}(u'')|$$

and, together with the inequality (5),

$$g_{1n}(u) \leqq \frac{1}{\sqrt{n+1}}\,g_{2n}(u)$$

where we may put

$$g_{2n}(u) = 2g_{1,n+1}(u) + \frac{2}{\sqrt{n+1}}\,|c_{n+3}(u')| + \frac{1}{n+1}\,|c_{n+2}(u'')|.$$

Obviously, $|c_n(u)| \leqq (n+1)^{-1}\,g_{2n}(u)$, and

$$\sum_{n=0}^{\infty} g_{2n}^2(u) \leqq C\,\|u\|^2_{\mathbf{W}_2^{(1)}(\mathbf{R})}.$$

Then, by induction,

$$\forall u \in \mathbf{W}_2^{(l)}(\mathbb{R}) : |c_n(u)| \leqq (n+1)^{-l/2}\,g_{ln}(u); \qquad \sum_{n=0}^{\infty} g_{ln}^2(u) \leqq C\,\|u\|^2_{\mathbf{W}_2^{(l)}(\mathbf{R})}. \tag{8}$$

9.2. Now let us estimate the rate of convergence of the Fourier series

$$u(x) = \sum_{n=0}^{\infty} c_n(u)\,\hat{H}_n(x)$$

for $u \in \mathbf{L}_2(\mathbb{R})$, where we shall use the $\mathbf{L}_2$ modulus of continuity

$$\omega^2(u, h) = \sup_{|t| \leqq h} \int_{-\infty}^{\infty} |u(x+t) - u(x)|^2\,\mathrm{d}x.$$

With the notations

$$u_h(x) = \frac{1}{h}\int_x^{x+h} u(t)\,\mathrm{d}t$$

and

$$P_n(u) = \sum_{k=0}^{n} c_k(u)\,\hat{H}_k(x)$$

we have

$$\|u - P_n u\| \leqq \|u - u_h - P_n(u) + P_n(u_h)\| + \|u_h - P_n(u_h)\|$$
$$= \|u - u_h - P_n(u - u_h)\| + \|u_h - P_n(u_h)\| \leqq \|u - u_h\| + \|u_h - P_n u_h\| \tag{9}$$

($\|\cdot\|$ means $\|\cdot\|_{\mathbf{L}_2(\mathbf{R})}$). Because of $u_h \in \mathbf{W}_2^{(1)}(\mathbb{R})$,

$$\|u_h - P_n(u_h)\| = \left\{\sum_{k=n+1}^{\infty} |c_k(u_h)|^2\right\}^{1/2} \leqq \left\{\sum_{k=n+1}^{\infty} \frac{g_{1k}^2(u_h)}{k+1}\right\}^{1/2}$$
$$\leqq \frac{1}{\sqrt{n+1}}\left\{\sum_{k=0}^{\infty} g_{1k}^2(u_h)\right\}^{1/2} \leqq \frac{C}{\sqrt{n+1}}\,\|u_h\|_{\mathbf{W}_2^{(1)}(\mathbf{R})}.$$

Furthermore,

$$\|u_h\|^2_{W_2^{(1)}(\mathbb{R})} = \|u_h\|^2_{L_2(\mathbb{R})} + \|u_h'\|^2_{L_2(\mathbb{R})} \leqq \|u\|^2_{L_2(\mathbb{R})} + \|u_h'\|^2_{L_2(\mathbb{R})}.$$

The identity $u_h'(x) = h^{-1}[u(x+h) - u(x)]$ gives

$$\|u_h'\|^2_{L_2(\mathbb{R})} = h^{-2} \int_{-\infty}^{\infty} |u(x+h) - u(x)|^2 \, dx \leqq h^{-2} \, \omega^2(u, h)$$

and thus

$$\|u_h - P_n(u_h)\| \leqq \frac{C}{\sqrt{n+1}} [\|u\| + h^{-1} \, \omega(u, h)].$$

Now we come to $\|u - u_h\|^2$,

$$\|u - u_h\|^2 = h^{-2} \int_{-\infty}^{\infty} \left| \int_0^h [u(x+t) - u(x)] \, dt \right|^2 dx \leqq h^{-1} \int_{-\infty}^{\infty} dx \int_0^h |u(x+t) - u(x)|^2 \, dt$$

$$= h^{-1} \int_0^h dt \int_{-\infty}^{\infty} |u(x+t) - u(x)|^2 \, dx = \omega^2(u, h).$$

Then, by the inequality (9),

$$\|u - P_n u\| \leqq \frac{C}{\sqrt{n+1}} \|u\| + \left(1 + \frac{C}{h\sqrt{n+1}}\right) \omega(u, h).$$

Thus, with $h := 1/\sqrt{n+1}$ finally we arrive at

$$\forall u \in L_2(\mathbb{R}): \|u - P_n u\|_{L_2(\mathbb{R})} \leqq \frac{C}{\sqrt{n+1}} \|u\|_{L_2(\mathbb{R})} + \left(1 + C\omega\left(u, \frac{1}{\sqrt{n+1}}\right)\right). \tag{10}$$

9.3. A similar estimate is true for functions u belonging to $W_2^{(l)}(\mathbb{R})$ with a positive integer l. Formula (8) gives

$$\|u - P_n(u)\|^2_{L_2(\mathbb{R})} = \sum_{k=n+1}^{\infty} |c_k(u)|^2 \leqq (n+1)^{-l/2} \sum_{k=n+1}^{\infty} g_{lk}^2(u).$$

It is easy to see that

$$g_{ln}(u) \leqq C \sum_{j=0}^{l} |c_{n+k_j}(u^{(j)})|$$

with non-negative integers k_j. But then

$$\sum_{k=n+1}^{\infty} g_{lk}^2(u) \leqq C \sum_{j=0}^{l} \sum_{k=n+1}^{\infty} |c_k(u^{(j)})|^2 = C \sum_{j=0}^{l} \|u^{(j)} - P_n(u^{(j)})\|^2_{L_2(\mathbb{R})}$$

and, by (10),

$$\sum_{k=n+1}^{\infty} g_{lk}^2(u) \leqq C \sum_{j=0}^{l} \left[\frac{1}{n+1} \|u^{(j)}\|_{L_2(\mathbb{R})} + \omega^2\left(u^{(j)}, \frac{1}{\sqrt{n+1}}\right)\right]$$

$$\leqq C \left[\frac{1}{n+1} \|u\|^2_{W_2^{(l)}(\mathbb{R})} + \omega_l^2\left(u, \frac{1}{\sqrt{n+1}}\right)\right],$$

where ω_l denotes the $W_2^{(l)}$ modulus of continuity. These relations lead to the estimate wanted,

$$\forall u \in W_2^{(l)}(\mathbb{R})\colon \|u - P_n(u)\|_{L_2(\mathbb{R})}$$
$$\leqq \frac{C}{(n+1)^{l/2}} \left[\frac{1}{\sqrt{n+1}} \|u\|_{W_2^{(l)}(\mathbb{R})} + \omega_l\left(u, \frac{1}{\sqrt{n+1}}\right)\right]. \tag{11}$$

9.4. Now we shall come to the case of a Euclidean space $\mathbb{R}^m$ of arbitrary dimension m. In the following we shall denote by x, y, t, h m-dimensional vectors, and by $\mathbf{k}, \mathbf{l}, \mathbf{n}$ m-dimensional multi-indices. By $\mathbf{1}$ resp. $\mathbf{0}$ we denote the m-dimensional vector or multi-index the components of which are all equal to 1 resp. 0. The relation $x \leqq y$ (or $x \geqq y$) for vectors or multi-indices means $x_k \leqq y_k$ (or $x_k \geqq y_k$) for every $k = 1, 2, \ldots, m$.

Let us introduce the functions

$$\hat{H}_{\mathbf{n}}(x) = \prod_{j=1}^{m} \hat{H}_{n_j}(x_j) = \frac{(-1)^{|\mathbf{n}|} e^{|x|^2/2}}{\sqrt{2^{|\mathbf{n}|}\mathbf{n}!\, \pi^{m/2}}} D^{\mathbf{n}} e^{-|x|^2}. \tag{12}$$

These functions will be called Hermite functions, too, and they form a complete orthonormal system in $L_2(\mathbb{R}^m)$. For the Fourier coefficients with respect to the system (12) and for the corresponding partial sums we preserve the notations

$$c_{\mathbf{n}}(u) = \int_{\mathbb{R}^m} u(x) \hat{H}_{\mathbf{n}}(x)\, dx, \qquad P_{\mathbf{n}}(u) = \sum_{\mathbf{k}=\mathbf{0}}^{\mathbf{n}} c_{\mathbf{k}}(u) \hat{H}_{\mathbf{k}}(x).$$

Then it is not difficult to obtain estimates similar to (8),

$$\forall u \in W_2^{(\mathbf{l})}(\mathbb{R}^m)\colon |c_{\mathbf{n}}(u)| \leqq \frac{1}{\sqrt{(\mathbf{n}+\mathbf{1})}} g_{\mathbf{l}\mathbf{n}}(u), \qquad \sum_{\mathbf{n}\geqq\mathbf{0}} g_{\mathbf{l}\mathbf{n}}^2(u) \leqq C \|u\|^2_{W_2^{(\mathbf{l})}(\mathbb{R}^m)}. \tag{13}$$

For the sake of notational simplicity we shall omit $\mathbb{R}^m$ in the notation of function spaces.

Defining the mollification $u_h(x)$ by

$$u_h(x) = \frac{1}{h^{\mathbf{1}}} \int_{x_1}^{x_1+h_1} \cdots \int_{x_m}^{x_m+h_m} u(t)\, dt$$

and taking into account the relation $D^{\mathbf{1}} u_h(x) = (h^{\mathbf{1}})^{-1}[u(x+h) - u(x)]$ we obtain as in the one-dimensional case the estimate similar to (11),

$$\forall u \in W_2^{(\mathbf{l})}\colon \|u - P_{\mathbf{n}}(u)\|_{L_2} \leqq \frac{C}{\sqrt{(\mathbf{n}+\mathbf{1})^{\mathbf{l}}}} \left[\frac{1}{\sqrt{(\mathbf{n}+\mathbf{1})^{\mathbf{1}}}} \|u\|_{W_2^{(\mathbf{l})}} + \omega_{\mathbf{l}}\left(u, \frac{1}{\sqrt{(\mathbf{n}+\mathbf{1})^{\mathbf{1}}}}\right)\right]. \tag{14}$$

9.2. Now let us consider the singular equation (4.5). We shall assume that the operator defined by the left-hand side of (4.5) is invertible, and the solution $u(x)$ of this equation to belong to $W_2^{(\mathbf{l})}$ for some multi-index $\mathbf{l}$. This is the case if e.g. the symbol of (4.5) belongs to the class $\mathfrak{R}_{\mathbf{l}+\mathbf{1},\lambda}$ for sufficiently large $\mathbf{l}$, $g \in W_2^{(\mathbf{l})}$, and the operator T is smoothing of sufficiently large order. We solve this singular equation by the least squares method with the coordinate functions (12) such that the approximate solutions are of the form

$$u_{\mathbf{n}}(x) = \sum_{\mathbf{k}=\mathbf{0}}^{\mathbf{n}} b_{\mathbf{k}} \hat{H}_{\mathbf{k}}(x), \qquad b_{\mathbf{k}} = \text{const.}$$

Then the formulae (3.9) and (14) yield the following error estimate for that approximate solution,

$$\|u - u_n\|_{\mathbf{L}_2} \leqq \frac{c_0 \|A\| \|A^{-1}\|}{\sqrt{(n+1)^l}} \left[\frac{1}{\sqrt{(n+1)^1}} \|u\|_{\mathbf{W}_2^{(l)}} + \omega_l \left(u, \frac{1}{\sqrt{(n+1)^1}}\right)\right]. \quad (15)$$

Here, $\|A\|$ and $\|A^{-1}\|$ are the norms of the corresponding operators in $\mathbb{R}^m$. Note that the constant c_0 can be easily estimated.

References

Abrahamse, M. B.

[1] The spectrum of a Toeplitz operator with a multiplicatively periodic symbol. J. Funct. Anal. **31**: 2, 224–233 (1979).

Agranovich, M. S.

[1] Elliptic singular integro-differential operators. Uspehi Mat. Nauk **20**: 5 (125), 3–120 (1965) (*Russian*); *also in*: Russ. Math. Surveys **20**, 1–121 (1965).

Agranovich, M. S., and M. I. Vishik

[1] Pseudodifferential operators. Izd. Moskov. Univ., Moscow 1968 (*Russian*).

Aleksidse, M. A.

[1] The solution of boundary value problems by the method of expansion with respect to non-orthogonal functions. Nauka, Moscow 1978 (*Russian*).

Alexandroff, P., and H. Hopf

[1] Topologie. I. Springer-Verlag, Berlin 1935.

Alexandrov, A. B.

[1] The norm of the Hilbert transform in the space of Hölder functions. Funkts. Anal. Prilozh. **9**: 2, 1–4 (1975) (*Russian*); *also in*: Funct. Anal. Appl. **9**, 94–96 (1975).

Alexandrov, P. S.

[1] Combinatorial topology. Goztekhizdat, Moscow and Leningrad 1947 (*Russian*).

Arnold, D. N., and W. L. Wendland

[1] Collocation versus Galerkin procedures for boundary integral methods. *In*: Boundary Element Methods in Engineering, ed. C. A. Brebbia, 18–33, Springer-Verlag, Berlin, Heidelberg, New York 1982.

[2] On the asymptotic convergence of collocation methods. Math. Comput. **41**, 349–381 (1983).

[3] The convergence of spline collocation for strongly elliptic equations on curves. Numer. Math. **47**, 313–341 (1985).

Aronszajn, N.

[1] Calculus of residues and general Cauchy formulas in $\mathbb{C}^n$. Bull. Sci. Math., II. Ser. **101**, 319–352 (1977).

Atiyah, M. F., and I. M. Singer

[1] The index of elliptic operators on compact manifolds. Bull. Amer. Math. Soc. **69**: 3, 422–433 (1963).

[2] The index of elliptic operators. I. Uspehi Mat. Nauk **23**: 5 (143), 99–142 (1968) (*Russian*).

Atkinson, F. V.

[1] Normal solvability of linear equations in normed spaces. Mat. Sb. **28**: 1, 3–14 (1951) (*Russian*).

Aubin, J. P.

[1] Approximation of elliptic boundary value problems. Wiley-Interscience, New York 1972.

Babuška, I., and A. K. Aziz

[1] Survey lectures on the mathematical foundation of the finite element method. *In*: The Mathematical Foundation of the Finite Element Method with Applications to Partial Differential Equations, ed. A. K. Aziz, 3–345, Academic Press, New York 1972.

Bakelman, Y. Ya., A. L. Werner, and B. E. Kantor
[1] Introduction to "global" differential geometry. Nauka, Moscow 1973 (*Russian*).
Bateman, H., and A. Erdélyi
[1] Higher transcendental functions. Vols. I–III. McGraw-Hill Company, Inc., New York 1953.
Baxter, G.
[1] A norm inequality for a "finite-section" Wiener-Hopf equation. Ill. J. Math. **7**, 97–103 (1963).
Besov, O. V., V. P. Ilyin, and P. I. Lizorkin
[1] L_p-estimates of a certain class of nonisotropic singular integrals. Dokl. Akad. Nauk SSSR **169**: 6, 1250–1253 (1966) (*Russian*).
Birman, M. S., and M. Z. Solomyak
[1] Double Stieltjes operator-integrals and problems on multipliers. Dokl. Akad. Nauk SSSR **171**: 5, 1251–1254 (1966) (*Russian*).
[2] Double Stieltjes operator-integrals. II. *In*: Probl. Mat. Fiz., Vyp. 2, 26–60, Izd. Leningrad. Univ., 1967 (*Russian*).
Bitsadze, A. V.
[1] Boundary value problems for elliptic equations of second order. Nauka, Moscow 1966 (*Russian*).
[2] Grundlagen der Theorie der analytischen Funktionen. Akademie-Verlag, Berlin 1973.
Bochner, S.
[1] Theta relations with spherical harmonics. Proc. Natl. Acad. Sci. USA **37**: 12, 804–808 (1951).
Boikov, I. B.
[1] On the approximate solution of singular integral equations. Dokl. Akad. Nauk SSSR **203**: 3, 511–514 (1972) (*Russian*); *also in*: Sov. Math. Dokl. **13**, 400–404 (1972).
Bojarski, B.
[1] A direct approach to the theory of systems of singular integral equations. *In the book*: N. I. Muskhelishvili, Singular integral equations, Third ed., 478–488, Nauka, Moscow 1968 (*Russian*).
[2] Theory of the generalized analytic vector. Ann. Pol. Math. XVII, 281–320 (1966) (*Russian*).
Bony, J.-M.
[1] Résolution des conjectures de Calderón et espaces de Hardy généralisés [d'après R. Coifman, G. David, A. McIntosh, Y. Meyer]. Astérisque **92–93**, 293–300 (1982).
Booss, B.
[1] Topologie und Analysis. Eine Einführung in die Atiyah-Singer-Indexformel. Springer-Verlag, Berlin, Heidelberg, New York 1977.
Böttcher, A.
[1] The finite section method for the Wiener-Hopf integral operator. Kand. Dissert., Rostov-on-Don State Univ., 1984 (*Russian*).
Böttcher, A., and B. Silbermann
[1] Über das Reduktionsverfahren für diskrete Wiener-Hopf-Gleichungen mit unstetigem Symbol. Z. Anal. Anw. **1**: 2, 1–5 (1982).
[2] Invertibility and asymptotics of Toeplitz matrices. Akademie-Verlag, Berlin 1983.
Boutet de Monvel, L.
[1] Boundary problems for pseudo-differential operators. Acta Math. **126**, 11–51 (1971).
Brackx, F., R. Delanghe, and F. Sommen
[1] Clifford analysis. Research Notes in Math. 76, Pitman, Boston, London, Melbourne 1982.
Breuer, M.
[1] Banachalgebren mit Anwendung auf Fredholmoperatoren und singuläre Integralgleichungen. Bonn. Math. Schr. 24, 1965.
Budyanu, M. S., and I. Gohberg
[1] General theorems on the factorization of matrix functions. I. The basic theorem. Mat. Issled. **III**: 2, 87–103 (1968) (*Russian*); *also in*: Amer. Math. Soc. Transl. (2) **102**, 5–26 (1973).
[2] General theorems on the factorization of matrix functions. II. Some tests and their consequences. Mat. Issled. **III**: 3, 3–18 (1968) (*Russian*); *also in*: Amer. Math. Soc. Transl. (2) **102**, 15–26 (1973).
Butzer, P. L., and R. J. Nessel
[1] Fourier analysis and approximation. I. Birkhäuser Verlag, Basel, Stuttgart 1971.

CALDERÓN, A. P.

[1] Lebesgue spaces of differentiable functions and distributions. Proc. of Symposia in Pure Mathematics IV, Amer. Math. Soc., 33–49 (1961).

[2] Singular integrals. Colloquium lectures given at Ithaka, New York at the seventieth summer meeting of the Amer. Math. Soc., 1–68 (1965).

[3] Commutators of singular integral operators. Proc. Natl. Acad. Sci. USA **53**: 5, 1092–1099 (1965).

[4] Algebras of singular integral operators. Proc. of Symposia in Pure Mathematics X, Amer. Math. Soc., 18–55 (1967).

[5] Cauchy integrals on Lipschitz curves and related operators. Proc. Natl. Acad. Sci. USA **74**: 4, 1324–1327 (1977).

[6] Lecture notes on pseudo-differential operators and elliptic boundary value problems. Publs. I.A.M., Ser. 2, N 1, 1976.

CALDERÓN, A. P., C. P. CALDERÓN, E. FABES, M. JODEIT, and N. M. RIVIERE

[1] Applications of the Cauchy integrals on Lipschitz curves. Bull. Amer. Math. Soc. **84**: 2, 287–290 (1978).

CALDERÓN, A. P., and A. ZYGMUND

[1] On the existence of certain singular integrals. Acta Math. **88**: 1–2, 85–139 (1952).

[2] Singular integral operators and differential equations. Amer. J. Math. **79**: 4, 901–921 (1957).

[3] On singular integrals. Amer. J. Math. **78**: 2, 289–309 (1956).

[4] On a problem of Mikhlin. Trans. Amer. Math. Soc. **78**: 1, 209–224 (1955).

[5] Addends to the paper "On a problem of Mikhlin". Trans. Amer. Math. Soc. **84**: 2, 559–560 (1957).

[6] On singular integrals with variable kernels. Applicable Analysis **7**, 221–238 (1978).

CHEBOTAREV, G. N.

[1] On a convolution type equation of first kind. Izv. Vyssh. Uchebn. Zaved., Mat. **2**, 80–92 (1967) (*Russian*).

[2] On an exceptional case of the Wiener-Hopf equation in the space of bounded functions. Izv. Vyssh. Uchebn. Zaved., Mat. **10**, 92–101 (1967) (*Russian*).

CHERSKI, YU. I.

[1] The general singular equation and equations of convolution type. Mat. Sb. **41**: 3 (83), 277–296 (1957) (*Russian*).

CHIKIN, L. A.

[1] Exceptional cases of the Riemann boundary value problem and of singular integral equations. Uchebn. Zap. Kazan. Univ. **113**: 10, 57–105 (1953) (*Russian*).

CLANCEY, K., and I. GOHBERG

[1] Localization of singular integral operators. Math. Z. **169**, 105–117 (1979).

[2] Factorization of matrix functions and singular integral operators. Birkhäuser Verlag, Basel, Boston, Stuttgart 1981.

COBURN, L. A.

[1] Weyl's Theorem for non-normal operators. Mich. Math. J. **13**, 285–286 (1966).

COBURN, L. A., and R. G. DOUGLAS

[1] Translation operators on the half-line. Proc. Natl. Acad. Sci. USA **62**, 1010–1013 (1969).

CODDINGTON, E. A., and N. LEVINSON

[1] Theory of ordinary differential equations. McGraw-Hill, New York 1955.

COIFMAN, R. R., and C. FEFFERMAN

[1] Weighted norm inequalities for maximal functions and singular integrals. Stud. Math. **51**, 241–250 (1974).

COIFMAN, R., A. MCINTOSH, and Y. MEYER

[1] L'integrale de Cauchy definit un opérateur borné sur L^2 pour les courbes Lipschitziennes. Ann. Math. **116**, 361–388 (1982).

COIFMAN, R. R., and Y. MEYER

[1] Au delà des opérateurs pseudo-différentiels. Astérisque **57**, 2–184 (1978).

COIFMAN, R. R., R. ROCHBERG, and G. WEISS

[1] Factorization theorems for Hardy spaces in several variables. Ann. Math. **103**, 611–635 (1976).

Costabel, M.

[1] Singular integral operators on curves with corners. Integral Equations Oper. Theory **3**: 3, 323–349 (1980).

[2] An inverse for the Gohberg-Krupnik symbol map. Proc. R. Soc. Edinb. **87 A**, 153–165 (1980).

[3] On the algebra generated by one-dimensional singular integral operators with piecewise continuous coefficients. *In*: Toeplitz Centennial, ed. I. Gohberg, 211–215, Oper. Theory: Advances and Applications, Vol. 4, Birkhäuser Verlag, Basel, Boston, Stuttgart 1982.

Cotlar, M.

[1] A unified theory of Hilbert transforms and ergodic theorems. Revista Mat. Cuyana **1**: 2, 105–167 (1955).

Dang, D., and D. Norrie

[1] A finite element method for the solution of singular integral equations. Comput. Math. Appl. **4**, 219–224 (1978).

Danilyuk, I. I.

[1] Nonregular boundary value problems in the plane. Nauka, Moscow 1975 (*Russian*).

Danilyuk, I. I., and Y. Vu. Shelepov

[1] On the boundedness in L_p of the singular operator with Cauchy kernel over a curve of bounded rotation. Dokl. Akad. Nauk SSSR **174**: 3, 514–517 (1967) (*Russian*).

David, G.

[1] Opérateurs intégraux singuliers sur certains courbes du plane complex. Ann. Sci. Ec. Norm. Sup. **17**, 1, 157–189 (1984).

Delanghe, R.

[1] On the singularities of functions with values in a Clifford algebra. Math. Ann. **196**, 293–319 (1972).

Dieudonné, J. A.

[1] Sur les homomorphismes d'espaces normés. Bull. Sci. Math., II. Ser. **67**, 72–84 (1943).

Douglas, R. G.

[1] Banach algebra techniques in operator theory. Academic Press, New York 1972.

Dow, M. L., and D. Elliott

[1] The numerical solution of singular integral equations over $(-1, +1)$. SIAM J. Numer. Anal. **16**, 115–134 (1979).

Duduchava, R. V.

[1] On singular integral operators on piecewise smooth lines. *In*: Function Theoretic Methods in Differential Equations, ed. R. P. Gilbert and R. J. Weinacht, 109–131. Pitman, London, San Francisco, Melbourne 1976.

[2] On bisingular integral operators with discontinuous coefficients. Mat. Sb. **101** (143), 584–609 (1976) (*Russian*).; *also in*: Math USSR Sb. **30**, 515–537 (1976).

[3] On integral operators of convolution type with discontinuous coefficients. Math. Nachr. **79**, 75–98 (1977) (*Russian*).

[4] On multidimensional singular integral operators. I: J. Oper. Theory **11**, 41–76 (1984), II: J. Oper. Theory **11**, 199–214 (1974).

[5] On multidimensional singular integral equations. The basic theorems. Soobshzh. Akad. Nauk Gruz. SSR **111**: 3, 465–468 (1983) (*Russian*).

Duistermaat, J. J.

[1] Fourier integral operators. Courant Inst. of Math. Sci., New York 1973.

Dunford, N., and J. T. Schwartz

[1] Linear operators. Part I: General theory. Interscience Publishers, Inc., New York 1958.

Dybin, V. B.

[1] The exceptional case of the paired integral equation of convolution type. Dokl. Akad. Nauk SSSR **176**: 2, 251–254 (1967) (*Russian*).

[2] Normalization of the Wiener-Hopf operator. Dokl. Akad. Nauk SSSR **191**: **4**, 759–762 (1970) (*Russian*); *also in*: Sov. Math. Dokl. **11**, 437–441 (1970).

[3] Normalization of the singular integral equation in the exceptional case. *In*: Mat. Analiz i Prilozh. VI, 46–61, Izd. Rostov. Univ., Rostov-on-Don 1974 (*Russian*).

[4] On the singular integral operator over the real line with almost periodic coefficients. *In*: Teoria Funkts., Differ. Uravn. i Prilozh., 98–108, Elista 1976 (*Russian*).

[5] Correctly posed Riemann boundary value problems for the case of standard almost periodic discontinuities in the coefficient. Math. Nachr. **122**, 301–338 (1985) (*Russian*).

[6] Singular integral equations with coefficients vanishing on countable sets. Math. Nachr. **124**, 65–84 (1985) (*Russian*).

[7] Correctly posed problems for singular integral equations. Izd. Rostov. Univ., Rostov-on-Don 1987 (*Russian*).

Dybin, V. B., and V. N. Gaponenko

[1] The Riemann boundary value problem with quasi-periodic degeneracy of the coefficient. Dokl. Akad. Nauk SSSR **212**: 5, 1046–1049 (1973) (*Russian*); *also in*: Sov. Math. Dokl. **14**: 5, 1516–1520 (1973).

Dybin, V. B., and N. K. Karapetyants

[1] Application of the method of normalization to a class of infinite systems of linear algebraic equations. Izv. Vyssh. Uchebn. Zaved., Mat. **10**, 39–49 (1967) (*Russian*).

Dynkin, E. M., and B. P. Osilenker

[1] Weighted estimations of singular integrals and their applications. Itogi Nauki i Tekhniki, Ser. Mathem. Analiz, Vol. 21, 42–129, VINITI, Moscow 1983 (*Russian*).

Dynin, A. S.

[1] Singular operators of arbitrary order on a manifold. Dokl. Akad. Nauk SSSR **141**: 1, 21–23 (1961) (*Russian*).

[2] Fredholmian elliptic operators on manifolds. Uspehi Mat. Nauk **17**: 2 (104), 194–195 (1962) (*Russian*).

Dzishkariani, A. V.

[1] On the solution of singular integral equations by approximate projection methods. Zh. Vychisl. Mat. Mat. Fiz. **19**, 1149–1161 (1979) (*Russian*); *also in*: U.S.S.R. Comput. Math. Math. Phys. **19**: 5, 61–74 (1980).

[2] On the solution of singular integral equations by collocation methods. Zh. Vychisl. Mat. Mat. Fiz. **21**, 355–362 (1981) (*Russian*).; *also in*: U.S.S.R. Comput. Math. Math. Phys. **21**: 2, 99–107 (1981).

Elliott, D.

[1] The classical collocation method for singular integral equations having a Cauchy kernel. SIAM J. Numer. Anal. **19**, 816–831 (1982).

Elschner, J.

[1] Über eine Klasse entarteter singulärer Integrodifferentialgleichungen. I. Math. Nachr. **90**, 197–211 (1979).

[2] Über die normale Auflösbarkeit von entarteten singulären Integrodifferentialgleichungen. Demonstr. Math. **X**: 3, 815–826 (1977).

[3] On a class of degenerate singular integro-differential equations. II. Math. Nachr. **103**, 255–281 (1981).

[4] Singular ordinary differential operators and pseudodifferential equations. Akademie-Verlag, Berlin 1985.

[5] Galerkin methods with splines for singular integral equations over (0, 1). Numer. Math. **43**, 265–281 (1984).

[6] A Galerkin method with finite elements for degenerate one-dimensional pseudodifferential equations. Math. Nachr. **111**, 111–126 (1983).

[7] On spline approximation for singular integral equations on an interval. J. Integral Equations (in print).

Elschner, J., and G. Schmidt

[1] On spline interpolation in periodic Sobolev spaces. Preprint P-MATH-01/81, Inst. Math. Akad. Wiss. DDR, Berlin 1983.

Elschner, J., and B. Silbermann

[1] Über eine Klasse entarteter gewöhnlicher Differentialgleichungen und das Kollokationsverfahren zu ihrer Lösung. Czech. Math. J. **29** (**104**), 551–563 (1979).

Erdogan, F., G. D. Gupta, and T. S. Cook

[1] Numerical solution of singular integral equations. *In*: Methods of Analysis and Solutions of Crack Problems, ed G. C. Sih, Vol. 1, 368–425, Leyden 1973.

Eskin, G. I.

[1] Boundary value problems for elliptic pseudodifferential equations. Nauka, Moscow 1973 (*Russian*); *English transl.*: Amer. Math. Soc. Transl. of Math. Monographs 52, Providence, R.I., 1981.

Fedosov, B. V.

[1] On the index of an elliptic system on a manifold. Funkts. Anal. Prilozh. **4**: 4, 57–67 (1970) (*Russian*); *also in*: Funct. Anal. Appl. **4** (1970), 312–320 (1971).

[2] A direct proof of the formula for the index of an elliptic system in Euclidean space. Funkts. Anal. Prilozh. **4**: 4, 83–84 (1970) (*Russian*); *also in*: Funct. Anal. Appl. **4** (1970), 339–341 (1971).

[3] Analytic index formulas for elliptic operators. Trudy Moskov. Mat. O.-va **30**, 159–242 (1974) (*Russian*); *also in*: Trans. Moscow Math. Soc. **30** (1974), 159–240 (1976).

Fefferman, C., and E. M. Stein

[1] H^p spaces of several variables. Acta Math. **129**, 137–193 (1972).

Fichera, G.

[1] Una introduzione alla teoria delle equazioni integrali singolari. Rend. Matem. e Applic. **17**, No 1–2, 82–191 (1958).

Fichtenholz, G. M.

[1] Differential- und Integralrechnung. Bd. I–III. VEB Deutscher Verlag der Wissenschaften, Berlin 1964.

Friedrichs, K. O.

[1] Pseudo-differential operators. An introduction. Lecture Notes, Courant Inst. Math. Sci., New York 1968.

Fueter, R.

[1] Über die analytische Darstellung der regulären Funktionen einer Quaternionenvariablen. Comment. Math. Helv. 8, 371–378 (1935).

Gabdulhaev, B. G.

[1] Approximate solution of singular integral equations by the method of mechanical quadratures. Dokl. Akad. Nauk SSSR **179**: 2, 260–263 (1968) (*Russian*).

[2] On a direct method for the solution of integral equations. Izv. Vyssh. Uchebn. Zaved., Mat. **3**, 51–60 (1965) (*Russian*).

Gabdulhaev, B. G., and P. N. Dushkov

[1] The method of mechanical quadratures for singular integral equations. Izv. Vyssh. Uchebn. Zaved., Mat. **12**, 3–14 (1974) (*Russian*).

Gadzhiev, A. D.

[1] Multipliers of Fourier series with respect to spherical functions and properties of the symbol of the multidimensional singular operator. Dokl. Akad. Nauk SSSR **266**: 2, 268–269 (1982) (*Russian*); *also in*: Sov. Math. Dokl. **26**, 304–305 (1982).

[2] Differentiability properties of the symbol of the singular operator in spaces of Bessel potentials on the sphere. Izv. Akad. Nauk Az. SSR, Ser. Fiz.-Tekh. Mat. Nauk **1**, 134–140 (1982) (*Russian*).

Gakhov, F. D.

[1] Boundary value problems. Fizmatgiz, Moscow 1963 (*Russian*); *English transl.*: Pergamon Press, Oxford 1966.

Gantmacher, F. R.

[1] Matrizenrechnung. I. VEB Deutscher Verlag der Wissenschaften, Berlin 1958.

Gegelia, T. G.

[1] On a generalization of G. Giraud's formula. Soobshzh. Akad. Nauk Gruz. SSR **16**: 9, 657–663 (1965) (*Russian*).

[2] Differentiability properties of the solutions of singular integral equations on surfaces. Trudy Gruz. Polytekh. Inst. **1** (**81**), 69–77 (1962) (*Russian*).

[3] On a property of the solutions of singular integral equations. Trudy Tbilis. Mat. Inst. **110**, 43–56 (1965) (*Russian*).

[4] On the boundedness of singular operators. Soobshzh. Akad. Nauk Gruz. SSR **20**: 5, 517–523 (1968) (*Russian*).

[5] Differentiability properties of certain integral transforms. Trudy Tbilis. Mat. Inst. **26**, 195–225 (1959) (*Russian*).

[6] On properties of multidimensional singular integrals in the space $L_p(S, \varrho)$. Dokl. Akad. Nauk SSSR **139**: 2, 279–282 (1961) (*Russian*).

[7] On the regularization of singular integral operators. Trudy Tbilis. Mat. Inst. **29**, 229–237 (1963) (*Russian*).

Gelfand, I. M., D. A. Raikov, and G. E. Shilov

[1] Kommutative normierte Algebren. VEB Deutscher Verlag der Wissenschaften, Berlin 1966.

Gelfand, I. M., and G. E. Shilov

[1] Verallgemeinerte Funktionen (Distributionen). Bd. I. VEB Deutscher Verlag der Wissenschaften Berlin 1960.

[2] Verallgemeinerte Funktionen (Distributionen). Bd. II. VEB Deutscher Verlag der Wissenschaften, Berlin 1962.

Gerasoulis, A.

[1] The use of piecewise quadratic polynomials for the solution of singular integral equations of Cauchy type. Comput. Math. Appl. 8, 15–22 (1982).

Gerasoulis, A., and R. Vichnevetsky (eds.)

[1] Numerical solution of singular integral equations. Proceedings of an IMACS International Symposium held at Lehigh University Bethlehem, Pennsylvania, USA, June 21–22, 1984.

Giraud, G.

[1] Sur une classe générale d'équations à intégrales principales. C. R. Acad. Sci. Française **202** (**26**), 2124–2126 (1936).

[2] Équations à intégrales principales. Ann. Sci. Ec. Norm. Super. **51,** fasc. 3 et 4, 251–372 (1934).

[3] Sur certaines opérations du type elliptique. C. R. Acad. Sci. Française **200**, 1651–1653 (1953).

[4] Sur une classe d'équations linéaires où figurent des valeurs principales d'intégrales simples. Ann. Sci. Ec. Norm. Super., 3 sér. **56,** 119–172 (1939).

Glushko, V. P.

[1] On operators of potential type and certain embedding theorems. Dokl. Akad. Nauk SSSR **126**: 3, 467–470 (1959) (*Russian*).

[2] Degenerate linear differential equations. Voronezh 1972 (*Russian*).

Gohberg, I.

[1] On the theory of multidimensional singular integral equations. Dokl. Akad. Nauk SSSR **133**: 6, 1279–1282 (1960) (*Russian*).

[2] Some questions on the theory of multidimensional singular integral equations. Izv. Mold. Fil. Akad. Nauk SSSR **10** (**76**), 39–50 (1960) (*Russian*).

[3] On linear equations in normed spaces. Dokl. Akad. Nauk SSSR **76**: 4, 477–480 (1951) (*Russian*).

[4] On an application of the theory of normed rings to singular integral equations. Uspehi Mat. Nauk **7**: 2, 149–156 (1952) (*Russian*).

[5] The problem of factorization in normed rings, functions of isometric and symmetric operators, and singular integral equations. Uspehi Mat. Nauk **19**: 1, 71–124 (1964) (*Russian*).

[6] On the number of solutions of the homogeneous singular integral equation with continuous coefficients. Dokl. Akad. Nauk SSSR **122**: 3, 327–330 (1958) (*Russian*).

Gohberg, I., and I. A. Feldman

[1] Convolution equations and projection methods for their solution. Nauka, Moscow 1971 (*Russian*); *German transl.*: Akademie-Verlag, Berlin 1974; *English transl.*: Amer. Math. Soc. Transl. of Math. Monographs 41, Providence, R.I. 1974.

Gohberg, I., and M. G. Krein

[1] The fundamentals on defect numbers, root numbers, and indices of linear operators. Uspehi Mat. Nauk **12**: 2, 44–118 (1957) (*Russian*); *also in*: Amer. Math. Soc. Transl. (2) **13**, 185–264 (1960).

[2] Systems of integral equations on a half-line with kernels depending upon the difference of the

arguments. Uspehi Mat. Nauk **13**: 2, 3–72 (1958) (*Russian*); *also in*: Amer. Math. Soc. Transl. (2) **14**, 217–287 (1960).
[3] Introduction to the theory of linear nonselfadjoint operators in Hilbert space. Nauka, Moscow 1965 (*Russian*); *English transl.*: Amer. Math. Soc. Transl. of Math. Monographs 18, Providence, R.I. 1969.

GOHBERG, I., and N. YA. KRUPNIK
[1] On the spectrum of singular integral operators in the spaces L_p. Stud. Math. **XXXI**, 347–362 (1968) (*Russian*).
[2] On the spectrum of singular integral operators in the spaces L_p with weight. Dokl. Akad. Nauk SSSR **185**: 4, 745–748 (1969) (*Russian*).
[3] Singular integral operators with piecewise continuous coefficients and their symbols. Izv. Akad. Nauk SSSR, Ser. Mat. **35**: 4, 940–964 (1971) (*Russian*); *also in*: Math. USSR Izv. **5** (1971), 955–979 (1972).
[4] Introduction to the theory of one-dimensional singular integral operators. Shtintsa, Kishinev 1973 (*Russian*); *German transl.*: Birkhäuser Verlag, Basel, Stuttgart 1979.
[5] On the norm of the Hilbert transform in the space L_p. Funkts. Anal. Prilozh. **2**: 2, 91–92 (1968) (*Russian*).
[6] Systems of singular integral equations in the spaces L_p with weight. Dokl. Akad. Nauk SSSR **186**: 5, 998–1001 (1969) (*Russian*); *also in*: Sov. Math. Dokl. **10**, 688–691 (1969).
[7] On the algebra generated by one-dimensional singular integral operators with piecewise continuous coefficients. Funkts. Anal. Prilozh. **4**: 3, 26–36 (1970) (*Russian*); *also in*: Funct. Anal. Appl. **4** (1970), 193–201 (1971).
[8] On one-dimensional singular operators with shift. Izv. Akad. Nauk Arm. SSR, Mat. **VIII**: 1, 3–12 (1973) (*Russian*).

GOHBERG, I., and V. I. LEVCHENKO
[1] On the projection method for the degenerate discrete Wiener-Hopf equation. Mat. Issled. **VII**: 3, 238–253 (1972) (*Russian*).

GOHBERG, I., A. S. MARKUS, and I. A. FELDMAN
[1] On normally solvable operators and ideals connected with them. Izv. Mold. Fil. Akad. Nauk SSSR **10 (76)**, 51–70 (1960) (*Russian*).

GOHBERG, I., and A. A. SEMENTSUL
[1] Toeplitz matrices composed by the Fourier coefficients of functions with discontinuities of almost periodic type. Mat. Issled. **V**: 4, 63–83 (1970) (*Russian*).

GOHBERG, I., L. LERER, and L. RODMAN
[1] Factorization indices for matrix polynomials. Bull. Amer. Math. Soc. **84**: 2, 275–277 (1978).

GOLBERG, M. A.
[1] The numerical solution of Cauchy singular integral equations with constant coefficients on $[-1, 1]$. J. Integral Equations **9**, 127–151 (1985).
[2] The convergence of a collocation method for a class of Cauchy singular integral equations. J. Integral Equations; to appear.

GOLBERG, M. A., and J. A. FROMME
[1] On the L_2-convergence of collocation for the generalized airfold equation. J. Math. Anal. Appl. **71**, 271–286 (1979).

GOLDBERG, S.
[1] Unbounded linear operators. Theory and applications. McGraw-Hill, New York 1966.

GORDADZE, E. G.
[1] On singular integrals with Cauchy kernel. Trudy Tbilis. Mat. Inst. **42**, 5–17 (1972) (*Russian*).
[2] On singular integrals on nonsmooth curves. *In*: Proceedings of a Symposium on Continuum Mechanics and Related Problems of Analysis, Sept. 23–29, 1971, Vol. II, 74–85, Metsniereba, Tbilisi 1974 (*Russian*).
[3] On the boundary value problem of linear conjugation for Radon curves. Soobshzh. Akad. Nauk Gruz. SSR **84**: 1, 29–32 (1976) (*Russian*).

GRUDSKI, S. M.
[1] Singular integral operators with infinite index and Blaschke products. Math. Nachr. **129** (1986). (*Russian*).

GRUDSKI, S. M., and V. B. DYBIN
[1] The Riemann boundary value problem in the space $L_p(\Gamma, \varrho)$ for the case of almost periodic discontinuities in its coefficient. *In*: Lin. Oper. (Mat. Issled., Vyp. 54), 36–49, Shtiintsa, Kishinev 1980 (*Russian*).

GÜRLEBECK, K.
[1] Über die optimale Interpolation verallgemeinert analytischer quaternionenwertiger Funktionen und ihre Anwendung zur näherungsweisen Lösung wichtiger räumlicher Randwertaufgaben der mathematischen Physik. Dissertation A, TH Karl-Marx-Stadt, 1984.

HAFTMANN, R.
[1] Numerische Behandlung einer singulären Integralgleichung aus der Strömungsmechanik. Dissertation A, TH Karl-Marx-Stadt, 1979.

HAIKIN, M. I.
[1] On the integral equation of convolution type of first kind. Izv. Vyssh. Uchebn. Zaved., Mat. **3**, 105–116 (1967) (*Russian*).
[2] The Wiener-Hopf equation in spaces of test functions and distributions. Izv. Vyssh. Uchebn. Zaved., Mat. **10**, 83–91 (1967) (*Russian*).
[3] On the regularization of operators with non-closed range. Izv. Vyssh. Uchebn. Zaved., Mat. **8**, 118–123 (1970) (*Russian*).
[4] On the exceptional case for the singular integral operator with piecewise continuous coefficients. Mat. Issled. **VII**: 3, 194–209 (1972) (*Russian*).
[5] On the zeros of the singular integral operator with piecewise continuous coefficients in the exceptional case. Mat. Issled. **VIII**: 1, 171–179 (1973) (*Russian*).
[6] The singular integral equation with continuous coefficients in the exceptional case. Trudy Semin. po Kraev. Zad. Kazan. Univ. **10**, 152–162 (1973) (*Russian*).
[7] On the range of the singular integral operator in the exceptional case. Izv. Vyssh. Uchebn. Zaved., Mat. **2**, 119–122 (1976) (*Russian*).
[8] On the exceptional case for singular integral operators with matrix coefficients. Izv. Vyssh. Uchebn. Zaved., Mat. **6**, 77–80 (1979) (*Russian*).
[9] The system of singular integral equations in the exceptional case as an ill-posed problem. Izv. Vyssh. Uchebn. Zaved., Mat. **7**, 79–83 (1979) (*Russian*).

HAIKIN, YU. E.
[1] On operators of convolution type in spaces with weight. Vestn. Leningrad. Univ. 13, 79–82 (1969) (*Russian*).
[2] On the boundedness of pseudo-differential operators in weighted spaces. Dokl. Akad. Nauk SSSR **190**: 2, 289–292 (1970) (*Russian*).
[3] Singular integral operators in the spaces L_p with weight. Sib. Mat. Zh. **14**: 2, 437–441 (1973) (*Russian*).

HALILOV, Z. I.
[1] Linear singular equations in a normed ring. Izv. Akad. Nauk SSSR, Ser. Mat. **13**: 2, 163–176 (1949) (*Russian*).

HALMOS, P.
[1] Measure theory. Van Nostrand, Inc., Princeton 1950.

HARDY, G. H., and J. E. LITTLEWOOD
[1] Some more theorems concerning Fourier series and Fourier power series. Duke Math. J. **2**: 2, 354–382 (1936).

HARDY, G. H., J. E. LITTLEWOOD, and G. POLYA
[1] Inequalities. University Press, London and Cambridge 1951.

HAUSDORFF, F.
[1] Zur Theorie der linearen metrischen Räume. J. Reine Angew. Math. **167**, 294–311 (1932).

Havin, V. P.
[1] Boundary properties of integrals of Cauchy type and of conjugate harmonic functions in regions with rectifiable boundary. Mat. Sb. **68**: 4 (110), 499–517 (1965) (*Russian*).
[2] On the continuity in L_p of the integral operator with Cauchy kernel. Vestn. Leningrad. Univ. 7, 103–108 (1967) (*Russian*).
Heinig, G., and B. Silbermann
[1] Factorization of matrix functions in algebras of bounded functions. *In*: Operator Theory: Advances and Applications, Vol. 14, 157–177, Birkhäuser Verlag, Basel 1984.
Heunemann, D.
[1] Über die normale Auflösbarkeit singulärer Integraloperatoren mit unstetigem Symbol. Math. Nachr. **80**, 157–163 (1977).
Hilbert, D.
[1] Grundzüge einer allgemeinen Theorie der linearen Integralgleichungen. B. G. Teubner, Leipzig und Berlin 1912.
Hörmander, L.
[1] Estimates for translation invariant operators. Acta Math. **104**, 93–140 (1960).
[2] Pseudo-differential operators. Commun. Pure Appl. Math. **18**: 3, 501–517 (1965).
[3] Linear partial differential operators. Springer-Verlag, Berlin, Göttingen, Heidelberg 1963.
[4] On the division of distributions by polynomials. Ark. Mat. **3**, 555–568 (1958).
[5] Fourier integral operators. I. Acta Math. **127**, 79–183 (1971).
Horváth, J.
[1] Sur les fonctions conjugées à plusieurs variables. Proc. K. Ned. Akad. Wet. **56**: 1, 17–19 (1952).
Hsiao, G. C., and W. L. Wendland
[1] A finite element method for some integral equations of the first kind. J. Math. Anal. Appl. **58**, 449–481 (1977).
[2] The Aubin-Nitsche lemma for integral equations. J. Integral Equations **3**, 299–315 (1981).
Hunt, R., B. Muckenhoupt, and R. Wheeden
[1] Weighted norm inequalities for the conjugate function and Hilbert transform. Trans. Amer. Math. Soc. **176**: 1, 227–251 (1973).
Hurewicz, W., and H. Wallman
[1] Dimension theory. Princeton Univ. Press, Princeton, New Jersey, 1948.

Ioakimidis, N. I., and P. S. Theocaris
[1] On the convergence of two direct methods for the solution of Cauchy type singular integral equations of the first kind. BIT **20**, 83–87 (1980).
Itskovich, I. A.
[1] The equivalence problem in the theory of two-dimensional singular integral equations. Uchebn. Zap. Kishinev. Univ. **11**, 7–11 (1954) (*Russian*).
[2] On Fredholm series. Dokl. Akad. Nauk SSSR **59**: 3, 423–425 (1948) (*Russian*).
[3] Integrals of Cauchy type as Hilbert space operators. Uchebn. Zap. Kishinev. Univ. **5**, 37–41 (1952) (*Russian*).
Ivanov, V. V.
[1] The theory of approximate methods and their application to the numerical solution of singular integral equations. Noordhoff Int. Publ., Leyden 1976.

Jen, E., and R. Srivastav
[1] Cubic splines and approximate solution of singular integral equations. Math. Comput. **37**, 417–423 (1981).
John, F., and L. Nirenberg
[1] On functions of bounded mean oscillation. Commun. Pure Appl. Math. **18**, 415–426 (1965).
Journé, J.-L.
[1] Calderón-Zygmund operators, pseudo-differential operators, and the Cauchy integral of Calderón. Lect. Notes Math., Vol. 994, Springer-Verlag, Berlin, Heidelberg, New York, Tokyo 1983.

JUNGHANNS, P.

[1] Kollokationsverfahren zur näherungsweisen Lösung singulärer Integralgleichungen mit unstetigen Koeffizienten. Math. Nachr. **102**, 17–24 (1981).

[2] Polynomiale Näherungsverfahren für singuläre Integralgleichungen auf beschränkten Intervallen. Dissertation B, TH Karl-Marx-Stadt, 1984.

JUNGHANNS, P., and B. SILBERMANN

[1] Zur Theorie der Näherungsverfahren für singuläre Integralgleichungen auf Intervallen. Math. Nachr. **103**, 199–244 (1981).

[2] Local theory of the collocation method for the approximate solution of singular integral equations. I. Integral Equations Oper. Theory **7**: 6, 791–807 (1984).

[3] Numerical analysis for one-dimensional Cauchy type singular integral equations. *In*: Probleme und Methoden der Mathematischen Physik, 8. Tagung in Karl-Marx-Stadt 1983, 122–129, Teubner-Texte zur Mathematik, Bd. 63, BSB B. G. Teubner Verlagsges., Leipzig 1984.

KANTOROVICH, L. V., and G. P. AKILOV

[1] Funktionalanalysis in normierten Räumen. Akademie-Verlag, Berlin 1964.

KANTOROVICH, L. V., and V. I. KRYLOV

[1] Approximate methods of higher analysis. Interscience, New York 1958.

KARPENKO, L. N.

[1] Approximate solution of a singular integral equation by means of Jacobi polynomials.. Prikl. Mat. Mekh. **30**, 564–569 (1966) (*Russian*).

KATO, T.

[1] Perturbation theory for linear operators. Springer-Verlag, Berlin, Heidelberg, New York 1966.

KHVEDELIDZE, B. V.

[1] Linear discontinuous boundary value problems of function theory, singular integral equations, and some of their applications. Trudy Tbilis. Mat. Inst. **23**, 3–158 (1956) (*Russian*).

[2] A remark on my paper "Linear discontinuous boundary value problems ...". Soobshzh. Akad. Nauk Gruz. SSR **21**: 2, 129–138 (1958) (*Russian*).

[3] The method of Cauchy type integrals for discontinuous boundary value problems in the theory of holomorphic functions of one complex variable. Itogi Nauk i Tekh., Ser. Sovrem. Probl. Mat. **7**, 5–162 (1975) (*Russian*); *also in*: J. Sov. Math. **7**: 3 (1977).

KHVOLES, A. A.

[1] On the boundedness of the multidimensional singular integral in Lipschitz spaces. Soobshzh. Akad. Nauk Gruz. SSR **74**: 2, 294–296 (1974) (*Russian*).

[2] Singular integral equations in Lipschitz spaces. Soobshzh. Akad. Nauk Gruz. SSR **76**: 1, 29–31 (1974) (*Russian*).

[3] On the smoothness of the solution of the singular integral equation in the space $L_p(M)$. Trudy Tbilis. Mat. Inst. **58**, 208–223 (1978) (*Russian*).

KINOSHITA, N., and T. MURA

[1] On boundary value problems of elasticity. Res. Rep. Fac. Eng. Meijy Univ. **8**, 56–82 (1956).

KÖHLER, U., and B. SILBERMANN

[1] Einige Ergebnisse über Φ_+-Operatoren in lokalkonvexen topologischen Vektorräumen. Math. Nachr. **56**, 145–153 (1973).

[2] Über algebraische Eigenschaften einer Klasse von Operatorenmatrizen und eine Anwendung auf singuläre Integraloperatoren. Math. Nachr. **57**, 245–258 (1973).

KOHN, J. J., and L. I. NIRENBERG

[1] An algebra of pseudo-differential operators. Commun. Pure Appl. Math. **18**, 269–305 (1965).

KOKILASHVILI, V. M.

[1] On singular integrals and maximal operators with Cauchy kernel. Dokl. Akad. Nauk SSSR **223**: 3, 555–558 (1975) (*Russian*); *also in*: Sov. Math. Dokl. **16** (1975), 941–945 (1976).

KOPP, P.

[1] Über eine Klasse von Randwertproblemen mit verletzter Lopatinski-Bedingung für elliptische Systeme erster Ordnung in der Ebene. Dissertation, TH Darmstadt, 1977.

KOSULIN, A. E.

[1] One-dimensional singular equations in generalized functions. Dokl. Akad. Nauk SSSR **163**: 5, 1054–1057 (1965) (*Russian*).

[2] One-dimensional singular equations in generalized functions. Mat. Sb. **70**: 2, 180–197 (1966) (*Russian*).

[3] One-dimensional singular equations in generalized functions. The case of a non-closed contour. Vestn. Leningrad. Univ. 7, 157–170 (1966) (*Russian*).

KÖTHE, G.

[1] Zur Theorie der kompakten Operatoren in lokalkonvexen Räumen. Port. Math. **13**, 97–104 (1954).

KOZAK, A. V.

[1] On the reduction method for multidimensional discrete convolutions. Mat. Issled. **VIII**: 3, 157–160 (1973) (*Russian*).

[2] A local principle in the theory of projection methods. Dokl. Akad. Nauk SSSR **212**: 6, 1287–1289 (1973) (*Russian*); *also in*: Sov. Math. Dokl. **14** (1973), 1580–1583 (1974).

[3] A local principle in the theory of projection methods. *In*: Integr. i Differ. Uravn. i Prilozh., 58–73, Elista 1983 (*Russian*).

KRACHKOVSKI, S. N., and A. S. DIKANSKI

[1] Fredholm operators and their generalizations. Itogi Nauki Tekh., Ser. Mat. Anal. 1968, 39–71, VINITI, Moscow 1969 (*Russian*).

KRASNOSELSKI, M. A., G. M. VAINIKKO, et al.

[1] Näherungsverfahren zur Lösung von Operatorgleichungen. Akademie-Verlag, Berlin 1973.

KRASNOSELSKI, M. A., P. P. ZABREIKO, et al.

[1] Integral operators in spaces of summable functions. Nauka, Moscow 1966 (*Russian*).

KRAVCHENKO, V. G., and G. S. LITVINCHUK

[1] On the symbol of the singular operator with Carleman shift. Ukr. Mat. Zh. **25**, 541–544 (1973) (*Russian*); *also in*: Ukr. Math. J. **25** (1973), 449–452 (1974).

KREIN, M. G.

[1] Integral equations on a half-line with kernels depending upon the difference of the arguments. Uspehi Mat. Nauk **13**: 5, 3–120 (1958) (*Russian*); *also in*: Amer. Math. Soc. Transl. (2) **22**, 163–288 (1962).

KREIN, S. G.

[1] Linear equations in Banach space. Nauka, Moscow 1971 (*Russian*).

KRUPNIK, N. YA.

[1] On multidimensional singular integral equations. Uspehi Mat. Nauk **20**: 6 (126), 119–123 (1965) (*Russian*).

[2] On multidimensional singular operators in spaces of test functions and distributions. Uchebn. Zap. Kishinev. Univ. **70**, 39–48 (1964) (*Russian*).

[3] On the question on the normal solvability and the index of singular integral equations. Uchebn. Zap. Kishinev. Univ. **82**, 3–7 (1965) (*Russian*).

[4] Some general questions on the theory of one-dimensional singular operators with matrix coefficients. *In*: Nesamosopryazh. Oper. (Mat. Issled., Vyp. 42), 91–112, Shtiintsa, Kishinev 1976 (*Russian*).

[5] On singular integral operators with matrix coefficients. *In*: Spectr. Svoistva Oper. (Mat. Issled., Vyp. 45), 93–100, Shtiintsa, Kishinev 1977 (*Russian*).

[6] Some consequences of the Hunt-Muckenhoupt-Wheeden theorem. *In*: Oper. v Banach. Prostr. (Mat. Issled., Vyp. 47), 64–70, Shtiintsa, Kishinev 1978 (*Russian*).

[7] Sufficient collections of n-dimensional representations of Banach algebras and the n-symbol. Funkts. Anal. Prilozh. **14**: 1, 63–64 (1980) (*Russian*); *also in*: Funct. Anal. Appl. **14**, 50–52 (1980).

[8] Conditions for the existence of an n-symbol and of a sufficient collection of n-dimensional representations of a Banach algebra. *In*: Lin. Oper. (Mat. Issled., Vyp. 54), 84–97, Shtiintsa, Kishinev 1980 (*Russian*).

[9] Banach algebras with symbol and singular integral operators. Shiintsa, Kishinev 1984 (*Russian*).

[10] Complete singular integral equations. *In*: Issled. po Funkts. Anal. i Differ. Uravn., 82–90, Shtiintsa, Kishinev 1984 (*Russian*).

KRUPNIK, N. YA., and V. I. NYAGA

[1] On singular integral operators in the case of a non-smooth contour. Mat. Issled. **X**: 1, 144–164 (1975) (*Russian*).

KUPRADZE, V. D.

[1] Boundary value problems of the theory of steady-state elastic vibrations. Uspehi Mat. Nauk **8**: 3 (55), 21–74 (1953) (*Russian*).

[2] Potential methods in the theory of elasticity. Fizmatgiz, Moscow 1963 (*Russian*).

KUPRADZE, V. D., T. G. GEGELIA, M. O. BASHELEISHVILI, and T. V. BURCHULADZE

[1] Three-dimensional problems of the mathematical theory of elasticity. Nauka, Moscow 1967 (*Russian*).

LAVRENTYEV, M. A.

[1] On the building up of the flow past an arc of given shape. Trudy ZAGI **118**, 3–56 (1932) (*Russian*).

LEITERER, J.

[1] Zur normalen Auflösbarkeit singulärer Integraloperatoren. Math. Nachr. **51**, 197–230 (1971).

[2] Über die normale Auflösbarkeit kompakt gestörter singulärer Integraloperatoren. Math. Nachr. **64**, 253–261 (1974).

LEITERER, J., and A. S. MARKUS

[1] On the normal solvability of singular integral operators in symmetric spaces. Mat. Issled. **VII**: 1, 72–82 (1972) (*Russian*).

LINDENSTRAUSS, J., and L. TZAFRIRI

[1] On the complemented subspace problem. Isr. J. Math. **9**, 263–269 (1971).

LITVINCHUK, G. S.

[1] Fredholm theory of systems of singular integral equations with Carleman shift and complex conjugate unknowns. Izv. Akad. Nauk SSSR, Ser. Mat. **31**: 3, 563–586 (1967), **39**: 6, 1414–1417 (1968) (*Russian*).

[2] Boundary value problems and singular integral equations with shift. Nauka, Moscow 1977 (*Russian*).

LIZORKIN, P. I.

[1] (L_p, L_q)-multipliers of Fourier integrals. Dokl. Akad. Nauk SSSR **152**: 4, 808–811 (1963) (*Russian*).

LORENZ, M.

[1] Über eine mehrdimensionale singuläre Integralgleichung mit ausgeartetem Symbol. Math. Nachr. **60**, 267–280 (1974).

[2] Über ein System mehrdimensionaler singulärer Integralgleichungen nicht normalen Typs. Wiss. Z. Tech. Hochsch. Karl-Marx-Stadt **16**: 3, 419–431 (1974).

[3] Über mehrdimensionale singuläre Integraloperatoren und Pseudomultiplikationsoperatoren nicht normalen Typs. Math. Nachr. **73**, 185–212 (1976).

[4] Nichtelliptische Pseudodifferentialoperatoren im R^n. Beitr. Anal. **14**, 63–71 (1979).

LOUNESTO, P.

[1] Spinor valued regular functions in hypercomplex analysis. Report HTKK-MAT A 154, Dissertation 433, Helsingin Teknillinen Korkeakoulu, 1979.

LUZIN, N. N.

[1] Integral and trigonometric series. Moscow 1915 (*Russian*).

MAGNARADZE, L. G.

[1] On a system of linear singular integro-differential equations and on the linear Riemann boundary value problem. Soobshzh. Akad. Nauk Gruz. SSR **4**: 1, 3–9 (1943) (*Russian*).

[2] The theory of a class of linear singular integro-differential equations and its applications to the problem of vibration of an aircraft wring of finite span, to the problem of the blow on a surface of water, and to analogous things. Soobshzh. Akad. Nauk Gruz. SSR **4**: 2, 103–110 (1943) (*Russian*).

MANDZHAVIDZE, G. F., and B. V. KHVEDELIDZE

[1] On the Riemann-Privalov problem with continuous coefficients. Dokl. Akad. Nauk SSSR **123**: 5, 791–794 (1958) (*Russian*).

MARCUS, M., and H. MINC

[1] A survey of matrix theory and matrix inequalities. Allyn and Bacon, Boston 1964.

MARKUS, A. S., and I. A. FELDMAN

[1] On the index of an operator matrix. Funkts. Anal. Prilozh. **11**: 2, 83–84 (1977) (*Russian*); *also in*: Funct. Anal. Appl. **11**, 149–151 (1977).

[2] On the relationship between certain properties of an operator matrix and its determinant. *In*: Lin. Oper. (Mat. Issled., Vyp. 54), 110–120, Shtiintsa, Kishinev 1980 (*Russian*).

MAZYA, V. G.

[1] On the degenerate problem with oblique derivative. Mat. Sb. **87**: 3 (129), 417–454 (1972) (*Russian*); *also in*: Math. USSR Sb. **16**, 429–469 (1972).

MAZYA, V. G., and YU. E. HAIKIN

[1] A remark on the continuity in L_2 of the singular integral operator. Vestn. Leningrad. Univ. 19, 156–159 (1969) (*Russian*).

[2] On the continuity of singular integral operators in normed spaces. Vestn. Leningrad. Univ. 1, 28–34 (1976) (*Russian*).

MAZYA, V. G., and B. P. PANEYAKH

[1] Degenerate elliptic pseudo-differential operators and the problem with oblique derivative. Trudy Moskov. Mat. O.-va **31**, 237–295 (1974). (*Russian*); *also in*: Trans. Moscow Math. Soc. **31**, 247–305 (1976).

MAZYA, V. G., and B. A. PLAMENEVSKI

[1] On singular equations with a symbol having zeros. Dokl. Akad. Nauk SSSR **160**: 6, 1250–1253 (1965) (*Russian*).

[2] On the Cauchy problem for singular hyperbolic integral equations of convolution type. Vestn. Leningrad. Univ. 19, 161–163 (1969) (*Russian*).

MAZYA, V. G., B. A. PLAMENEVSKI, and YU. E. HAIKIN

[1] On a correctly posed problem for singular integral equations with a symbol having zeros. Differ. Uravn. **13**: 8, 1479–1486 (1977) (*Russian*); *also in*: Differ. Equations **13**, 1028–1033 (1977).

MEISTER, E.

[1] Randwertaufgaben der Funktionentheorie. B. G. Teubner, Stuttgart 1983.

MEYER, CH.

[1] Über einige Klassen von singulären Operatoren und ihre Beziehung zu singulären Integraloperatoren. Dissertation, TH Karl-Marx-Stadt, 1975.

[2] Über einige Klassen von singulären Operatoren und ihre Beziehung zu singulären Integraloperatoren. Math. Nachr. **73**, 171–183 (1976).

MEYER, CH., and B. SILBERMANN

[1] Die Indexformel für eine Klasse von ausgearteten singulären Integraloperatoren mit Carlemanscher Verschiebung. Demonstr. Math. **X**: 1, 155–167 (1977).

MEYER, S.

[1] Über eine Klasse singulärer Integralgleichungen, deren Symbol auf einer Menge positiven Maßes entartet. Wiss. Z. Tech. Hochsch. Karl-Marx-Stadt **14**: 6, 681–695 (1972).

[2] Über singuläre Integralgleichungen, deren Symbol Entartungen vom logarithmischen Typ besitzt. Wiss. Z. Tech. Hochsch. Karl-Marx-Stadt **16**: 3, 399–418 (1974).

MIKHAILOV, L. G.

[1] On some one-dimensional integral operators with homogeneous kernels. Dokl. Akad. Nauk SSSR **176**: 2, 263–265 (1967) (*Russian*).

MIKHAILOVA-GUBENKO, N. M.

[1] Singular integral equations in Lipschitz spaces. Dokl. Akad. Nauk SSSR **159**: 3, 509–511 (1964) (*Russian*).

[2] Singular integral equations in Lipschitz spaces. I. Vestn. Leningrad. Univ. 1, 51–63 (1966) (*Russian*).

[3] Singular integral equations in Lipschitz spaces. II. Vestn. Leningrad. Univ. 7, 45–47 (1966) (*Russian*).

Mikhlin, S. G.

[1] Vorlesungen über lineare Integralgleichungen. VEB Deutscher Verlag der Wissenschaften, Berlin 1962.

[2] Concerning the theorem on the boundedness of the operator of singular integration. Uspehi Mat. Nauk **8**: 1 (53), 213–217 (1953) (*Russian*).

[3] The equivalence problem in the theory of singular integral equations. Mat. Sb. **3**: 1 (45), 121–140 (1938) (*Russian*).

[4] The composition of double singular integrals. Dokl. Akad. Nauk SSSR **2** (11): 1 (87), 3–6 (1936) (*Russian*).

[5] Singular integral equations with two independent variables. Mat. Sb. **1**: 4 (43), 535–550 (1936) (*Russian*).

[6] Addendum to the paper "Singular integral equations with two independent variables". Mat. Sb. **1**: 6 (43), 963–964 (1936) (*Russian*).

[7] Singular integral equations. Uspehi Mat. Nauk **3**: 3 (25), 29–112 (1948) (*Russian*).

[8] The composition of multidimensional singular integrals. Vestn. Leningrad. Univ. 2, 24–41 (1955) (*Russian*).

[9] On the theory of multidimensional singular integral equations. Vestn. Leningrad. Univ. 1, 3–24 (1956) (*Russian*).

[10] Extension of the operation of singular integration to the space L_2. Dokl. Akad. Nauk SSSR **19**: 5, 353–355 (1938) (*Russian*).

[11] Multidimensional singular integrals and integral equations. Fizmatgiz, Moscow 1962 (*Russian*); *English transl.*: Pergamon Press, New York 1965.

[12] Partielle Differentialgleichungen der Mathematischen Physik. Akademie-Verlag, Berlin 1977.

[13] On the differentiability of series with respect to spherical functions. Dokl. Akad. Nauk SSSR **126**: 2, 278–280 (1959) (*Russian*).

[14] The problem of the minimum of a quadratic functional. Gostekhizdat, Moscow 1952 (*Russian*).

[15] On a problem in the theory of singular integral equations. Dokl. Akad. Nauk SSSR **15**: 8, 429–432 (1937) (*Russian*).

[16] Reduction of a singular integral equation to an equivalent Fredholm equation. Dokl. Akad. Nauk SSSR **20**: 2–3, 93–96 (1938) (*Russian*).

[17] Singular integral equations in classes of Lipschitz functions. Dokl. Akad. Nauk SSSR **138**: 3, 541–544 (1961) (*Russian*).

[18] Singular integral equations in Sobolev spaces. Vestn. Leningrad. Univ. 13, 67–70 (1977) (*Russian*).

[19] Singular integral equations in Sobolev spaces. Vestn. Leningrad. Univ. 13, 67–70 (1977) (*Russian*).

[20] On the question on the index of a system of singular equations. Dokl. Akad. Nauk SSSR **152**: 3, 555–558 (1963) (*Russian*).

[21] On the index of a system of singular integral equations. *In*: Outlines of the Joint Soviet-American Symposium on Partial Differential Equations, 1–14, Novosibirsk 1963 (*Russian*).

[22] Two theorems on regularizers. Dokl. Akad. Nauk SSSR **125**: 4, 737–739 (1959) (*Russian*).

[23] On an inequality for the boundary values of harmonic functions. Uspehi Mat. Nauk **6**: 6 (46), 158–159 (1951) (*Russian*).

[24] Variationsmethoden der Mathematischen Physik. Akademie-Verlag, Berlin 1962.

[25] Numerische Realisierung von Variationsmethoden. Akademie-Verlag, Berlin 1969.

[26] The method of least squares in problems of mathematical physics. Uchebn. Zap. Leningrad. Univ. 16, 167–206 (1949). (*Russian*)

[27] On the method of least squares for multidimensional singular integral equations. *In*: Essays Dedicated to I. N. Vekua on the Occasion of His 70th Birthday, Nauka, Moscow 1977 (*Russian*).

[28] The spectrum of a family of operators of elasticity theory. Uspehi Mat. Nauk **28**: 3 (171), 43–82 (1973) (*Russian*); *also in*: Russ. Math. Surveys **28**: 3, 45–88 (1973).

[29] Singular integral equations with continuous coefficients. Dokl. Akad. Nauk SSSR **59**: 3, 435–438 (1948) (*Russian*).

[30] On the computation of the index of a system of one-dimensional singular equations. Dokl. Akad. Nauk SSSR **168**: 6, 1248–1250 (1966) (*Russian*); *also in*: Sov. Math. Dokl. **7**, 815–817 (1966).

[31] Approximation on a rectangular grid. Sijthoff & Noordhoff, Alphen aan den Rijn, Germantown 1979.

[32] On multipliers of Fourier integrals. Dokl. Akad. Nauk SSSR **109**: 4, 701–703 (1956) (*Russian*).

[33] Fourier integrals and multiple singular integrals. Vestn. Leningrad. Univ. 7, 143–155 (1957) (*Russian*).

MIKHLIN, S. G., and R. K. RADEVA

[1] On the approximate solution of singular integral equations. Izv. Vyssh. Uchebn. Zaved., Mat. **5**, 158–162 (1974) (*Russian*).

MIRANDA, C.

[1] Equazioni alle derivate parziali di tipo ellittico. Springer-Verlag, Berlin, Göttingen, Heidelberg 1955; *English transl.*: Partial differential equations of elliptic type, Springer-Verlag, Berlin, Heidelberg, New York 1970.

MOISIL, G., and N. THEODORESCU

[1] Functions holomorphes dans l'espace. Mathematica **V**, 142–159 (1931).

MUCKENHOUPT, B.

[1] Weighted norm inequalities for the Hardy maximal function. Trans. Amer. Math. Soc. **165**, 207–226 (1972).

[2] The equivalence of two conditions for weight functions. Stud. Math. **49**, 101–106 (1974).

MUHAMMED-NASER

[1] Hyperholomorphic functions. Sib. Mat. Zh. **12**: 6, 1327–1341 (1971) (*Russian*); *also in*: Sib. Math. J. **12** (1971), 959–968 (1972).

MÜLLER, CL.

[1] Neue Verfahren zur Lösung der elliptischen Randwertprobleme der Mathematischen Physik. Vortr. Rheinisch-Westfäl. Akad. Wiss. 288, 27–68 (1979).

MULTHOPP, H.

[1] Die Berechnung der Auftriebsverteilung von Tragflächen. Luftfahrt-Forschung **XV**: 4, 153–169 (1938).

MURAI, T.

[1] Boundedness of singular integral operators of Calderón type. Proc. Japan. Acad. **59** (A), 8, 364–367 (1983).

MUSHKELISHVILI, N. I.

[1] Singular integral equations. Noordhoff, Groningen 1953; *German transl.*: Akademie-Verlag, Berlin 1965.

MUSHKELISHVILI, N. I., and N. P. VEKUA

[1] The Riemann boundary value problem for several unknown functions and its applications to systems of singular integral equations. Trudy Tbilis. Mat. Inst. **12**, 1–46 (1943) (*Russian*).

NAIMARK, M. A.

[1] Normed rings. Gostekhizdat, Moscow 1956 (*Russian*); *German transl.*: VEB Deutscher Verlag der Wissenschaften, Berlin 1959; *English transl.*: Noordhoff, Groningen 1959.

NATANSON, I. P.

[1] Konstruktive Funktionentheorie. Akademie-Verlag, Berlin 1955.

NEDELEC, J. C.

[1] Approximation des equations intégrales en méchanique et en physique. Lecture Notes, Centre de Mathématiques Appliquées, Ecole Polytechnique, Palarseau 1977.

NIED, H. F.

[1] Numerical solution of a coupled system of singular integral equations with fracture mechanics applications. *In*: A. GERASOULIS and R. VICHNEVETSKY (eds.) [1] 73–79.

NIKOLSKI, S. M.
[1] Approximation of functions of several variables and embedding theorems. Nauka, Moscow 1969 (*Russian*); *English transl.*: Springer-Verlag, Berlin, New York 1976.
NIKOLSKI, YU. S.
[1] On singular integral operators in weighted spaces. Trudy Mat. Inst. Steklova **105**, 190–200 (1969) (*Russian*); *also in*: Proc. Steklov Inst. Math. **105** (1969), 233–245 (1971).
NOETHER, F.
[1] Über eine Klasse singulärer Integralgleichungen. Math. Ann. **82**, 42–63 (1921).
NYAGA, V. I.
[1] On the symbol of singular integral operators in the case of a piecewise Lyapunov contour. Mat. Issled. **IX**: 2, 109–125 (1974) (*Russian*).
[2] Conditions for the Fredholmness of singular integral operators with conjugation in the case of a piecewise Lyapunov contour. *In*: Issled. po Funkts. Anal. i Differ. Uravn., 90–102, Shtiintsa, Kishinev 1984 (*Russian*).

PALAIS, R. S.
[1] Seminar on the Atiyah-Singer index theorem. Princeton Univ. Press, Princeton, New Jersey, 1965.
PASCALI, D.
[1] Vecteurs analytiques généralisés. Rev. Roum. Math. Pures Appl. **X**: 6, 779–808 (1965).
[2] Sur la représentation de première espèce des vecteurs analytiques généralisés. Rev. Roum. Math. Pures Appl. **XII**: 5, 685–689 (1967).
PEETRE, J.
[1] On the differentiability of the solutions of quasilinear differential equations. Trans. Amer. Math. Soc. **104**: 3, 476–482 (1962).
PEREYRA, C. B.
[1] An extension of a theorem of E. Stein. Proc. Amer. Math. Soc. **19**: 6, 1396–1402 (1968).
PERLIN, P. I.
[1] A numerical method for the solution of the singular integral equations arising from the fundamental three-dimensional problems of elasticity theory. Mekh. Tverd. Tela **3**, 109–111 (1975) (*Russian*).
PETERHÄNSEL, W.
[1] Zur Theorie einer Klasse singulärer Integralgleichungen auf kompakten Mannigfaltigkeiten ohne und mit Rand. Dissertation, TH Darmstadt, 1980.
PETROVSKI, I. G.
[1] On some problems of the theory of partial differential equations. Uspehi Mat. Nauk **1**: 3–4, 44–70 (1946) (*Russian*).
PHAM THE LAI
[1] Potentiels élastiques; tenseurs de Green et de Neumann. J. Mec. **6**: 2, 211–242 (1967).
PIETSCH, A.
[1] Zur Theorie der σ-Transformationen in lokalkonvexen Vektorräumen. Math. Nachr. **21**, 347–369 (1960).
[2] Homomorphismen in lokalkonvexen Vektorräumen. Math. Nachr. **22**, 162–174 (1960).
[3] Nukleare lokalkonvexe Räume. Akademie-Verlag, Berlin 1965.
PILIDI, V. S.
[1] On the bisingular equation in the space L_p. Mat. Issled **VII**: 3, 167–175 (1972) (*Russian*).
PLAMENEVSKI, B. A.
[1] On singular equations in a cone. Dokl. Akad. Nauk SSSR **179**: 5, 1057–1059 (1968) (*Russian*).
[2] On the boundedness of singular integrals in spaces with weight. Mat. Sb. **76**: 4 (118), 573–592 (1968) (*Russian*).
[3] On algebras generated by pseudodifferential operators with isolated singularities of the symbols. Dokl. Akad. Nauk SSSR **248**: 2, 298–302 (1979) (*Russian*); *also in*: Sov. Math. Dokl. **20**, 1013–1017 (1979).
[4] On algebras generated by pseudodifferential operators with isolated singularities of the symbols. *In*: Spektr. Teoria i Volnov. Protsessy (Probl. Mat. Fiz., Vyp. 10), 209–241, Izd. Leningrad. Univ., 1981 (*Russian*).

PLAMENEVSKI, B. A., and V. N. SENICHKIN
[1] On the spectrum of C^*-algebras generated by pseudodifferential operators with isolated singularities of the symbols. Dokl. Akad. Nauk SSSR **261**: 6, 1304–1306 (1981) (*Russian*); *also in*: Sov. Math. Dokl. **24**, 686–689 (1981).
[2] On the spectrum of C^*-algebras of pseudodifferential operators with singularities in the symbols. Math. Nachr. **121**, 231–268 (1985) (*Russian*).
PLEMELJ, J.
[1] Ein Ergänzungssatz zur Cauchyschen Integraldarstellung analytischer Funktionen, Randwerte betreffend. Monatsh. Math. Phys. **76**, 205–210 (1908).
[2] Riemannsche Funktionenscharen mit gegebener Monodromiegruppe. Monatsh. Math. Phys. **76**, 211–245 (1908).
POGORZELSKI, W.
[1] Integral equations and their applications. Vol. I. Pergamon Press, PWN-Polish. Sci. Publ., Warsaw 1966.
POWER, S. C.
[1] Fredholm Toeplitz operators and slow oscillation. Can. J. Math. **32**: 5, 1058–1071 (1980).
PRIVALOV, I. I.
[1] The Cauchy integral. Izv. Saratov. Univ. Fiz.-Mat. Fak. II: 1, 1918.
[2] Randeigenschaften analytischer Funktionen. VEB Deutscher Verlag der Wissenschaften, Berlin 1956.
[3] Einführung in die Funktionentheorie. I–III. BSB B. G. Teubner Verlagsges., Leipzig 1958–1959.
PRÖSSDORF, S.
[1] Operators admitting an unbounded regularization. Vestn. Leningrad. Univ. 13, 59–67 (1965) (*Russian*).
[2] On the theory of systems of singular integral equations with degenerating symbol matrix. I: Vestn. Leningrad. Univ. 19, 58–73 (1965), II: Vestn. Leningrad. Univ. 7, 68–75 (1966) (*Russian*).
[3] On the stability of the index of the one-dimensional singular operator with degenerate symbol. *In*: Probl. Mat. Anal., 70–79, Izd. Leningrad. Univ., 1966 (*Russian*).
[4] On linear equations in spaces of test functions and distributions. Dokl. Akad. Nauk SSSR **166**: 4, 802–805 (1966) (*Russian*).
[5] The index of the one-dimensional singular operator with degenerate symbol in the space $C^\infty(\Gamma)$. Vestn. Leningrad. Univ. 7, 154–156 (1966) (*Russian*).
[6] Eindimensionale singuläre Integralgleichungen und Faltungsgleichungen nicht normalen Typs in lokalkonvexen Räumen. Habil.-Schrift, TH Karl-Marx-Stadt, 1967.
[7] Die Indexformel für ein System eindimensionaler singulärer Integralgleichungen nicht normalen Typs. Math. Z. **106**, 73–80 (1968).
[8] Über eine Verallgemeinerung des Ableitungsbegriffs und die Regularisierung singulärer Integralgleichungen. Wiss. Z. Tech. Hochsch. Karl-Marx-Stadt **10**: 5, 551–554 (1968).
[9] Ein Satz über die äquivalente Regularisierung abgeschlossener Operatoren. Wiss. Z. Tech. Hochsch. Karl-Marx-Stadt **10**: 5, 555–556 (1968).
[10] Über eine Klasse singulärer Integralgleichungen nicht normalen Typs. Math. Ann. **183**, 130–150 (1969).
[11] Zur Theorie der Faltungsgleichungen nicht normalen Typs. Math. Nachr. **42**, 102–131 (1969).
[12] Zur Lösung eines Systems singulärer Integralgleichungen mit entartetem Symbol. *In*: Elliptische Differentialgleichungen, Bd. I, 111–118, Akademie-Verlag, Berlin 1970.
[13] The singular integral equation with a symbol vanishing at a finite number of points. Mat. Issled. **VII**: 1, 116–132 (1972) (*Russian*).
[14] On systems of singular integral equations with degenerate symbol. Mat. Issled. **VII**: 2, 129–142 (1972) (*Russian*).
[15] Über koerzitive und nichtkoerzitive Probleme bei singulären Integralgleichungen. Mitt. Math. Ges. DDR, H. 3–4, 61–78 (1970).
[16] Systeme einiger singulärer Gleichungen vom nicht normalen Typ und Projektionsverfahren zu ihrer Lösung. Stud. Math. **LIII**, 225–252 (1975).

[17] Einige Klassen singulärer Gleichungen. Akademie-Verlag, Berlin 1974.

[18] Über eine Algebra von Pseudodifferentialoperatoren im Halbraum. Math. Nachr. **52**, 113–139 (1972).

[19] Zur Konvergenz der Fourierreihen hölderstetiger Funktionen. Math. Nachr. **69**, 7–14 (1975).

[20] Some classes of singular equations. North-Holland Publ. Comp., Amsterdam, New York, Oxford: 1978; *Russian transl.*: Mir, Moscow 1979.

[21] Allgemeine Konvergenztheoreme zu Projektionsverfahren für lineare Operatorgleichungen und einige Anwendungen. *In*: Theory of Nonlinear Operators, Proc. Summer School Berlin 1975, Abh. Akad. Wiss. DDR, Nr. 1 N (1977), 217–227, Akademie-Verlag, Berlin 1977.

[22] On approximation methods for the solution of one-dimensional singular integral equations. Appl. Anal. **7**, 259–270 (1978).

[23] On the degenerate Riemann-Hilbert problem. *In*: Partial Differential Equations, Banach Center Publ., Vol. 10, 301–310 (1983) (*Russian*).

[24] Zur Splinekollokation für lineare Operatoren in Sobolewräumen. *In*: Recent Trends in Mathematics, Reinhardsbrunn 1982, 251–262, Teubner-Texte zur Mathematik, Bd. 50, BSB B. G. Teubner Verlagsges., Leipzig 1983.

[25] Approximation methods for solving singular integral equations. *In*: Complex Analysis, Methods, Trends, and Applications, eds. E. Lanckau and W. Tutschke, 131–141, Akademie-Verlag, Berlin 1983.

[26] A localization principle in the theory of finite element methods. *In*: Probleme und Methoden der Mathematischen Physik, 8. Tagung in Karl-Marx-Stadt 1983, 169–177, Teubner-Texte zur Mathematik, Bd. 63, BSB B. G. Teubner Verlagsges., Leipzig 1984.

[27] Ein Lokalisierungsprinzip in der Theorie der Splineapproximation und einige Anwendungen. Math. Nachr. **119**, 239–255 (1984).

Prössdorf, S., and J. Elschner

[1] Finite element methods for singular integral equations on an interval. Engineering Analysis **1**: 2, 83–87 (1984).

Prössdorf, S., and A. Rathsfeld

[1] A spline collocation method for singular integral equations with piecewise continuous coefficients. Integral Equations Oper. Theory **7**: 4, 536–560 (1984).

[2] On spline Galerkin methods for singular integral equations with piecewise continuous coefficients. Numer. Math. **48**, 99–118 (1986).

[3] On strongly elliptic singular integral operators with piecewise continuous coefficients. Integral Equations Oper. Theory **8**: 6, 825–841 (1985).

Prössdorf, S., and G. Schmidt

[1] Notwendige und hinreichende Bedingungen für die Konvergenz des Kollokationsverfahrens bei singulären Integralgleichungen. Math. Nachr. **89**, 203–215 (1979).

[2] A finite element collocation method for singular integral equations. Math. Nachr. **100**, 33–60 (1981).

[3] A finite element collocation method for systems of singular integral equations. *In*: Complex Analysis and Applications '81, 428–439, Publishing House of the Bulgarian Acad. Sci., Sofia 1984.

Prössdorf, S., and B. Silbermann

[1] Über die normale Auflösbarkeit des singulären Integraloperators vom nicht normalen Typ. Math. Nachr. **55**, 73–88 (1973).

[2] Über die normale Auflösbarkeit von Systemen singulärer Integralgleichungen vom nicht normalen Typ. Math. Nachr. **56**, 131–144 (1973).

[3] Ein Projektionsverfahren zur Lösung singulärer Gleichungen vom nicht normalen Typ. Wiss. Z. Tech. Hochsch. Karl-Marx-Stadt **16**: 2, 367–376 (1974).

[4] Ein Projektionsverfahren zur Lösung abstrakter singulärer Gleichungen vom nicht normalen Typ und einige seiner Anwendungen. Math. Nachr. **61**, 133–155 (1974).

[5] Verallgemeinerte Projektionsverfahren zur Lösung singulärer Gleichungen vom nicht normalen Typ. Math. Nachr. **68**, 7–28 (1975).

[6] Projektionsverfahren zur Lösung von Systemen singulärer Gleichungen vom nicht normalen Typ. Rev. Roum. Math. Pures Appl. **XXII**: 7, 965–991 (1977).

[7] On the convergence of the reduction and collocation methods for systems of singular integral equations. Dokl. Akad. Nauk SSSR **226**: 3, 516–519 (1976) (*Russian*); *also in*: Sov. Math. Dokl. **17**, 140–143 (1976).

[8] General theorems on the convergence of projection methods for operator equations in Banach spaces. Dokl. Akad. Nauk SSSR **230**: 3, 527–529 (1976) (*Russian*); *also in*: Sov. Math. Dokl. **17**, 1347–1349 (1977).

[9] Einige allgemeine Sätze zur Theorie der Projektionsverfahren für lineare Operatorgleichungen in Banachräumen. Math. Nachr. **75**, 61–72 (1976).

[10] Zur Kollokations- und Reduktionsmethode für Systeme singulärer Integralgleichungen. *In*: VII. Internationaler Kongreß über Anwendungen der Mathematik in den Ingenieurwissenschaften, Weimar 1975, Berichte, 289–293, VEB Verlag für Bauwesen, Berlin 1975.

[11] Projektionsverfahren und die näherungsweise Lösung singulärer Gleichungen. Teubner-Texte zur Mathematik, BSB B. G. Teubner Verlagsges., Leipzig 1977.

[12] Über Näherungsverfahren zur Lösung singulärer Gleichungen. *In*: 7. Tagung über Probleme und Methoden der Mathematischen Physik in Karl-Marx-Stadt 1979, Tagungsberichte, Teil II, 95–114, Wiss. Schriftenreihe der TH Karl-Marx-Stadt, 1979.

PRÖSSDORF, S., and E. TEICHMANN

[1] Über die äquivalente Regularisierung einer singulären Integralgleichung nicht normalen Typs. Wiss. Z. Tech. Hochsch. Karl-Marx-Stadt **11**: 3, 361–366 (1969).

PRÖSSDORF, S., and G. UNGER

[1] Zur Faktorisierung von Matrixfunktionen in Algebren mit zwei Normen. Math. Nachr. **79**, 37–47 (1977).

PRÖSSDORF, S., and L. v. WOLFERSDORF

[1] Zur Theorie der Noetherschen Operatoren und einiger singulärer Integral- und Integrodifferentialgleichungen. Wiss. Z. Tech. Hochsch. Karl-Marx-Stadt **13**: 1, 103–116 (1971).

PRZEWORSKA-ROLEWICZ, D.

[1] Equations with transformed argument. Elsevier Sci. Publ. Comp., Amsterdam 1973.

PRZEWORSKA-ROLEWICZ, D., and S. ROLEWICZ

[1] Equations in linear spaces. PWN-Polish Sci. Publ., Warsaw 1968.

RATHSFELD, A.

[1] Über das Reduktionsverfahren für singuläre Integralgleichungen mit stückweise stetigen Koeffizienten. Math. Nachr. **127**, 125–143 (1986).

[2] Über die Stabilität von Näherungsverfahren für singuläre Integralgleichungen in $\mathbf{L}^p$ Z. Anal. Anw. (in print).

REMPEL, S., and B.-W. SCHULZE

[1] Index theory of elliptic boundary problems. Akademie-Verlag, Berlin 1982.

RIESZ, F., and B. SZ.-NAGY

[1] Vorlesungen über Funktionalanalysis. VEB Deutscher Verlag der Wissenschaften, Berlin 1956.

RIESZ, M.

[1] Sur les fonctions conjugées. Math. Z. **27**: 2, 218–244 (1927).

ROBERTSON, A. P., and W. ROBERTSON

[1] Topological vector spaces. Cambridge Univ. Press, Cambridge 1964.

RODE, L. O., and I. B. SIMONENKO

[1] Multidimensional singular integrals in classes of functions with general modulus of smoothness. Sib. Mat. Zh. **9**: 4, 928–936 (1968) (*Russian*).

ROGOZHIN, V. S.

[1] A general scheme for the solution of boundary value problems in the space of generalized functions. Dokl. Akad. Nauk SSSR **164**: 2, 277–280 (1965) (*Russian*).

ROZIN, A. L.

[1] Singular integrals and maximal functions in the space L_1. Soobshzh. Akad. Nauk Gruz. SSR **87**: 1, 29–32 (1977) (*Russian*).

[2] On the commutator of the singular integral operator and the multiplication operator. Vestn. Leningrad. Univ. 19, 59–65 (1977) (*Russian*).

RUZMETOV, E.
[1] Some properties of the multidimensional singular integral. Dokl. Akad. Nauk Tadzh. SSR **16**: 5, 12–15 (1978) (*Russian*).

SAAK, E. M.
[1] On the theory of multidimensional elliptic systems of first order. Dokl. Akad. Nauk SSSR **222**: 1, 43–46 (1975) (*Russian*); *also in*: Sov. Math. Dokl. **16**, 591–595 (1975).

SAGINASHVILI, A. I.
[1] Singular integral operators with semi-almost periodic discontinuities in the coefficient. Soobshzh. Akad. Nauk Gruz. SSR **95**: 3, 541–543 (1979) (*Russian*).

SALAEV, V. V.
[1] On a generalization of a theorem by S. G. Mikhlin concerning the behavior of the multidimensional singular operator in the classes $A_{\alpha,K}$. Uchebn. Zap. Az. Univ. **1**, 40–46 (1975) (*Russian*).

SAMKO, S. G.
[1] On the proof of the Babenko-Stein theorem. Izv. Vyssh. Uchebn. Zaved., Mat. **5**, 47–51 (1975) (*Russian*).

SARANEN, J., and W. L. WENDLAND
[1] On the asymptotic convergence of collocation methods with spline functions of even degree. Math. Comput. **45**, 41–108 (1985).

SARASON, D.
[1] Toeplitz operators with semi-almost periodic symbols. Duke Math. J. **44**: 2, 357–364 (1977).

SCHAEFER, H. H.
[1] Über singuläre Integralgleichungen und eine Klasse von Homomorphismen in lokalkonvexen Räumen. Math. Z. **66**, 147–163 (1956).
[2] Topological vector spaces. Springer-Verlag, New York, Heidelberg, Berlin 1971.

SCHÄFER, K.-J.
[1] Klassen von Pseudodifferentialoperatoren in Hilberträumen auf dem R^n mit invarianten und singulären Symbolen. Dissertation, Universität Kaiserslautern, 1980.

SCHECHTER, M.
[1] Principles of functional analysis. Academic Press, New York 1971.

SCHMIDT, G.
[1] On spline collocation for singular integral equations. Math. Nachr. **111**, 177–196 (1983).
[2] The convergence of Galerkin and collocation methods with splines for pseudodifferential equations on closed curves. Z. Anal. Anw. **3**: 3, 371–384 (1984).
[3] On spline collocation methods for boundary integral equations in the plane. Math. Methods Appl. Sci. **7**, 74–89 (1985).
[4] On ε-collocation for pseudodifferential equations on a closed curve. Math. Nachr. **126**, 183–196 (1986).
[5] Spline collocation for singular integro-differential equations over $(0, 1)$. Numer. Math. (in print).

SCHULZ, J.
[1] Ein Näherungsverfahren zur Lösung eindimensionaler singulärer Integralgleichungen nicht normalen Typs. Beitr. Numer. Math. **2**, 177–191 (1973).

SCHULZE, B.-W.
[1] Über eine Verheftung elliptischer Randwert-Probleme. Math. Nachr. **93**, 205–222 (1979).

SCHÜPPEL, B. M.
[1] Regularisierung singulärer Integralgleichungen vom nicht normalen Typ mit stückweise stetigen Koeffizienten. *In*: Function Theoretic Methods for Partial Differential Equations, Darmstadt 1976, Lect. Notes Math., Vol. 561, 430–442, Springer-Verlag, Berlin, Heidelberg, New York 1976.

SCHWARTZ, L.
[1] Théorie des distributions. I–II. Hermann, Paris 1950–1951.

SEELEY, R. T.
[1] Singular integrals on compact manifolds. Amer. J. Math. **81**: 3, 658–690 (1959).

[2] Regularisation of singular integral operators on compact manifolds. Amer. J. Math. **93**: 2, 265–275 (1961).
[3] Integro-differential operators on vector bundles. Trans. Amer. Math. Soc. **117**: 5, 167–204 (1965).
[4] The index of elliptic systems of singular integral operators. J. Math. Anal. Appl. **7**: 2, 289–309 (1963).

SEIFERT, H., and W. THRELFALL
[1] Lehrbuch der Topologie. B.G .Teubner, Leipzig 1934.

Séminaire Henri Cartan
[1] Théorème d'Atiyah-Singer sur l'indice d'un opérateur différential elliptique. Fasc. 1 et 2. Éc. Norm. Super., Paris 1965.

SEMENTSUL, A. A.
[1] On singular integral equations with coefficients having discontinuities of almost periodic type. Mat. Issled. **VI**: 3, 92–114 (1971) (*Russian*).

SHAMIR, E.
[1] The solution of Riemann-Hilbert systems with piecewise continuous coefficients. Dokl. Akad. Nauk SSSR **167**: 5, 1000–1003 (1966) (*Russian*).

SHEVCHENKO, V. I.
[1] On the Hölder continuity of the solutions of singular integral equations of normal type. *In*: Matem. Fizika, No. 15, 160–171, Naukova Dumka, Kiev 1974 (*Russian*).

SHILOV, G. E.
[1] On locally analytic functions. Uspehi Mat. Nauk **21**: 6, 177–182 (1966) (*Russian*).

SHUBIN, M. A.
[1] Pseudodifferential operators and spectral theory. Nauka, Moscow 1978 (*Russian*).

SILBERMANN, B.
[1] Über eine Klasse von singulären Integralgleichungen, deren Symbol nicht mehr als eine endliche Anzahl von Nullstellen ganzzahliger und gebrochener Ordnung aufweist. Math. Nachr. **47**, 245–260 (1970).
[2] Über eine Klasse einseitig invertierbarer singulärer Integraloperatoren nicht normalen Typs in gewissen Paaren von Banachräumen. I: Math. Nachr. **51**, 327–342 (1971), II: Math. Nachr. **52**, 297–313 (1972).
[3] On singular integral operators in spaces of infinitely differentiable and generalized functions. Mat. Issled. **VI**: 3, 168–179 (1971) (*Russian*).
[4] Über einen Zugang zu einer Klasse eindimensionaler singulärer Integraloperatoren nicht normalen Typs. Wiss. Z. Tech. Hochsch. Karl-Marx-Stadt **13**: 1, 135–142 (1971).
[5] Zur Theorie eindimensionaler singulärer Integraloperatoren nicht normalen Typs mit stückweise stetigen Koeffizienten. Math. Nachr. **57**, 371–384 (1973).
[6] Über paarige Operatoren nicht normalen Typs. Math. Nachr. **60**, 79–95 (1974).
[7] Über die einseitige Invertierbarkeit gewisser Klassen paariger Operatoren nicht normalen Typs. Wiss. Z. Tech. Hochsch. Karl-Marx-Stadt **16**: 3, 377–381 (1974).
[8] Singuläre Integralgleichungen vom nicht normalen Typ mit unstetigen Koeffizienten. Dissertation B, TH Karl-Marx-Stadt, 1974.
[9] Ein Projektionsverfahren für schwach ausgeartete singuläre Integralgleichungen. Z. Angew. Math. Mech. **55**, 525–527 (1975).
[10] Ein Projektionsverfahren für einen diskreten Wiener-Hopfschen Operator, dessen Koeffizientensymbole Nullstellen nicht ganzzahliger Ordnungen besitzen. Math. Nachr. **74**, 191–199 (1976).
[11] Lokale Theorie des Reduktionsverfahrens für Toeplitzoperatoren. Math. Nachr. **104**, 137–146 (1981).
[12] Lokale Theorie des Reduktionsverfahrens für singuläre Integraloperatoren. Z. Anal. Anw. **1**: 6, 45–56 (1982).

SIMONENKO, I. B.
[1] A new general method for the investigatoin of linear operator integral equations. I. Izv. Akad. Nauk SSSR, Ser. Mat. **29**: 3, 567–586 (19 65) (*Russian*).

[2] A new general method for the investigation of linear operator integral equations. II. Izv. Akad. Nauk SSSR, Ser. Mat. **29**: 4, 757–782 (1965) (*Russian*).

[3] The Riemann boundary value problem with measurable coefficients. Dokl. Akad. Nauk SSSR **135**: 3, 538–541 (1960) (*Russian*).

[4] The Riemann boundary value problem for n pairs of functions with measurable coefficients and its application to the study of singular integrals in L_p spaces with weight. Izv. Akad. Nauk SSSR, Ser. Mat. **28**: 2, 277–306 (1964) (*Russian*).

[5] The Riemann boundary value problem with continuous coefficients. Dokl. Akad. Nauk SSSR **142**: 2, 278–281 (1959) (*Russian*).

[6] The Riemann boundary value problem for n pairs of functions with continuous coefficients. Izv. Vyssh. Uchebn. Zaved., Mat. **20**: 1, 140–145 (1961) (*Russian*).

[7] Some general questions in the theory of the Riemann boundary value problem. Izv. Akad. Nauk SSSR, Ser. Mat. **32**: 5, 1138–1146 (1968) (*Russian*); *also in*: Math. USSR Izv. **2**, 1091–1099 (1968).

[8] On the problem of solvability of bisingular and of polysingular equations. Functs. Anal. Prilozh. **5**: 1, 93–94 (1971), (*Russian*).

Slobodetski, L. N.

[1] Generalized Sobolev spaces and their applications to boundary value problems for partial differential equations. Uchebn. Zap. Leningrad. Pedagog. Inst. im. Herzena **197**, 54–112 (1958) (*Russian*).

Smirnov, V. I.

[1] Lehrgang der höheren Mathematik. Bd. I–V. VEB Deutscher Verlag der Wissenschaften, Berlin 1964.

Sobolev, S. L.

[1] Equations of mathematical physics. Gostekhizdat, Moscow 1954 (*Russian*).

[2] Introduction to the theory of cubature formulas. Nauka, Moscow 1974 (*Russian*).

[3] Einige Anwendungen der Funktionalanalysis auf Gleichungen der mathematischen Physik. Akademie-Verlag, Berlin 1964.

Sofronov, I. D.

[1] On the approximate solution of singular integral equations. Dokl. Akad. Nauk SSSR **111**: 1, 37–39 (1956) (*Russian*).

Sokhozki, Yu. V.

[1] On definite integrals and functions utilized for expansions into series. St.-Petersburg 1873 (*Russian*).

Soldatov, A. P.

[1] On the Fredholm theory of operators. One-dimensional singular integral operators of classical type. Differ. Uravn. **XIV**: 7, 1285–1295 (1978) (*Russian*); *also in*: Differ. Equations **14**, 915–922 (1978).

[2] On the Fredholm theory of operators. One-dimensional singular integral operators of general form. Differ. Uravn. **XIV**: 4, 706–718 (1978) (*Russian*); *also in*: Differ. Equations **14**, 498–508 (1978).

Sprössig, W.

[1] Über ein System zweidimensionaler singulärer Integralgleichungen nicht normalen Typs. Wiss. Z. Tech. Hochsch. Karl-Marx-Stadt **16**: 3, 433–446 (1974).

[2] Über die Regularisierung eines Systems zweidimensionaler singulärer Integralgleichungen, dessen Symbol endlich viele Nullstellen ganzzahliger Ordnung besitzt. Dissertation, TH Karl-Marx-Stadt, 1974.

[3] Analoga zu funktionentheoretischen Sätzen im R^n. Beitr. Anal. **12**, 113–126 (1978).

[4] Räumliches Analogon zum komplexen T-Operator. Beitr. Anal. **12**, 127–137 (1978).

[5] Taylor- und Laurententwicklungen räumlich verallgemeinerter analytischer Funktionen. Wiss. Z. Tech. Hochsch. Karl-Marx-Stadt **21**: 7, 889–894 (1979).

[6] Anwendung der analytischen Theorie der Quaternionen zur Lösung räumlicher Probleme der linearen Elastizität in beschränkten Gebieten. Z. Angew. Math. Mech. **59**, 741–743 (1979).

[7] Über die Auflösung von Maxwell-Gleichungen in isotropen nichtlinearen Medien. Beitr. Anal. **15**, 71–76 (1981).

STEENROD, N.

[1] The topology of fibre bundles. Princeton, New Jersey, 1951.

STEIN, E. M.

[1] Singular integrals and differentiability properties of functions. Princeton Univ. Press, Princeton, New Jersey, 1970.

[2] Note on singular integrals. Proc. Amer. Math. Soc. **8**: 2, 250–254 (1957).

STEIN, E. M., and G. WEISS

[1] Introduction to Fourier analysis on Euclidean spaces. Princeton Univ. Press, Princeton, New Jersey, 1971.

[2] Generalization of the Cauchy-Riemann equations and representations of the rotation group. Amer. J. Math. **90**, 163–196 (1968).

STEINMÜLLER, J.

[1] Ein Projektionsverfahren zur Lösung einer Klasse singulärer Integralgleichungen, deren Symbol Nullstellen nicht ganzzahliger Ordnungen besitzt. Beitr. Anal. **13**, 127–141 (1979).

STERNBERG, S.

[1] Lectures on differential geometry. Prentice Hall, Inc., Englewood Cliffs, New Jersey, 1964.

STEPHAN, E., and W. L. WENDLAND

[1] Remarks to Galerkin and least squares methods with finite elements for general elliptic problems. Manuscr. Geod. **1**, 93–123 (1976).

STRANG, G., and G. FIX

[1] An analysis of the finite element method. Prentice-Hall, Englewood Cliffs, New Jersey, 1973.

SUDBERY, A.

[1] Quaternionic analysis. Math. Proc. Cambr. Philos. Soc. **85**, 199–225 (1979).

TAIBLESON, M.

[1] The preservation of Lipschitz spaces under singular integral operators. Stud. Math. **24**, 107–111 (1964).

THOMAS, K. S.

[1] Galerkin methods for singular integral equations. Math. Comput. **36**, 193–205 (1981).

TIMAN, A. F.

[1] Theory of approximation of functions of a real variable. Fizmatgiz., Moscow 1960 (*Russian*); *English transl.*: Pergamon Press, Oxford 1963.

TIMAN, T. A.

[1] On a problem of S. G. Mikhlin concerning the order of singular integrals at infinity. Mat. Zametki **3**: 4, 461–472 (1963) (*Russian*).

TITCHMARSH, E. C.

[1] Theory of Fourier integrals. Oxford 1957.

TOVMASYAN, N. E.

[1] On the theory of singular integral equations. Differ. Uravn. **3**: 1, 69–80 (1967) (*Russian*).

TREVES, F.

[1] Introduction to pseudodffferential and Fourier integral operators. Vol. I–II. Plenum Press, New York, London 1982.

TRICOMI, F.

[1] Formula d'inversione dell'ordine di due integrazioni doppio "con asterisco". Rend. d. R. Accad. Nazionale d. Lincei, t. **III**, ser. 6a, fasc. 9, 535–539 (1926).

[2] Equazioni integrale contenenti il valor principale di un integrale doppio. Math. Z. **27**, 87–133 (1928).

VAINIKKO, G. M.

[1] Some error estimates for the Bubnov-Galerkin method. I. Asymptotic estimates. Uch. Zap. Tartu. Gos. Univ. **150** (1964) (*Russian*).

VASILEVSKI, N. L.
[1] Banach algebras generated by certain two-dimensional integral operators. I: Math. Nachr. **96,** 245–255 (1980), II: Math. Nachr. **99**, 135–144 (1980) (*Russian*).
[2] On an algebra generated by two-dimensional integral operators with continuous coefficients in a subdomain of the unit disc. J. Integral Equations **2**, 111–116 (1980).
VEKUA, I. N.
[1] On a Riemann boundary value problem. Trudy Tbilis. Mat. Inst. **11**, 109–139 (1942) (*Russian*).
[2] New methods for the solution of elliptic equations. Gostekhizdat, Moscow, Leningrad 1948 (*Russian*).
[3] Verallgemeinerte analytische Funktionen. Akademie-Verlag, Berlin 1963.
VEKUA, N. P.
[1] Systems of singular integral equations. Nauka, Moscow 1970 (*Russian*).
[2] On a system of singular integro-differential equations and its applications to some boundary value problems of linear conjugation. Trudy Tbilis. Mat. Inst. **24,** 135–147 (1957) (*Russian*).
VERBITSKI, I. E.
[1] Projection methods for the solution of singular integral equations with piecewise continuous coefficients. *In*: Oper. v Banach. Prostr. (Mat. Issled., Vyp. 47), 12–24, Shtiintsa, Kishinev 1978 (*Russian*).
VERBITSKI, I. E., and N. YA. KRUPNIK
[1] Exact constants in the theorems of K. I. Babenko and B. V. Khvedelidze on the boundedness of the singular operator. Soobshzh. Akad. Nauk Gruz. SSR **85**: 1, 21–24 (1977) (*Russian*).
[2] Exact constants in theorems on the boundedness of singular operators in spaces L_p with weight and their applications. *In*: Lin. Oper. (Mat. Issled., Vyp. 54), 21–35, Shtiintsa, Kishinev 1980 (*Russian*).
VILENKIN, N. YA.
[1] Special functions and theory of group representations. Nauka, Moscow 1965 (*Russian*).
VISHIK, M. I.
[1] On strongly elliptic systems of differential equations. Mat. Sb. **29 (71)**, 615–676 (1951) (*Russian*).
[2] On an inequality for the boundary values of harmonic functions in the ball. Uspehi Mat. Nauk **6**: 2 (**42**), 165–166 (1951) (*Russian*).
VISHIK, M. I., and G. I. ESKIN
[1] Boundary value problems for general singular equations in a bounded region. Dokl. Akad. Nauk SSSR **155**: 1, 24–27 (1964) (*Russian*).
[2] Convolution equations in a bounded region. Uspehi Mat. Nauk **20**: 3 (**123**), 89–152 (1965) (*Russian*); *also in*: Russ. Math. Surveys **20**: 3, 86–151 (1965).
[3] Convolution equations in a bounded region in spaces with weighted norms. Mat. Sb. **69**: 1, 65–110 (1966) (*Russian*); *also in*: Amer. Math. Soc. Transl. (2) **67**, 33–82 (1968).
[4] Normally solvable problems for elliptic systems of convolution equations. Mat. Sb. **74**: 3, 326–356 (1967) (*Russian*); *also in*: Math. USSR Sb. **3**, 303–332 (1967).
[5] Convolution equations of variable order. Trudy Moskov. Mat. O.-va **16**, 25–50 (1967) (*Russian*); *also in*: Trans. Moscow Math. Soc. **16**, 27–52 (1967).
[6] Elliptic convolution equations in a bounded region and their applications. Uspehi Mat. Nauk **22**: 1, 15–76 (1967) (*Russian*); *also in*: Russ. Math. Surveys **22**: 1, 13–75 (1967).
VLADIMIROV, V. S.
[1] Methods of the theory of functions of several complex variables. Nauka, Moscow 1964 (*Russian*).
VOLPERT, A. I.
[1] On the index of a system of two-dimensional singular integral equations. Dokl. Akad. Nauk SSSR **142**: 4, 776–778 (1962) (*Russian*).
[2] Elliptic systems on the sphere and two-dimensional singular integral equations on the sphere. Mat. Sb. **59 (101)**, 195–214 (1962) (*Russian*).

WASHIZU, K., and M. IKEGAWA
[1] Finite element technique in lifting surface problems. *In*: International Symposium on Finite Element Methods in Flow Problems, 195–207, Univ. of Wales, Swansea 1974.

WATSON, G. N.

[1] A treatise on the theory of Bessel functions. 2nd ed., Cambridge Univ. Press, Cambridge 1944.

WENDLAND, W. L.

[1] Elliptic systems in the plane. Pitman, London, San Francisco, Melbourne 1979.

[2] Elliptic systems in the plane. Vortrag Jyväskylä "Colloquium on Mathematical Analysis" (1973).

[3] Asymptotic accuracy and convergence. *In*: Progress in Boundary Element Methods, ed. C. A. BREBBIA, Vol. 1, 289–313, Pentech Press, London, Plymouth 1981.

[4] Boundary element methods and their asymptotic convergence. *In*: Theoretical Acoustics and Numerical Treatments, ed. P. FILIPPI, CISM Courses and Lectures, No. 277, 135–216, Springer-Verlag, Wien, New York 1983.

[5] On the spline approximation of singular integral equations and one-dimensional pseudodifferential equations on closed curves. *In*: A. GERASOULIS and R. VICHNEVETSKY (eds.) [1], 113–119.

WIDOM, G.

[1] Singular integral equations in L_p. Trans. Amer. Math. Soc. **97**: 1, 131–160 (1960).

[2] On the spectrum of a Toeplitz operator. Pacific J. Math. **14**, 365–375 (1964).

YOOD, B.

[1] Properties of linear transformations preserved under addition of a completely continuous transformation. Duke Math. J. **18**: 3, 599–612 (1951).

YOSIDA, K.

[1] Functional analysis. Springer-Verlag, Berlin, Göttingen, Heidelberg 1965.

ZOLOTAREVSKY, V. A.

[1] On the convergence of the collocation method for systems of singular integral equations. Mat. Issled. **X**: 1, 56–69 (1974) (*Russian*).

[2] The solution of singular integral equations by the reduction method. Mat. Issled. **X**: 2, 38–52 (1974) (*Russian*).

[3] On the approximate solution of singular integral equations. Mat. Issled. **X**: 3, 82–94 (1974) (*Russian*).

[4] On the method of mechanical quadratures for systems of singular integral equations. Izv. Vyssh. Uchebn. Zaved., Mat. **4**, 47–55 (1976) (*Russian*).

ZOLOTAREVSKI, V. A., and V. N. SEICHUK

[1] The collocation method for the solution of singular integral equations over Lyapunov curves. Dokl. Akad. Nauk SSSR **258**: 1, 20–22 (1981) (*Russian*).

ZYGMUND, A.

[1] Sur le module de continuité de la somme de la série conjugée de la série de Fourier. Prace Matem. Fiz. **33**, 125–132 (1924).

[2] Intégrales singulières. Publ. Semin. Math. d'Orsay, Univ. de Paris, Faculté de Sci. d'Orsay, 1965.

[3] Trigonometric series. Vol. I–II. Cambridge Univ. Press, Cambridge 1959.

Notation index

A^*	adjoint of the operator A, 15
$\alpha(A)$	nullity of the operator A, 15
$A \mid X_0$	restriction of the operator A to the subspace X_0, 26
$A^{(-1)}$	generalized inverse of A, 22, 27
$\mathbf{A}_p$	Hunt-Muckenhoupt-Wheeden condition, 313
$[A, B]$	commutator $AB - BA$ of the operators A and B, 38
$\mathfrak{A}(\alpha, m)$	algebra of all functions in $\mathfrak{A}$ whose m-th Taylor derivative exists, 144
$A_{\alpha,k}$	class of functions satisfying the inequalities (5.1), (5.2), Chapter IX, 234
$A'_{\alpha,k}$	class of functions satisfying the inequalities (5.1), (5.3), Chapter IX, 234
$A \overset{x}{\sim} B$	the operators A and B are equivalent at a point x, 392
$\mathbf{BMO}(\mathbb{R}^m)$	space of functions of bounded mean oscillation, 316
$\beta(A)$	deficiency of the operator A, 15
Coker A	cokernel (defect space) of the operator A, 15
codim M	codimension of M, 16
$\mathbf{C}$	complex field, 31
$\mathbf{C}(\Gamma)$	space of continuous functions on Γ, 44
$\mathbf{C}^{\pm}(\Gamma)$	closure of $R^{\pm}(\Gamma)$ in $\mathbf{C}(\Gamma)$, 84
$\mathbf{C}^m(\Gamma)$	space of m times continuously differentiable functions, 66
$\mathbf{C}^{m,\lambda}(\Gamma)$	space of functions in $\mathbf{C}^m(\Gamma)$ whose m-th derivative satisfies a Hölder condition with exponent λ, 66, 188
$\mathbf{C}^{\infty}(\Gamma)$	space of infinitely differentiable functions, 33, 168
$\mathbf{C}^{-\infty}(\Gamma)$	dual space of $\mathbf{C}^{\infty}(\Gamma)$, 172
$\mathbf{C}(\Gamma; (t_j, m_j)_{j=1}^r)$	space of locally differentiable functions from $\mathbf{C}(\Gamma)$, 67
$D(A)$	domain of the operator A, 15
dim M	dimension of M, 15
δ_{ij}	Kronecker delta, 16
$D_{\Gamma}^{\pm}$	interior resp. exterior region with respect to the closed curve system Γ, 44
det A	determinant of the (operator) matrix A, 113
$\mathfrak{D}(A)$	linear dilation of the operator A, 135
$\mathscr{D}_{\gamma}$	collection of all γ-quasiuniform meshes, 463
$\partial/\partial\bar{z}$	Cauchy-Riemann operator, 187
$\mathbf{Di}(\bar{G})$	space of functions satisfying a Dini condition on $\bar{G}$, 207, 233
$\mathbf{DiL}(\bar{G})$	space of functions satisfying the inequality (4.1), Chapter IX, on $\bar{G}$, 233
δ	Beltrami operator, 215
(e, f)	value of the functional f at the element e, 16

$\hat{\in}$	belongs uniformly to, 263
$E_n(Y, f)$	$= \inf \{\|f - y_n\|_Y : y_n \in Y_n\}$, 429
$\Phi_\pm(X, Y)$	collection of $\Phi_\pm$-operators from X to Y, 22
$\Phi(X, Y)$	collection of Φ-operators from X to Y, 22
F	Fourier transform, 39
$f^{(n)}$	n-th Taylor (resp. mean) derivative of the function f, 144, 149
f^*	maximal function of Hardy and Littlewood, 313
$f^\#$	sharp function, 315
$G\mathbf{L}_\infty(\Gamma)$	group of invertible elements in $\mathbf{L}_\infty(\Gamma)$, 94
$G\mathscr{L}(X, Y)$	set of invertible operators in $\mathscr{L}(X, Y)$, 427
$\gamma_{n,m}$	$= \pi^{m/2} i^n \Gamma(n/2)/\Gamma((m + n)/2)$, 251
$\mathbf{H}^\mu(\Gamma)$	space of functions satisfying a Hölder condition with exponent μ, 44, 209
$\mathbf{H}^s$	periodic Sobolev space of order s, 459
H	singular integral operator with Hilbert kernel, 62
H_Γ	the operator defined in Theorem 5.1, Chapter II, 61
$\mathbf{H}^\lambda(\Gamma; (t_j, m_j)_{j=1}^r)$	space of locally differentiable functions from $\mathbf{H}^\lambda(\Gamma)$, 67
$\mathbf{H}(\mathbf{W}_2^l(S^{n-1}))$	the space defined in § 2 of Chapter XVI, 411
Im A	range (image space) of the operator A, 15
Ind A	index of the operator A, 15
ind a	index of the continuous function a, 76, 79
ind $a \mid \mathbf{L}_p(\Gamma, \varrho)$	index in $\mathbf{L}_p(\Gamma, \varrho)$ of the function $a \in \mathbf{L}_\infty(\Gamma)$, 100
$\mathrm{ind}_{p,\varrho}\, a$	p, ϱ-index of the function $a \in \mathbf{PC}(\Gamma)$, 104, 107
Ker A	kernel (null space) of the operator A, 15, 36
$\mathfrak{R}_n(\Gamma)$	collection of all product-sums of singular operators, 134
$k_{n,m}$	$= (2n + m - 2)\,(n + m - 3)!/((n - 2)!\, n!)$, number of linearly independent spherical functions of order n, 215
$\mathfrak{R}_{l,\lambda}$	class of symbols satisfying the condition (9.7), Chapter XI, 303
$\mathfrak{R}(A; \{X_n, Q_n\})$	convergence manifold of the operator A with respect to a projection method, 427
$\mathscr{K}(X, Y)$	collection of compact operators from $\mathscr{L}(X, Y)$, 427
$\mathscr{L}(X, Y)$	space of continuous linear operators from X to Y, 15
$\mathscr{L}(X)$	space of continuous linear operators on X, 15
$\mathfrak{L}(M)$	linear hull of M, 28
$\mathbf{Lip}_\mu(\Gamma)$	space of functions satisfying a Hölder (Lipschitz) condition with exponent μ, 44, 209
$\mathbf{L}_p(\Gamma)$	space of functions summable in the p-th power, 47
$\mathbf{L}_\infty(\Gamma)$	space of essentially bounded measurable functions, 47
$\mathbf{L}_p(\Gamma, \varrho)$	space $\mathbf{L}_p$ with weight ϱ, 53, 210, 284
$\mathbf{L}_p^+(\Gamma, \varrho)$	the subspace $P_\Gamma \mathbf{L}_p(\Gamma, \varrho)$, 63
$\mathring{\mathbf{L}}_p^-(\Gamma, \varrho)$	the subspace $Q_\Gamma \mathbf{L}_p(\Gamma, \varrho)$, 63
$\mathbf{L}_p(\Gamma; (t_j, m_j)_{j=1}^r)$	space of locally differentiable functions in $\mathbf{L}_p(\Gamma)$, 67
$\bar{\mathbf{L}}_p(\alpha, m; \beta, n), \tilde{\mathbf{L}}_p(\alpha, m; \beta, n)$	151, 152, 162
$\tilde{\mathbf{L}}_p^s(R_-, \alpha, \mu; R_+, \beta, \nu)$	166
$\tilde{\mathbf{L}}_p^s(\alpha, \mu, S_+; \beta, \nu, S_-)$	166
$\overline{M}$	closure of the set M, 15 32
$M_1 \dot{+} M_2$	direct sum of M_1 and M_2, 15
$\omega(u, t)$	modulus of continuity, 208

P_Γ	singular projection $\frac{1}{2}(I + S_\Gamma)$, 63
$\mathbf{PC}(\Gamma)$	algebra of piecewise continuous functions on Γ, 103
$\mathbf{PC}^{n\times n}(\Gamma)$	$= [\mathbf{PC}(\Gamma)]^{n\times n}$, 129
$\mathbf{PS}_d(\Delta)$	space of splines, 459
$P_n^{(\beta,\gamma)}$, $\tilde{P}_n^{(\beta,\gamma)}$	Jacobi polynomials, 447
P_M	the projection defined in § 1 of Chapter XV, 389
$\Pi\{X_n, Q_n\}$	class of operators for which a projection method converges, 427
Q_Γ	singular projection $\frac{1}{2}(I - S_\Gamma)$, 63
$\mathbb{R}$	real field, 31
$\mathbb{R}^m$	m-dimensional Euclidean space, 39
$\overline{\mathbb{R}}^m$	one-point compactification of the m-dimensional Euclidean space
$R(\Gamma)$, $R^\pm(\Gamma)$	collections of rational functions without poles on Γ, 44
$R(x)$	$= \pi^{-(m+1)/2}\,\Gamma\left(\frac{m+1}{2}\right)\frac{x}{\lvert x\rvert^{m+1}}$, the M. Riesz kernel, 291
$\mathcal{R}(\Gamma)$	class of bounded Riemann-integrable functions on Γ, 436
sp K	trace of the operator K, 22, 26
Smb A	symbol of the operator A, 36
S_Γ	singular integral operator with Cauchy kernel, 44
S_∞	Hilbert transform, 56
sing a	singular support of the function a, 94
$\mathring{S}$	multidimensional Cauchy integral operator, 419
$\mathbf{S}_d^n$, $\mathbf{S}_d(\Delta)$	spaces of splines, 458, 463
$u * v$	convolution of the functions u and v, 39
$\mathbf{W}_p^{(m)}$	Sobolev-Slobodetski space, 66, 209
$\mathbf{W}$	Wiener algebra, 84
$\mathbf{W}_p^{(l,\lambda)}$	collection of all functions in $\mathbf{W}_p^{(l)}$ whose derivative is Hölder continuous with exponent λ in the $\mathbf{L}_p$-norm, 436
X^*	conjugate (dual) space of X, 15, 32
X/M	quotient space (factor space), 16
$X\times Y$	product space, 30
$\lVert x\rVert$	norm (resp. seminorm) of x, 32
X^n	collection of n-dimensional vectors with components from X, 113
$X^{n\times n}$	collection of $n\times n$ matrices with entries from X, 113
$Y_{n,m}^{(k)}$	m-dimensional spherical functions of order n, 215

Author Index

Subject Index